ATOMIC PHYSICS
[MODERN PHYSICS]

Attoduekwe Samuel

ATOMIC PHYSICS
[MODERN PHYSICS]

(For M. Sc. & B. Sc. III Year Students)

S.N. GHOSHAL
*Khaira Professor of Physics (Retired),
Calcutta University*

S. CHAND
AN ISO 9001: 2000 COMPANY

S. CHAND & COMPANY LTD.
RAM NAGAR, NEW DELHI-110055

S. CHAND & COMPANY LTD.
(An ISO 9001 : 2000 Company)

Head Office : 7361, RAM NAGAR, NEW DELHI - 110 055
Phones : 23672080-81-82, 9899107446, 9911310888; Fax : 91-11-23677446
Shop at: **schandgroup.com**; e-mail: **info@schandgroup.com**

Branches :
- 1st Floor, Heritage, Near Gujarat Vidhyapeeth, Ashram Road,**Ahmedabad** - 380 014, Ph: 27541965, 27542369, ahmedabad@schandgroup.com
- No. 6, Ahuja Chambers, 1st Cross, Kumara Krupa Road, **Bengaluru** - 560 001, Ph: 22268048, 22354008, bangalore@schandgroup.com
- Bajaj Tower, Plot No. 243, Lala Lajpat Rai Colony, Raisen Road, **Bhopal** - 462 011, Ph: 4274723. bhopal@schandgroup.com
- 152, Anna Salai, **Chennai** - 600 002, Ph: 28460026, 28460027, chennai@schandgroup.com
- S.C.O. 2419-20, First Floor, Sector - 22-C (Near Aroma Hotel), **Chandigarh** -160 022, Ph: 2725443, 2725446, chandigarh@schandgroup.com
- Plot No. 5, Rajalakshmi Nagar, Peelamedu, **Coimbatore** -641004, (M) 09444228242, coimbatore@schandgroup.com **(Marketing Office)**
- 1st Floor, Bhartia Tower, Badambadi, **Cuttack** - 753 009, Ph: 2332580; 2332581, cuttack@schandgroup.com
- 1st Floor, 20, New Road, Near Dwarka Store, **Dehradun** - 248 001, Ph: 2711101, 2710861, dehradun@schandgroup.com
- Pan Bazar, **Guwahati** - 781 001, Ph: 2738811, 2735640 guwahati@schandgroup.com
- Padma Plaza, H.No. 3-4-630, Opp. Ratna College, Narayanaguda, **Hyderabad** - 500 029, Ph: 24651135, 24744815, hyderabad@schandgroup.com
- Mai Hiran Gate, **Jalandhar** - 144008, Ph: 2401630, 5000630, jalandhar@schandgroup.com
- 67/B, B-Block, Gandhi Nagar, **Jammu** - 180 004, (M) 09878651464 **(Marketing Office)**
- A-14, Janta Store Shopping Complex, University Marg, Bapu Nagar, **Jaipur** - 302 015, Ph: 2719126, jaipur@schandgroup.com
- Kachapilly Square, Mullassery Canal Road, Ernakulam, **Kochi** - 682 011, Ph: 2378207, cochin@schandgroup.com
- 285/J, Bipin Bihari Ganguli Street, **Kolkata** -700 012, Ph: 22367459, 22373914, kolkata@schandgroup.com
- Mahabeer Market, 25 Gwynne Road, Aminabad, **Lucknow** - 226 018, Ph: 2626801, 2284815, lucknow@schandgroup.com
- Blackie House, 103/5, Walchand Hirachand Marg, Opp. G.P.O., **Mumbai** - 400 001, Ph: 22690881, 22610885, mumbai@schandgroup.com
- Karnal Bag, Model Mill Chowk, Umrer Road, **Nagpur** - 440 032, Ph: 2723901, 2777666 nagpur@schandgroup.com
- 104, Citicentre Ashok, Govind Mitra Road, **Patna** - 800 004, Ph: 2300489, 2302100, patna@schandgroup.com
- 291/1, Ganesh Gayatri Complex, 1st Floor, Somwarpeth, Near Jain Mandir, **Pune** - 411 011, Ph: 64017298, pune@schandgroup.com **(Marketing Office)**
- Flat No. 104, Sri Draupadi Smriti Apartments, East of Jaipal Singh Stadium, Neel Ratan Street, Upper Bazar, **Ranchi** - 834 001, Ph: 2208761, ranchi@schandgroup.com **(Marketing Office)**
- Kailash Residency, Plot No. 4B, Bottle House Road, Shankar Nagar, **Raipur** - 492 007, Ph: 09981200834, raipur@schandgroup.com **(Marketing Office)**
- 122, Raja Ram Mohan Roy Road, East Vivekanandapally, P.O., **Siliguri**-734001, Dist., Jalpaiguri, (W.B.) Ph. 0353-2520750 **(Marketing Office)**
- Plot No. 7, 1st Floor, Allipuram Extension, Opp. Radhakrishna Towers, Seethammadhara North Extn., **Visakhapatnam** - 530 013, visakhapatnam@schandgroup.com (M) 09347580841 **(Marketing Office)**

© 1991, S.N. Ghoshal
All rights reserved. No part of this publication may be reproduced, stored in a retrieval system or transmitted, in any form or by any means, electronic, mechanical, photocopying, recording or otherwise, without the prior permission of the Publishers.
First Edition 1991
Subsequent Editions and Reprints 1996, 97, 2001, 2003, 2004, 2006, 2007, 2008, 2010
Reprint 2011
ISBN : 81-219-1095-1   **Code : 16 173**

PRINTED IN INDIA
By Rajendra Ravindra Printers (Pvt.) Ltd., Ram Nagar, New Delhi-110 055 and published by S. Chand & Company Ltd., 7361, Ram Nagar, New Delhi-110 055

Dedicated to
the memory of my teachers
Prof. M. N. SAHA, F.R.S.
Professor Emilio Segrè, N.L.

PREFACE TO THE SECOND EDITION

In this edition the book has been revised to include the postgraduate physics syllabi of Indian Universities in addition to the undergraduate honours syllabi covered in the previous edition. Apart from the new additions made in the existing chapters to fulfil this requirement, three new chapters have been added in this edition to deal with the quantum mechanical theories of atomic and molecular structures.

As a prerequisite for this, a brief introduction to the approximation methods of quantum mechanics is included in one of these new chapters (Ch. XII). In the next chapter, the theories of the fine structure of hydrogen-like atoms, Zeeman effect and the structures of the two-electron and many electron atoms are discussed. The explanation of the periodic classification of elements based on the vector model of the atom discussed in the existing book is further elaborated on the basis of quantum mechanics in this chapter. Both Thomas-Fermi statistical method and Hartree-Fock method of self-consistent fields have been included to provide the basis of the calculation of the energy-levels of complex atoms. Corrections to the central field approximation with special emphasis on spin-orbit coupling are discussed at length for both equivalent and non-equivalent electrons.

In chapter XIV, a brief introduction to the interaction of the radiation field with atomic systems has been included and the selection rules in certain special cases are derived.

The discussions on the molecular structure have been expanded to include the molecular orbital theory of the hydrogen molecular ion using LCAO approximation and of the hydrogen molecule. Besides, the Heitler-London theory of the hydrogen molecule is discussed briefly and the distinction between ionic and covalent bonding explained.

In the chapter on the special theory of relatively (Ch. XV) in the new edition, a new section of acceleration and force in relativistic mechanics has been added.

Some of the detailed calculations have been relegated to the appendices which also include brief introductions to Pauli's theory of spin and Dirac's theory of the relativistic electron.

I shall regard my efforts amply rewarded if the book is received with the same enthusiasm as the previous edition.

I wish to express my sincere thanks to the publishers, Messrs. S. Chand & Co., especially to Mr. Ravindra Gupta, the Director of the Company, for their continued interest in bringing out this edition. I also wish to thank Mr. R.M. Nath, the Manager of the Calcutta Office of the Company for his co-operation in various stages of the publication.

AUTHOR

PREFACE TO FIRST EDITION

The subject of Atomic and Nuclear Physics has grown tremendously in its scope during the last quarter of century. The subject was treated traditionally under the single banner of Modern Physics. However, at the present level of our understanding, it can best be presented in two distinct parts, viz. the Atomic Physics comprising the extra-nuclear part of the atom and the Nuclear Physics dealing with the structure of the atomic nucleus. In conformity with this approach, the present volume deals with the Atomic Physics part only. The Nuclear Physics part will be treated in a future volume.

The undergraduate curriculum on Atomic and Nuclear Physics of the Indian Universities has expanded considerably in recent years. However there is a dearth of books dealing with the subject, specially in the advanced undergraduate level. The present work was undertaken with a view to fill up this gap. My association with the teaching in the B.Sc. Honours classes at Presidency College, Calcutta for many years has been specially helpful to me in undertaking the work.

The present book is actually an outgrowth of a similar book in Bengali written by me under the Centrally sponsored scheme of production of books in regional languages at the University level. Since the publication of that book, many friends including my colleagues and students have urged me to bring out an English language edition of the book so that students at all-India level might benefit from it. The West Bengal State Book Board readily gave its permission to this project. However it must be emphasized that this is not a translation of the Bengali Book, though I have drawn freely form the latter.

Though chronological development of the subject has been followed, the traditional approach of discussing in detail some of the earlier work, both experimental and theoretical, has been avoided. Instead, the basic principles have been emphasized. The three main pillars on which the theoretical development of the subject rests, viz., Quantum Mechanics, Special Theory of Relatively and Statistical Mechanics have been treated in separate chapters.

In the first few chapters, the important landmarks of the Quantum Theory and its applications to explain the origin and nature of the spectra of the hydrogen-like atoms and of the alkali atoms have been discussed in detail. The doublet structure of the alkali spectra and of anomalous Zeeman effect have been explained by introducing the concept of the electron spin. The explanation of the periodic classification of elements on the basis of the electronic structure of the atom has been given in sufficient detail.

A brief chapter on laser has been included to expose the students to this new and exciting branch of physics. The production and properties of x-rays, the nature of the x-ray spectra and their explanation, Moseley's law and its importance have all been discussed in detail.

De Broglie's theory of matter wave and the experimental discovery of the wave property of the electron are discussed at length with emphasis on its impact on the subsequent development of the wave mechanical theory of subatomic systems.

After a systematic development of wave mechanics, the more complete explanations of the some of the atomic and subatomic phenomena based on the quantum mechanical theory have been given in some detail.

The application of the ideas of the electronic structure of the atom in the wider context of molecular structure, Raman effect and the properties of solids including metals and semi-conductors has been discussed at length in the last few chapters. A brief discussion on superconductivity is included in this connection to emphasize the exciting developments currently taking place in this field.

The students will find some interesting problems at the end of each chapter, the working out of which should be helpful in the proper understanding of the subject.

I am grateful to many friends and colleagues who have directly or indirectly helped me in completing the project. In particular my thanks are due to Dr. C. K. Majumdar and Dr. D. Roy Chowdhury for their many useful suggestions on the chapters on quantum mechanics, to Dr. J.N. Chakravarty of Narendrapur Ramkrishna Mission College for his continued interest in the project and to Dr. A. K. Saha for his help in preparing the manuscript.

Finally my special thanks are due to Messrs S. Chand Ltd., particularly to Shri Ravindra Kumar Gupta, Director and to Shri Ravi Gupta, Production Manager for their unflagging interest in bringing out this volume. I also wish to thank Mr. K. Mehra, Manager of the Calcutta office of S. Chand Ltd. for rendering all possible help in expediting the publication of the book. I hope the book will be found useful by the teachers and students all I shall consider my efforts amply rewarded if it benefits them.

AUTHOR

CONTENTS

Chapter	Pages
1. Atomic structure of matter	**1—10**
1.1 Dalton's atomic theory	1
1.2 Avogadro's hypothesis and Avogadro number	3
1.3 Determination of Avogadro number	5
1.4 Kinetic theory of matter	7
1.5 Atomic weight and the mass of the atoms	9
2. Cathode rays and positive rays	**11—32**
2.1 Discharge of electricity through gases	11
2.2 Cathode rays	13
2.3 Determination of the specific charge of the cathode rays	16
2.4 Dunnington's experiment	19
2.5 Determination of the charge of an electron; Millikan's oil drop experiment	21
2.6 Positive rays	25
2.7 Thomson's parabola method of analysis of the positive rays	25
2.8 Isotopes	29
2.9 Unit of atomic mass	30
3. Origin of quantum theory; Planck's theory of black body radiation	**33—45**
3.1 Black body radiation ; Wien's displacement law	33
3.2 Rayleigh–Jean's law	36
3.3 Planck's law of radiation ; Quantum hypothesis	41
3.4 Deductions from Plank's law	44
4. Structure of the hydrogen-like atoms; Bohr-Sommerfeld theory	**46—83**
4.1 Thomson model of the atom	46
4.2 Rutherford model of the atom	47
4.3 Bohr's postulates	48
4.4 Bohr's theory of the spectra of hydrogen-like atoms	50

(xi)

4.5	Origin of the spectral series	...	53
4.6	Energy levels	...	55
4.7	Ritz combination principle	...	59
4.8	Effect of finite mass of the nucleus on the spectrum	...	60
4.9	Discovery of heavy hydrogen	...	63
4.10	Bohr's correspondence principle	...	65
4.11	Wilson-Sommerfeld quantum condition	...	66
4.12	Sommerfeld's theory of elliptic orbits; relativity correction	...	69
4.13	Limitations of quantum theory	...	77
4.14	Resonance potentials ; Franck and Hertz experiment	...	77
4.15	Ionization potential	...	80
4.16	Fluorescence and phosphorescence	...	81
5.	**Emission of electrons from metal surfaces; Photoelectric and the thermionic effects**	...	**84—99**
5.1	Discovery of photoelectric effect	...	84
5.2	Lenard's experiment	...	84
5.3	Millikan's experiment	...	86
5.4	Failure of the electromagnetic theory	...	89
5.5	Einstein's light quantum hypothesis and photo-electric equation	...	90
5.6	Application of photoelectric equation	...	92
5.7	Photoelectric cells	...	93
5.8	Thermionic emission	...	94
5.9	Effect of temperature on thermionic emission	...	96
6.	**Alkali spectra, space quantization, electron spin; Periodic classification of elements**	...	**100—141**
6.1	Alkali spectra	...	100
6.2	Space quantization and normal Zeeman effect	...	103
6.3	Electron spin ; vector model of the atom	...	108
6.4	Doublet structure of the alkali spectral lines	...	116
6.5	Fine structure of the hydrogen spectral terms	...	118
6.6	Pauli's exclusion principle; Periodic classification of elements	...	120

6.7	Arrangement of electrons in atoms	...	123
6.8	Energy levels of complex atoms	...	130
6.9	Anomalous Zeeman effect	...	131
6.10	Paschen-Back effect	...	136
6.11	Stern and Gerlach's experiment	...	138
6.12	Stark effect	...	140

7. Laser ... 142—159

7.1.	Einstein's theory of atomic transitions	...	142
7.2	Laser principle	...	144
7.3	Experimental arrangement	...	149
7.4	Population inversion	...	150
7.5	Ruby-laser	...	153
7.6	Helium-neon laser	...	154
7.7	Uses of laser radiation	...	157

8. X-rays 160—209

8.1	Discovery of x-rays	...	160
8.2	Production of x-rays	...	161
8.3	Properties of x-rays	...	163
8.4	Measurement of the intensity of x-rays	...	164
8.5	Variation of x-ray intensity with wavelength	...	165
8.6	Origin of continuous x-rays	...	167
8.7	Origin of x-ray peaks	...	168
8.8	Fine structure of x-ray levels	...	171
8.9	Theoretical consideration of the fine structure of x-rays	...	173
8.10	Auger transition	...	175
8.11	Moseley's law	...	176
8.12	Absorption of x-rays	...	178
8.13	Absorption edges	...	182
8.14	Scattering of x-rays ; Thomson's theory	...	183
8.15	Determination of the number of electrons per atom	...	188
8.16	Compton scattering	...	189
8.17	Diffraction of x-rays	...	194
8.18	Determination of the wavelength of x-rays by crystal diffraction method	...	201

8.19	Curved crystal method	203
8.20	Refraction of x-rays	204
8.21	Determination of x-rays wavelength by ruled grating	205
8.22	Modification of Bragg equation due to refraction	206
8.23	Polarization of x-rays	207

9. Wave particle duality; Heisenberg's uncertainty principle ... 210—240

9.1	Particles and waves	210
9.2	Phase and group velocities	211
9.3	Particle wave	213
9.4	Relation between phase and group velocity of de Broglie waves	214
9.5	Discovery of matter waves; Davisson and Germer's experiment	216
9.6	G.P. Thomson's experiment	218
9.7	Effect of refraction of the electron beam	220
9.8	De Broglie wavelength of high energy electrons	222
9.9	Electron microscope	223
9.10	Need for a new mechanics for the sub-atomic particles	228
9.11	Particles and wave packets	229
9.12	Nature of matter waves	229
9.13	Uncertainty relation	231
9.14	Gamma-gay microscope experiment	235
9.15	Applicability of classical and quantum concepts	237
9.16	Principle of superposition	238

10. Introduction to wave mechanics ... 241—279

10.1	Introduction	241
10.2	Wave function ; Schrödinger wave equation	242
10.3	Operators in quantum mechanics	245
10.4	Physical interpretation of ψ ; Probability density	249
10.5	Normalization of the wave function	250
10.6	Probability current density; Conservation of probability	251
10.7	Separation of space and time in Schrödinger equation; Time-independent Schrödinger equation	253

10.8	Eigenfunctions and eigenvalues	...	254
10.9	Probability of stationary states	...	257
10.10	Degeneracy	...	259
10.11	Averages in quantum mechanics; Expectation values	...	259
10.12	Expectation values and correspondence principle; Ehrenfest's theorem	...	262
10.13	Formal proof of the uncertainly relation	...	264
10.14	Hermitian operators	...	266
10.15	Some properties of Hemitian operators	...	268
10.16	Reality of the eigenvalues of Hermitian operators	...	272
10.17	Equations of motion in quantum mechanics	...	273
10.18	Fundamental postulates of quantum mechanics	...	275

11. Solutions of Schrödinger equation in some simple cases ... **280—333**

11.1	Boundary conditions	...	280
11.2	Particle in a one-dimensional potential box with rigid walls	...	282
11.3	Step potential	...	285
11.4	One dimensional rectangular potential barrier	...	289
11.5	Explanation of α-decay	...	292
11.6	One dimensional rectangular potential well	...	294
11.7	Linear harmonic oscillator	...	298
11.8	Oscillator wave functions; Parity	...	301
11.9	Diatomic molecules	...	306
11.10	Free particle with one-dimensional motion; Momentum eigenfunctions	...	307
11.11	Angular momentum operators	...	309
11.12	Three dimensional motion in a central field; Problem of the hydrogen atom	...	310
11.13	ϕ-equation	...	312
11.14	θ-equation	...	312
11.15	Radial equation	...	314
11.16	The wave functions of the hydrogen-like atoms	...	319
11.17	Constants of motion in a central field	...	323
11.18	Parity operator	...	326

11.19	Dirac notations	...	328
11.20	Two particle system	...	329

12. Approximate methods in quantum mechanics ... 334—361

12.1	Stationary perturbation method	...	334
12.2	First order perturbation	...	335
12.3	Second order perturbation	...	337
12.4	Degenerate case	...	339
12.5	Time-dependent perturbation	...	341
12.6	Constant perturbation lasting for a finite time	...	344
12.7	Variational method	...	347
12.8	WKB approximation	...	349

13. Quantum mechanical theory of atomic structure ... 362—420

13.1	Quantum mechanical theory of the fine structure of hydrogen-like atoms	...	362
13.2	Fine structure of the hydrogen spectral terms	...	367
13.3	Hyperfine structure of spectral lines	...	370
13.4	Lamb and Retherford's experiment	...	370
13.5	Quantum mechanical theory of Zeeman effect	...	372
13.6	Strong field case (Paschen-back effect)	...	375
13.7	Two-electron atoms	...	376
13.8	Term scheme for the helium atom	...	380
13.9	Calculation of the ground state energy of the helium atom	...	381
13.10	Excited states of the helium atom	...	387
13.11	Identical particles	...	390
13.12	Many electron atoms	...	393
13.13	Periodic classification of elements	...	396
13.14	The nature of the central field; Thomas-Fermi model of the atom	...	397
13.15	Hartree-Fock method of self-consistent field	...	403
13.16	Application of Hartree-Fock method to many-electron atoms	...	405
13.17	Results of Hartree-Fock calculations	...	407
13.18	Corrections to the central field approximation	...	407

14. Atoms in radiation field ... 421—436

- 14.1 Interaction of electromagnetic field with atomic systems ... 421
- 14.2 Dipole selection rules for one electron atom ... 429
- 14.3 Forbidden transitions ... 431
- 14.4 Width of a spectral line ... 432
- 14.5 Selection rules for many electron atoms ... 434

15. Special theory of relativity ... 437—491

- 15.1 Frames of reference ; Newtonian relativity ... 437
- 15.2 Galilean transformation and electromagnetic theory ... 441
- 15.3 Michelson-Morley experiment ... 443
- 15.4 Lorentz- Fitzgerald contraction ... 447
- 15.5 Einstein's special theory of relativity ... 447
- 15.6 Lorentz transformation equations ... 450
- 15.7 Relativity of length measurement ... 453
- 15.8 Relativity of time measurement ... 454
- 15.9 Einstein's velocity addition theorem ... 456
- 15.10 Explanation of stellar abberation from the special theory of relativity ... 458
- 15.11 Fizean's experiment ... 460
- 15.12 Change of the mass of a body with velocity ... 461
- 15.13 Mass-energy equivalence ... 463
- 15.14 Some important mathematical relationships ... 465
- 15.15 Graphical representation of Lorentz transformation equations ... 466
- 15.16 Relativistic Doppler effect ... 471
- 15.17 Four vectors ... 477
- 15.18 Twin paradox ... 482
- 15.19 Einstein's general theory of relativity ... 484
- 15.20 Acceleration and force ... 485

16. Molecular spectra ... 492—537

- 16.1 Introduction ... 492
- 16.2 Origin of band spectra ... 494
- 16.3 Pure rotational spectrum of a diatomic molecule ... 497
- 16.4 Rotation-vibration spectra ... 501

16.5	Rational structure of the infrared bands of diatomic molecules	...	507
16.6	Electronic bands and their vibrational structure	...	510
16.7	Frank-Condon principle	...	511
16.8	Electronic states	...	513
16.9	Rotational structure of electronic bands	...	514
16.10	Raman effect	...	515
16.11	Characteristics of Raman spectrum	...	517
16.12	Classical theory of Raman effect	...	518
16.13	Quantum theory of Raman effect	...	519
16.14	Rotational structure of Raman spectrum	...	522
16.15	Experimental techniques of the study of Raman effect using laser source	...	525
16.16	Applications of Raman effect	...	527
16.17	Electronic structure of diatomic molecules	...	528
16.18	Hydrogen molecular ion; LCAO approximation	...	528
16.19	Hydrogen molecule; Heitler-London and molecular orbital theories	...	532

17. Structure and properties of solids   ... 538—587

17.1	Introduction	...	538
17.2	Interatomic forces and classification of solids	...	538
17.3	Lattice structure of crystals	...	544
17.4	Lattice energy	...	550
17.5	Miller indices	...	552
17.6	Laue diffraction equations	...	556
17.7	Atomic scattering factor and geometrical structure factor	...	559
17.8	Bragg's experiment on the structure of rock salt	...	561
17.9	Crystal structure determination	...	566
17.10	Arisotropy of the physical properties of single crystals	...	570
17.11	Specific heat of solids	...	571
17.12	Classical theory and its limitations	...	572
17.13	Einstein's theory of specific heat	...	573
17.14	Debye's theory of specific heat	...	575
17.15	Phonon	...	584

18. Metals and semi conductors ... 588—647

18.1	Specific heat of metals	...	588
18.2	Metallic conductivity; Free electron theory	...	591
18.3	Wiedemann-Franz law	...	597
18.4	Thermionic emission; Work function	...	599
18.5	Schottky effect	...	605
18.6	Field emission	...	607
18.7	Calculation of Fermi energy	...	608
18.8	Band theory of solids	...	611
18.9	Quantum mechanical theory of band structure	...	615
18.10	Motion of an electron according to band theory; Effective mass	...	620
18.11	Classification of solids according to band structure	...	622
18.12	Extrinsic semiconductors	...	625
18.13	Carrier concentration and positions of Fermi levels in semiconductors	...	628
18.14	Conductivity of semiconductors	...	633
18.15	Rectifying property of semiconductors	...	634
18.16	Hall effect	...	634
18.17	Superconductivity	...	636
18.18	Quantum nature of semiconductivity; Josephson effect	...	642

19. Magnetic properties of solids ... 648—674

19.1	Introduction	...	648
19.2	Origin of diamagnetism	...	652
19.3	Origin of paramagnetism	...	654
19.4	Origin of ferromagnetism	...	660
19.5	Einstein-de Haas experiment	...	664
19.6	Origin of molecular field	...	665
19.7	Domain structure of ferromagnetic materials	...	669
19.8	Antiferromagnetism ; Ferrimagnetism and ferrites	...	672

20. Statistical mechanics ... 675—694

20.1	Introduction	...	675
20.2	Statistical method for describing the state of a system	...	675

20.3	Equilibrium distribution ; Maxwell-Botzmann statistics	...	677
20.4	Calculation of the density of states	...	681
20.5	Quantum statistics	...	683
20.6	Exchange operator	...	685
20.7	Fermi-Dirac statistics	...	687
20.8	Bose-Einstein statistics	...	690
20.9	Derivation of Planck's law of radiation using Bose-Einstein statistics	...	693

Appendices ... 696—720

A I	Hermite polynomials	...	696
A II	Legendre polynomials and associated Legendre functions	...	697
A III	Spherical harmonics	...	699
A IV	Laguerre polynomials and associated Laguerre functions	...	700
A V	List of hydrogenic total wave functions	...	701
A VI	Table of important physical constants	...	702
A VII	Periodic table of elements	...	703
A VIII	Evaluation of the integrals in the calculation of the ground state energy of helium	...	705
A IX	Hamiltonian of a charged particle in an electromagnetic field	...	710
A X	Pauli's theory of spin	...	712
A XI	Dirac's theory of the relativistic electron	...	714

Index ... 721—725

1

ATOMIC STRUCTURE OF MATTER

1.1 Dalton's atomic theory

It is well-known to the students of modern science that all matter is made up of microscopically small particles known as *atoms*, which is a word derived from the Greek word ατομοζ, meaning *indivisible*. This atomic view of matter was known to the ancient philosophers in different parts of the world, including Democritus (400 B.C) in Greece and Kanada in India. However, the ancients had only vague qualitative idea of the ultimate constituents of matter, since their views were not established on any scientific observations as we know today.

The modern theory of the atomic constitution of matter was first propounded by the British chemist, John Dalton in 1803 on the basis of a number of well-known laws of chemical combination. These include the law of conservation of mass, the law of definite proportions and the law multiple proportions.

The first of these, was propounded by the French scientist Antoine Lavoisieur (1789). He showed that if a piece of pure tin is heated in a closed vessel so that it combines chemically with the oxygen of the air contained in the vessel, then the total weight of the substances inside the vessel plus the weight of the vessel remained the same before and after the chemical combination. Subsequently, it was established on the basis of many experiments that *the total weight of all the components taking part in a chemical reaction is equal to the total weight of all the products of the reaction*. This is what is known as the *law of conservation of mass*.

The *law of definite proportions* which was propounded by the French chemist J.L. Proust in 1799, states that *when two chemical elements A and B combine to produce a chemical compound C (say), then the elements A and B are always present in a definite proportion by weight in the compound C*. For example, the two elements hydrogen and oxygen are always present in the same ratio of 1.008 : 8 by weight in the chemical compound water irrespective of the method of production of water.

The *law of multiple proportions* was discovered by Dalton himself (1803). *When two elements A and B combine chemically to produce two or more chemical compounds, then the ratios of the weights of any one of* them (*say B*) which combines with a fixed weight of the other (*say A*) to form the different compounds $C_1, C_2...$ are equal to the ratios of small integers. As an example, we know that the elements carbon and oxygen combine to produce the two chemical compounds carbon monoxide (CO) and carbon dioxide (CO_2). In one mole of the former (*i.e.*, CO), 12×10^{-3} kg of carbon is in chemical combination with 16×10^{-3} kg of oxygen; in the latter (*i.e.*, CO_2), 12×10^{-3} kg of carbon is in chemical combination with 32×10^{-3} kg of oxygen. Thus the weights of oxygen which are in chemical combination with 12×10^{-3} kg of carbon in the above two compounds are in the ratio 16 : 32 or 1 : 2.

On the basis of the above laws, Dalton proposed his atomic hypothesis which consisted of the following postulates :

(*a*) All chemical elements are made up of extremely small particles, known as *atoms*. The atoms cannot be further subdivided by any chemical process. They retain their identity during chemical reaction.

(*b*) All atoms of the same element are identical *i.e.*, their weights and other properties are exactly the same. The atoms of different elements differ in their weights and other properties. Each element is characterised by the weight of its atoms.

(*c*) When different elements combine chemically, it is the atoms of these elements which combine together. This combination of the atoms take place in simple numerical ratios, *e.g.* 1 : 1, 1 : 2, 2 : 1, 2 : 3 *etc*.

With the help of these three postulates, it is possible to explain the laws of chemical combination stated above.

Since the weights and other properties of the atoms of different elements and their numbers taking part in a chemical reaction remain unchanged during the reaction and since the total weight of each element taking part in the reaction is equal to the combined weights of all the atoms present in it, we conclude that the total weight of all the elements taking part in the reaction must be equal to the total weight of all the substances produced in the reaction.

This is the law of conservation of mass.

Again since chemical compounds are produced by the combination of the atoms of two or more elements in simple numerical proportions and since all the atoms of the same element have equal weights, hence the elements taking part in a chemical reaction always combine together in definite proportions by weight.

This is the law of definite proportions.

Now consider the production of two types of chemical compounds C_1, C_2 by the combination of the two elements A and B. Suppose that in

Atomic Structure of Matter

each molecule of the first compound C_1, a_1 atoms of A are in chemical combination with b_1 atoms of B. If ω_A and ω_B are the weights of the atoms of A and B respectively, then the weights of A and B which are present in chemical combination in each molecule of C_1 are $a_1\omega_A$ and $b_1\omega_B$ respectively. Similarly if in each molecule of the second compound C_2, a_2 atoms of A are in chemical combination with b_2 atoms of B, then the respective weights of A and B in each molecule of C_2 are $a_2\omega_A$ and $b_2\omega_B$. Hence the weight of B which is in chemical combination with the unit weight of A in C_1 is $b_1\omega_B/a_1\omega_A$. Likewise, the weight of B which is in chemical combination with the unit weight of A in C_2 is $b_2\omega_B/a_2\omega_A$. So the ratio of the weights of B which are in chemical combination with unit weights of A in the two compounds C_1 and C_2 is

$$\frac{b_1\omega_B}{a_1\omega_A} : \frac{b_2\omega_B}{a_2\omega_A} \quad \text{or,} \quad a_2b_1 : a_1b_2$$

Since a_1, b_1, a_2 and b_2 are small integers, the ratio $a_2b_1 : a_1b_2$ is equal to the ratio of two small integers. This is the law of multiple proportions.

On the basis of the above postulates, Dalton was able to determine the relative masses of the atoms of a number of elements taking the atomic mass of hydrogen to be 1. However because of the inadequacy of the experimental data, particularly regarding the numbers of atoms of the different elements present in the molecules of the compounds, his results were in many cases incorrect. For instance, Dalton assumed the numbers of hydrogen and oxygen atoms in a molecule of water to be *one each*, *i.e.*, $a_H = 1$ and $a_O = 1$. From a knowledge of the ratio r of the weights of hydrogen and oxygen present in water determined experimentally (which is 1 : 8), he estimated the relative atomic mass of oxygen from the relation $r = a_H\omega_H/a_O\omega_O = 1/8$. Taking $\omega_H = 1$, he found $\omega_O = 8$, which, as we know today, is not correct.

1.2 Avogadro's hypothesis and Avogadro number

Soon after Dalton's atomic theory was proposed, it was found that certain experimental facts regarding the chemical combination of gases could not be properly reconciled with his theory.

It was observed by the French chemist Joseph Gay-Lussac that when two or more gases combined chemically to produce one or more gaseous substance, the volumes of the reacting gases and those of the product gases were always in simple proportions. Thus according to Gay-Lussac, two volumes of hydrogen gas combined with one volume of oxygen gas to produce two volumes of water vapour. This could not be reconciled with Dalton's idea that one atom of hydrogen combined with one atom of oxygen to produce one molecule of water.

In order to explain Gay-Lussac's observations, the Italian physicist L.R.A. Avogadro proposed his now famous *Avogadro hypothesis* (1811) which can be stated as follows :

Equal volumes of different gases contain equal numbers of molecules at the same temperature and pressure.

Thus if on the basis of Avogadro hypothesis it is assumed that there are n molecules in each one litre volume of a gas at a given temperature and pressure, then according to Gay-Lussac's observations, $2n$ molecules of hydrogen contained in two litres of hydrogen gas combine with n molecules of oxygen present in 1 litre of oxygen gas to produce $2n$ molecules of water contained in 2 litres of water vapour which is produced. Hence two molecules of hydrogen combine with one molecule of oxygen to produce two molecules of water vapour according to the following equation :

$$2H_2 + O_2 = 2H_2O$$

The above equation tells us that each molecule of hydrogen and oxygen contains two atoms of these elements respectively while each molecule of water vapour contains three atoms—two of hydrogen and one of oxygen. Experimentally it is found that 16 kg of oxygen gas combine with 2.016 kg of hydrogen gas to produce 18.016 kg of water vapour. Since in each molecule of water vapour, there are two atoms of hydrogen present in combination with one atom of oxygen, the ratio of the *atomic weights* of oxygen and hydrogen is

$$\left(16 : \frac{2 \cdot 016}{2}\right) \text{ or, } (16 : 1 \cdot 008)$$

Since there are two atoms of oxygen present in each molecule, the molecular wight of oxygen may be taken to be 32.00. Hence the molecular weight of hydrogen will be 2.016. Chemists define one *mole* of a substance as the weight of the substance equal to its molecular weight expressed in grams *i.e.*, 10^{-3} kg. The volume of one mole of a gas is known as the *molar volume* or *the gram-molecular volume*. Careful measurements have established that at standard temperature (0°C) and pressure (760 mm of Hg), one molar volume of a gas is $V = 22.4136$ litres. According to Avogadro hypothesis, the number of molecules in each mole of any gas is the same. This is therefore a *universal constant* and is known as Avogadro number, usually designated by the symbol N_0.

From the above discussion, it is obvious that there are N_0 molecules of oxygen in 32×10^{-3} kg or 32g of oxygen gas. So there are $2N_0$ oxygen atoms in it. Hence in 16×10^{-3} kg or 16 g of oxygen, which is the atomic weight of oxygen expressed in grams, there are N_0 atoms of oxygen. So in each gram-atomic weight of an element, the number of atoms is equal to the Avogadro number N_0. From this the mass of each atom of the element, expressed in gram, can be determined by dividing the gram-atomic weight by N_0. This shows clearly the importance of determining the Avogadro number very accurately.

1.3 Determination of Avogadro number

It was not until about 100 years after Avogadro's hypothesis had been proposed that the Avogadro number N_0 could be determined accurately. N_0 has been determined by various methods. These include methods based on different types of fluctuation phenomena and electrolytic method.

An example of the fluctuation phenomenon is the Brownian motion which is observed in the case of very small microscopically observable particles, *e.g.*, the suspended particles in the colloidal solutions or the smoke particles. The botanist Robert Brown discovered the phenomenon in 1827. The theory of Brownian motion was first developed by Albert Einstein and M. von Smoluchowski. According to them, there is an unbalanced force acting on the suspended microscopic particles inside a colloidal solution due to incessant bombardment of these particles by the molecules of the solution during their random thermal motion. This unbalanced force causes the suspended particles to move about inside the fluid along short zig-zag paths. Their directions of motion change abruptly and frequently with rapid jerks, each little movement being independent of the previous path. This motion is opposed by the viscous force in the fluid. It is possible to calculate theoretically the displacement of the particles as a function of time. In addition, it is also possible to determine theoretically the distribution of the particles in a given field of force, e.g., the gravitational field.

The French physicist Jean Perrin determined Avogadro number by the above two types of measurement. (For details see *A Treatise on Heat* by Saha and Srivastava).

Another example of the fluctuation phenomenon is the scattering of light in the atmosphere. Lord Rayleigh first pointed out that the origin of the blue colour of the sky is related to this scattering (1871). It can be shown that if the density of air is the same everywhere, then due to interference between the light waves scattered from different molecules in the air, the amplitude of the scattered light wave will become zero and no scattering will be observed. As a result the sky would appear black. Actually this does not happen due to the fluctuation of the atmospheric density within the different volume elements in the atmosphere. The smaller the size of the volume element, greater is the magnitude of this fluctuation (see Ch. XX). It is quite considerable within volume elements having sides of the order of the wavelength of light. Due to this fluctuation, the light waves scattered from different molecules are incoherent so that there is no destructive interference between them. As a result the total scattered intensity is equal to the sum of the intensities of the light scattered by the different molecules and depends on the number of air molecules per unit volume (n). On the basis of this assumption, Rayleigh proved that the intensity I of the scattered light is inversely proportional to the fourth power of the wavelength ($I \propto 1/\lambda^4$). It is possible to determine n by measuring I, using Rayleigh's theory, from which Avogadro number N_0 can be estimated. (For details see *Physical Optics* by Jenkins and White).

The most accurate method of determining N_0 is based on electrolysis. When an electric current is passed through an electrolyte, the latter is decomposed into its constituents carrying positive and negative electricity which are liberated at the two opposite electrodes. The famous British scientist Michael Faraday established the following two laws of electrolysis based on a variety of experiments :

(*i*) The amount of a substance liberated at an electrode is directly proportional to the total quantity of electric charge flowing through the electrolyte.

(*ii*) If the same quantity of electrical charge is passed through different electrolytes, then the amounts of the substances liberated at the respective electrodes are directly proportional to their chemical equivalents.

The electrical charge required for the liberation of that amount of a substance which has weight equal to its *equivalent weight* has been determined quite accurately by many experimenters. This is known as a *faraday* of electrical charge and has a value

$$F = 9 \cdot 64846 \times 10^4 \text{ coulombs per mole}$$

Evidently, for the liberation of each gram-atom (1 mole) of a monovalent substance, the quantity of electricity required is one faraday. As for example, for the liberation of $107 \cdot 88 \times 10^{-3}$ kg of silver or $1 \cdot 008 \times 10^{-3}$ kg of hydrogen or $35 \cdot 46 \times 10^{-3}$ kg of chlorine, the above quantity of electricity is required.

On the other hand, for liberating each gram-atom of a substance of valency v, the quantity of electricity required is vF.

From the above two laws of electrolysis, it is evident that at the time of electrolysis, each atom or group of atoms liberated at an electrode carries the same quantity of electricity. Such electrically charged atoms or groups of atoms are called *ions*. During electrolysis these ions move through the electrolyte resulting in the flow of electric current through it. The negative ions move towards the anode while the positive ions move towards the cathode and are ultimately liberated at the respective electrodes.

Atomic Structure of Matter

From the above discussions, it can be realized that there is a fundamental unit of electrical charge. Later researches have established that this is equal to the amount of electricity carried by an electron. An ion of a monovalent element carries this quantity of electricity. If we designate it by the symbol e, then we get the result

$$F = N_0 e$$

The electronic charge e has been determined accurately by various workers. The first to do this was R.A. Millikan of the U.S.A. (see § 2·5). Knowing e, the Avogadro number N_0 can be determined from the value of F given above.

Another very accurate method of determining N_0 is the x-ray diffraction studies on the structure of crystals. These studies give the spacings between the atoms in the crystal which in turn yield the value of the number of atoms per unit volume of the crystal, if the structure is known. From this, N_0 can be deduced, knowing the atomic mass and the density of the substance.

The currently accepted value of N_0 is

$$N_0 = 6 \cdot 02205 \times 10^{23} \text{ per mole}$$

1.4 Kinetic theory of matter

Following the development of the atomic theory of matter, discussed in the previous sections, James Clerk Maxwell of England and Ludwig Boltzmann of Germany developed the kinetic theory of matter, making use of the atomic view point. The theory was based on the following main assumptions:

(i) All gases are made up of a very large number of molecules. The molecules are considered rigid and perfectly elastic spheres. For a particular gas, all molecules are identical in all respects, *e.g.*, their mass, size *etc*.

(ii) The molecules are extremely small in size compared to the intermolecular distances.

(iii) The molecules are in incessant motion, moving with all possible velocities, the directions of their motion being entirely random. This state of the molecules in a gas is known as the state of *molecular chaos*.

(iv) During their motion, the molecules collide with one another and with the walls of the vessel. At each collision, the velocities of the molecules are changed in magnitude and direction.

(v) The collisions suffered by the molecules are perfectly elastic. As a result both momentum and energy conservations are satisfied.

(*vi*) There are no forces of attraction or repulsion between the molecules. As a result they do not influence the motion of one another except at the time of collisions and their energies are wholly kinetic.

(*vii*) Between successive collisions, the molecules move in straight lines with uniform velocities. The mean distance travelled by a molecule between two successive collisions is known as the *mean free path*.

It may be noted that since there is an enormously large number of molecules in a gas, it is not possible to analyse the motion of the individual molecules. There is no other way but to employ statistical method (see Ch XX). The basic postulate is the principle of molecular chaos mentioned above, according to which the probability of a molecule being at any point within a closed volume is independent of the position of the point. Further the probability of a molecule having any particular direction of its velocity is the same for all directions. In other words all directions are equally probable.

One of the consequences of the above postulate is that the density of the gas is the same everywhere within the enclosure. It is also possible to calculate the mean squared velocity of the gas molecules from which the temperature and pressure of the gas can be deduced. For this purpose it is necessary to know the nature of distribution of the molecular speeds.

The simplest assumption would be that all molecules move with the same speed which obviously cannot be true for such a large number of molecules. So Maxwell derived the *law of distribution of molecular velocities* towards the middle of the nineteenth century, which was later improved upon by Maxwell himself as also by Boltzmann and others. It is possible to deduce the different observable properties of the gas, e.g., its temperature, pressure, thermal conductivity, viscosity, etc. on the basis of the theory so developed. When these are compared with experimentally determined values of the different properties of the gas, the main assumptions underlying the kinetic theory of gases are found to be amply justified which in turn provide additional confirmation of atomic structure of matter. It may be noted that the fundamental laws governing the behaviour of the gases, *e.g.*, Boyle's law, Charles' law *etc.* are also derivable from the kinetic theory. (For details see *A Treatise on Heat* by Saha and Srivastava).

Since the volume occupied by a mole of a gas at standard temperature and pressure and the number of molecules in it are accurately known (see above), it is possible to find the mean volume available per gas molecule inside a closed volume. From this we can get the order of magnitude of the intermolecular distance which comes out to be of the order of 10^{-9} m. This gives us an idea about the smallness of the molecules. Assuming the gas molecules to be rigid spheres, it is possible to deduce their radius from the kinetic theory by measuring the thermal conductivity or the viscosity of the gases. These come out to be of the order of 1×10^{-10} m to 3×10^{-10} m. It is also possible to estimate the atomic radii for the solids by determining the molar volumes at their melting points. The values so derived are also of the same order of magnitude as above. In the table below, are listed the molecular radii of some common gases as also the atomic radii of a few elements:

Table 1.1

Gas	Molecular radius (10^{-10} m)	Element	Atomic radius (10^{-10} m)
Hydrogen	1·36	Carbon	1·00
Helium	1·09	Aluminium	2·02
Nitrogen	1·89	Sodium	2·21
Oxygen	1·81	Caesium	3·22
Air	1·87	Tin	1·61
Carbon dioxide	2·31	Bismuth	1·88

1.5 Atomic weight and the mass of the atoms

We have mentioned earlier about atomic and molecular weights. These are relative quantities. Since most of the elements react chemically with oxygen to produce chemical compounds, the atomic or molecular weights were defined in terms of oxygen as the standard before 1960.

Chemists used to take the atomic weight of natural oxygen to be exactly 16 and express the atomic weights of all other elements in the unit which is one-sixteenth the atomic weight of naturally occurring oxygen. We have seen that when expressed in this unit, the atomic weight of hydrogen is 1·008.

It should be noted that molecules are the smallest entities which can exist in the free state. The molecules of the inert gases like helium, neon *etc.* are monatomic. Hence their atomic and molecular weights are equal. On the other hand the molecules of hydrogen, nitrogen and oxygen are diatomic; so their molecular weights are twice their atomic weights. In general, if there are n atoms in the molecule of an element, then its molecular weight is n times its atomic weight.

The molecular weights of the compounds are obtained from the atomic weights of their constituent atoms, if the numbers of these atoms in a molecule are known.

The absolute mass of an atom may be obtained from a knowledge about the gram-atomic weight of the corresponding element. As an example, the gram-atomic weight of hydrogen is known to be 1.008×10^{-3} kg. Since there are N_0 (Avogadro number) atoms in this amount of hydrogen the absolute mass of each hydrogen atom in this scale is

$$M_H = \frac{1.008 \times 10^{-3}}{6.02205 \times 10^{23}} = 1.674 \times 10^{-27} \text{ kg}$$

Similarly the absolute masses of the atoms of other elements may be calculated.

It should be noted that the above absolute masses of the atoms are derived from the atomic weights determined by the chemists. These are different from the atomic masses determined by the physicists. This will be discussed further in § 2.9.

2

CATHODE RAYS AND POSITIVE RAYS

2.1 Discharge of electricity through gases

From the discussion in the previous chapter, it is clear that all matter is made up by the combination of a large number of very small entities known as atoms. Further from Faraday's laws of electrolysis, it is found that electrical charge has a fundamental unit. No charge smaller than this elementary unit charge can exist. All charges are integral multiples of this basic unit. The laws of electrolysis also tell us that there is in intimate connection between this elementary charge and the atomic view of matter. Later, various experiments on the discharge of electricity through gases have confirmed this.

We know that a dry gas is ordinarily a bad conductor of electricity. However under certain circumstances, electric current is found to flow through a gas. For instance, at the time of lightning or thunder clap, very high electric current passes through the air for a short duration. The sparks produced between the electrodes of an induction coil are also examples of momentary electric currents through the air. Other examples are the neon-signs used for advertisements or the fluorescent lamps used for household illumination. In all these, electric current passes through a gas or a vapour.

The basic arrangement for conducting experiments on the passage of electric current through a gas at different pressures is shown in Fig. 2·1.

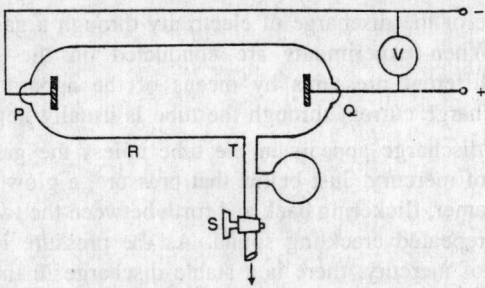

Fig. 2·1. Experimental arrangement for the study of electrical discharge in a gas.

11

PQ is a closed glass vessel which has two electrodes sealed into it at the two ends. There is a side tube T attached to PQ through which the air inside the vessel can be pumped out to reduce the pressure in it which can be regulated and measured by a manometer. It is possible to apply any desired potential difference between the electrodes which can be measured by means of an electrostatic voltmeter V. An electrometer can also be connected between the two electrodes externally to measure the current flowing through the gas between the electrodes.

When the gas pressure inside PQ is equal to atmospheric pressure, a very high electric field of the order of 30,000 volts or higher is necessary to produce an electric discharge through the gas. In this case, luminous electric spark is found to be produced between the electrodes from time to time attended with sharp crackling sound. This is known as *spark discharge* which proceeds along narrow and zig-zag paths. The sparks are momentary and repetitive. The different sparks proceed along different paths.

When the gas pressure is reduced, the potential difference required to produce the sparks (*sparking potential*) becomes less (see Fig. 2·2). However, when the pressure is very low, is becomes difficult to initiate a discharge through the gas.

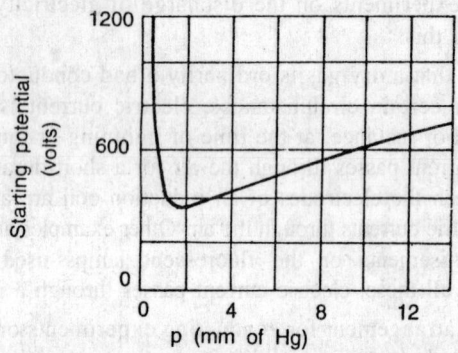

Fig. 2·2. Variation of starting potential for electrical discharge.

The nature of the discharge of electricity through a gas depends on its pressure. When experiments are conducted on the discharge of electricity at different pressures by means of the apparatus described above, the discharge current through the tube is usually kept constant.

No glow discharge appears in the tube unless the gas pressure is below 10 mm of mercury. Just below that pressure, a glow appears as a bright blue streamer, flickering back and forth between the two electrodes, attended with repeated crackling sound. As the pressure is reduced to about 0·5 mm of mercury, there is a stable discharge in the tube and a buzzing sound is heard. A pale purple luminous discharge, which broadens sidewise, spreads through the entire region between the cathode and the anode. This luminous column of light is known as the *positive column*. There is a narrow gap in this luminous column near the cathode known as *Faraday dark space*, named after Michael Faraday, who had first observed it. There is also a very narrow luminous region at the cathode itself which is known as the *negative glow*, so that the Faraday dark space lies between the positive column and negative glow.

As the gas pressure is reduced further, the negative glow is detached from the cathode and moves away from it towards the anode producing a new dark space between it and the cathode, known as the *Crooke's dark space*. Also under this condition, another luminous region known as the *cathode-glow*, appears at the cathode. During all these changes, the colours of the different luminous regions change.

At a pressure of about 0·3 mm of mercury, the positive column breaks up into a number of narrow bands of luminous regions, separated by intervening dark spaces. These are known as striations (see Fig. 2·3). As the pressure is reduced below about 0·1 mm of mercury, the Crooke's dark space widens and the negative glow becomes brighter and extends towards the anode. The positive column shrinks in size and ultimately disappears.

When the pressure is reduced below 0·01 mm of mercury, the whole tube is filled up by Crooke's dark space and no luminous discharge is observed. At this time, the electrical resistance between the anode and cathode becomes very high and it becomes difficult to maintain the discharge, unless the potential difference between the anode and cathode is very high. Under this condition, a faint fluorescent light is observed on the walls of the tube which may be bluish or greenish in colour, depending on the nature of the glass. Careful observations have shown that at this time a stream of high velocity charged particles is emitted from the cathode which falls on the anode as well as on the walls of the glass tube. As a result of this impact, the walls of the tube emit bluish or greenish *fluorescent radiation*. The streams of the particles emitted from the cathode are known as *cathode-rays*.

2.2 Cathode rays

J.J. Thomson, the famous British scientist, studied the properties of the cathode rays in detail. Following are some of the important properties of these rays.

(a) The cathode rays travel in straight lines. This can be demonstrated by an experimental arrangement shown in Fig. 2·4. A cross shaped obstacle O is placed facing the cathode in a discharge tube. When the discharge is started, a dark shadow S of the cross is formed on the wall of the tube opposite the cathode, the rest of the tube glowing with fluorescent light. Since the cathode rays, like light, cast shadows of obstacles put

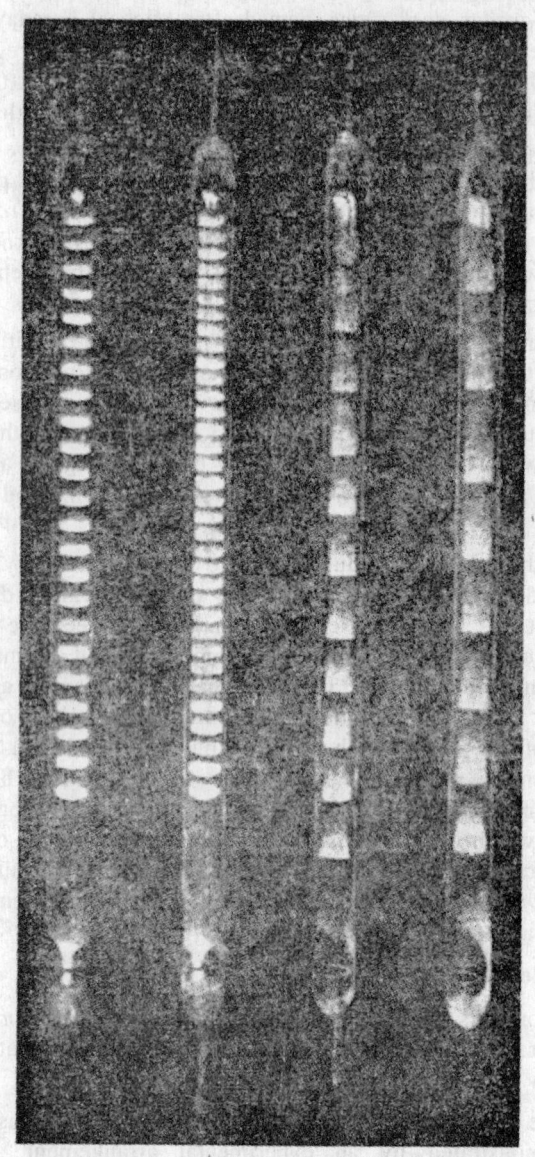

Fig. 2.3 Nature of electrical discharge in a gas at different pressures.

in their path, they must be moving in straight lines. This is further confirmed by the fact that if the cathode is concave in shape, then the cathode rays are focused at a point. If a small piece of platinum is kept at this point, it becomes heated by the impact of the cathode rays and ultimately begins to glow. This also proves that the cathode rays have very high kinetic energy. It is this kinetic energy which is converted into heat energy to heat up the platinum piece.

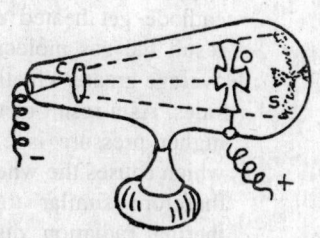

Fig. 2·4. Straight line paths of the cathode rays.

(b) The cathode rays are found to be deflected by electric and magnetic fields which prove that they are composed of electrically charged particles. From the directions of their deflections, it is found that these are *negatively charged particles*.

(c) Cathode rays produce fluorescence when they are incident on certain substances, like barium platino cyanide, zinc sulphide *etc.* As mentioned above, certain glasses also become fluorescent by their impact.

(d) When the cathode rays fall on some substance, they exert a pressure on it. This can be demonstrated by an experimental arrangement shown in Fig. 2·5. There is a small light wheel W

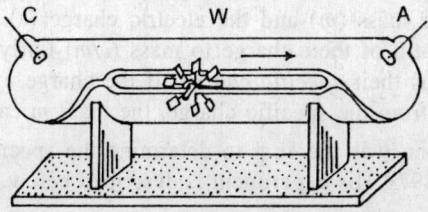

Fig. 2·5. Experiment to demonstrate pressure exerted by cathode rays.

placed inside a discharge tube which can roll on two parallel horizontal glass rails. A few mica vanes are attached to the wheel. When the cathode rays fall upon the mica vanes, the wheel is found to roll along the rails towards the anode A. According to Thomson, it is not merely the mechanical pressure due to the impact of the cathode rays on the mica vanes which is responsible for the rolling motion of the wheel, since this pressure cannot amount to much. Most probably due to the impact of the cathode rays, the sides of the vanes facing the cathode get heated up compared to the opposite sides. As a result the gas molecules near the heated sides acquire on the average greater kinetic energy than the molecules near the other sides. As a result of the impact of these molecules, a relatively higher pressure is exerted on the heated sides of the vanes which causes the wheel to roll towards the anode. This effect is therefore similar to the well-known *radiometer action* of thermal radiation, discovered by Crookes.

Thomson's experiments established that the cathode rays are made up of a beam of negatively charged particles. Due to the very large potential difference applied between the two electrodes in the discharge tube, they acquire very high kinetic energy in travelling from the cathode to the anode. As a result, they travel with very high velocity inside the discharge tube. It has been found that irrespective of the nature of the gas present in the tube, these negatively charged particles have always the same properties. This shows that they are universal constituents of all matter. These have since been named *electrons*, a name originally coined by Johnstone Stoney during his studies on electrolytic conduction.

Since all substances are made up of atoms and since the electrons are present in all substances we may conclude that they must be present within the atoms of all substances. Since the atoms are electrically neutral, the atoms must therefore be made up of two parts, one of which carries positive electricity while the other an equal amount of negative electricity.

2.3 Determination of the specific charge of the cathode rays

In order to determine the nature of the cathode rays, it is necessary to measure their mass (m) and the electric charge (e). It is possible to determine the ratio of their charge to mass (e/m) fairly accurately. This ratio is known as their *specific charge*. If the charge e is independently measured, then from the specific charge, the mass m can be determined.

J.J. Thomson was the first to determine the specific charge of the cathode rays (1897). His experiment is described below. In Fig. 2·6, T_1 is

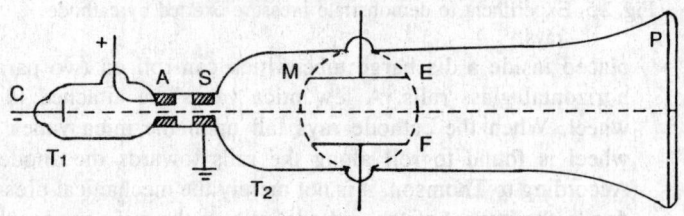

Fig. 2·6. Thomson's apparatus for measuring e/m for cathode rays.

a short discharge tube in which the cathode rays are produced by the application of a high potential difference between the cathode C and the anode A. The pressure within T_1 is kept very low. The cathode rays emitted from C are collimated in passing successively through the narrow holes in the anode A and in the slit S placed behind it. The collimated beam then enters the larger and wider tube T_2 in which also the pressure is very low. As it passes between the two parallel metal plates E, F between which an electric field is established perpendicular to the initial direction of the cathode rays the collimated beam suffers deflection from its original course opposite to the direction of the field. There is arrangement for applying a magnetic field also at the same place (M) in a direction perpendicular to the electric field and the direction of the cathode rays, due to which a magnetic force acts on the cathode rays in the same direction as the electric force. By properly choosing the direction of the magnetic field, the deflections due to the electric and magnetic fields can be made opposite to one another.

The cathode ray beam after emergence from the electric and magnetic field region falls upon a fluorescent screen P on which it produces a luminous spot. Due to the actions of the fields, the spot changes its position and its displacement from the undeflected position can be easily measured.

Let e and m be the charge and mass of the cathode rays and υ their velocity. Let X be the electric field and B the magnetic flux density through which the cathode rays pass. Then the force exerted on them by the electric field is

$$F_e = Xe \qquad (2\cdot3\text{--}1)$$

The magnetic force is

$$F_m = Be\upsilon \qquad (2\cdot3\text{--}2)$$

If F_e and F_m act in opposite directions, then the fields can be so adjusted that the deflection of the luminous spot on P produced by the cathode rays due to one field is exactly compensated by the action of the other. Under this condition the two forces are equal and opposite. It X_0 be the electric field in this case, then we can write

$$X_0 e = Be\upsilon \qquad (2\cdot3\text{--}3)$$

so that

$$\upsilon = X_0/B \qquad (2\cdot3\text{--}4)$$

Thus the velocity υ of the cathode rays can be determined from Eq. (2·3–4).

Knowing υ, the specific charge e/m can be determined by noting either the electric deflection or the magnetic deflection.

From Fig. (2·7a), we see that the cathode rays move along x before entering the electric field between the two plates E and F.

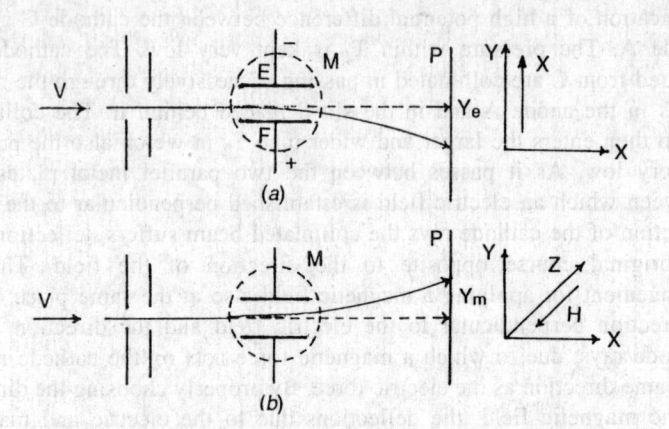

Fig. 2·7. Electric and magnetic deflections of cathode rays.

The action of the electric field is to deflect them along y. So we can write the equation of motion of the cathode rays along y as follows :

$$m\frac{d^2y}{dt^2} = Xe \qquad (2\cdot3-5)$$

As the cathode rays proceed along x towards the screen P, their velocity component dy/dt and displacement y increase which means that these are functions of x. So Eq. (2·3–5) can be rewritten as

$$\frac{d}{dx}\left(\frac{dy}{dt}\right)\frac{dx}{dt} = \frac{Xe}{m}$$

Since $dx/dt = \upsilon$ remains constant, we get after integration

$$\frac{dy}{dt} = \int \frac{Xe}{m\upsilon} dx = \frac{e}{m\upsilon} \int X dx \qquad (2\cdot3-6)$$

Since $dy/dt = 0$ at $x = 0$, the integration constant is zero in Eq. (2·3–6) which can be rewritten as

$$\frac{dy}{dx} \cdot \frac{dx}{dt} = \frac{e}{m\upsilon} \int X dx$$

Integrating once more, we get the deflection due to the electric field

$$y_e = \frac{e}{m\upsilon^2} \int_0^L dx \left(\int X dx \right) \qquad (2\cdot3-7)$$

It may be noted that in the above derivation, we have not considered the sign of the charge of the cathode rays.

We assume that the electric field extends from $x = 0$ to the position of the fluorescent screen where $x = L$.

Cathode Rays and Positive Rays

For a uniform electric field, X = constant between the plates and $X = 0$ outside. In this case the above two integrals can be evaluated easily. In practice, it is not possible to produce a strictly uniform electric field. The field fringes out at the edges of the plates E and F and falls to zero at some distance beyond the edges. So in this case, the field has to be *mapped* at different points in order to evaluate the integral.

The deflection produced by the magnetic field between the poles of the magnet M can similarly be found (see Fig. 2·7b). Using Eq. (2·3–2), we get the equation of motion of the cathode rays in this case as

$$m \frac{d^2 y}{dt^2} = Be\upsilon$$

which gives on integration

$$y_m = \frac{e}{m\upsilon} \int_0^L dx \left(\int B dx \right) \qquad (2\cdot3\text{–}8)$$

where y_m is the magnetic deflection. As before the integrals can be evaluated by mapping the magnetic field at different points along the path of the cathode rays.

Thus by measuring the electric *or* magnetic deflections, e/m for the cathode rays can be determined with the help of Eq. (2·3–7) or (2·3–8). Thomson showed that this had always the same value, irrespective of the nature of the residual gas in the discharge tube.

2.4 Dunnington's experiment

Thomson's experiment did not yield a very accurate value of e/m for the electrons. Later many improved methods have been devised to determine e/m for the electrons more accurately. We shall describe below the experiment of Dunnington.

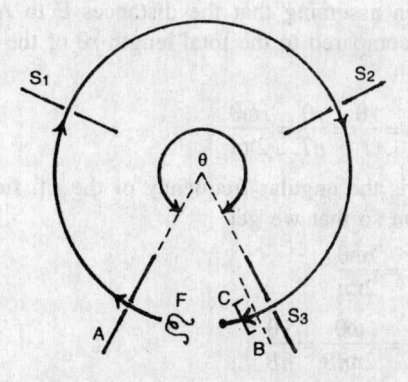

Fig. 2·8. Dunnington's experiment.

Dunnington's experimental arrangement is shown in Fig. 2·8. F is a metal filament which when heated emits electrons. There is a metal plate A with a narrow slit facing the filament F. A radio frequency alternating electric field $E = E_0 \sin \omega t$ is applied between F and A so that the electrons emitted by F are attracted towards A during the positive half cycle of the r.f. field and pass through the slit in A to emerge on the other side of it. The pressure inside the apparatus is kept very low. A steady and uniform magnetic field of flux density B is applied in a direction perpendicular to the plane of the diagram which causes the narrow beam of electrons emerging through A with a definite velocity υ to move along a circular trajectory defined by the slits A, S_1, S_2, S_3. This trajectory has a radius of curvature r defined by the relation

$$Be\upsilon = \frac{m\upsilon^2}{r} \qquad (2.4\text{--}1)$$

The electrons describing the circular trajectory reach the metallic collector C after passing through a grid B made of metal wire. Electrons having other velocities emerging through A will not be able to reach B, because they will follow trajectories of different radii of curvature.

The same r.f. field which acts between F and A is also applied between the grid B and the collector. If this electric field is such that when the electrons reach C they are not repelled by it, then only they will be collected by C and the current measuring device connected to C will register a current. It is so arranged that at every instant of time F and C are at the same potential; likewise A and B have always the same potential. Then if the time t taken by the electrons to travel from A to B is exactly equal to an integral multiple of the period T of the r.f. field, *i.e.*, if $t = nT$ where n is an integer, then the electrons will experience a repulsive field between B and C, so that they will not be able to reach C and the current recorded by the meter will show a minimum. If θ be the angle subtended by the electron trajectory between A and B at the centre of the circle, then assuming that the distances F to A and B to C are negligibly small compared to the total length $r\theta$ of the electron path, we can write

$$\upsilon = \frac{r\theta}{t} = \frac{r\theta}{nT} = \frac{r\omega\theta}{2\pi n}$$

where $\omega = 2\pi/T$ is the angular frequency of the r.f. field. But from Eq. (2·4–1), $\upsilon = Ber/m$ so that we get

$$\frac{Ber}{m} = \frac{r\omega\theta}{2\pi n}$$

or, $$\frac{e}{m} = \frac{\omega\theta}{2\pi nB} = \frac{\nu\theta}{nB} \qquad (2.4\text{--}2)$$

were $\nu = \omega/2\pi$ is the frequency of the r.f. field.

At the time of the experiment, the magnetic field is gradually changed till the current recorded becomes minimum. Then from the corresponding value of B, e/m can be calculated knowing v, n and θ. Usually n is not known. So it is necessary to note two successive minima of the current for n and $(n + 1)$ for which the magnetic flux densities are B_1 and B_2. Then from Eq. (2·4–2) one gets two equations from which n can be eliminated and e/m calculated.

Dunnington obtained a value of $e/m = 1.7597 \times 10^{11}$ coulombs per kilogram. The currently accepted value is

$$\frac{e}{m} = 1.7589 \times 10^{11} \text{ C/kg}$$

In the next section, we shall describe a method for the accurate determination of the electronic charge e. From the value of e so determined and that of e/m given above, the mass of the electron is found to be

$$m = 9.1084 \times 10^{-31} \text{ kg}$$

It may be noted that according to the special theory of relativity the mass of a particle increases with its velocity (see Ch. XV). The increase becomes appreciable only when the velocity becomes comparable to the velocity of light c in vacuum. For this reason, the value of the specific charge of the electrons does not appear to be a constant when its velocity is very high. This effect becomes particularly important when the experiments are done with the high energy β-rays which are electrons emitted by radioactive nuclei (see Vol. II) *. However in the experiments described above, the relativistic variation of mass may be entirely neglected and the value of e/m given above may be taken to be that for the electron at rest.

2.5 Determination of the charge of an electron, Millikan's oil drop experiment

The charge of an electron was first determined accurately by the famous American physicist R.A. Millikan (1909) who had improved upon the method devised by Thomson and Wilson for this purpose.

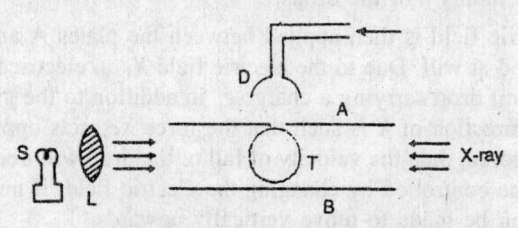

Fig. 2·9. Millikan's oil drop experiment.

---

* Whenever there is a reference to Vol. II, it will mean Volume II of this book.

Millikan's apparatus is shown in Fig. 2·9. A and B are two parallel horizontal circular metal plates between which an electric field upto several thousand volts per centimetre can be applied. At the centre of the upper plate, a few small holes are drilled, through which very small oil drops produced by an atomiser can be sprayed into the region between the plates. As the drops come out of the atomiser, they are charged by friction. If no electric field is applied between A and B, then the oil drops fall downwards by the action of gravity. During this fall, their motion is opposed by the viscosity of air.

It is known from hydrodynamics that when a spherical object of radius r falls through a fluid having the coefficient of viscosity η, then it attains a constant terminal velocity υ after falling through some distance. According to Stoke's law, the resistive force due to viscosity on the object is $6\pi \eta r \upsilon$ which is obviously equal to the force of gravity under this condition. If the density of oil is ρ and the density of air σ, then the force of gravity on the oil drop is

$$w = \frac{4}{3} \pi r^3 (\rho - \sigma) g \qquad (2\cdot5\text{--}1)$$

Hence from Stoke's law, we get

$$w = \frac{4}{3} \pi r^3 (\rho - \sigma) g = 6\pi \eta r \upsilon \qquad (2\cdot5\text{--}2)$$

So we get $\quad r^2 = \dfrac{9 \eta \upsilon}{2(\rho - \sigma) g} \qquad (2\cdot5\text{--}3)$

Hence $\quad w = \dfrac{4\pi}{3} \left(\dfrac{9\eta\upsilon}{2} \right)^{3/2} \dfrac{1}{\sqrt{(\rho - \sigma) g}} \qquad (2\cdot5\text{--}4)$

The space between the plates A and B is illuminated with the help of an electric bulb S and a lens L. The inside surfaces of the plates are painted black so that there is no reflection of light from them. The oil drops are observed through a long focus microscope T. A scale is attached to the eyepiece of the microscope with the help of which the velocities of the oil drops can be measured. Knowing the velocity, it is possible to determine the radius r of the drops.

An electric field is then applied between the plates A and B which can be changed at will. Due to the electric field X, an electrical force Xe_n acts upon an oil drop carrying a charge e_n in addition to the gravitational force. If the direction of X is such that the force Xe_n acts opposite to the gravitational force, then the velocity of fall of the drop will decrease. This velocity can be controlled by changing the electric field; if necessary the drop may even be made to move vertically upwards.

If the terminal velocity of an upward moving drop due to the application of the electric field be υ_1, then we get

$$Xe_n - w = 6\pi\eta r\upsilon_1 \qquad (2\cdot5\text{--}5)$$

From Eqs. (2·5–2) and (2·5–5) we get

$$\frac{Xe_n}{w} = \frac{\upsilon + \upsilon_1}{\upsilon} \qquad (2\cdot5\text{--}6)$$

Since w can be found from Eq. (2·5–4), we can determine e_n from Eq. (2·5–6).

Millikan then exposed the air between A and B to a beam of x-rays due to which some of the air molecules were ionised. Due to thermal collisions between these ionised air molecules and the oil drops, the charge on the oil drops may change. Suppose this happens to an oil drop of charge e_n which changes to e'_n. Due to this, the electrical force acting on it changes and hence its velocity also changes abruptly. If the terminal velocity of the drop now becomes υ_2, then we can write

$$Xe'_n - w = 6\pi\eta r\upsilon_2 \qquad (2\cdot5\text{--}7)$$

Subtracting Eq. (2·5–7) from Eq. (2·5–5), we have

$$X(e'_n - e_n) = 6\pi\eta r(\upsilon_2 - \upsilon_1)$$

So using Eq. (2·5–2), we get

$$\frac{X(e'_n - e_n)}{w} = \frac{\upsilon_2 - \upsilon_1}{\upsilon}$$

Using Eq. (2·5–4) we then get

$$e'_n - e_n = \frac{w}{X}\frac{\upsilon_2 - \upsilon_1}{\upsilon} = \frac{4\pi}{3X}\left(\frac{9}{2}\eta\right)^{3/2}\left\{\frac{\upsilon}{(\rho - \sigma)g}\right\}^{1/2}(\upsilon_2 - \upsilon_1) \qquad (2\cdot5\text{--}8)$$

In Eq. (2·5–8), the quantity $(\upsilon_2 - \upsilon_1)$ depends on the change of the charge on the oil drop. Millikan, by observations on a single drop found that its velocity changed repeatedly in the electric field which indicated repeated changes of its electrical charge. He could determine these changes with the help of Eq. (2·5–8) and found that these were always integral multiples of a minimum value. This minimum value of the charge was taken by him to be the electronic charge e.

Millikan also observed that for very small drops, the measured electronic charge came out to be somewhat higher. The reason for this is that in these cases, Stoke's law (Eq. (2·5–2) was not wholly correct. For very small drops, the impact of the air molecules on the drops due to their thermal motion causes the drops to fall downwards along zig-zag paths, much like the Brownian motion of extremely small sized particles suspended in a liquid. In this case the viscous resistance is decreased. Millikan assumed the viscous resistance to depend on the ratio (λ/r) between the mean free path λ of the air molecules and the radius of the drop. Since $\lambda \propto 1/p$ where p is the pressure of the air, we can modify Eq. (2·5–2) and write

$$w = \frac{4\pi}{3} r^3 (\rho - \sigma) g = \frac{6\pi \eta r \upsilon}{1 + b/pr} \qquad (2\cdot5\text{–}9)$$

where b is a constant. Assuming $b/pr \ll 1$, we then get the following modified equation in place of Eq. (2·5–8):

$$e'_n - e_n = \frac{4\pi}{3X} \left\{ \frac{9}{2} \frac{\eta}{1+b/pr} \right\}^{3/2} \times \left\{ \frac{\upsilon}{(\rho-\sigma)g} \right\}^{1/2} (\upsilon_2 - \upsilon_1) \qquad (2\cdot5\text{–}10)$$

From Eq. (2·5–8) we get the apparent value e' of the electronic charge. Then from Eqs. (2·5–8) and (2·5–10), we have

$$\left(\frac{e'}{e} \right)^{2/3} = 1 + \frac{b}{pr} \qquad (2\cdot5\text{–}11)$$

where e is the true value of the electronic charge. If e' is determined for the different drops and a graph is plotted with $(e')^{2/3}$ along the ordinate and $1/pr$ along the abscissa, then the graph should be a straight line. The intercept of this graph with the ordinate gives the true electronic charge e while its slope gives the constant b.

Millikan obtained a value $e = 1\cdot591 \times 10^{-19}$ coulomb from his measurements. Later experiments have yielded a somewhat higher value for e. One such experiment was by Laby and Hopper (1940) who used vertical plates to apply an electric field perpendicular to gravity. They measured the velocity of the drops by photographic method, taking the photographs of the drop under observation at every 1/25 second interval. Thus a continuous record of the position of the drop could be obtained from which its velocity could be determined. The value of e they obtained is close to the currently accepted value which is

$$e = 1\cdot60219 \times 10^{-19} \text{ coulomb}$$

The difference between this and Millikan's value for e is due to a number of errors in Millikan's experiment. Apart from the correction to Stoke's law for small drops mentioned above, these included errors due to:

(a) limitation in the applicability of Stoke's law for the drops falling in a medium of finite extent, Stoke's law being strictly valid for a medium of infinite extent;

(b) distortion of the electric field due to the presence of a few holes in the upper plate to introduce the oil drops;

(c) uncertainty in the value of η which was the chief source of error in Millikan's experiment.

If a more accurate value of η determined in later years is used, then Millikan's experimental data yields a value for e which is close to the currently accepted value.

It may be noted that fairly accurate value of e can also be deduced from a knowledge of F (Faraday) and the Avogadro number N_0 determined by methods other than the electrolytic method, e.g., the x-ray diffraction method. It is the comparison of the value of e so determined with the value found by Millikan's method that had pointed to the necessity of very accurate determination of the viscosity η of air which was undertaken by Bearden and others and which finally resolved the discrepency in the values of e obtained by the two methods.

2.6 Positive rays

We have seen earlier that both positive and negative ions are produced in the experiments on the discharge of electricity through gases under low pressure. Of these, the negative ions are the electrons which are the universal constituents of all substances. We have already discussed about their properties in the previous sections.

A German physicist Goldstein was the first to investigate the nature of the positive ions. He found that if there was a small hole (canal) in the cathode of the low pressure discharge tube then a luminous ray was observed behind the cathode (1886). He gave these the name 'canal rays'.

The canal rays are deflected by electric and magnetic fields which prove that, like cathode rays, they are made up of charged particles. From the directions of their deflections it is found that these particles carry positive electricity. For this reason, they were subsequently named 'positive rays'.

The origin of the positive rays can be understood easily. The positively charged ions produced in a discharge tube are attracted by the cathode and move towards it, away from the anode. Some of them pass through the hole in the cathode and suffer collisions with the low pressure residual gas molecules behind the cathode. As a result the molecules are excited to higher energy states and emit visible light during de-excitation. For this reason the path of the positive ions behind the cathode becomes luminous. These positive ion beams are known as the positive rays.

2.7 Thomson's parabola method of analyis of the positive rays

J.J. Thomson was the first to determine the specific charge of the positive rays. In Fig. 2·10 the apparatus used by Thomson is shown. B is a large hollow glass bulb which is evacuated by means of vacuum pump. If the gas pressure within the bulb is too low, the intensity of the positive rays becomes so weak that it become difficult to observe them. For this reason, a very narrow capillary tube is attached to one end of the bulb near the anode A through which the experimental gas is introduced at a very slow rate into the apparatus. This gas is ionised by bombardment

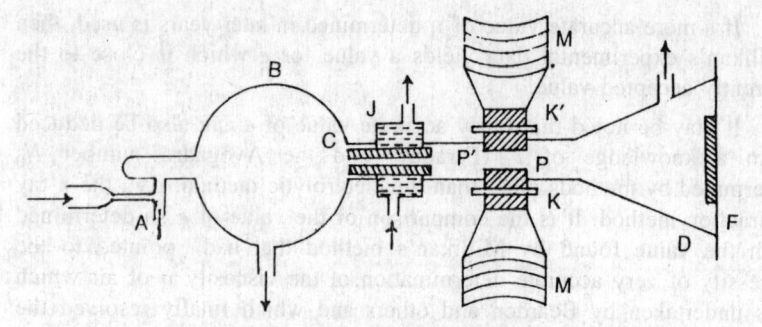

Fig. 2·10. Thomson's apparatus for measuring *e/m* for positive rays.

with a beam of electrons so that copious positive ions are produced near the anode.

At the other end of the bulb a glass tube is attached within which a cylindrical copper cathode C is situated. The end of the cathode facing the anode is made of aluminium to reduce its damage due to sputtering by continuous positive ion bombardment. There is a narrow hole, 0.1 mm in diameter, along the axis of the cathode from one end to the other. A fraction of the positive ions falling upon the cathode passes through this hole and forms a narrow collimated beam of positive rays on the other side of the cathode. Since the cathode gets heated by positive ion bombardment, there is an arrangement for cooling it by passing cold water through a glass jacket J surrounding it.

The collimated beam of positive rays is subjected to an electric and a magnetic field which are parallel and act at the same place. The magnetic field is produced by an electromagnet M. There are two parallel soft iron pieces P and P′ attached to the two poles of the magnet and insulated from the body of the latter by means of two mica sheets K, K′. A large potential difference is maintained between P and P′ to produce an electric field in the same direction as the magnetic field, also produced between P and P′ by passing electric current through the coils of the electromagnet. Thus parallel electric and magnetic fields are produced at the same place.

After emerging from the electric and magnetic fields, the positive ray beam enters a conical glass tube D at the far end of which a photographic plate F is placed with its plane perpendicular to the direction of the undeflected beam. The gas pressure within D can be reduced to about 10^{-4} mm of Hg with the help of charcoal cooled to liquid air temperature which s much lower than the pressure within the glass bulb B (10^{-3} mm of Hg. It is possible to maintain the pressure difference between B and D because of the presence of the cathode C with a long narrow hole separating them. Only a minute fraction of the gas within B leaks into D through the long narrow hole in C.

Cathode Rays and Positive Rays

The reason for maintaining very low pressure in D is to minimise loss of energy of the positive ions by collisions with the gas molecules within it during their passage to the photographic plate F. They would otherwise be lost to the beam due to scattering and loss of energy by such collisions and it would be difficult to measure the deflection of the beam by the electric and magnetic fields.

In the absence of electric and magnetic fields, the positive ray beam is not deflected and hits the photographic plate F near its middle, producing a black spot. When the fields are applied, the beam is deflected from its course. The deflection can be calculated in the following manner.

Let ε, M and v be respectively the charge, mass and velocity of the positive ions. Since the electric and magnetic fields both act along the same direction (y-axis), the deflections due to the two fields are mutually perpendicular. We have already calculated these deflections in the case of cathode rays (see § 2·3). From Eq. (2·3–7), we get the electric deflection

$$y = \frac{\varepsilon}{Mv^2} \int_0^L dx \left(\int X dx \right) \qquad (2\cdot7-1)$$

From Eq. (2·3–8) the magnetic deflection comes out to be

$$z = \frac{\varepsilon}{Mv} \int_0^L dx \left(\int B dx \right) \qquad (2\cdot7-2)$$

The integrals in Eqs. (2·7–1) and (2·7–2) depend on the electric field X and the magnetic flux density B and on their spatial variation. So for a given arrangement of the fields, the integrals may be taken to be constant and we can write

$$y = K_E \varepsilon / M v^2 \qquad (2\cdot7-3)$$

$$z = K_B \varepsilon / M v \qquad (2\cdot7-4)$$

where K_E and E_B are two constants, From Eqs. (2·7–3) and (2·7–4) we get

$$\frac{z^2}{y} = G \cdot \frac{\varepsilon}{M} \qquad (2\cdot7-5)$$

where G is a constant depending on the values of X and B and on the geometry of the apparatus. Note that the velocity of the ion v has disappeared from Eq. (2·7–5). So for given electric and magnetic fields, Eq. (2·7–5) gives us for ions of a particular type (ε/M = constant)

$$\frac{z^2}{y} = \text{constant} \qquad (2\cdot7-6)$$

This is an equation for a parabola. Thus ions of a given ε/M but travelling with different velocities are focused at different points on a parabola in the plane of the photographic plate (y-z plane). A few such parabolas are shown in Fig. 2·11. The different parabolas are produced by positive ions of different ε/M. The different points on a particular parabola are produced by ions of the same ε/M having different velocities.

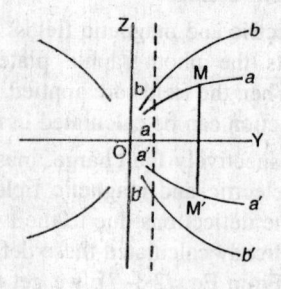

Fig. 2·11. Positive ray parabolas.

For two different types of ions having the same ε but different M, the parabola for the heavier ion lies closer to the y-axis (see Eq. (2·7–3). Again for ions of the same mass but carrying different charges, the parabola for those with a higher charge will be above the one with a lower charge. For example the parabola for the O^{++} ions lies above the parabola for the O^+ ions (see Fig. 2·12). Parabolas for the ions of the same element with higher charges are usually fainter.

Hydrogen being the lightest element, its parabola will be farthest from the y-axis.

If the field X and the flux density B are known, ε/M can be calculated from Eq. (2·7–5) by measuring y and z for any point on the parabola. To do this it is necessary to identify the y and z axes correctly. For this purpose, the photographic plate is exposed to the positive ray beam for some time with a given combination of the electric and magnetic fields. The fields are then successively reversed to get three other identical parabolas in three quadrants which are mirror images of the first one by the y and z axes. As seen in Fig. 2·11, distance between two corresponding points on the first parabola and its mirror image by the y-axis is $MM' = 2z$ which can be measured and ε/M calculated using Eq. (2·7–5).

Thomson's preliminary measurements, were made using hydrogen gas. He found for hydrogen $\varepsilon/M_H = 9.571 \times 10^7$ C/kg. This agrees well with electrolytic determination. Since the e.c.e of hydrogen is 1.0446×10^{-8} kg/C, the value of ε/M_H comes out to be 9.573×10^7 kg/C. So it can be concluded that the positive rays produced with hydrogen gas are nothing but singly charged positive hydrogen ions, i.e., hydrogen atoms from which an electron has been removed. Since in the hydrogen

Cathode Rays and Positive Rays

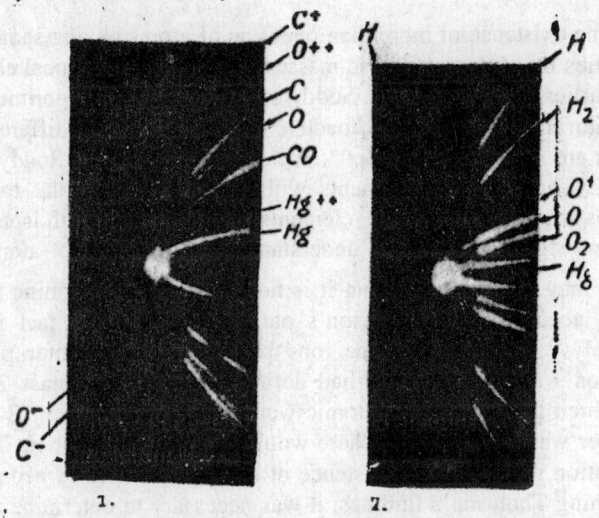

Fig. 2·12. Photographs of positive ray parabolas.

atom, there is only one electron revolving round the central positively charged nucleus (see Ch. IV), the removal of the electron leaves behind the positively charged hydrogen nucleus which is an elementary particle known as the *proton*. Like electrons, protons are also universal constituents of all atoms. It carries one electronic unit of positive electricity. Its mass is 1836·13 times the mass of the electrons.

2.8 Isotopes

While experimenting with neon gas, Thomson observed two close lying parabolas (1912). One of these was much brighter than the other. The atomic weight of the ions producing the brighter parabola was found to be 20 while that producing the fainter parabola was 22. The chemical atomic weight of neon, on the other hand, is found to be 20·2. Whatever the source of the neon gas might be and however pure it might be, there were always these two parabolas and their relative intensities were always the same. From this it was clear that the ions producing the fainter parabola could not be due to any impurity present in the experimental neon gas. So Thomson came to the conclusion that there are two types of atoms in neon gas, one of which has the atomic weight 20 while the other has the atomic weight 22. The two types of neon have the same chemical properties, only their atomic weights are different. If the abundance of the neon of atomic weight 20 in natural neon gas is 90% while that of the neon of atomic weight 22 is 10%, then the mean atomic weight of neon would be

$$M = 20 \times 0.9 + 22 \times 0.1 = 20.2$$

Thus it is possible to explain the observed atomic weight of naturally occurring neon on the above basis.

The existence of more than one type of atoms having same chemical properties but different atomic masses for the same chemical element had been earlier discovered by F. Soddy of England while experimenting with the naturally occurring radioactive elements. Such different atomic species are known, as *isotopes* *. We shall discuss about Soddy's work in Vol. II. Thomson's experiments with neon confirmed the existence of isotopes amongst the stable elements. The discovery of isotopes is of profound significance in the understanding of structure of atomic nuclei.

It may be mentioned that it is not possible to determine the atomic masses accurately by Thomson's parabola method. In fact it was not definitely established that the ions producing the brighter parabola in Thomson's neon experiment had definitely the atomic mass 20 and not 20·2 which is the chemical atomic weight of ordinary neon. Evidently, if the latter was the case, then there would be some doubt about Thomson's assumption regarding the existence of the two isotopes of neon. Thus for confirming Thomson's findings, it was necessary to determine the atomic masses more accurately than was possible by the parabola method. It was left to F.W. Aston, a British physicist, to determine the atomic masses very accurately by an instrument known as the mass spectrograph devised by him. We shall discuss about this and other types of mass spectrographs in Vol. II. Aston's measurements with neon confirmed Thomson's discovery of the isotopes beyond all doubts.

Subsequent researches have established the existence of more than one stable isotope of many of the elements in the periodic table. Thus hydrogen has been found to have two stable isotopes ^1H and ^2H. Oxygen has three stable isotopes ^{16}O, ^{17}O, and ^{18}O; chlorine has two stable isotopes ^{35}Cl and ^{37}Cl. Tin has the largest number of stable isotopes. The number is *ten*.

It may be noted that the superscript written to the left of the symbol for the isotope is the *mass number* which is the integer nearest to the atomic mass of the isotope. It is usually designated by the symbol A. This differs slightly from the actual atomic mass of the isotope. We shall revert to this subject in Vol. II.

2.9 Unit of the atomic mass

We have seen earlier that prior to 1960 the unit of atomic weight used by the chemists was one-sixteenth of the atomic weight of naturally occurring oxygen which was taken to be exactly 16 (§ 1·5).

After the discovery of isotopes, a different scale for the atomic masses was introduced by the physicists. They took the atomic mass of the ^{16}O isotope to be exactly 16 so that the unit of the atomic weight used by the physicists, known as the atomic mass unit (amu) was *one-sixteenth*

---

* The word *isos* in Greek means 'same' while *topos* means 'place'.

of the atomic weight of ^{16}O isotope. The ratio between the physical and chemical units was thus 1 : 1·00027.

A new unified scale of atomic weight was proposed by the German physicist Mattauch based on the atomic weight of ^{12}C isotope being taken as exactly 12. The unit of atomic weight or atomic mass in this scale is thus equal to *one-twelfth* of the atomic weight of ^{12}C isotope. The proposal was accepted by the International Union of Pure and Applied Physics (IUPAP) in 1960.

This unified atomic mass scale based on ^{12}C isotopic mass being exactly 12 is now in use by both physicists and chemists. In this scale, the *unified atomic mass unit* (u) is

$$1\,u = 1\cdot 66057 \times 10^{-27}\ \text{kilogram}$$

The relation between the new unit of atomic mass and the older unit used by the physicists based on the ^{16}O scale is

$$1\,u = 1\cdot 0003179\ \text{amu}$$

Problems

1. A stream of a cathode rays falls normally on a mica foil with a velocity 5×10^7 m/s and is totally absorbed by the foil. If the current is 10^{-4} A/m^2, calculate the pressure on the foil. $(2\cdot 85 \times 10^{-8}\ \text{N/m}^2)$

2. A beam of electrons of velocity υ enters into an electric field X between two parallel metal plates at right angles to the field direction. If the beam travels a distance l in the field show that its deflection due to the electric field will be $y_1 = Xel^2/2m\upsilon^2$ where e and m are the charge and mass of the electron. Neglect the effect of the fringing field at the edges of the plates.

3. In Prob. 2, if the angle between the initial and emergent directions of the electron beam is θ, prove that $\tan \theta = 2y_1/l$.

    If the electrons after emerging from the fields travels through a distance L outside the field, prove that the additional deflection of the beam is $y_2 = XelL/m\upsilon^2$. Calculate e/m of the electrons from the total deflection.

4. If the atomic weight of hydrogen is taken to be 1, calculate the mass of the hydrogen atom in kilogram, assuming the Avogadro number to be $6\cdot 025 \times 10^{23}$ per mole. $(1\cdot 66 \times 10^{-27}\ \text{kg})$

5. In an oil drop experiment, the charges on a number of oil drops were measured as (in units of 10^{-19} coulomb) 8·03, 11·2, 14·49, 6·36, 17·71, 9·72, 12·8, 3·16, 6·4 and 4·77 respectively. Determine the charge of an electron from the above data.

6. A charged oil drop falls vertically in air with a constant velocity of 2×10^{-4} m/s. Calculate its radius and mass from the following data :

$\rho = 800 \text{ kg/m}^3$; $\sigma = 1.13 \text{ kg/m}^3$; $\eta = 1.81 \times 10^{-5}$ N.s/m^2;
$e = 3.2 \times 10^{-19}$ C. (See Eq. 2.5–3).   (1.44×10^{-6} m ; 10^{-14} kg)

7. Calculate the atomic mass of ^{1}H, ^{4}He and ^{16}O in ^{12}C atomic mass scale. The masses of ^{1}H, ^{4}He and ^{12}C in ^{16}O scale are 1.00814, 4.003860 and 12.003807 amu respectively.
(1.0078203, 4.0025902, 15.994926)

3

ORIGIN OF THE QUANTUM THEORY ; PLANCK'S THEORY OF BLACKBODY RADIATION

3.1 Blackbody radiation ; Wien's displacement law

Quantum theory was first introduced in physics by Max Planck of Germany in the year 1900 while trying to explain the observed energy distribution of the electromagnetic radiation emitted by a blackbody. It is well-known that when a body is heated it emits electromagnetic radiation. Thus when a piece of iron is heated to a few hundred degrees centigrade, it emits e.m. radiation which is predominantly in the infra-red region. When the temperature is raised further to about 1000°C it begins to glow with reddish colour which shows that the radiation emitted by it has wavelengths shorter than in the previous case. Further heating makes it white-hot so that the radiation emitted is shifted towards blue which has still shorter wavelength in the visible spectrum. The above facts show that the nature of the radiation depends on the temperature of the emitter.

Apart from emitting e.m. radiation, a heated body also absorbs a part of the radiation falling on it. G.R. Kirchhoff in 1889 proposed the theorem that the ratio of the emissive power to the absorptive power of a body is a constant depending only on the temperature of the body and independent of its nature. By emissive power is meant the amount of radiant energy emitted by a body per unit area of its surface per unit time. On the other hand, the absorptive power measures the fraction absorbed of the radiant energy falling on the body. If a body absorbs all the radiant energy falling on it, then its absorptive power is unity. Such a body is called a blackbody. So according to Kirchhoff's theorem stated above, the emissive power of a black body is a function of temperature only.

An ideal blackbody can be realised in practice by heating to any desired temperature a hollow enclosure (cavity) C, with a very small orifice O as shown in Fig. 3·1. Its inner surface is coated with lampblack. Radiation entering the cavity through O is incident on its blackened inner surface and is partly absorbed and partly reflected. The reflected component is incident at another point on the inner surface of C and is again partly absorbed and partly reflected. The process goes on repeatedly as there is very little probability of any part of the radiation coming out of the cavity through the orifice again because of its smallness. At each reflection more than 98% of the incident beam is absorbed. Thus the cavity may be taken to have unit absorptive power so that it behaves like a blackbody.

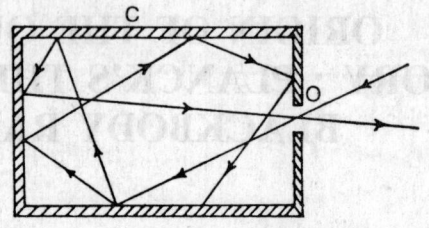

Fig. 3·1. A cavity as a black body.

The inner walls of the heated cavity also emit radiation, a part of which can come out through the orifice. This radiation obviously has the characteristics of blackbody radiation. Its spectrum can be analysed by an infra-red spectrometer, using a bolometer as a detector. Thus the emissive power of the blackbody for different wavelengths can be determined.

The variation of the intensity of the emitted radiation E_λ as a function of the wavelength is shown graphically in Fig. 3·2 for different temperatures of the blackbody. At a given temperature as λ increases

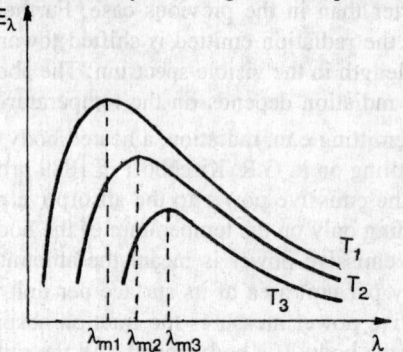

Fig. 3·2. Energy distribution of black body radiation ($T_1 > T_2 > T_3$).

E_λ at first increases with increasing λ at very short wavelengths, attains a maximum at some wavelength λ_m and then decreases again with further increase of λ. The value of λ_m depends only on the temperature T of the blackbody and decreases with increasing temperature. It is independent of the nature of the emitting body. The E_λ vs λ curves have the same nature at different temperatures. However at higher temperatures the intensity is higher at all wavelengths

Origin of the Quantum Theory ; Planck's Theory

The shift in the peak of the intensity distribution curves as the temperature is changed is found to obey the following empirical relationship, known as Wien's *displacement law* (1893) :

$$\lambda_m T = \text{constant} \tag{3.1-1}$$

The total power E radiated per unit area of a blackbody is found to depend on its absolute temperature. An empirical law connecting E and T was proposed by J. Stefan (1879) which was later deduced by L. Boltzmann from thermodynamics. This is known as *Stefan-Boltzmann law* which can be expressed mathematically as follows :

$$E = \sigma T^4 \tag{3.1-2}$$

where $\sigma = 5.67 \times 10^{-8}$ joules/sec/m^2 per deg^4 is known as Stefan's constant. E is obtained by integrating E_λ over all wavelengths :

$$E = E(T) = \int_0^\infty E_\lambda \, d\lambda \tag{3.1-3}$$

W. Wien, from thermodynamical considerations, proposed an empirical relationship between E_λ and λ for a given temperature T which is of the form

$$E_\lambda(T) \, d\lambda = \frac{A}{\lambda^5} f(\lambda T) \, d\lambda \tag{3.1-4}$$

where A is a constant and $f(\lambda T)$ is a function of the product λT.

Stefan-Boltzmann law and Wien's displacement law follow simply from Wien's distribution law (Eq. 3·1–4).

From Eq. (3·1–4), we get by writing $x = \lambda T$

$$E = \int_0^\infty E_\lambda \, d\lambda = A \int_0^\infty \frac{f(\lambda T)}{\lambda^5} \, d\lambda$$

$$= A T^4 \int_0^\infty \frac{f(\lambda T)}{(\lambda T)^5} \, d(\lambda T)$$

$$= A T^4 \int_0^\infty \frac{f(x)}{x^5} \, dx$$

The definite integral on the r.h.s. is a constant. Hence we get $E = \sigma T^4$ where σ is a constant. This is Stefan-Boltzmann law (Eq 3·1–2).

Again by differentiating Eq. (3·1–4) we get

$$\frac{dE_\lambda}{d\lambda} = -\frac{5A}{\lambda^6} f(\lambda T) + \frac{A T}{\lambda^5} f'(\lambda T) = 0$$

for maximum, for which $\lambda = \lambda_m$ (say). So we get

$$Tf'(\lambda_m T) - \frac{5f(\lambda_m T)}{\lambda_m} = 0$$

or, $\qquad x_m f'(x_m) - 5f(x_m) = 0$

where $x_m = \lambda_m T$. The above equation in a single variable x_m can have only one solution. Hence

$$x_m = \lambda_m T = \text{constant}$$

This is Wien's displacement law (Eq. 3·1–1).

3·2 Rayleigh-Jeans law

The functional form $f(\lambda T)$ cannot be deduced from thermodynamics. It is necessary to assume a suitable model for the radiating system in order to determine it.

Wien himself had proposed an expression for the functional form of $f(\lambda T)$ on the basis of some arbitrary assumptions regarding the mechanism of emission and absorption of radiation. Wien's law for the *energy density* u_λ of the blackbody radiation, based on these assumptions can be written as

$$u_\lambda \, d\lambda = \frac{a}{\lambda^5} \exp(-b/\lambda T) \, d\lambda \qquad (3\cdot2-1)$$

It can be shown that u_λ is simply related to E_λ. The constants a and b in Eq. (3·2–1) were chosen arbitrarily so as to fit the experimental energy distribution curves. Since the theory was not based on any plausible physical model, it proved to be quite unsatisfactory.

The simplest model is to regard the radiating system as composed of a collection of charged linear harmonic oscillators which according to the electromagnetic theory of light, radiate electromagnetic waves because of their accelerated motion. They can also absorb electromagnetic radiation. If we consider a cavity full of such radiation, then the atomic oscillators in the walls enclosing the cavity will continually exchange energy with the radiation in the cavity. Ultimately an equilibrium condition will be established when the energy density of the e.m. radiation will assume an equilibrium value determined by the temperature T of the cavity walls.

When the temperature of the walls is increased, the amplitudes of the existing modes of vibration of the oscillators are increased. Also new modes are excited for which the frequencies are higher.

Thus radiant energy density in the cavity is increased until a new equilibrium is established.

Now according to classical mechanics, the total energy of a linear harmonic oscillator of mass m oscillating along x is given by

Origin of the Quantum Theory ; Planck's Theory

$$E = E_k + V = \frac{p^2}{2m} + \frac{1}{2}\beta x^2 \tag{3.2-2}$$

where p is the momentum. β is a constant. It has thus two degrees of freedom. According to the law of equipartition of energy, the mean energy associated with each degree of freedom is $\frac{1}{2}kT$ and hence the mean energy of the oscillator is

$$<\epsilon> = kT \tag{3.2-3}$$

where k is Boltzmann constant.

To get the energy density of the radiation in the cavity for a given frequency $v = c/\lambda$, we have to find the number n_v of oscillators per unit volume having the frequency v and to multiply this by the mean energy $<\epsilon>$. n_v can be calculated by determining the number of modes of stationary vibrations which can be excited in a three dimensional box of specified dimensions with appropriate boundary conditions and can be calculated as follows.

Consider a cavity full of electromagnetic radiation with frequencies extending from 0 to ∞. The cavity is in thermal equilibrium with its walls which are at the temperature T. For simplicity, we assume the cavity to be cubical with the sides of length l each and the radiation is incident normally on the walls. Due to reflections from the walls standing waves are generated in the cavity with nodes at the walls.

The three dimensional equation for the propagation of electromagnetic waves can be written as

$$\nabla^2 \phi = \frac{\partial^2 \phi}{\partial x^2} + \frac{\partial^2 \phi}{\partial y^2} + \frac{\partial^2 \phi}{\partial z^2} = \frac{1}{c^2}\frac{\partial^2 \phi}{\partial t^2} \tag{3.2-4}$$

where ϕ is the electromagnetic field variable. Considering a sinusoidal wave with time dependence of the form $\phi \sim \exp(i\omega t)$ where $\omega = 2\pi v$ is the circular frequency, we get $\partial^2 \phi/\partial t^2 = -\omega^2 \phi$. Hence from Eq. (3.2-4) we have

$$\frac{\partial^2 \phi}{\partial x^2} + \frac{\partial^2 \phi}{\partial y^2} + \frac{\partial^2 \phi}{\partial z^2} + \frac{\omega^2}{c^2}\phi = 0 \tag{3.2-5}$$

where ϕ now represents the space part of the field variable. The above equation has to be solved with the boundary conditions

$\phi = 0$ at $x = 0, y = 0, z = 0$ and at $x = l, y = l, z = l$.

Using the method of separation of variables, we can write

$$\phi = \phi(x, y, z) = \phi_x(x)\phi_y(y)\phi_z(z) = \phi_x(x)\phi_{yz}(y, z) \tag{3.2-6}$$

Substitution in Eq. (3·2–5) then gives

$$\frac{1}{\phi_x}\frac{\partial^2 \phi_x}{\partial x^2}+\frac{\omega^2}{c^2}=-\frac{1}{\phi_{yz}}\left(\frac{\partial^2 \phi_{yz}}{\partial y^2}+\frac{\partial^2 \phi_{yz}}{\partial z^2}\right)=\frac{\omega^2_{23}}{c^2}=\text{constant} \quad (3\cdot2\text{–}7)$$

Since the first member in the above equation is a function of x alone while the second member is a function of (y, z), each must be equal to a constant which we have written as ω^2_{23}/c^2. So we get

$$\frac{1}{\phi_x}\frac{\partial^2 \phi_x}{\partial x^2}+\frac{\omega_1^2}{c^2}=0$$

where $\omega_1^2=\omega^2-\omega^2_{23}$. Hence we have

$$\frac{\partial^2 \phi_x}{\partial x^2}+\frac{\omega_1^2}{c^2}\phi_x=0 \quad (3\cdot2\text{–}8)$$

The solution of this equation is of the form

$$\phi_x=A_1\sin\frac{\omega_1 x}{c}+B_1\cos\frac{\omega_1 x}{c}$$

The boundary condition $\phi_x=0$ at $x=0$ requires us to put $B_1=0$ which gives

$$\phi_x=A_1\sin\frac{\omega_1 x}{c} \quad (3\cdot2\text{–}9)$$

Again since $\phi_x=0$ at $x=l$, we get

$$\frac{\omega_1 l}{c}=n_1\pi \quad\text{or}\quad \frac{\omega_1}{c}=\frac{n_1\pi}{l} \quad (3\cdot2\text{–}10)$$

where n_1 is an integer. So we get finally

$$\phi_x=A_1\sin\frac{n_1\pi x}{l} \quad (3\cdot2\text{–}11)$$

Similarly by the separation of the variables y and z and applying the appropriate boundary conditions we can solve for ϕ_y and ϕ_z:

$$\phi_y=A_2\sin\frac{n_2\pi y}{l} \quad\text{and}\quad \phi_z=A_3\sin\frac{n_3\pi z}{l} \quad (3\cdot2\text{–}12)$$

where n_2 and n_3 must be taken as integers to satisfy the boundary conditions $\phi=0$ at $y=l$ and $z=l$. Thus we get finally

$$\phi=\phi_0\sin\frac{n_1\pi x}{l}\sin\frac{n_2\pi y}{l}\sin\frac{n_3\pi z}{l} \quad (3\cdot2\text{–}13)$$

n_1, n_2 and n_3 are given by

$$n_1=\frac{\omega_1 l}{\pi c},\; n_2=\frac{\omega_2 l}{\pi c},\; n_3=\frac{\omega_3 l}{\pi c} \quad (3\cdot2\text{–}14)$$

Origin of the Quantum Theory ; Planck's Theory

where $\omega_1^2 + \omega_2^2 + \omega_3^2 = \omega^2$ so that

$$n_1^2 + n_2^2 + n_3^2 = \frac{\omega^2 l^2}{\pi^2 c^2} = \frac{4l^2 v^2}{c^2} = \frac{4l^2}{\lambda^2} \qquad (3\cdot2-15)$$

where $\lambda = c/v = \dfrac{2\pi c}{\omega}$ is the wavelength.

A set of values of n_1, n_2 and n_3 satisfying Eq. (3·12–15) represents a particular mode of vibration. To calculate the number of modes of vibration in the frequency interval v to $v + dv$ we represent the n_1, n_2, n_3 values in a three dimensional diagram with n_1 along the x-axis, n_2 along the y-axis and n_3 along the z-axis. Each combination of n_1, n_2, n_3 values is then represented by a point in this diagram whose coordinates are (n_1, n_2, n_3).

For simplicity, we first consider a two dimensional analogue in which n_1 is plotted along x and n_2 along y as shown in Fig. 3·3. The

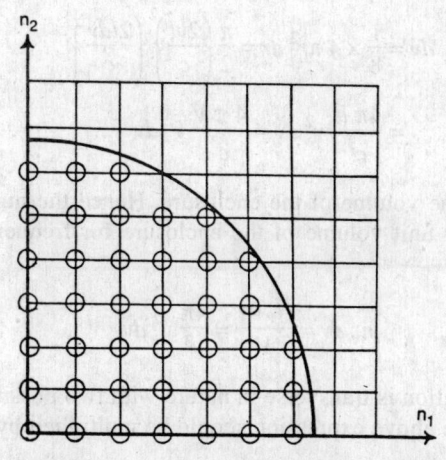

Fig. 3·3. Counting the number of modes of vibration with frequency less than v. This is equal to the number of lattice points within the quadrant.

representative points corresponding to the different modes of vibration are shown by the small circles at the points of intersection of the straight lines drawn parallel to the n_1 and n_2 axes, unit distance apart. For large values of n_1 and n_2, the number of representative points is nearly equal to the number of unit squares shown in the figure. Hence the number of modes of vibration between v and $v + dv$ can be determined by counting the number of unit squares in the annulus between the two circular arcs of radii $r = 2vl/c$ and $r + dr = 2(v + dv) \, l/c$ in the first quadrant. The first quadrant is chosen because n_1 and n_2 can assume positive values only.

This number is approximately equal to the area of the anulus divided by the area of each square which is unity.

In the actual three dimensional case, we have to draw straight lines parallel to the n_1, n_2 and n_3 axes unit distance apart. The points of intersection of these lines give the representative points corresponding to the different modes of vibration. As before, the number of modes of vibration for the frequencies lying between v and $v + dv$ is obtained by counting the number of unit cubes in the first octant of the three dimensional diagram between the two spheres of radii $r = 2vl/c$ and $r + dr = 2(v + dv)l/c$.

Since we can associate one representative point with each unit cube formed by the intersection of the three sets of mutually perpendicular straight lines parallel to the three axes, the number of such points $N_v\, dv$ is equal to the volume of the first octant of the spherical shell as above divided by the volume of each unit cube and is given by

$$N_v\, dv = \frac{1}{8} \times 4\pi r^2\, dr = \frac{\pi}{2}\left(\frac{2vl}{c}\right)^2\left(\frac{2l\,dv}{c}\right)$$

$$= \frac{4\pi l^3}{c^3} v^2\, dv = \frac{4\pi V}{c^3} v^2\, dv \qquad (3 \cdot 2\text{--}16)$$

Here $V = l^3$ is the volume of the enclosure. Hence the number of modes of vibration per unit volume of the enclosure for frequencies between v and $v + dv$ is

$$n_v\, dv = \frac{N_v\, dv}{V} = \frac{4\pi}{c^3} v^2\, dv$$

Since e.m. radiation is transverse in nature with two possible directions of polarization, the above expression should be multiplied by two. So we get finally

$$n_v\, dv = \frac{8\pi}{c^3} v^2\, dv \qquad (3 \cdot 2\text{--}17)$$

Transforming to wavelength λ, we get

$$n_\lambda\, d\lambda = \frac{8\pi}{\lambda^4}\, d\lambda \qquad (3 \cdot 2\text{--}18)$$

which gives the number of oscillators per unit volume emitting radiation of wavelengths lying between λ and $\lambda + d\lambda$. We thus get the energy density of the blackbody radiation in this wavelength range as given below by multiplying by $\langle\epsilon\rangle = kT$:

$$u_\lambda\, d\lambda = \frac{8\pi kT}{\lambda^4}\, d\lambda \qquad (3 \cdot 2\text{--}19$$

Eq. (3·2–19) is known as *Rayleigh-Jeans radiation law* (1900, 1909).

The energy density u_λ is simply related to intensity of the emitted radiation given below*

$$E_\lambda = \frac{c}{4} u_\lambda \qquad (3\cdot2\text{–}20)$$

Eq. (3·2–19) agrees well with the experimental results for longer wavelengths. However it fails completely at shorter wavelengths. According to Eq. (3·2–19), u_λ and hence E_λ approaches infinity as $\lambda \to 0$ whereas the experimental result shows that $E_\lambda \to 0$ as $\lambda \to 0$. This serious disagreement between theory and experiment, known as *ultraviolet catastrophe*, indicates the limitations of the classical mechanics on the basis of which the equipartition law is deduced and which is used in the above deduction.

3.3 Planck's law of radiation ; Quantum hypothesis

The failure of R-J formula to explain the observed energy distribution law of black body radiation showed that there was something wrong either with the equipartion law or with the classical electromagnetic theory or with both.

Max Planck in Germany examined the whole situation critically and put forward a bold new postulate (1900) regarding the nature of vibration of the linear harmonic oscillators which are in equilibrium with the electromagnetic radiation within a cavity. According to Planck, an oscillator can have a discrete set of energies which are integral multiples of a *finite quantum* of energy $\epsilon_0 = h\nu$ where h is a constant known as Planck's constant and ν is the frequency of the oscillator. Thus the energy of the oscillator can only have the values

$$\epsilon = n\,\epsilon_0 = nh\nu \qquad (3\cdot3\text{–}1)$$

where n is an integer or zero. Obviously the ground state energy, *i.e.*, energy in the lowest state of the oscillator will be zero.

Planck further assumed that the change in energy of the oscillator due to emission or absorption of radiation can also take place by a discrete amount $h\nu$. Planck estimated the value of h by fitting his theory to the experimental data (Fig. 3·2.). Planck's constant h is a universal constant and plays a crucial role in all quantum phenomena. Its value is $h = 6\cdot62618 \times 10^{-34}$ joule. second (Js).

Since radiation is emitted from the oscillators, and since according to Planck the change in energy of the oscillators can only take place by discrete amount, the energy carried by the emitted radiation will be $h\nu$ which is equal to the loss of energy of the oscillator. Obviously this is also

---
* See *A Treatise on Heat* by Saha and Srivastava.

the energy gain of the oscillator when it absorbs radiation. No absorption of energy by the oscillator can take place unless the energy of the radiation $h\nu'$ is equal to the possible energy change of the oscillator $h\nu$, i.e., unless $\nu' = \nu$.

According to the postulates of Planck, the oscillators can exist in a set of discrete energy states 0, $h\nu$, $2h\nu$, $3h\nu$ etc. (see Fig. 3·4.)

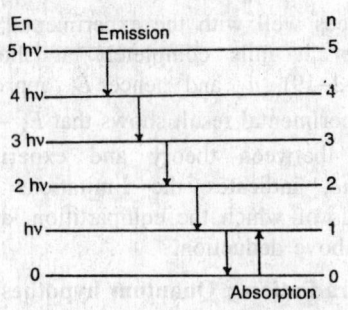

Fig. 3·4. Energy levels of an oscillator according to Planck.

The number of oscillators in an energy state $\epsilon_n = nh\nu$ is determined by the well-known Maxwell-Boltzmann distribution function :

$$N_n = N_0 \exp(-\epsilon_n/kT) = N_0 \exp(-nh\nu/kT) \tag{3.3–2}$$

For $\epsilon_n = 0$, $N_n = N_0$ so that N_0 is the number of oscillators in the ground state. The number N_n decreases exponentially with increasing energy ϵ_n.

We first calculate the mean energy $<\epsilon>$ of the oscillators according to the new ideas introduced by Planck. Since the energies of the oscillators can assume discrete values, we have to replace integrals by sums over all possible oscillator states to determine $<\epsilon>$. We get

$$<\epsilon> = \sum_{n=0}^{\infty} N_n \epsilon_n \Big/ \sum_{n=0}^{\infty} N_n$$

$$= \sum_{n=0}^{\infty} N_0 \epsilon_n \exp(-\epsilon_n/kT) \Big/ \sum_{n=0}^{\infty} N_0 \exp(-\epsilon_n/kT)$$

$$= \sum_{n=0}^{\infty} nh\nu \exp(-nh\nu/kT) \Big/ \sum_{n=0}^{\infty} \exp(-nh\nu/kT)$$

$$= \frac{h\nu x [1 + 2x + 3x^2 + 4x^3 + \ldots]}{1 + x + x^2 + x^3 + \ldots}$$

where $x = \exp(-h\nu/kT)$. The sums in the numerator and denominator in the above expression are easy to evaluate. We get

Origin of the Quantum Theory ; Planck's Theory

$$<\epsilon> = \frac{h\nu x (1-x)^{-2}}{(1-x)^{-1}} = \frac{h\nu x}{1-x} = \frac{h\nu}{x^{-1}-1}$$

$$\therefore \quad <\epsilon> = \frac{h\nu}{\exp(h\nu/kT) - 1} = \frac{\hbar\omega}{\exp(\hbar\omega/kT) - 1} \quad (3\cdot3-3)$$

where $\omega = 2\pi\nu$ is the circular frequency and $\hbar = h/2\pi$.

It can be easily seen that if $h\nu \ll kT$, the value of $<\epsilon>$ reduces to the classical limit kT. In this case we have, since $h\nu/kT \ll 1$

$$\exp(h\nu/kT) \approx 1 + \frac{h\nu}{kT}$$

so that $\quad <\epsilon> = \dfrac{h\nu}{1 + h\nu/kT - 1} = kT$

Obviously this corresponds to the continuous variation of the oscillator energy. Eq. (3·3–3) along with Eq. (3·2–17) gives Planck's radiation law

$$u_\nu \, d\nu = \frac{8\pi\nu^2 \, d\nu}{c^3} <\epsilon>$$

$$u_\nu \, d\nu = \frac{8\pi}{c^3} \frac{h\nu^3 \, d\nu}{\exp(h\nu/kT) - 1} \quad (3\cdot3-4)$$

or $\quad u_\lambda \, d\lambda = \dfrac{8\pi hc}{\lambda^5} \dfrac{d\lambda}{\exp(hc/\lambda kT) - 1} \quad (3\cdot3-5)$

Planck's formula (3·3–5) for the energy distribution of black body radiation agrees well with the experimental results both for the long wavelength and the short wavelength ends of the energy spectrum. It can be easily seen that it reduces to R-J law (Eq. 3·2–19) when $\lambda \to \infty$ and to Wien's law (Eq. 3·2–1) when $\lambda \to 0$

For very short wavelengths ($\lambda \to 0$), we have

$$\exp(hc/\lambda kT) - 1 \approx \exp(hc/\lambda kT)$$

Writing $b = hc/k$ which is a constant, we get

$$\lim_{\lambda \to 0} u_\lambda \, d\lambda = \frac{a}{\lambda^5} \exp(-b/\lambda T) \, d\lambda$$

which is the same as Eq. (3·2–1) proposed by Wien.

Again for very long wavelengths, $hc/\lambda \ll kT$. Hence

$$\exp(hc/\lambda kT) - 1 \approx 1 + \frac{hc}{\lambda kT} - 1 = \frac{hc}{\lambda kT}$$

So we get $\lim\limits_{\lambda \to \infty} u_\lambda \, d\lambda = \dfrac{8\pi hc}{\lambda^5} \dfrac{d\lambda}{hc/\lambda kT} = \dfrac{8\pi kT}{\lambda^4} d\lambda$

which is the same as Rayleigh-Jeans Law (Eq. 3·2–19).

3.4 Deductions from Planck's law

(a) Wien's Displacement law

Planck's formula (3·3–5) shows a maximum at $\lambda = \lambda_m$ when the denominator becomes minimum. The denominator can be written as

$$z = \lambda^5 \{\exp(hc/\lambda kT) - 1\}$$

Then $\dfrac{dz}{d\lambda} = 5\lambda^4 \{\exp(hc/\lambda kT) - 1\} - \lambda^5 \dfrac{hc}{\lambda^2 kT} \exp(hc/\lambda kT) = 0$ for $\lambda = \lambda_m$.

This gives

$$1 - \exp(-hc/\lambda_m kT) = hc/5\lambda_m kT$$

or, $\qquad 1 - \exp(-x) = x/5 \qquad (3.4\text{–}1)$

where $\qquad x = hc/\lambda_m kT$

This is a transcendental equation which cannot be solved analytically. It can be solved graphically. If we put $y = 1 - \exp(-x)$ and $y = x/5$ then the point of intersection of the two graphs given by the above two relations gives the solution which turns out to be $x = 4.9651$. This can be readily verified by substitution in Eq. (3.4–1). We thus get

$$\lambda_m T = \frac{hc}{kx} = \frac{hc}{4.9651\,k} = 0.0029 \text{ m.K} = \text{constant}$$

This is Wien's displacement law.

(b) Stefan-Boltzmann law

From Eq. (3·3–4) we obtain the total energy density of the radiation emitted by a blackbody as

$$u = \int_0^\infty u_\nu \, d\nu = \frac{8\pi h}{c^3} \int_0^\infty \frac{\nu^3 \, d\nu}{\exp(h\nu/kT) - 1}$$

$$= \frac{8\pi h}{c^3} \left(\frac{kT}{h}\right)^4 \int_0^\infty \frac{x^3 \, dx}{\exp x - 1}$$

Here $x = h\nu/kT$. To evaluate the integral, we note that

$$\frac{x^3}{\exp x - 1} = \frac{x^3}{\exp x \{1 - \exp(-x)\}}$$

$$= x^3 \exp(-x) \{1 - \exp(-x)\}^{-1}$$

$$= x^3 \{\exp(-x) + \exp(-2x) + \exp(-3x) \ldots\ldots\}$$

$$= \sum_p x^3 \exp(-px)$$

$$\therefore \int_0^\infty \frac{x^3 \, dx}{\exp x - 1} = \sum_p \int_0^\infty x^3 \exp(-px) \, dx = \sum_p \frac{3!}{p^4}$$

since $\int_0^\infty x^n \exp(-px) \, dx = n!/p^n$

Also since $\sum_p 1/p^4 = \pi^4/90$, we get

$$u = \frac{8\pi k^4 T^4}{c^3 h^3} \cdot \frac{\pi^4}{15} = aT^4 \qquad (3\cdot4\text{--}2)$$

where $\quad a = \dfrac{8\pi^5 k^4}{15 \, c^3 h^3} = \text{constant}$

As stated before, the intensity E of blackbody radiation is related to the energy density u (see Eq. (3·2–20)). We thus get Stefan-Boltzmann law (Eq. (3·1–2))

$$E = \frac{c}{4} u = \frac{ca}{4} T^4 = \sigma T^4$$

where the Stefan constant is given by

$$\sigma = \frac{ca}{4} = \frac{2\pi^5 k^4}{15 \, c^2 h^3} = 5\cdot 67 \times 10^{-8} \text{ joule/s/m}^2/\text{deg}^4 \qquad (3\cdot4\text{--}3)$$

Problems

1. The maximum in the energy distribution curve of the radiation emitted by the sun is 4820 Å. Calculate the surface temperature of the sun assuming it to be a black body. (6016 K)

2. Assuming each photon to carry an energy $\hbar\omega$, calculate from Planck's law the number of photons per unit volume having angular frequencies in the range ω to $\omega + d\omega$.

3. Using Planck's formula and the result of Prob. 2, calculate the mean angular frequency of the radiation emitted from a blackbody at the temperature T K.

4. The solar constant which is the amount of radiant energy received on the earth from the sun per second per unit area on a black surface perpendicular to the sun's rays at the mean distance $D = 1\cdot 496 \times 10^8$ km of the earth from the sun has a value $S = 1\cdot 3514 \times 10^3$ J/m². Assuming the sun to be a blackbody, calculate the surface temperature of the sun using Stefan-Boltzman law. (Radius of the sun $R = 6\cdot 96 \times 10^5$ km).

(5761 K)

4

STRUCTURE OF THE HYDROGEN-LIKE ATOMS ; BOHR-SOMMERFELD THEORY

4.1 Thomson model of the atom

In Chapter II, we have seen that atoms of all elements are made up of two parts with equal and opposite electricity in them, so that the atoms as a whole are electrically neutral. The negative electricity in the atom is carried by the electrons which are quite light compared to the total mass of an atom. If all electrons are removed from an atom, the positively charged part which is left behind will have almost the whole mass of the atom within it. We shall now investigate as to how the two parts are arranged inside the atom.

We know that positive and negative electricity attract each other according to Coulomb's law of force. If due to this attraction, they combine together, then there is possibility that each of the oppositely charged parts of the atom will lose its entire electrical charge. Obviously this does not happen and both types of charge retain their identity inside the atom.

The first proposal regarding the possible arrangement of the positively and negatively charged parts in the atom was made by J.J. Thomson in 1907 which is known as the *Thomson model of the atom*.

According to Thomson, the entire positive electricity of the atom is distributed within a sphere of atomic radius which is of the order of 10^{-10} m (see Table 1·1). Within this sphere of positive charge, the negatively charged point electrons are embedded at certain definite points, much like the plums in a pudding. The electrons are however not at rest, but oscillate with definite frequencies about their mean positions of rest.

The above atomic model of Thomson has some drawbacks. It is well-known that when the light emitted by different elements is analysed spectroscopically the emission spectrum is found to consist of light of certain definite wavelengths. These different wavelength components

Structure of the Hydrogen–Like Atom

appear as *spectral lines* of definite wavelengths or frequencies in the spectrogram (see Fig. (4·1). For each element, there are many such spectral lines of different frequencies in its emission spectrum.

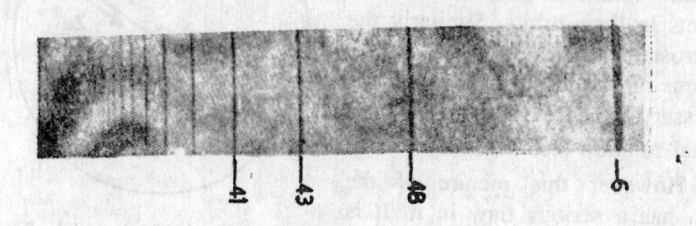

Fig. 4·1. Balmer series lines for hydrogen.

According to the electromagnetic theory of light, when an electron oscillates with a definite frequency, it mainly emits light of that particular frequency. In addition, light having frequencies twice, thrice etc. of the oscillation frequency may also be emitted, though with much reduced intensity. In the hydrogen atom, there is only one electron, which will oscillate with one particular frequency. So according to Thomson model, the emission spectrum of hydrogen will mainly consist of light of this particular frequency. However, even in this case, the emission spectrum is found to be made up of a large number of spectral lines of different frequencies having comparable intensities. The frequencies of these lines do not have any correspondence with the expected oscillation frequency of the electron in this case.

Later Ernest Rutherford showed (1911) by a series on experiments on the scattering of α-particles (*i.e.* nuclei of helium atoms) emitted by radioactive substances that the positive charge inside an atom is confined within a region having radius less than 10^{-14} m, *i.e.*, less than a ten-thousandth part of the radius of the atom. Further, it was found that almost the entire mass of the atom was concentrated within this positively charged core of the atom which has subsequently been named as the *nucleus of the atom*.

As a result of these findings of Rutherford, the Thomson model of the atom was totally rejected.

4.2 Rutherford model of the atom

If the positively charged heavy core of the atom has such a small radius then how are the electrons distributed in the atom? It is obvious that though they are attracted by the positive charge of the nucleus, they do not fall upon the nucleus, as otherwise they would not exist in the atom at all. So Rutherford proposed that the electrons revolve in orbits round the nucleus much like the planets revolving round the sun in their respective orbits. We know that in the case of the planets, the gravitational attraction of the sun provides the centripetal force mv^2/r of the rotational motion of the planets in their orbits. Similarly the electrostatic attractive force of the nucleus on the electrons supplies the necessary centripetal force for their orbital rotation (see Fig. 4·2).

However this picture of the atom has a serious flaw in it. It is known from the electromagnetic theory of light that an accelerated or decelerated electrically charged particle emits electromagnetic radiation and thereby loses its energy so that its velocity is decreased. Since the electron revolving in its orbit is acted upon by the centripetal force its motion is accelerated. Thus it will lose energy by the emission of electromagnetic radiation. So due to the attractive force of the nucleus it will move in orbits of continually decreasing radii and its path will be a spiral as shown in Fig. 4·3 (see Eq. 4·4–10) Finally it will fall upon the nucleus itself and will disappear. So in this model, there could not be any atom with the positively and negatively charged parts existing separately in it which is contrary to the observed facts (see Ch. II).

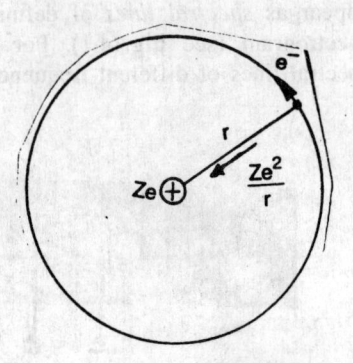

Fig. 4·2. Rutherford model of the atom

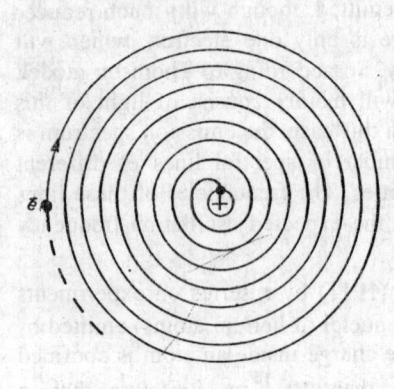

Fig. 4·3. Spiralling path of the electron.

In order to remove the above drawback of the Rutherford model, the famous Danish physicist Niels Bohr put forward some bold new proposals based on the quantum idea which had been proposed by Max Planck a few years earlier (see Ch. III).

4.3 Bohr's postulates

Bohr considered the simplest of all atoms, *viz.*, the hydrogen atom. According to the Rutherford model, there is only one electron revolving in the orbit of this atom round the nucleus which is a single proton. Bohr assumed that the electron revolves in a circular orbit. It may be noted that this simple picture of the hydrogen atom is also applicable to some ionized atoms. In the helium atom, there are two electrons. So the removal

Structure of the Hydrogen–Like Atom

of one electron from it leaves behind the singly charged helium ion, viz., He⁺ which has only one electron revolving in the orbit round the nucleus which carries two electronic units of positive electricity. Thus the He⁺ ion resembles the hydrogen atom. In general if an atom has Z electrons in its orbits and $(Z-1)$ of these are removed from it, then the $(Z-1)$ fold positively charged ion which is left behind resembles a hydrogen atom with one electron revolving in the orbit round the nucleus carrying Z units of positive electricity. All such ions are known as *hydrogen like atoms*. Besides He⁺, other examples are Li⁺⁺ ($Z = 3$), Be⁺⁺⁺ ($Z = 4$), B⁺⁺⁺⁺ ($Z = 5$) etc. In Fig. 4·4, some of these hydrogen-like atoms are shown diagrammatically.

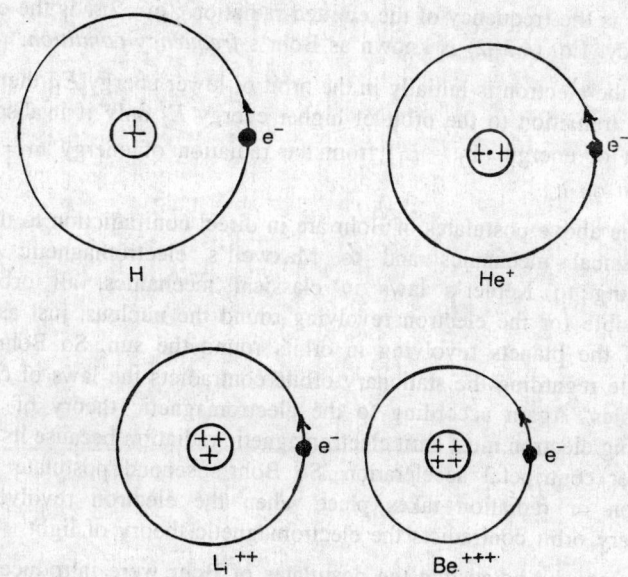

Fig. 4·4. Hydrogen-like atoms.

In trying to find a way out of the difficulties associated with the Rutherford model discussed in § 4·2, Bohr proposed the following three *postulates* (1912) :

(*a*) **Bohr's first postulate** : In the hydrogen atom, there is a single electron which can revolve round the nucleus in certain definite orbits known as the *stationary orbits*. In a stationary orbit, the electron has an angular momentum L which is an integral multiple of $\hbar = h/2\pi$ where h is Planck's constant, *i.e.*,

$$L = nh/2\pi = n\hbar \qquad (4 \cdot 3 - 1)$$

where n is known as the *quantum number*, and its possible values are $n = 1, 2, 3$, etc. Eq. (4·3–1) is known as Bohr's *quantum condition*. The electron in this case is in a stationary state (see § 10·7).

(b) **Bohr's second postulate** : When the electron revolves in a stationary orbit, it does not emit electromagnetic radiation as predicted by the electromagnetic theory of light.

(c) **Bohr's third postulate** : The emission of electromagnetic radiation from the hydrogen-like atom takes place when the electron makes a transition from one stationary orbit to another. If the transition is from an orbit of higher energy E_2 to an orbit of lower energy E_1, then the energy $h\nu$ of the emitted radiation, according to Planck's law, will be

$$h\nu = \hbar\omega = E_2 - E_1 \qquad (4\cdot3\text{-}2)$$

where ν is the frequency of the emitted radiation ; $\omega = 2\pi\nu$ is the circular frequency. Eq. (4·3-2) is known as Bohr's *frequency condition*.

If the electron is initially in the orbit of lower energy E_1, then it can make a transition to the orbit of higher energy E_2 only if it absorbs an amount of energy $(E_2 - E_1)$ from the radiation of energy $h\nu = E_2 - E$ incident on it.

The above postulates of Bohr are in direct contradiction to the laws of classical mechanics and to Maxwell's electromagnetic theory. According to Kepler's laws in classical mechanics, all orbits are permissible for the electron revolving round the nucleus, just as in the case of the planets revolving in orbits round the sun. So Bohr's first postulate regarding the stationary orbits contradicts the laws of classical mechanics. Again according to the electromagnetic theory of light, a revolving electron must emit electromagnetic radiation because its motion is under centripetal acceleration. So Bohr's second postulate that no emission or radiation takes place when the electron revolves in a stationary orbit contradicts the electromagnetic theory of light.

It may be noted that the postulates of Bohr were introduced as *ad hoc* hypotheses. The classical laws of mechanics were accepted by Bohr as the laws governing the motion of the electron. Only certain restrictions were imposed upon the permissible orbits through the quantum condition. Bohr made no attempts to propound any *new mechanics* to describe the motion of the electron.

The above postulates are also applicable to the other hydrogen–like atoms.

4.4 Bohr's theory of the spectra of hydrogen-like atoms

Consider a hydrogen-like atom in which an electron of mass m and charge $(-e)$ revolves in a circular orbit of radius r round a nucleus of charge $+Ze$. The electrostatic attractive force on the nucleus $Ze^2/4\pi\epsilon_0 r^2$ provides the centripetal force $m\upsilon^2/r$ to keep the electron in orbit so that we can write

Structure of the Hydrogen–Like Atom

$$\frac{1}{4\pi\epsilon_0} \cdot \frac{Ze^2}{r^2} = \frac{mv^2}{r}$$

which gives $\qquad v^2 = \frac{1}{4\pi\epsilon_0} \cdot \frac{Ze^2}{mr} \qquad$ (4·4–1)

Here ϵ_0 is the permittivity of vacuum and has the value $10^{-9}/36\pi$ F/m. From Bohr's quantum condition (4·4–1), we have

$$L = mr^2\omega = mvr = n\hbar \qquad (4\cdot4\text{–}2)$$

where ω is the angular velocity and v is the linear velocity of the electron in the orbit. From Eq. (4·4–2) we have

$$v = \frac{n\hbar}{mr} \qquad (4\cdot4\text{–}3)$$

Eliminating v from Eqs. (4·4–1) and (4·4–3) we get

$$\frac{1}{4\pi\epsilon_0} \cdot \frac{Ze^2}{mr} = \frac{n^2\hbar^2}{m^2 r^2}$$

which gives $\qquad r = \frac{4\pi\epsilon_0 n^2 \hbar^2}{mZe^2} \qquad$ (4·4–4)

Hence from Eq. (4·4–3) we get

$$v = \frac{Ze^2}{4\pi\epsilon_0 n\hbar} \qquad (4\cdot4\text{–}5)$$

Thus both the radius of the orbit and the electron velocity depend on the quantum number n. The orbit with $n = 1$ has the smallest radius. For hydrogen ($Z = 1$), this radius is known as the *Bohr radius* and is given by

$$a_0 = \frac{4\pi\epsilon_0 \hbar^2}{me^2} = 0.529 \times 10^{-10} \text{ m} = 0.529 \text{ Å} \qquad (4\cdot4\text{–}6)$$

where we have substituted the numerical values of ϵ_0, m, e and $\hbar$ given in appendix A–VI. The radii of the orbits are directly proportional to n^2 so that the radii of the successive orbits are in the ratios 1 : 4 : 9 : 16 : ......

Again the velocity of the electron is inversely proportional to n and is the highest in the smallest orbit having $n = 1$. The velocities become progressively less in the orbits of increasing radii. For hydrogen, the velocity of the electron in the Bohr orbit is

$$v_1 = \frac{e^2}{4\pi\epsilon_0 \hbar} = 2.18 \times 10^6 \text{ m/s} \qquad (4\cdot4\text{–}7)$$

This velocity is about 1/137 times the velocity of light c.

For the hydrogen-like atoms, the radii of the orbits become less as Z increases while velocity in the different orbits increases with increasing Z. It should be noted that the radius of the orbit as estimated above (Eq. (4·4–4) is of the same order of magnitude as the atomic radii estimated from the kinetic theory (see Table 1·1).

The total energy E of the electron in its orbit is equal to the sum of its kinetic energy E_k and potential energy V. We have, using Eq. (4·4–1)

$$E_k = \frac{1}{2}mv^2 = \frac{m}{8\pi\epsilon_0}\frac{Ze^2}{mr} = \frac{1}{8\pi\epsilon_0}\frac{Ze^2}{r} \qquad (4\cdot4\text{–}8)$$

The potential energy is given by

$$V = -\frac{1}{4\pi\epsilon_0}\int_r^\infty \frac{Ze^2}{r^2}\,dr = -\frac{Ze^2}{4\pi\epsilon_0 r} \qquad (4\cdot4\text{–}9)$$

So the total energy is

$$E = E_k = V = \frac{Ze^2}{8\pi\epsilon_0 r} - \frac{Ze^2}{4\pi\epsilon_0 r} = -\frac{Ze^2}{8\pi\epsilon_0 r} \qquad (4\cdot4\text{–}10)$$

Using Eq. (4·4–4) we then get the energy in the nth stationary orbit

$$E_n = -\frac{mZ^2 e^4}{32\pi^2\epsilon_0^2 n^2 \hbar^2} \qquad (4\cdot4\text{–}11)$$

Written in terms of h this becomes

$$E_n = -\frac{mZ^2 e^4}{8\epsilon_0^2 n^2 h^2} \qquad (4\cdot4\text{–}11a)$$

The energy of the electron given by Eq. (4·4–11) is negative. So as n increases i.e., as the electron goes to orbits of larger radii, E_n increases and converges to the limiting value $E_n = 0$ as $n \to \infty$. E_n is the lowest for the smallest orbit for which $n = 1$.

If the electron, initially in an orbit of quantum number n_i, makes a transition to a final orbit of quantum number n_f, then its energy is changed from E_i to E_f where we have from Eq. (4·4–11)

$$E_i = -\frac{mZ^2 e^4}{32\pi^2\epsilon_0^2 n_i^2 \hbar^2} \quad \text{and} \quad E_f = -\frac{mZ^2 e^4}{32\pi^2\epsilon_0^2 n_f^2 \hbar^2}$$

If $n_i > n_f$, then $E_i > E_f$. In this case, according to Bohr's third postulate, there will be emission of radiation of energy

$$\hbar\omega = E_i - E_f = \frac{mZ^2 e^4}{32\pi^2\epsilon_0^2 \hbar^2}\left(\frac{1}{n_f^2} - \frac{1}{n_i^2}\right) \qquad (4\cdot4\text{–}12)$$

Structure of the Hydrogen-Like Atom

so that the frequency of the emitted radiation is

$$v = \frac{\omega}{2\pi} = \frac{m Z^2 e^4}{64\pi^3 \epsilon_0^2 \hbar^3} \left(\frac{1}{n_f^2} - \frac{1}{n_i^2} \right) \quad (4.4\text{–}13)$$

If λ be the wavelength of the emitted radiation, then $v\lambda = c$. Since v full waves are contained in a path length c, the number of full waves per unit length which is known as the *wave number* is given by

$$\tilde{v} = \frac{v}{c} = \frac{1}{\lambda} = \frac{m Z^2 e^4}{64\pi^3 \epsilon_0^2 c\hbar^3} \left(\frac{1}{n_f^2} - \frac{1}{n_i^2} \right) \quad (4.4\text{–}14)$$

The unit of $\tilde{v}$ is m^{-1}. If we write

$$R_\infty = \frac{m e^4}{64\pi^3 \epsilon_0^2 c\hbar^3} = 1.09737 \times 10^7 \text{ m}^{-1} \quad (4.4\text{–}15)$$

then we get

$$\tilde{v} = R_\infty Z^2 \left(\frac{1}{n_f^2} - \frac{1}{n_i^2} \right) \quad (4.4\text{–}16)$$

R_∞ is known as the Rydberg constant. For hydrogen, we get

$$\tilde{v} = R_\infty \left(\frac{1}{n_f^2} - \frac{1}{n_i^2} \right) \quad (4.4\text{–}17)$$

In terms of R_∞, the energy of the electron in the nth orbit given by Eq. 4.4–16) can be written as

$$E_n = -\frac{2\pi c \hbar R_\infty Z^2}{n^2} \quad (4.4\text{–}18)$$

4.5 Origin of the spectral series

Long before Bohr's theory, Balmer in 1883 had shown that the experimentally determined wave number of the spectral lines of hydrogen in the visible region could be represented by a formula similar to Eq. (4.4–17):

$$\tilde{v} = A \left(\frac{1}{4} - \frac{1}{n^2} \right) = A \left(\frac{1}{2^2} - \frac{1}{n^2} \right)$$

where A is a constant and n takes on different integral values for the different lines of the spectrum : $n = 3, 4, 5, \ldots$ The value of A found from the measurement of the wavelength $\lambda = 1/\tilde{v}$ of the spectral lines was 1.09678×10^7 m^{-1} which is close to the value of R_∞ given above. This agreement establishes the essential correctness of Bohr's theory.

Since the wave numbers of the different spectral lines of hydrogen in the visible region can be expressed by a single mathematical formula as above, these lines are said to belong to a *spectral series* known as the *Balmer series*.

Later more such spectral series for hydrogen were discovered. The wave numbers of the lines of a particular series can be written as differences between two terms, one of which is a constant and the other is a running term depending on the cardinal numbers so that the lines have decreasing separation converging towards a *series limit*. These different series are :

Lyman series : $\tilde{\nu} = R_\infty \left(\dfrac{1}{1^2} - \dfrac{1}{n^2} \right)$, $n = 2, 3, 4, \ldots\ldots$

The lines of this series fall in the ultraviolet region.

Balmer series : $\tilde{\nu} = R_\infty \left(\dfrac{1}{2^2} - \dfrac{1}{n^2} \right)$, $n = 3, 4, 5, \ldots\ldots$

As seen above these lines fall in the visible region.

Paschen series : $\tilde{\nu} = R_\infty \left(\dfrac{1}{3^2} - \dfrac{1}{n^2} \right)$, $n = 4, 5, 6, \ldots\ldots$

The lines of this and the following series belong to the infra-red region.

Brackett series : $\tilde{\nu} = R_\infty \left(\dfrac{1}{4^2} - \dfrac{1}{n^2} \right)$, $n = 5, 6, 7, \ldots\ldots$

Pfound series : $\tilde{\nu} = R_\infty \left(\dfrac{1}{5^2} - \dfrac{1}{n^2} \right)$, $n = 6, 7, 8, \ldots\ldots$

In all cases R_∞ is the Rydberg constant.

The origin of the spectral series can easily be understood from Bohr's theory. If the electron in the hydrogen atom is initially in one of the stationary orbits with $n_i = 2, 3, 4, \ldots\ldots$ and makes transition to the final stationary orbit with $n_f = 1$, then the wave numbers of the spectral lines emitted due to these transitions given by (Eq. (4.4–17) are

$$\tilde{\nu} = R_\infty \left(\dfrac{1}{n_f^2} - \dfrac{1}{n_i^2} \right) = R_\infty \left(\dfrac{1}{1^2} - \dfrac{1}{n_i^2} \right)$$

which shows that these lines belong to the Lyman series.

Again if transitions take place from initial stationary orbits with $n_i = 3, 4, 5, \ldots\ldots$ to a final orbit with $n_f = 2$, then the wave numbers of the emitted spectral lines are given by

$$\tilde{\nu} = R_\infty \left(\dfrac{1}{n_f^2} - \dfrac{1}{n_i^2} \right) = R_\infty \left(\dfrac{1}{2^2} - \dfrac{1}{n_i^2} \right)$$

These lines belong to the Balmer series. Similarly the transitions from the initial orbits with $n_i = 4, 5, 6, \ldots\ldots$ to the final orbit with $n_f = 3$ give rise to the Paschen series ; the transitions from the initial orbits

Structure of the Hydrogen–Like Atom 55

with $n_i = 5, 6, 7, \ldots$ to the final orbit with $n_f = 4$ give rise to the Brackett series, and so on.

The origin of the different spectral series of hydrogen due to the transitions between the different orbits is shown in Fig. 4·5. It should be noted that the radii of the different orbits have not been drawn according to scale in this figure. The transitions are indicated by drawing arrows from the initial to the final orbits.

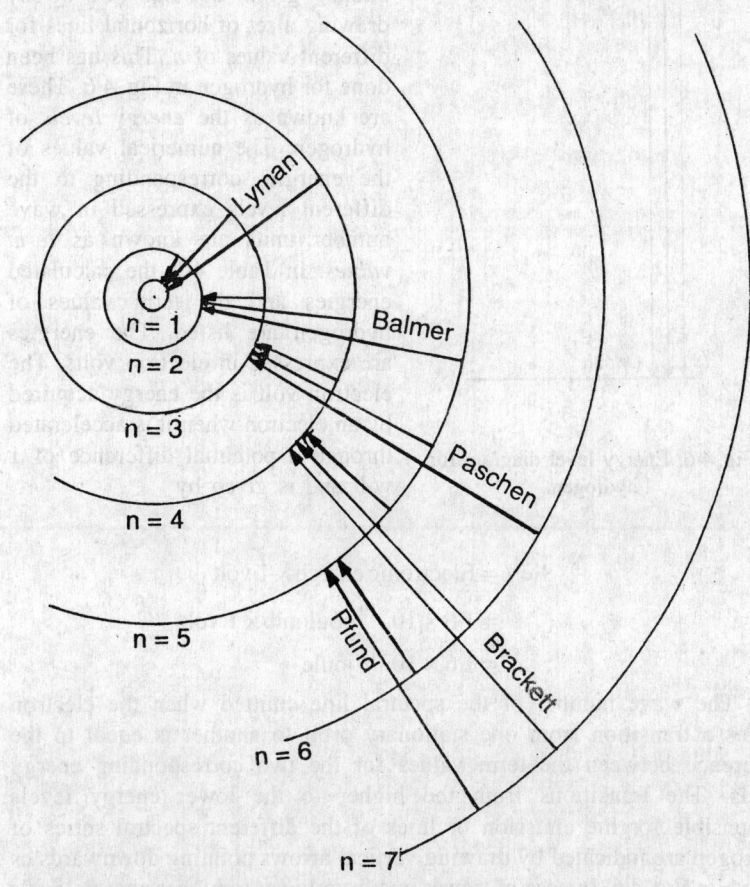

Fig. 4·5. Origin of the different spectral series of hydrogen according to Bohr.

4.6 Energy levels

According to Bohr's theory, the electron has definite energy in a stationary orbit. Eq. (4.4–18) gives this energy as

$$E_n = \frac{-2\pi c \hbar R_\infty Z^2}{n^2} = -\frac{ch R_\infty Z^2}{n^2} \qquad (4.6\text{–}1)$$

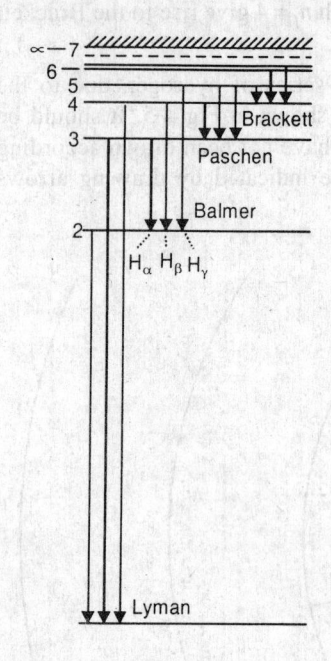

Fig. 4·6. Energy level diagram for hydrogen.

For hydrogen Z = 1, so that

$$E_n = -\frac{2\pi c\hbar R_\infty}{n^2} \quad (4\cdot 6\text{–}2)$$

We can represent the energies of the electron in the different orbits of hydrogen (or other hydrogen like atoms) given by Eq. (4·6–1) by drawing a set of horizontal lines for different values of n. This has been done for hydrogen in Fig. 4·6. These are known as the *energy levels* of hydrogen. The numerical values of the energies corresponding to the different levels expressed in wave number units are known as *term values*. In Table 4·1, the calculated energies and the term values of hydrogen are listed. The energies are expressed in electron volts. The electron volt is the energy acquired by an electron when it is accelerated through a potential difference of 1 volt and is given by

$$1 \text{ eV} = \text{Electronic charge} \times 1 \text{ volt}$$
$$= 1\cdot 6 \times 10^{-19} \text{ coulomb} \times 1 \text{ volt}$$
$$= 1\cdot 6 \times 10^{-19} \text{ joule}$$

The wave number of the spectral line emitted when the electron makes a transition from one stationary orbit to another is equal to the difference between the term values for the two corresponding energy levels. The transitions from the higher to the lower energy levels responsible for the emission of lines of the different spectral series of hydrogen are indicated by drawing vertical arrows pointing downwards as shown in Fig. 4·6. In case of transitions from lower to higher energy levels by the absorption of radiation, the arrows point vertically upwards. It will be seen from Fig. 4·6, that as the quantum number n increases, the energy levels become higher and are more closely spaced. When n is very large, the gap between the levels become almost indistinguishable. When $n \to \infty$, the energy becomes zero ($E_\infty = 0$). This corresponds to the electron being at an infinite distance from the nucleus ($r \to \infty$). It is no longer bound to the atom and is free to move away from it. The minimum energy

Structure of the Hydrogen–Like Atom

required to set free an electron originally bound in the lowest energy level ($n = 1$) of the hydrogen atom is given by (see Eq. (4·4–11)) :

$$I = E_\infty - E_1 = -E_1 = \frac{me^4}{32\pi^2 \epsilon_0^2 \hbar^2} = 2\pi c \hbar R_\infty \qquad (4.6\text{--}3)$$

It is possible to calculate I from the values of c, $\hbar$ and R_∞. It should be noted that normally the electron in an atom revolves in the smallest orbit for which $n = 1$ and the energy of the corresponding levels is the lowest. This is known as the *normal state* or the *ground state* of the atom. In Fig. 4·6, this is represented by the lowest horizontal line. In order that it may make a transition to a larger orbit *i.e.* to a higher energy state, it has to *absorb* energy. This energy may be supplied by exposing the atom to electromagnetic radiation or by hitting it with energetic electrons. If the energy so absorbed is equal to or greater than I then the electron will be released from the atom which is thus *ionized*. The energy I is known as the *ionization energy*. For hydrogen, the ionization energy is

$$I = 13.6 \text{ eV}$$

The corresponding *ionization potential* is 13·6 volts. This is equal to the potential difference through which an electron must be accelerated in order that it may gain an amount of energy equal to the ionization energy I. Obviously if such an electron from some external source strikes the electron bound in the hydrogen atom, the latter will be emitted and the atom will be ionized.

If an electron in the ground state of the hydrogen atom absorbs an energy E greater than I, then the extra energy $(E - I) = (E - 13.6)$ eV is equal to the kinetic energy after its release from the atom. For example, if $E = 20$ eV, kinetic energy of the electron set free from a hydrogen atom will be (20–13·6) or 6·4 eV. If the energy given to the electron is exactly equal to I, then it is just set free from the atom and has zero kinetic energy.

The electron set free from the atom can have any energy *i.e.*, it may have continuously varying energy. Its energy is no longer quantized so that the possible energy values in this case are no longer represented by a set of discrete levels as in Fig. 4·6 for the bound electron, but by the shaded portion above zero energy level ($n = \infty$). This energy region is known as a *continuum*.

In a gas discharge tube there are copious free electrons when the discharge is going on. These electrons acquire high energy under the influence of the large potential difference between the electrodes in the discharge tube. When they bombard the hydrogen atoms present in the tube, the electrons in the ground state in these atoms are raised to higher energy levels by absorbing energy from the incident electrons. They can stay in these *excited levels* (*i.e.*, levels of higher energy) for about

10^{-8} s and make transitions to lower energy levels emitting light corresponding to the different spectral lines in the process.

Table 4·1 Energy levels of hydrogen

Energy level (n)	Energy (eV)	Term values m^{-1}
1	−13·6	$1·09678 \times 10^7$
2	−3·4	$2·7420 \times 10^6$
3	−1·51	$1·2186 \times 10^6$
4	−0·85	$6·865 \times 10^5$
5	−0·544	$4·387 \times 10^5$
6	−0·378	$3·047 \times 10^5$
7	−0·277	$2·238 \times 10^5$
∞	0	0

In Table 4·2, the wave numbers and the wavelengths of some lines of the different spectral series of hydrogen are listed. It may be noted that the lines of the Balmer series are usually designated as $H_\alpha, H_\beta, H_\gamma$ etc., which have been indicated in the table.

Table 4·2

Spectral series	n_i	$\tilde{v}$ (m^{-1})	λ(Å)
Lyman	2	$8·2258 \times 10^6$	1216·0
($n_f = 1$)	3	$9·7491 \times 10^6$	1025·8
	4	$10·2823 \times 10^6$	972·5
	5	$10·5291 \times 10^6$	949·5
	∞	$10·9678 \times 10^6$	911·8
Balmer	3	$1·5233 \times 10^6$	6562·8 (H_α)
($n_f = 2$)	4	$2·0565 \times 10^6$	4861·3 (H_β)
	5	$2·3032 \times 10^6$	4340·5 (H_γ)
	6	$2·4373 \times 10^6$	4101·7 (H_δ)
	∞	$2·7420 \times 10^6$	3647·0
Paschen	4	$5·331 \times 10^5$	18,756
($n_f = 3$)	5	$7·799 \times 10^5$	12,821

Structure of the Hydrogen–Like Atom

Spectral series	n_i	$\tilde{\nu}$ (m^{-1})	λ(Å)
	6	$9\cdot139 \times 10^5$	10,939
	7	$9\cdot948 \times 10^5$	10,052
	∞	$12\cdot186 \times 10^5$	8806
Brackett	5	$2\cdot468 \times 10^5$	4·05 microns
($n_f = 4$)	6	$3\cdot808 \times 10^5$	2·63 "
	7	$4\cdot617 \times 10^5$	2·16 "
	8	$5\cdot141 \times 10^5$	1·94 "
	∞	$6\cdot855 \times 10^5$	1·46 "

4·7 Ritz Combination Principle

If an electron is initially in an excited energy level with $n = 3$, then it can come down first to the level with $n = 2$ and subsequently from the $n = 2$ level to the $n = 1$ level. Both these transitions will be accompanied by the emission of electromagnetic radiation. Alternatively the electron in the excited level $n = 3$ may make a transition directly to the $n = 1$ level. In the first case two different spectral lines having wave numbers $\tilde{\nu}_{32}$ and $\tilde{\nu}_{21}$ are emitted as shown in Fig. 4·7. In the second case, a single spectral line of wave number $\tilde{\nu}_{31}$ will be emitted. From Fig. 4·7. it is evident that

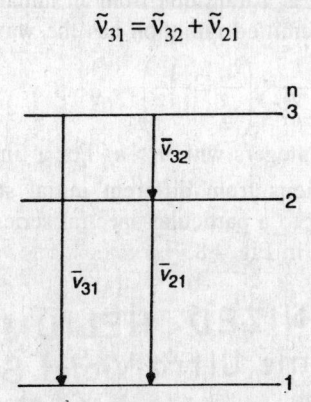

$$\tilde{\nu}_{31} = \tilde{\nu}_{32} + \tilde{\nu}_{21}$$

Fig. 4·7. Ritz combination principle.

Similarly, due to transition from an initial excited level with $n = 4$ to different lower levels with $n = 3, 2, 1$, a number of spectral lines may be emitted whose wave numbers are related to one another as follows:

$$\tilde{\nu}_{41} = \tilde{\nu}_{43} + \tilde{\nu}_{31} = \tilde{\nu}_{42} + \tilde{\nu}_{21} = \tilde{\nu}_{43} + \tilde{\nu}_{32} + \tilde{\nu}_{21}$$

Ritz was the first to observe that it is possible to express the wave number of a spectral line emitted by an atom as the sum of the wave numbers of two or more spectral lines. His discovery was made long before the advent of Bohr's theory and is known as *Ritz combination principle*. Bohr's theory which predicts the existence of discrete energy levels for the hydrogen-like atoms provides a proper explanation of the combination principle as discussed above.

The validity of Ritz combination principle can be checked with the help of Table 4·2. Thus the sum of the wave numbers of the first lines of the Paschen series ($n_i = 4$ to $n_f = 3$) and Balmer series ($n_i = 3$ to $n_f = 2$) is

$$\tilde{\nu}_{43} + \tilde{\nu}_{32} = 5 \cdot 331 \times 10^5 + 15 \cdot 233 \times 10^5 = 20 \cdot 564 \times 10^5 \text{ m}^{-1}.$$

This is in good agreement with the wave number of the second line of the Balmer series ($n_i = 4$ to $n_f = 2$) which is

$$\nu_{42} = 20 \cdot 565 \times 10^5 \text{ m}^{-1}$$

4·8 Effect of finite mass of the nucleus on the spectrum

We have seen that hydrogen-like atoms have spectra similar to the hydrogen spectrum. If we consider for instance He^+ which has a nuclear charge $+2e$ so that $Z = 2$ for it, then from Eq. (4·4–11) its term values are given by

$$E_n = -\frac{mZ^2 e^4}{32\pi^2 \epsilon_0^2 n^2 \hbar^2} = -\frac{me^4}{8\pi^2 \epsilon_0^2 n^2 \hbar^2}$$

So the *term values* of He^+ are four times larger than the corresponding hydrogen terms. If there is a transition from an initial energy level n_i to a final level n_f, then the emitted radiation has the wave number

$$\tilde{\nu} = \frac{me^4}{16\pi^3 \epsilon_0^2 c\hbar^3} \left(\frac{1}{n_f^2} - \frac{1}{n_i^2} \right)$$

n_i and n_f are two integers with $n_i > n_f$. For a final state with a given value of n_f the transitions from different initial states n_i give rise to spectral lines belonging to a particular spectral series for ionised helium. Such a series is shown in Fig. 4·8.

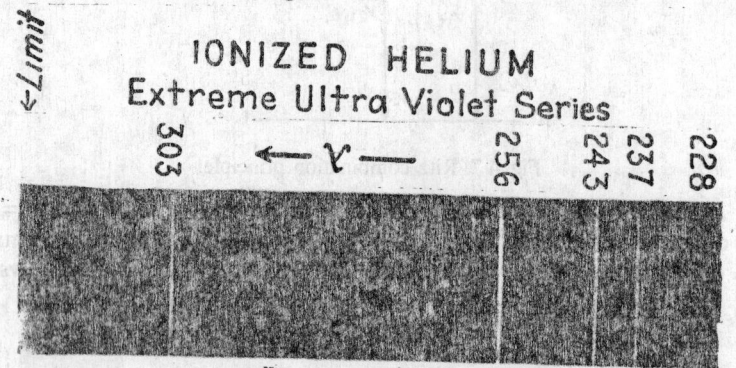

Fig. 4·8. Spectrum of ionised helium.

Structure of the Hydrogen–Like Atom

Consider transitions from the $n_i = 5, 6, 7, 8$ etc. levels of He^+ to the final level with $n_f = 4$. The corresponding spectral series is known as the Pickering series. Its lines can be represented by the formula

$$\tilde{\nu} = \frac{m\,e^4}{16\pi^3 \epsilon_0^2 c\hbar^3}\left(\frac{1}{4^2} - \frac{1}{n_i^2}\right), n_i = 5, 6, 7, 8, \ldots$$

$$= \frac{m\,e^4}{64\pi^3 \epsilon_0^2 c\hbar^3}\left(\frac{1}{2^2} - \frac{1}{n^2}\right), n = \frac{n_i}{2}$$

$$= R_\infty\left(\frac{1}{2^2} - \frac{1}{n^2}\right)$$

Thus the alternate lines of the Pickering series with n_i even ($n_i = 6, 8, \ldots$) of He^+ should have the same wavelengths as the spectral lines of the Balmer series of hydrogen (see page 54).

Careful measurements have shown that though the lines of the two series have approximately the same wave numbers, the lines of the Pickering series of He^+ have slightly higher wave numbers than the corresponding lines of the Balmer series which is not expected from Bohr's theory.

The reason for the above difference is due to the finite masses of the nuclei of hydrogen and helium. In the original Bohr theory, the nucleus was assumed to be at rest which happens only if it is infinitely heavy. However a nucleus of finite mass cannot be at rest. The combined electron-nucleus system will revolve around their common centre of mass with the same angular velocity. The mass of the hydrogen nucleus is 1836 times heavier than the electron mass. The helium nucleus is four times heavier than the hydrogen nucleus. Even though these masses are much higher than the electronic mass m, they cannot be considered infinitely heavy.

The revolution of the electron and a nucleus of finite mass $M \gg m$ about the common centre of mass C is shown if Fig. 4.9. If r_1 and r_2 are the radii of the orbits of the electron and the nucleus, assumed circular, then we can write

$$mr_1 = Mr_2$$

so that

$$r_2 = \frac{m}{M} r_1$$

If the electron-nucleus distance be r, then

$$r = r_1 + r_2 = r_1\left(1 + \frac{m}{M}\right)$$

The combined angular momentum of the electron and the nucleus is

$$L = mr_1^2 \omega + Mr_2^2 \omega$$

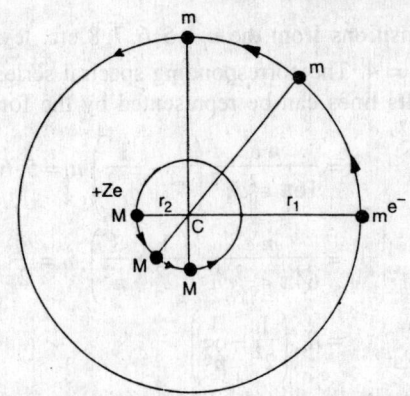

Fig. 4·9. Motion of the electron and the nucleus around the centre of mass.

$$= mr_1^2\omega\left(1 + \frac{m}{M}\right)$$

$$= \frac{mr^2\omega}{1 + m/M}$$

Suppose $\quad \mu = \dfrac{mM}{m+M} = \dfrac{m}{1 + m/M}$

μ is known as the *reduced mass*. Then we get

$$L = \mu r^2\omega$$

Thus we may take into account the motion of the nucleus by replacing the electron-nucleus system by a particle of mass μ revolving round an in infinitely heavy nucleus at rest in an orbit of radius r with an angular velocity ω. Bohr's quantum condition (Eq. (4·3–1)) then gives us

$$\mu r^2 \omega = n\hbar \qquad (4\cdot 8\text{–}1)$$

where $n = 1, 2, 3, \ldots\ldots$ The electrostatic force between the electron and the nucleus is

$$F_e = \frac{Ze^2}{4\pi\epsilon_0 r^2} = m\cdot\omega^2 r_1 = M\omega^2 r_2 = \mu\omega^2 r \qquad (4\cdot 8\text{–}2)$$

The sum of the kinetic energies of the electron and the nucleus is

$$E_k = \frac{1}{2} mr_1^2\omega^2 + \frac{1}{2} Mr_2^2\omega^2 = \frac{1}{2}\mu r^2 \omega^2 = \frac{Ze^2}{8\pi\epsilon_0 r}$$

The potential energy is

$$V = -\frac{Ze^2}{4\pi\epsilon_0 r}$$

So the total energy is

$$E = E_k + V = -\frac{Ze^2}{8\pi\epsilon_0 r} \qquad (4\cdot 8\text{–}3)$$

Structure of the Hydrogen–Like Atom

Then using Eqs. (4·8–1), (4·8–2) and (4·8–3), we get

$$E_n = -\frac{\mu Z^2 e^4}{32\pi^2 \epsilon_0^2 n^2 \hbar^2} \qquad (4\cdot 8\text{–}4)$$

If we compare the above equation with Eq. (4·4–11), we see that the only difference between the two is that the electronic mass m is replaced by the reduced mass μ when the motion of the nucleus due to its finite mass is considered. The Rydberg constant will now be

$$R_M = \frac{\mu e^4}{64\pi^3 \epsilon_0^2 c \hbar^3} = \frac{R_\infty}{1 + m/M} \qquad (4\cdot 8\text{–}5)$$

where R_∞ is the Rydberg constant for an infinitely heavy nucleus given by Eq. 4·4-15.

The transition between two energy levels results in the emission of electromagnetic radiation of wave number

$$\tilde{\nu} = R_M Z^2 \left(\frac{1}{n_f^2} - \frac{1}{n_i^2} \right) = \frac{R_\infty Z^2}{1 + m/M} \left(\frac{1}{n_f^2} - \frac{1}{n_i^2} \right) \qquad (4\cdot 8\text{–}6)$$

The Rydberg constant depends on the nuclear mass M. Since $M \gg m$, the difference between R_M and R_∞ is small. For hydrogen and helium, we get

$$R_H = 1\cdot 09677 \times 10^7 \text{ m}^{-1}, \quad R_{He} = 1\cdot 09722 \times 10^7 \text{ m}^{-1}$$

The above value of R_H agrees very well with the experimental value derived from the study of the hydrogen spectrum ($1\cdot 09678 \times 10^7 \text{ m}^{-1}$) discussed in § 4.5. Similarly in the case of helium too, the agreement between the experimental value and the theoretical value given above is very good.

The reduced mass $\mu < m$. It increases with increasing nuclear mass M and becomes equal to the electronic mass as $M \to \infty$. So R_M increases as M increases. As $M \to \infty$, $R_M \to R_\infty$. Since $R_{He} > R_H$ the spectral lines of the Pickering series of He^+ should have slightly higher wave numbers or shorter wavelengths than the corresponding lines of the Balmer series of hydrogen. We have seen above that this is actually observed experimentally.

4·9 Discovery of heavy hydrogen

Naturally occurring hydrogen has two isotopes having mass numbers 1 and 2. The latter is known as *heavy hydrogen or deuterium*. Its nucleus is known as the *deutron*. The two types of hydrogen have the same chemical properties. The atoms of both types of hydrogen have only one electron revolving in their orbits. Their nuclei have the same charge but different masses : $M_D \approx 2M_H$. So according to the discussion in § 4·8, the Rydberg constant for hydrogen and deuterium are slightly different :

$$R_H = \frac{R_\infty}{1 + M/M_H} \text{ and } R_D = \frac{R_\infty}{1 + m/M_D}$$

Since $M_D > M_H$, R_D is slightly greater than R_H. Calculation gives $R_D = 1·09707 \times 10^7 \text{ m}^{-7}$. Hence the lines of the deuterium spectra will have slightly higher wave numbers than the corresponding lines of the hydrogen spectra.

Three American scientists H. C. Urey, F.G. Brickwedde and G.M. Murphy were the first to observe (1931) the presence of faint companion lines on the short wavelength sides of each of the hydrogen spectral lines. From this observation, they concluded that these fainter lines were due to a hitherto undiscovered isotope of hydrogen. From the measured wave number differences between the faint companion lines of D and the corresponding much brighter parent lines of H the atomic mass of the isotope was estimated to be *twice* than that of hydrogen. From the faintness of the lines, it is evident, that its abundance in naturally occurring hydrogen is quite low. Later, improved methods of measurement have yielded a relative abundance of only about 1 part in 7000 (H — 99·985 %; D — 0·015%) for deuterium.

The wavelength difference between the corresponding spectral lines of hydrogen and deuterium can be determined as follows :

$$\lambda_H = \frac{1}{\tilde{\nu}_H} = \frac{1}{R_H}\left(\frac{1}{n_2^2} - \frac{1}{n_1^2}\right)^{-1} = \frac{1 + m/M_H}{R_\infty}\left(\frac{1}{n_2^2} - \frac{1}{n_1^2}\right)^{-1}$$

$$\lambda_D = \frac{1}{\tilde{\nu}_D} = \frac{1}{R_D}\left(\frac{1}{n_2^2} - \frac{1}{n_1^2}\right)^{-1} = \frac{1 + m/M_D}{R_\infty}\left(\frac{1}{n_2^2} - \frac{1}{n_1^2}\right)^{-1}$$

$$\therefore \quad \Delta\lambda = \lambda_H - \lambda_D = \frac{m/M_H - m/M_D}{R_\infty\left(\frac{1}{n_2^2} - \frac{1}{n_1^2}\right)}$$

Since both m/M_H and m/M_D are very small numbers, we can write the denominator in the above expression to be approximately equal to $1/\lambda_H$. Also since $M_D \approx 2M_H$ we get finally

$$\Delta\lambda = \lambda_H \cdot \frac{m}{2M_H} = \frac{\lambda_H}{3672}$$

The H_β line of the Balmer series ($n_1 = 4 \to n_2 = 2$) of hydrogen has the wavelength $\lambda_H = 4681$ Å. So from the above expression, we get $\Delta\lambda = 1·28$ Å for the wavelength difference of the H_β lines of hydrogen and

4·10 Bohr's Correspondence Principle

We have stated earlier that Bohr's theory of the hydrogen spectra was *ad hoc* in nature (§ 4.4). Planck's quantum theory proposed a few years earlier and Einstein's light quantum hypothesis to explain photoelectron emission (see later) were also *ad hoc* in nature. In none of these, was any attempt made to devise a new theory of atomic mechanics.

The laws of classical mechanics are applicable to explain the motion of all macroscopic bodies, starting from the very large heavenly bodies like the galaxies, stars and planets down to the very small objects like the atoms and molecules as a whole. The kinetic theory of matter which describes the motion of the molecules in a gas is based on classical mechanics.

Bohr pointed out that the new laws of atomic mechanics required to describe the motion of the subatomic particles (like the electrons inside the atoms) have to be developed in such a way that these laws must go over to the laws of classical mechanics in the limit of larger macroscopic bodies. This is known as *Bohr's correspondence principle* and forms an important guide line in formulating the laws of atomic mechanics.

Since the orbital radius of the electron becomes very large when the quantum number n is very large, the laws of classical mechanics should be applicable in these cases. In other words, according to the correspondence principle, the results derived from Bohr's theory should go over to those deduced from classical physics in the limit of very large quantum numbers.

According to the laws of classical electromagnetic theory, the frequency of the radiation (f) emitted by a circulating charged particle is equal to the frequency of rotation of the charge or an integral multiple of it. For an electron in the hydrogen atom rotating in its orbit, the frequency of rotation is, according to Eqs. (4·4-4) and (4·4-5).

$$f = \frac{v}{2\pi r} = \frac{e^2}{4\pi\epsilon_0 n\hbar} \cdot \frac{me^2}{8\pi^2\epsilon_0 n^2\hbar^2} = \frac{me^4}{32\pi^3\epsilon_0^2 n^3\hbar^3}$$

Again from Eq. (4·4-13), the frequency of the radiation emitted due to transition from the $(n+1)$th to the nth orbit of the hydrogen atom is

$$v = \frac{\omega}{2\pi} = \frac{me^4}{64\pi^3\epsilon_0^2\hbar^3}\left\{\frac{1}{n^2} - \frac{1}{(n+1)^2}\right\}$$

$$= \frac{me^4}{64\pi^3\epsilon_0^2\hbar^3} \cdot \frac{2n+1}{n^2(n+1)^2}$$

If $n \gg 1$, then we get

$$v = \frac{me^4}{64\pi^3 \epsilon_0^2 \hbar^3} \cdot \frac{2n}{n^4} = \frac{me^4}{32\pi^3 \epsilon_0^2 n^3 \hbar^3} = f$$

Thus in the limit of very large quantum numbers, the results of Bohr's theory are the same as those obtained from classical electromagnetic theory, as required by the correspondence principle.

In the above discussion, we have assumed that the transition takes place from the $(n + 1)$th to the nth orbit so that the change of quantum number is by 1 unit. If on the other hand, the change of quantum number is 2, 3, ... units, *i.e.*, if the transition is from the $(n + 2)$th, $(n + 3)$th etc. orbits to the nth orbit, then for large n, the frequencies of the emitted radiation can be shown to be twice, three times etc. of the frequency of rotation. As we have seen above, the classical electromagnetic theory predicts the emission of radiation of these frequencies also.

4·11 Wilson-Sommerfeld quantum conditions

In Bohr's theory, the electron orbit is assumed to be circular. Actually however, the orbit of a particle acted upon by an attractive force directed towards a fixed point (*central force*) is an ellipse with the centre of force at one of the foci of the ellipse, if the force $F \propto 1/r^2$ and the energy E of the particle is negative. The orbit is circular only as a special case. For example, we know from Kepler's laws that the planets revolve round the sun in elliptic orbits with the sun at one of the foci. The force acting on the planets is the gravitational attraction of the sun which is of the inverse-square type. Since the electrons in the atom are acted upon by the Coulomb attractive force of the nucleus which is also of the inverse-square type, these will in general describe elliptic orbits as long as they are bound in the atom ($E < 0$).

The position of a particle describing an elliptic orbit in a plane can be represented by the two polar coordinates r and θ as shown in Fig. 4·10. In general, both r and θ change periodically. In one complete

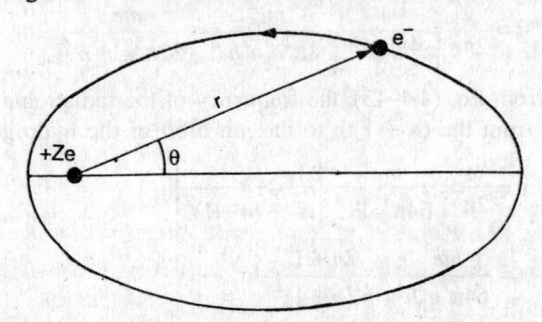

Fig. 4·10. Elliptic orbit of the electron.

revolution θ changes from 0 to 2π while r changes from a minimum value at one end of the ellipse (*perihelion*) to a maximum value at the other end (*aphelion*) and back again to the minimum value.

The British physicist W. Wilson and the German scientist Arnold Sommerfeld independently discovered (1916) the generalized rules for the quantization for the periodic motion of a particle having one or more than one degree of freedom.

Let us first consider the motion of the linear harmonic oscillator of mass m. If x be the displacement of the particle from the mean position of rest, the equation of motion will be

$$m\ddot{x} + m\omega^2 x = 0$$

where $-m\omega^2 x$ is the force of restitution, always directed towards the mean position of rest. $\omega = 2\pi \nu$ is the circular frequency. Since $p = m\dot{x}$ is the momentum of the particle, its total energy which is the sum of its kinetic and potential energies is given by

$$E = E_k + V = \frac{1}{2} m\dot{x}^2 + \frac{1}{2} m\omega^2 x^2 = \frac{p^2}{2m} + \frac{x^2}{2/m\omega^2}$$

The above equation can be rewritten as

$$\frac{p^2}{2mE} + \frac{x^2}{2E/m\omega^2} = 1 \tag{4.11-1}$$

Eq. (4·11–1) is the equation of an ellipse with the two semi-axes $a = \sqrt{2mE}$ and $b = \sqrt{2E/m\omega^2}$. According to this equation, the position x and the momentum p of the particle at any instant is determined by the coordinates of a point on this ellipse (see Fig. 4·11). When the oscillator completes one full oscillation, the point describes one complete revolution on the ellipse.

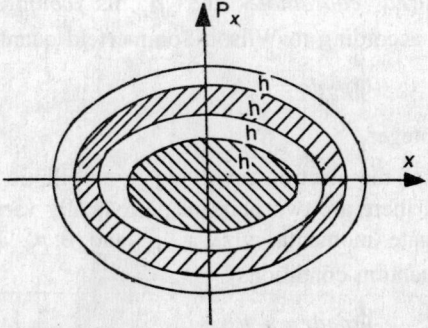

Fig. 4·11. Plot of p_x vs. x for a linear harmonic oscillator.

It should be borne in mind that the ellipse given by Eq. (4·11–1) is not the elliptic orbit we had been talking about earlier. The change of position of the point on the ellipse as discussed above does not represent the change of position of any material particle. It simply gives the mathematical relationship between the momentum p and the position x of the harmonic oscillator at different instants of time.

The area of the ellipse given by Eq. (4·11–1) is given by

$$J = \oint p\,dx = \pi ab = \pi\sqrt{2mE}\sqrt{\frac{2E}{m\omega^2}} = \frac{2\pi E}{\omega} = \frac{E}{v} \qquad (4\cdot11\text{–}2)$$

Here the symbol $\oint$ denotes integration over a complete cycle. According to Planck, the energy of an atomic oscillator should be $E = nh\nu$ where n is an integer. Substituting this quantum condition of Planck in Eq. (4·11–1), we get

$$J = \oint p\,dx = nh \qquad (4\cdot11\text{–}3)$$

where $n = 1, 2, 3, \ldots$ Thus the harmonic oscillator oscillates in such a way that the ellipse given by Eq. (4·11–1) which is a graphical representation of a mathematical relationship between the coordinate and momentum of the oscillator at different instants of time can have an area which is an integral multiple of Planck's constant h. Hence the area of the strips between the successive ellipses in Fig. 4·11 are h each.

The two dimensional space with x and p as two rectangular coordinates in known as the two-dimensional *phase-space* and the integral given by Eq. (4·11–3) is called the *phase integral* or the *action-integral*. A point in the phase space represents the state of motion of the particle at a particular instant of time. The ellipse represented by Eq (4·11–1) is thus the path of the *phase-point*. Eq. (4·11–3) gives us the rule for the quantization of the action integral. According to Wilson and Sommerfeld this rule of quantization is applicable for any pair of periodically varying coordinate and its *conjugate* momentum. If q represents a periodically varying *generalized coordinate* and p_q its *conjugate generalized momentum*, then according to Wilson-Sommerfeld quantum condition

$$\oint p_q\,dq = n_q h \qquad (4\cdot11\text{–}4)$$

where n_q is an integer.

In the case of the electron revolving in an elliptic orbit (cyclically periodic motion), there are two pairs of periodically varying coordinates and their conjugate momenta, viz., r, p_r and θ, p_θ. Hence we have following two quantum conditions

$$\oint p_r\,dr = n_r h \qquad (4\cdot11\text{–}5)$$

$$\oint p_\theta\,d\theta = n_\theta h \qquad (4\cdot11\text{–}6)$$

The *radial quantum number* n_r and the *orbital* or *azimuthal quantum number* n_θ can only have integral values. So in place of the single

Structure of the Hydrogen–Like Atom

quantum number n of Bohr's theory we now have the two quantum numbers n_r and n_θ. Both Planck's quantum conditions (3·3–1) and Bohr's quantum condition (4·3–1) follow as special cases of the general quantum condition (4·11–4).

For a *cyclically periodic motion*, if the radial distance r is a constant (circular orbit), then the only variable is θ which changes from 0 to 2π in a complete period. The conjugate momentum p_θ is actually the angular momentum of the electron which remains constant under the action of a *central force*. So we have

$$\oint p_\theta \, d\theta = p_\theta \int_0^{2\pi} d\theta = 2\pi \, p_\theta = nh$$

where we have written $n_\theta = n$ which is an integer. So we get

$$p_\theta = \frac{nh}{2\pi} = n\hbar$$

which is Bohr's quantum condition.

4.12 Sommerfeld's theory of elliptic orbits ; Relativity correction

Sommerfeld used the two quantum conditions (4·11–5) and (4·11–6) to calculate the energy of the electron revolving in an elliptic orbit. Since the angular momentum p_θ = constant under the action of a central force, we have

$$\oint p_\theta \, d\theta = 2\pi p_\theta = kh$$

where we have written $n_\theta = k$. So we get

$$p_\theta = kh \qquad (4 \cdot 12-1)$$

k is an integer.

It is known from coordinate geometry that the following relation holds between the coordinates r and θ for an elliptic orbit.

$$r = \frac{a(1-\epsilon^2)}{1+\epsilon \cos \theta} \qquad (4 \cdot 12-2)$$

where ϵ is the eccentricity of the ellipse given by

$$\sqrt{1-\epsilon^2} = \frac{b}{a} \qquad (4 \cdot 12-3)$$

Here a and b are the semi-major and semi-minor axes of ellipse. Eq. (4·12–2) shows that as θ changes by 2π, r goes through one complete cycle of change, coming back to its original value since $\cos(2\pi + \theta) = \cos \theta$.

The radial component of the momentum is

$$p_r = m\dot{r} = m \frac{dr}{d\theta} \cdot \frac{d\theta}{dt}$$

$$= mr^2 \dot{\theta} \left(\frac{1}{r^2} \frac{dr}{d\theta} \right) = \frac{p_\theta}{r^2} \frac{dr}{d\theta} \qquad (4\cdot12\text{–}4)$$

From Eqs. (4·11–5) and (4·12–4) we get

$$n_r h = \oint p_r \, dr = \oint \frac{p_\theta}{r^2} \frac{dr}{d\theta} \cdot \frac{dr}{d\theta} d\theta$$

$$= p_\theta \oint \left(\frac{1}{r} \frac{dr}{d\theta} \right)^2 d\theta \qquad (4\cdot12\text{–}5)$$

Since p_θ is a constant it has been taken out of the integral sign.

From Eq. (4·12–2) we get

$$\frac{dr}{d\theta} = \frac{a(1-\epsilon^2)\epsilon \sin\theta}{(1+\epsilon\cos\theta)^2}$$

so that

$$\frac{1}{r} \frac{dr}{d\theta} = \frac{\epsilon \sin\theta}{1+\epsilon\cos\theta} \qquad (4\cdot12\text{–}6)$$

Then from Eq. (4·12–5) we get

$$n_r h = p_\theta \oint \frac{\epsilon^2 \sin^2\theta \, d\theta}{(1+\epsilon\cos\theta)^2} = p_\theta \times I \text{ (say)}$$

Integration by parts gives

$$I = \frac{\epsilon \sin\theta}{1+\epsilon\cos\theta} \Big]_0^{2\pi} - \int_0^{2\pi} \frac{\epsilon \cos\theta \, d\theta}{1+\epsilon\cos\theta}$$

$$= \int_0^{2\pi} \left(\frac{1}{1+\epsilon\cos\theta} - 1 \right) d\theta = 2\pi \left\{ \frac{1}{\sqrt{1-\epsilon^2}} - 1 \right\}$$

The value of the first of the above integrals may be found from any standard table of integrals. We then have since $p_\theta = k\hbar$

$$n_r h = 2\pi k \hbar \left\{ \frac{1}{\sqrt{1-\epsilon^2}} - 1 \right\} \qquad (4\cdot12\text{–}7)$$

Then we get

$$\frac{1}{\sqrt{1-\epsilon^2}} = 1 + \frac{n_r}{k} = \frac{k+n_r}{k} = \frac{n}{k} \text{ (say)} \qquad (4\cdot12\text{–}8)$$

Since n_r and k are both integers, their sum $n = n_r + k$ is also an integer. It is called the *principal quantum number*.

Structure of the Hydrogen–Like Atom

We thus get using Eq. (4·12–3)

$$\frac{b}{a} = \sqrt{1-\epsilon^2} = \frac{k}{n} \qquad (4\cdot12\text{–}9)$$

Since $b \leq a$, we must have $k \leq n$. For a circular orbit, $b = a$ and $\epsilon = 0$. So in this case $k = n$. Thus the maximum value of the azimuthal quantum number is $k = n$. In this case the radial quantum number $n_r = 0$. Again since k is a positive integer, its minimum value is $k = 1$. For $k = 0$ the angular momentum of the electron $p_\theta = 0$ which means that the electron would execute simple harmonic motion through the nucleus instead of moving in an elliptic orbit around the latter. Since this is not possible, we have to rule out the value $k = 0$ so that the minimum value is $k = 1$ as stated above. Thus the permissible values of k are

$$k = 1, 2, 3, \ldots n$$

Since $n_r = n - k$, the corresponding values of n_r are

$$n_r = (n-1), (n-2), (n-3), \ldots 0$$

The possible values of b/a are

$$\frac{b}{a} = \frac{1}{n}, \frac{2}{n}, \frac{3}{n}, \ldots 1$$

The energy of the electron in the elliptic orbit is

$$E = E_k + V = \frac{p_r^2}{2m} + \frac{p_\theta^2}{2mr^2} - \frac{1}{4\pi\epsilon_0}\frac{Ze^2}{r}$$

Using Eq. (4·12–4) we get

$$E = \frac{p_\theta^2}{2mr^2}\left\{\left(\frac{1}{r}\frac{dr}{d\theta}\right)^2 + 1\right\} - \frac{1}{4\pi\epsilon_0}\frac{Ze^2}{r}$$

or,

$$\left(\frac{1}{r}\frac{dr}{d\theta}\right)^2 = \frac{2mE}{p_\theta^2}r^2 + \frac{mZe^2}{2\pi\epsilon_0 p_\theta^2}r - 1 \qquad (4\cdot12\text{–}10)$$

Again from Eq. (4·12–2) we have

$$\epsilon^2 \sin^2\theta = \epsilon^2 - \epsilon^2 \cos^2\theta$$

$$= \epsilon^2 - \left\{\frac{a(1-\epsilon^2)}{r} - 1\right\}^2$$

$$= (\epsilon^2 - 1) + \frac{2a(1-\epsilon^2)}{r} - \frac{a^2(1-\epsilon^2)^2}{r^2}$$

So from Eqs. (4·12–2) and (4·12–6) we get

$$\left(\frac{1}{r}\frac{dr}{d\theta}\right)^2 = \frac{\epsilon^2 \sin^2\theta}{(1+\epsilon\cos\theta)^2}$$

$$= \frac{r^2}{a^2(1-\epsilon^2)^2}\left\{(\epsilon^2-1)+\frac{2a(1-\epsilon^2)}{r}-\frac{a^2(1-\epsilon^2)^2}{r^2}\right\}$$

$$= -\frac{r^2}{a^2(1-\epsilon^2)}+\frac{2r}{a(1-\epsilon^2)}-1 \tag{4.12-11}$$

Eqs. (4·12–10) and (4·12–11) must be valid for all values of r on the ellipse. This is possible if the coefficients of each power of r are separately equal in both equations. Hence we get

$$\frac{2mE}{p_\theta^2} = -\frac{1}{a^2(1-\epsilon^2)} \tag{4.12-12}$$

$$\frac{mZe^2}{2\pi\epsilon_0 p_\theta^2} = \frac{2}{a(1-\epsilon^2)} \tag{4.12-13}$$

From Eqs. (4·12–12) and (4·12–13) we get

$$\frac{4\pi\epsilon_0 E}{Ze^2} = -\frac{1}{2a} \tag{4.12-14}$$

or
$$E = -\frac{Ze^2}{8\pi\epsilon_0 a} \tag{4.12-15}$$

Using Eqs. (4·12–1), (4·12–8) and (4·12–13), we have

$$a = \frac{4\pi\epsilon_0 p_\theta^2}{mZe^2(1-\epsilon^2)} = \frac{4\pi\epsilon_0 k^2\hbar^2}{mZe^2}\cdot\frac{n^2}{k^2} = \frac{4\pi\epsilon_0 n^2\hbar^2}{mZe^2} \tag{4.12-16}$$

$$\therefore \qquad E = -\frac{mZ^2 e^4}{32\pi^2\epsilon_0^2 n^2\hbar^2} \tag{4.12-17}$$

Thus the total energy does not depend on the azimuthal quantum number k; it depends on the principal quantum number n only. The value of E as given above is the same as the found from Bohr's theory.

Eq. (4·12–16) shows that the semi-major axis a is also independent of k, but depends only on n. The above expression for a is the same as the expression for radius of the circular orbit in Bohr's theory.

Since for a given principal quantum number there are n different elliptic orbits with different values of k, the electron has the same energy while revolving in any one of these orbits. Such a system is said to be degenerate. All these elliptic orbits for a given n but different k have the same value of the semi-major axis a, but different values of the semi-minor axis b (see Fig. 4·12).

The degeneracy, i.e., equality of the energy values for the different elliptic orbits of the electron for a given n but different k arises because of the equality of the period of variation of the coordinates r and θ. As a result, the two quantum conditions (4·11–5) and (4·11–6) are really not

Structure of the Hydrogen–Like Atom

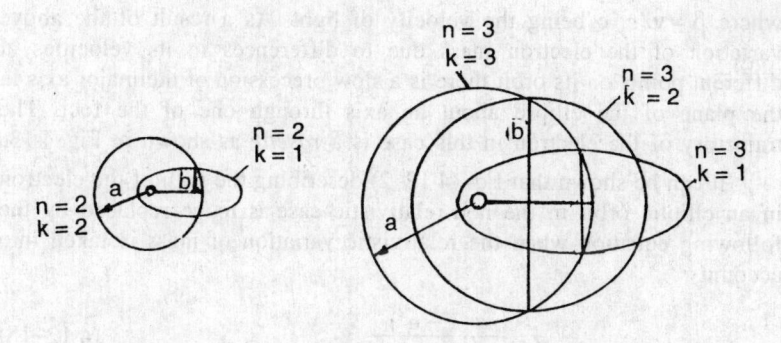

Fig. 4·12. Different possible orbits for definite principal quantum numbers n.

two independent conditions. This is a general result which holds when the ratios of the periods of variation of two or more coordinates describing the motion of a system are rational numbers *i.e.*, when the periods are commensurable. If however, the ratios are irrational numbers so that they are not commensurable, then only the different quantum conditions of the type Eq. (4·11-4) for the different pairs of the coordinates and their conjugate momenta will be independent. The system is said to be non-degenerate in this case.

From the above discussion it is clear that the energy levels of the electron in the hydrogen-like atoms as found from Sommerfeld's theory are identical with those found from Bohr's theory. So the predicted energy spectra will also be identical is both cases. However the experimentally observed spectra for these atoms do not agree with those predicted by these theories. Experiments with high resolution spectroscopes reveal a fine structure of the spectral lines of these atoms i.e., each spectral line as predicted by these theories is found to be split up into a number of very closely lying component lines which cannot be explained on the basis of these theories.

Sommerfeld pointed out that origin of the fine structure of the spectral lines of the hydrogen-like atoms was due to the relativistic variation of the mass of the electron. When an electron in the course of its revolution in an elliptic orbit comes closer to the nucleus, its velocity becomes relatively higher than when it is farther away. Due to this, its mass is relatively larger when it is closer to the nucleus as predicted by the special theory of relativity. According to this theory, the mass m of a particle moving with a velocity v is related to its rest mass m_0 (which is its mass when the velocity is 0) by the following formula (see § 15·12).

$$m = \frac{m_0}{\sqrt{1 - \beta^2}}$$

where $\beta = v/c$, c being the velocity of light. As a result of the above variation of the electron mass due to differences in its velocities at different points on its orbit there is a slow precession of the major axis in the plane of the ellipse about an axis through one of the foci. The trajectory of the electron in this case is a *rosette* as shown in Fig. 4·13.

It can be shown that Eq. (4·12–2) describing the path of the electron in an elliptic orbit in the non-relativistic case is to be replaced by the following equation when the relativistic variation of mass is taken into account :

$$r = \frac{a(1-\epsilon^2)}{1+\epsilon \cos \gamma \theta} \qquad (4\cdot12\text{–}18)$$

where

$$\gamma^2 = 1 - \frac{Z^2 e^4}{(4\pi\epsilon_0)^2 p^2 c^2} \qquad (4\cdot12\text{–}19)$$

Here we have written $p_\theta = p$ which is a constant.

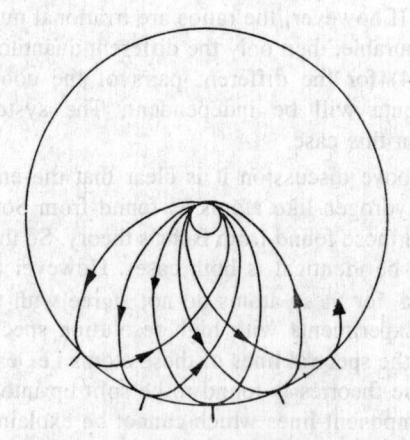

Fig. 4·13. Precession of the elliptic orbit.

For the non-relativistic case ($c \to \infty$), $\gamma = 1$ and r returns to the same value when θ changes by 2π so that it has the same periodicity as θ. However when $\gamma < 1$ as in the relativistic case, r does not return to the same value when θ changes by 2π and it has now a different periodicity from θ. In fact, in this case the value of r becomes the same only when θ changes by $2\pi/\gamma > 2\pi$ since

$$\cos \gamma (2\pi/\gamma + \theta) = \cos (2\pi + \gamma \theta) = \cos \gamma \theta$$

So the axis of the ellipse precesses through an angle

$$\Delta\theta = \frac{2\pi}{\gamma} - 2\pi$$

when r returns to its original value.

Structure of the Hydrogen–Like Atom

Taking the effect of relativistic precession into account, Somemerfeld calculated the total energy to the electron to be (see *Atomic Structure and Spectral Lines* by A. Sommerfeld)

$$E_{nk} = -\mu c^2 \left[1 + \frac{\alpha^2 Z^2}{(n-k+\sqrt{k^2-\alpha^2 Z^2})^2}\right]^{-\frac{1}{2}} - \mu c^2$$

where μ is the reduced mass and α is called Sommerfeld's *fine structure constant* given by

$$\alpha = \frac{e^2}{4\pi\epsilon_0 c\hbar} = \frac{1}{137 \cdot 03605} \quad (4 \cdot 12\text{--}20)$$

α is equal to the ratio of the velocity of the electron in the first Bohr orbit of hydrogen to the velocity of light c.

The above expression for E_{nk} can be expanded in ascending powers of α^2 as follows:

$$E_{nk} = -\frac{\mu e^4 Z^2}{32\pi^2 \epsilon_0^2 n^2 \hbar^2}\left[1 + \frac{Z^2\alpha^2}{n^2}\left(\frac{n}{k}-\frac{3}{4}\right) + \frac{Z^4\alpha^4}{n^4}\left(\frac{n^3}{4k^3}+\frac{3n^2}{4k^2}-\frac{3n}{2k}+\frac{5}{8}\right)\right.$$
$$\left. +\frac{Z^6\alpha^6}{n^6}\left(\frac{n^5}{8k^5}+\frac{3n^4}{8k^4}+\frac{n^3}{8k^3}-\frac{15n^2}{8k^2}+\frac{15n}{8k}-\frac{35}{64}\right)+\ldots\ldots\right] \quad (4 \cdot 12\text{--}21)$$

Since the successive terms diminish rapidly, we get, retaining terms up to α^2

$$E_{nk} = -\frac{\mu Z^2 e^4}{32\pi^2 \epsilon_0^2 n^2 \hbar^2}\left[1 + \frac{Z^2\alpha^2}{n^2}\left(\frac{n}{k}-\frac{3}{4}\right)\right] \quad (4 \cdot 12\text{--}22)$$

Since $\alpha \ll 1$, the difference in energy from Bohr's theory as given by the second term within the bracket is quite small. This correction term due to the relativistic variation of mass depends on k. So an energy level for a given n now splits up into k closely lying sub-levels of slightly different term values. This fine structure splitting of the energy levels of the hydrogen-like atoms is illustrated in Fig. 4·14. where the fine structure splitting has been shown very much magnified. For a given n the sub-levels with higher k lie above those with lower k.

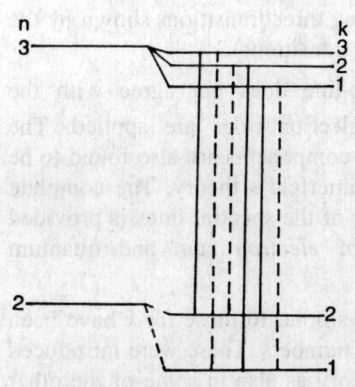

Fig. 4·14. Splitting of the energy levels of hydrogen due to relativity effect. The splittings are shown highly magnified.

The transitions between the sub-levels of different k for two different values of n give rise to the splitting into a number of closely lying components of a single spectral line expected from Bohr's theory due to transition between these two levels of different n. This is illustrated in Fig. 4·14 where the expected fine structure splitting of the H_α line of the Balmer series is shown. This line is due to the transition from the $n_1 = 3$ to the $n_2 = 2$ level. According to Sommerfeld's relativistic theory, the upper level with $n_1 = 3$ splits into three sub-levels with $k = 3, 2$ and 1 while the lower level with $n_2 = 2$ splits into two sub-levels with $k = 2$ and 1. As will be seen from the figure, if all possible transitions between the sub-levels of the upper state and those of the lower state are considered, the H_α line should split up into *six* components. However the number of components actually observed is less. The reason for this is that some of the transitions between the upper and lower sub-levels are forbidden. Only those transitions can occur for which the azimuthal quantum number k changes by 1 unit only. This is known as the *selection rule* for the transition which in the present case is

$$\Delta k = k_f - k_i = \pm 1 \qquad (4\cdot12\text{–}23)$$

Thus the possible transitions consistent with the above selection rule are :

$$k_i = 1 \text{ to } k_f = 2 \qquad (\Delta k = +1)$$
$$k_i = 2 \text{ to } k_f = 1 \qquad (\Delta k = -1)$$
$$k_i = 3 \text{ to } k_f = 2 \qquad (\Delta k = -1)$$

So only three of the transitions are allowed which have been shown by the full lines in Fig. 4·14. The remaining three transitions shown by the broken lines are not allowed (forbidden transitions).

The observed splitting of the H_α-line does not agree with the expected splitting even when these selection rules are applied. The wavelength differences between the split components are also found to be different from those expected from Sommerfeld's theory. The complete explanation of the fine structure splitting of the spectral lines is provided by the introduction of the concept of *electron spin* and quantum electrodynamics (see § 6·5).

It may be noted that selection rules similar to those for k have been discovered in the cases of other quantum numbers. These were introduced on *ad hoc* basis in Bohr-Sommerfeld theory as also in some of the other theories in the pre-quantum mechanics days. In some cases, Bohr's correspondence principle provides explanation for the origin of the selection rules. However a full and satisfactory explanation can only be provided by quantum mechanics (see Ch. XI).

Structure of the Hydrogen–Like Atom

It should be mentioned here that the azimuthal quantum number k of Bohr-Sommerfeld theory has been replaced by the new azimuthal quantum number $l = (k - 1)$ in quantum mechanics (Ch. XI). As we shall see later that for a given value of n, the possible values of l are 0, 1, 2, 3, ... $(n - 1)$. The orbital angular momentum of the electron has a value $v_l = \sqrt{l(l + 1)} \cdot \hbar$.

4·13 Limitations of the old quantum theory

The quantum theory developed before the discovery of quantum mechanics is known as the old quantum theory. Bohr-Sommerfield theory and its extension by the introduction of the electron spin (see Ch. VI) provided successful explanation of the spectra not only of the hydrogen-like atoms, but also of some complex atoms, *e.g.*, the alkali atoms. In these atoms, a single valence electron revolves in an external orbit so that their outer structure somewhat resembles the hydrogen-like atoms. However where such resemblance is lacking, the old quantum theory fails completely in explaining the observed spectra. Even in the case of the two-electron helium atom, the theory is unable to provide a satisfactory explanation of the observed spectra. Further it is not possible to explain the intensities or the polarizations of the spectral lines by the old quantum theory.

All these limitations of the old quantum theory were overcome with the discovery of the new mechanics of the subatomic particles (*quantum mechanics*) by W. Heisenberg (1925) and E. Schroedinger (1926) and its further extension by P.A.M. Dirac (1928, 1930). Quantum mechanics is now the most powerful tool in the hands of scientists to explain various phenomena in the sub-atomic domain including the explanation of the atomic spectra (see Ch. XI and XIII).

4·14 Resonance potentials ; Franck and Hertz experiment

The study of the spectra of hydrogen-like atoms gave convincing support to Bohr's predictions about the existence of discrete energy levels. Further confirmation of these predictions were provided by a series of entirely different type of experiments by J. Franck and G. Hertz in Germany within a short time after the formulation of Bohr's theory (1916).

Their experimental arrangement in shown in Fig. 4·15. T is a sealed tube within which a metal filament F can be heated by passing an electric current through it so that there is electron emission (thermionic emission) from F. P is a collector plate at the other end of the tube and G is a wire-grid placed just in front of P. Thus the electrons from F have to pass through the grid G to reach P. The pressure within T which contains a small amount of the experimental gas or vapour is kept quite low. Franck and Hertz used mercury vapour in their experiment.

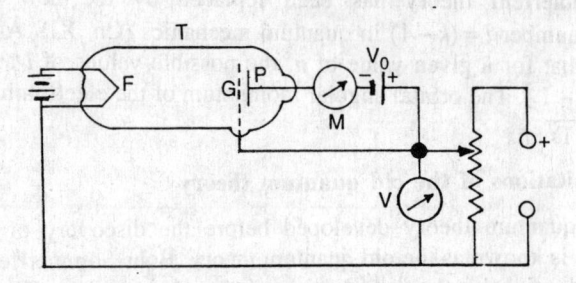

Fig. 4·15 Franck and Hertz experiment.

The grid G is kept at a positive potential with respect to F so that the electrons gain kinetic energy in travelling from F to G which becomes maximum when they reach G. The collector P is kept at a slightly negative potential with respect to G. This *retarding potential* V_0 is usually about 0·5 volt, so that if the electron energy is less than eV_0 while passing through G, they are unable to reach P.

If an electron travelling from F to G collides with a gaseous atom then the atom may receive some energy from the colliding electron. If the collision is elastic, then the energy transfer is negligibly small since the atoms are much heavier than the electrons. On the other hand if the collision is with an electron bound in the atom, then there may be considerable energy transfer from the impinging electron to the atomic electron.

According to Bohr's theory, an atomic electron can exist in different discrete energy levels. Normally it is in the ground level. If the energy of the impinging electron is not sufficient to raise the electron bound in the atom from the ground level to the first excited energy level by collision, then there will not be any energy transfer between the two in spite of the collision. Thus the electrons coming from F will reach the grid G with the full energy gained by them and will be able to overcome the retarding potential V and fall on P. The meter M will then record an electric current. As the potential difference between F and G is increased, the energy gained by the electrons coming from F increases and the current recorded by M also increases. Finally when the energy gained by these electrons as they reach G becomes equal to the energy difference between the ground level and the first excited level of atomic electrons, then the latter may be transferred to the higher level by gaining sufficient energy from the incident electrons. When this happens, the entire energy of an incident electron is transferred to the atomic electron. As these *inelastic collisions* take place just behind the grid G, the incident electron loses its entire kinetic energy and is unable to reach P by overcoming the retarding potential V_0. When this happens, there is a sudden drop in the current

Structure of the Hydrogen–Like Atom

recorded by M. The corresponding potential difference between F and G is known as the *resonance potential*.

As the potential difference between F and G is further increased the current recorded by M begins to rise again. The collisions between the on-coming electrons and the atomic electrons now take place some distance behind G towards F. Thus even after losing their entire energies by collision, the on-coming electrons again acquire sufficient energy when they reach G and are able to overcome the retarding potential V_0 to fall on P so that M again records some current.

The nature of variation of the current recorded by M with the change of the potential difference between F and G is shown graphically in Fig. 4·16. The current after its sudden drop at the resonance potential begins to rise again. When the potential difference between F and G becomes twice the resonance potential the current suddenly drops again to a low value. The alternate rise and fall of the electron current with the increase of the potential difference between F and G occur repeatedly as shown in Fig. 4·16.

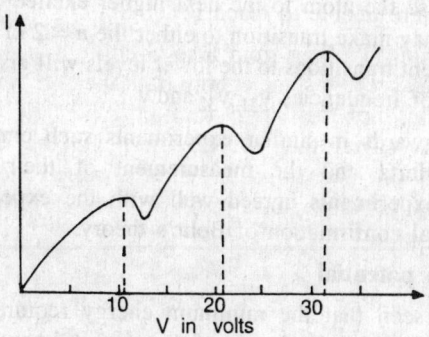

Fig. 4·16. Variation of the plate current with voltage in Franck and Hertz experiment.

The different maxima of the current occur at potentials which are integral multiples of the resonance potential so that the difference in the potentials between two consecutive peaks is equal to the resonance potential.

The alternate rise and fall of the electron current recorded by M can be understood in the following manner. When the potential difference between F and G is twice the resonance potential, the electrons coming from F gain energy equal to the energy difference between the two levels of the atom when it has travelled half way from F to G. So if it collides with an atom at this point, it causes transition of the atom from the ground level to the excited level and loses its entire energy in the process. It then starts afresh with zero kinetic energy towards G and again acquires the same amount of energy on reaching G. Collision with a second atom at this point results in the transition of this atom to the upper level while the incident electron again loses its entire kinetic energy. As a result it is unable to overcome the retarding potential so that the current recorded by M suddenly drops which gives rise to the second peak.

Similarly the other peaks arise as the electron coming from F collides with larger numbers of atoms at different points along its path from F to G to cause transitions between the levels of the corresponding atoms. These take place at potentials which are integral multiples of the resonance potentials.

It may be noted that according to Bohr's theory, the ground state energy of hydrogen is -13.6 eV while that of the first excited state is -3.4 eV. So the resonance potential of hydrogen is 10·2 volts which has been confirmed experimentally. Once the electron is raised to the first excited energy level with $n = 2$ due to bombardment by an incident electron accelerated to the resonance potential, it remains in that level for about 10^{-8} s. It then returns to the ground level by the emission of light of frequency v_{21} (see § 4·5). If the energy of the incident electron is sufficient to raise the atom to the next higher excited level with $n = 3$, then the atom may make transition to either the $n = 2$ or the $n = 3$ excited levels. Subsequent transitions to the lower levels will give rise to the three emission lines of frequencies v_{31}, v_{32} and v_{21}.

Soon afterwards in similar experiments such emission lines were observed by Hertz and the measurement of their wavelengths by spectroscopic experiments agreed well with the expected wavelengths giving additional confirmation of Bohr's theory.

4·15 Ionization potential

We have seen that the minimum energy required to set free an electron bound in the ground state of the atom is known as its ionization energy (§ 4·6). This energy may be supplied to the atom by exposing it to photons or by bombarding it with electrons from outside. In Fig. 4·17, the experimental arrangement for measuring the ionization energy of an atom is shown. F is a metal filament within the sealed tube T which can be

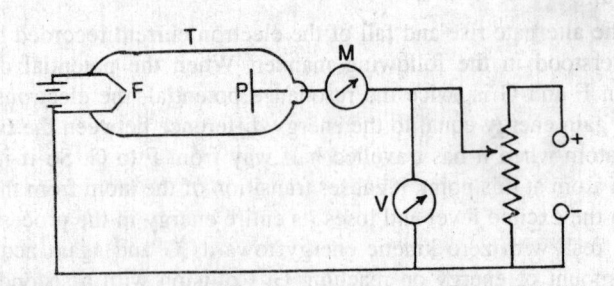

Fig. 4·17. Measurement of ionization potential.

Structure of the Hydrogen–Like Atom

heated by passing an electric current through it. The thermionic electrons emitted from F are accelerated towards the anode P which is maintained at a positive potential with respect to F. When they reach P, an electric current is recorded by the meter M connected to P.

The tube T which is maintained at a low pressure contains a small amount of the experimental gas or vapour. As the potential difference V between F and P is increased, the current I increases according to Child-Langmuir law ($I \propto V^{3/2}$). At a definite value of $V = V_i$ the rate of change of I shows a sudden increase as shown in Fig. 4·18. The reason for this is that when $V > V_i$, the energy gained by the electrons coming from F becomes greater than the ionization energy of the gaseous atoms within T. As a result they are able to ionize some of these atoms by collision. The atomic electrons which are thereby emitted add to the swarm of electrons coming from F so that the current recorded by M begins to rise more rapidly with increasing V when the potential is greater than V_i.

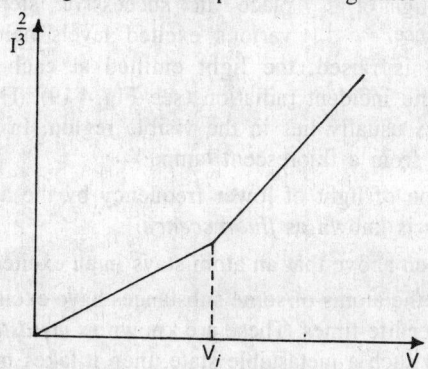

Fig. 4·18. Change of plate current with electron accelerating potential.

The potential V_i is known as the *ionization potential*. An electron accelerated to this potential gains energy equal to the ionization energy of the particular atom. The ionization potential and the resonance potential discussed in § 4·14 are both known as *critical potentials* of the atom.

The value of the ionization potential for hydrogen determined by the above method comes out to be 13·6 volts which agrees well with that determined by spectroscopic methods.

4·16 Fluorescence and phosphorescence

The use of fluorescent lamps for illumination purposes is quite common. The principle of action of these lamps depends on the phenomenon of *fluorescence*. In a fluorescent lamp there is a small amount of mercury which becomes vaporised by a suitable arrangement when the lamp is switched on. Subsequently an electric discharge is passed through the mercury vapour which causes the mercury atoms to emit electromagnetic radiation. Both visible radiation and ultraviolet radiation are emitted. The inner surface of the glass tube of the lamp is coated with some fluorescent material, *e.g.*, barium platino-cyanide. These materials are known as phosphors. The ultraviolet light emitted by the mercury atoms is absorbed by the fluorescent material. As a result, the atoms within this material are raised to higher excited levels from which they make transitions to the lower levels within $\sim 10^{-8}$ s. If the transition is directly to the ground state then the radiation of the same frequency is re-emitted. If however, the transition takes place in successive steps through the various excited levels below the level to which the atom is raised, the light emitted at each step has lower frequency than the incident radiation (see Fig. 4·19). The light of these lower frequencies usually lies in the visible region. In this way visible light is obtained from a fluorescent lamp.

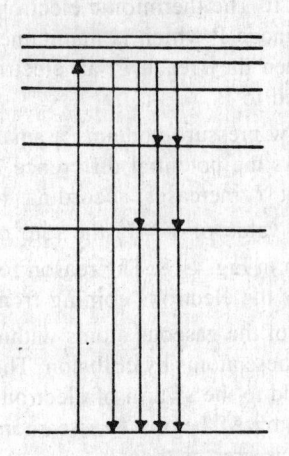

Fig. 4·19. Origin of fluorescence.

The emission of light of lower frequency by the action of light of higher frequency is known as *fluorescence*.

We have seen above that an atom stays in an excited level for about 10^{-8} s. However the atoms of some substances have excited energy levels which have longer life-times. These are known as *metastable states*. If an atom is raised to such a metastable state, then it takes much longer time to return to the lower levels. As a result, these substances continue to emit light for a long time after the removal of the source of the exciting radiation. These are known as phosphorescent substances and the phenomenon as *phosphorescence*. Phosphorescence is widely utilized in illuminating various objects in the dark, *e.g.*, the buttons of alarm clocks, switches of household electric light, painted signboards etc.

Problems

1. Calculate the orbital radius and the velocity of the electron in the first Bohr orbit of hydrogen. (0·529 Å ; 2.2×10^6 m/s)

2. Calculate the energies in eV and the term values in m^{-1} of the electron in the first three orbits of hydrogen. Hence find the wavelengths of the spectral lines due to the transitions between them.

3. Calculate the Rydberg constant for an infinitely heavy atom.

4. The Rydberg constants for hydrogen and He^+ ion are respectively 1.0967758×10^7 m^{-1} and 1.0972227×10^7 m^{-1}. If the ratio of the

Structure of the Hydrogen–Like Atom

nuclear masses of hydrogen and helium is 3·9726, calculate the proton to electron mass ratio. (1836)

5. Hydrogen has three isotopes of mass numbers 1, 2 and 3. If the ratios of the masses of the nuclei of these are equal to the ratios of their mass numbers, calculate the wavelength differences of the H_α lines of the second and the third isotopes with respect to the first.

(1·79 Å ; 2·38 Å)

6. Calculate the gravitational force between the proton and the electron in the first Bohr orbit of hydrogen and determine the ratio of this force to the Coulomb force between the two. (3.63×10^{-47} N ; 4.4×10^{-40})

7. Calculate the ratios of the relativistic energy splitting of the two different k sub-levels of the $n = 2$ level of hydrogen (using Sommerfeld's expression) from the Bohr energy of the level to the original energy. (3.32×10^{-6} ; 16.6×10^{-6})

8. A hydrogen atom in the normal state is bombarded with electrons of kinetic energies (a) 11·8 eV and (b) 5·5 eV respectively. What are the minimum kinetic energies of the electron in the two cases after the bombardment? (1·6 eV ; 5·5 eV)

9. A hydrogen atom in its ground state is bombarded with electrons to excite it to a higher level so that during the de-excitation of the atom the first Balmer line is produced. What should be the minimum energy of the bombarding electron. What other spectral lines may be produced in this case? (12·09 eV)

10. The system consisting of an electron and positron revolving about their centre of mass is known as a positronium. Calculate the distance between the two particles in the ground state. Also calculate the Rydberg constant and ionization potential.

(1·06 Å ; 0.54893×10^7 m^{-1} ; 6·8 eV)

11. In a muonic atom, a muon of charge $-e$ and mass $m = 207\, m_e$ takes the place of the orbital electron. If the nucleus is a proton, find the distance of the muon from the nucleus in the ground state of the atom. Also find the binding energy of the muon. What is the energy of the radiation emitted for the transition $n = 2$ to $n = 1$?

(2.85×10^{-13} m ; 2·53 keV ; 1·8975 keV)

12. The wavelength of the resonance line of hydrogen arising out of the transition $n = 2$ to $n = 1$ is found to be shifted by 1 Å when observed in the direction of motion of the hydrogen atom due to Doppler effect. What is the velocity of the emitting atom? (2.467×10^5 m/s)

5

EMISSION OF ELECTRONS FROM METAL SURFACES ; PHOTO-ELECTRIC AND THERMIONIC EFFECTS

5·1 Discovery of photoelectric effect

Planck's postulate regarding the discrete nature of the possible energy states of an oscillator marked a radical departure from the ideas of classical physics (see Ch. III). According to the laws of classical mechanics, the energy of a linear harmonic oscillator can vary continuously, depending on the amplitude of its vibrations. At the beginning Planck's ideas faced stiff opposition. Gradually, however, these new ideas were found to be quite useful in explaining many hitherto unexplained physical phenomena.

One of these was the *photoelectric effect*, discovered by Heinrich Hertz (1888) in Germany and studied in detail by Hallwachs, Elster and Geitel. It was observed that a plate of a metal such as zinc when exposed to ultraviolet radiation became positively charged which showed that it had lost negative charges from its surface. These negatively charged particles were later identified as electrons by P. Lenard in Germany (1899). The phenomenon is known as *photoelectric* effect and the emitted electrons are known as photoelectrons.

5·2 Lenard's experiment

Lenard's apparatus for the measurement of e/m of the negatively charged particles emitted in photoelectric process is shown diagrammatically in Fig. 5·1. The whole apparatus is enclosed in the tube B in which a very low pressure is maintained with the help of a pump. C is a plane metal plate kept at a negative potential while A is another plane metal plate parallel to C at some distance from it which is kept at zero potential. Thus A is positive w.r.t. C. There is a small hole at the centre of A through which the beam of the negative ions emitted from C can pass. There is a side tube attached to B which has a quartz window G covering its open end. Ultraviolet light can pass through G to fall upon the

Emssion of Electrons from Metal Surfaces

plate C due to which negative ions are emitted from C by photoelectric process. These are attracted towards A and a fraction of them passes through the hole in it to fall on the collector plate D. The resulting feeble electric current is detected with the help of an electrometer.

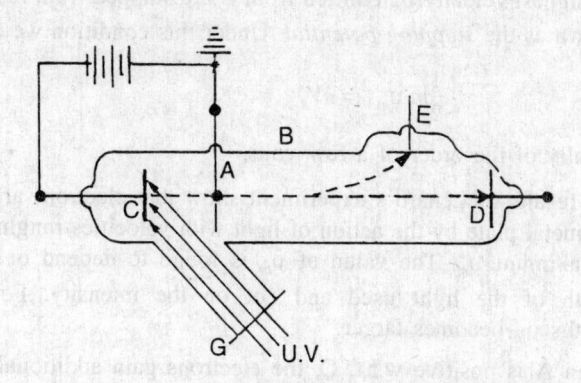

Fig. 5·1. Lenard's experiment on photoelectricity.

The enclosure B is placed between the poles of a magnet. Suppose the magnetic field due to this magnet acts perpendicular to the plane of the diagram. By the action of the magnetic field, the negative ions coming out through the hole in A describe a circular arc and fall upon the collector E on one side instead of upon D. The resulting electric current can be recorded by another electrometer connected to E.

From the geometry of the arrangement, it is possible to determine the radius of curvature R of the negative ions. From the values of R and the magnetic induction B, it is possible to determine the e/m for the negative ions emitted from C. If V be the potential difference between C and A, then the kinetic energy gained by these negative ions will be

$$\frac{1}{2} m v^2 = eV$$

Equating the centripetal to the magnetic force we then get

$$Bev = \frac{mv^2}{R}$$

so that
$$\frac{e}{m} = \frac{2V}{B^2 R^2} \quad (5 \cdot 2 - 1)$$

Lenard's measurement of the e/m proved conclusively that the negative ions emitted from C by the action of ultraviolet light were electrons.

Lenard also observed that if a positive potential was applied to C instead of a negative potential, the electric current recorded at C decreased. This is due to the fact that the negatively charged electrons emitted from C then experience a repulsive force in going towards A which is now negative w.r.t. C. As the positive potential on C is increased, the current recorded at D ultimately becomes zero. If V_s is the potential difference between C and A when this happens, electrons of all velocities upto the highest velocity v_m emitted from C are stopped from reaching A. V_s is known as the *stopping potential*. Under this condition we can write

$$\frac{1}{2}m v_m^2 = eV_s \qquad (5\cdot2\text{–}2)$$

V_s is usually of the order of a few volts.

The results of Lenard's experiment show that electrons are emitted from the metal plate by the action of light with velocities ranging from 0 upto a maximum v_m. The value of v_m is found to depend only on the wavelength of the light used and not on the intensity. For shorter wavelengths v_m becomes larger.

When A is positive w.r.t. C, the electrons gain additional velocity over and above their emission velocity. In Lenard's experiments the velocity thus acquired was much greater than the emission velocity of the electrons which was therefore neglected in deriving Eq. (5·2–1) for e/m.

Another very important conclusion drawn from Lenard's experiment is that the photoelectric current is independent of the wavelength of light; it depends only on the intensity of the light used.

Later similar experiments were performed by O.W. Richardson, K.T. Compton and others which confirmed Lenard's findings. Albert Einstein in 1905 proposed a theory of photoelectric emission based on the ideas underlying the newly discovered quantum theory of Max Planck. Einstein's theory was a radical departure from the classical electromagnetic theory of light. Considering the importance of the matter, R.A. Millikan in 1916 performed a series of experiments which gave conclusive support to the new theory of Einstein.

5·3 Millikan's experiments

Millikan's experimental arrangement is shown in Fig. 5·2. He used alkali metals since these emit photoelectrons under the action of both visible and u.v. light.

P is a highly evacuated bulb fitted with two side tubes. One of these has a quartz window O at its end through which visible or u.v. radiation may enter the bulb and fall upon the cylindrical pieces of the

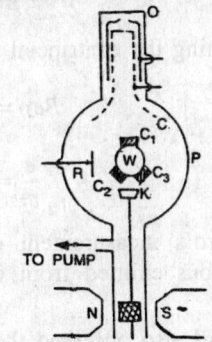

Fig. 5·2 Millikan's experiment on photoelectron emission

alkali metals C_1, C_2 and C_3 mounted on the periphery of a wheel W. The wheel can be rotated by some external arrangement so that the faces of the metal pieces may be scraped by the sharp knife-edge K and fresh clean surfaces of these may be exposed to the radiation entering through O.

C is a collector grid placed just below O. It is made of oxidised copper wire so that photoelectrons are not emitted from its surface when the incident light passes through it. An electrometer connected with the collector measures the photoelectric current. Potential difference can be applied between C and W which can be varied as necessary.

Millikan used monochromatic radiation of different wavelengths and measured the photoelectric current emitted from the surfaces of C_1, C_2, C_3 under the action of the incident radiation. Keeping the wavelength of the light fixed, he first measured the photoelectric current for different intensities I of the incident light by varying the potential difference V between C and W. The results of such measurements are shown in Fig. 5·3. When the potential of C is positive w.r.t. W, the current is almost

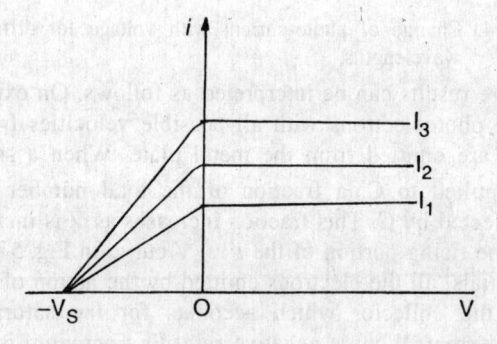

Fig. 5·3. Change of photo-current with voltage for different light intensities.

constant (saturated) for a given intensity of the light. If the potential of C is made negative w.r.t. W, the current decreases and goes to zero at a particular potential $-V_s$ which is the same for the different intensities. The saturation current increases at higher intensities. The negative potential V_s is known ass the *stopping potential*.

Millikan next measured the variation of the photoelectric current with the p.d. between C and W by using light of different wavelengths λ. The results of these measurements are shown in Fig. 5·4. As before, the current attains saturation when the potential of C is positive and decreases at negative potentials. For a particular wavelength λ_1, the current goes to zero for a stopping potential $-V_1$. When the wavelength is changed to λ_2, the stopping potential is $-V_2$ and so on. For the different wavelengths $\lambda_1, \lambda_2, \lambda_3$ etc., the saturation currents were adjusted to have the same value by adjusting the intensity of the light used. As will be seen from

Fig. 5·4, the stopping potential has larger negative value for shorter wavelength of the light used, *i.e.*, for higher frequency. It may be noted that while measuring the potential V, an error due to contact potential creeps in. By using an additional electrode R of platinum, Millikan corrected for this error in the measurement of the potential.

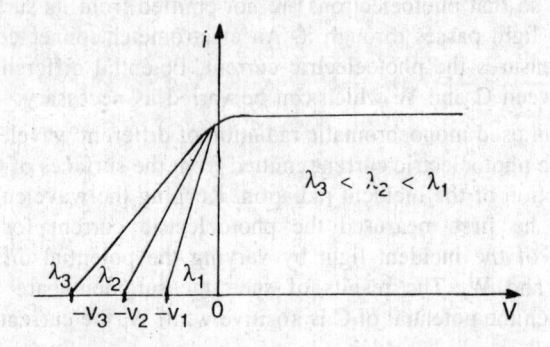

Fig. 5·4. Change of photo-current with voltage for different wavelengths.

The above results can be interpreted as follows. On exposure to the incident light, photoelectrons with all possible velocities from 0 upto a maximum v_m are emitted from the metal plate. When a small positive potential is applied to C, a fraction of the total number of electrons emitted is collected by C. This fraction increases as V is increased which is shown by the rising portion of the *i* vs. V curve in Fig. 5·4. For higher positive potentials, all the electrons emitted by the action of the light are collected by the collector which accounts for the saturation of the photoelectric current. When a negative *retarding potential* is applied, the lower energy electrons are unable to reach the collector so that *i* gradually decreases with increasing negative potential. Finally for a potential $-V_s$, the photoelectrons of all velocities upto the maximum v_m are prevented from reaching the collector. Obviously at this point, we can equate the maximum kinetic energy of the emitted electrons to the energy required to overcome the effect of the retarding potential, so that if *e* be the charge and *m* the mass of the electrons, we can write

$$\frac{1}{2} m v_m^2 = e V_s$$

We can then summarize the important conclusions drawn from the results of the above experiments on photoelectric effect :

(a) The photoelectric current depends upon the intensity of the light used. It is independent of the wavelength of the light.

(b) The photoelectrons are emitted with all possible velocities from 0 upto a maximum v_m which is independent of the intensity of the incident light but depends only on its wavelength (or frequency). It is found that if v be the frequency of the light used, then the maximum kinetic energy of the emitted photoelectrons increases linearly with v (see Fig. 5·5).

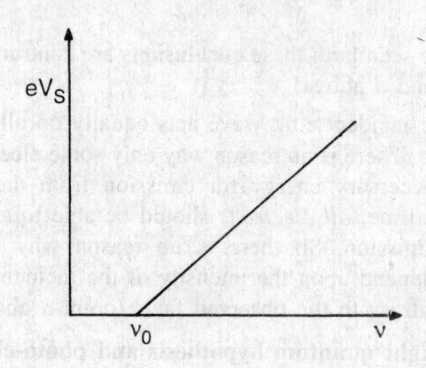

Fig. 5·5. Variation of the velocity of the photoelectrons with the frequency of the incident light.

(c) Photoelectron emission is an instantaneous effect. There is no time gap between the incidence of the light and the emission of the photoelectrons.

(d) The straight line graph showing the variation of the maximum kinetic energy of the emitted electrons with the frequency v of the light intersects the abscissa at some point v_0. No photoelectron emission takes place if the frequency $v < v_0$. This minimum frequency v_0 is known as the *threshold frequency*. Its value depends on the nature of the emitting material.

5·4 Failure of the electrommagnetic theory

The above experimental facts cannot be explained on the basis of classical electromagnetic theory according to which light consists of mutually perpendicular oscillating electric and magnetic fields propagated as a transverse wave with finite velocity which is a characteristic of the medium. The intensity of light is determined by the amplitudes of these electromagnetic oscillations. When light falls on an electron in a metal, it is acted upon by the oscillating electric field and gains energy from the latter. Larger the amplitude of the electric vector, larger is the quantity of energy gained by the electron. So according to this theory, the energy of the emitted electron should depend on the intensity of the incident light which is contrary to the observed fact (point *b* above).

Further, according to the e.m. theory, the velocity of the emitted electrons should not depend on the frequency of light. Whatever may be the frequency of the incident light, the electron would be emitted if it gets sufficient time to collect the necessary energy for emission. As the e.m. waves pass by an electron, the latter gets a small amount of energy from each passing wave. Ultimately when it is able to collect sufficient energy from the succession of passing waves to break away from the metal, it is emitted. The time necessary for this purpose may be of the order of some seconds.

As we have seen both these conclusions are contrary to the observed facts (points c and d above).

Finally, the incident e.m. wave acts equally on all the electrons on the metal surface. There is no reason why only some electrons will be able to collect the necessary energy for emission from the passing waves. Given sufficient time, *all electrons* should be able to collect the energy necessary for emission. So there is no reason why the photoelectric current should depend upon the intensity of the incident light. However, this again is contrary to the observed facts (point a above).

5·5 Einstein's light quantum hypothesis and photo-electric equation

We have seen above that the maximum kinetic energy of the emitted photoelectrons increases linearly with the frequency of the incident light. We can express this by an equation of the form

$$\frac{1}{2} m v_m^2 = eV_s = a\nu - \epsilon \qquad (5.5-1)$$

where a and ϵ are two constants. The constant a was determined accurately by Millikan (1916) as also by other workers and was found to be equal to Planck's constant h, so that we may rewrite Eq. (5·5–1) as follows:

$$\frac{1}{2} m v_m^2 = eV_s = h\nu - \epsilon \qquad (5.5-2)$$

If we put $\nu = \nu_0 = \epsilon/h$, then from Eq. (5·5–2) we see that the kinetic energy of the emitted photoelectron becomes zero. So there will be no photoelectron emission if $\nu < \nu_0$, ν_0 being the threshold frequency mentioned above. Eq. (5·5–2) is known as *Einstein's photoelectric equation*.

Einstein proposed that the quantum hypothesis of Planck could be used to explain all the observed characteristics of the photoelectric phenomena mentioned above (1905) and was able to deduce the empirical photoelectric equation on theoretical grounds. According to Planck the atomic oscillators in the blackbody can only have discrete energy values $nh\nu$ (n is an integer) and emission or absorption of energy by them takes place in discrete steps by the amount $h\nu$. Einstein went a step further and postulated that light is emitted from a source in the form of bundles of energy of the amount $h\nu$ known as light *quantum or photon*, much like bullets coming out from a gun. h is again Planck's constant. This is known as Einstein's *light quantum hypothesis*.

When a photon of energy $h\nu$ falls on a metal surface, the electron absorbs the energy $h\nu$ and is emitted from the metal provided $h\nu$ is greater than the work ϵ required to remove the electron from the metal. ϵ is known as the *work function* of the metal. The surplus of energy $(h\nu - \epsilon)$ is taken away by the electron as its kinetic energy E_k. Thus the photoelectric equation finds a simple explanation. Obviously if $h\nu < \epsilon$ i.e., if $\nu < \nu_0$ no photoelectron emission can take place. This explains the existence of the threshold frequency. Further, according to this theory, larger the number of photons falling on the metal, greater is the probability of their encounter with the atomic electrons and hence greater is the photoelectric current. So the increase of photoelectric current with the increasing light intensity finds an easy explanation in this theory.

Finally as soon as a photon of energy $h\nu > \epsilon$ falls on an electron, the latter absorbs it and is emitted *instantaneously*. Thus the instantaneous emission is also easily explained in this theory.

Einstein's light-quantum hypothesis postulates *corpuscular nature* for light in contrast to the wave nature postulated in the electromagnetic theory of light. The postulate of wave nature has found support through such phenomena as the interference, diffraction and polarisation of light. On the other hand, the photoelectric effect provides a direct evidence in favour of the corpuscular nature of light.

It should be noted that Planck's constant h can be determined accurately with the help of Eq. (5·5–2) which can be rewritten as

$$V_s = \left(\frac{h}{e}\right)\nu - \left(\frac{\epsilon}{\varepsilon}\right) \qquad (5\cdot5-3)$$

If we plot the stopping potential V_s against the frequency of the incident light, then the resulting graph will be a straight line. The slope of this graph is equal to (h/e). Millikan determined Planck's constant h by this method by using the value of the electronic charge e determined by him. He found $h = 6\cdot55 \times 10^{-34}$ J.s. Later determination by Parker, Langenberg, Denenstein and Taylor (1969) gave $h = 6\cdot6262 \times 10^{-34}$ J.s. (See § 18·18).

The work function ϵ of the emitting material can be determined from the intercept $\nu_0 = \epsilon/h$ with the abscissa of the straight line graph in Fig. 5·5. Its numerical value is of the order of a few elctron-volts for most conductors and is equal to their thermionic work function (see Ch. XVIII). The measured numerical values are strongly influenced by the presence of impurities in the experimental material and the degree of cleanliness of the emitting surface.

Photoelectron emission takes place not only when light falls on a metal surface, but also when other materials in solid, liquid or gaseous state are irradiated with electromagnetic radiation of different wavelengths starting from γ-rays and x-rays on the short wavelength end of the spectrum to ultraviolet and visible radiation in the intermediate wavelength region. It is also observed in some cases with infra-red radiation which has longer wavelengths. When photoelectrons are emitted from the isolated atoms, as in gases, the minimum energy needed for their emission is equal to the energy of binding of the electron in the atomic orbit. The binding energies of the electrons in the outermost orbits of the atoms are usually of the order of a few electron volts. So these can be emitted by the action of visible or ultraviolet light. On the other hand, the electrons in the inner orbits of the heavier atoms are bound much more strongly and require for their emission irradiation by x-rays or γ-rays.

5·6 Application of photoelectric equation

We can illustrate the application of Eq. (5·5–2) by the following simple example.

Suppose visible light of wavelength 4500 Å falls on a piece of sodium. If the work function for sodium be 2·3 volts, then what is the maximum kinetic energy of the emitted photoelectrons? What is the value of the stopping potential?

We first calculate the energy of the incident light. We know that the energy of the photon is

$$E = h\nu = \frac{hc}{\lambda} = \frac{6 \cdot 626 \times 10^{-34} \times 3 \times 10^8}{1 \cdot 6 \times 10^{-19} \lambda(\text{Å}) \times 10^{-10}} \text{ eV}$$

$$= \frac{12412 \cdot 5}{\lambda(\text{Å})} \text{ eV}*$$

Since $\lambda = 4500$ Å in the present case we get

$$E = \frac{12412 \cdot 5}{4500} = 2 \cdot 76 \text{ eV}$$

The frequency of the light is

$$\nu = \frac{c}{\lambda} = \frac{3 \times 10^8}{4500 \times 10^{-10}} = 6 \cdot 67 \times 10^{14} \text{ s}^{-1}$$

The maximum energy of the emitted photoelectrons is

$$E_m = E - \epsilon = 2 \cdot 76 - 2 \cdot 23 = 0 \cdot 46 \text{ eV}$$

The stopping potential is

$$V_s = \frac{E_m}{e} = 0 \cdot 46 \text{ volts}$$

---

* For easily remembering the numerical expression, we can write the approximate form: $E(\text{eV}) = 12345/\lambda$ Å.

The velocity of these electrons is given by $mv_m^2/2 = eV_s$.
So we get

$$v_m = \sqrt{\frac{2eV_s}{m}} = \sqrt{\frac{2 \times 1.6 \times 10^{-19} \times 0.46}{9.11 \times 10^{-31}}} = 4 \times 10^5 \text{ m/s}$$

5·7 Photoelectric cells

Different types of photocells have been developed based on photoelectric effect. These can be used for the measurement of light intensity as also in various control devices.

Two types of phototubes are generally used, the vacuum cells and the gas filled cells. In both, a cathode C made of some light sensitive material and an anode A facing it are enclosed within a sealed tube made of glass or quartz (see Fig. 5·6). Electrical connections are provided with external circuits by means of electrodes sealed through the enclosing tube. For operation with ultraviolet, quartz tubes are preferred. The cathode surface is usually made much larger in order to increase photo emission.

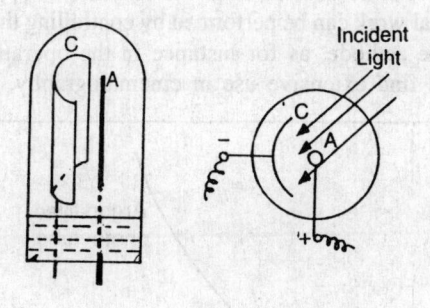

Fig. 5·6. Photoelectric cell.

In the *vacuum cell*, the tube is highly evacuated. At low anode voltages, the photocurrent increases with the applied voltage according to the $V^{3/2}$ law. At higher voltages, the current reaches saturation. In this region, the current increases linearly with increasing light intensity. Hence such tubes are useful for comparison of light intensities. By using cathodes made of different light sensitive materials, it is possible to construct phototubes which are sensitive in different wavelength regions. For example in the case of white light, the cathode is prepared by depositing metallic caesium on a silver surface. This is then coated with caesium oxide. Such a cathode is designated as Cs-O-Ag. In spectrophotometers, the phototubes which are used have usually K-O-Ag cathodes. The high frequency response of the vacuum tubes are usually quite good.

The photoelectric current in vacuum tubes is generally quite small. It is of the order of a few microamperes for an anode voltage of about 100 volts. The sensitivity of the tube is greatly increased if a small amount of some inert gas, *e.g.*, argon is introduced in the tube. The photoelectrons emitted from the cathode are attracted by the anode potential and gain sufficient energy while moving towards the anode to ionize the molecules of the gas within the tube. This gives rise to an increase of the photocurrent by a factor of 10 or more depending on the anode potential.

In these *gas filled* cells, there is no definite relationship between the photocurrent and the light intensity, so that they are not very useful for the purpose of comparing light intensities.

Inert gasses are used in the gas filled cells to prevent chemical reaction with the cathode material. The gas chosen should have low ionization potential so that the anode potential required for the operation of the tube may not be too high. Finally the atomic weight of the gas should be relatively low to keep the transit time of the photoelectrons from the cathode to the anode as low as possible in order to get better high frequency response of the phototube.

The characteristics of both vacuum cells and gas-filled cells are shown in Fig. 5·7. The internal resistance of the cells are usually quite high, so that they can be used as amplifiers in certain applications Various types of mechanical work can be performed by controlling the intensity of the light falling on the cathode, as for instance in the operation of a burglar alarm. Photocells find extensive use in cinematography, television *etc*.

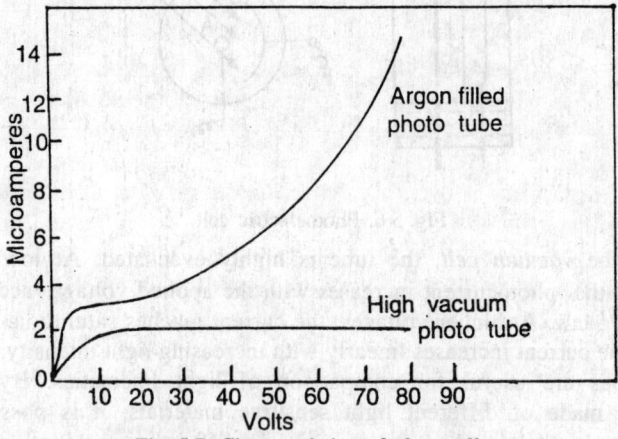

Fig. 5·7. Characteristics of photocells.

5·8 Thermionic emission

It has been known for nearly two hundred years that air and other gases in contact with a heated metal become electrically conducting. Towards the end of the nineteenth century, Elster and Geitel observed that a white hot metal lost negative electricity. The effect becomes more pronounced if the metal is initially charged with negative electricity. The British physicist O.W. Richardson performed a series of experiments to

Emssion of Electrons from Metal Surfaces

show that the loss of negative electricity from a white hot metal was due to the emission of negatively charged particles from its surface. This phenomenon is known as *thermionic emission*. The e/m for these particles can be determined by an experiment described below.

In Fig. 5·8, CD is a very thin filament of some pure metal stretched along the axis of a metal cylinder AB enclosed within a highly evacuated closed glass tube. The filament is heated by passing an electric current through it from a battery E. There in an arrangement to apply a potential difference between the filament and the cylinder as shown in the figure.

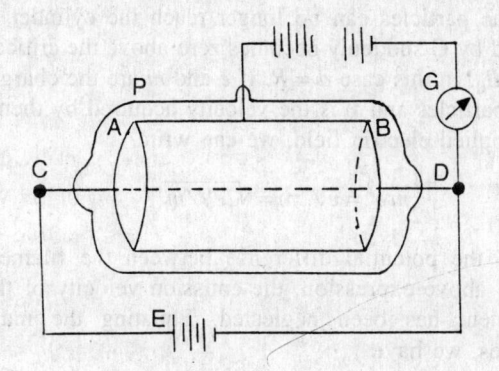

Fig. 5·8. Apparatus for the study of thermionic emission.

G is a meter for measuring very low currents. If the filament is heated to a very high temperature and is kept at a negative potential with respect to the cylinder AB, then the meter G records a small electric current. If the filament is made positive with respect to the cylinder, no current is indicated by G. This shows that negatively charged particles are emitted from the metal filament.

In Fig. 5·9, the cross section of the apparatus in a perpendicular plane is shown. The circle A represents the cylinder AB while the dot C at its centre is the section of the filament CD. Due to the potential difference between C and A, a radial electric field is established between them. The negatively charged particles emitted from the heated filament are accelerated radially towards the cylinder due to the applied electric field and produce the current as recorded by G. If now a magnetic field is applied parallel to the filament, then the paths of the charged particles emitted from the

Fig. 5·9. Sectional view of the apparatus shown in Fig. 5·8.

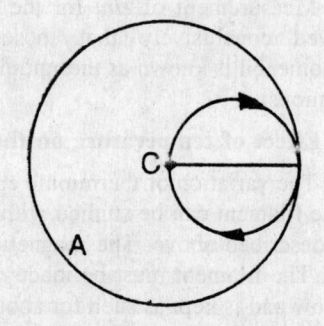

cylinder are no longer radially straight but are curved. If the radius of the filament CD is small compared to that of the cylinder AB, then these curved paths are almost circular, since in this case the emitted particles acquire the maximum possible velocity under the action of the radial electric field within a very short distance from the filament. They then proceed practically with a constant velocity towards the cylinder and hence describe circular paths under the action of the magnetic field.

As the magnetic induction field B is increased, the diameter d of the curved trajectory of the particles decreases and ultimately becomes equal to the radius R of the cylinder AB when $B = B_0$ (say). When d becomes less than R, the particles can no longer reach the cylinder so that the current recorded by G suddenly becomes zero above the critical magnetic induction field B_0. In this case $d = R$. If e and m are the charge and mass of the emitted particles and v is the velocity acquired by them under the action of the applied electric field, we can write

$$\frac{1}{2}mv^2 = eV, \quad v = \sqrt{2eV/m}$$

Here V is the potential difference between the filament and the cylinder. In the above expression, the emission velocity of the particles from the filament has been neglected. Equating the magnetic and centripetal forces, we have

$$Bev = \frac{mv^2}{d/2} = \frac{2mv^2}{d}$$

$$\therefore \quad d = \frac{2mv}{Be} = \frac{2m}{Be}\sqrt{\frac{2eV}{m}} = \sqrt{\frac{8mV}{B^2 e}}$$

For $B = B_0$, $d = R$ so that we get

$$\frac{e}{m} = \frac{8V}{B_0^2 R^2} \tag{5.8-1}$$

Measurement of e/m for the particles emitted from a heated metal showed conclusively that these were nothing but electrons. The phenomenon is known as thermionic emission and the emitted electrons as thermions.

5.9 Effect of temperature on thermionic emission

The variation of thermionic emission with the change of temperature of the filament can be studied with the help of an apparatus similar to the one described above. The magnetic field of course is not applied in this case. The filament must be made of a highly pure metal which is heated to glow and is kept as such for about a week to remove all impurities from its surface by evaporation. The enclosing tube P is continuously pumped to keep it at a very low pressure. After this preliminary treatment of the filament, a potential difference is applied between the filament and the cylinder and the current recorded by the meter is measured.

As the positive potential V on the anode is increased, more and more of the emitted electrons are drawn to the latter so that the thermionic current increases with increasing V. When V is sufficiently high, all the electrons emitted by the filament are drawn to the anode which makes the recorded thermionic current to become saturated as shown in Fig. 5·10.

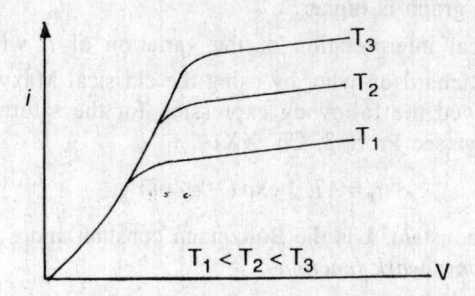

Fig. 5·10. Variation of thermionic current with voltage at different temperatures.

As long as the filament current remains constant, the temperature of the filament also remains constant. From measurement of the filament current and the potential difference between the two ends of the filament, the resistance of the heated filament can be determined. If its resistance at 0°C is known then from a knowledge of its temperature coefficient of resistance, the temperature of the heated filament can be determined. The temperature can also be determined with the help of an optical pyrometer or by using Stefan-Boltzmann formula. In this way the saturation thermionic current i_s at a particular filament temperature can be determined.

If now the filament current in increased, the temperature of the filament also increases. The saturation thermionic current at this increased filament temperature can then be determined by the same method as described above. The nature of the variation of the saturated thermionic current i_s with change of temperature is shown in Fig. 5·11.

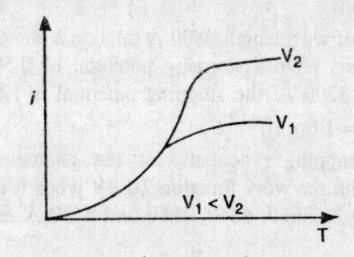

Fig. 5·11. Variation of saturated thermionic current with temperature for different voltages.

At low temperature, all the electrons emitted by the filament are drawn to the anode even at a low potential V. So the effect of change of V is not noticeable at lower temperatures. At higher temperature, the emission increases. V must be sufficiently high in this case to draw all the emitted electrons to the anode. Otherwise space charge accumulates between the filament and the anode which limits the current drawn to the latter as shown by the flat portions of the graph. At higher voltages, larger fractions of the emitted electrons are drawn to the anode so that the flat portion of the graph is higher.

Theoretical interpretation of the variation of i_s with T was first provided by Richardson who, by using the classical Maxwell-Boltzmann statistics, derived the following expression for the saturated thermionic current density (see Prob. 3, Ch. XX):

$$i_s = AT^{1/2} \exp(-\epsilon/kT) \qquad (5.9-1)$$

where A is a constant, k is the Boltzmann constant and $\epsilon = e\phi$ is known as the *thermionic work function*.

Later by using Fermi-Dirac statistics, the above formula was modified to yield the Richardson-Dushman equation (see §18.4, Ch. XVIII).

$$i_s = AT^2 \exp(-\epsilon/kT) \qquad (5.9-2)$$

As we shall see in Chapter XVIII it is the latter formula which gives a correct representation of the phenomenon. It may be noted that the above formula applies when there is no accumulation of space charge between the filament an the anode. The thermionic work function ϵ referred to above is the same as the work function discussed in connection with photoelectron emission. We shall discuss about it in detail it in Ch. XVIII.

Problems

1. The work function of caesium is 1·99 eV. What is the maximum wavelength of the light which can emit photoelectrons from a photo-cathode of caesium. If light of wavelength 4100 Å falls on the Cs photocathode, what is the maximum kinetic energy of the emitted photoelectrons. (6237·4 Å ; 1·037 eV)

2. When light of wavelength 4800 Å falls on a photo-cathode, the emitted photoelectrons have a stopping potential of 0·59 volt. With light of wavelength 3200 Å, the stopping potential is 1·87 volts. Calculate h, assuming $e = 1.6 \times 10^{-19}$ C. (6.56×10^{-34} J.s)

3. Find the stopping potentials for the photoelectrons emitted from rubidium with the work function 2·1 eV when it is irradiated with light of wavelengths 3650, 4340, 4860 and 6230 Å respectively.
(1·3, 0·76, 0·45 eV)

4. The temperature of a metal wire is increased from 1200 K to 1800 K due to which the thermionic emission is increased by a factor 10^3.

Calculate the thermionic work function of the metal using Richardson-Dushman equation. (1·89 eV)

5. A photocell has a cathode made of zinc which has the work function $\epsilon = 3.74$ eV. When light of wavelength 2620 Å falls on the photocathode, the photocurrent is found to be stopped by a retarding potential of 1·5 V. Calculate the magnitude of the contact potential difference of the cell and indicate its polarity. (Hint : The contact p.d. is to be added to the stopping potential). (− 0·5 V)

6. What is the maximum potential to which a caesium metal plate will be raised if it is exposed to light of wavelength 4500 Å? (1·8 volts)

7. Find the maximum kinetic energy of the photoelectrons released from the surface of caesium exposed to radiation with electric vector $X = X_0 (1 + \cos wt) \cos w_0 t$ where $w = 6 \times 10^{14}$ s^{-1} and $w_0 = 3.6 \times 10^{15}$ s^{-1} and the work function of caesium is 1·89 eV.
(0·88 eV)

6

ALKALI SPECTRA ; SPACE QUANTIZATION ; ELECTRON SPIN ; PERIODIC CLASSIFICATION OF ELEMENTS

6·1 Alkali Spectra

The spectra of the alkali atoms, *e.g.*, lithium, sodium, potassium *etc.* show considerable similarity to the spectra of the hydrogen atoms. Actually there are more series in them than in hydrogen. However the lines in a particular series are found to have a regularly decreasing separation and converge towards a limit as in the hydrogen spectra (§ 4·5) For this reason, their wave numbers can be represented as the differences between the members of two term series of the type

$$\bar{v} = T_1(n_1) - T_2(n_2)$$

The term values $T(n)$ are different from the hydrogen terms as discussed in Ch. IV. For a particular series, one of the terms is a constant while the other which is the *running term* can be written as

$$T(n) = \frac{R}{(n+p)^2}$$

where R is the Rydberg constant. n, as before, is an integer while p is a constant for a particular series known as the *quantum defect* ; $p < 1$. The quantity $n^* = n + p$ is called the *effective quantum number*. Obviously n^* is non-integral.

The reason for the similarity between the spectra of the alkali atoms and the hydrogen spectra is that all alkali metals have one easily removable *valence electron* outside a tightly bound core of electrons in their atoms. For instance, lithium for which $Z = 3$ has three electrons in its orbits, two of which revolve in orbits close to the nucleus and make up the core while the third revolves in an outer orbit. Thus the two core electrons screen the nuclear charge of $+ 3e$ units and hence the electrostatic potential felt by the outermost electron is that due to an effective charge of $+e$ units as in the case of the hydrogen atom for which $Z = 1$.

The screening of the nuclear charge $+Ze$ by the core electrons of charge $-(Z-1)e$ is however not complete. The outermost valence electron revolves in an elliptic orbit with the nucleus at one focus of the ellipse. During the course of its revolutions, the electron is at varying distances from the nucleus and may penetrate into the core region. When it is close to the nucleus, the screening is less effective due to this penetration and the valence electron is acted upon by an effective positive charge greater than $+e$. The degree of penetration depends upon the eccentricity of the ellipse. We have seen from the theory of elliptic orbits (§ 4·12) that the eccentricity ϵ depends upon the azimuthal quantum number $l = k - 1$. Smaller l is more eccentric is the ellipse. In such cases, the valence electron penetrates deeply into the core region and comes so close to the nucleus during its revolution that the full nuclear charge $+Ze$ may act upon it, the screening being practically nil. Obviously, in this case, the alkali spectral terms will depart considerably from the hydrogen terms. On the other hand, for the less eccentric orbits *i.e.*, for larger l, the screening will be much more effective and the spectral terms of the alkali atoms will resemble more closely the hydrogen terms.

Spectroscopists designate the term values of the alkali atoms of different l by special symbols. Thus $l = 0$ terms are called S terms, $l = 1$ terms are known as P terms, $l = 2$ as D terms, $l = 3$ as F terms, and so on. In the case of the hydrogen atom, the terms with different l-values for a given principal quantum number n have only slightly different energy values due to relativity correction and electron spin (see § 6·5). However, because of the different degrees of penetration into the core region by the valence electrons in the alkali atoms, the energy values for the terms with different l for a given n are widely different. The quantum defect p decreases with increasing azimuthal quantum number l for a given n. It is the highest for the S terms (most eccentric ellipse) and very low for the F terms (eccentricity very small). The value of p also depends on n.

In Fig. 6·1 the energy levels for the lithium atom and the possible transitions between them are shown. As can be seen, the energy levels fall into different groups according to the values of l. Thus all S terms ($l = 0$) are grouped in one vertical column. Similarly the P terms ($l = 1$), D terms ($l = 2$) and F terms ($l = 3$) are placed in different vertical columns. The numerals by the sides of the levels represent the n-values. Thus for 2S levels, $n = 2$, $l = 0$, for 3P level, $n = 3$, $l = 1$ and so on. The transitions between the levels are governed by the selection rule

$$\Delta l = \pm 1. \tag{6.1-1}$$

Thus transitions can take place from the S levels to the P levels only ; from the P levels to the S and D levels and so on. The spectral series originating from some of these transitions are given the following special names :

$nP \to n_0S$: Principal series. $nS \to n_0P$: Sharp series.
$nD \to n_0P$: Diffuse series. $nF \to n_0D$: Fundamental series.

In Fig. 6·1, the values of the effective quantum numbers n^* are shown on the right hand side. The departure of n^* from the integral values for the lithium terms is obvious from the figure. The integers shown in the figure on the right side correspond to the positions of the hydrogen terms. The terms of lithium depart most from the hydrogen terms for $l = 0$. The lack of agreement with the hydrogen terms becomes progressively less pronounced as l increases. From the figure, it is obvious that it is the least for the F terms ($l = 3$).

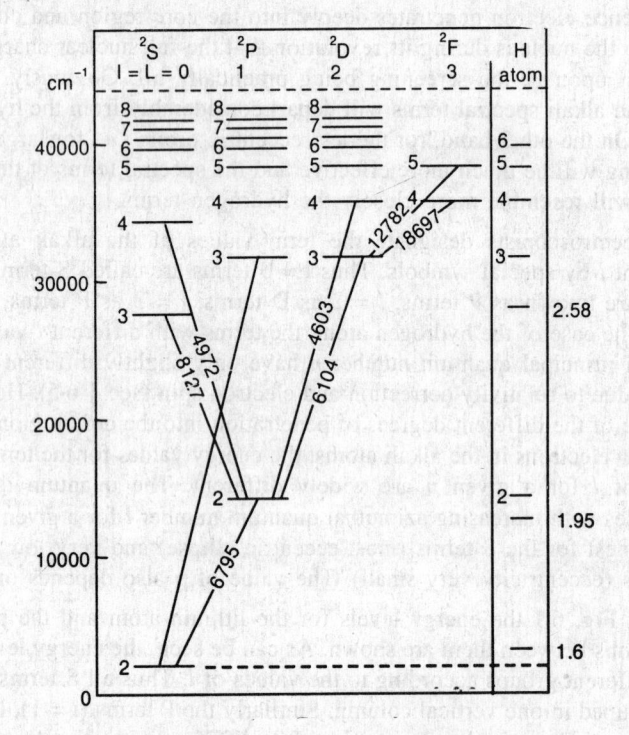

Fig. 6·1. Energy levels for lithium atoms.

The well-known D lines of sodium belong to the principal series and originate from the transition 3P to 3S (see Fig. 6·12). From what has been stated above, each spectral line of any one of the alkali spectral series should be single. So the sodium D-line should also be single. However it is actually found to consist of two closely lying components. These components can easily be resolved by means of an ordinary spectrometer.

They are known as D_1 and D_2 lines having wavelengths 5896 Å and 5890 Å respectively. Observations show that such *doublet structure* of the spectral lines is a characteristic feature of the spectra of all alkali atoms. The origin of the doublet structure can be explained by introducing the concept of electron spin and will be discussed in § 6·4.

6·2 Space quantization and normal Zeeman effect

According to Bohr-Sommerfeld theory, two quantum numbers are required to explain the motion of the electron in an atom. These are the principal quantum number n and the azimuthal quantum number k. In quantum mechanics, k is replaced by the quantum number $l = k - 1$. Description of the electron motion in terms of these two quantum numbers essentially implies that the motion is confined to the orbital plane, there being only two degrees of freedom r and θ. The periodicity of the radial motion is described in terms of the radial quantum number $n_r = (n - k)$ which is thus not an independent quantum number.

The orbital plane of the electron may have different orientations in space. In Fig. 6·2 a, an electron orbit is shown with the orbital angular momentum vector $\mathbf{p}_k$ perpendicular to the orbital plane. In order to study the orientation of the vector $\mathbf{p}_k$ in space, we consider a special direction

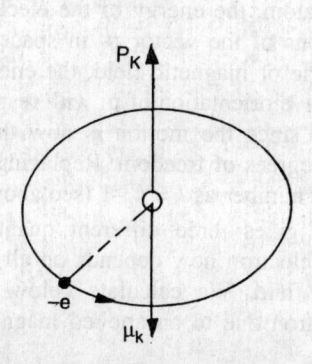

Fig. 6·2 (a) Angular momentum and magnetic moment of the electron due to orbital motion. In quantum mechanics, $\mathbf{p}_k$ and μ_k are replaced by $\mathbf{p}_l$ and μ_l.

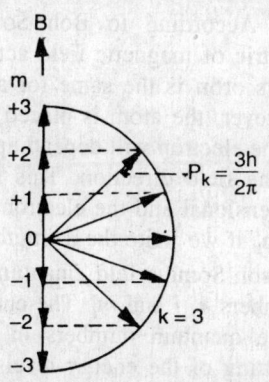

Fig. 6·2 (b) Space quantization of the orbital angular momentum.

fixed in space. This may be the direction of an applied electric or magnetic field. According to classical mechanics, the orbital plane of the electron and hence the vector $\mathbf{p}_k$ may be oriented in any arbitrary direction with respect to a fixed direction in space. However the situation is different in the case of the atomic orbit.

According to Sommerfeld, the orbital angular momentum $\mathbf{p}_k$ of the electron in the atom can be oriented only in some limited number of special directions with respect to a direction fixed in place. Then according to the *rules of space quantization* proposed by Sommerfeld, only those orientations of the vector $\mathbf{p}_k$ are possible for which its components along the fixed direction can have the values

$$p_k \cos \phi = m_k \hbar \qquad (6\cdot2\text{-}1)$$

where ϕ is the angle between the vector $\mathbf{p}_k$ and the fixed direction in space. m_k is known as the *magnetic quantum number* for orbital motion. Its possible values are

$$m_k = 0, \pm 1, \pm 2 \ldots \pm k \qquad (6\cdot2\text{-}2)$$

i.e., m_k can have $(2k + 1)$ values. Since $p_k = k\hbar$, Eq. 6·2-1) gives

$$\cos \phi = \frac{m_k}{k} \qquad (6\cdot2\text{-}3)$$

So the angle of orientation θ between $\mathbf{p}_k$ and the fixed direction can also have $(2k + 1)$ possible values as shown in Fig. 6·2b. (See however, § 6·3).

According to Bohr-Sommerfeld theory, if there is no external electric of magnetic field acting on the atom, the energy of the electron in its orbit is the same for all orientations of the vector $\mathbf{p}_k$ in space. If however, the atom is placed in an electric or magnetic field, the energy of the electron will depend upon the relative orientation of $\mathbf{p}_k$ with respect to the field direction. This is expected, since the motion is now three dimensional and the electron has three degrees of freedom. Replacing $\mathbf{p}_k$ by $\mathbf{p}_l$, if we write the azimuthal quantum number as $l = k - 1$ (see above), Wilson-Sommerfeld quantum condition gives three different quantum numbers n, l and m_l. The energy of the electron now depends on all the three quantum numbers in an applied field. We calculate below the splitting of the energy levels of an electron due to an applied magnetic field.

The rotation of the electron in the orbit constitutes an electric current. Such a current loop behaves like a magnetic shell having a magnetic moment $\mu_l = Ai$ where A is the area of the current loop and i is the current strength. If the orbit is assumed circular of radius a, then the area of the loop is $A = \pi a^2$ so that $\mu_l = \pi a^2 i$. Let the electron revolve in its orbit with a circular frequency $\omega = 2\pi\nu$, where ν is the frequency of revolution. Then since in each revolution a quantity of charge e crosses a section of the orbit, e being the electronic charge, the total charge crossing any section of the orbit in unit time, which is equal to the current, is

$$i = e\nu = e\omega/2\pi$$

Hence the magnetic moment associated with the orbital rotation of the electron becomes

$$\mu_l = \pi a^2 \cdot \frac{e\omega}{2\pi} = \frac{e\omega a^2}{2} \qquad (6\text{-}3)$$

The magnetic moment has a direction at right angles to the plane of the electron orbit. Actually it is opposite to the direction of the orbital angular momentum $\mathbf{p}_l$ of the electron, since the electron carries a negative charge. This is shown in Fig. 6·2a.

Now the magnitude of $\mathbf{p}_l$ is given by

$$p_l = m_e a^2 \omega \qquad (6\cdot2\text{-}5)$$

where we have written the electron mass as m_e to avoid confusion. So the ratio of μ_l to p_l is

$$\frac{\mu_l}{p_l} = \frac{e}{2m_e} \qquad (6\cdot2\text{-}6)$$

This ratio is known as the *gyromagnetic ratio*. The same expression for the gyromagnetic ratio is obtained if the electron is assumed to rotate in an elliptic orbit.

Since the orbital angular momentum is given by $p_l = l\hbar$ where $l = 0$, 1, 2, 3, ... we get

$$\mu_l = \frac{e}{2m_e} p_l = l \frac{e\hbar}{2m_e} = l\,\mu_B \qquad (6\cdot2\text{-}7)$$

where
$$\mu_B = \frac{e\hbar}{2m_e} = 9\cdot2741 \times 10^{-24} \text{ joule/tesla*} \qquad (6\cdot2\text{-}8)$$

μ_B is known as the Bohr magneton which is the basic unit of atomic magnetic moments.

The current loop due to the electron circulating in its orbit thus behaves like a tiny magnet of atomic size. We know that if a magnet is placed in an external magnetic field, it tends to align in the field direction. Before that the magnet will oscillate back and forth about the field direction. The atomic magnet mentioned above however does not oscillate about the field direction in this way. The magnetic moment vector μ_l actually precesses about the field direction, being aligned at a definite angle with respect to the latter, just like the axis of a spinning top precessing about the earth's gravitational field if it is inclined to the vertical. This behaviour is characteristic of all gyroscopic motions when an external torque acts upon them. In the present case, the electron revolving in its orbit is similar to a spinning top and hence its axis of

---

* In S.I. system the unit of magnetic flux density is tesla : 1 tesla = 10^4 gauss.

revolution precesses about the direction of the magnetic field which exerts a torque upon it.

If B is the magnetic flux density, the potential energy of the electron due to the magnetic interaction between its orbital magnetic moment μ_l and the magnetic field is given by

$$\varepsilon_B = -\vec{\mu_l} \cdot \mathbf{B} \tag{6·2-9}$$

If θ is the angle between $\mathbf{p}_l$ and $\mathbf{B}$, then since μ_l has a direction opposite to $\mathbf{p}_l$, we get

$$\varepsilon_B = \mu_l B \cos\theta \tag{6·2-10}$$

Using Eqs. (6·2-3) and (6·2-7), with l replacing k we get

$$\varepsilon_B = l\,\mu_B B\,\frac{m_l}{l} = m_l\,\mu_B\,B \tag{6·2-11}$$

Hence the total energy of the electron in a magnetic field becomes

$$E_{nlm} = E_{nl} + m_l\,\mu_B\,B \tag{6·2-12}$$

where E_{nl} is the energy of the electronic energy level in the absence of the magnetic field with k replaced by l. Writing μ_B explicitly from Eq. (6·2-8), we get

$$E_{nlm} = E_{nl} + \frac{e\hbar}{2m_e}\,m_l\,B$$

$$= E_{nl} + m_l\,\hbar\,\frac{eB}{2m_e}$$

$$= E_{nl} + m_l\,\hbar\,\omega_L \tag{6·2-13}$$

where
$$\omega_L = \frac{eB}{2m_e} = 8\cdot782 \times 10^{10} B \text{ s}^{-1} \tag{6·2-14}$$

$\nu_L = \omega_L/2\pi$ is known as *Larmor precessional frequency*.

Eq. (6·2-13) shows that an electronic energy level with given values n and l splits up into $(2l+1)$ close lying sublevels, each with a slightly different energy determined by the magnetic quantum number m_l. The state of lowest energy has its angular momentum antiparallel to the field direction ($m_l < 0$). As a result of this splitting of the level, a spectral line originating from the transition between two energy levels with given pairs of n and l values splits up into a number of components by the action of the magnetic field. This occurs due to the transitions from the sublevels of the upper states with different values of m_l to those of the lower state. Such transitions between the sublevels are governed by the selection rule

$$\Delta m_l = 0,\ \pm 1 \tag{6·2-15}$$

Alkali Spectra ; Space Quantization ; Periodic....

In Fig. 6·3, is shown the magnetic splitting of a spectral line originating from the transition between two levels with $l = 2$ and $l = 1$ respectively governed by the above selection rule. The wave number of the resultant spectral line due to such transition in the magnetic field is given by (see Eq. 6·2-12)

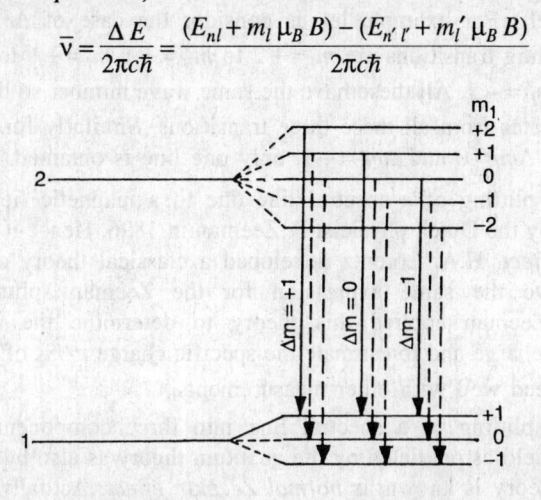

$$\bar{\nu} = \frac{\Delta E}{2\pi c \hbar} = \frac{(E_{nl} + m_l \mu_B B) - (E_{n'l'} + m_l' \mu_B B)}{2\pi c \hbar}$$

Fig. 6·3. Normal Zeeman effect. Zeeman splittings are shown highly exaggerated.

$$= \frac{(E_{nl} - E_{n'l'})}{2\pi c \hbar} + \frac{\mu_B B}{2\pi c \hbar} (m_l - m_l')$$

so that
$$\bar{\nu} = \bar{\nu}_0 + \frac{eB}{4\pi m_e c} \Delta m_l \qquad (6 \cdot 2\text{-}16)$$

where $\bar{\nu}_0 = (E_{nl} - E_{n'l'})/2\pi c\hbar$ is the original wave number of the unsplit spectral line. Since $\Delta m_l = 0, \pm 1$, the original line splits up into *three components* by the application of the magnetic field. The wave numbers of the three components are :

$$\bar{\nu}_1 = \bar{\nu}_0 \qquad (\Delta m_l = 0)$$

$$\bar{\nu}_2 = \bar{\nu}_0 + \frac{eB}{4\pi m_e c} = \bar{\nu}_0 + \frac{\omega_L}{2\pi c} \qquad (\Delta m_l = +1)$$

$$\bar{\nu}_3 = \bar{\nu}_0 - \frac{eB}{4\pi m_e c} = \bar{\nu}_0 - \frac{\omega_L}{2\pi c} \qquad (\Delta m_l = -1)$$

First of these has the same wave number as the original line. The other two have wave numbers lying symmetrically on either side of the original line. The wave number differences of these from the original line are $\Delta \bar{\nu} = \pm \omega_L/2\pi c$ which are independent of the original wave number.

From Fig. 6·3 it may appear that the original line splits up into nine components due to the magnetic field. Actually this is not so. All the component lines originating from a given change Δm_l of the magnetic quantum number have the same wave number. This is due to the fact that the splitting between the adjacent sub-levels is the same for the upper and lower levels. For example let us consider the case of $\Delta m_l = +1$. The corresponding transitions are $m_l = +2$ to $m_l' = +1$, $m_l = +1$ to $m_l' = 0$ and $m_l = 0$ to $m_l' = -1$. All these have the same wave number so that only one line originates from all these three transitions. Similarly for each of the transitions $\Delta m_l = 0$ and $\Delta m_l = -1$, only one line is obtained.

The splitting of a spectral line due to a magnetic field was first observed by the Dutch physicist P. Zeeman in 1896. Hence it is known as *Zeeman effect*. H.A. Lorentz developed a classical theory of the effect which gave the same expression for the Zeeman splitting as Eq. (6·2-16). Zeeman applied this theory to determine the sign of the electronic charge and to estimate the specific charge e/m_e of the electron which agreed well with other measurements.

The splitting of a spectral line into three components due to a magnetic field as predicted by the quantum theory as also by the Lorentz electron theory is known as *normal Zeeman effect*. Actually the normal Zeeman effect is not commonly observed. In most cases, a spectral line is found to split up into more than three components under the influence of the magnetic field upto several tesla of flux density (*i.e.*, several ten thousand gauss of the magnetic field strength in c.g.s. unit). This type of splitting is known as *anomalous Zeeman effect* which can be explained only when the existence of the electron spin is taken into account (see § 6·7).

6·3 Electron spin ; Vector model of the atom

In order to account for the doublet structure of the alkali spectral lines as also to provide an explanation for the anomalous Zeeman effect, two Dutch physicists G.E. Uhlenbeck and S.A. Goudsmit proposed the hypothesis of the electron spin (1925).* According to them the electron has a spinning motion like a top. We may draw an analogy with the motion of the earth which while revolving round the sun also spins about its own axis once in 24 hours. The electron, while revolving in its orbit round the nucleus, has a spinning motion with an intrinsic spin angular momentum

$$p_s = s\,\hbar \qquad (6\cdot3\text{-}1)$$

---

*Actually it was Wolfgang Pauli who first pointed out (1924) that a fourth quantum number was needed to explain the alkali doublets. He called it Zweideutigkeit meaning two-valuedness, but failed to recognise its connection with any dynamical property of the electron. This was subsequently done by Goudsmit and Uhlenbeck.

where $s = \frac{1}{2}$ is known as the spin quantum number. The spin angular momentum $\mathbf{p}_s$ is a vector. Just as the orbital angular momentum $\mathbf{p}_l$ of the electron is associated with an orbital magnetic moment so also the spin angular momentum $\mathbf{p}_s$ is associated with an intrinsic magnetic moment μ_s. However the ratio of μ_s/p_s is different from the gyromagnetic ratio for the orbital motion given by Eq. (6·2-6) and can be written as

$$\frac{\mu_s}{p_s} = g_s \frac{e}{2m_e} \qquad (6\cdot3\text{-}2)$$

where one must assume that $g_s = 2$ in order to obtain agreement with the experimental results. Hence we get by using Eq. (6·3-1)

$$\mu_s = 2 \cdot \frac{e}{2m_e} \cdot p_s = \frac{e}{m_e} \cdot \frac{\hbar}{2} = \frac{e\hbar}{2m_e} = \mu_B \qquad (6\cdot3\text{-}3)$$

Thus the intrinsic magnetic moment of the electron is equal to the Bohr magneton μ_B. It should be noted that the above values of μ_s and p_s are also obtained from the relativistic quantum mechanical theory of the electron proposed by P. A. M. Dirac. It is not possible to visualize the spin of a point particle like an electron classically. Uhlenbeck and Goudsmit introduced the idea of the electron spin as an *ad hoc* hypothesis. However in Dirac's theory, spin appears quite naturally in solving the Dirac equations for the relativistic electron. (See App. A-XI).

As stated at the end of § 4·12, the quantum mechanical value of the orbital angular momentum of the electron is

$$p_l = \sqrt{l(l+1)}\ \hbar \qquad (6\cdot3\text{-}4)$$

where $l = 0, 1, 2,... (n-1)$ is the azimuthal quantum number which replaces the quantum number k of Bohr-Sommerfeld theory. n is the principal quantum number. A rigorous derivation of Eq. (6·3-4) will be given in Ch. XI.

Similarly the spin angular momentum of the electron. according to quantum mechanics is given by

$$p_s = \sqrt{s(s+1)}\ \hbar \qquad (6\cdot3\text{-}5)$$

with $s = 1/2$.

In an external magnetic field, the vector $\mathbf{p}_l$ can be oriented in such a manner that its component in the field direction can assume values determined by Sommerfeld's space quantization rule (Eq. 6·2-1) :

$$p_l \cos \phi = m_l \hbar$$

where m_l as before, can take up the $(2l+1)$ integral (including zero) values :

$$m_l = 0, \pm 1, \pm 2,... \pm l$$

θ is the angle between p_l and the field direction.

The largest component of p_l along the field direction is thus $l\hbar$ when $m_l = l$. This value is less than the magnitude of p_l which means that even in this case the vector p_l cannot align in the direction of magnetic field B (see Fig. 6.4). This result is different from what is expected from the old quantum theory according to which the maximum value of the component of the vector p_k (or p_l) along the field direction is equal to the magnitude of the vector itself. The vector is actually aligned in the direction of the field in this case as shown in Fig. 6.2 b.

The value of $\cos \phi$ according to the quantum mechanical theory is given by

$$\cos \phi = \frac{m_l}{\sqrt{l(l+1)}} \qquad (6.3\text{-}6)$$

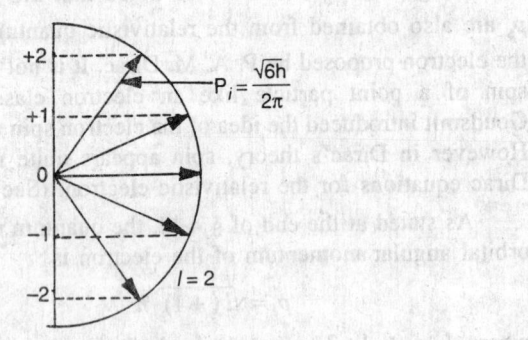

Fig. 6.4. Components of the vector p_l in the field direction according to quantum mechanics.

so that its maximum value is $\sqrt{l/(l+1)}$.

The spin angular momentum p_s of the electron can align either parallel or antiparallel to an applied magnetic field so that its components in the field direction are $m_s \hbar$, where $m_s = +\frac{1}{2}$ (parallel) and $m_s = -\frac{1}{2}$ (antiparallel) as shown is Fig. 6.5.

In the absence of any external field, the motion of the electron in an atom can be fully described in terms of the three quantum numbers n, l and s.

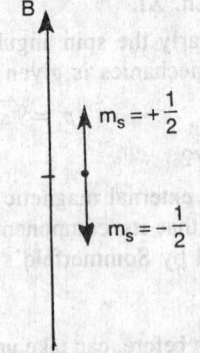

Fig. 6.5 Orientation of the spin angular momentum of the electron with respect to the field direction.

Alkali Spectra ; Space Quantization ; Periodic....

In addition, the magnetic quantum number must be introduced when the atom is under the influence of the magnetic field.

The above considerations apply to the case of a single electron atom, *i.e.*, hydrogen-like atom, or to the case when there is a single electron revolving in the outermost orbit, as in the alkali atoms. When there are more than one electrons in an atom, the l vectors for the different electrons must be added up vectorially to give the resultant orbital angular momentum L of the atom :

$$\mathbf{L} = \sum_i \mathbf{l}_i = \mathbf{l}_1 + \mathbf{l}_2 + \mathbf{l}_3 + \ldots \qquad (6\cdot3\text{-}7)$$

where $\mathbf{l}_1, \mathbf{l}_2, \mathbf{l}_3$ etc. are the orbital angular momenta of the different electrons. Notice that here we have substituted the vector $\mathbf{l}$ in place of $\mathbf{p}_l = l\hbar$.

The value of the total orbital quantum number L of the atom must be an integer. For example in the case of two electrons, if $l_1 = 2$ and $l_2 = 1$, then L can have any one of the values 3, 2 or 1.

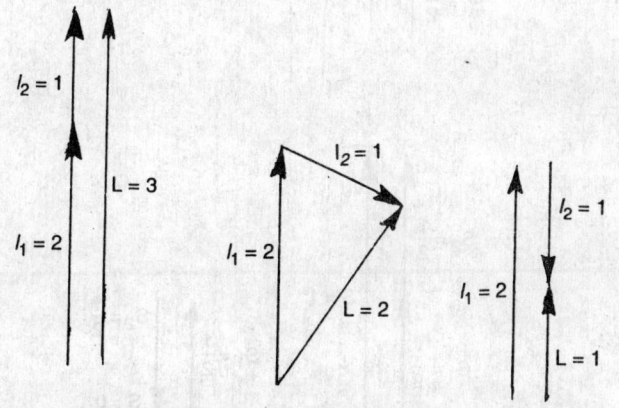

Fig. 6·6. Composition of the orbital angular momentum vectors for two electrons.

This is illustrated in Fig. 6·6. According to quantum mechanics, the magnitude of the resultant orbital angular momentum of the atom will be $p_L = \sqrt{L(L+1)}\, \hbar$. In the above example if $L = 2$, then $p_L = \sqrt{6}\, \hbar$.

Likewise, for an atom with more than one electrons in it, the resultant spin angular momentum is given by

$$\mathbf{S} = \sum_i \mathbf{s}_i = \mathbf{s}_1 + \mathbf{s}_2 + \mathbf{s}_3 + \ldots \qquad (6\cdot3\text{-}8)$$

Here $\mathbf{s}_1, \mathbf{s}_2, \mathbf{s}_3 \ldots$ are the spins of the individual electrons in the atom. Each of these is equal to 1/2. According to quantum mechanics the resultant spin angular momentum will have the value

$$p_S = \sqrt{S(S+1)}\, \hbar$$

The rule for vector addition of the spins of the different electrons requires that these can either be parallel or antiparallel to one another which is illustrated in Fig. 6·7. Thus if an atom has an even number of electrons, the above rule requires that the value of S for it can be either zero or an integer. On the other hand, if the number of electrons is odd, then the value of S will be half-odd integral.

From Fig. 6·7, we can see that for 3 electrons, $S = \frac{3}{2}$ or $\frac{1}{2}$, while for 2 electrons, $S = 1$ or 0.

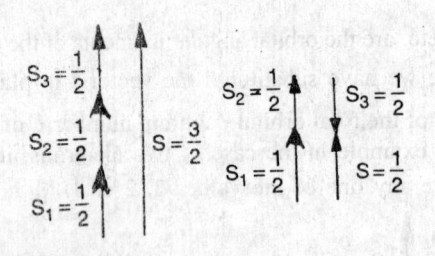

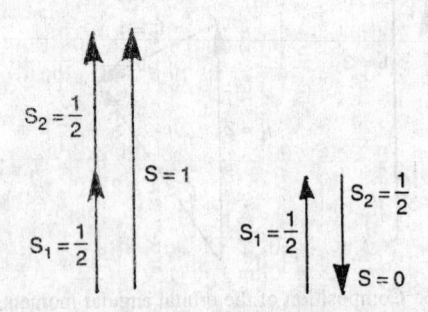

Fig. 6·7. Composition of the spin vectors for (a) three electrons and (b) two electrons.

Since an electron in an atom has two different angular momenta $\mathbf{p}_l$ and $\mathbf{p}_s$, its resultant angular momentum $\mathbf{p}_j$ is obtained by the vector addition of $\mathbf{p}_l$ and $\mathbf{p}_s$. Writing $\mathbf{p}_j = \mathbf{j}\hbar$ where j is called the *total* or *inner quantum number*, we get

$$\mathbf{p}_j = \mathbf{p}_l + \mathbf{p}_s$$

or
$$\mathbf{j} = \mathbf{l} + \mathbf{s} \qquad (6 \cdot 3\text{-}9)$$

For a single electron, $s = 1/2$ and hence j can have two values: $j = l + 1/2$ or $j = l - 1/2$ corresponding to parallel or antiparallel alignments of s w.r.t. l. For example if $l = 1$, then j can be 3/2 or 1/2 (see Fig. 6·8). For the case of $l = 0$, j can have only one value *viz.*, $j = 1/2$.

Alkali Spectra ; Space Quantization ; Periodic....

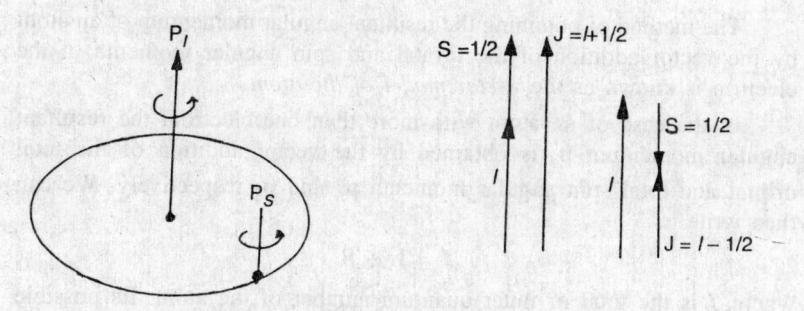

Fig. 6·8. Composition of **l** and **s** vectors to form **j** vector for a single electron.

The above method of finding the resultant angular momentum **j** by the vector addition of **l** and **s** is somewhat modified in quantum mechanics. As in the case of p_l and p_s, the quantum mechanical value of p_j is given by

$$p_j = \sqrt{j(j+1)}\,\hbar \qquad (6\cdot3\text{-}10)$$

where the possible values of j are the same as those given above. The vector $\mathbf{p}_s$ of magnitude $\sqrt{s(s+1)}\,\hbar$ cannot be aligned parallel or antiparallel to the vector $\mathbf{p}_l$ which has the magnitude $\sqrt{l(l+1)}\,\hbar$ to yield the resultant $\mathbf{p}_j$ given by Eq. (6·3-10) The vectors $\mathbf{p}_l$ and $\mathbf{p}_s$ must be inclined suitably

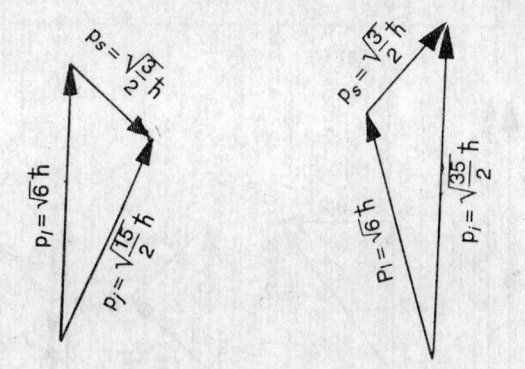

Fig. 6·9. Composition of $\mathbf{p}_l$ and $\mathbf{p}_s$ vectors according to quantum mechanics for $l = 2$ and $s = 1/2$.

to yield the resultant $\mathbf{p}_j$. This is illustrated in Fig. 6·9 for the case of $l = 2$, $s = 1/2$. The angle between $\mathbf{p}_l$ and $\mathbf{p}_s$ is given by

$$\cos(\mathbf{l},\mathbf{s}) = \frac{j(j+1) - l(l+1) - s(s+1)}{2\sqrt{l(l+1)}\,\sqrt{s(s+1)}} \qquad (6\cdot3\text{-}11)$$

The method of obtaining the resultant angular momentum of an atom by the vector addition of the orbital and spin angular momenta of the electron is known as the *vector model of the atom*.

In the case of an atom with more than one electron, the resultant angular momentum p_J is obtained by the vector addition of the total orbital and total spin angular momenta p_L and p_S respectively. We can then write

$$J = L + S$$

where J is the total or inner quantum number of the atom. Its possible values are

$$J = (L+S), (L+S-1), (L+S-2), \ldots |L-S| \qquad (6\cdot3\text{-}12)$$

If $L > S$ the number of possible J values is $(2S + 1)$ while it is $(2L + 1)$ if $L < S$. For $L = 0$, there is only one value of $J = S$. Similarly for $S = 0$, the only possible value is $J = L$. The origin of the vector J by the vector addition of L and S is illustrated in Fig. 6·10.

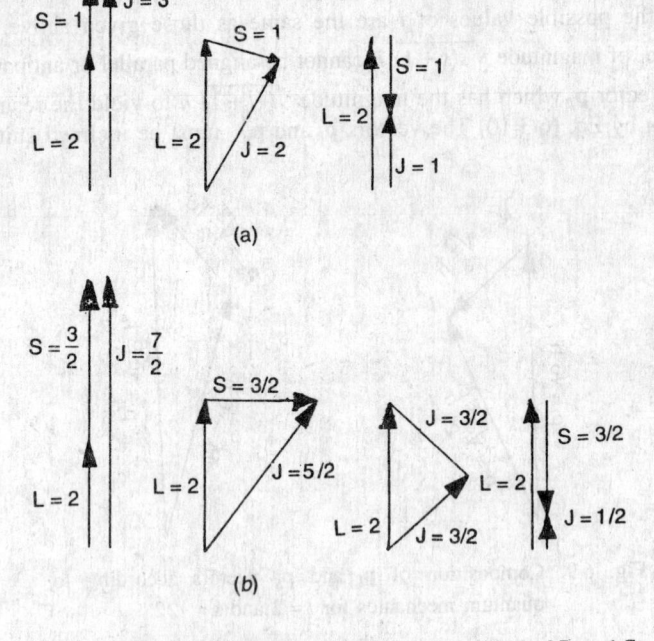

Fig. 6·10. Origin of the vector **J** by the composition of **L** and **S** vectors for more than one electron for (a) $L = 2$ and $S = 1$; (b) $L = 2$ and $S = 3/2$.

Alkali Spectra ; Space Quantization ; Periodic....

It may be noted that the method of forming the resultant angular momentum **J** as discussed above is due to the spin-orbit interaction between the total **L** and **S** vectors of the atom. This type of couping between **L** and **S** is known as Russel-Saunders or *L-S* coupling and is found to be prevalent in majority of the atoms.

The interaction giving rise to *L-S* coupling is actually magnetic in nature. The magnetic moment μ_S due to the electron spin is influenced by the magnetic moment μ_L due to its orbital motion, the strength of the interaction depending upon the relative orientation of the two which is determined by the rules of space quantization. Due to the spin-orbit interaction, the vectors **L** and **S** precess about the direction of **J**. (See § 13·18).

There is another type of coupling known as *j-j* coupling for forming the resultant angular momentum vector **J** of an atom. In this case the **l** and **s** vectors of the individual electrons combine to produce the resultant vector $\mathbf{j} = \mathbf{l} + \mathbf{s}$ of the different electrons. The vectors $\mathbf{j}_i$ of the different electrons are then vectorially combined to yield the resultant angular momentum vector $\mathbf{J} = \Sigma_i \mathbf{j}_i$ of the atom. This type of coupling is observed in the case of the excited states of some heavy atoms. In some cases, mixture of *L-S* and *j-j* couplings is also observed.

If an atom is placed in a magnetic field then its resultant angular momentum **J** precesses about the field direction in such a way that the components of **J** along the magnetic field are determined by Sommerfeld's rule of space quantization. The vector **J** can align itself in $(2J+1)$ directions w.r.t. the field such that the components of the total angular momentum along the field direction are given by

$$p_J \cos \theta = m_J \hbar \qquad (6 \cdot 3\text{-}13)$$

where the magnetic quantum number m_j can have the following $(2j+1)$ values :

$$m_J = J, J-1, J-2, \ldots -J \qquad (6 \cdot 3\text{-}14)$$

Notice that though *J* can be both integral and half-integral, difference between the consecutive values of m_J is 1.

The above result is somewhat modified in quantum mechanics according to which the resultant angular momentum has a value $p_J = \sqrt{J(J+1)}\, \hbar$. Thus the largest component of p_J along the field direction which has a value $J\hbar$ is less than the magnitude of the vector $\mathbf{p}_J$. This is shown is Fig. 6·11.

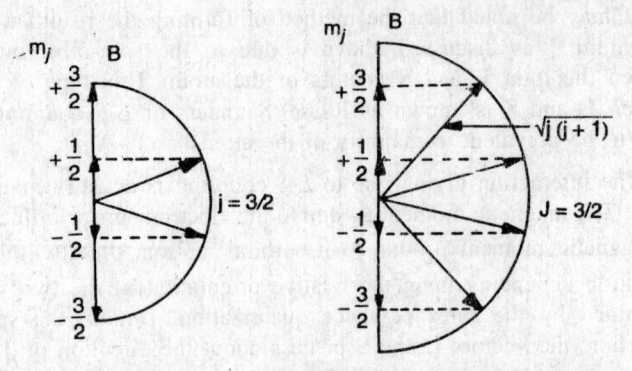

Fig. 6·11 Orientation of the **J** vector in an external field according to (a) old quantum theory ; (b) according to quantum mechanics.

6·4 Doublet structure of the alkali spectral lines

It was noted at the end of § 6·1 that the alkali spectral lines have doublet structure. This arises due to the existence of electron spin. Since only the single valence electron outside the core region is responsible for the origin of the alkali spectra, we have to consider the magnetic interaction between the orbital magnetic moment μ_l and the spin magnetic moment μ_s of this electron. This magnetic interaction causes a difference in energy between the two terms of the same l having two different values of $j = l \pm \frac{1}{2}$. Such splitting of the energy levels of the same l but different j is known as the *multiplicity* of the term or level. Obviously for the alkali atom, the multiplicity is 2. In general, if an atom having more than one electron has a resultant spin S, then a term having a definite L splits up into $(2S+1)$ different terms, each having a different J if $L > S$. Thus the multiplicity is $(2S+1)$. For the alkali atoms, $S = s = \frac{1}{2}$ so that the multiplicity is $2s + 1 = 2$ as mentioned above.

In Fig. 6·12, the doublet structure of the energy levels of sodium is shown. As can be seen, the P, D, F etc. terms all have this doublet structure. For example, since $l = 1$ for the P terms, $j = \frac{3}{2}$ or $\frac{1}{2}$. These two terms are designated as $^2P_{3/2}$ and $^2P_{1/2}$ respectively. Similarly the D terms with $l = 2$ have $j = \frac{5}{2}$ or $\frac{3}{2}$ and are designated as $^2D_{5/2}$ and $^2D_{3/2}$ respectively. *In the above notations, the superscript 2 to the left of the letters P,D etc. indicating the l-values of the terms denote the multiplicity of the term, while the subscripts ½, ³⁄₂ etc. to the right of the letters denote the j-values of the terms.* The $l = 0$ or S term is single since j can have only one value in this case, viz., $j = \frac{1}{2}$. Even so, since $s = \frac{1}{2}$ the multiplicity is 2 for it so that it is written as $^2S_{1/2}$.

Alkali Spectra ; Space Quantization ; Periodic....

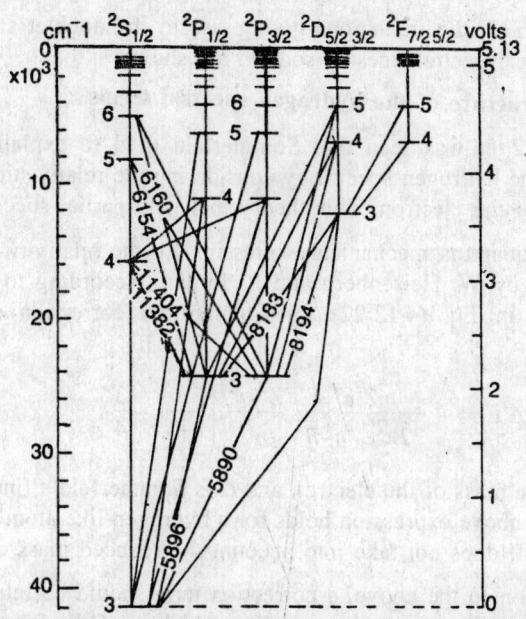

Fig. 6·12. Doublet structure of the energy levels of sodium. The wave lengths are in angstroms. The numbers opposite the levels are values of n.

Since all the energy levels are now split into two due to spin-orbit interaction, the transitions between them give rise to a number of components in place of a single spectral line. These transitions are governed by the selection rule

$$\Delta j = 0, \pm 1 \qquad (6\cdot4\text{-}1)$$

It is possible to explain the fine structure of the alkali spectral lines with the help of the two selection rules (6·1-1) and (6·4-1). For example in the case of the sodium lines of the principal series $n\,P \longrightarrow n_0 S$ with $n_0 = 3$, the following transitions are possible :

$$n\,^2P_{1/2} \longrightarrow 3^2S_{1/2} \;(\Delta j = 0)$$

$$n\,^2P_{3/2} \longrightarrow 3^2S_{1/2} \;(\Delta j = 1)$$

Thus these lines will each have a doublet structure. The two sodium D lines originate due to the following transitions :

$$3^2P_{1/2} \longrightarrow 3^2S_{1/2} \;(D_1 \text{ line})$$

$$3^2P_{3/2} \longrightarrow 3^2S_{1/2} \;(D_2 \text{ line})$$

In Fig. 6·12, the transitions giving rise to the doublet structure of these and other spectral lines of sodium are shown.

6·5 Fine structure of the hydrogen spectral terms

In § 4·12, it was seen that Sommerfeld tried to explain the fine structure of the hydrogen spectra by considering the relativistic variation of the mass of the electron. This theory met with partial success only.

Later a quantum mechanical expression for the relativity correction was deduced by W. Heisenberg and P. Jordan. According to them, the second term in Eq. (4·12-22) for Sommerfeld correction should be replaced by

$$\Delta E_r = -\frac{mZ^2 e^4}{32\pi^2 \varepsilon_0^2 n^2 \hbar^2} \cdot \frac{Z^2 \alpha^2}{n^2} \left(\frac{n}{l+\frac{1}{2}} - \frac{3}{4} \right) \tag{6·5-1}$$

where m is the mass of the electron and α is Sommerfeld's fine structure constant. The above expression holds for a hydrogen-like atom of nuclear charge $+Ze$. It does not take into account the reduced mass correction.

In addition to the above, a correction term should be introduced to take into account the spin-orbit coupling which is of the form

$$\Delta E_{LS} = \frac{mZ^2 e^4}{32\pi^2 \varepsilon_0^2 n^2 \hbar^2} \cdot \frac{Z^2 \alpha^2}{n} \cdot \frac{j(j+1) - l(l+1) - s(s+1)}{2l(l+\frac{1}{2})(l+1)} \tag{6·5-2}$$

For hydrogen like atoms, there is only one orbital electron for which $s=\frac{1}{2}$ and j can have two values : $j = l + \frac{1}{2}$ or $j = l - \frac{1}{2}$. Combining the above two expressions, we get for hydrogen ($Z = 1$)

$$\Delta E = \Delta E_r + \Delta E_{LS}$$

$$= -\frac{mZ^2 e^4}{32\pi^2 \varepsilon_0^2 n^2 \hbar^2} \cdot \frac{Z^2 \alpha^2}{n^2} \left[\frac{n}{j+\frac{1}{2}} - \frac{3}{4} \right] \tag{6·5-3}$$

The above equation shows that the correction term depends on j only and is the same for two consecutive l values for which j is the same, viz., $l_1 = j - \frac{1}{2}$ and $l_2 = j + \frac{1}{2}$. Thus $^2S_{1/2}$ and $^2P_{1/2}$ terms for the same n have the same energy according to this theory. Similarly $^2P_{3/2}$ and $^2D_{3/2}$ terms for the same n have the same energy ; and so forth. In Fig. 6·13 are shown the possible transitions for the origin of the fine structure splitting of the H_α line of hydrogen according to the above considerations. However even with these corrections there is slight disagreement between the theoretical predictions and the experimental results. (see § 13·1 and § 13·2 in Ch. XIII).

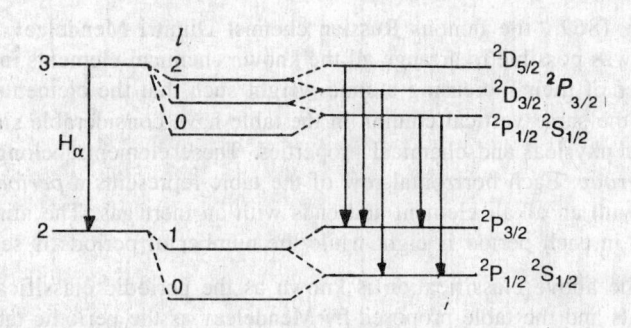

Fig. 6·13. Origin of the fine structure of the Balmer line of hydrogen. The levels in the middle are due to relativity correction. The levels on the extreme right are due to spin-relativity correction. The splittings are not according to scale.

The problem was finally solved by considering the interaction between the electron and the radiation field. According to this quantum electrodynamical theory, levels with the same j but different l do not have the same energy as shown in Fig. 6·13. Thus $^2S_{1/2}$ and $^2P_{1/2}$ levels of hydrogen for the same n should have slightly different energies ; similarly $^2P_{3/2}$ and $^2D_{3/2}$ levels for the same n have different energies. These predictions of the theory were fully confirmed by a very elegant experiment performed by two American physicists W.E. Lamb and R.E. Retherford in 1950. Using micro-wave technique they showed that there was a difference of energy between the $2\ ^2S_{1/2}$ and $2\ ^2P_{1/2}$ levels of hydrogen corresponding to a frequency difference of $\Delta \nu = \Delta E/h = 1058$ mega-hertz (MHz) which agrees very well with the theoretical prediction. (see § 13·4).

Another consequence of the above theory is the so-called anomalous value of the gyromagnetic ratio for the electron spin. According to the spin hypothesis of Goudsmit and Uhlenbeck as also from Dirac electron theory, the value of this ratio in the unit of Bohr magneton should be $g_s = 2$ (see Eq. 6·3-2). However according to the above theory it is given by

$$g_s = 2 \times 1 \cdot 001160$$

which has been confirmed experimentally. This gives the electron magnetic moment

$$\mu_e = 2 \times 1 \cdot 001160\ \mu_B = -9 \cdot 2849 \times 10^{-24}\ \text{J/T}$$

6.6 Pauli's exclusion principle ; Periodic classification of elements

In 1869, the famous Russian chemist Dimitri Mendeleev showed that it was possible to arrange all the known chemical elements in a table in order of their increasing atomic weight such that the elements which fall in the same vertical column in the table have considerable similarity in their physical and chemical properties. These elements belong to the same *group*. Each horizontal row of the table represents a *period* which starts with an alkali element and ends with an inert gas. The number of *groups* in each period is eight while the number of periods is seven.

The above classification is known as the periodic classification of elements and the table proposed by Mendeleev as the periodic table (see Appendix A-VII). It is possible to explain the periodic classification on the basis of the arrangement of the electrons in the atoms of the different elements.

We have seen that the energy of an electron in an atom is determined by the three quantum numbers, n, l and j. In addition, when a magnetic field is present the magnetic quantum number m_j which determines the component of **j** along the field direction has to be considered. Since m_j can have $(2j + 1)$ values, the electronic state is $(2j + 1)$-*fold degenerate* which means that the $(2j + 1)$ sub-levels of the same n, l and j but different m_j have the same energy. In a magnetic field however these $(2j + 1)$ sublevels split up and have different energies. Since there is always some magnetic field present in an atom because of the magnetic moments μ_l and μ_s associated respectively with the orbital and spin motions of the electron, these $(2j + 1)$ sublevels have different energies. Hence the energy state of an electron requires the four quantum numbers n, l, j and m_j for its complete specification.

In a magnetic field, the total angular momentum **j** formed by the vector addition of the orbital and spin angular momenta **l** and **s** precesses about the field direction. However, if the magnetic field is very strong, the coupling between **l** and **s** may be broken and these two vectors will independently precess about the field direction, their components along the field direction being determined by the quantum numbers m_l and m_s respectively. j loses its significance in this case. Thus an alternative way of characterising an electronic state in an atom is to specify the four quantum numbers n, l, m_l and m_s.

To understand how to specify the above four numbers for an electron in an atom, Wolfgang Pauli, the famous Austrian physicist, proposed the 'exclusion principle' in 1925. According to Pauli's exclusion principle, no two electrons in an atom can have all the four quantum numbers n, l, m_l and m_s the same. This means that if two electrons in an atom have the same values of the three quantum numbers n, l and m_l, the fourth quantum

Alkali Spectra ; Space Quantization ; Periodic....

number m_s must be different for them. As an example if $n = 2$, $l = 1$ and $m_l = -1$ for two electrons in an atom, then $m_s = +\frac{1}{2}$ for one of them and $m_s = -\frac{1}{2}$ for the other. Both of them cannot have $m_s = +\frac{1}{2}$ or $-\frac{1}{2}$.

We shall now see how, with the help of Pauli's exclusion principle, it is possible to understand the arrangement of the electrons in an atom.

According to Bohr-Sommerfeld picture, n different elliptic orbits are possible in an atom for a given value of the principal quantum number n. These are characterised by different values of the azimuthal quantum number l which can take up the values 0, 1, 2,...... $(n - 1)$. Electrons revolving in these orbits have slightly different energies. The energies are higher for the orbits of larger l. The different values of the principal quantum number n determine the broad division of the energy levels into groups or *shells*. Thus $n = 1$ is known as the K-shell, $n = 2$ as the L-shell, $n = 3$ the M-shell and so on. Shells with larger n have higher energies. The energy differences between the shells are large compared to the energy splitting between the orbits or the *subshells* of different l in a shell.

The subshells with different l are also given special designations. Thus $l = 0, 1, 2, 3, 4$ etc. subshells are designated by the letters s, p, d, f, g etc. For example $3d$ means the d subshell with $l = 2$ in the M shell ($n = 3$).

Each of the energy levels corresponding to subshells with different l further splits up due to magnetic ineraction into $(2l + 1)$ components with different m_l which can take up the values $m_l = 0, \pm 1, \pm 2, ... \pm l$. These have slightly different energy values. Finally each of these components can have two different values of $m_s = \pm \frac{1}{2}$ for which the energies are again slightly different. So for a definite value of l the total number of sublevels is $2(2l + 1)$ for which the energies are different. The m_l and m_s values for these are different. Because of Pauli's exclusion principle there can be only one electron in each of these sublevels with definite values of l, m_l and m_s so that the maximum number of electrons that can be accommodated in a subshell with a definite l is $2(2l + 1)$. We can now easily calculate the total number of electronic energy levels for a given n. This is

$$2 \sum_{l=0}^{n-1} (2l + 1) = 2n^2 \qquad (6 \cdot 6\text{--}1)$$

Thus the maximum number of electrons that can be placed in a given atomic shell with a definite n consistent with the exclusion principle is $2n^2$. These are listed in Table 6·1 for different values of n.

Table 6·1

Shell	K	L		M			N			
n	1	2		3			4			
Sublevel	s	s	p	s	p	d	s	p	d	f
l	0	0	1	0	1	2	0	1	2	3
Maximum no. of electrons in the sublevel.	2	2	6	2	6	10	2	6	10	14
Maximum no. of electrons in the shell.	2	8		18			32			

As an example, consider the 1s subshell for which $n = 1$ and $l = 0$. So $m_l = 0$. Since m_s can be $+\frac{1}{2}$ or $-\frac{1}{2}$, the maximum possible number of electrons in this subshell is 2. If both these electrons are present in an atom, these are designated as $1s^2$. The values of m_s are $+\frac{1}{2}$ for one of these and $-\frac{1}{2}$ for the other. If however, only one electron is present in this subshell it is designated by the symbol $1s^1$. Its m_s value can be either $+\frac{1}{2}$ or $-\frac{1}{2}$. Not more than two electrons can occupy this subshell because, as seen above, n, l and m_l are the same for it. So m_s must be different for the different electrons placed in it. Since there are only two possible values of m_s, there is no place for a third electron in this subshell because that would violate Pauli's exclusion principle. The third electron must go to the next subshell of higher energy which is in the L shell with $n = 2$.

In the L shell, there are two subshells with $l = 0$ and 1. Of these $l = 0$ or 2s subshell can contain a maximum of 2 electrons. The other subshell 2p with $l = 1$ can have three possible values of m_l given by $m_l = 0, \pm 1$. For each of these m_s can have two values, $+\frac{1}{2}$ or $-\frac{1}{2}$. The possible quantum numbers for the electrons in the 2p subshell are shown in Table 6·2.

Table 6·2

n	l	m_l	m_s
2	1	+1	$+\frac{1}{2}$
2	1	+1	$-\frac{1}{2}$
2	1	0	$+\frac{1}{2}$
2	1	0	$-\frac{1}{2}$
2	1	−1	$+\frac{1}{2}$
2	1	−1	$-\frac{1}{2}$

It is evident that no two states amongst the six listed in Table 6·2 have the same set of the four quantum numbers n, l, m_l, m_s. So each of these states can be occupied by an electron, consistent with the exclusion principle. Thus the maximum number of electrons that can be accommodated in the $2p$ subshell is $2(2l+1) = 2(2 \times 1 + 1) = 6$ as shown in the table. The electrons in the $2p$ subshell can be designated as $2p^i$ where i is the number of electrons which can have any value from 1 to 6. (see § 13·13 in Ch. XIII)

6·7 Arrangement of the electrons in the atoms

To consider the arrangement of the electrons in the atoms, we shall assume that the number of electrons in an atom is equal to the serial number of the element in the periodic table which is known as its atomic number Z.

The very first element in the periodic table is hydrogen ($Z = 1$) having only one electron in it which can be placed in the $1s$ subshell. This is because the $n = 1$ or K shell is closest to the nucleus so that an electron in this shell is most strongly bound in the atom and its energy is the lowest. Since it is a general law of dynamics that all systems tend to attain the state of minimum potential energy, the electrons while occupying the states in the atoms also obey this law. The electron in the hydrogen atom is designated as $1s^1$.

The next element in the periodic table is helium with $Z = 2$ so that there are two electrons in its atom. Since a maximum of two electrons can be in the $1s$ subshell both the electrons in the He atom can be placed in it. The electronic configuration of He is designated as $1s^2$. The K shell is completely filled up with two electrons, so that the first period in the periodic table ends at helium.

The second period starts with the element lithium for which $Z = 3$. This is an alkali element. Of the three electrons in the Li atom, two go to the $1s$ subshell in the K shell which is thereby completed. The third electron goes to the $2s$ subshell in the L shell. Thus the electronic configuration of lithium can be designated as $1s^2 2s^1$.

In the next element beryllium with $Z = 4$, there are four electrons, of which two go into $1s$ subshell and completes the K shell while the other two go into the $2s$ subshell of the L shell. The electronic configuration is $1s^2 2s^2$. The $2s$ subshell is filled up completely at Be.

There are five electrons in the next element boron so that $Z = 5$ for it. Of the five electrons in this atom, two go to $1s$ and two to $2s$ subshells. The fifth electron must to the $2p$ subshell ; the corresponding energy level has slightly higher energy than for the $2s$ subshell. The electronic configuration of boron is $1s^2 2s^2 2p^1$.

In the subsequent five elements carbon, nitrogen, oxygen fluorine and neon with Z = 6, 7, 8, 9 and 10 respectively, the $2p$ subshell is gradually filled up with increasing number of electrons. It becomes completely filled at neon with six electrons and with that the L shell is completed to that the second period ends with neon. Its electronic configuration is $1s^2\ 2s^2\ 2p^6$.

Both the elements helium and neon with which end the first and the second periods respectively are inert gases. Both of them are placed in the last (eighth) column of the periodic table.

The third period begins with the second alkali element sodium (Z = 11). After completely filling up of the K and L shells with 2 and 8 electrons respectively, the eleventh electron goes to the M shell with $n = 3$. Actually this electron goes to the $3s$ subshell so that the electronic configuration of Na is $1s^2\ 2s^2\ 2p^6\ 3s^1$. In the periodic table sodium is placed just below lithium in the first column, both having a single electron in the respective outermost s subshells with completed inner shells. In lithium, the outermost electron in $2s^1$ while in sodium it is $3s^1$. The similarity of the electronic configuration in the outermost shells of these two elements is responsible for the similarities of their physical and chemical properties. From this it can be inferred that the positions of the outermost *valence electrons* in the atoms determine their physical and chemical properties.

In the next element magnesium (Z = 12), the twelfth electron is placed in the $3s$ subshell so that its electron configuration is $1s^2\ 2s^2\ 2p^6\ 3s^2$. This configuration is similar to that of beryllium with $Z = 4$ which has the configuration $1s^2\ 2s^2$. In both cases, there are two valence electrons in the s subshell. Both the elements belong to the *alkaline earth group* and are placed in the second column of the periodic table. There are many similarities in their physical and chemical properties.

The arrangement of the electrons in the atoms of the different elements of the periodic table is shown in Table 6·3. In the table, the elements are arranged in order of their increasing atomic number Z.

From the table it will be seen that as stated before the K shell is completely filled up at helium. The $l = 1$ or p subshells of the L, M, N, O and P shells are filled up respectively at neon (Z = 10), argon (Z = 18), krypton (Z = 36), xenon (Z = 54) and radon (Z = 86) which are all inert gases. They are all placed in the last column of the periodic table. When the outermost orbit of a given l is completely filled with electrons, the electrons in the atom are very strongly bound. It is normally difficult for such atoms to take up one more electron or to give up one of its electrons. It is known that in chemical reaction, electrons rearrange their positions in neighbouring atoms. Since it is difficult for the inert gases to do this, they do not enter into chemical combination with any other element. That is the reason for their nomenclature. Their ground states are $^1S_0$.

In each of the elements immediately following the inert gases in the periodic table, there is a single valence electron in the outermost s subshell ($l = 0$) of the L, M, N, O or P shell outside the completely filled s or p subshell of the preceding shell. Each of these is an alkali element and they are all placed in the first column of the periodic table. Besides Li ($Z = 3$) and Na ($Z = 11$), these include K ($Z = 19$), Rb ($Z = 37$), Cs ($Z = 55$) and Fr ($Z = 87$). They have very high electro-positivity. In chemical reactions, they can easily exchange the losely bound outermost valence electron (s electron) with an electro-negative element. They are all mono-valent elements, having similar physical and chemical properties. Their spectra show doublet structure. The ground state of each of these elements is $^2S_{1/2}$.

In the second column of the periodic table are the elements of the alkaline earth group. They have two valence electrons in the outermost s ($l = 0$) subshells of their atoms. Apart from Be ($Z = 4$) and Mg ($Z = 12$), they include Ca ($Z = 20$), Sr ($Z = 38$), Ba ($Z = 56$) and Ra ($Z = 88$). Their chemical properties are similar and they are all divalent elements. Their ground states are $^1S_0$.

The elements in the following columns (3 to 7) of the periodic table have respectively a total of 3 to 7 electrons in the outermost s and p subshells of their atoms. The differences in the physical and chemical properties of these elements in the different columns in any period of the periodic table are due to the differences in the number of electrons in the outermost subshells of their atoms. Their valencies increase by one unit from one column to the next. The elements in the different periods falling in the same column have similar physical and chemical properties due to the similarly in the electronic configuration in the outermost subshells of their atoms. They have the same valency. For example the elements B, Al, Ga, In and Tl with $Z = 5, 13, 31, 49$ and 81 respectively, falling in the third column of the periods 2 to 6 are all trivalent and are similar chemically. Their ground states are $^2P_{1/2}$.

The elements fluorine, chlorine, bromine, iodine and astatine in the seventh column with $Z = 9, 17, 35, 53$ and 85 respectively belonging to the halogen group have five electrons each in the outermost p subshells of their atoms, viz., $2p, 3p, 4p, 5p$ and $6p$ respectively, i.e., their outermost electronic configuration is np^5. Thus they are short of one electron each for attaining the inert gas electron configuration. These elements are highly electro-negative. During chemical reactions, they can easily take an electron from an electropositive element (e.g., an alkali element) and complete their outermost p subshells. They have similar physical and chemical properties. They have the ground states $^2P_{1/2}$.

Table 6·3 shows that the $3p$ orbit is completed at argon ($Z = 18$). So it might be expected that the nineteenth electron in the next element potassium ($Z = 19$) with which the fourth period begins will go to the $3d$

subshell. But actually that does not happen. It goes to the 4s subshell. This is due to the fact that this electron is more strongly bound in the 4s subshell than in the 3d subshell. In the next element calcium (Z = 20), the twentieth electron also goes to the 4s subshell. However in the succeeding elements starting from Sc (Z = 21) to Zn (Z = 30), the additional electrons fill up the 3d subshell which lies inside the 4s subshell. This 3d subshell

Table 6.3

Element	Z	K (n = 1)	L (n = 2)		M (n = 3)			N (n = 4)	
		$l = 0$ (1s)	$l = 0$ (2s)	1 (2p)	$l = 0$ (3s)	1 (3p)	2 (3d)	$l = 0$ (4s)	1 (4p)
H	1	1							
He	2	2							
Li	3	2	1						
Be	4	2	2						
B	5	2	2	1					
C	6	2	2	2					
N	7	2	2	3					
O	8	2	2	4					
F	9	2	2	5					
Ne	10	2	2	6					
Na	11	There are 10 electrons in 1s to 2p shells			1				
Mg	12				2				
Al	13				2	1			
Si	14				2	2			
P	15				2	3			
S	16				2	4			
Cl	17				2	5			
Ar	18				2	6			
K	19	There are 18 electrons in 1s to 3p shells					...	1	
Ca	20						...	2	
Sc	21						1	2	
Ti	22						2	2	
V	23						3	2	
Cr	24						5	1	
Mn	25						5	2	
Fe	26						6	2	
Co	27						7	2	
Ni	28						8	2	
Cu	29						10	1	
Zn	30						10	2	

(Contd.)

Alkali Spectra ; Space Quantization ; Periodic....

Table 6.3 (Contd.)

Element	Z	K (n = 1) l = 0 (1s)	L (n = 2) l = 0 (2s)	L (n = 2) 1 (2p)	M (n = 3) l = 0 (3s)	M (n = 3) 1 (3p)	M (n = 3) 2 (3d)	N (n = 4) l = 0 (4s)	N (n = 4) 1 (4p)
Ga	31						10	2	1
Ge	32						10	2	2
As	33						10	2	3
Se	34						10	2	4
Br	35						10	2	5
Kr	36						10	2	6

Element	Z	Inner electron configuration	N (n = 4) l = 2 (4d)	N (n = 4) 3 (4f)	O (n = 5) l = 0 (5s)	O (n = 5) 1 (5p)	O (n = 5) 2 (5d)	P (n = 6) l = 0 (6s)
Rb	37	There are 36 electrons in 1s to 4p shells	...	...	1			
Sr	38		...	...	2			
Y	39		1	...	2			
Zr	40		2	...	2			
Nb	41		4	...	1			
Mo	42		5	...	1			
Tc	43		6	...	1			
Ru	44		7	...	1			
Rh	45		8	...	1			
Pd	46		10	...	...			
Ag	47	There are 46 electrons in 1s to 4d shells		...	1			
Cd	48			...	2			
In	49			...	2	1		
Sn	50			...	2	2		
Sb	51			...	2	3		
Te	52			...	2	4		
I	53			...	2	5		
Xe	54			...	2	6		
Cs	55	There are 46 electrons in 1s to 4d shells		...	2	6		1
Ba	56			...	2	6		2
La	57				2	6	1	2
Ce	58			2	2	6	...	2
Pr	59			3	2	6	...	2
Nd	60			4	2	6	...	2
Pm	61			5	2	6	...	2
Sm	62			6	2	6	...	2
Eu	63			7	2	6	...	2
Gd	64			7	2	6	1	2

(Contd.)

Element	Z	Inner electron configuration	N (n = 4)		O (n = 5)			P (n = 6)
			l = 2 (4d)	3 (4f)	l = 0 (5s)	1 (5p)	2 (5d)	l = 0 (6s)
Tb	65	There are 46 electrons in 1s to 4d shells		9	2	6	...	2
Dy	66			10	2	6	...	2
Ho	67			11	2	6	...	2
Er	68			12	2	6	...	2
Tm	69			13	2	6	...	2
Yb	70			14	2	6	...	2
Lu	71			14	2	6	1	2

Element	Z	Inner electron configuration	O (n = 5)		P (n = 6)			Q (n = 7)
			l = 2 (5d)	3 (5f)	l = 0 (6s)	1 (6p)	2 (6d)	l = 0 (7s)
Hf	72	There are 68 electrons in 1s to 5p shells	2	...	2			
Ta	73		3	...	2			
W	74		4	...	2			
Re	75		5	...	2			
Os	76		6	...	2			
Ir	77		9	...	...			
Pt	78		9	...	1			
Au	79	There are 78 electrons in 1s to 5d shells		...	1			
Hg	80			...	2			
Tl	81			...	2	1		
Pb	82			...	2	2		
Bi	83			...	2	3		
Po	84			...	2	4		
At	85			...	2	5		
Rn	86			...	2	6		
Fr	87			...	2	6	...	1
Ra	88			...	2	6	...	2
Ac	89			...	2	6	1	2
Th	90			...	2	6	2	2
Pa	91			2	2	6	1	2
U	92			3	2	6	1	2
Np	93			4	2	6	1	2
Pu	94			6	2	6	...	2
Am	95			7	2	6	...	2
Cm	96			7	2	6	1	2
Bk	97			8	2	6	1	2
Cf	98			10	2	6	...	2
E	99			11	2	6	...	2
Fm	100			12	2	6	...	2
Mv	101			13	2	6	...	2

(Contd.)

No	102		14	2	6	...	2
Lw	103		14	2	6	1	2
Ku	104		14	2	6	2	2
Ha	105		14	2	6	3	2

is completed with 10 electrons at zinc. In all these elements, the electronic structure in the outermost $4s$ subshell is almost the same (1 or 2) so that there is no regular increase in the valencies of the elements form Sc to Ni ($Z = 28$). They constitute the *first transition group of elements*. Of these the last three, *viz.* Fe, Co and Ni with $Z = 26$, 27 and 28 respectively which are placed in the eighth column of the periodic table, show ferromagnetic behaviour at room temperature.

The next element copper ($Z = 29$) has a completed $3d$ subshell with ten electrons and a single valence electron in the $4s$ subshell. So it is monovalent and is placed in the first column of the same (*i.e.*, fourth) period. In the next element Zn with $Z = 30$, there are two $4s$ electrons outside the completed $3d$ subshell. Thus starting from copper, we have a core of closed K, L and M shells outside which the electrons are successively placed in the $4s$ and $4p$ subshells of the N shell upto Kr ($Z = 36$) which is an inert gas. With this the fourth period is completed.

It may be noted that though copper and zinc are placed in the first and second columns of the fourth period they do not belong to the alkaline and alkaline earth groups respectively, since they have completed $3d$ inner subshell each unlike the former which have completed p subshells. For this reason, they are placed slightly shifted with respect to the alkaline and alkaline earth group elements in the periodic table.

After krypton ($Z = 36$), the fifth period begins. In the first two elements Rb ($Z = 37$) and Sr ($Z = 38$), the additional electrons go to the $5s$ subshell in the O shell outside the completed $4p$ subshell of the N shell. After this, starting from Y ($Z = 39$), the inner $4d$ subshell begins to fill up which is completed at Pd ($Z = 46$) with 10 electrons. The situation is similar to the case of the elements from Sc to Ni in the previous shell. These elements from Y to Pd belong to the *second transition group*. The elements Ru, Rh and Pd with $Z = 44$, 45 and 46 respectively in this group are collectively placed in the last column of the periodic table below Fe, Co, Ni in the previous period.

After this, starting from Ag ($Z = 47$), the additional electrons are placed in the $5s$ and $5p$ subshells of the O shell upto Xe ($Z = 54$) which is an inert gas. The fifth period is completed at xenon.

In the next two elements Cs ($Z = 55$) which is an alkali element and Ba ($Z = 56$) which belongs to the alkaline earth group, the additional electrons go the $6s$ subshell of the P shell. In La ($Z = 57$), the additional electron goes to the 5d subshell. However from the next element Ce ($Z = 58$) upto Lu ($Z = 71$), the 4f subshell begins to fill up. These 14 elements belong to the *rare-earth* group. Since their external electron configuration is almost the same ($5s^2\ 5p^6\ 6s^2$) there is remarkable similarity in their chemical properties. In a few of these, there is also a single 5d electron (Gd, Lu with $Z = 64$ and 71). The rare-earths, as also the transition group elements, have high paramagnetic subsceptibilities.

The third transistion group of elements starts from Hf ($Z = 72$) when the inner 5d subshell is gradually filled up, The last three elements of this group Os, Ir and Pt ($Z = 76, 77$ and 78) are placed in the eighth column of the periodic table below Ru, Rh and Pd of the previous period.

From Au ($Z = 79$) upto Rn ($Z = 86$), the 6s and 6p subshells of the P shell are gradually filled up and the sixth period ends with radon ($Z = 86$) which is an inert gas.

In the next two elements Fr ($Z = 87$) and Ra ($Z = 88$), the extra electrons go to the 7s subshell of the Q shell. After this the *actinide series* begins with Ac ($Z = 89$) and ends at the transuranic element (see Vol. II) nobelium (No; $Z = 102$) when the inner 5f subshell is filled up gradually. These constitute the second rare-earth group.

6·8 Energy levels of complex atoms

With the help of the vector model of the atom, it is possible to determine the energy levels of complex atoms with more than one valence electrons in the outermost orbits.

Consider the case of an atom with two non-equivalent valence electrons, *i.e.*, electrons with different values of n or l (or both). We can apply the methods discussed in § 6·3 to get the possible states in this case. As an example, for two non-equivalent p electrons, the possible values of the total spin is $S = \frac{1}{2} + \frac{1}{2} = 1$ or $S = \frac{1}{2} - \frac{1}{2} = 0$. Since $l = 1$ for each electron, the possible values of L are 0, 1, 2. So the following energy levels or spectroscopic states are possible ; for $S = 1$: ^3S, ^3P, ^3D; for $S = 0$: ^1S, ^1P, ^1D.

If the values of both n and l are the same for the two electrons (*equivalent electrons*), the situation is more complicated. In this case Pauli's exclusion principle is to be considered which excludes certain states. All orientations of l and s are no longer allowed. Since n and l are the same, m_l or m_s must be different for the electrons to satisfy Pauli principle. For example in the case of two equivalent p electrons, if l is in the same direction for both, we get $L = 2$ or a D term. In this case m_l is the same for both ($+1$, 0 or -1). So m_s must be different for the two which means $m_s = +\frac{1}{2}$ for one and $m_s = -\frac{1}{2}$ for the other so that $m_s = 0$ and $S = 0$ and *i.e.*, only the ^1D term can occur. ^3D term is not possible due to Pauli

Alkali Spectra ; Space Quantization ; Periodic.... 131

principle. More detailed considerations show that only the terms ^1S, ^3P and ^1D are permissible in this case. (see § 13·18)

We give below a table of the energy levels for two equivalent (e) and two non-equivalent (n) electrons in some cases.

Table 6·4

Electrons		States
ss	e	^1S
	n	^1S ^3S
pp	e	^1S ^3P ^1D
	n	^1S ^3S ^1P ^3P ^1D ^3D
dd	e	^1S ^3P ^1D ^3F ^1G
	n	^1S ^3S ^1P ^3P ^1D ^3D ^1F ^3F ^1G ^3G

For three electrons, the l and s values of the third should be combined with those of the other two vectorially to get the possible states, subject to Pauli's exclusion principle. Similar procedure can be followed for larger number of electrons.

In the case of closed shells, there is the maximum possible number of electrons in these shells. In order to fulfil Pauli principle, all electron spins must be in antiparallel pairs giving $m_S = 0$ and $S = 0$. Further $m_L = 0$ and $L = 0$ for them so that they always form $^1S_0$ ground state as in the case of inert gases. For a complex atom, we can therefore leave out the closed shells altogether and consider only the valence electrons in the outermost shells.

The ground states of the electron configuration in some typical cases by the above considerations are given below.

Single s valence electron (H, Na, K *etc.*) : $^2S_{1/2}$
Single p valence electron (B, Al, Ga *etc.*) : $^2P_{1/2}$
Two s valence electrons (Be, Mg, Ca *etc.*) : $^1S_0$
Two p valence electrons (C, Si, Ge *etc.*) : $^3P_0$

See also Ch. XIII

6·9 Anomalous Zeeman effect

In § 6·2, it was mentioned that anomalous Zeeman effect is more commonly observed than normal Zeeman effect. Its origin can be explained by considering the spin of the electron in addition to its orbital motion. If **L** and **S** represent the orbital and spin angular momentum vectors (in units of $\hbar$), then their resultant **J** will precess about the direction of an applied magnetic field. There are $(2J + 1)$ possible orientations of J w.r.t. the field direction, its components in the field direction being given by $m_J \hbar$ where $m_J = J, J, -1, J-2, \ldots -J$. In a given magnetic field, an atomic energy level with a definite set of values for L and J splits up into $(2J + 1)$ closely lying component levels. The splitting between the adjacent levels depends on the value of J. This can be calculated in the following manner.

We know that the vectors **L** and **S** precess about the direction of **J**. Since the magnetic moments $\vec{\mu_L}$ and $\vec{\mu_S}$ due to the orbital and spin motions are aligned antiparalled to the vectors **L** and **S** respectively, these two vectors also precess about the direction of **J**. However the resultant magnetic moment $\vec{\mu} = \vec{\mu_L} + \vec{\mu_S}$ has a direction different from that of **J**. The reason for this can be understood with reference to Fig. 6·14.

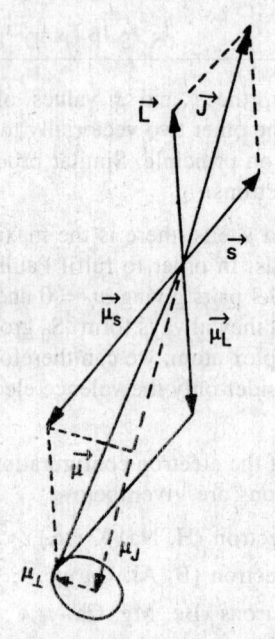

Fig. 6·14. Origin of Lande g-factor.

In Fig. 6·14, the vector $\vec{\mu_L}$ has been drawn opposite to the vector L with length shown equal to the length of L. Since the gyromagnetic ratio $\mu_S/p_S = \dfrac{e}{m_e}$ for the spin motion is twice that for the orbital motion $\left(\mu_L/p_L = \dfrac{e}{2m_e}\right)$, the vector $\vec{\mu_S}$ should be drawn with its length *twice* that of S, having the direction opposite to S. This is what has been done in Fig. 6·14. This means that the resultant magnetic moment vector $\vec{\mu}$ will

Alkali Spectra ; Space Quantization ; Periodic....

have a direction different from the resultant angular momentum $J = L + S$. Hence as $\vec{\mu_L}$ and $\vec{\mu_S}$ precess about the direction of J, their resultant $\vec{\mu}$ also precesses about J. So the component $\mu_\perp$ becomes zero. On the other hand the component μ_J of $\vec{\mu}$ along J is equal to the sum of the components of $\vec{\mu_L}$ and $\vec{\mu_S}$ in this direction and is given by

$$\mu_J = \mu_L \cos(L, J) + \mu_S \cos(S, J) \qquad (6.9-1)$$

Here $\cos(L, J)$ and $\cos(S, J)$ are the cosines of the angles between J and the vectors L and S respectively. Remembering that the squares of these vectors are $J(J+1)$, $L(L+1)$ and $S(S+1)$ respectively, it can be easily proved with the help of Fig. 6·15. that

$$\cos(L, J) = \frac{L(L+1) + J(J+1) - S(S+1)}{2\sqrt{L(L+1)} \times \sqrt{J(J+1)}}$$

$$\cos(S, J) = \frac{S(S+1) + J(J+1) - L(L+1)}{2\sqrt{S(S+1)} \times \sqrt{J(J+1)}}$$

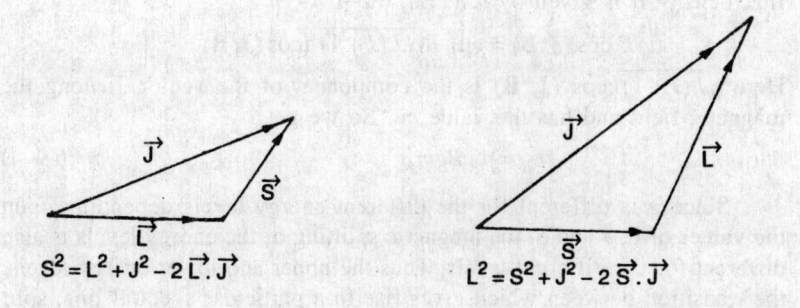

$S^2 = L^2 + J^2 - 2\vec{L}\cdot\vec{J}$

$L^2 = S^2 + J^2 - 2\vec{S}\cdot\vec{J}$

Fig. 6·15. Orientations of vectors L and S w.r.t. J.

From the discussions in § 6·2 and § 6·3 we can write the quantum mechanical values of μ_L and μ_S as

$$\mu_L = \frac{e}{2m_e} p_L = \frac{e\hbar}{2m_e} \cdot \sqrt{L(L+1)}$$

$$\mu_S = 2 \times \frac{e}{2m_e} p_S = \frac{e\hbar}{2m_e} \cdot 2\sqrt{S(S+1)}$$

Hence from Eq. (6·9–1) we get

$$\mu_J = \frac{e\hbar}{2m_e} \left\{ \sqrt{L(L+1)} \cdot \frac{L(L+1) + J(J+1) - S(S+1)}{2\sqrt{L(L+1)} \cdot \sqrt{J(J+1)}} \right.$$

$$\left. + 2\sqrt{S(S+1)} \cdot \frac{S(S+1) + J(J+1) - L(L+1)}{2\sqrt{S(S+1)} \cdot \sqrt{J(J+1)}} \right\}$$

$$= \frac{e\hbar}{2m_e} \left\{ \frac{L(L+1) + J(J+1) - S(S+1)}{2\sqrt{J(J+1)}} \right.$$

$$\left. + 2 \cdot \frac{S(S+1) + J(J+1) - L(L+1)}{2\sqrt{J(J+1)}} \right\}$$

$$= \frac{e\hbar}{2m_e} \sqrt{J(J+1)} \left\{ 1 + \frac{J(J+1) + S(S+1) - L(L+1)}{2J(J+1)} \right\}$$

This can be written as

$$\mu_J = \sqrt{J(J+1)} \, g \, \mu_B \qquad (6\cdot 9\text{--}2)$$

where μ_B is the Bohr magneton and g is the Landé g-factor or splitting factor given by

$$g = 1 + \frac{J(J+1) + S(S+1) - L(L+1)}{2J(J+1)} \qquad (6\cdot 9\text{--}3)$$

Thus the splitting of the atomic energy levels in a magnetic field of flux density B is given by (c.f. Eq. 6·2–10)

$$\varepsilon_B = \mu_J B \cos(\mathbf{J}, \mathbf{B}) = g\mu_B B \sqrt{J(J+1)} \, (\cos(\mathbf{J}, \mathbf{B}))$$

Here $\sqrt{J(J+1)} \cos(\mathbf{J}, \mathbf{B})$ is the component of the vector $\mathbf{J}$ along the magnetic field and has the value m_J. So we get

$$\varepsilon_B = \mu_B B g m_J \qquad (6\cdot 9\text{--}4)$$

Since g is different for the different energy levels depending upon the values of L, J and S, the magnetic splitting of the energy levels is also different for the different levels. Thus the upper and lower energy levels, the transition between which gives rise to a particular spectral line, split up by different amounts in a magnetic field. The transitions between these split levels are governed by the selection rule

$$\Delta m_j = 0, \pm 1 \qquad (6\cdot 9\text{--}5)$$

They give rise to the magnetic splitting of the spectral line. The number of components into which a line splits up is in general different from that expected (two or three) when the electron spin is not considered (normal Zeeman effect). Hence it is known as the anomalous Zeeman effect.

As an example, we may consider the magnetic splitting of the sodium D-lines. When a sodium vapour lamp is placed in a magnetic field and the D-lines emitted from the lamp are analysed by a high resolution spectrometer, it is found that the D_1 line splits into four components while the D_2 line splits into six components. These splittings can be easily explained with the help of the theory developed above.

Alkali Spectra ; Space Quantization ; Periodic....

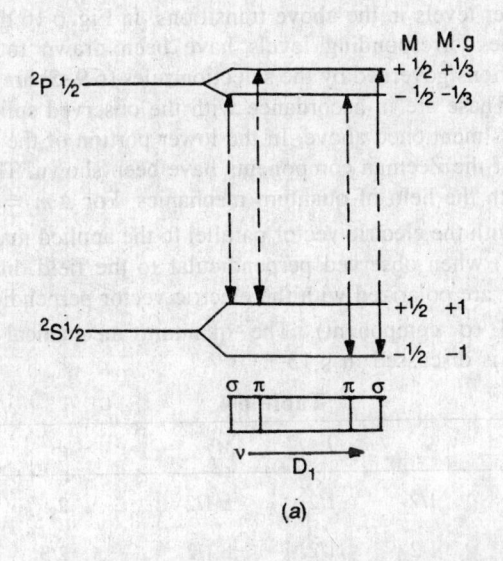

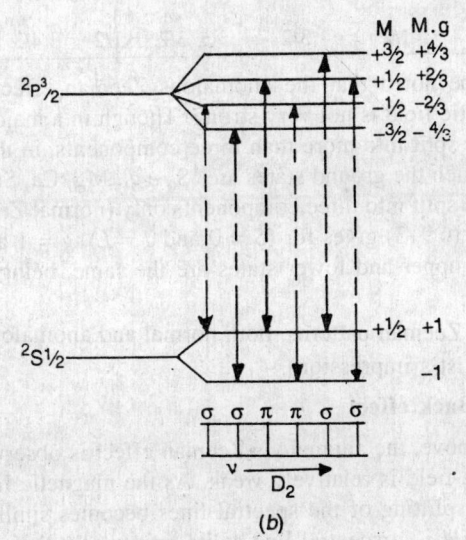

Fig. 6·16. Origin of anomalous Zeeman effect for sodium D_1 and D_2 lines.

We know that the D_1 and D_2 lines originate due to the transitions $3^2P_{1/2} \longrightarrow 3^2S_{1/2}$ and $3^2P_{3/2} \longrightarrow 3^2S_{1/2}$ respectively. The values of the Lande' g-factors for the above three levels are listed in Table 6·4 along with the corresponding quantum numbers.

It is evident from the table that the values of gm_J is different for the upper and lower levels in the above transitions. In Fig. 6·16 the magnetic splitting of the corresponding levels have been drawn to scale. The possible transitions governed by the selection rules (6·9–5) are also shown in the figure. These are in accordance with the observed splitting of the D_1 and D_2 lines mentioned above. In the lower portion of the figures, the polarizations of the Zeeman components have been shown. These can be determined with the help of quantum mechanics. For $\Delta m_J = 0$, the lines are polarized with the electric vector parallel to the applied magnetic field (π-component) when observed perpendicular to the field direction. For $\Delta m_J = \pm 1$, they are polarized with the electric vector perpendicular to the magnetic field (σ component). The quantum mechanical theory of Zeeman effect is discussed in § 13·5.

Table 6·4

Level	L	S	J	M_J	g	gM_J
$3^2S_{1/2}$	0	1/2	1/2	± 1/2	2	± 1
$3^2P_{1/2}$	1	1/2	1/2	± 1/2	2/3	± 1/3
$3^2P_{3/2}$	1	1/2	3/2	± 3/2, ± 1/2	4/3	±2, ±2/3

It should be noted that the anomalous Zeeman effect is observed when the magnetic field is not very strong. Though in a majority of cases, a spectral line is split into more than three components, in the case of the elements, for which the ground states are $^1S_0$ e.g., Mg, Ca, Sr, Cd etc., the spectral lines are split into three components only (normal Zeeman effect). In this case, Eq. (6·9–3) gives for ($S = 0$ and $J = L$), $g = 1$ and hence the splittings of the upper and lower states are the same, being independent of L and J.

Pictures of Zeeman patterns, both normal and anomalous, are shown in Fig. 6·17 (Artist's impression).

6·10 Paschen–Back effect

As stated above, the anomalous Zeeman effect is observed when the applied magnetic field is relatively weak. As the magnetic field becomes very strong, the splitting of the spectral lines becomes similar to normal Zeeman splitting, i.e., a spectral line splits into two or three components only. This is known as Paschen–Back effect.

In a very strong magnetic field, the coupling between the orbital and the spin angular momentum vectors **L** and **S** to form the resultant angular momentum vector **J** is broken. The two vectors **L** and **S** now independently precess about the field direction. There are $(2L + 1)$ possible components of **L** in the field direction given by $m_L = L, L-1, L-2, \ldots\ldots -L$. Similarly the components of **S** in the field

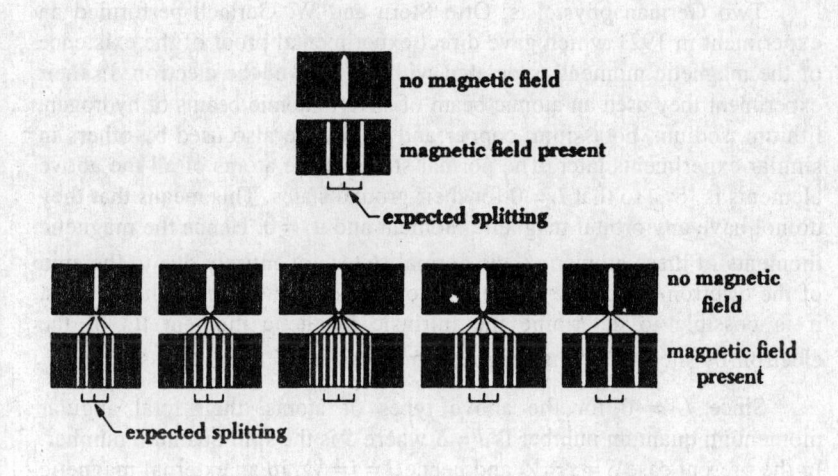

Fig. 6·17. Normal and anomalous Zeeman patterns. (Courtesy : Introduction to Atomic Spectra by W.E. White published by McGraw Hill Book Co. Inc. New York).

direction are $(2S + 1)$ in number, given by $m_S = S, S-1, \ldots\ldots\ldots -S$. The magnitude of the splitting of an energy level in a magnetic field of flux density B is approximately given by

$$\varepsilon_B = \mu_B B m_L + 2\mu_B B m_S = \mu_B B (m_L + 2m_S) \qquad (6\cdot10\text{--}1)$$

where μ_B is the Bohr magneton. The factor 2 in the second term is due to the factor 2 in the gyromagnetic ratio for the spin moment. So we get normal splitting as given in § 6·2 (see Eq. 6·2–11). The selection rules for the transition are $\Delta m_L = 0, \pm 1$ and $\Delta m_S = 0$.

There is a residual interaction between **L** and **S**, which has been neglected here. When this is considered, the splitting is more complicated, given by (for hydrogen)

$$\Delta E = \lambda_{nl} m_l m_s \quad (l \neq 0) \qquad (6\cdot10\text{--}2)$$

where λ_{nl} depends on the energy E_n of the unsplit level. For $l = 0$, $\Delta E = 0$. (see § 13·6).

6·11 Stern and Gerlach's experiment

Two German physicists, Otto Stern and W. Gerlach performed an experiment in 1921 which gave direct experimental proof of the existence of the magnetic moment associated with the spin of the electron. In their experiment they used an atomic beam of silver. Atomic beams of hydrogen, lithium, sodium, potassium, copper and gold were also used by others in similar experiments later. The normal states of the atoms of all the above elements is $^2S_{1/2}$ so that $L = 0$ for their ground states. This means that they do not have any orbital magnetic moment and $\mu_L = 0$. Hence the magnetic moments of these atoms in their normal states are entirely due to the spin of the electron. Since they have only one valence electron in their atoms, it is possible to determine the intrinsic magnetic moment μ_B of the electron by measuring the magnetic moment μ_S of these atoms.

Since $L = 0$ for the above types of atoms, their total angular momentum quantum number is $J = S$ where S is the spin quantum number. In the present case $S = s = ½$ and hence $J = j = ½$. In an external magnetic field the vector j can have two possible orientations so that the magnetic quantum number can have the two values $m_j = + ½$ or $m_j = - ½$.

When these atoms behaving as tiny magnets of atomic size are subjected to a homogeneous magnetic field, a torque acts upon them just like the torque acting on a small magnetic needle which arises because of the equal and opposite forces acting on the two poles of the magnet. Due to this torque, these atomic magnets precess about the field direction. The reason for this is the same as was discussed in § 6·2 following Eq. (6·2–8).

If, however, the atomic magnet is placed in an *inhomogeneous* magnetic field, then a net translatory force acts upon it in the direction of the field. The forces acting on the two poles of the magnet are now different as can be understood from Fig. 6·18. The energy of the atom in the external magnetic field is given by [see Eq. (6·9–4)].

$$W = \varepsilon_B = g\mu_B \, m_j \, B$$

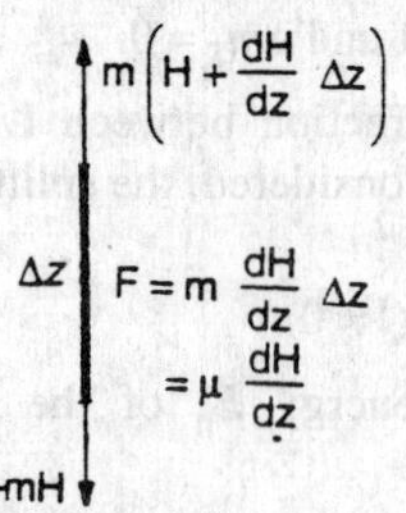

$$m\left(H + \frac{dH}{dz} \Delta z\right)$$

$$\Delta z \quad F = m \frac{dH}{dz} \Delta z$$

$$= \mu \frac{dH}{dz}$$

$-mH$

Fig. 6·18. Force on an atomic magnet in an inhomogeneous magnetic field.

Hence the translatory force acting on the atom is

$$F = -\frac{dW}{dz} = -g\,\mu_B\,m_j\,\frac{dB}{dz} \qquad (6\cdot11\text{--}1)$$

Here the inhomogeneity of the magnetic field has been assumed to be in the z-direction so that the translatory force is also along z. The above equation shows that the force increases as the inhomogeneity dB/dz increases.

The experimental arrangement of Stern and Gerlach is shown in Fig. 6·19. O is an oven in which metallic silver is heated to produce a neutral *atomic beam* of silver which comes out of the oven through a

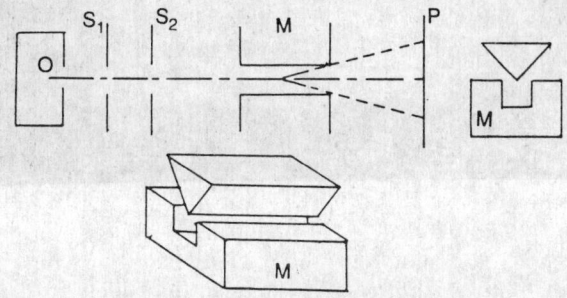

Fig. 6·19. Stern and Gerlach's experiment.

small orifice. The beam is finely collimated by a slit system and the collimated beam is allowed to pass through a highly inhomogeneous magnetic field at M and falls upon a photographic plate P.

In order to produce the inhomogeneous field, a magnet with specially shaped pole pieces was used. As shown in the figure, one of the pole pieces was in the shape of *knife-edge* while the other had a plane face with a channel cut on its surface parallel to the knife-edge and facing the latter. Since the magnetic flux lines are highly converging towards the knife-edge, the inhomogeneity near the latter is so high, that the field changes appreciably within atomic dimensions in its vicinity.

When there is no magnetic field, the atomic beam travels undeflected towards the photographic plate P. When the inhomogeneous field acts, each silver atom in the beam is acted upon by a translatory force given by Eq. (6·11–1) at right angles to the path of the beam due to which these atoms are now incident at a point on P slightly off the axis. Since there are two possible values of m_j ($=\pm\,\frac{1}{2}$), the beam splits up into two and produces two short lines on the photographic plate as shown in Fig. 6·20. From the distance between the two lines, it is possible to determine the magnetic moment of the silver atoms and hence the electronic magnetic moment.

Stern and Gerlach's experiment also provides a direct proof of the idea of space quantization. Since $J=j=\frac{1}{2}$ in the present case, there are

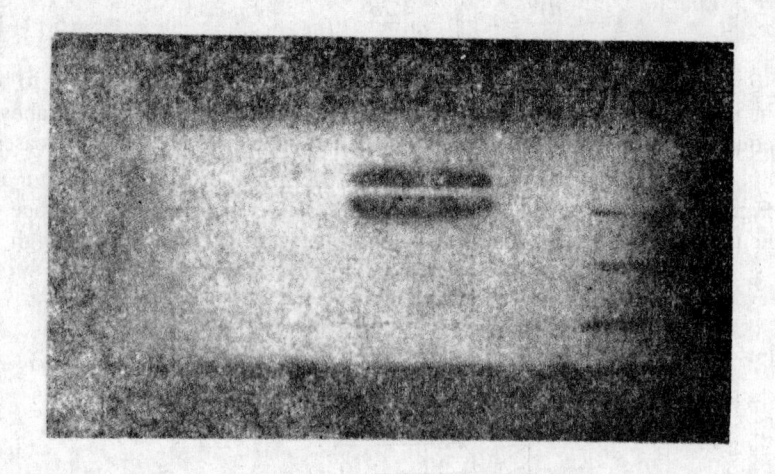

Fig. 6·20. Result of Stern and Gerlach's experiment.

$(2J + 1)$ or two possible orientations of J in the magnetic field. Hence the beam splits into two in the magnetic field.

6·12 Stark Effect

An effect similar to Zeeman effect in which a spectral line is split into several components by the action of an electric field was discovered by J. Stark in 1913 in Germany. This is known as Stark effect.

Stark observed that the canal rays produced behind a perforated cathode in a discharge tube (see § 2·6) emitted spectral lines which were split up into a number of components in an intense electric field. As in Zeeman effect the components show various degrees of polarization.

The splitting is, in the first instance, proportional to the electric field. However in an intense field the pattern is quite different, as in Paschen–Back effect.

The complete theory of Stark effect, based on quantum, mechanics gives good agreement with the observed results (see *Physics of Atoms and Molecules* by Bransden and Joachain).

Alkali Spectra ; Space Quantization ; Periodic.... 141

Problems

1. Determine the angles between the applied magnetic field and the orbital angular momentum vectors for $k = 1, 2, 3$ where k is the azimuthal quantum number of the old quantum theory.

2. Calculate the Larmor precessional frequency for a magnetic induction field of 0·5 T. Hence calculate the splitting in wave numbers of a spectral line due to normal Zeeman effect for the same field.
   If the wavelength of a spectral line is 5893 Å, what is the percentage splitting of the wavelength of the line in the above case ?
   $$(0.7 \times 10^{10}\, s^{-1}\,;\, 23.33\, m^{-1}\,;\, 1.37 \times 10^{-3}\, \%)$$

3. Calculate the angles between the applied magnetic field and the p_l vectors according to quantum mechanics for $l = 1, 2, 3$. Calculate the ratio of the largest component of p_l along the field direction and the magnitude of p_l in the above cases.

4. For a many electron atom with $L = 2$, $S = 1$ and $J = 2$, calculate the angle between the **L** and **S** vectors both according to old quantum theory and quantum mechanics.

5. Apply the rules of space quantization to find the possible values of J obtained by the vector addition of **L** and **S** for the following combinations, using old quantum theoretical results : $L = 1, S = 1$; $L = 2, S = 1, L = 1, S = 3/2$; $L = 2, S = 3/2$. Also draw the quantum mechanical vector diagrams in the above cases.

6. Calculate the Landé g-factor for $^3S_1$ and $^3P_1$ levels. Show the anomalous Zeeman splitting of the spectral line due to transition between these levels. (2 ; 3/2)

7. Calculate the energy splitting of the two levels in Prob. 6 if an induction field of 1 T is applied. $(8.7 \times 10^{-5}\, eV\,;\, 11.6 \times 10^{-5}\, eV)$

8. In a Stern-Gerlach type experiment, a collimated beam of lithium atoms (mass M) in the ground state $^2S_{1/2}$ moves along the x-axis with a velocity υ through an inhomogeneous magnetic field of inhomogeneity $\partial B/\partial z$ along the z-axis. If the length of the path in the field is a and the distance from the far end of the field to a photographic plate placed perpendicular to x is b, then show that the deflection of the atoms with a given m_j on the plate is
   $$\delta = \frac{g\, \mu_B m_J}{2M\upsilon^2} \frac{\partial B}{\partial z} a(a + 2b)$$
   Calculate the distance between the two components of the split beam on a screen if $a = 10$ cm, $b = 20$ cm, $\partial B/\partial z = 20$ T/m and the kinetic energy of the atoms is 0·04 eV. (0·72 mm)

9. Calculate the magnetic moment of the muon which has a mass 207 m_e.
   $(4.48 \times 10^{-26}\, J/T)$

10. Calculate the magnetic field for which the splitting between the successive Zeeman components of the terms $3^2P_{1/2}$ and $3^2P_{3/2}$ of sodium is equal to 1/20 of the fine structure splitting of these terms.
    (2·775 T ; 1·388 T)

7

LASER

7·1 Einstein's theory of atomic transitions

Transitions between the atomic energy states is a statistical process. It is not possible to predict which particular atom will make a transition from one state to another at a particular instant. However in an assembly of a very large number of atoms, it is possible to calculate the rate of radiative transitions between two states, based on the laws of probability. Albert Einstein was the first to calculate the probability of such transition, assuming the atomic system to be in equilibrium with electromagnetic radiation (1917).

Consider an assembly of atoms at an absolute temperature T in which the atoms may be in different energy states. If n_0 be number of atoms per unit volume in the ground state ($E = 0$), then the number of atoms n per unit volume in an excited state of energy E is given by the Boltzmann distribution law:

$$n = n_0 \exp(-E/kT) \qquad (7\cdot1\text{-}1)$$

where $k = 1\cdot38 \times 10^{-23}$ joule/kelvin is the Boltzmann constant. The maximum number of atoms is in the ground state, the number decreasing with increasing excitation. It n_1 and n_2 be the numbers of atoms per unit volume in the states of energies E_1 and E_2 respectively, then from Eq. (7·1-1) we get

$$\frac{n_2}{n_1} = \exp\left\{-\frac{(E_2 - E_1)}{kT}\right\} \qquad (7\cdot1\text{-}2)$$

If $E_2 > E_1$, then $n_2 < n_1$.

Now the atoms in the higher energy state E_2 make *spontaneous transitions* to the lower energy state E_1 at a certain rate, determined by n_2 and the probability of spontaneous transition A_{21} from E_2 to E_1. So the number of such transitions per unit volume per second is $A_{21} n_2$. These transitions lead to the emission of e.m. radiation of energy $h\nu = E_2 - E_1$. It should be noted that $1/A_{21}$ is a measure of the life-time of the upper state against spontaneous decay to the lower state.

Now suppose that the atoms are in equilibrium with e.m. radiation of frequency $v = (E_2 - E_1)/h$ having energy density u_v. Due to absorption of energy from this radiation, some of the atoms in the lower energy state E_1 make upward transitions to the higher energy state E_2. The rate of such transitions is determined by the number n_1 and the energy density u_v of the radiation and is given by $B_{12} n_1 u_v$ per unit volume per second. Here B_{12} is the constant of proportionality determining the probability of such *induced transitions*.

In addition to the above two types of transitions, there may also be downward induced transitions of some atoms from the energy state E_2 to the state E_1 due to the action of the e.m. field of the radiation, resulting in *stimulated emission* of radiation of frequency $v = (E_2 - E_1)/h$. The rate of such induced or stimulated transitions will be $B_{21} n_2 u_v$, where B_{21} is the probability of such transitions.

The coefficients A_{12}, B_{12} and B_{21} are known as Einstein's A and B coefficients. Under equilibrium the numbers of upward and downward transitions per unit volume per second are equal. So we can write

$$A_{21} n_2 + B_{21} n_2 u_v = B_{12} n_1 u_v$$

From this we get

$$u_v = \frac{A_{21} n_2}{B_{12} n_1 - B_{21} n_2} = \frac{A_{21}}{B_{21}} \cdot \frac{1}{\left(\dfrac{B_{12}}{B_{21}}\right) \dfrac{n_1}{n_2} - 1}$$

$$= \frac{A_{21}}{B_{21}} \cdot \frac{1}{\left(\dfrac{B_{12}}{B_{21}}\right) \exp(hv/kT) - 1} \qquad (7 \cdot 1\text{-}3)$$

Eq. (7·1-3) must agree with Planck's radiation formula (Eq. 3·3-4). Comparing the two we get

$$B_{12} = B_{21} \qquad (7 \cdot 1\text{-}4)$$

and

$$\frac{A_{21}}{B_{21}} = \frac{8\pi h v^3}{c^3} \qquad (7 \cdot 1\text{-}5)$$

so that we get

$$u_v = \frac{8\pi h v^3}{c^3} \cdot \frac{1}{\exp(hv/kT) - 1}$$

which is Planck's radiation formula.

Einstein's A and B coefficients cannot be determined from classical e.m. theory. It is possible to calculate the B coefficient from quantum mechanics by using Dirac's theory of time-dependent perturbation. (see § 14.1)

7·2 Laser principle

Laser is one of the most important discoveries in physics in recent years. The word *laser* is an acronym coined from the words *L*ight *A*mplification by *S*timulated *E*mission of *R*adiation which is the full descriptive name of the phenomenon.

Two Russian physicists N. Basov and A.M. Prokhorov and the American physicist C.H. Townes independently discovered a similar phenomenon in the field of microwaves known as Maser (*M*icrowave *A*mplification by *S*timulated *E*mission of *R*adiation) in 1954. Subsequently Townes and A.L. Shawlow suggested the possibility of similar stimulated emission in the case of visible radiation. Finally optical maser or laser was discovered by T.H. Maiman in 1960.

The principle of laser is based on the phenomenon of stimulated emission of radiation, the theory of which had been worked out by Einstein in 1917 and was discussed in the previous section. We know that light is emitted from a source when the atoms in the source make transitions from an excited to a lower energy state spontaneously. Normally the atoms exist in the excited state for about 10^{-8} s. If the energy of the excited state is E_2 and that of the lower energy state is E_1, then the energy of the emitted photon in this spontaneous transition is

$$h\nu_{12} = E_2 - E_1$$

On the other hand, if a photon of energy $h\nu_{12}$ falls upon an atom in the state of energy E_1, then it will make a transition to the higher energy state E_2 by the absorption of the photon.

Due to spontaneous transition, the light photons are emitted in all possible directions. Besides, there is no definite phase relationship between the different photons so that the emitted light is incoherent in nature.

Besides the above two types, a third type of transition from an upper energy state E_2 to a lower energy state E_1 may be induced by an incident photon of energy $h\nu_{12} = E_2 - E_1$ giving rise to the *stimulated emission of radiation*. All the above three types of transition are shown diagramatically in Fig. 7·1.

The photon emitted during stimulated emission has the same energy as the incident photon. It is emitted in the same direction and has the same phase as the latter. Thus we get two coherent photons in this case. If these two photons are now incident on two other atoms in the state E_2, then it will result in the induced emission of two more photons so that there will be four coherent photons of the same energy. These four photons may then induce transitions in four other atoms in the energy state E_2, thereby giving rise to the stimulated emission of four fresh coherent photons of the same energy so that the number of coherent photons of energy $h\nu_{12}$ is

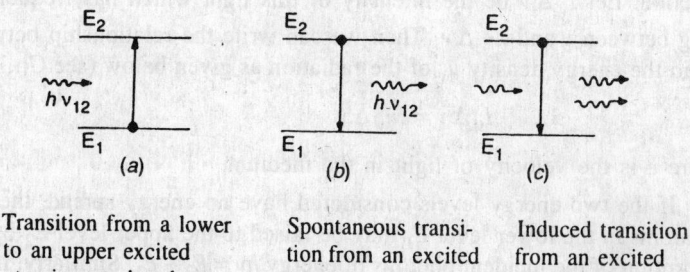

Laser

Fig. 7·1

(a) Transition from a lower to an upper excited atomic state by absorption of a photon.

(b) Spontaneous transition from an excited to a lower state.

(c) Induced transition from an excited to a lower state.

now increased to eight. If the process can be made to go on in a chain, we may ultimately be able to increase the intensity of the coherent radiation enormously. In Fig. 7·2, such amplification of the number of coherent photons due to stimulated emission is shown.

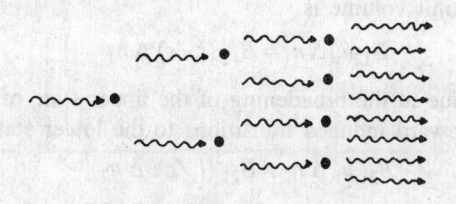

Fig. 7·2. Amplification due to stimulated emission of radiation.

The necessary condition for this type of amplification of the light intensity by the stimulated emission of radiation is that the number of atoms in the upper energy state E_2 must be sufficiently increased.

Normally the number of atoms in the upper energy state is much lower than in the lower energy state ($n_2 << n_1$) due to the Boltzmann factor (see Eq. 7·1-2 ; see also Prob. 5, Ch XVI). As a result the rate of downward induced transitions $B_{21} n_2 u_v$ is much less than that of the upward transitions $B_{12} n_1 u_v$ between the same two levels due to absorption of photons. To increase the rate of stimulated emission from E_2 to E_1, it is necessary to make $n_2 > n_1$. This is known as *population-inversion*. We shall see later how this can achieved.

Suppose by some means, such population inversion between the states E_2 and E_1 has been achieved in a medium so that $n_2 > n_1$. Let a collimated beam of light be propagated through the medium in a particular direction. Let $I_\nu \Delta \nu$ be the intensity of this light which has frequencies lying between ν and $\nu + \Delta \nu$. Then we can write the relationship between I_ν and the energy density u_ν of the radiation as given below (see Ch. III):

$$I_\nu \Delta \nu = c u_\nu \Delta \nu$$

where c is the velocity of light in the medium.

If the two energy levels considered have no energy spread, then all the atoms in the lower level E_1 may be raised to the upper level E_2 by the absorption of the incident photons of energy $h\nu = E_2 - E_1$. Similarly, in the reverse process, all the atoms in the level E_2 may undergo induced transition to the lower level E_1 by the action of the incident light. Actually however, due to Doppler effect and other causes, the energy levels do not have absolutely well-defined energies, but have finite energy spread. So all the atoms in the level E_1 or E_2 will not undergo transitions to the other level by the action of the light of definite frequency ν. Due to such broadening of the level E_1, out of n_1 atoms per unit volume in this level, only the number Δn_1 will make transitions to the upper state by the absorption of light of frequency ν. The number of such upward transitions per second per unit volume is

$$B_{12} u_\nu \Delta n_1 = B_{12} (I_\nu / c) \Delta n_1$$

Similarly due to the broadening of the upper state of energy E_2 the number of downward induced transitions to the lower state is

$$B_{21} u_\nu \Delta n_2 = B_{21} (I_\nu / c) \Delta n_2$$

Here B_{12} and B_{21} denote the B-coefficient of Einstein (see § 7·1). For each upward transition, a quantity of energy $h\nu$ is absorbed from the incident light beam. Similarly for each downward induced transition, an equal amount of energy is added to the beam. So the rate of change of energy density of the beam is

$$\frac{d}{dt}(u_\nu \Delta \nu) = (B_{21} u_\nu \Delta n_2) h\nu - (B_{12} u_\nu \Delta n_1) h\nu$$

$$= h\nu (B_{21} \Delta n_2 - B_{12} \Delta n_1) u_\nu \qquad (7\cdot2\text{-}1)$$

Since the light wave is propagated through a distance $dx = cdt$, we get ($\because B_{12} = B_{21}$)

$$c \frac{d}{dx}(u_\nu \Delta \nu) = \frac{d}{dx}(I_\nu \Delta \nu) = \frac{h\nu}{c}(\Delta n_2 - \Delta n_1) B_{12} I_\nu$$

so that

$$\frac{dI_\nu}{dx} = \frac{h\nu}{c}\left(\frac{\Delta n_2}{\Delta \nu} - \frac{\Delta n_1}{\Delta \nu}\right) B_{12} I_\nu \qquad (7\cdot2\text{-}2)$$

Hence

$$\frac{dI_\nu}{I_\nu} = \frac{h\nu}{c}\left(\frac{\Delta n_2}{\Delta \nu} - \frac{\Delta n_1}{\Delta \nu}\right) B_{12} dx = \beta dx \qquad (7\cdot2\text{-}3)$$

where
$$\beta = \frac{h\nu}{c}\left(\frac{\Delta n_2}{\Delta \nu} - \frac{\Delta n_1}{\Delta \nu}\right) B_{12} \qquad (7\cdot2\text{-}4)$$

$\beta = \beta_\nu$ is known as the *gain-constant*. It depends on the frequency ν of the light. On integration, we get

$$I_\nu = I_{\nu 0} \exp(\beta_\nu x) \qquad (7\cdot2\text{-}5)$$

Since the levels E_1 and E_2 have finite widths, the transitions between different levels within them give rise to lines of different frequencies (see Fig. 7·3). So the spectral line due to transition between two levels has a finite spread of energy or frequency. Let this frequency spread be $\Delta \nu$ about the mean frequency ν_0. It is possible to estimate the value of β at this mean frequency ν_0. If we assume $\Delta n = n$ for each energy, we get for $\nu = \nu_0$

$$\beta = \beta_0 = \frac{h\nu_0}{c \Delta \nu}(n_2 - n_1) B_{12} \qquad (7\cdot2\text{-}6)$$

Using Eq. (7·1-5) we get

$$\beta = \beta_0 = \frac{\lambda_0^2}{8\pi \Delta \nu}(n_2 - n_1) A_{12} \qquad (7\cdot2\text{-}7)$$

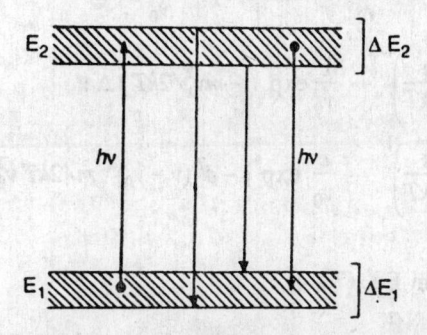

Fig. 7·3. Transitions between levels with finite widths.

If $n_2 > n_1$ as a result of population inversions, $\beta_0 > 0$ so that according to Eq. (7·2-5), I_ν increases as x increases. This means that as the beam of light progresses through the medium, its intensity I_ν increases exponentially; *i.e.*, there is amplification of the intensity of the beam.

On the other hand if $n_2 < n_1$ as is to be normally expected from the Boltzmann formula, $\beta_0 < 0$ and I_ν decreases exponentially as x increases, so that there is no amplification in this case.

The above discussion shows that in order to produce light amplification by stimulated emission of radiation, population inversion is an essential criterion.

In order to determine the nature of variation of β with v, it is necessary to know the nature of the spectral line-broadening. If this broadening is Gaussian, then the gain constant β is found to be a Gaussian function of the frequency v. Its value is the highest at the centre of the spectral line.

If the line broadening is due to Doppler effect produced by the random thermal motion of the atoms, we can estimate the gain constant by considering the fraction of atoms whose x-component of velocity lies between v_x and $v_x + \Delta v_x$. This is given by (see Ch. XX, Prob. 3)

$$\frac{\Delta n}{n} = \left(\frac{m}{2\pi kT}\right)^{1/2} \exp(-mv/2kT)\, \Delta v_x \qquad (7\cdot2\text{-}8)$$

The change in frequency due to Doppler effect is given by (see § 15·16)

$$\frac{v - v_0}{v_0} = \frac{v_x}{c} \qquad (7\cdot2\text{-}9)$$

This gives

$$v_x = \frac{c(v - v_0)}{v_0} \quad \text{and} \quad \Delta v_x = \frac{c\,\Delta v}{v_0}$$

$$\therefore\ \Delta n = n\left(\frac{m}{2\pi kT}\right)^{1/2} \cdot \frac{c}{v_0} \exp(-mv_x^2/2kT)\, \Delta v$$

$$= n\left(\frac{m}{2\pi kT}\right)^{1/2} \cdot \frac{c}{v_0} \exp[-c^2(v-v_0)^2\, m/2kT\, v_0^2]\, \Delta v \qquad (7\cdot2\text{-}10)$$

We then get from Eq. (7·2-4),

$$\beta = h\left(\frac{m}{2\pi kT}\right)^{1/2} (n_2 - n_1) \exp\{-mc^2(v-v_0)^2/2kT\, v_0^2\}\, B_{12} \qquad (7\cdot2\text{-}11)$$

Thus for a Gaussian line shape, the gain constant has Gaussian variation as stated above.

At the line centre $v = v_0$, we have, using Eq. (7·1-5)

$$\beta_0 = h\left(\frac{m}{2\pi kT}\right)^{1/2} (n_2 - n_1)\, B_{12}$$

$$= \frac{A_{12}\lambda_0^3}{8\pi}\left(\frac{m}{2\pi kT}\right)^{1/2} (n_2 - n_1) \qquad (7\cdot2\text{-}12)$$

This gives the maximum value of the gain constant.

7·3 Experimental arrangement

The basic experimental arrangement for producing a laser beam is shown in Fig. 7·4. The lasing medium is enclosed within a closed vessel

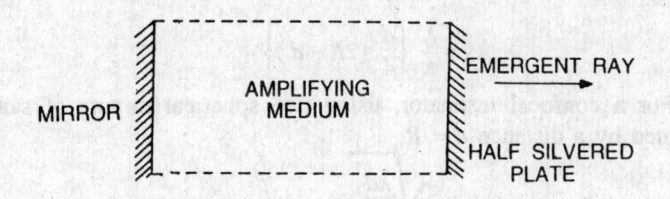

Fig. 7·4 Basic experimental arrangement for producing a laser beam.

with two reflectors at its two ends. If by some means, sufficient population inversion is achieved in the atoms of the medium, then there is amplification of the light intensity by the laser process. Due to repeated reflections from the two reflectors, the beam of light gets sufficient chance to interact with a large number of the excited atoms in the metastable states to cause stimulated emission. There is a small opening at the centre of one of the reflectors through which a fraction of the laser beam comes out. In this way a coherent beam of light, highly collimated and of very high intensity is obtained.

Actually the lasing medium in the closed vessel acts as a cavity resonator within which stationary waves of definite wavelengths are excited so that there is an integral number of half waves between the two reflectors which act as two nodes. Since the wavelength of the light wave is small compared to the distance d between the reflectors, the number of half waves between the reflectors is very large. The difference in the frequencies between two consequtive modes of vibration is given by $v_{n+1} - v_n = c/2d$ where c is the velocity of light. This shows that there is a spread in the frequency of the laser beam which can be reduced by controlling the quality factor (Q) of the resonator. Usually, the spread in frequency is of the order of a few thousand hertz. Since the frequency of the laser radiation is in the range of $10^{14} - 10^{15}$ hertz, this spread in frequency in negligibly small which accounts for the very high degree of monochromaticity of this radiation.

The energy of the laser beam in the cavity resonator between the two mirrors falls off laterally from the axis. If the lateral distribution is assumed to be a Gaussian function $\exp(-\rho^2/w^2)$, ρ being the lateral distance from the axis, then the parameter w may be regarded as the *spot size* which is a function of the longitudinal distance z measured from the *midpoint* between the two spherical mirrors and the wavelength λ and is given by

$$w^2 = w_0^2 + \frac{\lambda^2 z^2}{\pi^2 w_0^2}$$

where w_0 is the spot size at the centre. Its value is determined by the radii of curvature of the mirrors and their separation d. For a symmetrical cavity in which both the mirrors have the same radius of curvature R, w_0 is given by

$$w_0^2 = \frac{\lambda}{\pi} \left\{ \frac{d}{4} (2R - d) \right\}^{1/2}$$

For a confocal resonator, using two spherical minors of same R separated by a distance $d = R$

$$w_0 = \sqrt{\frac{\lambda d}{2\pi}}$$

This gives $\quad w^2 = \frac{\lambda d}{2\pi} + \frac{2\lambda z^2}{\pi d}$

So the spot size at the mirrors $(z = \pm d/2)$ is

$$w_{d/2} = \sqrt{\frac{\lambda d}{\pi}}$$

This shows that the beam coming out of the resonator has very little divergence.

7.4 Population inversion

There are different methods of achieving population inversion in the atomic states which is an essential requirement for producing a laser beam.

Normally most of the atoms in a medium are in the ground state of energy E_0. There are broadly four different methods of raising them to an excited state of energy E_1 which are discussed below.

(i) *Excitation with the help of photons* : If the atoms are exposed to electromagnetic radiation of frequency $v = (E_1 - E_0)/h$, then there is selective absorption of this radiation due to which the atoms are raised to the excited state E_1. This method, known as *optical pumping*, is used in the ruby laser.

(ii) *Excitation by electrons* : This method is used in some gas lasers. Electrons are released from the atoms due to high voltage electric discharge through a gas. These electrons are accelerated to high velocities due to the strong electric field inside the discharge tube. When they collide with the neutral gas atoms, a fraction of these atoms are raised to the excited state : $(e + X \longrightarrow X^* + e)$.

(iii) *Inelastic collision between atoms* : If a gas contains two different kinds of atoms X and Y, then during an electric discharge through the gas, some of these atoms are raised to excited states X^* and Y^*. Suppose one of these excited atoms, say X^*, is in a metastable state which has a mean life long compared to the normal mean life of an excited atom which is $\sim 10^{-8}$ s (see Ch. IV). If the excitation energies of X^* and Y^* are nearly equal, then some of the atoms of Y in their ground state may be raised to the excited state due to inelastic collision with X^* :

$$X^* + Y \longrightarrow X + Y^*$$

As a result the number of the excited atoms Y^* goes on increasing. This type of collision is known as collision of second kind.

This method is employed in helium-neon laser. Helium has an excited state which is metastable. As the excited He^* atoms in the metastable state collide with the Ne atoms in the ground state, the latter are raised to the excited state. The number of such excited Ne^* atoms goes on increasing continually because of the metastability of the generating He^* state. Methods (ii) and (iii) are known as *electrical pumping*.

(iv) *Excitation by chemical method* : Sometimes an atom or a molecule which is a product of a chemical reaction is produced in an excited state. As an example hydrogen and fluorine combine to produce HF molecule in the excited state. Thus the number of such excited atoms or molecules may be considerably greater than those in the ground state giving rise to the population inversion.

Optical pumping is more suitable for solid state lasers or liquid dye lasers. Light from a powerful source is absorbed by the active material and the atoms are thereby pumped into the *pump-level* (or rather pump-band, since the line broadening mechanisms in these materials produce very large broadening). As a result these bands can absorb a fairly large fraction of the light emitted from the exciting lamp.

The exciting lamp is usually in the form of a helix surrounding the laser rod or a cylinder placed alongside the cylindrical laser rod enclosed in an elliptical or circular reflecting cylinders. The energy stored in a capacitor bank is discharged into the exciting flash lamp.

The pumping efficiency can be split up into three factors :

(i) transfer efficiency (η_t) ;

(ii) lamp radiative efficiency (η_r) ;

(iii) pump quantum efficiency (η_q).

The transfer efficiency η_t is the ratio of the energy pumped into the laser rod to the power emitted by the exciting lamp. η_r is the efficiency of conversion of the electrical power into light entering the laser rod in the wavelength range of the effective pumping band λ_1 to λ_2. η_q is the ratio of the number of atoms decaying from the upper laser level to the number of atoms actually raised to the upper laser level by the incident exciting radiation.

η_t for a helical exciting lamp is usually much smaller than that of a linear lamp in an elliptical enclosure. However the former provides more uniform pumping.

A quantity of considerable importance is the distribution of light within the laser rod. For a helical exciting lamp (or for $R_0 < R_l$ for cylindrical lamps) the pump energy density is non-uniform outside a central core of radius $r < R_l/n$ where n is the refractive index of the laser rod (l) which is not desirable. This is usually overcome by surrounding the rod with a suitable cladding or by grinding the outer lateral surface of the rod. Similar devices are also helpful when $R_{ex} > R_l$

The pumping rate (w_p) is the rate at which the upper laser level is made populated by the exciting radiation and depends on the properties of the laser e.g. quantum efficiency $\eta_q(\lambda)$, absorption cross section $\sigma(\lambda)$, concentration of the active ions, rod radius R_l and the ratio r/R_l at a given point.

Gas lasers usually employ electrical pumping which is also used in the case of semiconductor lasers.

In this method, a current in passed through the gas producing ions and free electrons. They acquire additional kinetic energy due to the electric field and excite some neutral atoms by collision. This type of impact excitation is caused more readily by the electrons which have much higher average energy than the ions. After a while an equilibrium condition is established.

Both elastic and inelastic collisions are involved in electron impact excitation. The *pumping rate* depends on the total collision cross section (σ_e), ground state population of the atoms (N_g) and the electron flux (ϕ_e) which depends on the electron velocity ($\phi_e = n_e \upsilon_e$). σ_e depends on the electron beam energy and has a maximum at an electron energy of few eV higher than the threshold value to cause the transition to the laser level from the ground state.

Actually since the electrons have an energy distribution within the discharge, one has to integrate the expression for the pumping rate for monoenergetic electrons assuming a suitable energy distribution for the electrons. The pumping rate is found to depend on the ratio X/p where X is the electric field and p is the gas pressure. A change in the pumping rate can be achieved by changing the current density in the gas discharge.

Electrical pumping of gas lasers is a very complicated process and a closed expression for the pumping rate cannot be obtained unlike in the case of optical pumping.

7.5 Ruby Laser

The first laser invented by Maiman was a ruby laser in which a single crystal of ruby in the shape of a cylindrical rod 8 cm long and 5 mm in diameter, served as the lasing medium (see Fig. 7.5). The two plane faces A and B of the rod were highly polished. The face A was silvered so that the laser rays produced within the ruby cylinder were fully reflected from it and proceeded backward towards the other face. The face B was half-silvered so that even though the major portion of the light falling on it was reflected back into the crystal, a small fraction (about 1%) could come out through it as the laser beam.

The ruby crystal was enclosed within an electronic flash lamp L in the shape of a coil. Light of wavelength 5500 Å was emitted from it. The ruby laser is a pulsed laser, usually operated by high power pulses of short duration.

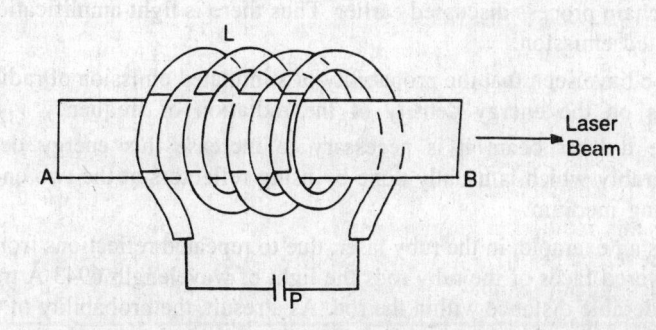

Fig. 7.5 Ruby laser.

Ruby is actually a crystal of aluminium oxide. In place of some of the aluminium atoms, there are chromium atoms. The energy levels of the chromium atoms are shown in Fig 7.6. Of the three levels shown, the first excited state E_1 is a metastable state having a mean life of about 3×10^{-3} s which is about 10^5 times longer than the normal mean life of an excited atomic state. The higher excited state E_2 is a normal excited state having a mean life of the order of 10^{-8} s.

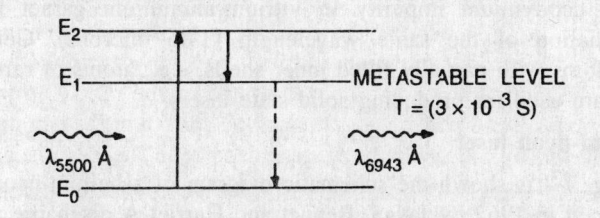

Fig. 7.6 Energy levels of the chromium atoms.

When the lamp L is switched on, a large fraction of the chromium atoms in the ground state E_0 are raised to the excited state E_2 by the absorption of light of wavelength 5500 Å. These make transitions to the lower excited state E_1 within about 10^{-8} s and are trapped in this metastable state. Thus the number of atoms in the excited state E_1 becomes ultimately greater than the number in the ground state so that there is population inversion between them by the method of optical pumping.

When the chromium atoms in the metastable state E_1 make spontaneous transition to the ground state, red light of wavelength 6943 Å is emitted. A photon of this wavelength emitted from one of the chromium atoms can then induce transition in another chromium atom in the excited state E_1 so that two coherent photons of the above wavelength are obtained. These two coherent photons then induce further transitions to give rise to a large number of coherent photons of the same wavelength by the chain process discussed earlier. Thus there is light amplification by stimulated emission.

We have seen that the probability of stimulated emission of radiation depends on the energy density of the radiation of frequency v_{12}. To produce a laser beam it is necessary to increase this energy density considerably which is usually done by using reflectors at the two ends of the lasing medium.

As an example, in the ruby laser, due to repeated reflections from the two silvered faces of the ruby rods the light of wavelength 6943 Å travels a considerable distance within the rod. As a result, the probability of these light photons to interact with other chromium atoms raised to the metastable level E_1 increases appreciably. Obviously only the beam of light incident normally upon the two end faces can undergo such repeated reflections so that the intensity of this beam which travels parallel to the axis of the ruby cylinder is considerably amplified. Out of this, about 1% comes out through the half silvered face as the laser beam.

Various other types of solid state lasers were invented after Maiman's original ruby laser, *e.g.*, by using calcium tungstate ($CaWO_4$) or glass with neodymium as impurity which gives laser beam in the infra-red region of wavelength 1·06 microns. A high efficiency solid state laser uses neodymium impurity in yttrium-aluminium garnet (YAG) giving radiation of the same wavelength (1·06 microns). Generally impurity atoms with partially filled inner shells, *e.g.*, atoms of rare earth elements, are used for producing solid state lasers.

7·6 Helium-neon laser

In Fig. 7·7 is shown the schematic diagram of a helium-neon laser first produced in 1962 by Javan, Bennet and Harriet A discharge tube is filled with a mixture of helium and neon gases in the ratio of 7 : 1 and at a total pressure of 1 torr. There are two curved mirrors fitted at the two ends of the tube. Discharge is produced by the application of a high voltage between two electrodes sealed into the discharge tube. The *Brewster windows* used at the two ends are inclined at the Brewster angle $\theta = \tan^{-1} n$ in the plane of incidence to be transmitted through the two windows and this allows light polarised in a particular plane to come out. The He-Ne laser operates continuously, usually at low power.

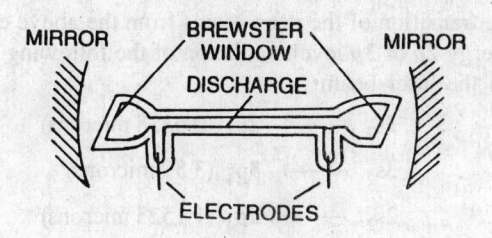

Fig. 7·7 Helium-neon laser.

The energy levels of helium and neon are shown in Fig. 7·8. Due to the electric discharge in the laser tube, some of the helium atoms are raised to the 2^3S or 2^1S metastable states. Their numbers build up since

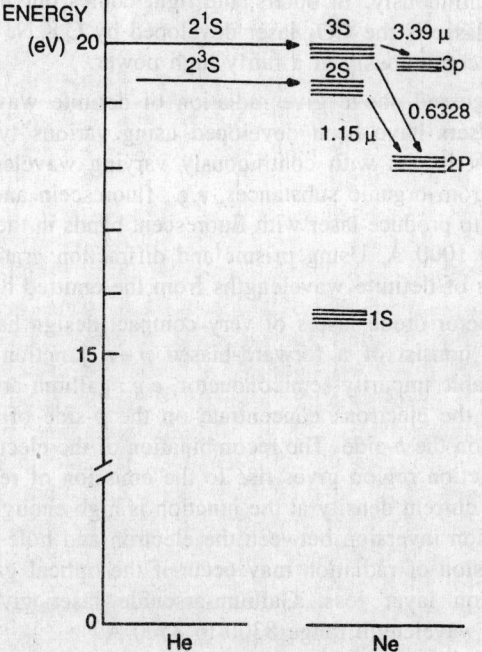

Fig. 7·8 Energy levels of helium and neon.

they cannot make optical transitions to the lower states of helium. The energies of the two excited states 3s and 2s of neon are slightly lower than those of the above metastable states of heium. Due to collisions between the helium atoms in these metastable states, the neon atoms in the ground state are raised to the above mentioned 3s or 2s excited states, leading to population inversion in the neon atoms :

$$He^* \, (^3S) + Ne \longrightarrow He + Ne^* \, (2s)$$
$$He^* \, (^1S) + Ne \longrightarrow He + Ne^* \, (3s)$$

Due to the transition of the neon atoms from the above excited states to the lower energy $2p$ or $3p$ levels radiation of the following wavelengths are obtained in the laser beam :

$$3s_2 \longrightarrow 2p_4 \; (0\cdot6328 \text{ microns})$$
$$3s_2 \longrightarrow 3p_4 \; (3\cdot39 \text{ microns})$$
$$2s_2 \longrightarrow 2p_4 \; (1\cdot1523 \text{ microns})$$

Several other weaker laser lines are also obtained from the He-Ne laser.

Lasers have been designed using solids, liquids and gases to give radiation ranging from the microwave to the ultraviolet regions. Some lasers, *e.g.*, He-Ne laser or the YAG (yttrium-aluminium garnate) laser can be used continuously. In others, the light comes out in pulses. An interesting gas laser is the CO_2 laser developed by C.K.N. Patel (1964) which operates continuously at a fairly high power.

Lasers discussed above give radiation of definite wavelengths. In recent years, lasers have been developed using various types of dyes which give light beams with continuously varying wavelengths. Some dyes prepared from organic substances, *e.g.*, fluorescein and rhodamine have been used to produce laser with fluorescent bands in the wavelength range 500 Å to 1000 Å. Using prisms and diffraction grating, one can select radiations of definite wavelengths from the emitted band.

Semiconductor diode lasers of very compact design have come up recently. These consist of a forward-biased $p-n$ junction in a single crystal of a suitable impurity semiconductor, *e.g.*, gallium arsenide. With a forward bias, the electrons concentrate on the p-side of the junction while the holes on the n-side. The recombination of the electrons and the holes in the junction region gives rise to the emission of recombination radiation. If the current density at the junction is high enough, then there may be population inversion between the electron and hole levels. Thus stimulated emission of radiation may occur if the optical gain is higher than the junction layer loss. Gallium-arsenide laser gives infra-red radiation in the wavelength range 8300 to 8500 Å.

Table 7·1
(Table of some common laser sources)

Laser Source	Wavelength (nm)	Type
Ruby (Cr)	694·3	p
He-Ne	633, 1152	Cw
N_2	337	p
CO_2-N_2-He	10568-10629	Cw, p
Ar II	514, 502, 496, 488, 476	Cw
He-Cd	325, 442	Cw
H_2	156-161, 116-124	p
Nd-doped YAG	1060	p, Cw
Nd-doped glass	1060	p, Cw
Ga As	904	p
Ga As P	860	p

p - pulsed; Cw - continuous working

7·7 Uses of laser radiation

The main characteristics of the laser radiation are : (i) the intensity of this radiation is very high ; (ii) the laser beam is a perfectly parallel beam ; (iii) the radiation from the laser tube is almost completely monochromatic and (iv) the laser beam is spatially coherent.

Due to the last two properties of laser, it is possible to perform different types of interference experiments with it. For instance, in performing Young's experiment on interference with laser beam, it is not necessary to first pass the beam through a pin hole and then split it into two coherent beams with two subsequent pin holes. Interference fringes are observed if the laser beam is allowed to fall on the latter two pin holes directly.

Coherence can be either spatial or temporal. Spatial coherence occurs if the phase difference between two points on a wave front is zero at all instants of time. If the lasing medium is homogeneous the output beam is coherent over its whole cross section. Temporal coherence depends on the duration of the output wave. The interval over which the output is represented by the same plane wave is known as the *coherence time*. It can be much longer than the period of oscillation of the light vector and can be as high as 10^{-3} s. Thus it is possible to obtain interference between two beams of light coming from two different laser sources.

The concepts of spatial and temporal coherence are independent.

Experiments with Michelson's interferometer have shown that interference fringes are obtained even with a path difference of 9 m between the two rays. This proves the temporal coherence of the laser radiation. It may be noted that with an ordinary source of light, interference fringes are observed upto a maximum of 3 m path difference between the two rays in the Michelson's interferometer.

Laser beams have been used for very accurate alignment of objects. The two mile long linear accelerator of the Stanford University (SLAC) was aligned with the help of laser beam. Since the rays in the laser beam are almost perfectly parallel, it is possible to focus them within an extremely small area. If a laser beam with a cross section of radius a is focused by a lens of focal length f, then the lateral spread of the beam is $f\lambda/a$ where λ is the wavelength. Since f and a are of the same order of magnitude, the lateral spread is of the order of λ. Hence the area of cross section of the focused beam is of the order of λ^2 which is $\sim 10^{-14}\,m^2$. This is much smaller than the cross section of the focused beam in the case of ordinary light.

Since the laser beam can be focused into such a small spot, a large amount of light energy can be concentrated in an extermely small area. For this reason, it is possible to cut minute holes in different materials as also to cut sheets of metals of high melting point with the help of laser beam. It can also be used for medical purposes, *e.g.*, in repairing retinal detachment in the eye.

Because of the highly directional property of the laser rays, it is possible to measure distances very accurately with the help of laser beam. The American astronauts landing on the moon had placed reflectors on the surface of the moon. Laser beams sent from the earth were reflected back by these. By measuring the time required for the light signal to travel from the earth to the moon and back, the earth to moon distance has been measured with an error of only 0.3 m. This is much less than the error of about 80 km in the measurements made by other methods.

Application of laser beams in diverse fields of science and technology is now quite common. These include determination and control of the motion of moving objects like aircrafts or rockets. The method thus makes possible the control of the motion of missiles to hit distant tragets during warfare. The field of communication technology has been revolutionized by the use of laser in conjuction with optical fibres. Because of the very high frequency of the carrier light waves (10^{14} to 10^{15} Hz), it is possible to transmit a very large number of speech signals (10^{11} to 10^{12}) over the same channel. Another interesting application is the production of three dimensional pictures by the method of holography based on the coherence property of the laser rays. Attempts are being made to use laser rays for the development of fusion reactor (see Vol. II) and for isotope separation.

The very high intensity of the laser beam implies that the amplitude of the corresponding electromagnetic wave is very large. So it is possible to investigate the *nonlinear* optical properties of different materials with the help of laser rays.*

Problems

1. What is the ratio of stimulated to spontaneous emission rates for the sodium D line at 200°C
2. Calculate the gain constant β of a laser having the following parameters : Inversion density $(n_2 - n_1) = 5 \times 10^{22}/m^3$; Wavelength = 650 nm ; Life time for spontaneous emission $= 2 \times 10^{-4}$ s ; Line width $\Delta\lambda = 15$ Å. (3·95)
3. Calculate the inversion density for He- Ne laser operating at 633 nm if the temperature of the discharge is 150°C and the gain constant is 3% per metre. The life time of the upper state for spontaneous transition is 10^{-7} s. $(3.15 \times 10^{14}/m^3)$
4. Calculate the spot size of a He- Ne laser beam at the centre of a confocal cavity if the length of the cavity is 1·24m. The wave length is 633·0 nm. (0·5 mm)
5. Calculate the frequency difference between two adjacent modes in the above case if the length of the cavity is 1 m. (150 MHz)
6. What is the Doppler broadening (full width at half maximum, FWHM) of the neon line in He-Ne laser at $T = 300$ K if $\lambda = 633.0$ nm? (1·7 GHz)
7. How many different modes of frequencies fall within the FWHM of the Ne line in the He-Ne laser of Prob. 5 ? (23)

---

* See Principles of Lasers by O. Svelto, Plenum Press (1982)

8
X-RAYS

8·1 Discovery of x-rays

X-rays were discovered by W.C. Röntgen in Germany in 1895 almost accidentally while investigating the properties of cathode rays. The discharge tube used by Röntgen is shown in Fig. 8·1. A very high potential difference (~ 40,000 volts) was applied between the anode A and the cathode C of the tube which was kept at a very low pressure. A large induction coil was used to generate the required high voltage.

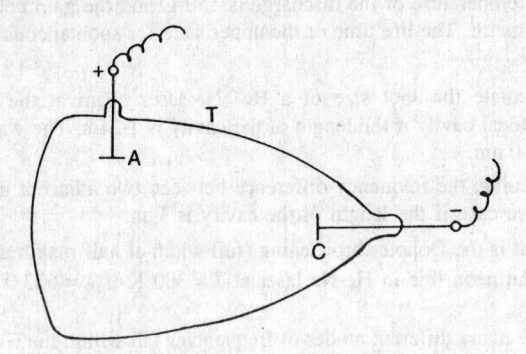

Fig. 8·1. Röntgen's apparatus for the discovery of x-rays.

The discharge tube was completely enclosed within a black card-board box so that no visible or ultraviolet light could come out of it. Under this condition, it was observed that if a piece of paper coated with a fluorescent substance like barium-platino-cyanide was held outside the cardboard box it began to glow with fluorescent light when the discharge was going on in the discharge tube inside the box. Such fluorescence was observed even when the paper was held 2 m away from the box. Obviously the fluorescence could not be caused by the cathode rays produced in the discharge tube since these could not penetrate the thick glass wall of the tube and travel a distance of 2 m in air outside the cardboard box. Their penetrability is much less. So Röntgen concluded that the fluorescence was produced by some hitherto unknown radiation of very high penetrating power which was coming out of the discharge tube. It was noticed that this radiation came mainly from those regions on the wall of the discharge tube where the cathode rays were incident. He named this radiation x-rays though the name Röntgen rays was used by many workers in later years.

Röntgen investigated the properties of the radiation in detail. He found that many substances which were opaque to visible or ultraviolet rays were transparent to the X-rays. The degree of transparency however differed for different substances. Thus paper was found to be completely transparent; the radiation could penetrate a thick volume of a book of about 1000 pages. The intensity of the radiation coming out on the other side of the book was not reduced appreciably. The rays could penetrate through a thin foil of tin, or a wooden board 2 to 3 cm thick. An aluminium sheet, about 15 mm thick, reduced the intensity considerably; however it did not completely absorb the radiation. On the other hand, x-rays were found to be easily absorbed by foils of metals of high atomic number (Z), such as gold ($Z = 79$) or lead ($Z = 82$).

Röntgen also noticed that x-rays could penetrate through the skin and flesh of the body of an animal. However it could not penetrate through the bone easily. Thus if some organ of the body of an animal, such as the hand of a human being, was placed on a fluorescent screen and x-rays were allowed to fall on the organ, the shadow of the bone could be observed on the screen. Because of this differential penetrating power of the radiation, different hospitals in Europe began to use it as a valuable new aid in the field of surgery very soon after its discovery. In fact there are few instances in the history of physics where a new fundamental discovery found application in the practical field so soon after the discovery.

8·2 Production of x-rays

After his initial experiments, Röntgen designed a new type of discharge tube for the generation of x-rays which is illustrated in Fig. 8·2. The pressure within the tube had to be maintained at about 10^{-3} mm of Hg below which the production of x-rays became difficult. At

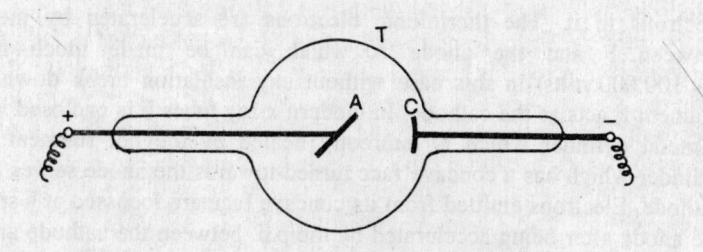

Fig. 8·2 X-ray tube developed by Röntgen.

the pressure of 10^{-3} mm, some air molecules become ionised under the influence of the very high electric field within the tube. The heavier positive ions are attracted to the cathode C and due to their impact electrons are emitted from C. These electrons are then attracted towards the anode A and hit the latter with very high velocity gained by acceleration under the influence of the high p.d. between the cathode and the anode. Due to the impact of the electrons on the anode, x-rays are emitted from it. The cathode C is usually concave in shape with its focus on the anode so that the electrons emitted from C are focused on a small region of A near the focus and x-rays are emitted from this small region. A p.d. of about 30,000 to 50,000 volts is required to be applied between C and A. This type of x-ray tube is known as the *gas-tube*. In these tubes, the electron current falling on the anode, the potential difference between C and A and the gas pressure inside the tube are interdependent.

For many years, the x-ray tube designed by Röntgen was the only type of x-ray source. In 1913, the American scientist W.D. Coolidge devised a new type of x-ray tube, known as the Coolidge tube, which is illustrated in Fig. 8·3. F is a tungsten filament which can be heated electrically so that it may emit thermionic electrons. The gas pressure in the tube can be kept much lower in this tube compared to that in the gas tube, since ion impact on the cathode is no longer the main source of

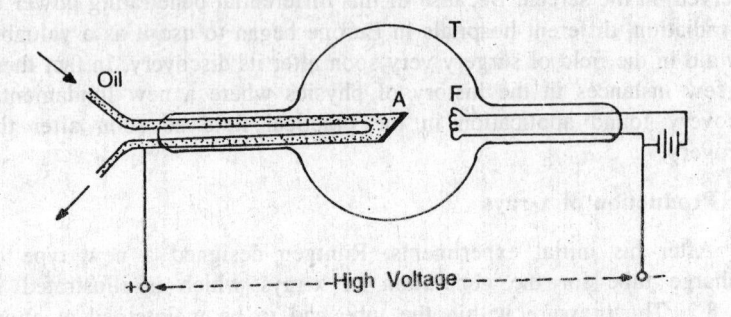

Fig. 8·3. Coolidge tube.

electrons in it. The thermionic electrons are accelerated by the p.d. between F and the anode A which can be made much higher (~ 100,000 volts) in this case without any insulation break down. The filament F acts as the cathode. In modern x-ray tubes F is enclosed within a metal cylinder which is indirectly heated by the hot filament. This cylinder which has a concave face turned towards the anode serves as the cathode. Electrons emitted from its concave face are focussed at a spot on the anode after being accelerated by the p.d. between the cathode and the anode which can be varied at will. As a result, x-rays are emitted from the anode, the intensity of the beam being regulated by regulating the filament current. The electron beam hitting the anode heats up the latter. To cool it, it is usually mounted at one end of a hollow copper cylinder through which water or oil is circulated. This prevents melting of the anode by excessive heating.

In modern x-ray tubes, a high alternating p.d. produced by means of a step-up transformer is applied between the cathode and the anode. During the half cycle of this a.c. potential in which the anode is positive w.r.t. to the cathode, the electrons are attracted to the former and generate x-rays by impact upon the latter. If necessary the a c. potential can be rectified and filtered by using inductors and capacitors before applying it to the x-ray tube.

In 1941, the American physicist D.W. Kerst devised an electron accelerating machine known as the *betatron* by which electrons can be accelerated up to energies of the order of 10^8 electron-volts. X-rays produced by such high energy electron beam are of extremely high energy and have been used to investigate the structure of atomic nuclei. Besides, the betatron is being increasingly used in the field of medicine and surgery for diagonstic and therapeutic purposes. The principle of working of the betatron will be described in Vol. II.

8·3 Properties of x-rays

The following are the chief properties of the x-rays :

(a) X-rays can penetrate through most substances. However their penetrability in different substances is different. They are able to penetrate more in low density substances and less in high density substances. For example, ordinary glass is quite transparent to x-rays, but *lead-glass* is almost completely opaque to them. We have already discussed about the penetrating power of x-rays in § 8·1.

(b) X-rays can produce fluorescence in different substances, *e.g.*, barium plantino-cyanide, uranium glass, rock salt, different compounds of calcium, *etc*.

(c) X-rays can blacken photographic plates. The degree of blackening depends upon the intensity of the x-rays incident upon the plate. Thus x-ray intensity can be measured with the help of photographic plates.

(d) X-rays ionize the gas through which they travel. The ionizing power depends on the intensity of the x-ray beam. Thus x-ray intensity can also be determined by measuring their ionizing power.

(e) X-rays are not deflected by electric or magnetic fields. This proves that unlike cathode rays or positive rays they are not beams of charged particles.

(f) X-ray travel in straight lines like ordinary light.

(g) Though Röntgen and other early investigators did not observe reflection or refraction of x-rays, later workers proved that x-rays are both reflected and refracted.

(h) X-rays can be diffracted with the help of crystaline substances. They can also be polarized.

From the above characteristics of the x-rays, it is clear that they are electromagnetic radiation like ordinary light or ultraviolet radiation. Their wavelengths can be measured by diffraction experiments using crystals. It has been found that these wavelengths are much shorter than those of visible or u.v. radiation as the following list shows :

Wavelength of visible radiation : ~ 4000 to 8000 Å

Wavelength of u.v. radiation : ~ 100 to 1000 Å

Wavelength of X-rays : ~ 10^{-2} to 100 Å

8·4 Measurement of the intensity of x-rays

Both photographic and ionization methods can be used to determine the intensity of X-rays. Ionization method gives the intensity more accurately. For such measurements, an instrument known as the *ionization chamber* is usually employed. This is illustrated in Fig. 8·4.

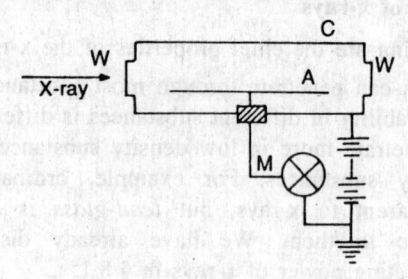

Fig. 8·4. Ionization chamber.

C is a cylindrical chamber made of brass, the two open ends of which are closed by two thin sheets (W, W) of some low Z metal, *e.g.*, aluminium. These are known as the windows of the chamber. X-ray beam can enter the chamber through one of the windows and come out through the other. The chamber is usually 20 to 100 cm long. It is filled with some gas or vapour (*e.g.*, air, methyl bromide *etc.*).

The x-ray beam passing through the chamber ionizes the gas within it. As a result both positive ions and negative electrons are produced in the gas. A metal rod A insulated from the wall of the chamber is placed inside it, parallel to the axis of the cylinder. A potential difference of few hundred volts is maintained between A and C. The positive ions are drawn to the cathode while the negative ions move to the anode so that an ionization current is produced in the chamber. The rod A is connected to an electrometer M which can measure the very small ionization current produced by the x-rays. The ionization current is found to be linearly proportional to the intensity of the x-ray beam passing through the chamber. Thus the measurement of the ionization current gives a measure of the intensity of the x-ray beam. (See Vol. II).

8.5 Variation of the x-ray intensity with wavelength

It is found that x-rays of different wavelengths are emitted from the x-ray tube. If the intensity I of the x-rays is measured as a function of the wavelength λ and the variation of I is plotted graphically, then the graphs of the nature shown in Figs. 8·5 and 8·6 are obtained. When the target (*i.e.*, the anode) from which the x-rays are emitted is made of some high Z metal. *e.g.*, tungsten ($Z = 74$), the I vs. λ graph has the appearance shown in Fig. 8·5. It is found that I varies continuously with λ upto potential differences of the order of 50,000 volts between the anode and the cathode. For a given potential difference, there is a minimum wavelength λ_m below which no x-rays are emitted. With increasing wavelength, the intensity at first increases and after attaining a maximum it begins to decrease again.

In Fig. 8·5, the variation of I with λ for several different values of the p.d. is shown. It is found that the minimum wavelength λ_m decreases with increasing p.d. For a given wavelength, the intensity is higher when the p.d. is higher.

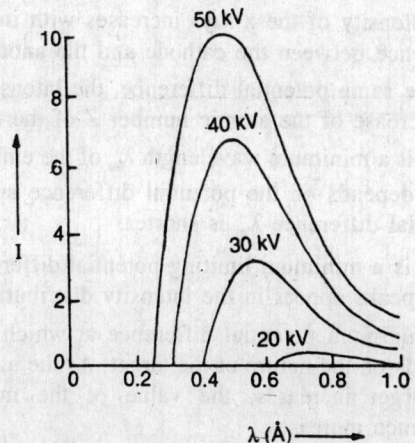

Fig. 8·5. Intensity vs. wavelength of x-rays emitted from tungsten at different voltages.

With a low Z target, such as molybdenum ($Z = 42$), the variation of I with λ has the appearance shown in Fig. 8·6. In this case also, there is continuous variation of I with λ for $V < 20,000$ volts. However at higher potential differences ($V \sim 25,000$ volts or higher) several discrete peaks appear, superimposed upon the continuous background at certain definite wavelengths. For targets with still lower, Z, *e.g.*, copper ($Z = 29$), such peaks begin to appear at still lower p.d. ($V \geq 8000$ volts). The wavelengths at which the peaks appear are different for the copper target from those for the molybdenum target. From the experimental results shown in Figs. 8·5 and 8·6, the following conclusions can be drawn :

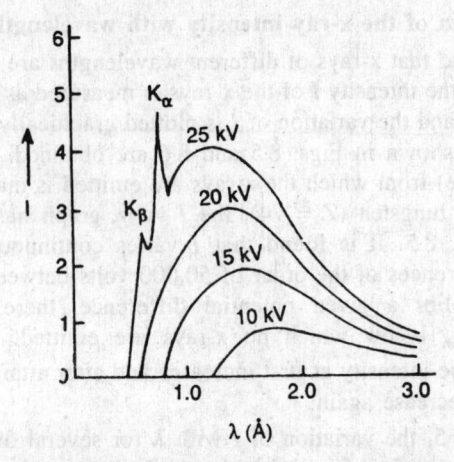

Fig. 8·6. Intensity vs. wavelength of x-rays emitted from molybdeum for different voltages. Notice the peaks at higher voltages.

(a) The intensity of the x-rays increases with increasing potential difference between the cathode and the anode.

(b) For the same potential difference, the intensity increases with the increase of the atomic number Z of the target.

(c) There is a minimum wavelength λ_m of the emitted x-rays whose value depends on the potential difference applied. For higher potential difference λ_m is shorter.

(d) There is a minimum limiting potential difference above which sharp peaks appear in the intensity distribution curve.

(e) The minimum potential difference at which the peaks appear depends on the nature of the target. As the atomic number Z of the target increases, the value of the minimum potential difference increases.

(f) The positions of the discrete peaks (i.e. their wavelengths) in the intensity distribution curve are characteristic of the nature of the target. They do not depend on the potential difference applied. These are known as *characteristic* x-rays. For targets with higher values of Z the peaks appear at relatively shorter wavelengths.

(g) The heights of the peaks increase with increasing potential difference.

8·6 Origin of the continuous x-rays

It is known from the electromagnetic theory of light that when light charged particles *e.g.*, electrons or positrons travel with accelerated or decelrated motion, they emit electromagnetic radiation of different frequencies (v). A part of their kinetic energy is transformed into the energy of the emitted radiation. When the electrons accelerated within the x-ray tube hit the target (*i.e.*, the anode) their motion becomes decelerated and their velocities decrease. As a result they emit electromagnetic radiation with a continuous distribution of wavelengths starting from a minimum.

We know that the energy of a photon of wavelength λ is given by

$$E_v = hv = hc/\lambda \qquad (8\cdot6-1)$$

where h is Planck's constant and c is the velocity of light in vacuum. Evidently, the maximum energy of the emitted photon will be equal to the initial energy of the incident electron :

$$E_m = hv_m = hc/\lambda_m = eV \qquad (8\cdot6-2)$$

where V is potential difference between the cathode and the anode within the x-ray tube. Due to this the electrons emitted from the cathode acquire an energy eV when they reach the anode. The wavelength of the emitted radiation depends on the fraction of incident electron energy (eV) that is transformed into x-radiation. Eq. (8·6–2) shows that when the entire energy eV is transformed into x-radiation, the frequency is the maximum (v_m) and the wavelength minimum (λ_m) given by

$$\lambda_m = hc/eV \qquad (8\cdot6-3)$$

If V is measured in volts we get

$$\lambda_m = \frac{12413}{V \text{(volts)}} \text{ Å}$$

Thus if $V = 10{,}000$ volts, $\lambda_m = 1\cdot24$ Å; on the other hand, if $V = 50{,}000$ volts, $\lambda_m = 0\cdot248$ Å.

The above mechanism of the emission of electromagnetic radiation from an accelerated or decelerated electron is known as *bremsstrahlung* * process and the radiation so emitted as bremsstrahlung. The continuous distribution of wavelengths emitted in this process is analogous to the continuous distribution of wavelengths in visible white light. Hence this type of x-radiation is also sometime called *white radiation*.

Usually only a fraction of the energy carried by the electron beam in the x-ray tube is converted into x-rays. The rest goes to heat the target.

---

* The word *brems* in German means 'brake' while *strahlung* means radiation so that the word bremsstrahlung means literally brake-radiation, *i.e.*... radiation due to deceleration.

Sometimes, the X-rays are classified according to their penetrating power. Thus the most penetrating radiation is called *hard radiation*. The radiation with relatively less penetrating power is known as *medium radiation* while that with very low penetrating power is known as *soft radiation*.

Explanation of the spectral distribution of the continuous X-rays :

The nature of the spectral distribution of the continuous x-rays from an x-ray tube (see Fig. 8.5) can be understood on the basis of the theory of bremsstrahlung based on electromagnetic theory of light.*

Because of the deceleration of the electron beam in the direction of its motion within the target inside the x-rays tube, one has to take the spectral energy distribution of the bremsstrahlung x-ray for electrons of different energies and assign different weights to these curves. These are based on the variation of reflecting power of the crystal grating for different wavelengths, the absorption of x-rays within the x-ray tube, in air and in the window of the ionization chamber and the different ionizing powers of the beams of different wavelengths.

The spectral energy distribution of the bremsstrahlung radiation is found to be independent of the frequency and hence can be written as

$$I_\lambda = I_\nu \, (\nu^2/c^2) \propto \frac{1}{\lambda^2}$$

Taking into account the factors outlined above, it is possible to explain, the observed spectral distribution shown in Fig. 8.5. (See *X-rays in Theory and Practice* by Compton and Allison).

8.7 Origin of the x-ray peaks

The peaks observed in the x-ray wavelength distribution curves (Fig. 8.6) are actually spectral lines in the x-ray region. Their origin lies in the interior electronic structure of the atoms.

We have seen (Ch. IV) that if an electron in the outermost orbit of an atom is excited to a higher energy level, it returns to the lower energy levels within $\sim 10^{-8}$ sec. During this transition radiation of definite wavelengths lying in the visible or ultraviolet regions are emitted giving rise to the discrete spectral lines in these regions. These electrons in the outermost orbits of the atoms are rather loosely bound (see Ch. IV). Their binding energies are of the order of a few electron volts which is comparable to the energies of the emitted photons. Actually the binding energy is somewhat greater than the energy of the emitted photon. For

---

*[1] See *Classical Electricity and Magnetism* by Panofsky and Philips.

X-Rays

instance, the mean energy of the photons corresponding to the sodium D-lines ($\lambda = 5893$ Å) is

$$E_v = \frac{hc}{\lambda} = \frac{12413}{5893} = 2 \cdot 1 \text{ eV}$$

On the other hand the binding energy of the outermost valence electron in sodium is 5·1 eV.

The wavelengths of the x-ray lines are usually of the order of a few angstrom units or even shorter. This means that the energies of the x-ray photons are of the order of 10^4 eV or higher. Hence the atomic electrons whose transitions are responsible for the origin of the x-ray lines must have binding energies of the order of 10^4 eV or higher. So these are bound to atoms much more strongly than the valence electrons in the outermost orbits. This can happen only if they revolve in orbits which lie close to the nucleus, *i. e.*, in the inner atomic orbits.

It was noted in Ch. VI that the electrons in the atoms are arranged in different atomic shells. Of these, two electrons can exist in the innermost K-shell which are the most strongly bound, since they are nearest to the atomic nucleus. Outside this, there may be a maximum of eight electrons in the L-shell which are much less strongly bound than the K electrons. Outside the L-shell there is the M-shell in which there can be a maximum of eighteen electrons which are still less strongly bound than the L-electrons. When an electron accelerated to very high energy in the x-ray tube strikes the target, it may collide inelastically with an inner shell electron (*e.g.*, a K-electron) in a target atom due to which this electron may be emitted from the atom. If the kinetic energy of the incident electron is $E_e = eV$ and the binding energy of the electron in the K-shell of the atom is E_K, then the electrons emitted from the K-shell will have a kinetic energy.

$$\frac{1}{2}mv^2 = eV - E_K \qquad (8\cdot7\text{--}1)$$

Obviously the minimum kinetic energy of the incident electron required to eject an electron from the K-shell is equal to its binding energy E_K.

The vacancy in the K-shell which is created as a result of the emission of a K electron can be filled up by the transition of a relatively loosely bound electron from an outer shell (*e.g.* L, M, N etc.) into the K-shell. The excess energy of this electron is taken away by a photon which is emitted due to this transition. If such transition takes place from an L-shell and the binding energy of the L-electron is E_L, then the energy of the emitted x-ray photon will be

$$h\nu_{KL} = E_K - E_L \qquad (8\cdot7\text{--}2)$$

Thus the energy $h\nu_{KL}$ of the emitted photon depends upon the binding energies of the electrons in the K and L shells. These binding energies are characteristics of the target atom. Hence the x-ray spectral lines have definite frequencies and wavelengths which are characteristics of the target atom.

For a given target, more than one x-ray spectral line is usually observed. These have different wavelengths. The x-ray lines originating out of the transitions from L, M, N etc. shells to the K-shell are known as the lines of the K-series of the simply K-lines. Similarly, the x-ray lines originating due to the transitions from M, N etc. shells to the L-shell are known as the L-lines ; and so on. The origin of the lines of the different x-ray series is best understood with the help of *Kossel diagram*, shown in Fig. 8·7. The corresponding *energy level diagram* is shown in Fig. 8·8.

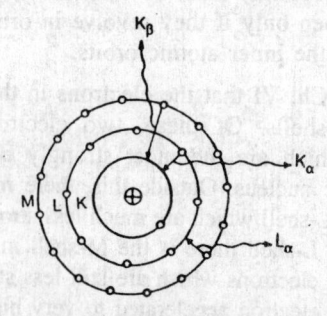

Fig. 8·7. Kossel diagram.

When all the electrons are present in the atom in their respective shells, the energy of the atom is taken to be zero. If now a K-electron is released from the atom due to absorption of the energy E_K by it, the total energy of the system made up of ionized atom thus produced and the electron emitted with zero kinetic energy increases to E_K. Similarly, the system composed of an atom from which an L-electron is released with zero kinetic energy and the released electron has a total energy which is

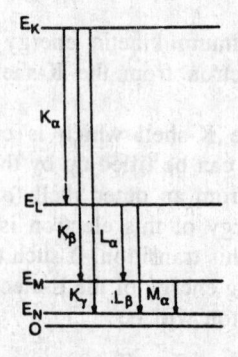

Fig. 8·8. Origin of the x-ray lines.

greater than that of the unionized atom by an amount E_L. In Fig. 8·8, these different energy levels of the atom are represented by a number of horizontal lines. Since $E_K > E_L > E_M$ etc., the K-level is shown highest with L, M etc. levels coming successively below the former. When there is transition of an L-electron to a vacant K-shell, the corresponding transition in the energy level diagram is from the K-level to the L-level. As a result, the energy of the emitted x-ray photon becomes equal to the energy-difference (E_K-E_L) between the K and L-levels.

The x-ray lines originating from the transitions between the different x-ray levels are usually designated by the symbols α, β γ etc. as below :

K-level to L-level transition produces $K_α$-line ;

K-level to M-level transition produces $K_β$-line ;

L-level to M-level transition produces $L_α$-line ;

L-level to N-level transition produces $L_β$-line ;

and so on. The frequencies of these lines are given by

$$v(K_α) = \frac{E_K - E_L}{h}$$

$$v(K_β) = \frac{E_K - E_M}{h}$$

$$v(L_α) = \frac{E_L - E_M}{h}$$

$$v(L_β) = \frac{E_L - E_N}{h}$$

As we have seen above, the x-ray peaks (*i.e.*; the x-ray spectral lines) are characteristics of the target elements. So they are known as *characteristic radiation*. These may also be produced by allowing the *continuous* x-*rays* generated by the bremsstrahlung process in the x-ray tube (see § 8·6) to be scattered from different elements outside the tube. These x-ray photons can eject electrons from the inner K, L etc. orbits of the atom of the scatterer if their energy hv is greater than the binding energies of the electrons in these orbits. The vacancy thus created in an inner orbit is subsequently filled up by the transition of an electron from an outer orbit. The surplus energy due to such transition is carried away by a characteristic x-ray photon. The characteristic radiation is also sometimes termed as fluorescent radiation (see Ch. IV).

8·8 Fine structure of x-ray levels

Careful measurements show that the characteristic x-ray lines are not single ; they have fine structures of their own. Each of them is composed of a number of closely lying components. Their origin can be explained by considering the three quantum numbers n, l, j associated with each electronic energy level (see Ch. VI). For a given principal quantum number n which broadly determines the positions of the different electronic shells e.g., K ($n = 1$), L ($n = 2$), M ($n = 3$) etc., the azimuthal quantum number l can assume the n different values : $l = 0, 1, 2, ...(n-1)$. Again for a given l, the total quantum number j can assume the two values : $j = l + \frac{1}{2}$ or $j = l - \frac{1}{2}$. For $l = 0$ however, j can have only one value, viz., $j = \frac{1}{2}$.

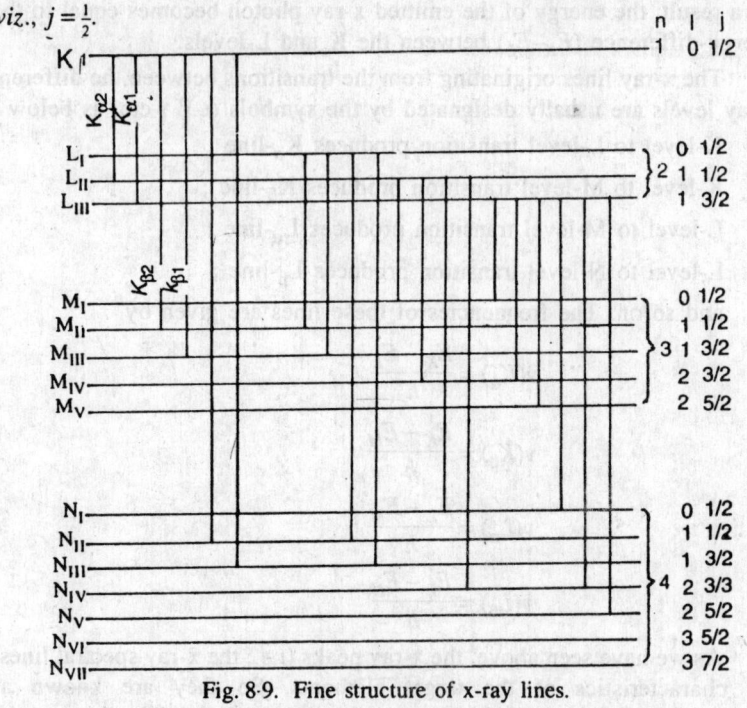

Fig. 8·9. Fine structure of x-ray lines.

On the basis of the above quantum numbers, the different x-ray levels can be classified in the manner shown in Table 8·1.

The transitions between the different x-ray levels are governed by the following two selection rules :

$$\Delta l = \pm 1 \text{ and } \Delta j = 0, \pm 1$$

These are analogous to the selection rules (6·1-1) and (6·4-1) applicable in the case of the transitions for the electron in the outermost atomic orbits.

The fine structure of the x-ray levels and the possible transitions between them are shown in Fig. 8·9. From the figure, it will be seen that each K-line actually splits up into two component lines. For instance, the K_α line is composed of the two components $K_{\alpha 1}$ and $K_{\alpha 2}$; K_β-line is made up of the two components $K_{\beta 1}$ and $K_{\beta 2}$. The component $K_{\alpha 1}$ originates due to the transition from the K-level to the L_{III}-level for which

$n = 2$, $l = 1$ and $j = 3/2$. On the other hand the $K_{\alpha 2}$ component originates out of the transition from the K-level to the L_{II} level with $n = 2$, $l = 1$ and $j = \frac{1}{2}$. The splittings of the L, M etc. lines are more complex.

Table 8·1

Shell	n	l	j	Energy Level	Multiplicity of the Level
K	1	0	1/2	K_I	1
L	2	0	1/2	L_I	
		1	1/2, 3/2	L_{II}, L_{III}	3
M	3	0	1/2	M_I	
		1	1/2, 3/2	M_{II}, M_{III}	5
		2	3/2, 5/2	M_{IV}, M_V	
N	4	0	1/2	N_I	
		1	1/2, 3/2	N_{II}, N_{III}	
		2	3/2, 5/2	N_{IV}, N_V	7
		3	5/2, 7/2	N_{VI}, N_{VII}	

Since the binding energy of an electron in the K-shell is much higher than that of an electron in the L-shell, the energies and hence the frequencies of the K-lines are much higher than those of the L-lines so that the wavelengths of the K-lines are much shorter than those of the L-lines. Similarly the wavelengths of the L-lines are much shorter than those of the M-lines which are again shorter than the wavelengths of the N-lines ; and so on. *i.e.*,

$$\lambda_K < \lambda_L < \lambda_M < \lambda_N \ldots$$

or, $\quad \nu_K > \nu_L > \nu_M > \nu_N \ldots$

8·9 Theoretical considerations of the fine structure of x-ray levels

The splitting of the x-ray levels discussed above is known as the fine structure of the x-ray levels. As we saw in § 4·4., the radii of the orbits of the hydrogen-like atoms are inversely proportional to Z. So in the heavy atoms of the target materials of the x-ray tubes, these radii will have quite small values. The electrons in the L-shell of such an atom may be regarded as being in the Coulomb field of the nuclear charge reduced by two units due to the screening of the former by the two K-electrons. This is known as *internal screening*. In addition, there will be *external screening* due to the electrons in the orbits outside the L-orbit, which is usually much smaller. Both these combine to reduce the effective charge of the nucleus to a value $Z_e e = (Z - \sigma) e$ where σ is the combined screening constant which is close to 2 for the L-electrons. For the M-electrons, the two inner K and L shells have 2 and 8 electrons so that for these the screening constant should be close to 10. Applying Bohr formula (see § 4·12), we then get the term value (see Eq. 4·12-21)

$$T_n = -\frac{E_n}{2\pi c \hbar} = \frac{R_N Z_e^2}{n^2} \qquad (8\cdot9\text{-}1)$$

Applying Sommerfeld's fine structure correction to the above formula (see § 4·12) we get the term value (see Eq. 4·12-22)

$$T_n = R_N \frac{(Z-\sigma)^2}{n^2} + \frac{R_N \alpha^2 (Z-\sigma')^4}{n^4} \left(\frac{n}{k} - \frac{3}{4}\right)$$
$$+ R_N \frac{\alpha^4 (Z-\sigma')^6}{n^6} \left(\frac{n^3}{4 k^3} + \frac{3n^2}{4 k^2} - \frac{3n}{2k} + \frac{5}{8}\right) \qquad (8\cdot9\text{-}2)$$

The application of the above equation to the observed energy levels shows that the screening constants σ and σ' are different, though a priori, they might be expected to be the same. α is Sommerfeld's fine structure constant.

The above expression was later derived by Gordon on the basis of Dirac electron theory in which the azimuthal quantum number k of Sommerfeld was replaced by $j + 1/2$ where j is the total quantum number $(j = l \pm 1/2)$. We then get

$$T_n = \frac{R_N (Z-\sigma)^2}{n^2} + \frac{R_N \alpha^2}{n^4} (Z-\sigma')^4 \left(\frac{n}{j+1/2} - \frac{3}{4}\right)$$
$$+ \frac{R_N \alpha^4}{n^6} (Z-\sigma')^6 \left[\frac{1}{4} \frac{n^3}{(j+1/2)^3} + \frac{3}{4} \frac{n^2}{(j+1/2)^2} - \frac{3}{2} \frac{n}{j+1/2} + \frac{5}{8}\right]$$
$$(8\cdot9\text{-}3)$$

The correction terms in Eq. (8·9-3) involving j are due to spin-orbit interaction as well as from relativistic effects. They are known as *spin-relativity correction*. Eq. (8·9-3) can be used to calculate the wave number difference between $L_{II} - L_{III}$ and $M_{II} - M_{III}$ levels for which n, s and l are the same, but j is different (see Fig. 8·9). For $L_{II} - L_{III}$, $n = 2$, $s = 1/2$, $l = 1$ and $j = 1/2, 3/2$. So we get

$$\Delta \bar{\nu} (L_{II} - L_{III}) = \frac{R_N \alpha^2}{16} (Z - \sigma_L')^4 \left[1 + \frac{5}{8} \alpha^2 (Z - \sigma_L')^2\right] \qquad (8\cdot9\text{-}4)$$

X-Rays

Similarly for the $M_{II} - M_{III}$, $n = 3$, $s = {}^1/_2$, $l = 1$ and $j = {}^1/_2, {}^3/_2$ so that the wave number separation is

$$\Delta \bar{v} (M_{II} - M_{III}) = \frac{R_N \alpha^2}{81} (Z - \sigma_M')^4 \left[\frac{3}{2} + \frac{31}{32} \alpha^2 (Z - \sigma_M')^2 \right] \quad (8.9-5)$$

Regular and irregular doublet laws :

Comparison with experimental data gives $\sigma_L' \approx 3.5$ and $\sigma_M' \approx 8.5$ which does not depend on Z. These values agree with those calculated by Pauling from purely theoretical considerations. Eqs. (8.9–2) and (8.9–3) show that the spin-relativity doublet wave number separation is proportional to the fourth power of the effective atomic number : $\Delta \bar{v} \propto (Z - \sigma')^4$. This is known as the *regular doublet law*.

If we retain only the first term on the r.h.s. of Eq. (8.9-3), we see that the difference in the term values for the doublets such as $L_I - L_{II}$, $M_I - M_{II}$, $M_{III} - M_{IV}$ etc. which have the same n, S and J but different L (see Fig. 8.9) is proportional to the difference in the values of the screening constants σ of the two component levels of the doublet and is independent of Z. This is known as the *irregular double law* of Hertz. It is found to agree fairly well with the observed separations between the term values of such doublets which are known as *screening doublets*.

8·10 Auger transition

We have seen that when a K-electron is ejected from an atom either by electron impact or by the absorption of an x-ray photon, the resulting vacancy in the K-shell is filled by the transition of an electron from an outer shell, *e.g.*, the L-shell which corresponds to a transition from the K-level to the L-level of the atom. This transition is usually associated with the emission of the characteristic x-ray photon of energy $h \nu_{KL} = E_K - E_L$. However in some cases, the surplus energy in such transition may be directly absorbed by another L-electron which is thereby emitted. Since there are now two vacancies in the L-shell, the resulting energy level of the atom will be different from the L-level and may be called an LL-level, its energy being given by E_{LL}. The kinetic energy of the ejected second electron will then be

$$\frac{1}{2} mu^2 = E_K - E_{LL}$$

Such a transition resulting in the emission of two electrons from the same atom is known as Auger transition which had first been observed by the French physicist Pierre Auger in 1925 in a cloud chamber photograph. The effect is known as Auger effect. Since no electromagnetic radiation is emitted in this type of transition, it is also known as *radiationless transition*.

It should be noted that the surplus energy in the transition of the first L-electron to the vacant K-shell may also be absorbed by an M-electron instead of a second L-electron as mentioned above. In this case, the second electron would be an M-electron, its energy being slightly different from that in the previous case. There may be similar ejection from other outer shells also.

The phenomenon of radiationless transition was at first thought to be due to the K_α x-ray photon emitted during the transition of the first L-electron to the vacant K-shell being absorbed by a second L-electron *in the same atom*. For this reason it was called *inner photo-effect*. However, later it was realized that actually no photon is at all emitted in the process. The entire surplus energy of the first transition is taken by the second L-electron which is thereby emitted.

Radiationless transitions are also observed in the transitions of electrons in the outermost shells of atoms, as also in the case of nuclear transitions giving rise to the emission of internal conversion electrons (see Vol. II).

8·11 Moseley's law

It was noticed by H.G.J. Moseley* of England in 1913 that the frequencies of the characteristic x-ray lines varied in a regular manner with the atomic number Z of the target atom. He found that if the square root of the frequency $v_{K\alpha}$ of the K_α-radiation was plotted as a function of Z, then a straight line graph was obtained, showing a linear relationship $\sqrt{v_{K\alpha}} \propto Z$. This is known as Moseley's law. Similarly the graph of $\sqrt{v_{k\beta}}$, the square root of the frequency of the K_β-radiation, against Z is found to be a straight line, whose slope is different from that of the previous graph. In Fig. 8·10. such Moseley diagrams are illustrated for the K_α and K_β radiations.

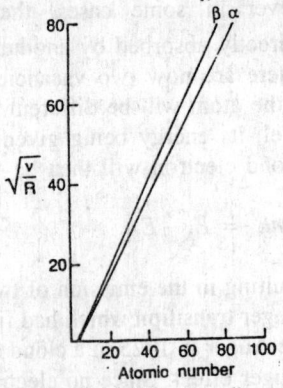

Fig. 8·10. Moseley diagram.

---

* Moseley was killed in action during the first world-war.

These straight line graphs can be represented by an equation of the type

$$\sqrt{v} = C_1 (Z - a) \qquad (8\cdot11\text{-}1)$$

so that

$$v = C (Z - a)^2 \qquad (8\cdot11\text{-}2)$$

C_1, C and a are three constants for a particular radiation. For K_α radiation, it is found that $C = \frac{3}{4} Rc$ where R is the Rydberg constant and c is the velocity of light in vacuum. $a \approx 1$ for K_α radiation so that we can write

$$v_{K\alpha} = \frac{3}{4} Rc (Z-1)^2 = Rc (Z-1)^2 \left(\frac{1}{1^2} - \frac{1}{2^2} \right) \qquad (8\cdot11\text{-}3)$$

Eq. (8·11-3) is similar to the expression for the first line of the Lyman series (see § 4·5) of hydrogen, the only difference being the presence of the factor $(Z-1)^2$ in the present case in place of $Z^2 = 1$ in the case of the hydrogen atom.

We have seen earlier that Mendeleev had arranged the elements in the periodic table in accordance with their increasing atomic weight (Ch VI). However, some anomalies were noticed in this arrangement. For example, according to Mendeleev, the three elements iron, cobalt and nickel in the first transition group should appear in the order Fe, Ni and Co, *i.e.*, nickel would come before cobalt if arranged in order of the increasing atomic weight. However, the chemical properties of these elements suggest that nickel should come *after* cobalt and not before it. Moseley diagram for the K_α- lines shows that the frequency of K_α- line for cobalt is lower than that of nickel which means that according to Eq. (8·11-3), Z for cobalt is less than that of nickel. Hence according to Moseley diagram cobalt should come before nickel and not after it.

Before Moseley, the real significance of the atomic number of the elements had not been realized. His work showed that the elements in the periodic table should be arranged according to their increasing serial numbers in the Moseley diagram instead of their increasing atomic weights. This serial number Z in the Moseley diagram is the atomic number of the element.

Moseley found that some elements seem to be missing in the Moseley diagram, *e.g.*, the element with $Z = 43$. These elements were discovered later.

Moseley equation (8·11-3) can be explained from the theory of the origin of the x-ray line. We have seen that the K_α-lines originate due to transition of an electron from the L-shell ($n = 2$) to the K-shell ($n = 1$). According to the Bohr theory, the transition from $n = 2$ to $n = 1$ level of a hydrogen-like atom gives rise to the emission of radiation of frequency

$$\nu = RcZ^2 \left(\frac{1}{1^2} - \frac{1}{2^2}\right)$$

In this case there is only one orbital electron on which the Coulomb attractive force due to the entire nuclear charge $+Ze$ acts. The factor Z^2 in the above expression appears due to this reason. In the case of the emission of the K_α x-rays, one out of the two K-electrons is first emitted from the atom leaving a single electron in the K-shell. The negative charge of this electron partially screens the nuclear charge $+Ze$ of the atom so that the effective Coulomb force acting on the L-electron is that due approximately to a net positive charge $+(Z-1)e$. Since the emission of the K_α radiation is due to the transition of an L-electron to the K-shell, the appearance of the factor $(Z-1)^2$ in Eq. (8.9-3) can thus be understood.

It should be noted that the K_α, K_β etc., lines have doublet structure. This was not known at the time of Moseley's discovery. So in the original Moseley diagrams, the frequencies plotted were the mean frequencies of the two components of the doublet. More precise measurements in later years have shown the Moseley diagram for the K_α-lines actually consists of two closely lying straight lines with slightly different slopes.

As in the case of the K-radiation, the X-ray spectral lines of the L-radiation can be represented by a formula of the type (8.11-2) in which $C = 5/36$ and $a = 7.4$ for L_β lines, so that we can write

$$\nu_{L\beta} = R c (Z - 7.4)^2 \left(\frac{1}{2^2} - \frac{1}{3^2}\right) \qquad (8.11\text{-}4)$$

The above formula shows that the L_β radiation originates due to the transition of an M electron ($n = 3$) to the L orbit ($n = 2$).

8.12 Absorption of x-rays

When a parallel beam of x-rays is allowed to pass through a slab of some material, the intensity of the beam emerging from the other side of the slab is found to decrease. If a monochromatic beam of x-rays of initial intensity I falls normally on a thin slab of some material of thickness Δx, as shown in Fig. 8.11, the diminution in the intensity ΔI is found to be proportional both to I and to Δx so that we can write

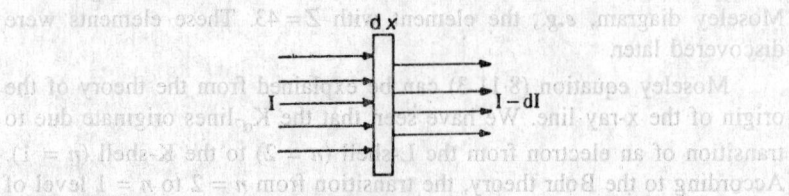

Fig. 8.11 Absorption of x-rays.

$$\Delta I \propto I \Delta x$$
or, $\quad\quad\quad\quad \Delta I = -\mu I \Delta x \quad\quad\quad\quad$ (8·12-1)

Here μ is a contant known as the *absorption coefficient* of the material. The negative sign on the r.h.s. of Eq. (8·12-1) shows that the intensity decreases as x increases.. In the limit of an infinitesimally small thickness dx, the above equation reduces to

$$\frac{dI}{I} = -\mu \, dx \quad\quad\quad (8 \cdot 12\text{-}2)$$

Integrating Eq. (8·12-2) we get

$$I = I_0 \exp(-\mu x) \quad\quad\quad (8\cdot 12\text{-}3)$$

Here I_0 is the intensity of the incident beam and I is the intensity after the beam has passed through a finite thickness x of the substance.

If the thickness x is measured in the unit of length then μ is called the *linear absorption coefficient*. If the intensity is reduced to half its initial value for a thickness $x_{1/2}$ of the substance, then we get

$$I/I_0 = \frac{1}{2} = \exp(-\mu x_{1/2})$$

which gives $\quad\quad \mu = \dfrac{\ln 2}{x_{1/2}} = \dfrac{0\cdot 693}{x_{1/2}} \quad\quad\quad (8\cdot 12\text{-}4)$

$x_{1/2}$ is known as the half-value thickness of the substance. μ can be determined by measuring $x_{1/2}$. Obviously μ has the dimension of the reciprocal of $x_{1/2}$. If the latter is measured in metres μ has the demension m^{-1}. Eq. (8·12-3) gives

$$\ln I = \ln I_0 - \mu x$$

The plot of ($\ln I$) against x is a straight line of slope μ (Fig. 8·12)

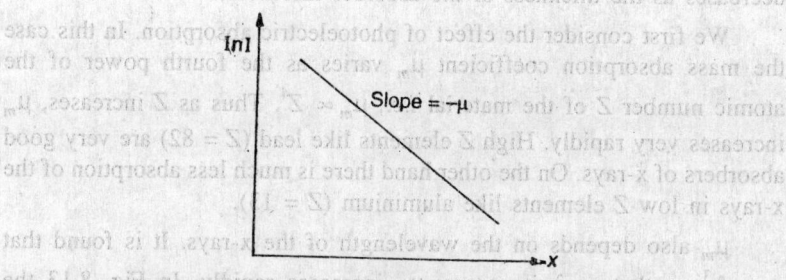

Fig. 8·12. Log plot of x-ray intensity vs. absorber thickness.

From a measurement of this slope also μ can be determined. Since substances in different physical states (*e.g.*, and solid, liquid and gas) have widely different densities the linear absorption coefficients may vary over wide limits due to the density factor. For the comparison of the absorbing powers of different substances it is therefore more appropriate to use the concept of mass absorption coefficient. Eq. (8·12-2) can be written as

$$\frac{dI}{I} = -\frac{\mu}{\rho}\rho\,dx = -\frac{\mu}{\rho}\,dm = -\mu_m dm \qquad (8\cdot12\text{-}5)$$

where ρ is the density of the absorber so that $dm = \rho dx$ is the mass of a thin slab of the absorber of unit cross sectional area and thickness dx. Thus in the above equation, the thickness of the absorber is expressed in terms of its mass per unit area (kg/m^2) instead of the unit of length. If $dm = 1$, then

$$dI/I = -\mu_m$$

Here $\mu_m = \mu/\rho$ is called the *mass absorption coefficient* of the absorbing substance. Integrating Eq. (8·12-5), we get

$$I = I_0 \exp(-\mu_m m) \qquad (8\cdot12\text{-}6)$$

where $m = \rho x$ is the mass per unit area of an absorber of finite thickness x.

The diminution in the intensity of a collimated beam of x-rays while passing through matter may be due to several reasons : (*a*) Firstly there may be photo-electric absorption of the x-ray photons by the electrons in the inner shells (*e.g.*, K, L, M, *etc.*) of the atoms which are thereby ejected. This is usually accompanied by the emission of characteristic fluorescent radiation due to the transitions of electrons from the outer shells. The net effect is the gradual diminution of the energy of the beam as it travels through the absorber. (*b*) Secondly there may be scattering of the x-rays by the atomic electrons. (*c*) At higher energies, there may be photo-nuclear and other effects due to which the energy of the x-ray beam decreases as the thickness of the absorber increases.

We first consider the effect of photoelectric absorption. In this case the mass absorption coefficient μ_m varies as the fourth power of the atomic number Z of the material *i.e.*, $\mu_m \propto Z^4$. Thus as Z increases, μ_m increases very rapidly. High Z elements like lead (Z = 82) are very good absorbers of x-rays. On the other hand there is much less absorption of the x-rays in low Z elements like aluminium (Z = 13).

μ_m also depends on the wavelength of the x-rays. It is found that $\mu_m \propto \lambda^3$ so that as λ increases, μ_m increases rapidly. In Fig. 8·13 the variation of μ_m with λ is shown graphically. As λ increases, μ_m at first increases continuously. Then at a definite wavelength $\lambda = \lambda_K$, there is a sudden decrease in the value of μ_m. After this discontinuity, μ_m again rises continuously as λ increases till there is again a sudden decrease in its value at another definite wavelength $\lambda = \lambda_{LI}$. Immediately after this there are two further discontinuities at two closely lying wavelengths λ_{LII} and λ_{LIII}.

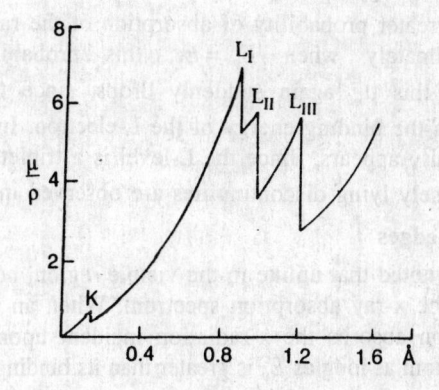

Fig. 8·13 Variation of μ_m with λ.

The wavelength λ_K at which the discontinuity appears is known as the K-*absorption edge*. The wavelengths λ_{LI}, λ_{LII} and λ_{LIII} at which the next three closely lying discontinuities appear are known as L_I, L_{II} and L_{III} absorption edges respectively. The wavelengths for the absorption edges depend on the nature of the absorber. The K-absorption edge has a shorter wavelength than the L-absorption edges. The wavelength λ_K for the K-absorption edge of a given absorber is slightly less than the minimum wavelength of the characteristic x-rays emitted by the absorber. As an example, in the case of copper (Z = 29) K_α lines, $\lambda(K_{\alpha 2}) = 1\cdot 5443$ Å, $\lambda(K_{\alpha 1}) = 1\cdot 5405$ Å ; for the $K_{\beta 1}$ and K_γ lines, we have $\lambda(K_{\beta 1}) = 1\cdot 3922$ Å and $\lambda(K_\gamma) = 1\cdot 3810$ Å and so on. On the other hand, the wavelength for the K-absorption edge of copper is found to be $1\cdot 3802$ Å which is shorter than the wavelengths of all the charactertstic K x-ray lines of copper.

The appearance of the discontinuities in Fig. 8·13 can be understood in the following manner. As long as the energy E_ν of the incident x-radiation is greater than the binding energy E_K of the electron in the K-shell of the absorber atom, the K-electron will be emitted by absorbing the incident x-ray photon. Its kinetic energy will be $mv^2/2 = E_V - E_K$. The process is actually an instance of photoelectric emission from the inner shell (K-shell) of the atom. As the wavelength λ of the incident radiation increases, i.e., as E_ν decreases, μ_m increases continuously till $E_\nu = E_K$. When E_ν becomes less than E_K, the incident radiation is no longer able to eject the K-electron by photoelectric process. So μ_m suddenly decreases and a discontinuity appears at $E_\nu = E_K$.

For $E_\nu < E_K$, the incident radiation can still eject an L-electron for which the binding energy $E_L < E_K$. So as λ is further increased, μ_m again rises due to the greater probability of absorption of the radiation by the L-electrons. Ultimately when $E_\nu = E_L$, this probability becomes maximum. After this μ_m again suddenly drops, since the energy E_ν becomes less than the binding energy of the L-electron. In this way, the second discontinuity appears. Since the L-level is a triplet (L_I, L_{II}, L_{III}), actually three closely lying discontinuities are observed in this case.

8·13 Absorption edges

It should be noted that unlike in the visible region, no discrete lines are observed in the x-ray absorption spectrum. When an inner electron, like the K-electron absorbs the x-radiation incident upon it, it will be ejected from the atom as long as E_ν is greater than its binding energy. There cannot be any transition of this electron to an outer shell, since outer shells are all filled with electrons. So incident radiation of all energies E_ν greater than the binding energy of the electron will be absorbed, which means that x-rays of all wavelengths less than a definite limiting value will be absorbed. The absorption spectrum therefore shows a continuum with a sharp cut off on the long wavelength side (see Fig. 8·14). This cut off corresponds to an *absorption edge*.

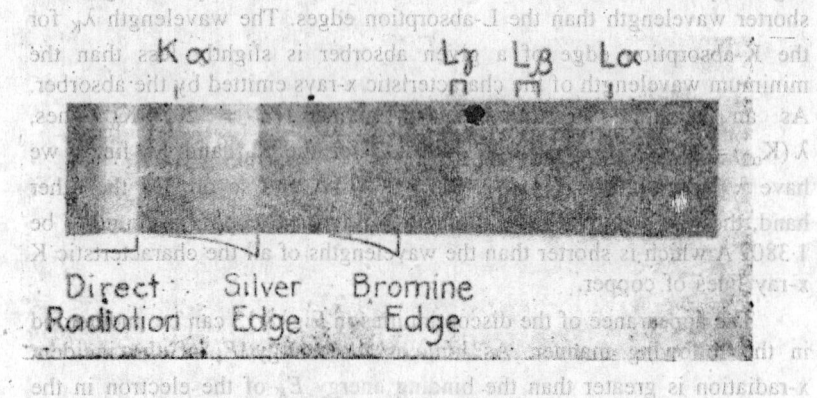

Fig. 8·14 X-ray absorption edges. (Courtesy : Introduction to Atomic Spectra by H.E. White, published by McGraw-Hill Book Inc. (New York).

Since the energies of the K, L, M etc. absorption edges are equal to the binding energies E_K, E_L, E_M etc. of the electrons in the corresponding shells, it is possible to obtain the energy differences of the x-ray levels by measuring the energy differences between the different absorption edges. Actually this method is utilized in measuring the energies of the levels involved in x-ray emissions.

8·14 Scattering of x-rays : Thomson's theory

We have seen that when an x-ray beam passes through some material, the radiation interacts with the atoms of the material mainly in two different ways, viz. by photo-electric absorption and by scattering. The scattering process can be explained on the basis of the e.m. theory of light by assuming that the atomic electrons are set into forced vibrations by the action of the electric vector of the incident electromagnetic wave. Due to this they emit radiation in different directions. This occurs at the expense of the energy of the incident beam which is thus reduced increasingly as the beam passes through greater thicknesses of the material. The atomic electrons thus act as *scattering centres*. The intensity of a collimated beam of x-rays in the direction of incidence on the material is reduced both due to absorption and scattering processes (see Eq. 8·12–1). The reduction in intensity is usually measured in terms of the extinction coefficient τ which is a sum of two terms :

$$\tau = \mu + \sigma$$

Here μ is the absorption coefficient mentioned earlier while σ is known as the *scattering coefficient*. We can also write the *mass extinction coefficient* by dividing τ by the density ρ of the material :

$$\tau_m = \frac{\tau}{\rho} = \mu_m + \sigma_m$$

where $\mu_m = \frac{\mu}{\rho}$ is the mass absorption coefficient and $\sigma_m = \frac{\sigma}{\rho}$ is the mass scattering coefficient.

The scattering process actually takes place from the individual atoms. If M be the atomic weight of the scatterer, then the number of atoms per unit mass of the scatterer is $\frac{N_0}{M}$ where N_0 is the Avogadro number. Hence the number density of the atoms is $n_a = \frac{N_0 \rho}{M}$. Let us consider the passage of the x-rays through a slab of the scatterer of thickness dx and unit cross sectional area. Then the probability of the x-ray beam being scattered by the atoms within this slab is proportional to the number of atoms in the slab which is equal to $n_a\, dx = N_0\, \rho\, \frac{dx}{M}$. So the reduction in the intensity I of the incident beam can be written as

$$dI \propto \frac{N_0 \rho}{M} I\, dx$$

or

$$\frac{dI}{I} = -\sigma_a \frac{N_0 \rho}{M} dx \qquad (8 \cdot 14\text{-}1)$$

Here σ_a is a constant. We can also write in analogy with Eq. (8·12-2)

$$\frac{dI}{I} = -\sigma \, dx \tag{8·14-2}$$

where σ is the *linear scattering coefficient per atom*. Hence we have

$$\sigma = \sigma_a \frac{N_0 \rho}{M} = \sigma_a n_a$$

so that the *atomic scattering coefficient* is given by

$$\sigma_a = \frac{\sigma M}{N_0 \rho} = \frac{\sigma}{n_a} \tag{8·14-3}$$

Since σ has the dimension of the reciprocal of length and n_a has the dimension of the reciprocal of volume, σ_a has the dimension of an area.

If E be the electric field of the incident e.m. wave, then the force acting on an electron of charge e and mass m is Ee. So the acceleration of the electron is

$$f = \frac{eE}{m} \tag{8·14-4}$$

The electron is set into forced vibration with the frequency of the incident radiation. According to the e.m. theory of light such an electron emits electromagnetic waves of the same frequency. This radiation is actually the scattered radiation and is emitted in different directions from the scattering centre. If the velocity of the electron oscillating under the influence of the incident electric vector is small compared to the velocity of light c, then it can be proved on the basis of the e.m. theory of light that the electric field intensity of the scattered wave at a distance r from the scattering centre in a direction making an angle β with respect to the incident electric vector E is given by (see Fig. 8·15)*

$$E'_\beta = \frac{ef \sin \beta}{4\pi \epsilon_0 c^2 r} = \frac{Ee^2 \sin \beta}{4\pi \epsilon_0 mc^2 r} \tag{8·14-5}$$

where $\epsilon_0 = 10^{-9}/36\pi$ F/m is the permittivity of empty space.

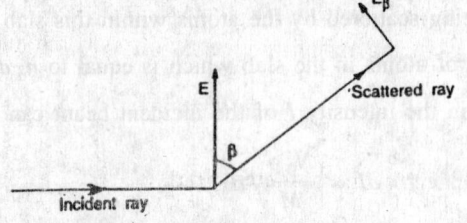

Fig. 8·15. Scattering of x-rays.

---

* see *ibid*.

According to Poynting's theorem, the intensity of the incident beam is

$$I = \epsilon_0 c E^2 \tag{8.14-6}$$

Similarly the intensity of the scattered beam is

$$I'_\beta = \epsilon_0 c E'^2_\beta = \frac{\epsilon_0 c E^2 e^4 \sin^2 \beta}{16\pi^2 \epsilon_0^2 m^2 c^4 r^2}$$

$$= \frac{e^4 E^2}{16\pi^2 m^2 c^3 \epsilon_0} \cdot \frac{\sin^2 \beta}{r^2} \tag{8.14-7}$$

Hence we have

$$\frac{I'_\beta}{I} = \frac{e^4 \sin^2 \beta}{16\pi^2 \epsilon_0^2 m^2 c^4 r^2} \tag{8.14-8}$$

This gives the intensity of the scattered beam in terms of the incident intensity in a direction β with respect to the direction of the incident electric vector. Actually the incident radiation is unpolarized. So we have to take the average of $\sin^2 \beta$ in Eq. (8.14-8).

To do this we consider an incident beam along the x-axis (Fig. 8.16) so that the electric vector E is in the y–z plane. We resolve E into the two components E_y and E_z. If the scattering centre is at O and the direction of the scattered beam OP which is assumed to be in the x-y plane makes an angle θ with the direction of the incident beam, then the electric field intensities E'_1 and E'_2 in the scattered beam associated with the two components E_y and E_z of the incident electric vector E will be

$$E'_1 = \frac{E_y e^2}{4\pi\epsilon_0 mc^2 r} \sin\left(\frac{\pi}{2} - \theta\right) = \frac{E_y e^2 \cos\theta}{4\pi\epsilon_0 mc^2 r} \tag{8.14-9}$$

$$E'_2 = \frac{E_z e^2}{4\pi\epsilon_0 mc^2 r} \tag{8.14-10}$$

In the above expression, we have made use of the fact that the direction of the scattered ray OP makes an angle $\beta = (\pi/2 - \theta)$ with the y-axis while it is perpendicular to the z-axis. E'_1, E'_2 are mutually perpendicular. So the resultant electric intensity of the scattered wave E'_θ is given by

$$E'^2_\theta = E'^2_1 + E'^2_2$$

The intensity of the scattered beam is then

$$I'_\theta = \epsilon_0 c\, E'^2_\theta = \epsilon_0 c \left(\frac{e^2}{4\pi\epsilon_0 mc^2 r}\right)^2 (E_y^2 \cos^2\theta + E_z^2) \tag{8.14-11}$$

Its time-averaged value is given by

$$\overline{I'_\theta} = \frac{e^4}{16\pi^2 \epsilon_0 m^2 c^3 r^2} \left(\overline{E_y^2} \cos^2\theta + \overline{E_z^2} \right) \quad (8\cdot14\text{-}12)$$

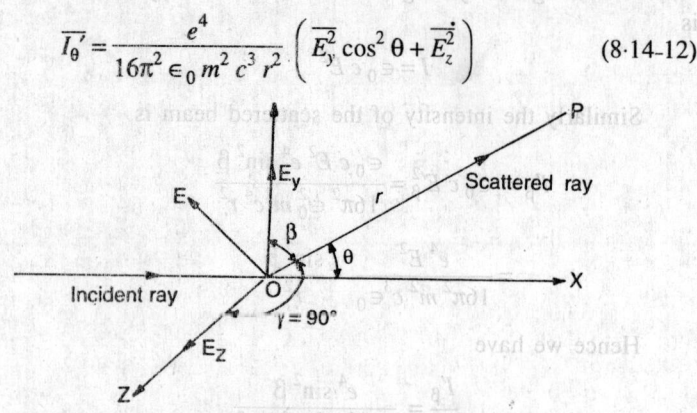

Fig. 8·16. Resolution of the electric vector of the incident beam in x-ray scattering.

The time-averaged value of the incident intensity is given by

$$\overline{I} = \epsilon_0 c\, \overline{E^2} = \epsilon_0 c\, (\overline{E_y^2} + \overline{E_z^2}) \quad (8\cdot14\text{-}13)$$

Because of the random orientation of the electric vector in the incident unpolarized beam, we have

$$\overline{E_y^2} = \overline{E_z^2} = \frac{1}{2\epsilon_0 c} \overline{I} \quad (8\cdot14\text{-}14)$$

Hence we get $\quad \overline{I'_\theta} = \dfrac{e^4}{16\pi^2 \epsilon_0 m^2 c^3 r^2} \cdot \dfrac{\overline{I}}{2\epsilon_0 c} (1 + \cos^2\theta)$

Omitting the bars, we then have

$$I'_\theta = I \cdot \frac{e^4}{32\pi^2 \epsilon_0^2 m^2 c^4 r^2} (1 + \cos^2\theta) \quad (8\cdot14\text{-}15)$$

This gives the scattered intensity due to a single electron. If there are n_e electrons per unit volume of the scatterer and if it is assumed that the scattering from the different electrons takes place independently of one another (incoherent scattering), then the resultant intensity of the scattered beam for a scatterer of thickness Δx and unit cross section is given by

$$I_S = I \cdot \frac{n_e e^4 \Delta x}{32\pi^2 \epsilon_0^2 m^2 c^4 r^2} (1 + \cos^2\theta) \quad (8\cdot14\text{-}16)$$

It should be noted that in the above derivation it has been assumed that the electrons are completely free ; *i.e.*, the binding of the electrons in the atoms has been neglected.

Eq. (8·14-16) shows that the intensity of the scattered radiation is different in different directions. It is minimum for $\theta = \pi/2$, i.e., in a direction perpendicular to the incident direction. Further, in this case, the scattered radiation is plane polarized with the electric vector along the z-axis. We can integrate Eq. (8·12-16) by assuming r to be large compared to the dimensions of the scatterer. In this case, if we assume the scatterer to be concentrated at the origin of the co-ordinate system, and draw a sphere of radius r about it then the area of the ring shaped strip on this sphere between the two cones of semi-vertical angles θ and $\theta + d\theta$ (see Fig. 8·17) is given by

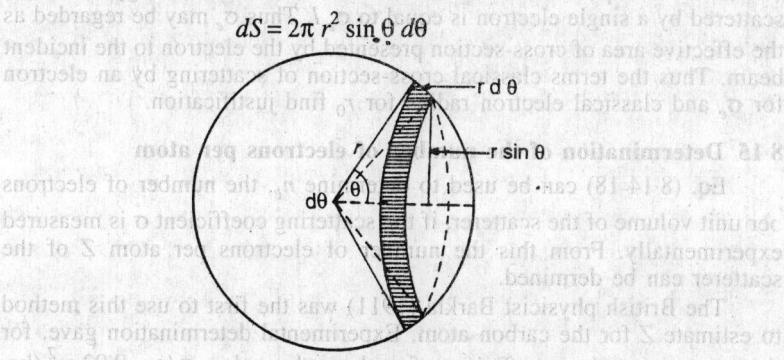

$$dS = 2\pi\, r^2 \sin\theta\, d\theta$$

Fig. 8·17. Calculation of solid angle.

Hence the total scattered energy is

$$W_S = \int I_S\, dS = I\, \frac{n_e\, e^4\, \Delta x}{32\, \pi^2\, r^2\, m^2\, c^4\, \epsilon_0^2} \cdot 2\pi\, r^2 \int_0^\pi (1+\cos^2\theta)\sin\theta\, d\theta$$

$$= \frac{8\pi}{3} n_e \left(\frac{e^2}{4\pi\epsilon_0\, mc^2}\right)^2 I\, \Delta x \qquad (8\cdot14\text{-}17)$$

Then from Eq. (18·14-2) we get (neglecting the negative sign)

$$\sigma = \frac{1}{I}\frac{\Delta I}{\Delta x} = \frac{W_s}{I\,\Delta x} = \frac{8\pi}{3} n_e \left(\frac{e^2}{4\pi\epsilon_0\, mc^2}\right)^2 \qquad (8\cdot14\text{-}18)$$

σ is the scattering coefficient. Eq. (18·14-18) is known as Thomson scattering formula, first deduced by J.J. Thomson. This type of scattering is known as Thomson scattering. Since there is no change of wavelength associated with this type of scattering, it is sometimes called *unmodified scattering*.

Eq. (8·14-18) gives the *classical scattering cross-section* for a free electron :

$$\sigma_e = \frac{\sigma}{n_e} = \frac{8\pi}{3} r_0^2 \rightarrow 6\cdot63 \times 10^{-29}\, \text{m}^2 \qquad (8\cdot14\text{-}19)$$

where
$$r_0 = \frac{e^2}{4\pi \epsilon_0 mc^2} = 2.81 \times 10^{-15} \text{ m} \qquad (8.14\text{-}20)$$

r_0 is known as the *classical electron radius*. Both σ_e and r_0 are universal constants, depending only on the three fundamental constants e, m and c. Since σ has the dimension of the reciprocal of length and n_e (electron number density) has the dimension of the reciprocal of volume, σ_e has the dimension of an area and r_0 the dimension of a length.

If a collimated beam of radiation of intensity I is incident on a surface placed normal to the beam, then the energy falling on a unit area of the surface per second is I. The fraction of this energy which is scattered by a single electron is equal to $\sigma_e I$. Thus σ_e may be regarded as the effective area of cross-section presented by the electron to the incident beam. Thus the terms classical cross-section of scattering by an electron for σ_e and classical electron radius for r_0 find justification.

8.15 Determination of the number of electrons per atom

Eq. (8.14-18) can be used to determine n_e, the number of electrons per unit volume of the scatterer, if the scattering coefficient σ is measured experimentally. From this the number of electrons per atom Z of the scatterer can be dermined.

The British physicist Barkla (1911) was the first to use this method to estimate Z for the carbon atom. Experimental determination gave, for the *mass scattering coefficient* of carbon, the value $\sigma/\rho = 0.02$ m²/kg. Here σ is expressed in m^{-1} and ρ in kg/m³.

From Eq. (8.14-19) we get

$$n_e = \frac{3\sigma}{8\pi r_0^2} = \frac{\sigma}{6.63 \times 10^{-29}} = \frac{0.02 \rho}{6.63 \times 10^{-29}} = 3 \times 10^{26} \rho$$

The number of carbon atoms per unit volume is given by

$$n_a = \frac{N_0 \rho}{12} = \frac{6.025 \times 10^{23} \rho \times 10^3}{12} = 5.02 \times 10^{25} \rho$$

So the number of electrons per carbon atom is

$$Z = \frac{n_e}{n_a} = \frac{3 \times 10^{26}}{5.02 \times 10^{25}} \approx 6$$

This is equal to the serial number (*i.e.*, the atomic number) of carbon in the periodic table of elements. So it is obvious the atomic number of carbon is equal to the number of electrons in the carbon atom. This conclusion has been proved to be true for other elements also. Experiments on the scattering of the α-particles from radioactive substances by James Chadwick in England has shown that the positive charge carried by the nucleus of an atom is also equal to the atomic number of the element (see Vol II). Hence we may conclude that the nuclear charge of an atom is equal to the total negative charge of the electrons in the atom.

8·16 Compton scattering

In 1924, the famous American scientist A.H. Compton, while experimenting on the scattering of monochromatic x-rays from carbon, observed that the radiation scattered in a particular direction consisted of x-rays of two different wavelengths. One of these had the same wavelength as the incident radiation, while the wavelength of the other was somewhat longer, so that its frequency was lower (see Fig. 8·20). The first of these was the same as the unmodified scattering discussed in the previous section. As we have seen this scattering can be accounted for by the classical e.m. theory. The other scattering in which the frequency of this radiation changes is a type of inelastic scattering of the incident x-ray photon and is known as *Compton scattering* and the effect known as *Compton effect*.

The theory of the effect was worked out by Compton assuming photon character of radiation. According to him, a photon not only carries a quantity of energy $h\nu$, but it also has a certain amount of momentum like a material particle. According to Maxwell's theory, electromagnetic radiation has a momentum which is in the direction in which the energy of the radiation flows, *i.e.*, in the direction of the Poynting vector N and it has a magnitude N/c^2 where c is the velocity of light. Similar result can be deduced from the special theory of relativity according to which the total energy of a particle of rest mass m_0 moving with a velocity v is given by (see Eq. 15·13-3)

$$E = mc^2 = m_0 c^2 / \sqrt{1 - v^2/c^2}$$

Here $m = m_0/\sqrt{1 - v^2/c^2}$ is the mass of the moving particle.

From the above equation this is given by

$$m = E/c^2$$

The photon as conceived by Einstein (see Ch. V), has no rest mass, but it has an energy $E = h\nu$ and it moves with the velocity c. If we assume that the above relation between the mass of a moving particle and its total energy is applicable in the case of the photon, then we may assign the mass $m = h\nu/c^2$ to the moving photon so that its momentum will be $mc = h\nu/c$.

Now consider the collision of a photon of energy $h\nu$ with an electron of rest mass m_0 which is initially at rest. As a result of this collision, the electron gets some kinetic energy and momentum at the expense of the corresponding quantities of the photon which are thereby reduced. Suppose the electron moves with a velocity v in a direction making an angle θ with the direction of the incident photon which is scattered in the direction ϕ w.r.t. to the incident direction (see Fig. (8·18). If the frequency of the scattered photon is ν', then its energy is $h\nu'$. Obviously $\nu' < \nu$, since the photon loses some energy by collision. The momenta of the incident and scattered photons are $h\nu/c$ and $h\nu'/c$ respectively. The kinetic energy gained by the electron is $E_k = (mc^2 - m_0 c^2)$ while its momentum is $mv = m\beta c$ where $\beta = v/c$ and m is the mass of the electron when it is moving with the velocity v.

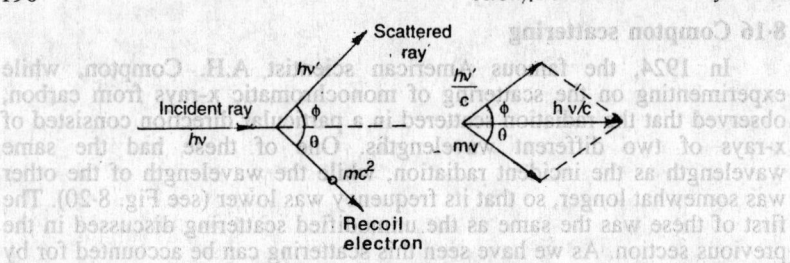

Fig. 8.18 Compton scattering

We now apply the laws of conservation of energy and momentum to the above collision process. From the law of conservation of energy we have

$$h\nu = h\nu' + E_k = h\nu' + \frac{m_0 c^2}{\sqrt{1-\beta^2}} - m_0 c^2 \qquad (8\cdot16\text{-}1)$$

Applying the law of conservation of momentum along and prependicular to the direction of motion of the incident photon, (see inset of Fig. 8·18) we get

$$\frac{h\nu}{c} = \frac{h\nu'}{c}\cos\phi + \frac{m_0 \beta c}{\sqrt{1-\beta^2}}\cos\theta \qquad (8\cdot16\text{-}2)$$

$$0 = \frac{h\nu'}{c}\sin\phi - \frac{m_0 \beta c}{\sqrt{1-\beta^2}}\sin\theta \qquad (8\cdot16\text{-}3)$$

The above two equations can be rewritten as

$$\frac{m_0 \beta c}{\sqrt{1-\beta^2}}\cos\theta = \frac{h\nu}{c} - \frac{h\nu'}{c}\cos\phi \qquad (8\cdot16\text{-}2a)$$

$$\frac{m_0 \beta c}{\sqrt{1-\beta^2}}\sin\theta = \frac{h\nu'}{c}\sin\phi \qquad (8\cdot16\text{-}3a)$$

If we square and add the above two equations, we get

$$\frac{m_0^2 \beta^2 c^2}{1-\beta^2} = \frac{h^2 \nu^2}{c^2} + \frac{h^2 \nu'^2}{c^2} - \frac{2h^2 \nu\nu'}{c^2}\cos\phi \qquad (8\cdot16\text{-}4)$$

Rearranging the terms in Eq. (8·16-1) and squaring we get

$$\left(\frac{m_0 c^2}{\sqrt{1-\beta^2}}\right)^2 = (h\nu - h\nu' + m_0 c^2)^2$$

$$= h^2 \nu^2 + h^2 \nu'^2 - 2h^2 \nu\nu' + m_0^2 c^4 + 2hm_0 c^2 (\nu - \nu')$$

Dividing by c^2 and rearranging we get

$$\frac{m_0^2 c^2}{1-\beta^2} - m_0^2 c^2 = \frac{m_0^2 \beta^2 c^2}{1-\beta^2}$$

$$= \frac{h^2 v^2}{c^2} + \frac{h^2 v'^2}{c^2} - \frac{2h^2 vv'}{c^2} + 2m_0 h (v - v') \quad (8\cdot16\text{-}5)$$

From Eqs. (8·16-4) and (8·16-5) we get after subtraction

$$0 = 2m_0 h (v - v') - \frac{2h^2 vv'}{c^2} (1 - \cos \phi)$$

which gives

$$\frac{v - v'}{vv'} = \frac{h}{m_0 c^2} (1 - \cos \phi)$$

or

$$\frac{c}{v'} - \frac{c}{v} = \frac{h}{m_0 c} (1 - \cos \phi) \quad (8\cdot16\text{-}6)$$

But the wavelength of the incident radiation is $\lambda = c/v$ while that of the scattered radiation is $\lambda' = c/v'$. If we define Compton wavelength of the electron as

$$\lambda_c = \frac{h}{m_0 c} \quad (8\cdot16\text{-}7)$$

then we get from Eq. (8·16-6) the change in wavelength due to Compton scattering at the angle ϕ

$$\Delta \lambda = \lambda' - \lambda = \lambda_c (1 - \cos \phi) = 2 \lambda_c \sin^2 \frac{\phi}{2} \quad (8\cdot16\text{-}8)$$

Eq. (8·16-8) shows that $\Delta \lambda$ depends only on the angle of scattering ϕ; it is independent of the wavelength of the incident radiation and of the scattering material.

The Compton wavelength λ_c is a universal constant depending on the values of h, m_0 and c. Substituting for these, we get

$$\lambda_c = 24\cdot2 \times 10^{-13} \text{ m} = 0\cdot024 \text{ Å}$$

For scattering at 90°, we get

$$\Delta \lambda = 2\lambda_c \sin^2 45° = \lambda_c = 2\cdot42 \times 10^{-2} \text{ Å}$$

The above results were verified by Compton using x-rays of different wavelengths (see below). Compton effect has also been observed for γ-rays. (see Vol II).

The energy of the recoil electron E_k during Compton scattering can easily be calculated with the help of Eqs. (8·16-2a) and (8·16-3a). Taking the ratio of these two equations, we get

$$\tan \theta = \frac{v' \sin \phi}{v - v' \cos \phi} = \frac{\sin \phi}{v/v' - \cos \phi} \quad (8\cdot16\text{-}9)$$

Writing $\alpha = hv/m_0 c^2$, we have from Eq. (8·16-6)

$$\frac{1}{v'} = \frac{1}{v} + \frac{h}{m_0 c^2}(1-\cos\phi) = \frac{1}{v}\left(1 + 2\alpha \sin^2\frac{\phi}{2}\right)$$

which gives
$$v' = \frac{v}{1 + 2\alpha \sin^2\frac{\phi}{2}} \qquad (8\cdot16\text{-}10)$$

Substituting in Eq. (8·16-9), we have

$$\tan\theta = \frac{2\sin\frac{\phi}{2}\cos\frac{\phi}{2}}{1 + 2\alpha \sin^2\frac{\phi}{2} - \cos\phi} = \frac{2\sin\frac{\phi}{2}\cos\frac{\phi}{2}}{2\sin^2\frac{\phi}{2} + 2\alpha \sin^2\frac{\phi}{2}}$$

or, $\qquad \cot\theta = (1+\alpha)\tan\dfrac{\phi}{2} \qquad (8\cdot16\text{-}11)$

This gives the relation between the angle of scattering of the photon (ϕ) and the angle of emission of the recoil electron (θ). The kinetic energy of the recoil electron is

$$E_k = hv - hv' = hv\left[1 - \frac{1}{1 + 2\alpha \sin^2\frac{\phi}{2}}\right]$$

$$= hv \frac{2\alpha \sin^2\frac{\phi}{2}}{1 + 2\alpha \sin^2\frac{\phi}{2}}$$

or, $\qquad E_k = hv \dfrac{\alpha(1-\cos\phi)}{1 + \alpha(1-\cos\phi)} \qquad (8\cdot16\text{-}12)$

When expressed in terms of the electron recoil angle θ, the above equation becomes

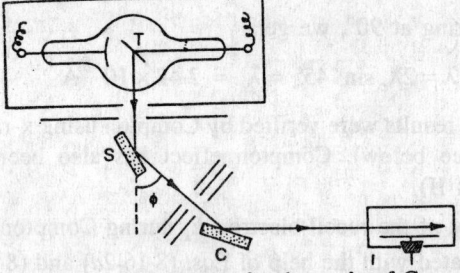

Fig. 8·19. Experimental arrangement for the study of Compton scattering.

$$E_k = hv \cdot \frac{2\alpha \cos^2\theta}{(1+\alpha)^2 - \alpha^2 \cos^2\theta} \qquad (8\cdot16\text{-}12a)$$

X-Rays

Experimental study of Compton effect

In Fig. 8·19 an experimental arrangement for the study of Compton effect is shown. T is a molybdenum target in an x-ray tube. The monochromatic K_α radiation from T, after scattering from a carbon scatterer S through a known angle passes through a number of slits to fall on the crystal C in a Bragg spectrometer. (see § 8·14). The x-rays diffracted by C enter the ionization chamber I which measures the intensity of the diffracted beam.

By measuring the angle of diffraction at which the intensity maximum is observed, it is possible to determine the wavelength of the x-rays scattered by S at a given angle from Bragg equation. (Eq. 8·17-1).

Compton observed two lines of two different wavelengths in the scattered radiation. At 90° scattering angle, one of these was found to have the wavelength of the molybdenum K_α x-rays (0·708 Å), while the other had the wavelength 0·731 Å (see Fig. 8·20) The wavelength difference 0·023 Å between the two agreed fairly well with the value of $\Delta \lambda$ expected from the Compton equation (8·16-8) for $\phi = 90°$ ($\Delta \lambda = 0.024$ Å). This wavelength difference has been determined using monochromatic x-rays of different wavelengths. In all cases it is found to have the value given above.

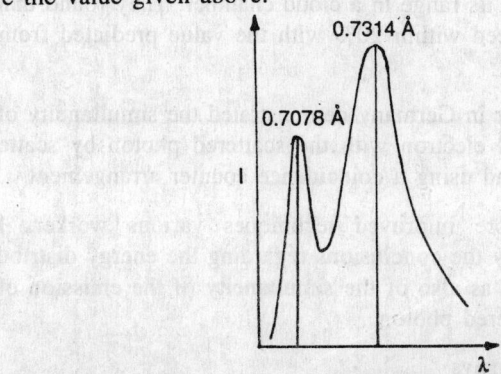

Fig. 8·20 Intensity *vs.* wavelength plot in Compton scattering.

Observations at other angles of scattering have also confirmed the theoretical predictions of Compton.

It should be noted that the appearance of the peak at the longer wavelength 0·731 Å in Fig. 8·20 is due to Compton scattering from the electron which may be considered free, since its energy of binding in the atom is small compared to the energy $h\nu$ of the photon. The other peak at the wavelength of the incident radiation is due to scattering from a bound electron. In this case the recoil momentum is taken up by the entire atom which being much heavier compared to the electron produces negligible wavelength shift and hence the scattered photon has the same energy and same wavelength as the incident photon.

Compton effect gives conclusive evidence in support of corpuscular character of electromagnetic radiation.

The energy distribution of the recoil electrons has been determined experimentally and the simultaneity of their emission with the scattered photons has also been establisted.

From Eq. (8·16-11) we see that as the angle of scattering ϕ varies between 0 and π, the angle of recoil electron emission θ varies from $\pi/2$ to 0 i.e. the electrons are always ejected at angles less than 90°. E_k is maximum for $\phi = \pi$ while $E_k = 0$ for $\phi = 0$, i.e., for $\theta = 0°$ and $\theta = 90°$ respectively (see Eq. 8·16-12).

The maximum energy of the recoil electron (for $\theta = 0°$) is (see Eq. 8·16-12a)

$$(E_k)_{max} = h\nu \frac{2\alpha}{1 + 2\alpha} \qquad (8 \cdot 16\text{-}13)$$

Compton and Simon (1925) determined the energy of the recoil electron by measuring its range in a cloud chamber. They found that the maximum energy agreed within 20% with the value predicted from the theory.

Bothe and Geiger in Germany demonstrated the simultaneity of the emission of the recoil electron with the scattered photon by scattering x-rays by hydrogen and using a coincidence counter arrangement.

Later, using more improved techniques various workers have verified more precisely the conclusions regarding the energy distribution of the recoil electrons as also of the simultaneity of the emission of the electron and the scattered photon.

8·17 Diffraction of x-rays

Shortly after the discovery of x-rays, it was realized that this new radiation was a kind of electromagnetic radiation with wavelength much shorter than that of visible radiation. We know that the wavelength of ordinary light can be measured with the help of a diffraction grating in which the grating elements must be comparable to the wavelength to be measured. In the initial stages, attempts were made to produce diffraction of x-rays using ordinary ruled gratings. Though it was not possible to make any accurate determination of the wavelength of the x-rays by this method, it was realized that these wavelengths were of the order of an angstrom.

X-Rays

The German physicist von Laue (1912) was the first to point out that diffraction of x-rays could be produced more easily with the help of crystals. Laue estimated the distance between the atoms in a crystal from a knowledge about the number of atoms per unit volume in the crystal. He showed that this was of the order of 10^{-9} to 10^{-10} m which is comparable the wavelength of the x-rays. From the observed symmetry in the structure of crystals it is evident that the entire crystal is made up by the regular repetitive arrangement of a basic unit or pattern of atomic dimensions. In other words, the atoms or molecules in a crystal are arranged in regular array, layer by layer, inside the crystal. The distance between the successive layers is comparable to the x-ray wavelength. This regularity in the arrangement of the atomic layers inside the crystal led Laue to conclude that crystals are most suitable for producing diffraction of x-rays.

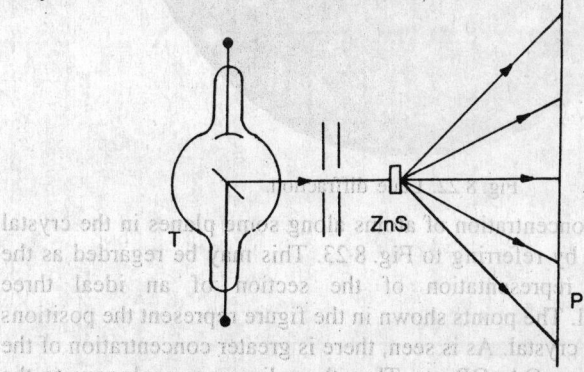

Fig. 8·21. Friedrich and Knipping's experiment.

At his suggestion, Friedrich and Knipping performed the following experiment to produce x-ray diffraction. Their experimental arrangement is shown in Fig. 8·21. A collimated beam of continuous x-rays from an x-ray tube was passed through a zinc sulphide crystal (ZnS) and the radiation coming out on the other side was allowed to fall on a photographic plate P. When the plate was developed, it was found that there was an intense black spot at the centre, *i.e.*, at the point where the incident beam hit the plate normally. Surrounding this central spot, there were a number of additional black spots arranged in a regular pattern. In Fig. 8·22, such a Laue diffraction pattern produced by a lithium fluoride (LiF) crystal is shown.

The famous British physicist, W.L. Bragg suggested a simple explanation for the production of Laue diffraction pattern. According to Bragg there are certain special planes known as *cleavage planes* within a crystal along which the atoms in the crystal are more densely concentrated. If the crystal is hit with a sharp pointed end, it splits along these cleavage planes.

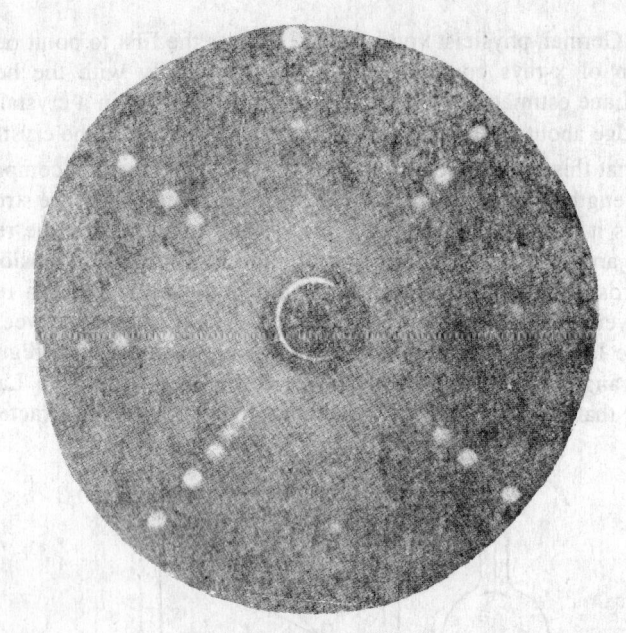

Fig. 8·22. Laue diffraction.

The greater concentration of atoms along some planes in the crystal can be understood by referring to Fig. 8·23. This may be regarded as the two dimensional representation of the section of an ideal three dimensional crystal. The points shown in the figure represent the positions of the atoms in the crystal. As is seen, there is greater concentration of the points along the lines OA, OB *etc*. Thus these lines are analogous to the cleavage planes in an actual crystal. According to Bragg, the incident x-rays are preferentially reflected from the cleavage planes which are rich in atoms. For this reason, the rays are reflected preferentially in certain directions giving rise to the production of the black spots on the photographic plate in these directions (see Ch. XVII for further discussion).

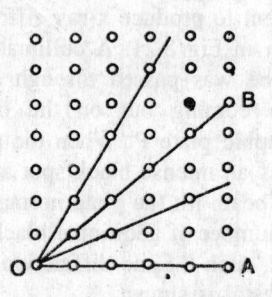

Fig. 8·23. Two dimensional lattice.

Bragg's experiment

To justify the above assumptions, Bragg performed an experiment in which he allowed a collimated beam of monochromatic x-rays to be incident on a cleavage plane of a crystal. He observed that a black spot was produced on a photographic plate in the expected direction of the reflected ray. Later he replaced the photographic plate by an ionization chamber which was placed on one of the arms of a spectrometer while the crystal was placed on the prism table (see Fig. 8·24). By this arrangement it was possible to turn the crystal C and the ionization chamber I independently through any desired angle. The x-rays emitted by the target

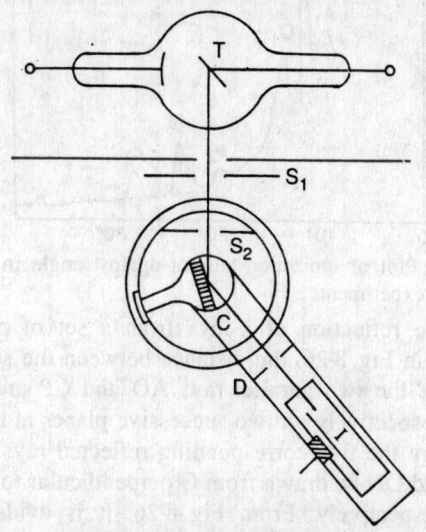

Fig. 8·24. Bragg spectrometer.

T in an x-ray tube were collimated by a system of slits S_1, S_2. The rays reflected by C were collimated by the slit D and entered the ionization chamber I. If now the prism table was turned through a given angle, the reflected ray would still enter I if the spectrometer arm on which it rested was turned through twice that angle. Bragg measured the ion current I for different angles of incidence θ and obtained a graph of I vs. θ. Such a graph is shown in Fig. 8·25. As will be seen from the figure, the intensity of the reflected ray shows sudden increases at certain definite angles of incidence. The origin of these intensity peaks can be understood as follows. Suppose a finely collimated beam of monochromatic x-rays falls upon the cleavage planes. The angles of incidence and reflection in these cases are usually measured with respect to these planes and are known as as *glancing angles*. According to Huyghens' theory, each atom in the cleavage plane acts as a secondary source under the action of the incident x-rays. Though these emit secondary wavelets in different directions, their intensities become appreciable only in those directions for which the glancing angles of incidence and reflection are equal. This is due to the constructive interference between the secondary wavelets given out by the atoms on the different parallel layers of the cleavage planes in the direction of the reflected ray. In the case of ordinary light, reflection takes place from the upper surface of the reflector. However x-rays penetrate some distance in the crystal and reflection takes place from the atoms in the different adjacent parallel layers.

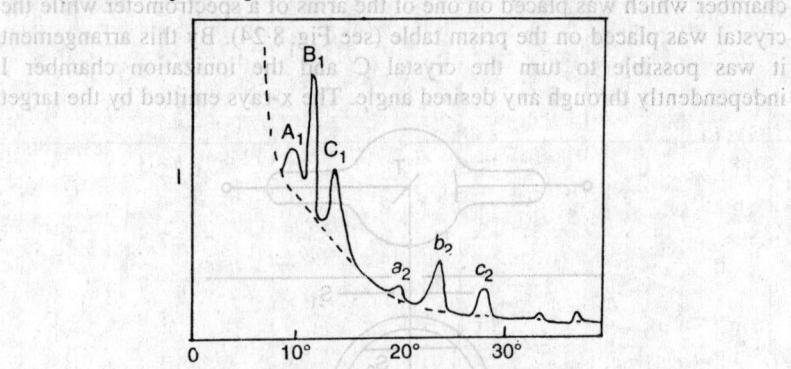

Fig. 8·25. Plot of ionization current against angle in Bragg's experiment.

Consider the reflection of x-rays from a set of parallel cleavage planes as shown in Fig. 8·26, the distance between the successive planes being d. Suppose the two parallel rays AO and CP are incident at the points O and P respectively on two successive planes at the glacing angle θ. OB and PD are the two corresponding reflected rays parallel to each other. Let OM and ON be drawn from O perpendicular to the incident and reflected rays respectively. From Fig. 8·26, it is evident that the ray reflected from the second plane travels an extra distance (MP + PN) w.r.t. the ray reflected from the first plane. If this path difference is equal to an integral multiple of the wavelength λ of the x-rays, then there will be constructive interference between the rays reflected from the two planes and there will be an intensity maximum in this direction. The condition for such an intensity maximum to occur is thus given by

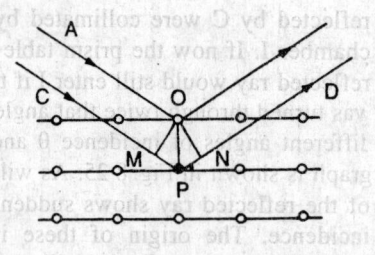

Fig. 8·26. Bragg reflection.

$$MP + PN = n\lambda$$

where n is an integer. But MP = PN = OP sin θ = d sin θ so that the condition for the intensity maximum becomes

$$2d \sin \theta = n\lambda \qquad (8\cdot17\text{--}1)$$

Eq. (8·17–1) is known as *Bragg equation*. It should be noted that even though the above phenomenon has been termed reflection, it is actually an instance of the diffraction of x-rays from different parallel atomic layers in the crystal.

If $n = 1$ in Eq. (8·17–1), then we get *first order reflection*. If the corresponding glacing angle of reflection is θ_1, then $2d \sin \theta_1 = \lambda$ and we get an intensity peak at $\theta = \theta_1$. Similarly for the second order reflection $n = 2$. If $\theta = \theta_2$ in this case, then we have $2d \sin \theta_2 = 2\lambda$; a second intensity peak appears at $\theta = \theta_2$. In Fig. 8·25, the peaks in the ionization current for different orders of reflection are shown. As will be seen, there are three peaks of the ionization current A_1, B_1 and C_1 in the first order. These correspond to three different wavelengths. Similarly there are three peaks A_2, B_2 and C_2 in the second order ; the relative heights of these peaks are similar to the peaks in the first order. The peaks in Fig. 8·25 actually represent the x-ray spectrum in different orders of the characteristic radiation emitted by the target in the x-ray tube. If θ_1, θ_2 and θ_3 are the angles of incidence (or reflection) for the different orders of reflection with $n = 1, 2$ and 3 respectively for a given wavelength λ, then from Eq. (8·17–1) we can write

$$\sin \theta_1 : \sin \theta_2 : \sin \theta_3 = 1 : 2 : 3$$

For example, in an experiment on the diffraction of rhodium x-rays from rock salt crystal, it is found that $\theta_1 = 11·8°, \theta_2 = 23·5°$ and $\theta_3 = 36°$ so that

$$\sin \theta_1 : \sin \theta_2 : \sin \theta_3 = 0·204 : 0·40 : 0·63 \simeq 1 : 2 : 3$$

Eq. (8·17–1) can be used to determine the wavelength of x-rays if the distance d between the successive planes in the crystal is known. On the other hand, if λ is known, then d can be determined.

Explanation of the origin of Laue spots

It is possible to understand the origin of the Laue spots on the basis of the above discussions. In Fig. 8·27, the different groups of parallel cleavage planes in a crystal are represented by drawing sets of parallel lines with relatively larger concentration of the black dots on them representing the crystal atoms. In the figure, these lines have been marked 1, 2, 3 *etc.* and the corresponding glancing angles on them by $\theta_1, \theta_2, \theta_3$, *etc.* for a parallel beam of x-rays incident from a particular direction. Suppose the interplanar distances for the different sets are d_1, d_2, d_3 *etc.* These are different for the different sets, though they are interrelated.

For a given direction of the incident x-rays, there will be constructive interference between the waves reflected from the different layers for a particular set of cleavage planes (*i.e.*, parallel lines) provided Bragg equation (8·17–1) is satisfied for the wavelength of the incident radiation i.e., if the incident beam contains x-radiation of this particular wavelength, the intensity of the beam reflected from this particular set of planes becomes quite strong and a Laue spot is observed in the direction of the reflected ray. Fig. 8·27 also shows that there is only a limited number of sets of planes with rich concentration of atoms. For this reason, even when the incident beam has a continuous distribution of wavelengths extending over a wide range, the number of Laue spots is not very large.

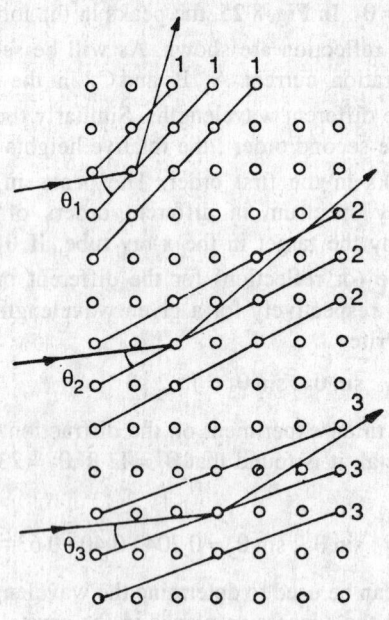

Fig. 8·27. X-ray reflection from different planes.

Since the positions of the Laue spots depend upon the distance between the successive cleavage planes, it is possible in principle to determine the structure of a crystal from the Laue diffraction pattern. However, in practice, it is quite difficult to determine the crystal structure except in some very simple cases. The diffraction pattern obtained by Bragg method, on the other hand, is simpler to analyse. This is because, monochromatic x-rays are used in this method and the incident rays are reflected from a single set of cleavage planes only. For this reason, only a single order of the spectrum is observed at a time. In Laue method, continuous x-rays are used. Out of these the crystal picks up rays of definite wavelengths and diffract them from the different sets of cleavage planes to produce the different orders of the spectrum at a time. For this reason, Laue method is more complex. However Bragg method is very time consuming (see also § 17·9).

X-Rays

The Bragg equation (8·17-1) was deduced above by considering reflection from two successive layers of atoms only, so that this method is analogous to the determination of the diffraction equation for a two dimensional diffraction grating. For a crystal on the other hand, diffraction from a three dimensional grating should be considered. This was done by von Laue by considering the superposition of the secondary waves from any two atoms in the crystal who obtained a set of equations (see Ch. XVII) from which Bragg equation was obtained as a special case.

8·18 Determination of the wavelength of x-rays by crystal diffraction method

As we have seen (§ 8·17) if the distance d (grating space) between the successive crystal planes, is known, then the wavelength λ of the x-rays can be determined by using Bragg equation. To determine the grating space in the crystal, a knowledge about the structure of the crystal is necessary. Bragg made a precise determination of the structure of rock salt (NaCl) crystal (see § 17·8. It has a simple cubic structure as shown in Fig. 8·28, with the ionized sodium and chlorine atoms at the alternate corners of each of the elementary cubes. The entire crystal is built up by

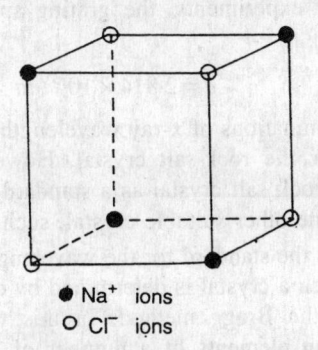

● Na^+ ions
○ Cl^- ions

Fig. 8·28 Cubic crystal of rock salt.

the repeated face to face arrangement of a very large number of such elementary cubes in a three dimensional pattern. The above regularity in the arrangement of a basic structure in a crystal is similar to the regularly repeated arrangement of a basic pattern in a lattice and hence such a crystal structure is known as a lattice structure (see § 8·17).

To determine the distances between the successive layers of atoms in the NaCl crystal (i.e., the length of sides of the basic unit cube) we proceed as follows. In the cubic crystal like the rock salt crystal, there are eight atoms, one each at the eight corners of the basic cube out of which the entire crystal is built up. It is easily seen that each atom actually lies at the common corner of eight adjacent cubes in such an arrangement so that on the average one atom may be assigned to each cube, since there are eight atoms per cube. Thus when we consider a very large number of such cubes, we may assume that the number of cubes is equal to the number of atoms in the crystal.

If there are n molecules of NaCl per unit volume of the crystal, the total number of atoms in it is $2n$. If M be the molecular weight and ρ the density of the rock salt crystal, we can write

$$n = N_0 \rho / M$$

where N_0 is the Avogadro number. Since the number of *unit cubes* per unit volume is $2n$, when n is very large, the volume of each such unit cube is

$$v = \frac{1}{2n} = \frac{M}{2N_0 \rho}$$

If d be the side of each unit cube, we get

$$d = v^{1/3} = (M/2N_0 \rho)^{1/3}$$

For the rock salt crystal, $M = 58.45$ and $\rho = 2.164 \times 10^3$ kg/m^3. Using the value of the Avogadro number N_0 which was known in the early years of crystal diffraction experiments, the grating space of rock salt was estimated to be

$$d = 2.814 \times 10^{-10} \text{ m}$$

All earlier determinations of x-ray wavelengths were done using the above value of d for the rock-salt crystal. However it is not always advantageous to use rock salt crystal as a standard for x-ray wavelength determination. So some other suitable crystal, such as calcite ($CaCO_3$) is used in many cases as the standard for the wavelength determination. The grating element of such a crystal is determined by comparing it with that of NaCl crystal by the Bragg method. In this way Siegbahn (1931) determined the grating elements of a number of crystals. *e.g.*, calcite, quartz, gypsum and mica. His values were however later shown to be somewhat erroneous due to the error in the value of N_0.

Later Bearden (1935) corrected these values from observations of the diffraction by a calcite crystal of x-rays whose wavelength had been measured with a ruled grating. The corrected values of d for rock salt and calcite crystals are :

$$d(\text{NaCl}) = 2.8197 \times 10^{-10} \text{ m}$$

$$d(\text{calcite}) = 3.03560 \times 10^{-10} \text{ m}$$

Using the above values of d, the wavelength of x-rays can be determined from the Bragg equation (8.17–1).

Bragg method of the measurement of wavelengths using a single crystal can also be applied in the case of γ-rays which are usually of shorter wavelengths compared to the x-rays from x-ray tubes. The shortest wavelength measured by this method is that of the γ-ray line of wavelength 0·016 Å.

8·19 Curved crystal method

When the x-ray wavelengths are very short, the Bragg angle of reflection (θ) is very small which is rather difficult to measure accurately. Hence accurate determination of very short wavelengths is difficult by this method. A new method for the determination of very short wavelengths of x-rays and γ-rays was devised by Mlle Cauchois (1932) using a curved crystal at the suggestion of Du Mond.

Fig. 8·29. shows a schematic diagram of a curved crystal x-ray spectrometer. A piece of flat quartz crystal is placed between two cylindrical steel surfaces to bend it into a curved crystal of radius of curvature 2 m. There are openings on both sides of the steel plates to allow the radiation to pass through them. The atomic planes in the crystal are prependicular to the curved surface and when produced meet at some point O. A circle of diameter equal to the radius of curvature of the crystal is drawn through O and the crystal. It is known as the focal circle. γ-rays from a small radioactive source placed at some point R on this circle are incident on the crystal planes and are reflected so as to pass to the other side of the crystal and are detected by a scintillation counter (see Vol. II) after passing through the lead collimator which prevents direct rays from R to enter the counter. From the geometry it can be seen that when the reflected rays from the different crystal planes are produced backwards they converge to a point O′ on the focal circle which thus forms the virtual image of the source. Further if the beam is narrow, the γ-rays are incident at the same angle θ on the different crystal planes. The Bragg angle θ is determined by the position of the source.

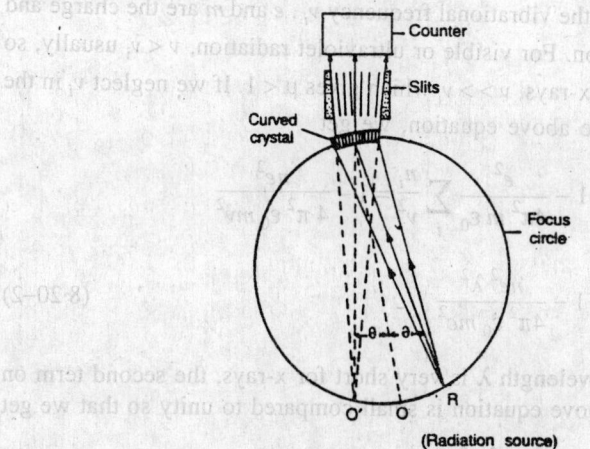

Fig. 8·29. Curved crystal spectrometer.

Wavelengths smaller than 0·01 Å have been measured with an accuracy of 10^{-5} Å by the curved crystal spectrometer.

8·20 Refraction of x-rays

Soon after discovery of x-rays, Röntgen, Barkla and others tried to observe the refraction of these rays. However their attempts did not meet with success. It was later noticed by Stenström (1919) that the wavelength of x-rays determined with the help of Bragg equation was slightly different when the measurement was made with the reflected rays in the different orders. This was attributed to the refraction of x-rays while entering the crystal. Obviously, Bragg equation (8·17–1) needs modification to take the effect of refraction into account. Later observations have shown that for all materials the refractive index $\mu < 1$ for x-rays. This means that when a collimated x-ray beam enters into a material medium from vacuum, it will be refracted *away* from the normal to the surface of separation so that the glancing angle of refraction $\theta' < \theta$, the glancing angle of incidence. This is just the opposite of what happens in the case of ordinary light for which $\theta' > \theta$. We know that the frequency of x-rays is much greater than that of visible radiation. It is thus much greater than the vibrational frequencies of the classical oscillators which are assumed to be responsible for the emission of light according to e.m. theory. According to Sellmeier equation, the refractive index μ of a medium for radiation of frequency v is given by

$$\mu^2 = 1 + \frac{e^2}{4\pi^2 \varepsilon_0 m} \sum_i \frac{n_i}{(v_i^2 - v^2)} \qquad (8\cdot 20\text{--}1)$$

where $\varepsilon_0 = 10^7/4\pi c^2$ F/m. n_i is the density of the classical oscillators (electrons) having the vibrational frequency v_i. e and m are the charge and mass of the electron. For visible or ultraviolet radiation, $v < v_i$ usually, so that $\mu > 1$. But for x-rays, $v >> v_i$ which gives $\mu < 1$. If we neglect v_i in the denominator of the above equation, we get

$$\mu^2 = 1 - \frac{e^2}{4\pi^2 m \varepsilon_0} \sum_i \frac{n_i}{v^2} = 1 - \frac{ne^2}{4\pi^2 \varepsilon_0 mv^2}$$

$$= 1 - \frac{ne^2 \lambda^2}{4\pi^2 \varepsilon_0 mc^2} \qquad (8\cdot 20\text{--}2)$$

Since the wavelength λ is very short for x-rays, the second term on the r.h.s. of the above equation is small compared to unity so that we get

$$\mu = 1 - \frac{ne^2 \lambda^2}{8\pi^2 \varepsilon_0 mc^2} \qquad (8\cdot 20\text{--}3)$$

Here $n = \sum_i n_i$ is the electron density in the material. From Eq. (8·20–3) we get for calcite, for $\lambda = 0.7$ Å

$$\mu = 1 - 1.84 \times 10^{-6}$$

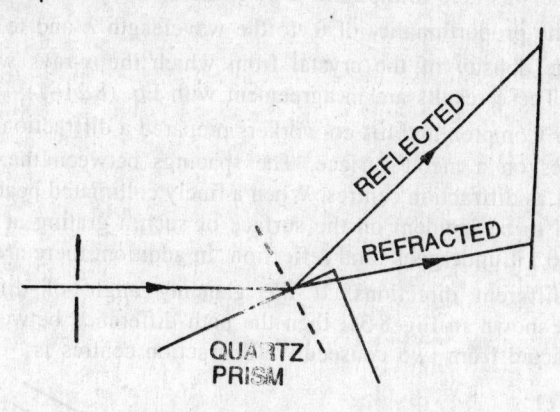

Fig. 8·30. Refraction of x-rays.

This shows that μ is only very slightly less than unity. This means that for x-rays, there is very little difference in the refractive indices of solids and gases e.g., air. Hence it is very difficult to measure the refractive indices of solids in this case. M. Siegbahn and his coworkers were able to measure the deviation of an x-ray beam due to refraction using a quartz prism. Their experimental arrangement is shown in Fig. 8·30. A collimated beam of monochromatic x-rays was allowed to be incident on one of the faces of a right angled quartz prism at a very small glancing angle. They were able to demonstrate the deviation of the refracted ray by photographic method and verify the applicability of dispersion theory to the case of x-rays. Bearden has used this method in an attempt to measure the refractive indices with sufficient accuray and obtain independent values of the wavelength of x-ray lines.

8·21 Determination of x-ray wavelengths by ruled grating

Since for x-rays the refractive index $\mu < 1$ for all materials, an x-ray beam will undergo total external reflection when it is refracted from air into a solid medium if the glancing angle of incidence $\theta < \theta_c$, the critical glancing angle. Hence we can write $\mu = \cos \theta_c$. Since θ_c is a very small angle, we can write

$$\mu = \cos \theta_c = 1 - \theta_c^2/2$$

We then get, using Eq. (8·20–3):

$$\theta_c = \sqrt{2(1-\mu)} = \sqrt{\frac{ne^2}{4\pi^2 \varepsilon_0 mc^2}} \cdot \lambda \qquad (8\cdot21-1)$$

For x-rays of wavelength $\lambda = 0.7$ Å, we get for calcite, $\theta_c \approx 0.1°$. A.H. Compton (1922) was the first to produce total reflection of a finely collimated beam of x-rays from a glass plate. By measuring the critical angle θ_c, he was able to measure μ of glass for x-rays. He was also able to prove the proportionality of θ_c to the wavelength λ and to the square root of the density of the crystal from which the x-rays were totally reflected. These results are in agreement with Eq. (8·21-1).

Later Compton and his co-workers prepared a diffraction grating by ruling lines on a metal surface. The spacings between the successive rulings act as diffraction centres. When a finely collimated beam of x-rays is allowed to be incident on the surface of such a grating at a glancing angle $\theta > \theta_c$, it undergoes total reflection. In addition, there are diffracted rays in different directions. If the glancing angle of diffraction is $(\theta + \alpha)$ as shown in Fig. 8·31, then the path difference between the two rays diffracted from two consecutive diffraction centres is

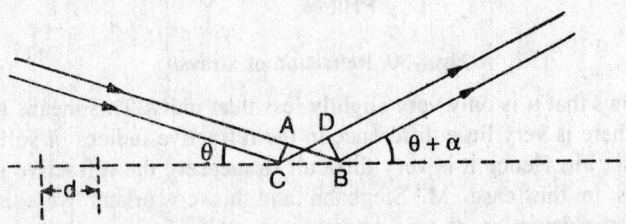

Fig. 8·31. Ruled grating method for x-ray wavelength determination.

$$AB - CD = d\{\cos\theta - \cos(\theta + \alpha)\} = n\lambda \quad (8·21\text{-}2)$$

Here n can be a positive or a negative integer or zero. The grating space d can be measured quite accurately and is independent of the crystal structure. Usually a grating with about 50000 lines per m is used. The wavelength of x-rays can be determined quite accurately by this method.

8·22 Modification of Bragg equation due to refraction

Consider a parallel beam of monochromatic x-rays of wavelength λ incident on a crystal at a glancing angle θ. As the beam enters the crystal the wavelength is changed to λ' due to refraction. Let the glancing angle of refraction be θ'. Inside the crystal the glancing angles of incidence and reflection on the crystal planes is θ' (see Fig. 8·32). The refractive index of the crystal is given by

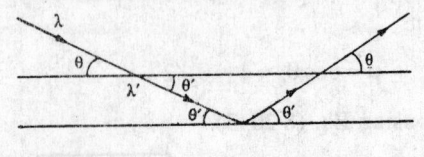

Fig. 8·32. X-ray reflection within a crystal.

X-Rays

$$\mu = \frac{\lambda}{\lambda'} = \frac{\cos \theta}{\cos \theta'}$$

The Bragg equation inside the crystal will be

$$2d \sin \theta' = n\lambda'$$

But
$$\sin \theta' = \sqrt{1 - \cos^2 \theta'} = \sqrt{1 - \cos^2 \theta / \mu^2}$$

$$= \frac{\sqrt{\mu^2 - \cos^2 \theta}}{\mu}$$

So we get
$$n\lambda' = \frac{n\lambda}{\mu} = \frac{2d}{\mu} \sqrt{\mu^2 - \cos^2 \theta}$$

So that
$$n\lambda = 2d \sqrt{\mu^2 - \cos^2 \theta}$$

$$= 2d \sin \theta \sqrt{1 - \frac{1 - \mu^2}{\sin^2 \theta}} \qquad (8 \cdot 22{-}1)$$

Since $1 - \mu^2 << 1$, we can write

$$1 - \mu^2 = (1 + \mu)(1 - \mu) \approx 2(1 - \mu)$$

So we have
$$n\lambda = 2d \sin \theta \sqrt{1 - \frac{2(1 - \mu)}{\sin^2 \theta}}$$

$$\approx 2d \sin \theta \left(1 - \frac{1 - \mu}{\sin^2 \theta}\right) \qquad (8 \cdot 22{-}2)$$

Eq. $(8 \cdot 22{-}2)$ is the modified Bragg equation. The correction term in this equation depends on the quantity $\delta = (1 - \mu)$. Since δ is a very small quantity, we can substitute for $\sin^2 \theta$ in Eq. $(8 \cdot 22{-}2)$ its value obtained from the Bragg equation $(8 \cdot 17{-}1)$

$$\sin^2 \theta = n^2 \lambda^2 / 4d^2$$

So we get finally $n\lambda = 2d \sin \theta \left(1 - \dfrac{4d^2 \delta}{n^2 \lambda^2}\right) \qquad (8 \cdot 22{-}3)$

For higher orders (n large), the correction term becomes smaller so that the unmodified Bragg equation $(8 \cdot 17{-}1)$ can be used to calculate λ for higher orders of diffraction.

8·23 Polarization of x-rays

According to the electromagnetic theory of light e.m. radiation scattered at an angle $\theta = 90°$ w.r.t. to the incident direction becomes polarized. From (Eq. $8 \cdot 14{-}9$), it can be seen that the electric intensity E_1' associated with the y-component of the incident electric vector E is zero at $\theta = 90°$: $E_1'(\theta = 90°) = 0$. On the other hand, the electric intensity E_2' associated with the z-component of E is independent of the scattering angle. Hence the radiation scattered along y is polarized with only the z-component of the electric vector being present (see Fig. 8·16).

Using the above result, an experimental arrangement can be devised to produce polarization of x-rays.

The experimental arrangement is shown in Fig. 8·33 schematically. A collimated x-ray beam from a tungsten target P is scattered by the first

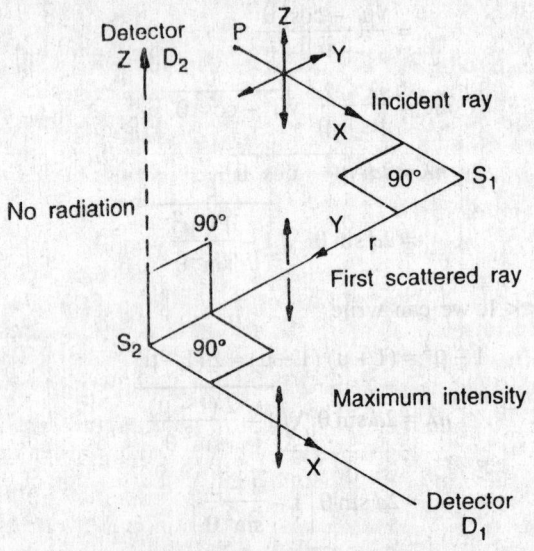

Fig. 8.33. Polarization of x-rays.

scatterer S_1. The beam scattered at 90° is allowed to suffer a second scattering from another scatterer S_2. The intensity of the twice scattered beam proceeding at right angles to $S_1 S_2$ is measured in two mutually perpendicular directions. It is found that the intensity is maximum along the direction $S_2 X$ which lies in the plane of PS_1 and $S_1 S_2$ while it is almost zero along the direction $S_2 Z$ perpendicular to this plane. This is in agreement with the theoretical prediction. The rays scattered by S_1 along $S_1 S_2$ are polarised along the z-direction, perpendicular to the plane of PS_1 and $S_1 S_2$. Since e.m. radiation is transverse in nature, there cannot be any z-component of the twice scattered beam along the z-direction, as that would make the vibrations longitudinal.

Problems

1. If a potential difference of 30,000 volts is applied between the anode and cathode of an x-ray tube, what is the minimum wavelength of the emitted x-rays. (0·4136 Å)

X-Rays

2. The wavelengths of the K and L_{II} absorption edges of copper are 1·3774 Å and 12·9 Å respectively. What is the wavelength of the $K\alpha_2$ line of copper? (1·542 Å)

3. The copper (Z = 29) target in an x-ray tube has a small amount of impurity within it. The x-rays emitted from the tube show a line of wavelength 0·53832 Å in addition to the copper K_{α_1} line of wavelength 1·541232Å. Using Moseley's law calculate the atomic number of the impurity. (48)

4. A potential difference of 60 kilovolts is applied between the anode and the cathode in an x-ray tube. The emitted x-rays are allowed to fall on thin foils made of cobalt, molybdenum and palladium respectively. If the K absorption edges of these metals have the wavelengths 1·6040, 0·61848 and 0·50795 Å respectively, what will be the maximum kinetic energies of the photoelectrons ejected from the K-shells of their atoms? (52·261 : 39·931 ; 35·563 kilovols)

5. The mass absorption coefficients of nickel ($\rho = 8.60 \times 10^3$ kg/m^3) for the wavelengths 0·880, 1·00, 1·235 and 1·389 Å are 8·13, 11·85, 20·8 and 28·6 kg/m^2 respectively. Calculate the half-value thickness in each case. Draw a graph of the mass absorption coefficient vs. wavelength. Represent the same by an approximate formula.

6. Using the relation between the angle of recoil electron emission θ and the angle of Compton scattering of photon φ (Eq. 8·16–9), express the energy of emission of the recoil electrons as a function of θ.

7. KBr crystal has a cubic structure. If its density is 2.75×10^3 kg/m^3 and its molecular weight is 119·01, calculate its lattice constant. If palladium $K_{\alpha 2}$ x-rays ($\lambda = 0.58863$ Å) are diffracted from this crystal, what will be the glancing angle of the first order of diffraction? (3·3 Å ; 5° 7′)

8. Use equation (8·20–3) to calculate the refractive index of calcite crystal (CaCO$_3$) for x-rays of wavelength 0·7 Å. The molecular weight of calcite is 100·09 and its density is 2.93×10^3 kg/m^3. $(1 - 1.8 \times 10^{-6})$

9. Calculate the energy of the γ-rays which suffer Compton scattering from electrons at rest so that the angle of scattering of the photon and the angle of recoil electron emission are both 40°. What is the energy of the scattered photon? (1·16 MeV ; 0·757 MeV)

10. Prove that if during Compton scattering the recoil electron has very low the energy, then the change in the wavelength of the photon scattered at φ is

$$\Delta \lambda = \frac{\lambda_0}{2}\left(\frac{\lambda'}{\lambda} + \frac{\lambda}{\lambda'} - 2 \cos \phi\right).$$

11. If the spectral distribution of bremstratilung radiation is of the form $I_v = C(v_m - v)$ where C is a constant and v_m is the maximum frequency of the radiation emitted when a beam of electrons of energy eV hits a target in the x-ray tube, prove that the most probable wavelength of the emitted radiation as

$$\lambda_{pr} = 3hc/2eV$$

9

WAVE PARTICLE DUALITY ; HEISENBERG'S UNCERTAINTY PRINCIPLE

9·1 Particles and waves

Photoelectric effect and Compton scattering are evidences of the corpuscular character of light. On the other hand, interference, diffraction and polarisation reveal the wave nature of light. Louis de Broglie of France proposed that the dual character exhibited by light may also be a characteristic of the subatomic material particles, such as electrons (1924). In his attempt to discover the underlying significance of Bohr's quantum condition, de Broglie tried to fit into the Bohr orbit an exact number of standing waves in analogy with the integral number of full waves in a stretched string.

In order that an integral number of waves may be fitted into the nth Bohr orbit of radius r we have to write $2\pi r = n\lambda$ where n is an integer. Comparing this with Bohr's quantum condition we get

$$m\upsilon r = m\upsilon \cdot \frac{n\lambda}{2\pi} = n\hbar$$

so that
$$\lambda = \frac{2\pi\hbar}{m\upsilon} = \frac{h}{p} \qquad (9\cdot1\text{--}1)$$

where $p = m\upsilon$

is the momentum of the electron.

So de Broglie proposed that a material particle of energy E and momentum p may exhibit wave characteristics having wavelength determined by Eq. (9·1–1). Its frequency is determined by the Planck formula

$$E = h\nu \qquad (9\cdot1\text{--}2)$$

Eqs. (9·1–1) and (9·1–2) may be rewritten as

$$p = \hbar k \text{ and } E = \hbar\omega \qquad (9\cdot1\text{--}3)$$

where $k = 2\pi/\lambda$ is the magnitude of the propagation vector ; $\omega = 2\pi\nu$ is the circular frequency.

Classically a wave of definite wavelength and frequency is of infinite extent in space and is of infinite duration. On the other hand a corpuscle is localized at a definite point in space at a given instant of time and has a definite momentum p and energy E. Thus the basic characteristics of wave and corpuscle are incompatible according to classical mechanics.

To reconcile the wave and corpuscular view points, it should be possible to localize the wave in a finite region of space. This can be done by the superposition of waves of different wavelengths upon one another by the method of Fourier transform.

9·2 Phase and group velocities

The phase velocity u of a monochromatic wave is the velocity with which a definite phase of the wave, such as its crest or trough is propagated in a medium. The phase velocity cannot be measured experimentally. It is not possible to distinguish between the successive waves for a monochromatic wave since they are all exactly similar. To do this it is necessary to affix distinguishing marks on the waves. We know that during the propagation of radio waves, signals are transmitted after *modulation*. Modulated signals are produced by the superposition of many plane waves of different frequencies, centred about a mean frequency. The energy carried by such modulated signals is transmitted with a velocity known as the *group velocity* (v_g). This is different from the *phase velocity* of the component waves. Production of the modulated signals is nothing but putting a distinguishing mark on the wave train. Measuring devices can then easily follow the progress of this distinguishing mark with time.

The simplest method to signal a wave train is by producing beats by the superposition of two waves of slightly different frequencies. These mutually reinforce to produce maxima at certain points at regular intervals with much weaker disturbances in between (see Fig. 9·1).

If the number of superposed waves is increased, the regions with prominent humps will be narrower while the intervening regions of weaker disturbances will become broader. In the limit, if an infinite number of waves of continuously varying frequencies extending over a finite range is superposed we get a single hump within a narrow region with no disturbances at any other point. This is known as a *wave packet*. This is the method of Fourier integral. It may be noted that if the frequencies of the component waves extend over an infinite range, the wave packet will shrink to a single point (δ-function).

There is definite relationship between the group velocity v_g and the phase velocity u. To determine this we consider the superposition of two waves of slightly different frequencies v_1 and v_2, having the same amplitude a (see Fig. 9·1). If λ_1 and λ_2 are the wavelengths of the two waves, we can write for the two waves

$$y_1 = a \sin(\omega_1 t - k_1 x) \quad (9\cdot1\text{–}1)$$

$$y_2 = a \sin(\omega_2 t - k_2 x) \quad (9\cdot2\text{–}2)$$

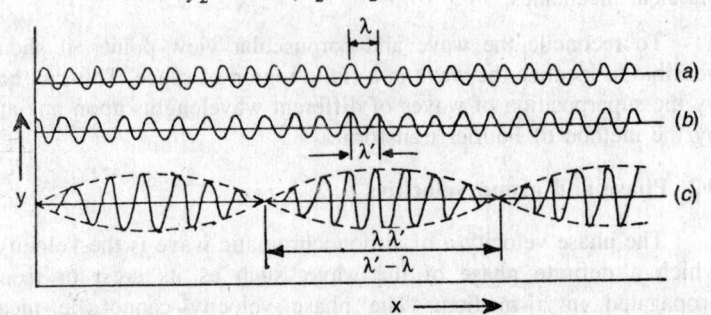

Fig. 9·1. Superposition of two waves of slightly different wavelengths.

Here $\omega_1 = 2\pi v_1$ and $\omega_2 = 2\pi v_2$ are the circular frequencies while $k_1 = 2\pi/\lambda_1$ and $k_2 = 2\pi/\lambda_2$ are the magnitudes of the propagation vectors of the two waves.

The resultant wave obtained by the superposition of the two waves is given by

$$y(x, t) = y_1 + y_2$$
$$= a \sin(k_1 x - \omega_1 t) + a \sin(k_2 x - \omega_2 t)$$
$$= 2a \cos\left(\frac{k_1 - k_2}{2} x - \frac{\omega_1 - \omega_2}{2} t\right) \sin\left(\frac{k_1 + k_2}{2} x - \frac{\omega_1 + \omega_2}{2} t\right) \quad (9\cdot2\text{–}3)$$

The resultant wave has the circular frequency $(\omega_1 + \omega_2)/2$ and propagation vector of magnitude $(k_1 + k_2)/2$. Its amplitude varies slowly according to the cosine term with the circular frequency $(\omega_1 - \omega_2)/2$. The velocity of propagation of the maxima in the amplitudes is given by $(\omega_1 - \omega_2)/(k_1 - k_2)$. The phase of the wave is propagated with the velocity $(\omega_1 + \omega_2)/(k_1 + k_2)$. Since ω_1 and ω_2 are nearly equal, the phase velocity is given by

$$u = \lim_{\omega_2 \to \omega_1} \frac{\omega_1 + \omega_2}{k_1 + k_2} = \frac{\omega_1}{k_1} \quad (9\cdot2\text{–}4)$$

On the other hand, the group velocity is

$$\upsilon_g = \lim_{\omega_2 \to \omega_1} \frac{\omega_1 - \omega_2}{k_1 - k_2} = \frac{d\omega}{dk} \qquad (9\cdot2\text{–}5)$$

The above expression for υ_g also holds in the general case of superposition of an infinite number of monochromatic waves of continuously varying frequencies.

From Eq. (9·2–4) we get by writing ω and k for ω_1 and k_1 respectively

$$\frac{du}{dk} = \frac{1}{k} \frac{d\omega}{dk} - \frac{\omega^2}{k^2} = \frac{\upsilon_g - u}{k}$$

This gives

$$\upsilon_g = u + k \frac{du}{dk} = u - \lambda \frac{du}{d\lambda} \qquad (9\cdot2\text{–}6)$$

For light waves in vacuum, there is no dispersion. Hence $du/dk = 0$, so that $\upsilon_g = u = c$. This is also true for elastic waves in a homogeneous medium.

9·3 Particle wave

If a particle, such as an electron, sometime behaves like a wave, then we may ask as to whether it is the phase velocity or the group velocity of the wave that should be identified with the particle velocity υ. In actual practice no wave can be of infinite extent. A train of light waves is of finite extent since the emission of light from excited atoms takes place in times of the order of 10^{-8} s. During this interval, the emitted wave train extends over about 3 m. We have seen that such a wave train can be obtained by the superposition of monochromatic waves of different wavelengths. So it must be propagated with the group velocity and not with the phase velocity. The energy carried by the wave is also transported with the group velocity. Thus all physically measureable quantities associated with the wave packet are propagated with the group velocity. The phase velocity, on the other hand, is the velocity of propagation of the phase of the disturbance. In the case of light wave it may even exceed the velocity of light in vacuum (c). But υ_g must be always less than c, since according to the special theory of relativity any physically measurable quantity must have a velocity less than c.

De Broglie proposed that the particle velocity υ should be put equal to the group velocity υ_g of the associated wave :

$$\upsilon = \upsilon_g \qquad (9\cdot3\text{–}1)$$

We can write the energy of a particle of rest mass m_0 as

$$E = \hbar\omega = \frac{m_0 c^2}{\sqrt{1-\beta^2}} \qquad (9\cdot3-2)$$

$$\omega = \frac{m_0 c^2}{\hbar\sqrt{1-\beta^3}} \qquad (9\cdot3-3)$$

The phase velocity

$$u = \frac{\omega}{k} = \frac{m_0 c^2}{\hbar k \sqrt{1-\beta^2}} \qquad (9\cdot3-4)$$

Then

$$\frac{du}{dk} = -\frac{m_0 c^2}{\hbar k^2 \sqrt{1-\beta^2}} + \frac{m_0 c^2 \beta}{\hbar k(1-\beta^2)^{3/2}} \cdot \frac{d\beta}{dk} \qquad (9\cdot3-5)$$

So by using Eqs. (9·2–6), (9·3–4) and (9·3–5) we get

$$\upsilon = \beta c = \upsilon_g = u + k\frac{du}{dk}$$

$$= \frac{m_0 c^2 \beta}{\hbar(1-\beta^2)^{3/2}} \cdot \frac{d\beta}{dk}$$

Hence

$$dk = \frac{m_0 c}{\hbar} \cdot \frac{d\beta}{(1-\beta^2)^{3/2}}$$

Integration gives

$$k = \frac{m_0 c}{\hbar} \frac{\beta}{\sqrt{1-\beta^2}} + C$$

C is the integration constant. Assuming $k = 0$ when $\upsilon = 0$ we get $C = 0$.

So we have

$$k = \frac{2\pi}{\lambda} = \frac{m_0 \upsilon}{\hbar\sqrt{1-\beta^2}} = \frac{m\upsilon}{\hbar} = \frac{p}{\hbar}$$

or,

$$p = \hbar k = \frac{h}{\lambda} \qquad (9\cdot3-6)$$

Eq. (9·3–6) which is the same as Eq. (9·1–1) is the famous de Broglie equation which gives the mathematical relationship between the momentum p of a particle which is a dynamical variable characteristic of a corpuscle and the wavelength which is a characteristic of the associated wave.

9·4 Relation between phase and group velocities of de Broglie waves

The phase velocity of the de Broglie waves is (see Eq. 9·2–4)

$$u = \lambda v = \frac{\omega}{k} \qquad (9.4-1)$$

For an electron we have (see Eq. 9·1–3)

$$E = \hbar\omega = \frac{m_0 c^2}{\sqrt{1-\beta^2}}$$

and $$p = \hbar k = \frac{m_0 \beta c}{\sqrt{1-\beta^2}}$$

So we have

$$\frac{E}{p} = \frac{c^2}{v} = \frac{\omega}{k} \qquad (9.4-2)$$

Hence we get

$$u = \frac{\omega}{k} = \frac{c^2}{v} \qquad (9.4-3)$$

where v is the velocity of the electrons and c is the velocity of light in vacuum.

According to de Broglie $v = v_g$. Hence we have

$$v = \frac{c^2}{v_g} \qquad (9.4-4)$$

or, $$u v_g = c^2 \qquad (9.4-5)$$

Eq. (9·4–5) is the relation between the phase and group velocities of the de Broglie waves.

Now the relativistic expression for the total energy E of the electron gives

$$E^2 = p^2 c^2 + m_0^2 c^4 \qquad (9.4-6)$$

Thus using Eq. (9·1–3), we get

$$\hbar^2 \omega^2 = \hbar^2 c^2 k^2 + m_0^2 c^4 \qquad (9.4-7)$$

Since $\omega = uk$ we have from Eq. (9·4–7)

$$\hbar^2 u^2 k^2 = \hbar^2 c^2 k^2 + m_0^2 c^4$$

Hence $$u = c\sqrt{1 + \frac{m_0^2 c^2}{\hbar^2 k^2}}$$

$$= c\sqrt{1 + \frac{m_0^2 c^2 \lambda^2}{4\pi^2 \hbar^2}} \qquad (9.4-8)$$

Eq. (9·4–8) gives the phase velocity u of the de Broglie waves. It shows that u is greater than c.

This is also consistent with the relation between u and v_g (Eq. 9·4–5), since $v_g = v$ must always be less than c.

Eq. (9·4–8) also shows that the phase velocity of the de Broglie waves depends on the wavelength λ even in vacuum. This behaviour the of de Broglie waves is different from that of the light wave for which the phase velocity is independent of λ in vacuum.

9·5. Discovery of matter waves : Davisson and Germer's experiment

When de Broglie proposed the hypothesis of matter waves, there was no experimental evidence of its existence. Soon afterwards, two American physicists J. Davisson and L.H. Germer performed an experiment (1927) which for the first time demonstrated the wave nature of electrons.

Their experimental arrangement is shown in Fig. 9·2. A beam of electrons from a heated filament F was collimated by a system of slits and was then accelerated by the anode A having a small hole in it. The emergent electron beam from this hole was then incident on a single crystal of nickel S. The electrons were scattered by S in different directions. The intensity of the scattered beam was measured with the help of a Faraday cup C and a sensitive galvanometer G. The collector cup was able to detect only those electrons which had the same energy as the incident beam so that only the electrons elastically scattered by S were detected. By changing the position of the collector along an arc of a circle about a point in the crystal as its centre, it was possible to determine the scattered intensity in different directions.

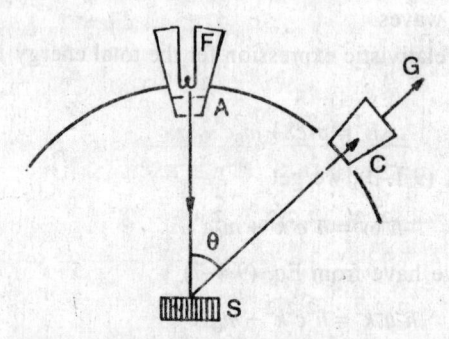

Fig. 9·2. Davisson and Germer's experiment.

The variation of the scattered intensity is plotted as a function of the scattering angle in a *polar graph* in Fig. 9·3. The radius vector is proportional to the scattered intensity. The angle between the radius vector and the ordinate gives the scattering angle θ. The different curves in Fig. 9·3 correspond to electrons accelerated to different energies by changing the accelerating potential between F and A.

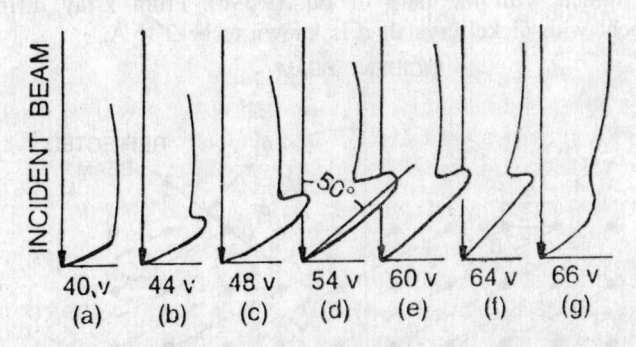

Wave Particle Duality : Heisenberg's Uncertainty

Fig. 9·3. Results of Davisson and Germer's experiment.

For an accelerating potential of 40 volts, a smooth curve as shown in Fig. 9.3a was obtained. At 44 volts, a small but distinct hump appears at an angle of about 60° ((Fig. 9·3b.) The hump increases at higher accelerating potentials and the angle at which it appears also changes. The hump becomes most prominent at 54 volts and appears at the angle of 50°. At still higher potentials, the hump decreases until it disappears completely above about 68 volts.

The above results show preferential 'reflection' of 54 eV electrons at $\theta = 50°$ and is an evidence of the diffraction of electrons from the single crystal of nickel. As discussed in Ch. VIII, in a crystal lattice there are atoms or groups of atoms arranged in a regular array. For this reason, a crystal can be used to diffract x-rays which have wavelengths comparable to the spacing between the atomic planes. The Bragg equation $2d \sin \theta = n\lambda$ for the maxima gives the wavelength if the lattice constant d is known. θ is the angle of diffraction at which the maximum occurs. Consider such a regular array of atoms arranged in a lattice. Sets of parallel planes having larger concentration of atoms can be drawn through the array. In two dimensions, these appear as sets of parallel straight lines as shown in Fig. 9·4.

Consider a beam of electrons incident normally upon the crystal face. The beam makes an angle θ with the normal to the set of planes. Then according to Bragg's law, there will be reinforcement of electron waves if the following condition is satisfied :

$$n\lambda = 2D \sin \psi = 2D \cos \phi \qquad (9 \cdot 5-1)$$

where D is the spacing between the atomic layers under consideration and $\psi = \pi/2 - \phi$. If d be the distance between the atoms in the surface layer then $D = d \sin \phi$ so that we get

$$n\lambda = 2d \sin \phi \cos \phi = d \sin 2\phi = d \sin \theta \qquad (9 \cdot 5-2)$$

where $\theta = 2\phi$ is the angle between the incident and reflected beams. $n = 1, 2, 3....$ and λ is the de Broglie wavelength of the electrons which can be determined with the help of Eq. (9·3–6). From x-ray diffraction experiment with nickel crystal, d is known to be ·2·15 Å.

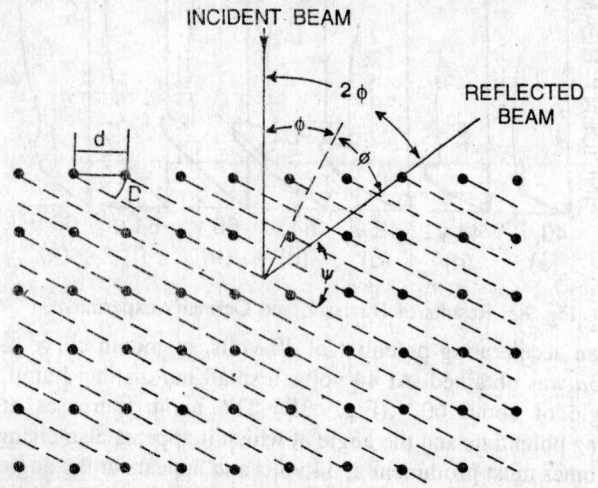

Fig. 9·4. Bragg reflection of electron waves for normal incidence.

In Davisson and Germer's experiment, there was a diffraction maxima at $\theta = 50°$ when the electron accelerating potential was 54 volts. Then from Eq. (9·5–2) we get for $n = 1$

$$\lambda = 2·15 \sin 50° = 1·65 \text{ Å}$$

On the other hand from de Broglie equation, we have

$$\lambda = \frac{h}{mv} = \frac{h}{\sqrt{2mE}} = \frac{h}{\sqrt{2meV}}$$

$$= \frac{6·62 \times 10^{-34}}{(2 \times 9·11 \times 10^{-31} \times 1·6 \times 10^{-19} V)^{1/2}}$$

$$= \frac{12·26}{\sqrt{V}} \text{ Å} \qquad (9·5–3)$$

where V is in volts. In the present case $V = 54$ volts.

Hence $\lambda = \dfrac{12·26}{\sqrt{54}} = 1·67 \text{ Å}$

Thus the wavelength determined from the diffraction measurement agrees with that obtained from de Broglie relation very well. This provides the justification of de Broglie's hypothesis of matter waves.

9·6 G. P. Thomson's experiment

An independent confirmation of de Broglie's theory was provided by the experiment of G.P. Thomson[*] (1928) in which a narrow beam of

---

[*] G.P. Thomson was the son of the great physicist J.J. Thomson, the discoverer of the electron.

Wave Particle Duality : Heisenberg's Uncertainty 219

electrons accelerated to potentials varying between 10,000 to 60,000 volts was passed through very thin films of metals, *e.g.*, gold or silver. The transmitted electrons were detected photographically by the plate P (see Figs. 9·5 and 9·6). The pattern observed consisted of a series of concentric

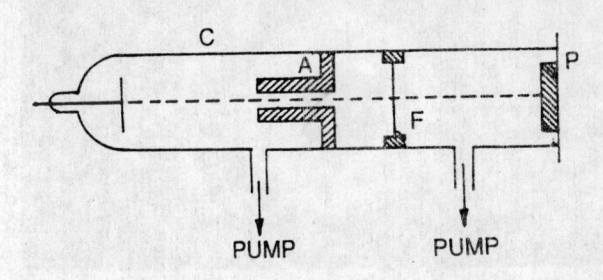

Fig. 9·5. G.P. Thomson's experimental arrangement for electron diffraction.

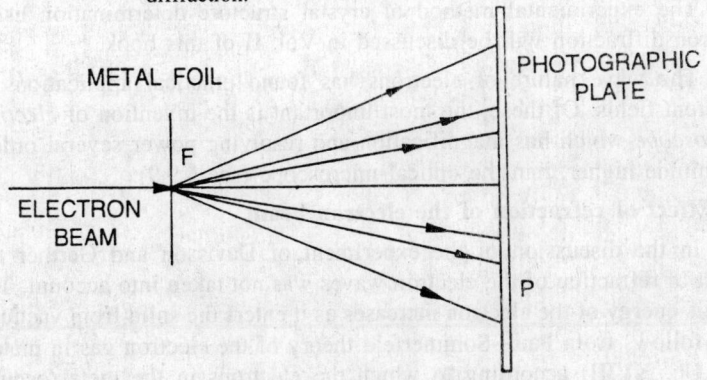

Fig. 9·6. Schematic diagram of Thomson's experiments.

rings about a central spot as shown in Fig. 9·7. Assuming that these are similar to Debye-Scherrer rings produced by x-rays diffracted by a powdered crystal (see Ch. XVII), the spacings between the atomic planes could be determined by Thomson assuming the validity of de Broglie relation $\lambda = h/p$. These agreed quite well with similar determination by x-ray diffraction experiments for different metals.

Wave nature of particles other than electrons, *e.g.*, protons, neutrons *etc.*, have also been discovered. Beams of atoms and molecules have also been found to exhibit wave characteristics. Of these, the neutron waves are specially suitable for crystal structure determination. Neutrons, being uncharged, are not affected by the Coulomb field of the atomic nucleus. Being much heavier than electrons, their de Broglie wavelengths are much shorter than those of electrons of the same energy. For example, *thermal neutrons* at room temperature (300 K) having the mean kinetic energy $kT = 0.026$ eV have de Broglie wavelength of about 1·8 Å which is comparable to the wavelengths of x-rays used in crystal structure determination. Electrons of the same energy have $\lambda = 77$ Å.

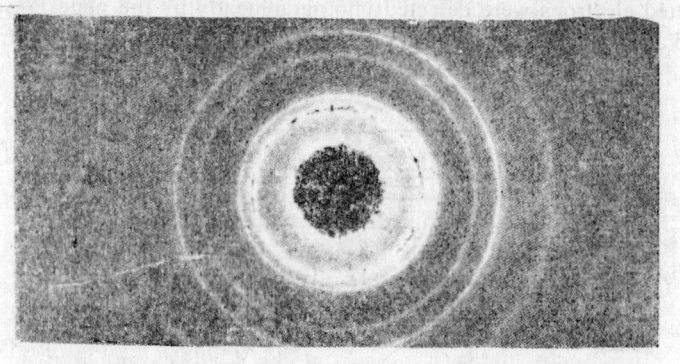

Fig. 9·7. Photograph of electron diffraction pattern.

The experimental method of crystal structure determination using neutron diffraction will be discussed in Vol. II of this book.

The wave nature of electrons has found practical applications in different fields. Of these, the most important is the invention of *electron microscope* which has magnification and resolving power several orders magnitude higher than the optical microscope (see § 9·9).

9·7 Effect of refraction of the electron beam

In the discussion of the experiment of Davisson and Germer the effect of refraction of the electron waves was not taken into account. The kinetic energy of the electron increases as it enters the solid from vacuum. This follows from Pauli-Sommerfeld theory of the electron gas in metals (see Ch. XVIII) according to which the electrons in the metal occupy successive energy levels starting from the bottom of a potential well upto the Fermi level ϵ_f. If the depth of the potential well is U_0 then we can write

$$\frac{1}{2} m v_1^2 = \frac{1}{2} m v_2^2 - U_0$$

where v_1 and v_2 are velocities of the electron in vacuum and in the solid medium respectively. If λ_1 and λ_2 be the wavelengths of the electron wave in vacuum and in the medium respectively, the refractive index of the medium will be given by

$$\mu = \frac{\lambda_1}{\lambda_2} = \frac{v_2}{v_1} \qquad (9 \cdot 7-1)$$

Here we have assumed $\lambda = h/p \propto \dfrac{1}{v}$. Obviously the above relationship is different from that valid in the case of light wave. In the latter case μ is the ratio of the velocity of light wave in vacuum to that in the medium. Since

$$v_2^2 = v_1^2 + 2U_0/m$$

we have
$$\mu = \frac{1}{v_1}\sqrt{v_1^2 + 2U_0/m} = \sqrt{1 + U_0/E_{in}} \qquad (9.7\text{-}2)$$

where $E_{in} = \frac{1}{2}mv_1^2$ is the kinetic energy of the electron in the first medium, *i.e.* vacuum. If this energy is acquired by acceleration through a potential difference of V, then $E_{in} = eV$. Also if we write $U_0 = e\,\Delta V$, ΔV being the depth of the potential well in the solid in volts, we get

$$\mu = \sqrt{1 + \Delta V/V} \qquad (9.7\text{-}3)$$

By measuring the position of the diffraction maxima, it is possible to determine the refractive index μ (see below), and so knowing V, the depth of the potential well ΔV can be estimated.

At the time of thermionic emission, the electrons have to overcome the attractive potential due to the image force of the emitting surface, the same kind of force which acts on the electron as it passes from the first to the second medium discussed above. As we have seen this results in the change of potential energy of the electron, which is in conformity with the Pauli-Sommerfeld theory of metallic solids (see Ch XVIII).

Careful measurement by Davisson and Germer on single crystals of nickel gave $\Delta V \approx 18$ volts. On the other hand measurements on thermionic effect gives a value of about 5 volts for the thermionic work function ϕ. The difference between ΔV and ϕ is of the order of magnitude of Fermi energy of nickel calculated on the basis of Fermi-Dirac statistics (see Ch. XVIII).

We now consider the effect of refraction of the electron waves at the time of electron diffraction from crystals.

As shown in Fig. 9.8 suppose a collimated beam of mono-energetic electrons is incident normally on a crystal face so that it enters undeviated

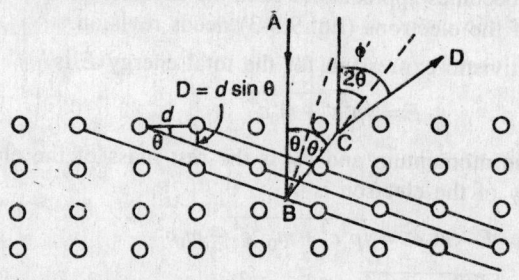

Fig. 9.8. Effect of refraction on electron diffraction.

inside the crystal. If the beam is incident at an angle θ on a particular atomic plane inside the crystal and is reflected at the same angle, then the angle between the incident ray AB and the reflected ray BC is 2θ. Then as the reflected beam emerges from the crystal, it is refracted along CD which makes an angle ϕ' with the normal to the crystal face at C.

We can write the Bragg equation inside the crystal as $n\lambda_2 = 2D \cos\theta$ where $D = d \sin\theta$ is the spacing between the atomic planes from which the reflection of the electron beam takes place. d is the spacing between the atoms in the crystal. We thus get

$$n\lambda_2 = 2d \sin\theta \cos\theta = d \sin 2\theta = d \sin\phi \qquad (9\cdot7\text{--}4)$$

where $\phi = 2\theta$ is the angle between the incident and the reflected beams inside the crystal. This beam then emerges from the crystal after being refracted at the angle ϕ' at the crystal face, so that we can write

$$\mu = \frac{\lambda_1}{\lambda_2} = \frac{\sin\phi'}{\sin\phi} \qquad (9\cdot7\text{--}5)$$

We then get from Eq. (9·7–4)

$$n\lambda_1 = n\lambda_2 \frac{\lambda_1}{\lambda_2} = n\lambda_2 \frac{\sin\phi'}{\sin\phi}$$

$$= d \sin\phi \frac{\sin\phi'}{\sin\phi} = d \sin\phi' \qquad (9\cdot7\text{--}6)$$

Here λ_1 is the de Broglie wavelength of the electrons outside the crystal and ϕ' is the measured angle between the incident and reflected beams outside the crystal. Eq. (9·7–6) is the modified Bragg equation which should be used in determining the de Broglie wavelength in place of Eq. (9·5–2) to get the correct result. (Note the change of notation for the angles in the two equations). Actually Davisson and Germer used this equation to determine λ in their first experiment.

9·8 De Broglie wavelength of high energy electrons

For very high energy electrons the relativistic variation of its mass with velocity becomes appreciable. Then the expression for the de Broglie wavelength of the electrons (Eq. 9.5-3) needs revision.

The relativistic expression for the total energy E is

$$E = \sqrt{p^2 c^2 + m_0^2 c^4}$$

where p is the momentum and m_0 is the rest mass of the electron. The kinetic energy of the electron is

$$E_k = E - m_0 c^2 = \sqrt{p^2 c^2 + m_0^2 c^4} - m_0 c^2$$

so that $\sqrt{p^2 c^2 + m_0^2 c^4} = E_k + m_0 c^2$

Squaring, we get

$$p^2c^2 + m_0^2c^4 = E_k^2 + 2m_0c^2 E_k + m_0^2c^4$$

or, $\quad p^2c^2 = E_k(E_k + 2m_0c^2)$

Hence $\quad p = \dfrac{1}{c}\sqrt{E_k(E_k + 2m_oc^2)} \quad$ (9.8–1)

So the de Broglie wavelength of the electron is

$$\lambda = \frac{h}{p} = \frac{ch}{\sqrt{E_k(E_k + 2m_0c^2)}} \quad (9.8\text{–}2)$$

For an electron accelerating potential of V, we have $E_k = eV$, so that

$$\lambda = \frac{ch}{\sqrt{eV(eV + 2m_0c^2)}} = \frac{ch}{e\sqrt{V(V + 2m_0c^2/e)}} \quad (9.8\text{–}3)$$

The rest energy of the electron is $m_0c^2 = 0.51$ MeV $= 0.51 \times 10^6$ eV so that $2m_0c^2/e = 1.02 \times 10^6$ volts. If V is expressed in volts, we get

$$\lambda = \frac{ch}{e\sqrt{V(V + 1.02 \times 10^6)}}$$

$$= \frac{3 \times 10^8 \times 6.62 \times 10^{-34}}{1.6 \times 10^{-19}\sqrt{V(V + 1.02 \times 10^6)}}$$

$$= \frac{12.41 \times 10^{-7}}{\sqrt{V(V + 1.02 \times 10^6)}}$$

$$= \frac{12.41 \times 10^3}{\sqrt{V(V + 1.02 \times 10^6)}} \text{ Å} \quad (9.8\text{–}4)$$

It should be noted that if the electron kinetic energy is small so that $E_k << m_0c^2$, then Eq. (9.8–2) reduces to the nonrelativistic expression for λ given by Eq. (9.5–3).

A few typical values of λ are : for $E_k = 0.1 \times 10^6$ eV, $\lambda = 0.0371$ Å ; for $E_k = 1$ MeV, $\lambda = 0.00873$ Å. If we use Eq. (9.5–3), we get $\lambda = 0.0388$ Å and $\lambda = 0.0123$ Å for the above two cases.

9.9 Electron microscope

The wave nature of the electron has found wide practical applications, the most important of which is the invention of the electron microscope. Microscopes are used to produce magnified images of very small objects. Besides, the detailed structures of such objects are also revealed with the help of microscopes. Higher the resolving power of a microscope, more minutely can the detailed structure of an object be studied with the microscope.

In an optical microscope the object is illuminated with visible light. The limit of resolution of a microscope depends on the wavelength of the light used and is given by

$$\Delta x = \frac{0.61 \lambda}{\text{numerical aperture}}$$

where Δx is the minimum separation between two point objects which can be just resolved. The numerical aperture depends on the angle subtended by the objective lens at the object and the refractive index of the object space. Smaller Δx is, better is the limit of resolution, which means that the object should be illuminated with light of the shortest possible wavelength for getting better resolution. Thus in an optical microscope, the limit of resolution is of the order of the wavelength of visible radiation which is a few thousand angstroms. If however, the object is illuminated with electron waves, much better resolution can be obtained, since as we have seen, the de Broglie wavelength of the electron waves can be made much shorter than that of visible radiation. Thus for electrons accelerated by a potential difference of 10^5 volts, the energy of the electron beam is 10^5 eV so that the de Broglie wavelength is

$$\lambda = \frac{ch}{\sqrt{eV(eV + 2m_0 c^2)}}$$

$$= \frac{3 \times 10^8 \times 6.62 \times 10^{-34}}{\{10^5 (10^5 + 2 \times 0.51 \times 10^6)\}^{1/2} \times 1.6 \times 10^{-19}}$$

$$= 0.0371 \times 10^{-10} \text{ m} = 0.0371 \text{ Å}$$

With higher energy electrons, the wavelength is even shorter. So it is possible to construct electron microscopes with a limit of resolution of the order of a fraction of an angstrom quite easily.

In Fig. 9.9. the schematic diagram of the different parts of an electron microscope is shown. In place of the lens systems used for the different parts in an optical microscope, electric and magnetic fields are used in an electron microscope. It is possible to focus the electron beam with the help of the magnetic field produced by a current carrying coil which thus acts as a lens in an electron microscope.

In Fig. 9.9. the thermionic electrons emitted by a heated filament F are first accelerated through a large potential difference (~50 to 100 kilo-volts) towards the anode A. Before reaching the anode, the electrons pass through the grid G. By adjusting the grid potential, the intensity of the electron beam can be controlled. The accelerated electron beam passes through a small hole in the anode and is condensed with the help of the magnetic field produced by the coil C which acts as the condenser lens. The condensed beam falls on the specimen S to illuminate it. The specimen must be very thin so that the illuminating electron beam may pass through it without losing any appreciable energy. The emergent beam then passes through another magnetic field produced by the coil O which acts as the objective lens. The objective lens produces a magnified intermediate image I_1. The electrons coming from I_1 then pass through another magnetic field produced by the coil E which acts as the eye-piece. The final image I produced by the eye-piece is either projected by E on a fluorescent screen or can be photographed by a suitable arrangement.

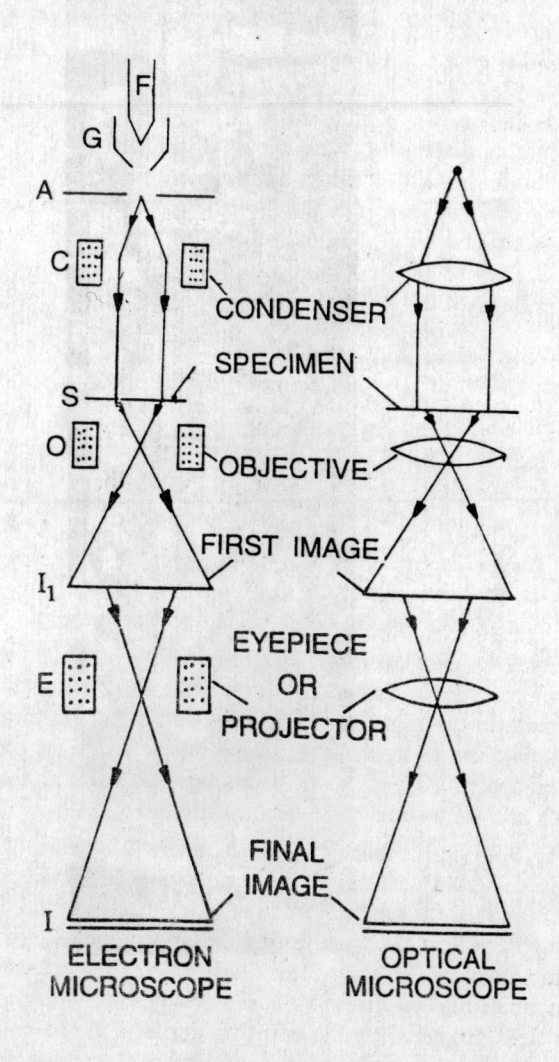

Fig. 9·9 Electron microscope (schematic diagram).

Fig. 9·10. Photograph of an electron microscope. Courtesy : Saha Institute of Nuclear Physics, Calcutta.

Fig. 9·11. Image formed by an electron microscope. Courtesy : Saha Institute of Nuclear Physics, Calcutta.

In Fig. 9·9, side by side with the parts of an electron microscope, the corresponding parts of an optical microscope are shown for comparison.

The photograph of an electron microscope is shown in Fig. 9·10.

It may be noted that the pressure within an electron microscope is kept very low ($<10^{-5}$ mm of Hg) so that the electron beam may not suffer scattering or absorption within the instrument.

In Fig. 9·11 is shown the photograph of the magnified image of a specimen examined by an electron microscope.

9·10 Need for a new mechanics for subatomic particles

The idea of quantization of a physical quantity was first introduced in Physics by Max Planck in 1900 to explain the observed energy distribution of black-body radiation.

A few years later, Einstein introduced the idea of light quantum or photon (1905) in explaining the experimental results on photoelectric emission. A subsequent landmark in the development of quantum ideas was the formulation of Bohr's theory (1912) to explain the wavelengths of the spectral lines of hydrogen and hydrogen-like atoms, based on the three famous postulates of Bohr.

However in formulating his theory of the motion of the electron in the hydrogen atom Bohr did not develop any new dynamical theory. He based his theory on the Newtonian equations of motion, the same equations which explain the Kepler's laws of planetary motion. However he made the new *ad hoc* proposition that out of the infinite number of orbits classically possible only those are permissible for which the angular momentum is an integral multiple of $h/2\pi$ where h is Planck's constant. The more broad-based quantum condition proposed by Sommerfeld and Wilson on the phase integral of motion ($\oint p\,dq = nh$, p and q denoting the generalised momentum and generalised co-ordinate) was also not based on any new dynamical theory.

Thus neither Bohr's quantum condition nor that of Sommerfeld and Wilson was based on any new dynamics of motion of the atomic electrons. The quantum conditions were introduced as *ad hoc* hypotheses to restrict certain dynamical quantities (like the angular momentum or the phase-integral) of classical mechanics to a set of quantized values. It is not possible to explain the quantization of these dynamical quantities on the basis of classical mechanics. Whatever the law of forces one might assume, classical mechanics always predicts continuous variation of the dynamical quantities like the angular momentum, energy *etc*. Further, the dual character of light and material particles (wave and corpuscular) as revealed through diverse experiments was a puzzle which could not be reconciled on the basis of the existing theories. All these called for the urgent development of a new mechanics for the microscopic systems.

9·11 Particles and wave packets

We have seen that the wave nature of matter was first proposed by de Broglie. Later the existence of matter waves was discovered experimentally by Davisson, Germer and G.P. Thomson. In order to develop a consistent theory of the motion of the particles in an atomic system the wave nature of the subatomic particles may be regarded as a basic assumption. Thus the representation of a particle by a *complex wave function* must be one of the fundamental postulates of the new mechanics applicable in the subatomic domain, the wave particle dualism being reconciled through the relations :

$$E = \hbar \omega \quad \text{and} \quad p = \hbar k$$

ω is the angular frequency and k the magnitude of the wave vector. In the classical limit, the particle velocity υ is taken to be equal to the group velocity of the wave packet which is an example of the "correspondence principle" according to which the quantum mechanical results must approach the results of classical mechanics in the classical limit.

In experiments which depend on the corpuscular characteristics, the behaviour of the wave packet as a whole is manifested. On the other hand, in interference or diffraction experiments, the wave characteristics of the individual wavelength components of the packet are manifested.

9·12 Nature of matter waves

Electron diffraction experiments show that a beam of electrons behaves very much like a beam of photons.

Consider the double slit experiment with electrons similar to Young's experiments in optics (Fig. 9·12a). Let a parallel beam of monoenergetic electrons be incident on two parallel slits each of width a and separated by b where a and b have the dimensions of the de Broglie wavelength of the electrons, $\lambda = h/p$.

If any one slit is shut off, a single slit diffraction pattern as shown in Fig. 9·12b. will be observed. When both the slits are open, the pattern has the appearance shown Fig. 9·12c. The interference effect between the two slits shows up by the appearance of the interference fringes within the envelope of the single slit diffraction pattern.

If the beam intensity is gradually reduced, the same pattern is observed after a sufficiently long time. Even if the intensity is such that essentially a single electron falls on the slit plate at a time, the same interference pattern is observed if a sufficiently large number of electrons is aimed at the slits over a long period of time. So we may be tempted to conclude that each electron interacts with both slits at a time, though according to the classical view a single electron would be expected to pass through only one of the slits at a time. Thus if the electron is regarded as a particle, its motion at one of the slits is affected by the conditions at the other slit. This is a violation of the *condition of locations* (valid strictly in classical physics) according to which the behaviour of a particle at a

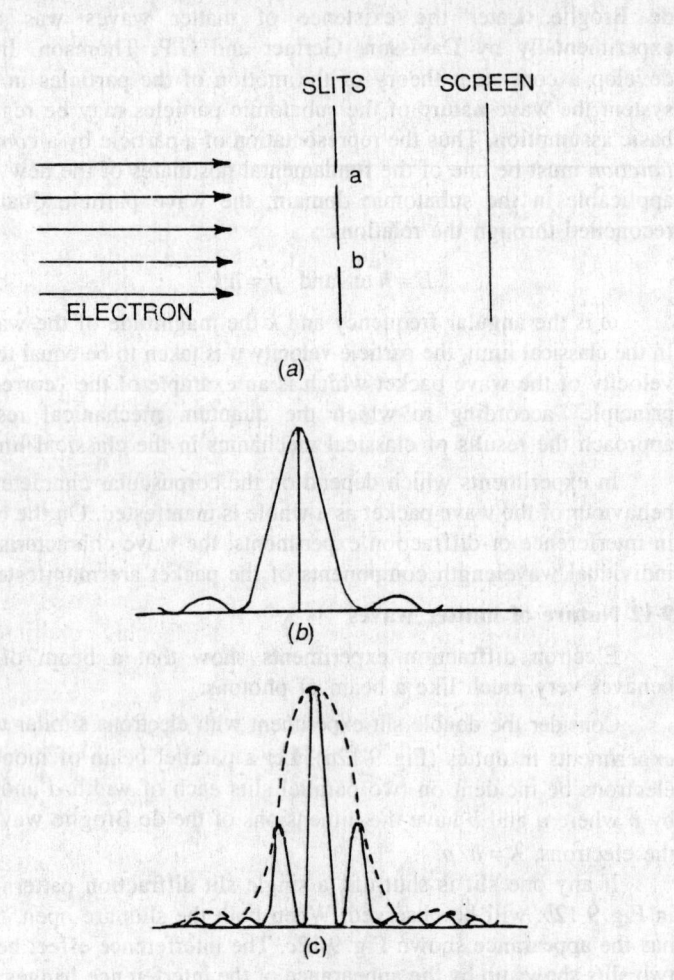

Fig. 9·12. Double slit experiment with electrons.
(a) Experimental arrangement ;
(b) Diffraction pattern with one slit closed ;
(c) Diffraction pattern with both slits open.

point should be influenced by the conditions (*e.g.*, fields) at that point only.

The above argument also applies to the case of the two slit interference pattern produced by light when considered from the photon point of view : each photon would appear to interact with both the slits simultaneously.

Now there are strong reasons to believe that neither the electrons nor the photons can be split into smaller fragments. Hence the results of the double slit interference experiments with electrons or photons can only be interpreted by assuming wave characteristics associated with them. However, these waves have no classical analogue. Instead we have to interpret *the modulus squared of the wave amplitude as the probability density and its integral over a finite region of space as the probability of finding the particle in that region.**

A classical wave, like the e.m. wave, measures certain physical quantities. The amplitude squared of such a wave is a measure of a physically measurable quantity like the energy density or the charge density. For instance, if it measures the charge density, then for a single incident electron in a diffraction experiment, the corresponding wave packet is so split by the diffracting slits that at the positions of the maxima, the detectors detect a fraction of the (amplitude)^2 each. Thus each detector should detect only a fraction of the electronic charge. This may be true on the average if a large number of incident electrons are considered, but it is certainly not true if a single electron is incident on the diffracting plate at a time. This will be detected as a whole by either one or the other detector since the charge on the individual electrons does not split.

So, in quantum mechanics, even though we regard the incident electron wave as divided into two parts in a diffraction experiment, it is the intensity of the wave in a given direction which is proportional to the absolute square of the wave amplitude. So the intensity has a 'probabilistic' interpretation. A quantity which depends quadratically on the amplitude always represents the probability of a physical event to happen.

Thus the probability interpretation of intensities is a distinguishing feature of quantum mechanics in contrast to the classical wave theory.

9·13 Uncertainty relation

The fact that a particle exhibits wave characteristics means that there should be a suitable wave equation to describe its motion. The solution of this equation gives a wave function Ψ (**r**. t) which would describe the state of motion of the particle everywhere in space at every instant of time. In particular, the probability interpretation discussed above should enable us to determine the probability of certain physical events to happen if the wave function is known.

Actually the wave function Ψ will not normally be a monochromatic wave of definite wavelength, since for such a wave, the modulus squared has the same value everywhere in space. So according to the probability interpretation, the particle will have the same probability of

---

* For further details see *Introductory Quantum Mechanics* by the author.

being found anywhere in space. To be able to localize the particle in a finite region, the wave should be such that it has finite amplitude only in that region of space and zero everywhere else. This would mean that the particle would have finite probability of being found in that particular region of space and zero probability outside that region. Such wave functions are shown in Fig. 9·13 for one dimensional motion.

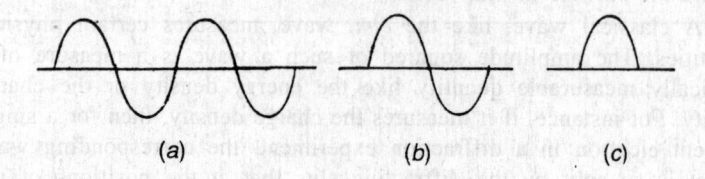

Fig. 9·13. Localized wave packets.

Such limited wave trains can be built up by the superposition of waves of different wavelengths which vary continuously by the method of Fourier integral theorem. So λ for such a wave train does not have a definite value, but a range of values. The amptitudes of the component waves of different wavelengths making up the wave train will also be usually different.

The above discussion shows that the representation of a particle by a wave and its probability interpretation rules out the possibility of locating the particle exactly at a particular point in space and of assigning it a definite momentum *at the same time*. The limited wave train which represents the particle shows that it can be anywhere in the given limited region of space. The range of wavelengths of the waves which must be superposed to obtain such a wave train determines the range in which the momentum of the particle must lie, since the wavelength and momentum are related by de Broglie relation. In order to reduce the extent of the region of location (as in Fig. 9·13) a larger range of wavelengths have to be chosen for superposition, which means a larger range of the momentum.

Thus the wave representation of the particle implies some uncertainty Δx of the position x of the particle and a corresponding uncertainty Δp in specifying its momentum p simultaneously. The more exactly we want to localize the particle (smaller uncertainty Δx in the position x) the less exactly specified will be the momentum (greater uncertainty Δp in momentum p). Heisenberg was the first to point out this inherent limitation in specifying the position and momentum of a particle regarded as a wave which is known as the *uncertainty principle*. He proposed the following relation between the uncertainties Δx and Δp :

$$\Delta x \cdot \Delta p \geq \hbar \tag{9-13-1}$$

where $\hbar = h/2\pi$; h is Planck's constant. This relation is known as Heisenberg's *uncertainty relation*.

The uncertainty relation can be derived in an elementary way as follows :

If there are n waves in a given length Δx, we can write

$$\Delta x = n\lambda \qquad (9\cdot 13\text{-}2)$$

where λ is the mean wavelength. For another wavelength $\lambda + \Delta \lambda$ we can write

$$\Delta x = (n - \Delta n)(\lambda + \Delta \lambda)$$

So we have $\qquad n\lambda + n\,\Delta\lambda - \lambda\,\Delta n - \Delta n \cdot \Delta\lambda = n\lambda$

Neglecting the term $\Delta n \cdot \Delta \lambda$ we then get,

$$\frac{\Delta \lambda}{\lambda} = \frac{\Delta n}{n}$$

If the uncertainty in the number of waves contained in Δx is taken as $\Delta n = 1$, we get $\Delta\lambda/\lambda = 1/n$. But since $p = h/\lambda$, $|\Delta p| = (h/\lambda^2)\,\Delta\lambda$ so that

$$\left|\frac{\Delta p}{p}\right| = \left|\frac{\Delta \lambda}{\lambda}\right| = \frac{1}{n}$$

or, $\qquad \Delta p = p/n \qquad (9\cdot 13\text{-}3)$

Combining Eqs. (9·13-2) and (9·13-3) we get

$$\Delta x \cdot \Delta p = n(h/p) \cdot (p/n) = h$$

Actually the above equality should be replaced by an inequality since Δn may be greater than the assumed value 1. Hence

$$\Delta x \cdot \Delta p \geq h$$

The above one-dimensional consideration can be extended to 3-dimensions. For the different co-ordinate directions, we have

$$\Delta x \cdot \Delta p_x \geq h, \quad \Delta y \cdot \Delta p_y \geq h, \quad \Delta z \cdot \Delta p_z \geq h \qquad (9\cdot 13\text{-}4)$$

It may be noted that the uncertainty in x will have no effect on the precision with which y to z can be measured and vice-versa, *i.e.*, the uncertainty relation holds between the conjugate quantities like the co-ordinate x and its conjugate momentum p_x.

The uncertainty relation can be expressed in other forms. If E is the energy of a quantum system at the time t, then the uncertainties ΔE and Δt are related by the equation

$$\Delta E \cdot \Delta t \geq h \qquad (9\cdot 13\text{-}5)$$

It may be mentioned that the magnitude of the product of the uncertainties is itself somewhat vague. It depends on the way the uncertainties are defined. However the vagueness never exceeds a factor of 10.

A more precise definition of the uncertainties in x and p can be given in terms of the standard deviations or the root mean squared deviations of these quantities as in statistics. Thus the mean squared deviation of x is

$$\Delta x^2 = \langle(x - \langle x \rangle)^2\rangle = \langle x^2 - 2x\langle x \rangle + \langle x \rangle^2\rangle$$
$$= \langle x^2 \rangle - 2\langle x \rangle \langle x \rangle + \langle x \rangle^2 = \langle x^2 \rangle - \langle x \rangle^2 \quad (9\cdot13\text{-}6)$$

where the angular brackets < > denote the mean values which are known as the expectation values in quantum mechanics.

Similarly the mean squared deviation of the momentum p_x is give by

$$\Delta p_x^2 = \langle(p_x - \langle p_x \rangle)^2\rangle = \langle p_x^2 \rangle - \langle p_x \rangle^2 \quad (9\cdot13\text{-}7)$$

The uncertainties in x and p_x are then

$$\Delta x = \{\langle x^2 \rangle - \langle x \rangle^2\}^{1/2} \quad (9\cdot13\text{-}8)$$

$$\Delta p_x = \{\langle p_x^2 \rangle - \langle p_x \rangle^2\}^{1/2} \quad (9\cdot13\text{-}9)$$

Using the definitions of the quantum mechanical expectation values given in § 10·12, it can then be proved that

$$\Delta x \cdot \Delta p_x > \hbar/2 \quad (9\cdot13\text{-}10)$$

A formal proof of this will be given in §. (10·13).

The uncertainty principle points out the basic limitation of our experimental capabilities when dealing with micro-systems. The laws of nature are consistent only if we accept this basic limitation.

The entire fabric of quantum mechanics is based on the validity of the uncertainty principle which "protects" quantum mechanics, so to say.*§§

If it were possible to measure both x and p_x silmultaneously with accuracies better than the limit imposed by the uncertainty principle, then quantum mechanics would collapse (Heisenberg). Uptil now, nobody has been able to figure out ways of achieving such accuracies. So 'quantum mechanics maintains its perilous but still correct existence'.

In general if q and p denote two canonically conjugate variables, e.g., the generalized co-ordinate and generalized momentum, then the uncertainty relation, using the above definition, become

$$\Delta q \cdot \Delta p \geq \frac{\hbar}{2} \quad (9\cdot13\text{-}11)$$

It should be noted that due to the uncertainty principle, our idea regarding *causality* needs revision. The classical concept of *cause and effect* applies to systems which are left undisturbed. Since a small system

---

* §§ See *Feynman lectures*, Vol. III by Richard P. Feynman, Robert B. Leighton and Mathews Sands.

Wave Particle Duality : Heisenberg's Uncertainty

like the atomic system cannot be observed without producing serious disturbance in it, we cannot expect to find strict casual connection between the results of our observations. There is an unavoidable indeterminancy in the calculation of the observational results. Quantum mechanical theory only enables us to calculate the probability of our obtaining a particular result when we make the observation.

9·14 γ-ray microscope experiment

Heisenberg's uncertainty relation can be illustrated by considering a simple hypothetical experiment. This is the experiment on the detection of an electron by a microscope proposed by Niels Bohr. To locate it as exactly as possible, Bohr assumed that very short wavelength radiation (γ-rays) may be used to 'illuminate' the electron and this radiation scattered from the electron may then be observed by means of a so called γ-ray, microscope. Since ordinary optical parts cannot focus γ-rays, such an experiment is actually hypothetical.

Now the limit of resolution of a microscope is given by

$$\Delta x \approx \lambda/\sin\alpha \qquad (9\cdot14\text{-}1)$$

where Δx is the minimum separation between two objects which can be resolved as separated by the microscope. 2α is the angle which the objective lens subtends at the position of the electron (see Fig. 9·14).

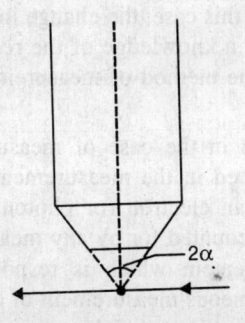

Fig. 9·14. Gama ray microscope experiment.

$\sin\alpha$ measures the numerical aperture of the microscope. The microscope cannot locate the position of the electron more precisely than Δx which is thus a measure of inexactitude in the position of the electron.

γ-rays of energy $h\nu$ have momentum $p = h\nu/c = h/\lambda$. Assuming the scattered γ-ray to have the same momentum p which it has originally when it enters the microscope, the x-component of this momentum p_x can have any value between 0 and $p\sin\alpha$, since it can be scattered in any direction between $(\pi/2 - \alpha)$ and $(\pi/2 + \alpha)$ to enter the microscope. Thus, there will be an uncertainty in determining the *x-component* of its momentum of $p\sin\alpha$ which is also the uncertainty in defining the

x-component of the momentum of the electron since the latter gains recoil momentum from the photon. Hence

$$\Delta p_x = p \sin \alpha = (h/\lambda) \sin \alpha \qquad (9\cdot14\text{-}2)$$

Combining Eqs. (9·14-1) and (9·14-2) we get,

$$\Delta x \cdot \Delta p_x = (\lambda/\sin \alpha) \cdot (h/\lambda) \sin \alpha = h$$

which is Heisenberg's uncertainty relation.

Uncertainty principle has the consequence that the position and momentum of a physical system cannot be simultaneously determined with arbitrary precision. Trying to improve the precision of measurement of position reduces the precision of measurement of momentum and vice-versa. These uncertainties in the simultaneous measurement of position and momentum are, however, not due to any defect in the method of measurement. They arise as a result of an inherent limitation of nature. However precise may be the method of measurement, there is no escape from these uncertainties.

The situation is different in classical physics. It is well known that even in this case, the method of measurement of any physical quantity changes the value of the measured quantity. For instance, in measuring the current in a circuit by an ammeter, the presence of the ammeter itself reduces the value of the current because of the finite resistance of the ammeter coil. However, in this case, the change in the value of the current can be accounted for from a knowledge of the resistance of the ammeter coil. So the influence of the method of measurement can be corrected in this case.

However, this not so in the case of measurements on the atomic scale. The change introduced in the measurement of the position (or its conjugate momentum) of an electron (or photon) due to the method of measurement cannot be accounted for by any means. It is this change due to the method of measurement which is responsible for the inherent uncertainties in the simultaneous measurement of conjugate quantities like position and momentum as expressed by the uncertainly relation. The entire fabric of quantam mechanics is based on the validity of the uncertainty principle.

Bohr enunciated a principle known as the *complementarity principle* to express this inherent limitation in the case of atomic and subatomic particles.

As we have seen, the particle and wave characters of corpuscles or photons are *complementary*. However they are mutually exclusive, *i.e.*, if in some experiment the particle character is revealed, the wave character cannot be simultaneously revealed in the same experiment and vice-versa. This mutual exclusiveness is known as the *principle of complementarity*.*

---

* For further details see *Introductory Quantum Mechanics* by the author.

9·15 Applicability of classical and quantum concepts

The uncertainty relation can be used to decide about the applicability of classical and quantum concepts.

Let us examine the motion of an electron in a Wilson cloud chamber. If it produces a track of width 0·01 cm and its energy is 10^5 eV, then the uncertainty in its position at right angles to the track which is along x is $\Delta y = 0·01 = 10^{-2}$ cm $= 10^{-4}$ m. Hence the minimum uncertainty in its momentum is $\Delta p \approx \Delta p_y = \hbar / \Delta y = 1·05 \times 10^{-34} / 10^{-4} = 1·05 \times 10^{-30}$ kg. m/s On the other hand the momentum of the electron is

$$p = p_x = \sqrt{2mE} = (2 \times 9·1 \times 10^{-31} \times 1·6 \times 10^{-19} \times 10^5)^{1/2}$$

$$= 1·70 \times 10^{-22} \text{ kg m/s}$$

Hence $\Delta p \ll p$. Since the motion of the electron along its track in a cloud chamber is governed by the laws of classical mechanics, we conclude that classical mechanics is applicable in those cases for which $\Delta p \ll p$, i.e., the uncertainty in momentum measurement is small compared to the total momentum. So we can analyse its motion on corpuscular theory.

On the other hand, for an electron in the $n = 1$ Bohr orbit of the hydrogen atom, the radius of the orbit is

$$a_0 = \epsilon_0 h^2 / \pi me^2 = 4\pi \epsilon_0 \hbar^2 / me^2$$

where ϵ_0 is the dielectric constant of empty space. Since the electron can be anywhere along the orbit, we may say that the uncertainly in its position will be equal to the circumference of the orbit which is

$$2\pi a_0 = 8\pi^2 \epsilon_0 \hbar^2 / me^2$$

Hence the uncertainty in its momentum is

$$\Delta p = \hbar / 2\pi a_0 = me^2 / 8\pi^2 \hbar \epsilon_0$$

The velocity of the electron is $v = e^2 / 4\pi \epsilon_0 \hbar$. Hence its momentum is $mv = \dfrac{me^2}{4\pi \epsilon_0 \hbar}$ i.e., $\Delta p \sim p$. We know that classical mechanics is not applicable in the case of the motion of the electron in the atomic orbit. So we conclude that quantum considerations are necessary when the uncertainty in the momentum is comparable to or greater than the total momentum. We have to consider the wave character of the electron in this case.

It may be noted that the circumference of the electron orbit is equal to its de Broglie wavelength for $n = 1$

$$2\pi a_0 = 8\pi^2 \epsilon_0 \hbar^2 / me^2 = \lambda$$

The reason for this agreement can be understood from de Broglie's theory. Since we have for the electron, the angular momentum $mvr = nh/2\pi$, we get from the de Broglie relation (Eq. 9·3–6)

$$2\pi r = nh/mv = n\lambda$$

Thus there must be an integral number of full waves in the orbit of atomic radius. So we may say that only those orbits are permitted which contain an integral number of full de Broglie waves. This gives an "explanation" of the quantum condition on the basis of de Broglie's theory (see Fig. 9·15).

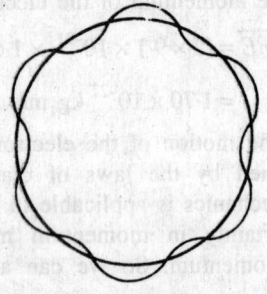

Fig. 9·15. De Broglie wave in atomic orbit.

The situation is analogous to the case of stationary waves in a string with both ends fixed. Only those modes are excited which have nodes at the ends so that for each mode, an integral number of half waves are contained within the length of the string, satisfying the conditions $L = n\lambda/2$.

9·16 Principle of superposition

The interference effect discussed in § 9·13 requires us to assume the validity of the *principle of superposition*. We are familiar with this principle in the case of superposition of optical disturbances which give rise to the well-known phenomena of interference and diffraction of light. In the case of coherent light, the amplitudes at a point are added with proper phases to get the resultant disturbance. The modulus squared of this resultant disturbance then gives the intensity at the point.

Let us now examine the principle of superposition from the corpuscular point of view. If we apply this principle in the case of the double slit interference experiment described in § 9·12, we have to add up the wave amplitudes with proper phases representing the *states of the individual electron* (*or photon*) to get the resultant amplitude. However the superposition principle in this case is basically different from that applied in classical physics.. As we had stated earlier, the interference effect is observed even if only one electron (or photon) passes through the slits at a time if we allow sufficiently long time to elapse so that a large number of electrons ultimately passes through the slits, one after another.

To apply the principle of superposition in this case, we have to consider the different states of the individual particles. In the case under consideration, these will correspond to the two possible paths of the particle through the two slits. The resultant state is to be regarded as a result of the superposition of these two states.

In classical physics, a complete set of numerical values for all the coordinates and velocities at a given instant of time is specified to represent a state. Because of the limitation in the precision of observation as enunciated by the uncertainly relation, a quantum mechanical state, on the other hand, is specified by a fewer or more indefinite data which are theoretically possible without mutual contradiction.

A state formed by the superposition of two or more states in quantum mechanics is thus intermediate in character. The result of measurement of some property of the system will approach more closely to one of the superposed states if the latter has greater probability of occurrence in the process of superposition. Thus when an observation is made of an atomic system the result in general will not be determinate. If the experiment is repeated several times under identical conditions, several different results corresponding to the different superposed states will generally be found so that some kind of average result may be assigned to the composite state obtained by the superposition principle.

Thus the superposition principle in quantum mechanics demands indeterminacy in the results of observations.. In this respect it is basically different from the superposition principle in classical physics.

It may be noted that the principle of superposition is one of the basic postulates of quantum mechanics.

Problems

1. Calculate the de Broglie wavelength of electrons of total energies 2 MeV and 20 MeV. $(6.4 \times 10^{-13}$ m ; 6.2×10^{-14} m).
2. Calculate the de Broglie wavelength of protons of kinetic energy 20 MeV and compare it with the radius of a medium heavy nucleus.
$(6.4 \times 10^{-15}$ m)
3. Assuming the relation between the phase velocity and group velocity, prove that for low energy electrons, the group velocity is equal to the velocity of the electrons. Use non-relativistic expression.
4. In an electron microscope electrons accelerated to 10^4 volts are used. What is the maximum possible resolving power of the microscope ? If x-rays of the same energy are used in an imaginary x-ray microscope, what will be the resolving power ? $(8.16 \times 10^8$; $8.06 \times 10^7)$
5. Calculate the de Broglie wavelength of thermal neutrons at 0°C and compare it with that of electrons of same average energy.
$(\lambda_n = 1.52$ Å ; $\lambda_e = 65.3$ Å)

6. A beam of electrons of energy 400 eV is incident on a nickel surface at an angle of 20° to the normal. If the gain in energy of the electrons in going into nickel is 18 eV, calculate the angle of refraction of the beam at the surface. (18·35°)

7. How much kinetic energy is required to localize a neutron within an atomic spacing in a nickel crystal ($d = 2.15$ Å) ? (4.5×10^{-4} eV)

8. Show that if a proton is observed by an electron microscope the limitation on the shortest distance for the localization of the proton is set by the de Broglie wavelength of the incident electron. (Use non-relativistic expressions).

9. Electrons having an initial de Broglie wavelength λ_e are observed by scattering protons of de Broglie wavelength λ_p from them. If $\lambda_e << m_p \lambda_p / m_e$ then prove that the shortest distance within which the electron can be localized is λ_e. (Use non-relativistic expressions).

10. A particle is confined within a potential box with rigid walls of width l. If the uncertainty in localizing the particle is taken to be l, calculate its ground state energy, assuming its momentum to be equal to the uncertainty in its momentum. Compare this result with that obtained in § 11·2. Can you account for the difference in the two values ?

($\hbar^2 / m l^2$)

10
INTRODUCTION TO WAVE MECHANICS

10·1 Introduction

We have seen that the origin of quantized motion cannot be explained by classical mechanics. The old quantum theory is also inadequate to give a logical explanation for the quantization of certain dynamical quantities in the atomic or subatomic domain.

According to Heisenberg the reason for this failure of the old quantum theory is due to the formulation of the theory on the basis of classical concepts. Heisenberg argued that we should try to develop an atomic mechanics on the basis of quantities which are actually observable in the atomic domain, *e.g.*, the energies of the atomic states found by measuring the wavelengths of the spectral lines, intensities of the spectral lines *etc.* Quantities *e.g.*, the positions or the velocities of the electrons which appear in the old quantum theory cannot be precisely determined because of the uncertainty principle and hence cannot describe the behaviour of atomic systems.

It was on this basis that Heisenberg (1925), for the first time, invented the *matrix mechanics* which was further developed by Born and Jordan (1925). Soon afterwards, Erwin Schrödinger (1926) independently developed *wave mechanics* on the basis of de Broglie's hypothesis of wave-particle duality and proposed a wave equation for describing the motion of atomic systems. The Schrödinger wave equation has become the foundation stone of non-relativistic quantum mechanics or wave mechanics.

Later Schrödinger (1926) showed that though the matrix mechanics of Heisenberg and his wave mechanics were entirely different in form, in content they were identical.

Subsequently P.A.M. Dirac (1930) proposed a general formalism of quantum mechanics which is a unifying concept of the above two formulations. Besides, the theory of interaction of the radiation field with atomic particles was developed by Dirac (1927), Jordan and Pauli (1928) and others. The foundations of relativistic quantum mechanics were also laid about this time by Dirac (1928).

10·2 Wave function ; Schrödinger wave equation

The space time behaviour of an atomic system is determined by the laws of probability. Just as the intensity of light at different points in interference or diffraction experiments is determined by the amplitude squared of light wave, so also in the case of an atomic electron, the probability of its being found at a particular point is determined by the square of the amplitude of the corresponding wave function.

Thus we may say that for any quantum system the wave function $\Psi(\mathbf{r}, t)$ determines the entire space-time behaviour of the system. Our first step is to develop a wave equation satisfied by $\Psi(\mathbf{r}, t)$.

This equation cannot be 'deduced' on theoretical considerations. We have to guess the equation. Its success will depend on its ability to explain various observed phenomena.

Some fundamental characteristics of the equation can be inferred :

(a) The differential equation satisfied by the wave function must be *first order in time*, so that if Ψ is known at some point $\mathbf{r}$ at the initial instant t_0, then at any subsequent instant t, the function $\Psi(\mathbf{r}, t)$ at the same point will be uniquely determined. This may be regarded as the basic postulate of quantum mechanics.

(b) The wave equation must be *linear* so that if $\Psi_1(\mathbf{r}, t)$ and $\Psi_2(\mathbf{r}, t)$ are two solutions of the equation, then any linear combination $(C_1\Psi_1 + C_2\Psi_2)$ will also be a solution. This ensures the validity of the superposition principle discussed earlier.

(c) The wave equation must be consistent with de Broglie's hypothesis and correspondence principle, *i.e.*, the conclusions of the new theory should be valid even in those cases for which classical mechanics is applicable. Just as the laws of geometrical optics are obtained as approximations from the wave theory of light, so also the laws governing the motion of the wave-packets as a whole in de Broglie's theory should be derivable as approximations from the new wave mechanics. This may be termed "correspondence principle"

Schrödinger equation for one-dimensional motion

The wave-packet representing an electron can be built up by the superposition of harmonic waves of continuously varying wave numbers $k = 2\pi/\lambda$ of appropriate amplitudes. Let us consider one of the component waves of a particular wave number k and a particular frequency $\nu = \omega/2\pi$ which by superposition with the others make up the wave-packet. We consider one-dimensional wave propogation, say along x. If $\Psi(x, t)$ represents the wave function we write

Introduction to Wave Mechanics

$$\Psi(x, t) = A \exp i(kx - \omega t) \tag{10-2-1}$$

If λ be the wavelength, we have from de Broglie theory

$$p = \frac{h}{\lambda} = \hbar k \tag{10-2-2}$$

Here p is the momentum of the particle. p and k are both along x. For a free particle (no force) of mass m, the energy is

$$E = p^2/2m = \hbar^2 k^2/2m = h\nu = \hbar\omega \tag{10-2-3}$$

Differentiating $\Psi(x, t)$ *once* with respect to t, we get

$$\frac{\partial \Psi}{\partial t} = -i\omega \Psi \tag{10-2-4}$$

On the other hand, differentiating $\Psi(x, t)$ twice with respect to x gives

$$\frac{\partial^2 \Psi}{\partial x^2} = -k^2 \Psi \tag{10-2-5}$$

From equations (10-2-4) and (10-2-5), we get

$$i\hbar\, \partial \Psi/\partial t = i\hbar (-i\omega) \Psi = \hbar\omega \Psi$$

and $-(\hbar^2/2m)\, \partial^2 \Psi/\partial x^2 = -(\hbar^2/2m)(-k^2 \Psi) = (\hbar^2 k^2/2m) \Psi$

Then, by the use of Eq. (10-2-3) we get,

$$i\hbar \frac{\partial \Psi}{\partial t} = -\left(\frac{\hbar^2}{2m}\right) \frac{\partial^2 \Psi}{\partial x^2} \tag{10-2-6}$$

Equation (10-2-6) is the one-dimensional time-dependent Schrödinger wave equation for a free particle of mass m. Note that Eq. (10-2-6) does not contain any of the parameters of the component waves of the wave packet representing the particle, *e.g.*, its wave number or frequency. Thus any component wave with a given set of values of these dynamical parameters obey the same wave equation. So the motion of the wave-packet formed by the superposition of such component waves is also determined by this wave equation.

Eq. (10-2-6) is the quantum mechanical analogue of the classical relationship (Eq. 10-2-3) between energy and momentum.

If the particle is acted upon by a force field derivable from a potential $V(x)$, then its total energy will be

$$E = \frac{p^2}{2m} + V(x)$$

In this case, in place of Eq. (10-2-3) we get, since $E = \hbar\omega$ and $p = \hbar k$,

$$\hbar\omega = \frac{\hbar^2 k^2}{2m} + V(x)$$

We then get using Eq. (10·2-4) and (10·2-5)

$$i\hbar \frac{\partial \Psi}{\partial t} = \hbar \omega \Psi$$

and $\quad -\dfrac{\hbar^2}{2m} \dfrac{\partial^2 \Psi}{\partial x^2} + V(x) \Psi = \left\{ \dfrac{\hbar^2 k^2}{2m} + V(x) \right\} \Psi$

So in place of Eq. (10·2-6) we get the one dimensional time-dependent Schrödinger wave equation as

$$i\hbar \frac{\partial \Psi}{\partial t} = \left\{ -\frac{\hbar^2}{2m} \frac{\partial^2}{\partial x^2} + V(x) \right\} \Psi \qquad (10\cdot2\text{-}7)$$

Schrödinger equation for three-dimensional motion

For three dimensional motion, the wave function can be written as

$$\Psi(r, t) = A \exp\{i(\mathbf{k} \cdot \mathbf{r} - \omega t)\} \qquad (10\cdot2\text{-}8)$$

where $\qquad \mathbf{k} \cdot \mathbf{r} = k_x x + k_y y + k_z z \qquad (10\cdot2\text{-}9)$

Let us operate Ψ by the Laplacian operator

$$\nabla^2 = \frac{\partial^2}{\partial x^2} + \frac{\partial^2}{\partial y^2} + \frac{\partial^2}{\partial z^2}$$

From Eqs. (10·2-8) and (10·2-9) we have

$$\frac{\partial \Psi}{\partial x} = ik_x \Psi \text{ and } \frac{\partial^2 \Psi}{\partial x^2} = -k_x^2 \Psi$$

Similarly $\quad \dfrac{\partial^2 \Psi}{\partial y^2} = -k_y^2 \Psi, \; \dfrac{\partial^2 \Psi}{\partial z^2} = -k_z^2 \Psi$

Hence $\quad \nabla^2 \Psi = -(k_x^2 + k_y^2 + k_z^2) \Psi = -k^2 \Psi$

so that $\quad -\dfrac{\hbar^2}{2m} \nabla^2 \Psi = \dfrac{\hbar^2 k^2}{2m} \Psi = \dfrac{p^2}{2m} \Psi$

Hence using Eq. (10·2-3) we get

$$i\hbar \frac{\partial \Psi}{\partial t} = \hbar \omega \Psi = \frac{\hbar^2 k^2}{2m} \Psi = \frac{p^2}{2m} \Psi = -\frac{\hbar^2}{2m} \nabla^2 \Psi$$

So the three-dimensional time-dependent Schrödinger wave equation for a free particle becomes

$$i\hbar \frac{\partial}{\partial t} \Psi(\mathbf{r}, t) = -\frac{\hbar^2}{2m} \nabla^2 \Psi(\mathbf{r}, t) \qquad (10\cdot2\text{-}10)$$

As in the case of one-dimensional motion, the Schrödinger equation for three-dimensional motion in the presence of a potential field $V(\mathbf{r})$ becomes

$$i\hbar \frac{\partial}{\partial t} \Psi(\mathbf{r},t) = \left\{ -\frac{\hbar^2}{2m} \nabla^2 + V(\mathbf{r}) \right\} \Psi(\mathbf{r},t) \quad (10\cdot2\text{-}11)$$

Hamiltonian :

Eq. (10·2-7) can be written as

$$i\hbar \frac{\partial}{\partial t} \Psi(x,t) = \hat{H}\Psi(x,t) \quad (10\cdot2\text{-}12)$$

where $\hat{H}$ is known as the *Hamiltonian operator* or simply the *Hamiltonian* (see § 10·3) for one-dimensional motion

$$\hat{H} = -\frac{\hbar^2}{2m} \frac{\partial^2}{\partial x^2} + V(x) \quad (10\cdot2\text{-}13)$$

Similarly Eq. (10·2-11) can be writen as

$$i\hbar \frac{\partial}{\partial t} \Psi(\mathbf{r},t) = \hat{H}\Psi(\mathbf{r},t)$$

where the three-dimensional Hamiltonian is given by

$$\hat{H} = -\frac{\hbar^2}{2m} \nabla^2 + V(\mathbf{r}) \quad (10\cdot2\text{-}14)$$

10·3 Operators in quantum mechanics

In mathematics, operators provide us with tools for obtaining new functions from a given function. An operator $\hat{\alpha}$ operating on the function $f(x)$ generates a new function $g(x)$: $\hat{\alpha} f(x) = g(x)$.

As an example let $\hat{\alpha} = d/dx$ and $f(x) = x^3$. Then

$$\hat{\alpha} f(x) = \frac{d}{dx} x^3 = 3x^2$$

Thus the new function generated is $g(x) = 3x^2$

In quantum mechanics, each dynamical variable is represented by an operator and provides a link with the latter through the correspondence principle.*

Operators in quantum mechanics are *linear operators*. A linear operator commutes with constants and obeys distributive and associative laws. The sum of two linear operators is commutative, but their product may or may not be commutative.

---

* Throughout this book we shall represent the operator corresponding to the dynamical variable α by putting the *caret* symbol (^) above α, *i.e.*, by the notation $\hat{\alpha}$.

Commutators :

Thus if $\hat{\alpha}$ and $\hat{\beta}$ are two linear operators, then these may or may not *commute*, *i.e.*, *the* product operator $\hat{\alpha}\hat{\beta}$ may or may not be equal to the operator $\hat{\beta}\hat{\alpha}$. The *commutator* or the *commutator bracket* of two operators $\hat{\alpha}$ and $\hat{\beta}$ is defined as

$$[\hat{\alpha}, \hat{\beta}] = \hat{\alpha}\hat{\beta} - \hat{\beta}\hat{\alpha} \qquad (10\cdot3\text{-}1)$$

When $\hat{\alpha}$ and $\hat{\beta}$ commute, the commutator vanishes ; *i.e.*,

$$[\hat{\alpha}, \hat{\beta}] = 0 \qquad (10\cdot3\text{-}2)$$

If $\hat{\alpha}$ and $\hat{\beta}$ do not commute, then the commutator $[\hat{\alpha}, \hat{\beta}] \neq 0$.

Examples :

(a) Let $\hat{\alpha} = \partial/\partial x$ and $\hat{\beta} = \partial^2/\partial x^2$; let $f(x)$ be a function on which they operate. Then

$$\hat{\alpha}\hat{\beta} f(x) = \frac{\partial}{\partial x} \frac{\partial^2}{\partial x^2} f(x) = \frac{\partial^3 f}{\partial x^3}$$

$$\hat{\beta}\hat{\alpha} f(x) = \frac{\partial^2}{\partial x^2} \frac{\partial}{\partial x} f(x) = \frac{\partial^3 f}{\partial x^3}$$

Hence $\quad (\hat{\alpha}\hat{\beta} - \hat{\beta}\hat{\alpha}) f(x) = \left(\dfrac{\partial}{\partial x} \dfrac{\partial^2}{\partial x^2} - \dfrac{\partial^2}{\partial x^2} \dfrac{\partial}{\partial x} \right) f(x)$

$$= \frac{\partial^3}{\partial x^3} f(x) - \frac{\partial^3}{\partial x^3} f(x) = 0$$

i.e., $\quad \left[\dfrac{\partial}{dx}, \dfrac{\partial}{\partial x^2} \right] \equiv \dfrac{\partial}{\partial x} \dfrac{\partial^2}{\partial x^2} - \dfrac{\partial^2}{dx^2} \dfrac{\partial}{\partial x} = 0 \qquad (10\cdot3\text{-}3)$

So the operators $\partial/\partial x$ and $\partial^2/\partial x^2$ commute.

(b) Let $\hat{\alpha} = x$, $\hat{\beta} = \partial/\partial x$; let $f(x)$ be a function. Then

$$\hat{\alpha}\hat{\beta} f(x) = x \frac{\partial}{\partial x} f(x)$$

$$\hat{\beta}\hat{\alpha} f(x) = \frac{\partial}{\partial x} x f(x) = x \frac{\partial}{\partial x} f(x) + f(x)$$

Hence $\quad (\hat{\alpha}\hat{\beta} - \hat{\beta}\hat{\alpha}) = \left(x \dfrac{\partial}{\partial x} - \dfrac{\partial}{\partial x} x \right) f(x)$

$$= x \frac{\partial f}{\partial x} - x \frac{\partial f}{\partial x} - f(x) = -\mathbf{1} \cdot f(x)$$

This relation holds for any arbitrary differentiable function $f(x)$. Removing $f(x)$ from both sides, we get

Introduction to Wave Mechanics

$$\left[\hat{x}, \frac{\partial}{\partial x}\right] = x\frac{\partial}{\partial x} - \frac{\partial}{\partial x}x = -1 \tag{10.3-4}$$

So the operators $\hat{x}$ and $\partial/\partial x$ do not commute.

Quantum mechanical operators :

To understand the role of operators in quantum mechanics, let us differentiate the wave function $\Psi(x, t)$ for a free particle given by Eq. (10·2-1) with respect to x and t. We get

$$\frac{\partial}{\partial x}\Psi(x,t) \frac{\partial}{\partial x} A \exp i(kx - \omega t) = ik\Psi(x,t) \tag{10.3-5}$$

$$\frac{\partial}{\partial t}\Psi(x,t) = \frac{\partial}{\partial t} A \exp i(kx - \omega t) = -i\omega \Psi(x,t) \tag{10.3-6}$$

Then using Eqs. (10·2-2) and (10·2-3), we can write

$$\frac{\hbar}{i}\frac{\partial}{\partial x}\Psi(x,t) = \hbar k \Psi(x,t) = p_x \Psi(x,t) \tag{10.3-7}$$

$$i\hbar\frac{\partial}{\partial t}\Psi(x,t) = \hbar\omega \Psi(x,t) = E\Psi(x,t) \tag{10.3-8}$$

Thus the dynamical variables momentum and energy of a particle can be represented by the two mathematical operators $\frac{\hbar}{i}\frac{\partial}{\partial x}$ and $i\hbar\frac{\partial}{\partial t}$ respectively.

In addition since the coordinate x is a multiplying operator, we get the following results :

$$x \longrightarrow \hat{x} = x, \quad p_x \longrightarrow \hat{p}_x = \frac{\hbar}{i}\frac{\partial}{\partial x} = -i\hbar\frac{\partial}{\partial x} \tag{10.3-9}$$

$$E \longrightarrow \hat{E} = i\hbar\,\partial t \tag{10.3-10}$$

In the above equations, corresponding to the dynamical variables written to the left of the arrows, the quantum mechanical operators are given on the right. These operator representations (Eq. (10·3-9) are known as Schrödinger representations. Thus if $\alpha(x, p_x)$ is a dynamical variable which is a function of x and p_x, then the operator representation of α will be

$$\hat{\alpha}(x, p_x) = \alpha(\hat{x}, \hat{p}_x) = \alpha\left(\hat{x}, \frac{\hbar}{i}\frac{\partial}{\partial x}\right) \tag{10.3-11}$$

Examples :

Kinetic energy of a particle : For one dimensional motion, this is given

$$E_k = \frac{p_x^2}{2m}$$

Then applying the rule given above, we have

$$\hat{E}_k = \frac{1}{2m}\hat{p}_x^2 = \frac{1}{2m}\left(\frac{\hbar}{i}\frac{\partial}{\partial x}\right)^2 = -\frac{\hbar^2}{2m}\frac{\partial^2}{\partial x^2} \qquad (10\cdot 3\text{-}12)$$

Hamiltonian for one dimensional motion: The classical expression for the Hamiltonian function for one dimensional motion which is equal to the total energy of the particle is given by

$$H(x, p_x) = \frac{p_x^2}{2m} + V(x)$$

The *Hamiltonian operator* or simply the *Hamiltonian* for the particle is then given by

$$\hat{H} = \frac{1}{2m}\hat{p}_x^2 + V(\hat{x}) = \frac{1}{2m}\left(\frac{\hbar}{i}\frac{\partial}{\partial x}\right)^2 + V(x) = -\frac{\hbar^2}{2m}\frac{\partial^2}{\partial x^2} + V(x)$$
$$(10\cdot 3\text{-}13)$$

This form of $\hat{H}$ is identical with Eq. (10·2-13).

It may be noted that the quantum mechanical operators for x and p_x satisfy the following commutation relations (see Eq. 10·3-4)]:

$$\left[\hat{x}, \hat{p}_x\right] = \left[\hat{x}, \frac{\hbar}{i}\frac{\partial}{\partial x}\right] = -i\hbar\left(x\frac{\partial}{\partial x} - \frac{\partial}{\partial x}x\right) = i\hbar \qquad (10\cdot 3\text{-}14)$$

i.e., the commutator $\left[\hat{x}, \hat{p}_x\right] \neq 0$. As will be seen later, the commutation relation satisfied by two operators in quantum mechanics is intimately related to the possibility of simultaneous measurement of the two corresponding dynamical variables in the same state.

Dirac has shown that Eq. (10·3-14) giving the commutation relation $\left[\hat{x}, \hat{p}_x\right] = i\hbar$ is the basic quantization relation in quantum mechanics for one dimensional motion.

The above quantization relation is satisfied not only in Schrödinger or coordinate representation but also in other representations (e.g. momentum representation or matrix representation). For greater details, see *Introductory Quantum Mechanics* by S.N. Ghoshal (2nd Ed.).

As in the case of motion along x, it can be shown that for motions along y and z, the corresponding quantum mechanical operators in Schrödinger representation for the coordinates y and z and the momentum components p_y and p_z are given by:

$$\hat{y} = y, \quad \hat{p}_y = \frac{\hbar}{i}\frac{\partial}{\partial y} \qquad (10\cdot 3\text{-}15)$$

$$\hat{z} = z, \quad \hat{p_z} = \frac{\hbar}{i}\frac{\partial}{\partial z} \qquad (10\cdot3\text{-}16)$$

and
$$[\hat{y}, \hat{p_y}] = i\hbar, \quad [\hat{z}, \hat{p_z}] = i\hbar \qquad (10\cdot3\text{-}17)$$

The set of relations (10·3-9), (10·3-15) and (10·3-16) gives us the operator representations of the coordinate and momentum vectors:

$$\hat{\mathbf{r}} = \mathbf{r}, \quad \hat{\mathbf{p}} = \frac{\hbar}{i}\nabla \qquad (10\cdot3\text{-}18)$$

where ∇ is the differential operator gradient vector, with the cartesian components $\partial/\partial x, \partial/\partial y, \partial/\partial z$.

Three dimensional motion:

As in the one dimensional case, the operator representation of a dynamical variable $\alpha(\mathbf{r}, \mathbf{p})$ is obtained by substituting for $\mathbf{r}$ and $\mathbf{p}$ the operators $\hat{\mathbf{r}}$ and $\hat{\mathbf{p}}$ as given above:

$$\hat{\alpha}(\mathbf{r}, \mathbf{p}) = \alpha(\hat{\mathbf{r}}, \hat{\mathbf{p}}) = \alpha\left(\mathbf{r}, \frac{\hbar}{i}\nabla\right) \qquad (10\cdot3\text{-}19)$$

As an example, the Hamiltonian operator $\hat{H}$ for three dimensional motion can be obtained from the classical expression for Hamiltonian:

$$H = \frac{p^2}{2m} + V(\mathbf{r})$$

Hence $\hat{H} = \frac{1}{2m}\hat{p}^2 + V(\hat{\mathbf{r}}) = -\frac{\hbar^2}{2m}\nabla^2 + V(\mathbf{r}) \qquad (10\cdot3\text{-}20)$

which is the same as Eq. (10·2-14).

Another important operator is the angular momentum operator $\hat{L}$ with the components $\hat{L_x}, \hat{L_y}, \hat{L_z}$ which will be considered in detail in § 11·11.

10·4 Physical interpretation of Ψ; Probability density

We have seen that the wave function $\Psi(x, t)$ which describes the complete space-time behaviour of a particle in one dimensional motion has appreciable amplitudes in those regions where the particle is likely to be found with greater probability. As discussed in § 9·13, quantities which are quadratic in the amplitude of the wave function are to be interpreted as the probability of finding a particle in a given region of space.

We shall assume that the quantity

$$|\Psi(x, t)|^2 dx = \Psi^*(x, t)\Psi(x, t)\, dx$$

is proportional to the probability of finding the particle in the interval x to $x + dx$ at the time t, where Ψ^* is the complex conjugate of Ψ. The total probability of finding the particle anywhere in space is then

$$P \propto \int_{-\infty}^{\infty} |\Psi(x,t)|^2 \, dx$$

We define the position *probability density* as

$$\rho(x,t) = |\Psi(x,t)|^2 \Big/ \int_{-\infty}^{\infty} |\Psi(x,t)|^2 \, dx \qquad (10\text{-}4\text{-}1)$$

Hence the total probability will be

$$P = \int_{-\infty}^{\infty} \rho(x,t) \, dx = \int_{-\infty}^{\infty} |\Psi(x,t)|^2 \, dx \Big/ \int_{-\infty}^{\infty} |\Psi(x,t)|^2 \, dx = 1 \qquad (10\text{-}4\text{-}2)$$

This is as it should be, since the total probability must be unity.

Since $|\Psi(x,t)|^2$ is necessarily positive, Eq. (10·4-2) shows that the probability density $\rho(x,t)$ is always positive which is consistent with the expected behaviour of probability.

10·5 Normalization of the wave function

If $\Psi(x,t)$ is multiplied by a constant C such that $\Psi_N(x,t) = C\Psi(x,t)$ where $\Psi_N(x,t)$ satisfies the relation

$$\int_{-\infty}^{\infty} |\Psi_N(x,t)|^2 \, dx = |C|^2 \int_{-\infty}^{\infty} |\Psi(x,t)|^2 \, dx = 1 \qquad (10\text{-}5\text{-}1)$$

then $\Psi_N(x,t)$ is said to be the *normalized* wave function. From Eq. (10·5-1) we have

$$|C|^2 = 1 \Big/ \int_{-\infty}^{\infty} |\Psi(x,t)|^2 \, dx \qquad (10\text{-}5\text{-}2)$$

C is called the normalization constant. Obviously a wave function is normalizable if $\int_{-\infty}^{\infty} |\Psi(x,t)|^2 dx$ or more generally $\int_{\tau} |\Psi(\mathbf{r},t)| \, d\tau$ over all space remains finite. This is known as the *square-integrability* of the wave function. It should be noted that since the modulus squared $|C|^2$ is determined by Eq. (10·5-2), the normalization constant C remains undefined to the extent of a phase factor.

Using Eqs. (10·4-1), (10·5-1) and (10·5-2), we get the probability density

$$\rho(x,t) = |\Psi_N(x,t)|^2 = \Psi_N^*(x,t)\Psi_N(x,t) \qquad (10\text{-}5\text{-}3)$$

Obviously the probability of finding the particle in an interval x to $x + dx$ will be

Introduction to Wave Mechanics 251

$$\rho(x,t)\,dx = |\Psi_N(x,t)|^2 dx \qquad (10\cdot5\text{-}4)$$

The above relations can be generalized for the case of three dimensional motion to give

$$\rho(\mathbf{r},t) = |\Psi(\mathbf{r},t)|^2 = \Psi^*(\mathbf{r},t)\,\Psi(\mathbf{r},t) \qquad (10\cdot5\text{-}5)$$

$$P = \int_\tau |\Psi(\mathbf{r},t)|^2 d\tau = 1 \qquad (10\cdot5\text{-}6)$$

where $\Psi(\mathbf{r},t)$ is here regarded as normalized and the integration is carried out over the entire three dimensional space.

10·6 Probability current density ; Conservation of probability

Since the total probability

$$\int_\tau \rho(\mathbf{r},t)\,d\tau = \int_\tau |\Psi(\mathbf{r},t)|^2 d\tau = 1 = \text{constant}$$

at every instant of time, any decrease of probability in a given volume element $d\tau$ must be associated with the corresponding increase of probabiliy in some other element. The situation is similar to the flow of charge from a given volume. The change in the total quantity of charge contained in it is equal to the net outward flow of charge through the surfaces enclosing the given volume and is given by the *equation of continuity* :

$$\frac{\partial \rho}{\partial t} + \nabla \cdot \mathbf{j} = 0$$

where ρ is the charge density and $\mathbf{j}$ is the current density. An analogous relation can be deduced in the case of probability density.

Considering a finite volume τ enclosed by the surface S, the rate of change of the probability of findng the particle is given by

$$\frac{\partial}{\partial t}\int_\tau \rho(r,t)\,d\tau = \frac{\partial}{\partial t}\int_\tau \Psi^*\Psi\,d\tau = \int_\tau \left(\frac{\partial \Psi^*}{\partial t}\Psi + \Psi^*\frac{\partial \Psi}{\partial t}\right) d\tau \qquad (10.6\text{-}1)$$

Ψ satisfies the Schrödinger equation

$$\hat{H}\,\Psi(r,t) = i\hbar\frac{\partial}{\partial t}\Psi(r,t) \qquad (10.6\text{-}2)$$

Writing $\hat{H}$ explicitly, we have

$$\left\{-\frac{\hbar^2}{2m}\nabla^2 + V(r,t)\right\}\Psi(r,t) = i\hbar\frac{\partial}{\partial t}\Psi(r,t)$$

Assuming V to be real, we get the complex conjugate of the above equation as

$$\left\{-\frac{\hbar^2}{2m}\nabla^2 + V(r,t)\right\}\Psi^*(r,t) = -i\hbar\frac{\partial}{\partial t}\Psi^*(r,t)$$

So we have
$$\frac{\partial \Psi}{\partial t} = \frac{1}{i\hbar}\left\{-\frac{\hbar^2}{2m}\nabla^2 + V\right\}\Psi \qquad (10\cdot6\text{-}3)$$

$$\frac{\partial \Psi^*}{\partial t} = -\frac{1}{i\hbar}\left\{-\frac{\hbar^2}{2m}\nabla^2 + V\right\}\Psi^* \qquad (10\cdot6\text{-}4)$$

Then
$$\frac{\partial \Psi^*}{\partial t}\Psi + \Psi^*\frac{\partial \Psi}{\partial t} = \frac{i}{\hbar}\left\{\left(-\frac{\hbar^2}{2m}\nabla^2\Psi^* + V\Psi^*\right)\Psi\right.$$
$$\left. -\Psi^*\left(-\frac{\hbar^2}{2m}\nabla^2\Psi + V\Psi\right)\right\}$$
$$= -\frac{i\hbar}{2m}(\Psi\nabla^2\Psi^* - \Psi^*\nabla^2\Psi)$$
$$= -\frac{i\hbar}{2m}\nabla\cdot(\Psi\nabla\Psi^* - \Psi^*\nabla\Psi) \qquad (10\cdot6\text{-}5)$$

Then we have from Eqs. (10·6-1) and (10·6-5)

$$\frac{\partial}{\partial t}\int_\tau \rho\, d\tau = -\frac{i\hbar}{2m}\int_\tau \nabla\cdot(\Psi\nabla\Psi^* - \Psi^*\nabla\Psi)\, d\tau \qquad (10\cdot6\text{-}6)$$

We define probability current density as

$$\mathbf{j} = \frac{i\hbar}{2m}(\Psi\nabla\Psi^* - \Psi^*\nabla\Psi) \qquad (10\cdot6\text{-}7)$$

Then from Eq. (10·6-7) we get by the application of vector divergence theorem

$$\frac{\partial}{\partial t}\int_\tau \rho\, d\tau = -\int \nabla\mathbf{j}\, d\tau = -\oint_S \mathbf{j}\cdot\partial S = -\oint_S j_n\, \partial S \qquad (10\cdot6\text{-}8)$$

wnere **n** denotes the unit vector normal to the surface element dS. and j_n denotes the normal component of j.

Since Eq. (10·6-8) must hold for any arbitrary volume element we have

$$\frac{\partial \rho}{\partial t} + \nabla\cdot\mathbf{j} = 0 \qquad (10\cdot6\text{-}9)$$

This is an equation of continuity which expresses the *conservation of probability density*.

Eq. (10.6-7) can also be written as

$$\mathbf{j} = -\frac{\hbar}{m}\text{Im}(\Psi\nabla\Psi^*) = \frac{\hbar}{m}\text{Re}(i\Psi\nabla\Psi^*) \qquad (10\cdot6\text{-}10)$$

Introduction to Wave Mechanics 253

where Im and Re represent the imaginary and real parts respectively. For one dimensional motion, the probability current density is given by

$$j_x = \frac{i\hbar}{2m}\left(\Psi\frac{\partial \Psi^*}{\partial x} - \Psi^*\frac{\partial \Psi}{\partial x}\right) \tag{10-6-11}$$

10·7 Separation of space and time in Schrödinger equation : Time independent Schrödinger equation

In quantum mechanics, the classical Hamiltonian function H has the operator representation (Eq. 10·2-14)

$$\hat{H} = -\frac{\hbar^2}{2m}\nabla^2 + V(\mathbf{r})$$

Now in classical mechanics, if the potential energy $V(\mathbf{r})$ is not an explicit function of time, then the energy is a constant of motion and $H = E$. We shall assume that this relation is valid in quantum mechanics also. So from the time-dependent Schrödinger equation (10·2-12), we get

$$i\hbar\frac{\partial}{\partial t}\Psi(\mathbf{r}, t) = \hat{H}\Psi(\mathbf{r}, t) = E\Psi(\mathbf{r}, t) \tag{10-7-1}$$

where E = constant. Integration of Eq. (10·7-1) gives

$$\Psi(\mathbf{r}, t) = \psi(\mathbf{r})\exp(-iEt/\hbar) \tag{10-7-2}$$

From Eq. (10·7-1) we then get

$$\hat{H}\psi(\mathbf{r})\exp(-iEt/\hbar) = E\psi(\mathbf{r})\exp(-iEt/\hbar) \tag{10-7-3}$$

Since $\hat{H}$ does not contain time explicitly, it does not operate on the expotential. Hence we get

$$\hat{H}\psi(\mathbf{r}) = E\psi(\mathbf{r}) \tag{10-7-4}$$

If $\hat{H}$ is written explicitly, we get

$$\left\{-\frac{\hbar^2}{2m}\nabla^2 + V(\mathbf{r})\right\}\psi(\mathbf{r}) = E\psi(\mathbf{r}) \tag{10-7-5}$$

or

$$\nabla^2\psi(\mathbf{r}) + \frac{2m}{\hbar^2}\{E - V(\mathbf{r})\}\psi(\mathbf{r}) = 0 \tag{10-7-6}$$

Eq. (10·7-5) or (10·7-6) is the time-independent Schrödinger equation.

For one dimensional motion, the time-independent Schrödinger equation is

$$\left\{-\frac{\hbar^2}{2m}\frac{d^2}{dx^2} + V(x)\right\}\psi(x) = E\psi(x) \tag{10-7-7}$$

or

$$\frac{d^2\psi(x)}{dx^2} + \frac{2m}{\hbar^2}\{E - V(x)\}\psi(x) = 0 \tag{10-7-8}$$

Each solution of the time-independent Schrödinger equation (10·7-6) or (10·7-8) corresponds to a definite energy. If we write $\psi = \psi_n(\mathbf{r})$ as the solution for $E = E_n$, then a particular solution of Eq. (10·2-11) can be written as

$$\Psi_n(\mathbf{r}, t) = \psi_n(\mathbf{r}) \exp(-i E_n t/\hbar) \qquad (10\cdot7\text{-}9)$$

The wave function $\Psi_n(\mathbf{r}, t)$ therefore belongs to a definite energy E_n. Obviously, the probability density $|\Psi_n(\mathbf{r}, t)|^2$ for this state is independent of time. Such states are known as *stationary states*.

Eq. (10·6-9) tell us that if $\nabla \cdot \mathbf{j} = 0$, then the probability current density ρ is constant in time. This corresponds to stationary state.

10·8 Eigenfunctions and eigenvalues

We have seen above that for the stationary states of a system the spatial part of the wave function $\psi(\mathbf{r})$ obeys the time-independent wave equation (10·7-6). This equation actually belongs to a class known as *eigenvalue equations*.

An equation of the type (10·7-6) can be solved only for a special set of values of E consistent with the boundary conditions. These are known as energy *eigenvalues* and the corresponding wave-functions as energy *eigenfunctions*.

In general if we consider an operator $\hat{\alpha}$ which operating on a function $\phi(x)$ multiplies the latter by a constant a, then $\phi(x)$ is called an *eigenfunction* of $\hat{\alpha}$ belonging to the eigenvalue a. To each operator $\hat{\alpha}$, there belongs in general a set of eigenvalues a_n and a set of eigenfunctions ϕ_n defined by the equation

$$\hat{\alpha} \phi_n(x) = a_n \phi_n(x) \qquad (10\cdot8\text{-}1)$$

The eigenfunctions are thus a special set of functions which when operated upon by an operator $\hat{\alpha}$ remain unchanged, being simply multiplied by the corresponding eigenvalues.

To be an eigenfunction, $\phi(x)$ must be well-behaved i.e., it must be single-valued and square-integrable so that the integral of its modulus-squared must be finite. Actually the nature of the operator determines the criteria to be fulfilled by a well-behaved function. Functions which are continuous, single-valued and remain finite or vanish at infinity are usually suitable as well-behaved functions. In the case of eigenvalue equations involving differential operators, the standard continuity conditions require that both $\phi(x)$ and its derivative remain continuous at all points.

Example :

Let $\hat{\alpha} \phi(x) = (d^2/dx^2) \phi(x) = a \phi(x) \qquad (10\cdot8\text{-}2)$

If $a < 0$, we get by writing $a = -k^2$,

$$d^2\phi/dx^2 + k^2\phi = 0 \tag{10.8-3}$$

The solution is

$$\phi(x) = A\exp(ikx) + B\exp(-ikx) \tag{10.8-4}$$

$\phi(x)$ is well-behaved in the entire range $-\infty \leq x \leq \infty$. But if $a > 0$, then we get by writing $a = k^2 (k\text{ real})$

$$\frac{d^2\phi}{dx^2} - k^2\phi = 0 \tag{10.8-5}$$

The solution is $\quad \phi(x) = A\exp(kx) + B\exp(-kx) \tag{10.8-6}$

For $x > 0$, $\exp(kx) \to \infty$ as $x \to \infty$. Hence for the range $0 < x \leq \infty$, we must put $A = 0$ to get a well-behaved solution which is

$$\phi(x) = B\exp(-kx) \tag{10.8-7}$$

For $x < 0$, $\exp(-kx) \to \infty$ as $x \to -\infty$. We must have $B = 0$ in this case and the well-behaved solution is

$$\phi(x) = A\exp(+kx) \tag{10.8-8}$$

In quantum mechanics, we are concerned with the solution of the eigenvalue problems under given sets of boundary conditions. Eigenvalue problems are also frequently encountered in classical physics, as for examples in the case of the transverse vibration of a string fixed at its two ends.

For a string parallel to the x axis fixed rigidly at the two ends and vibrating along y, only a discrete set of frequencies is permissible, consistent with the boundary conditions. The corresponding discrete set of the functions $y_n(x, t)$ constitutes the set of eigenfunctions for the different frequencies.

The general solution is obtained by a linear superposition of the eigenfunctions.

$$y(x, t) = \sum_{n=1}^{\infty} y_n(x, t) \tag{10.8-9}$$

We have seen earlier (§ 10.2) that one of the basic assumptions of quantum mechanics is that the wave equation satisfied by ψ must be linear so that the principle of superposition of states discussed in § 9.17 may be valid. The stationary state solutions of the time-independent Schrödinger equation may thus be superposed to give a general solution as in the classical string problem. These stationary state solutions represent the energy eigenfunctions belonging to different possible energy eigenvalues consistent with the boundary conditions. They form a *complete orthogonal* set, apart from being well behaved.

If a set of functions $u_n(x)$ has the property that

$$\int_a^b u_m^*(x)\, u_n(x)\, dx = 0 \quad (m \neq n) \tag{10.8-10}$$

then the functions $u_n(x)$ are said to be orthogonal in the interval a to b. It may be noted that this property of the functions is analogous to the condition of orthogonality of two vectors.

In quantum mechanics if $\hat{\alpha}$ is an operator representing a dynamical variable α such that the eigenvalue equation

$$\hat{\alpha}\, u_n(x) = a_n\, u_n(x)$$

is satified then the above equation is interpreted by saying that the result of measurement of α in the state $u_n(x)$ is certainly a_n. Here $u_n(x)$ is the eigenfunction of $\hat{\alpha}$ belonging to the eigenvalue a_n. In general if there is set of eigenfunctions $u_n(x)$ (with different n) of $\hat{\alpha}$ belonging to the different eigenvalues a_n, then it is assumed that any arbitrary continuous function $\psi(x)$ satisfying the same boundary conditions as $u_n(x)$ can be expanded in terms of the eigenfunctions $u_n(x)$:

$$\psi(x) = \sum_n C_n u_n(x) \tag{10.8-11}$$

This property of the eigenfunctions is known as *completeness*. Mathematical proof of completeness of the eigenfunctions can be provided for a wide range of operators, *e.g.*, the energy and momentum eigenfunctions.* For momentum eigenfunctions (see § 11·10) the above series representation is identical with Fourier series expansion.

The coefficients C_n are determined by utilizing the orthogonality property of the eigenfunctions (see later).

To illustrate the idea of completeness of the eigenfunctions, we note for example that in the Fourier expansion of the general solution of the string problem given above, the set of sine and cosine functions form a complete set. However an expansion in terms of either the sine functions or the cosine functions would not in general give the required functions. These, by themselves, do not form a complete set. It is only when all possible solutions of the eigenvalue equation are included in the expansion of the given arbitrary function that the requirement of completeness is fulfilled.

Using Eq. (10·7-9), we can write down the general solution of the Schrödinger equation for one dimensional motion as

$$\Psi(x, t) = \sum_n C_n u_n(x) \exp(-iE_n t/\hbar) \tag{10.8-12}$$

The above expansion is possible if the general solution
_____
* See *Quantum Mechanics* by L.I. Schiff.

Introduction to Wave Mechanics 257

$\Psi(x, 0) = \psi(x)$ for $t = 0$ can be expanded in terms of the stationary state solutions of the time-independent Schrödinger equation.

It may be noted that instead of a discrete spectrum, the stationary states may form a continuum. In this case, the sum in Eq. (10·8-12) is to be replaced by integral over all such states for different energies. There may also be the case in which both discrete and continuum of states may constitute the stationary states.

The expansion coefficients C_n in Eq. (10·8-12) can be evaluated by writing the solution (Eq. 10·8-11) for $t = 0$. Multiplying both sides of Eq. (10·8-12) by $u_m(x)$ and integrating we get

$$\int_a^b \psi(x) u_m^*(x) \, dx = \sum_n C_n \int_a^b u_m^*(x) u_n(x) \, dx$$

$$= C_m \int_a^b u_m^*(x) u_m(x) \, dx$$

$$= C_m \int_a^b |u_m(x)|^2 \, dx$$

since the only non-vanishing term on the r.h.s. is that for $m = n$. Hence

$$C_m = \int_a^b \psi(x) u_m^*(x) \, dx \bigg/ \int_a^b |u_m(x)|^2 \, dx \qquad (10\cdot 8\text{-}13)$$

If the functions $u_m(x)$ are normalized in the interval a to b, then,

$$\int_a^b |u_m(x)|^2 \, dx = 1$$

So writing n in place of m, we get

$$C_n = \int_a^b \psi(x) u_n^*(x) \, dx \qquad (10\cdot 8\text{-}14)$$

In this case the eigenfunctions $u_n(x)$ form an *orthonormal* set.

10·9 Probability of stationary states

The physical interpretation of the coefficients C_n is that $|C_n|^2$ gives the probability P_n of the occurrence of the state $u_n(x)$. According to the probability interpretation of the wave function, $\int_{-\infty}^{\infty} |\psi(x)|^2 \, dx$ is a measure of the total probability (at $t = 0$) of finding the particle in any one of the complete set of stationary states $u_n(x)$ while $\int_{-\infty}^{\infty} |C_n u_n(x)|^2 dx$ gives the probability P_n of the occurrence of the state $u_n(x)$. Then from definition, we get

$$P_n = \int_{-\infty}^{\infty} |C_n u_n(x)|^2 dx \Big/ \int_{-\infty}^{\infty} |\psi(x)|^2 dx \qquad (10\cdot 9\text{-}1)$$

Now $\int_{-\infty}^{\infty} |\psi(x)|^2 dx = \int_{-\infty}^{\infty} \psi^*(x) \psi(x) dx = \sum_n |C_n|^2$ § $\qquad (10\cdot 9\text{-}2)$

This is a constant. So we have from Eq. (10·9-1), assuming the u_n's to be normalized, the probability of the state $u_n(x)$

$$P_n = \frac{|C_n|^2}{\sum_n |C_n|^2} \int_{-\infty}^{\infty} |u_n(x)|^2 dx = \frac{|C_n|^2}{\sum_n |C_n|^2} \qquad (10\cdot 9\text{-}3)$$

The total probability of any of the states occurring is thus

$$\sum_n P_n = \sum_n |C_n|^2 \Big/ \sum_n |C_n|^2 = 1$$

If $\psi(x)$ is normalized, then

$$\int_{-\infty}^{\infty} |\psi(x)|^2 dx = \sum_n |C_n|^2 = 1$$

So the probability of the occurrence of the state $u_n(x)$ becomes

$$P_n = \int_{-\infty}^{\infty} |C_n u_n(x)|^2 dx = |C_n|^2 \qquad (10\cdot 9\text{-}4)$$

It may be noted that in most problems of interest in quantum mechanics, the operators belong to a class known as *Hermitian operators* (see § 10·13). The eigenvalues of Hermitian operators are real. It can be shown on general consideration that the eigenfunctions of Hermitian operators are orthogonal

---

§ See *Introductory Quantum Mechanics* by the author.

Introduction to Wave Mechanics

10·10 Degeneracy

It should be noted that the general solution of the Schrödinger equation may include more than one linearly independent eigenfunctions for the same energy eigenvalue. In this case E is said to be *degenerate*. Specific instances of degeneracy will be discussed in connection with the one dimensional motion of a free particle (§ 11·10) and the hydrogen problem (§ 11·15).

In general if there are n linearly independent solutions of an eigenvalue equation for a single eigenvalue then that eigenvalue is said to be *n-fold degenerate* and the *degree of degeneracy* is n.

10·11 Averages in quantum mechanics ; Expectation values

We have seen that for a microscopic system, it is usually not possible to make precise determination of the canonically conjugate quantities like the position and momentum. Quantum mechanical methods permit us to calculate the probability of the occurrence of a given value of any of these quantities during a measurement. This means that the result of two successive measurements of say, the position of a particle, may not yield the same value. We define the *expected average* or the *expectation value* of a dynamical variable such as the position co-ordinate by considering either a large number of independent measurements on the same system or single measurements on each of a large number of identical systems.

Consider the measurement of the dynamical variable α represented by the quantum mechanical operator $\hat{\alpha}$. We want to find the average of α in a state given by the wave function

$$\psi(x) = \sum_n C_n u_n(x) \qquad (10\cdot11\text{-}1)$$

where $\psi(x)$ is obtained by the superposition of the set of normalized eigenfunctions $u_n(x)$. These satisfy the relation $\hat{\alpha} u_n(x) = a_n u_n(x)$. This means that the result of measurement of α in the state $u_n(x)$ is definitely a_n. So by definition, the average of the quantity α is

$$<\alpha> = \sum_n a_n P_n = \sum_n a_n |C_n|^2 / \sum_n |C_n|^2$$

since the system can exist in different states $u_n(x)$ with different possible values of a_n. Here P_n is the probability of finding the value a_n as a result of measurement of α. Eq. (10·8-14) gives

$$C_n = \int_{-\infty}^{\infty} \psi(x) u_n^*(x) \, dx \qquad (10\cdot11\text{-}2)$$

so that
$$C_n^* = \int_{-\infty}^{\infty} \psi^*(x) u_n(x) dx \qquad (10\cdot11\text{-}3)$$

Hence $\sum_n a_n |C_n|^2 = \sum_n a_n C_n \int_{-\infty}^{\infty} \psi^*(x) u_n(x) dx$

Since the integral and the summation are interchangeable and since $\hat{\alpha}$ operates on $u_n(x)$ only we can write

$$\sum_n a_n |C_n|^2 = \int_{-\infty}^{\infty} dx\, \psi^*(x) \left\{ \sum_n C_n a_n u_n(x) \right\}$$

$$= \int_{-\infty}^{\infty} dx\, \psi^*(x) \left\{ \sum_n C_n \hat{\alpha}\, u_n(x) \right\}$$

$$= \int_{-\infty}^{\infty} dx\, \psi^*(x)\, \hat{\alpha} \left\{ \sum_n C_n u_n(x) \right\}$$

$$= \int_{-\infty}^{\infty} dx\, \psi^*\, \hat{\alpha}\, \psi(x) \qquad (10\cdot11\text{-}4)$$

Further from Eq. (10·9-2)

$$\int_{-\infty}^{\infty} |\psi(x)|^2 dx = \sum_n |C_n|^2$$

So we have

$$<\alpha> = \frac{\int_{-\infty}^{\infty} \psi^*(x) \hat{\alpha}\, \psi(x) dx}{\int_{-\infty}^{\infty} \psi^*(x) \psi(x) dx} \qquad (10\cdot11\text{-}5)$$

If $\psi(x)$ is normalized, the denominator in the above equation is equal to 1. So we get in this case

$$<\alpha> = \int_{-\infty}^{\infty} \psi^*(x)\, \hat{\alpha}\, \psi(x) dx \qquad (10\cdot11\text{-}6)$$

We can thus enunciate the following rule for calculating the expectation values of dynamical variables in quantum mechanics :

Place the operator representing the dynamical variable to act on the

Introduction to Wave Mechanics

wave function $\psi(x)$ and multiply the result by $\psi^*(x)$ and integrate over the spatial coordinates. Here $\psi(x)$ is assumed normalized.

For example the expectation values of the x-coordinate and the linear momentum p_x for one dimensional motion are

$$<x> = \int_{-\infty}^{\infty} \psi^*(x) \, x \, \psi(x) \, dx = \int_{-\infty}^{\infty} x \, |\psi(x)|^2 \, dx$$

$$<p_x> = \int_{-\infty}^{\infty} \psi^*(x) \left(\frac{\hbar}{i} \frac{\partial}{\partial x}\right) \psi(x) \, dx = -i\hbar \int_{-\infty}^{\infty} \psi^*(x) \left|\frac{\partial \psi(x)}{\partial x}\right| dx$$

It may be noted that the same expression for the expectation value is obtained if $\psi(x)$ is replaced by $\Psi(x, t)$

For three-dimensional motion, the expection value is given by

$$<\alpha> = \frac{\int_\tau \psi^*(\mathbf{r}) \hat{\alpha} \, \psi(\mathbf{r}) \, d\tau}{\int_\tau \psi^*(\mathbf{r}) \, \psi(\mathbf{r}) \, d\tau} \tag{10.11-7}$$

If $\psi(\mathbf{r})$ is normalized, we get

$$<\alpha> = \int_\tau \psi^*(\mathbf{r}) \, \hat{\alpha} \, \psi(\mathbf{r}) \, d\tau \tag{10.11-8}$$

Using the above results we can calculate the expectation value of energy for an energy eigenstate $\Psi_n(\mathbf{r}, t)$.

$$<E> = \int_\tau \Psi_n^*(\mathbf{r}, t) \, \hat{E} \, \Psi_n(r, t) \, dx$$

$$= \int_\tau \psi_n^*(\mathbf{r}) \exp(iE_n t/\hbar) \left(i\hbar \frac{\partial}{\partial t}\right) \psi_n(\mathbf{r}) \exp(-iE_n t/\hbar) \, d\tau$$

$$= i\hbar \int_\tau \psi_n^*(\mathbf{r}) \exp(iE_n t/\hbar) \left(-\frac{iE_n}{\hbar}\right) \psi_n(\mathbf{r}) \exp(-iE_n t/\hbar) d\tau$$

$$= E_n \int_\tau \psi_n^*(\mathbf{r}) \, \psi_n(\mathbf{r}) \, d\tau$$

Since the $\psi_n(\mathbf{r})$'s are normalized eigenfunctions, we get

$$<E> = E_n$$

Here we have assumed that the energy eigenvalues E_n are real (see §10.14).

The above result shows that the expectation value of energy for the stationary state n is certainly E_n which remains constant in time. This is a special case of the general result that the expectation value of any observable in quantum mechanics whose operator does not depend explicitly on time is independent of time for a stationary state.

10·12 Expectation values and correspondence principle ; Ehrenfest's theorem

According to correspondence principle (§ 10·1) we may expect that the classical relationships between different dynamical variables *e.g.*, coordinate and momentum of a particle will also hold good between the expectation values of these quantities for the quantum mechanical wave packet associated with the particle, a result known as *Ehrenfest's theorem*.

Since the wave packet representing a particle has a finite extension in space, one has to take the average values of the dynamical variables over this extension to bring out the correspondence between the laws governing the motion of a particle in classical physics and that of a wave-packet in quantum mechanics.

The expectation value of the coordinate x for one dimensional motion, is

$$<x> = \int_{-\infty}^{\infty} x |\Psi|^2 \, dx = \int_{-\infty}^{\infty} x \rho \, dx \qquad (10\cdot12\text{-}1)$$

where $\rho(x, t) = |\Psi(x, t)|^2$ is the probability density. Differentiating with respect to time, we have

$$\frac{d}{dt}<x> = \frac{d}{dt} \int_{-\infty}^{\infty} x \rho \, dx = \int_{-\infty}^{\infty} x \frac{\partial \rho}{\partial t} \, dx$$

$$= -\int_{-\infty}^{\infty} x \frac{\partial j_x}{\partial x} \, dx \qquad (10\cdot12\text{-}2)$$

where we have used the equation of continuity (Eq. 10·6-9) for one dimensional motion, j_x being the probability current density along x, given by (Eq. 10·6-11).

Multiplying by the particle mass m and integrating by parts, we get

$$m \frac{d}{dt}<x> = -m \int_{-\infty}^{\infty} x \frac{\partial j_x}{\partial x} \, dx = -m x j_x \Big]_{-\infty}^{\infty} + m \int_{-\infty}^{\infty} j_x \, dx$$

$$= m \int_{-\infty}^{\infty} j_x \, dx \qquad (10\cdot12\text{-}3)$$

where the first term becomes zero at $\pm \infty$ since ψ and hence j_x goes to zero as $x \to \pm \infty$. Substituting for j_x from Eq. (10·6-11) and integrating by parts we get

$$m \frac{d}{dt} <x> = m \frac{i\hbar}{2m} \left\{ \int_{-\infty}^{\infty} \Psi \frac{\partial \Psi^*}{\partial x} dx - \int_{-\infty}^{\infty} \Psi^* \frac{\partial \Psi}{\partial x} dx \right\}$$

$$= \frac{i\hbar}{2} \left\{ \Psi \Psi^* \Big]_{-\infty}^{\infty} - \int_{-\infty}^{\infty} \Psi^* \frac{\partial \Psi}{\partial x} dx - \int_{-\infty}^{\infty} \Psi^* \frac{\partial \Psi}{\partial x} dx \right\}$$

$$= -i\hbar \int_{-\infty}^{\infty} \Psi^* \frac{\partial \Psi}{\partial x} dx = \int_{-\infty}^{\infty} \Psi^* \left(-i\hbar \frac{\partial}{\partial x} \right) \Psi \, dx$$

$$= \int_{-\infty}^{\infty} \Psi^* \hat{p}_x \Psi \, dx \tag{10·12-4}$$

where we have used the operator representation of the x-component of the momentum. The r.h.s., by definition, is the expectation value of p_x. So we get

$$m \frac{d}{dt} <x> = <p_x> \tag{10·12-5}$$

Equation (10·12-5) is identical with the relationship between the coordinate and the momentum of a particle in classical physics. We thus see that the same relation also holds between the expectation values of the corresponding observables in quantum mechanics which is in conformity with the correspondence principle.

The above result can easily be generalized in the case of three dimensional motion.

Similarly the classical relationship between the rate of change of the momentum and the force acting on a particle is found to hold in quantum mechanics when the expectation values of these quanties are considered.

$$\frac{d}{dt} <p_x> = <F_x> \tag{10·12-6}$$

The above results provide justification for the operator representation of dynamical variables in quantum mechanics. For, by assuming the correspondence principle as the basic postulate, we could interpret Eq. (10·12-5) by stating that the rate of change of $m<x>$ with time should be equal to the expectation value of the x-component of the momentum. Hence the operator $(-i\hbar \partial/\partial x)$ operating on ψ on the r.h.s. of Eq. (10·12-4) should be interpreted as the operator representation $\hat{p}_x$ of the momentum p_x.

10.13. Formal proof of the uncertainty relation

We saw in § 9.13 that two canonically conjugate quantities, e.g., the coordinate (x) and the momentum (p_x) cannot be simultaneously measured with infinite precision. The inexactitudes in the simultaneous measurement of these quantities are related to each other through Heisenberg's uncertainty relations (Eq. 9·13-10).

We shall, in this section, give a formal proof of the uncertainty relation, using the definitions of the inexactitudes given at the end of §9.13.

The root mean squared deviation (or the standard deviation) for the coordinate x and the momentum p_x are given by (see Eqs. 9·13-8 and 9·13-9).

$$\Delta x = <(x - <x>)^2>^{1/2} = \{<x^2> - <x>^2\}^{1/2} \qquad (10 \cdot 13\text{-}1)$$

$$\Delta p_x = <(p_x - <p_x>)^2>^{1/2} = \{<p_x^2> - <p_x>^2\}^{1/2} \qquad (10 \cdot 13\text{-}2)$$

Now the quantum mechanical expectation values of x, x^2, p_x and p_x^2 are given by

$$<x> = \int \psi^* x \psi \, dx, \quad <x^2> = \int \psi^* x^2 \psi \, dx \qquad (10 \cdot 13\text{-}3)$$

$$<p_x> = \int \psi^* \left(-i\hbar \frac{\partial}{\partial x}\right) \psi \, dx = -i\hbar \int \psi^* \frac{\partial \psi}{\partial x} dx \qquad (10 \cdot 13\text{-}4)$$

$$<p_x^2> = \int \psi^* \left(-i\hbar \frac{\partial}{\partial x}\right)^2 \psi \, dx = -\hbar^2 \int \psi^* \frac{\partial^2 \psi}{\partial x^2} dx \qquad (10 \cdot 13\text{-}5)$$

Let $f(x)$ and $g(x)$ be two functions of x which are well-behaved. We define

$$F = \int f^* f \, dx, \qquad G = \int g^* g \, dx \qquad (10 \cdot 13\text{-}6)$$

$$J = \int f^* g \, dx \qquad J^* = \int g^* f \, dx \qquad (10 \cdot 13\text{-}7)$$

Consider the integral

$$I(\alpha) = \int (\alpha f^* + g^*)(\alpha f + g) \, dx$$

$$= \alpha^2 \int f^* f \, dx + \alpha \int (f^* g + g^* f) \, dx + \int g^* g \, dx$$

$$= F \alpha^2 + H \alpha + G \qquad (10 \cdot 13\text{-}8)$$

where we have written $J + J^* = H$; α is a real parameter.

The integrand in Eq. (10·13-8) being a modulus squared cannot be negative. Hence

$$I(\alpha) = F \alpha^2 + H \alpha + G \geq 0 \qquad (10 \cdot 13\text{-}9)$$

for all values of α. This condition imposes certain restrictions on the values of the coefficients F, G and H. Considering the minimum value of

Introduction to Wave Mechanics

the function $I(\alpha)$ when $\alpha = \alpha_0$, we note that α_0 can be found from the condition

$$I(\alpha_0) = F\alpha_0^2 + H\alpha_0 + G = 0 \tag{10.13-10}$$

In order that the above equation may have real roots, the following condition must be satisfied:

$$H^2 - 4FG \leq 0$$

or,
$$FG \geq \frac{H^2}{4} \tag{10.13-11}$$

i.e.,
$$FG \geq \frac{1}{4}(J + J^*)^2 \tag{10.13-12}$$

Let us choose the wave function $\psi(x)$ such that

$$<x> = \int \psi^* x \psi \, dx = 0$$

$$<p_x> = \int \psi^* \left(-i\hbar \frac{\partial}{\partial x}\right) \psi \, dx = 0$$

Then Eqs. (10.13-1) and (10.13-2) give

$$\Delta x^2 = <x^2>, \Delta p_x^2 = <p_x^2> \tag{10.13-13}$$

Let us write

$$f(x) = \hat{p}_x \psi = -i\hbar \frac{\partial \psi}{\partial x} \tag{10.13-14}$$

$$g(x) = ix\psi \tag{10.13-15}$$

Then
$$F = \int f^* f \, dx = \int \left(i\hbar \frac{\partial \psi^*}{\partial x}\right)\left(-i\hbar \frac{\partial \psi}{\partial x}\right) dx$$

$$= \hbar^2 \int \frac{\partial \psi^*}{\partial x} \frac{\partial \psi}{\partial x} dx$$

$$= \hbar^2 \left[\psi^* \frac{\partial \psi}{\partial x}\right]_{-\infty}^{\infty} - \hbar^2 \int_{-\infty}^{\infty} \psi^* \frac{\partial^2 \psi}{\partial x^2} dx$$

The first term on the r.h.s. of the above equation in the last step is zero because ψ must be well-behaved (both ψ and $d\psi/dx$ must vanish as $x \to \pm \infty$). Hence

$$F = -\hbar^2 \int \psi^* \frac{\partial^2 \psi}{\partial x^2} dx = \int \psi^* \left(-i\hbar \frac{\partial}{\partial x}\right)^2 \psi \, dx$$

$$= \int \psi^* \hat{p}_x^2 \psi \, dx = <p_x^2> \tag{10.13-16}$$

$$G = \int g^* g \, dx = \int (-ix\psi^*)(ix\psi) dx$$

$$= \int \psi^* x^2 \psi \, dx = <x^2> \qquad (10\cdot 13\text{-}17)$$

Also $\quad J = \int f^* g \, dx = \int \left(i\hbar \frac{\partial \psi^*}{\partial x} \right)(ix \psi) \, dx$

$$= -\hbar \int x \psi \frac{\partial \psi^*}{\partial x} dx \qquad (10\cdot 13\text{-}18)$$

$$J^* = -\hbar \int x \psi^* \frac{\partial \psi}{\partial x} dx \qquad (10\cdot 13\text{-}19)$$

$$\therefore \quad H = J + J^* = -\hbar \int x \left(\psi \frac{\partial \psi^*}{\partial x} + \psi^* \frac{\partial \psi}{\partial x} \right) dx$$

$$= -\hbar \int x \frac{\partial}{\partial x}(\psi^* \psi) \, dx = -\hbar [x \psi^* \psi]_{-\infty}^{\infty} + \hbar \int_{-\infty}^{\infty} \psi^* \psi \, dx$$

$$(10\cdot 13\text{-}20)$$

The first term on the r.h.s. of the square bracket in the above equation is zero due to the requirement of well-behaviour of $\psi(x)$. Then assuming $\psi(x)$ to be normalized, we get

$$H = J + J^* = \hbar \int \psi^* \psi \, dx = \hbar \qquad (10\cdot 13\text{-}21)$$

From Eqs. (10·13-16) and (10·13-17), we have by using Eqs. (10·13-13)

$$FG = \Delta x^2 \, \Delta p_x^2 \qquad (10\cdot 13\text{-}22)$$

Hence from Eq. (10·13-11) we get by using Eqs. (10·13-21) and (10·13-22)

$$\Delta x^2 \cdot \Delta p_x^2 \geq \frac{\hbar^2}{4} \qquad (10\cdot 13\text{-}23)$$

or, $\qquad\qquad\qquad\qquad \Delta x \cdot \Delta p_x \geq \dfrac{\hbar}{2} \qquad (10\cdot 13\text{-}24)$

10·14 Hermitian operators

We know that the expectation value of a dynamical variable gives the expected average of the result of measurement of the dynamical variable. So it must be a real number since it is a physically measureable quantity. The operators representing dynamical variables whose expectation values are real are said to be Hermitian operators. If $\hat{\alpha}$ is the operator representation of a dynamical variable α, then the expectation value of α, assuming one dimensional motion, is given by

$$<\alpha> = \int \psi^* \hat{\alpha} \psi \, dx \qquad (10\cdot 14\text{-}1)$$

where $\psi = \psi(x)$ is the normalized wavefunction specifying the state of the system.

Introduction to Wave Mechanics

The complex conjugate of the expectation value of α is

$$<\alpha>^* = \left(\int \psi^* \hat{\alpha} \psi \, dx\right)^* = \int \psi \, \hat{\alpha}^* \psi^* \, dx \qquad (10\cdot 14\text{-}2)$$

For the expectation value to be real we must have

$$<\alpha> = <\alpha>^*$$

or,
$$\int \psi^* \hat{\alpha} \psi \, dx = \int \psi \, \hat{\alpha}^* \psi^* \, dx \qquad (10\cdot 14\text{-}3)$$

Eq. (10·14-3) is the condition for the operator $\hat{\alpha}$ to be *Hermitian*.

In the three dimensional case, the above condition becomes

$$\int \Psi^* \hat{\alpha} \psi \, d\tau = \int \psi \, \hat{\alpha}^* \psi^* \, d\tau \qquad (10\cdot 14\text{-}3a)$$

where ψ is now a function of the three coordinates x, y, z.

A more general condition for an operator $\hat{\alpha}$ to be Hermitian is that

$$\int f^* \hat{\alpha} g \, dx = \int g \, \hat{\alpha}^* f^* \, dx \qquad (10\cdot 14\text{-}4)$$

where $f(x)$ and $g(x)$ are two normalizable functions § §.

In the three dimensional case, we can write

$$\int f^* \hat{\alpha} g \, d\tau = \int g \, \hat{\alpha}^* f^* \, d\tau \qquad (10\cdot 14\text{-}5)$$

As we shall see below, the eigenvalues of Hermitian operators are real and their eigenfunctions are orthogonal.

Examples :

We can easily prove the Hermitian character of the operators representing some of the dynamical variables, *e.g.*, the angular momentum, linear momentum *etc*.

As will be seen later in Chapter XI, the z-component of the orbital angular momentum L_z has the following operator representation

$$\hat{L}_z = -i\hbar \frac{\partial}{\partial \phi}$$

where ϕ is the azimuthal angle.

If f and g are two functions, we have

$$\int_0^{2\pi} f^* \hat{L}_z g \, d\phi = \int_0^{2\pi} f^* \left(-i\hbar \frac{\partial}{\partial \phi}\right) g \, d\phi = -i\hbar \int_0^{2\pi} f^* \frac{\partial g}{\partial \phi} d\phi$$

---

§ § For a proof of this, see *Introductory Quantum Mechanics* by the author.

Integrating by parts we get

$$\int_0^{2\pi} f^* \hat{L}_z g \, d\phi = -i\hbar \left\{ f^* g \Big|_0^{2\pi} - \int_0^{2\pi} g \frac{\partial f^*}{\partial \phi} d\phi \right\}$$

The first term on the r.h.s. of the above equation is zero because of the single-valuedness of the wave functions

$$f^*(2\pi) = f^*(0), \, g(2\pi) = g(0)$$

So we get

$$\int_0^{2\pi} f^* \hat{L}_z g \, d\phi = \int_0^{2\pi} g \left(i\hbar \frac{\partial}{\partial \phi} \right) f^* d\phi = \int_0^{2\pi} g \hat{L}_z^* f^* d\phi$$

This is in accordance with the general requirement for an operator to be Hermitian (Eq. 10·14-4). So $\hat{L}_z$ is Hermitian. We next consider the momentum operator $\hat{p}_x$ for one dimensional motion. We have

$$\int_{-\infty}^{\infty} f^* \hat{p}_x g \, dx = \int_{-\infty}^{\infty} f^* \left(-i\hbar \frac{\partial}{\partial x} \right) g \, dx = -i\hbar \int_{-\infty}^{\infty} f^* \frac{\partial g}{\partial x} dx$$

$$= -i\hbar \left\{ f^* g \Big|_{-\infty}^{\infty} - \int_{-\infty}^{\infty} g \frac{\partial f^*}{\partial x} dx \right\}$$

In view of the requirement that the wave functions must vanish at infinity the first term on the r.h.s. becomes zero. So we get

$$\int_{-\infty}^{\infty} f^* \hat{p}_x g \, dx = \int_{-\infty}^{\infty} g \left(i\hbar \frac{\partial}{\partial x} \right) f^* dx = \int_{-\infty}^{\infty} g \hat{p}_x^* f^* dx$$

So $\hat{p}_x$ is Hermitian. The proof of the Hermitian character of the Hamiltonian operator is left to the reader as a problem.

10.15 Some properties of Hemitian operators

(A) We first prove the following two theorems :

(i) The sum (or difference) of two Hermitian operators is Hermitian.
(ii) The product of two Hermitian operators is Hermitan if and only if they commute.

The proof of the first is very simple and is left to the reader as an exercise.

We prove the second.

Introduction to Wave Mechanics

Let $\hat{\alpha}$ and $\hat{\beta}$ be two Hermitian operators. Then if $u(x)$ and $v(x)$ are two functions of x, we have from Eq. (10·14-4),

$$\int u^* \hat{\alpha} v \, dx = \int v \hat{\alpha}^* u^* \, dx \qquad (10\cdot15\text{-}1)$$

$$\int v^* \hat{\alpha} u \, dx = \int u \hat{\alpha}^* v^* \, dx \qquad (10\cdot15\text{-}2)$$

Let the operator $\hat{O} = \hat{\alpha}\hat{\beta}$. Then if $f(x)$ and $g(x)$ are two functions, we can write

$$\int f^* \hat{O} g \, dx = \int f^* \hat{\alpha} \hat{\beta} g \, dx = \int f^* \hat{\alpha} (\hat{\beta} g) \, dx$$
$$= \int (\hat{\beta} g) \hat{\alpha}^* f^* \, dx$$

in view of Eq. (10·14-4). This can be written as

$$\int f^* \hat{O} g \, dx = \int (\hat{\beta} g)(\hat{\alpha} f)^* \, dx = \int (\hat{\alpha} f)^* \hat{\beta} g \, dx$$
$$= \int g \hat{\beta}^* (\hat{\alpha} f)^* \, dx = \int g (\hat{\beta} \hat{\alpha})^* f^* \, dx$$

in view of Eq. (10·14-4) again. So we get

$$\int f^* (\hat{\alpha} \hat{\beta}) g \, dx = \int g (\hat{\beta} \hat{\alpha})^* f^* \, dx \qquad (10\cdot15\text{-}3)$$

But for $\hat{O} = \hat{\alpha}\hat{\beta}$ to be Hermitian, we must have

$$\int f^* \hat{O} g \, dx = \int g \hat{O}^* f^* \, dx$$

or, $$\int f^* (\hat{\alpha} \hat{\beta}) g \, dx = \int g (\hat{\alpha} \hat{\beta})^* f^* \, dx \qquad (10\cdot15\text{-}4)$$

Comparing Eq. (10·15-3) and (10·15-4) we get

$$(\hat{\beta} \hat{\alpha})^* = (\hat{\alpha} \hat{\beta})^*$$

or, $$\hat{\beta} \hat{\alpha} = \hat{\alpha} \hat{\beta}$$

Hence $\hat{\alpha}$ and $\hat{\beta}$ must commute in order that the product $\hat{\alpha}\hat{\beta}$ of the two Hermitian operators $\hat{\alpha}$ and $\hat{\beta}$ may be Hermitian.

Example : We know that the operators $\hat{x} = x$ and $\hat{p}_x = -i\hbar \dfrac{\partial}{\partial x}$ do not commute (see Eq. 10·3-14)

$$\hat{x}\hat{p}_x - \hat{p}_x\hat{x} = i\hbar$$

From the theorem just proved it follows that the operator $\hat{x}\hat{p}_x$ is not Hermitian.

According to Dirac, the term dynamical variable in quantum mechanics is applicable to those quantities whose operators are Hermitian.

Thus even though x and p_x are dynamical variables, their operators being Hermitian, the product xp_x is not a dynamical variable, because the operators $\hat{x}$ end $\hat{p}_x$ do not commute.

(B) *Hermitian conjugate* :

The Hermitian conjugate $\hat{\alpha}^\dagger$ (alpha dagger) of an operator $\hat{\alpha}$ (not necessarily Hermitian) is defined by the relation

$$\int f^* \hat{\alpha}^\dagger g\, dx = \int g\, \hat{\alpha}^* f^*\, dx \tag{10.15-5}$$

where $f(x)$ and $g(x)$ are two functions of x. $\hat{\alpha}^\dagger$ is also called the Hermitian adjoint of $\hat{\alpha}$.

We prove the following important properties of the Hermitian conjugate $\hat{\alpha}^\dagger$.

(a) If $\hat{\alpha} = \hat{\alpha}^\dagger$, then $\hat{\alpha}$ is a Hermitian operator. $\hat{\alpha}$ is said to be self-adjoint in this case. We have in this case

$$\int f^* \hat{\alpha}^\dagger g\, dx = \int f^* \hat{\alpha} g\, dx \tag{10.15-6}$$

But from the defining relation (10.15-5), we have

$$\int f^* \hat{\alpha}^\dagger g\, dx = \int g\, \hat{\alpha}^* f^*\, dx$$

$$\therefore \quad \int f^* \hat{\alpha}\, g\, dx = \int g\, \hat{\alpha}^* f^*\, dx \tag{10.15-7}$$

in view of Eq. (10.15-6). The above relation (10.15-7) shows that $\hat{\alpha}$ is Hermitian.

If $\hat{\alpha}^\dagger = -\hat{\alpha}$, then $\hat{\alpha}$ is said to be anti-Hermitian.

(b) If $\hat{\alpha}$ is a non-Hermitian operator then $(\hat{\alpha} + \hat{\alpha}^\dagger)$ and $i(\hat{\alpha} - \hat{\alpha}^\dagger)$ are Hermitian.

(*i*) Let $\hat{O} = \hat{\alpha} + \hat{\alpha}^\dagger$. Then

$$\int f^* \hat{O} g\, dx = \int f^* (\hat{\alpha} + \hat{\alpha}^\dagger)\, g\, dx$$

$$= \int f^* \hat{\alpha}\, g\, dx + \int f^* \hat{\alpha}^\dagger g\, dx$$

$$= \int f^* \alpha\, g\, dx + \int g\, \hat{\alpha}^* f^*\, dx \tag{10.15-8}$$

But for $\hat{\alpha}^\dagger$ we can write

$$\int g^* \hat{\alpha}^\dagger f\, dx = \int f \hat{\alpha}^* g^*\, dx$$

Taking the complex conjugate, we get

$$\left(\int g^* \hat{\alpha}^\dagger f\, dx \right)^* = \left(\int f \hat{\alpha}^* g^*\, dx \right)^*$$

Introduction to Wave Mechanics 271

or, $$\int g(\hat{\alpha}^\dagger)^* f^* \, dx = \int f^* \hat{\alpha} g \, dx \tag{10.15-9}$$

Substituting in Eq. (10·15-8), we get
$$\int f^* \hat{O} g \, dx = \int g(\hat{\alpha}^\dagger)^* f^* \, dx + \int g \hat{\alpha}^* f^* \, dx$$
$$= \int g(\hat{\alpha} + \hat{\alpha}^\dagger)^* f^* \, dx = \int g \hat{O}^* f^* \, dx \tag{10.15-10}$$

So $\hat{O} = \hat{\alpha} + \hat{\alpha}^\dagger$ is a Hermitian operator.

(ii) Let $\hat{O} = i(\hat{\alpha} - \hat{\alpha}^\dagger)$. Then as before.
$$\int f^* \hat{O} g \, dx = i \int f^* \hat{\alpha} g \, dx - i \int g \hat{\alpha}^* f^* \, dx \tag{10.15-11}$$

But $$\int f^* \hat{\alpha} g \, dx = \int g(\hat{\alpha}^\dagger)^* f^* \, dx$$

from Eq. (10·15-9). Then
$$\int f^* \hat{O} g \, dx = i \int g (\hat{\alpha}^\dagger)^* f^* \, dx - i \int g \alpha^* f^* \, dx$$
$$= \int g i (\hat{\alpha}^\dagger - \hat{\alpha})^* f^* \, dx$$

But $$\hat{O}^* = -i \{\alpha^* - (\alpha^\dagger)^*\}$$
$$= i(\hat{\alpha}^\dagger - \hat{\alpha})^*$$
$$\therefore \quad \int f^* \hat{O} g \, dx = \int g \hat{O}^* f^* \, dx$$

Hence $\hat{O} = i(\hat{\alpha} - \hat{\alpha}^\dagger)$ is Hermitian.

(c) *Hermitian adjoint of $\hat{\alpha}^\dagger$ is $\hat{\alpha}$.*
$$\int f^* (\hat{\alpha}^\dagger)^\dagger g \, dx = \int g(\hat{\alpha}^\dagger)^* f^* \, dx$$

But $$\left(\int g^* \hat{\alpha}^\dagger f \, dx \right)^* = \left(\int f \hat{\alpha}^* g^* \, dx \right)^*$$

or, $$\int g(\hat{\alpha}^\dagger)^* f^* \, dx = \int f^* \hat{\alpha} g \, dx$$

$$\therefore \quad \int f^* (\hat{\alpha}^\dagger)^\dagger g \, dx = \int f^* \hat{\alpha} g \, dx$$

$$(\hat{\alpha}^\dagger)^\dagger = \hat{\alpha} \tag{10.15-12}$$

(D) *Hermitian adjoint of the sum and product of two arbitrary operators* :

(i) Let $\hat{\alpha}$ and $\hat{\beta}$ be two operators. Then
$$(\hat{\alpha} + \hat{\beta})^\dagger = \hat{\alpha}^\dagger + \hat{\beta}^\dagger \tag{10.15-13}$$

The proof is very simple and is left to the reader as an exercise.

(ii) For two operators $\hat{\alpha}$ and $\hat{\beta}$,

$$(\hat{\alpha}\hat{\beta})^{\dagger} = \hat{\beta}^{\dagger}\hat{\alpha}^{\dagger} \qquad (10 \cdot 15\text{-}14)$$

Let $\hat{O} = \hat{\alpha}\hat{\beta}$.

Then $\int f^* \hat{O}^{\dagger} g\, dx = \int g \hat{O}^* f^* dx = \int g (\hat{\alpha}\hat{\beta})^* f^* dx$

$$= \int g \hat{\alpha}^* (\hat{\beta} f)^* dx = \int (\hat{\beta} f)^* \hat{\alpha}^{\dagger} g\, dx$$

from the defining relation (10·15-5) for the Hermitian conjugate. Then

$$\int f^* \hat{O}^{\dagger} g\, dx = \int (\hat{\alpha}^{\dagger} g)\, \hat{\beta}^* f^* dx$$

$$= \int f^* \hat{\beta}^{\dagger} (\hat{\alpha}^{\dagger} g)\, dx = \int f^* (\hat{\beta}^{\dagger} \hat{\alpha}^{\dagger})\, g\, dx$$

$$\therefore \quad (\hat{\alpha}\hat{\beta})^{\dagger} = \hat{\beta}^{\dagger}\hat{\alpha}^{\dagger}$$

10·16 (a) Reality of the eigenvalues of a Hermitian operator

If $u(x)$ is an eigenfunction of a Hermitian operator belonging to the eigenvalue a, then we can write

$$\hat{\alpha}\, u(x) = a u(x)$$

Taking the complex conjugate, we get

$$\hat{\alpha}^* u^*(x) = a^* u^*(x)$$

So we have

$$<\alpha> = \int u^* \hat{\alpha}\, u\, dx = a \int u^* u\, dx \qquad (10 \cdot 16\text{-}1)$$

and $<\alpha>^* = \int u \hat{\alpha}^* u^* dx = \int u a^* u^* dx = a^* \int u^* u\, dx \qquad (10 \cdot 16\text{-}2)$

Since $<\alpha> = <\alpha>^*$, we have

$$a \int u^* u\, dx = a^* \int u^* u\, dx$$

which gives $a = a^*$

Thus a must be real which proves that the eigenvalue of a Hermitian operator is real.

As stated in § 10·14, the Hamiltonian is a Hermitian operator. Hence the eigenvalues of the Hamiltonian operator, which are the energy values of the system in its stationary states must be real.

(b) Orthogonality of the eigenfunctions of Hermitian operators

We have seen earlier (§ 10·8) that the eigenfunctions of a quantum mechanical operator must be orthogonal. This can easily be proved by considering the Hermitian character of these operators.

Let $\hat{\alpha}$ be the operator representation of the dynamical variable α. Let $u_m(x)$ and $u_n(x)$ be two eigenfunctions of $\hat{\alpha}$ belonging to the eigenvalues a_m and a_n respectively. Then we can write

$$\hat{\alpha}\, u_m(x) = a_m u_m(x) \qquad (10 \cdot 16\text{-}3)$$

Introduction to Wave Mechanics

$$\hat{\alpha}\, u_n(x) = a_n u_n(x) \tag{10.16-4}$$

Taking the complex conjugate of the first equation we have

$$\hat{\alpha}^*\, u_m^*(x) = a_m^* u_m^*(x) = a_m u_m^*(x) \tag{10.16-5}$$

Since the eigenvalues are real we have put $a_m^* = a_m$. Using the above equations, we have from the general requirement of hermiticity of operators (Eq. 10.14-4)

$$\int u_m^* \hat{\alpha}\, u_n\, dx - \int u_n \hat{\alpha}\, u_m^*\, dx$$

$$= \int u_m^* a_n u_n\, dx - \int u_n a_m u_m^*\, dx$$

$$= (a_n - a_m) \int u_m^* u_n\, dx = 0$$

If the eigenvalues a_m and a_n are different, *i.e.*, if $a_m \neq a_n$, then from the above equation we get the orthogonality condition

$$\int u_m^* u_n\, dx = 0$$

For the three dimensional case we can use Eq. (10.16-5) to get

$$\int u_m^* u_n\, d\tau = 0 \ (m \neq n)$$

10.17 Equations of motion in quantum mechanics

Dynamical variables like the coordinate x or the momentum p are usually time varying in classical mechanics. However their operator representations in quantum mechanics do not change with time. The time variation of the quantum mechanical states comes through the dependence of the wave function $\psi(x, t)$ on time. In quantum mechanics, the time derivatives of the expectation values of the dynamical variables are of physical significance.

Consider an operator $\hat{\alpha}$ representing the dynamical variable α. The expectation value of α is given by

$$<\alpha> = \int \psi^*(x, t)\, \hat{\alpha}\, \psi(x, t)\, dx$$

Here we assume one dimensional motion. The time rate of change of $<\alpha>$ is then

$$\frac{d}{dt}<\alpha> = \frac{d}{dt} \int \psi^* \hat{\alpha}\, \psi\, dx$$

$$= \int \left(\frac{\partial \psi^*}{\partial t} \hat{\alpha}\, \psi + \psi^* \frac{\partial \hat{\alpha}}{\partial t} \psi + \psi^* \hat{\alpha} \frac{\partial \psi}{\partial t} \right) dx \tag{10.17-1}$$

In the above expression we have provided for the time variation of the operator $\hat{\alpha}$ through the second term when the operator $\hat{\alpha}$ contains time explicitly. When this is lacking (e.g., for $\hat{x}$ or $\hat{p}$) this term is zero.

Schrödinger equation gives us

$$i\hbar \frac{\partial \psi}{\partial t} = \hat{H} \psi, \quad \text{or} \quad \frac{\partial \psi}{\partial t} = -\frac{i}{\hbar} \hat{H} \psi$$

Taking the complex conjugate, we get

$$\frac{\partial \psi^*}{\partial t} = \frac{i}{\hbar} \hat{H}^* \psi^*$$

Then

$$\frac{d}{dt} <\alpha> = \frac{i}{\hbar} \int \{(H^* \psi^*) \hat{\alpha} \psi - \psi^* \hat{\alpha}(\hat{H} \psi)\} dx + \int \psi^* \frac{\partial \hat{\alpha}}{\partial t} \psi \, dx$$

Since $\hat{H}$ is a Hermitian operator, we have

$$\int (H^* \psi^*) \hat{\alpha} \psi \, dx = \int (\hat{\alpha} \psi) \hat{H}^* \psi^* \, dx = \int \psi^* \hat{H} \hat{\alpha} \psi \, dx$$

Hence

$$\frac{d}{dt} <\alpha> = \frac{i}{\hbar} \int \psi^* (\hat{H}\hat{\alpha} - \hat{\alpha}\hat{H}) \psi \, dx + \int \psi^* \frac{\partial \hat{\alpha}}{\partial t} \psi \, dx \quad (10\cdot17\text{-}2)$$

The above equation gives us the time rate of change of the expectation value of $\hat{\alpha}$ if the commutator $[\hat{H}, \hat{\alpha}]$ is known. In particular if $\hat{\alpha}$ does not depend on time explicitly, we get

$$\frac{d}{dt} <\alpha> = \frac{i}{\hbar} \int \psi^* (\hat{H}\hat{\alpha} - \hat{\alpha}\hat{H}) \psi \, dx \quad (10\cdot17\text{-}3)$$

or,

$$\frac{d}{dt} <\alpha> = \frac{i}{\hbar} <\hat{H}\hat{\alpha} - \hat{\alpha}\hat{H}> \quad (10\cdot17\text{-}4)$$

The quantity $<\hat{H}\hat{\alpha} - \hat{\alpha}\hat{H}>$ on the right hand side is the expectation value of the commutator of $\hat{H}$ and $\hat{\alpha}$. We can symbolically write

$$\dot{\alpha} = \frac{d\alpha}{dt} = \frac{i}{\hbar} [\hat{H}, \hat{\alpha}] \quad (10\cdot17\text{-}5)$$

Eq. (10·17-5) is known as the equation of motion in quantum mechanics. If $\hat{H}$ and $\hat{\alpha}$ commute we get

$$\dot{\alpha} = \frac{d\alpha}{dt} = 0 \quad (10\cdot17\text{-}6)$$

In this case we say that the dynamical variable α is a constant of motion.

We know from classical mechanics that for a conservative system the Hamiltonian H is independent of time so that its operator representation $\hat{H}$ in quantum mechanics does not contain time explicitly. Hence

$$\dot{H} = \frac{dH}{dt} = \frac{i}{\hbar} [\hat{H}, \hat{H}] = 0 \quad (10\cdot17\text{-}7)$$

This shows that for a conservative system, the energy is a constant of motion in quantum mechanics.

It can be shown by using Hamilton's canonical equations and the classical Poisson bracket that the above results are analogous to the results of classical mechanics*.

This analogy is again a clear evidence of the validity of correspondence principle.

10·18 Fundamental postulates of quantum mechanics

In the theory developed so far we have tried to make plausible various new ideas on the basis of the underlying physical principles involved. There is however an alternative approach in which the formal mathematical structure of quantum mechanics is based on a number of fundamental postulates. The theory developed on the basis of these postulates, apart from being logically more satisfying, derives justification from its success in accounting for a wide variety of atomic and subatomic phenomena.

Though we shall not go into the development of the formal structure mentioned above, we shall in this section enumerate and briefly discuss the fundamental postulates all of which have been introduced in the course of our earlier discussions.

The fundamental postulates of quantum mechanics which we give here are five in number. These are applicable in general to a *quantum mechanical system* though in our discussion, we shall restrict ourselves to a *single particle system* for the sake of simplicity.

Postulate I

There is a wave function $\psi(\mathbf{r}, t) = \psi(x, y, z, t)$ which completely describes the space-time behaviour of the particle, consistent with the uncertainty principle.

Postulate II

Dynamical variables or observables which are the physically measureable properties of the particle are represented by mathematical operators in quantum mechanics.

These operators are linear and Hermitian. The most important of the operators are those representing the position vector $\mathbf{r}$ and the momentum $\mathbf{p}$.

In the table below we list the operators associated with these and certain other important dynamical variables in the Schrödinger representation.

* See *Introductory Quantum Mechanics* by the author.

Table IV-I

Dynamical variable	Symbol	Operator representation
Position vector	$\mathbf{r}$	$\hat{\mathbf{r}} = \mathbf{r}$
Cartesian components of $\mathbf{r}$	x, y, z	$\hat{x} = x, \hat{y} = y, \hat{z} = z$
Linear momentum	$\mathbf{p}$	$\hat{\mathbf{p}} = \dfrac{\hbar}{i} \nabla$
Cartesian components of $\mathbf{p}$	p_x, p_y, p_z	$\hat{p}_x = \dfrac{\hbar}{i} \dfrac{\partial}{\partial x}$ $\hat{p}_y = \dfrac{\hbar}{i} \dfrac{\partial}{\partial y}$ $\hat{p}_z = \dfrac{\hbar}{i} \dfrac{\partial}{\partial z}$
Angular momentum	$\mathbf{L} = \mathbf{r} \times \mathbf{p}$	$\hat{\mathbf{L}} = \dfrac{\hbar}{i} (\mathbf{r} \times \nabla)$
Cartesian components of $\mathbf{L}$		$\hat{L}_x = -i\hbar \left(y \dfrac{\partial}{\partial z} - z \dfrac{\partial}{\partial y} \right)$ $\hat{L}_y = -i\hbar \left(z \dfrac{\partial}{\partial x} - x \dfrac{\partial}{\partial z} \right)$ $\hat{L}_z = -i\hbar \left(x \dfrac{\partial}{\partial y} - y \dfrac{\partial}{\partial x} \right)$
Hamiltonian	H	$\hat{H} = -\dfrac{\hbar^2}{2m} \nabla^2 + V(\mathbf{r})$ $= -\dfrac{\hbar^2}{2m} \left(\dfrac{\partial^2}{\partial x^2} + \dfrac{\partial^2}{\partial y^2} + \dfrac{\partial^2}{\partial z^2} \right) + V(\mathbf{r})$

The rules for obtaining the operator representations of various dynamical variables were discussed in § 10·3.

Postulate III

The only possible result of measurement of a dynamical variable α are the eigenvalues of the operator $\hat{\alpha}$ satisfying the eigenvalue equation

$$\hat{\alpha} u_n = a_n u_n$$

where u_n is the eigenfunction of the operator $\hat{\alpha}$ belonging to the eigenvalue a_n.

Introduction to Wave Mechanics

The eigenfunctions u_n are well-behaved, i.e., they must be single-valued and square-integrable. They also form a complete set and are orthogonal (see § 10·8).

The eigenvalue equation satisfied by the Hamiltonian operator associated with the Hamiltonian function H of classical mechanics, known as the Schrödinger wave equation, can be written as

$$\hat{H}\psi_n = E_n\psi_n$$

where E_n is the energy of the system.

Postulates IV

The probability $\rho d\tau$ of finding a particle in the volume element $d\tau$ is given by

$$\rho\, d\tau = \psi^*\psi\, d\tau = |\psi|^2\, d\tau$$

where $\rho = |\psi|^2$ is known as the probability density. In a finite region of space the probability of finding the particle is obtained by integrating the above expression over the volume under consideration

$$\int_\tau \rho\, d\tau = \int_\tau |\psi|^2\, d\tau$$

The integral on the r.h.s. must always remain finite since the probability must be finite for any physically admissible state. In particular if we multiply ψ by a suitable complex number, we can make the total probability of finding the particle somewhere in space unity, so that we get

$$\int_\tau \rho\, d\tau = \int_\tau |\psi|^2\, d\tau = 1$$

where the integration is carried out over the entire available space. ψ is said to be normalized in this case.

Postulate V

The *expectation value* or the *expected average* of the results of a large number of measurements of a physical property α of the particle is given by

$$<\alpha> = \int_\tau \psi^*\hat{\alpha}\psi\, d\tau$$

where $\hat{\alpha}$ is the operator representing the dynamical variable α. Here we assume ψ to be normalized. If this is not so, then we have to write :

$$<\alpha> = \frac{\int_\tau \psi^*\hat{\alpha}\psi\, d\tau}{\int_\tau \psi^*\psi\, d\tau}$$

It may be noted that the state ψ of the system can be built up by applying the *principle of superposition* (see § 9·17)

$$\psi = \sum_n C_n u_n$$

where the u_n's are the solutions of the eigenvalue equation (see Postulate III). C_n's are complex numbers such that $|C_n|^2 = C_n^* C_n$ gives the probability of finding the particle in the state represented by the eigenfunction u_n. They can be evaluated by utilising the orthogonality property of the eigenfunctions.

Problems

1. Find the eigenvalues and the eigenfunctions of the operator $\left(x + \dfrac{d}{dx}\right)$.

   [$A \exp\{-(x-a)^2/2\}$]

2. Are the following pairs of operators commuting or non-commuting?

   $\hat{x}, \hat{p}_x\,; \hat{y}, \hat{p}_y\,; \hat{z}, \hat{p}_z\,; \hat{x}, \hat{p}_y\,;\ \hat{p}_y, \hat{p}_z\,; \hat{L}_x, \hat{L}_y\,; \hat{L}^2, \hat{L}_z$.

3. (a) Justify the statement that the probability current density cannot be directly measured.

   (b) Using the expression for the probability current density prove that $\psi_n(\mathbf{r}, t)$ cannot be real.

4. Prove that for one-dimensional motion, probability current density for a stationary state is independent of x and t.

5. Prove the expressions for the probability current density given in Eq. (10·6-10).

6. Prove the expression for the one-dimensional probability current density given in Eq. (10·6-11).

7. Prove that tthe classical relationship between the time rate of change of the momentum p_x and the force F_x acting on a particle holds in quantum mechanics if the expectation values of these quantities are considered i.e.,

   $$\frac{d}{dt}<p_x> = <F_x>$$

8. If we define the uncertainties Δx and Δp_x as

   $$\Delta x = \{<x^2> - <x>^2\}^{1/2},\ \Delta p_x = \{<p_x^2> - <p_x>^2\}$$

   then using the results of Prob. 14, prove that $\Delta x\, e \Delta p_x = \hbar/2$ for a normalised Gaussian packet at $t = 0$ (see Prob. 13).

9. Prove that if $\hat{\alpha}$ and $\hat{\beta}$ are two linerar operators, then $\hat{\alpha} + \hat{\beta}$ and $\hat{\alpha}\hat{\beta}$ are also linear operators.

10. If $\hat{\alpha}$ is an operator representing a dynamical variable α and u is an eigenfunction of $\hat{\alpha}$ belonging to the eigenvalue a so that $\hat{\alpha} u = au$, prove that the expectation values $<\alpha> = a$ and $<\alpha^n> = a^n$.

11. (i) Prove that the sum of two Hermitian operators is Hermitian.

    (ii) Prove that the indentity operattor 1, which operating on a functiton gives the function itself, is Hermitian.

Introduction to Wave Mechanics

12. Prove that the product of two Hermitian operators is Hermitian if and only if they commute.

13. Calculate the normalization constant for the Gaussian wave-packet $\psi(x) = A \exp(-\sigma^2 x^2/2) \exp(ikx)$ at $t = 0$. $\qquad (\sigma/\sqrt{\pi})^{1/2}$

14. Calculatte the expectation values $<x>$, $<x^2>$, $<p_x>$ and $<p_x^2>$ for the Gaussian wave-packet given in Prob. 13. $(0, 1/2\,\sigma^2, \hbar k; \hbar^2 k^2 + +\hbar\,\sigma^2/2)$

11

SOLUTIONS OF SCHRÖDINGER EQUATION IN SOME SIMPLE CASES

In the present chapter, we shall illustrate the methods of the solution of the Schrödinger equation for a few simple systems under one dimensional motion and for a system under three dimensional motion. In particular we shall see how the quanntum conditions introduced on ad hoc basis in the old quantum theory follow quite naturally in solving the Schrödinger equation with appropriate boundary conditions.

(A) One dimensional motion

11.1 Boundary conditions

(a) The time-independentt Schrödinger equation is a linear second order differential equation. If the potential V is finite everywhere, though it may not be continuous at all points, it can be shown that the wave function ψ and its gradient must remain continuous, finite and single-valued everywhere for the behaviour of a physical system to be described uniquely by the wave function.

Consider the one-dimensional Schrödinger equation which can be written in the form

$$\frac{d^2\psi}{dx^2} = \frac{2m}{\hbar^2}(V-E)\psi$$

If the potential $V(x)$ is continuous everywhere, then $d\psi/dx$ must be continuous everywhere. Now consider some point $x = a$ where the potential has a finite discontinuity (see Fig. 11.1). Integrating the above equation from $a-\delta$ to $a+\delta$ where δ is small, we get

$$\int_{a-\delta}^{a+\delta} \frac{d}{dx}\left(\frac{d\psi}{dx}\right) dx = \frac{2m}{\hbar^2} \int_{a-\delta}^{a+\delta} (V-E)\psi\, dx$$

$$\therefore \left(\frac{d\psi}{dx}\right)_{a+\delta} - \left(\frac{d\psi}{dx}\right)_{a-\delta} = \frac{2m}{\hbar^2} \int_{a-\delta}^{a+\delta} (V-E)\psi\, dx \qquad (11 \cdot 1\text{-}1)$$

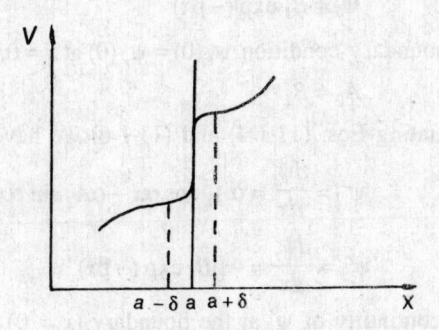

Fig. 11·1. Potential discontinuity.

Since $V(x)$ has a finite discontinuity at $x = a$, the integral on the r.h.s.-vanishes in the limit $\delta \to 0$, so that we have

$$\lim_{\delta \to 0} \left(\frac{d\psi}{dx}\right)_{a+\delta} = \lim_{\delta \to 0} \left(\frac{d\psi}{dx}\right)_{a-\delta}$$

Thus $d\psi/dx$ remains continuous everywhere including the points of finite potential discontinunity.

The finiteness of $\psi(x)$ everywhere follows from the probability interpretation of the wave function. The continuity of $\psi(x)$ follows from the fact that $d\psi/dx$ exists everywhere.

(b) If the potential $V \to \infty$ at some point, then the boundary conditions at the potential discontinuity can be derived in the following way. We assume that there is a potential discontinuity at $x = 0$ (see Fig. 11·5). To the left of $x = 0$, $V = 0$ while to its right $V = V_0 =$ constant. We let $V_0 \to \infty$ in the limit. The Schrödinger equations in the two regions can be written as (assuming $E < V_0$).

$$-\frac{\hbar^2}{2m}\frac{d^2\psi_l}{dx^2} = E\psi_l \qquad (11\cdot1\text{-}2)$$

$$-\frac{\hbar^2}{2m}\frac{d^2\psi_r}{dx^2} + V_0\psi_r = E\psi_r \qquad (11\cdot1\text{-}3)$$

Then writing $\alpha^2 = 2mE/\hbar^2$ and $\beta^2 = 2m(V_0 - E)/\hbar^2$, we get the solutions (see § 11·3) as :

$$\psi_l = A_1 \sin \alpha x + A_2 \cos \alpha x \qquad (11\cdot1\text{-}4)$$

$$\psi_r = B_1 \exp(-\beta x) + B_2 \exp(\beta x) \qquad (11\cdot1\text{-}5)$$

Since ψ_r must remain finite at all points $x > 0$, it must remain finite as $x \to \infty$. This is possible only if we put $B_2 = 0$. Then

$$\psi_r = B_1 \exp(-\beta x) \qquad (11 \cdot 1\text{-}6)$$

Applying the boundary condition $\psi_l(0) = \psi_r(0)$ at $x = 0$, we get

$$A_2 = B_1$$

Again differentiating Eqs. (11·1-4) and (11·1-6) we have

$$\psi'_l = \frac{d\psi_l}{dx} = \alpha A_1 \cos \alpha x - \alpha A_2 \sin \alpha x$$

$$\psi'_r = \frac{d\psi_r}{dx} = -\beta B_1 \exp(-\beta x)$$

Then the continuity of ψ' at the boundary ($x = 0$) gives us

$$\alpha A_1 = -\beta B_1 = -\beta A_2$$

or, $\qquad A_2 = -\alpha A_1/\beta \qquad (11 \cdot 1\text{-}7)$

If now we let $V_0 \to \infty$, then $\beta \to \infty$ and we get $A_2 = 0$ so that $\psi = \psi_l = A_1 \sin \alpha x$ and $\psi = \psi_r = 0$ so that the wave function vanishes for $x \geq 0$.

It may be noted that the constant A_1 is not determined from these relations so that $d\psi/dx$ remains undetermined at the boundary $x = 0$. A_1 can only be fixed by normalization.

11·2 Particle in a one-dimensional potential box with rigid walls

Consider a particle of energy E confined within a region 0 to l on the x-axis. At $x = 0$ and $x = l$, there are two absolutely rigid, impenetrable potential walls of infinite height (Fig. 11·2). This means

$V = \infty$ for $x \leq 0$ and $x \geq l$

$V = 0$ for $0 < x < l$

As we have seen in the previous section, the boundary conditions in this case require that ψ must vanish at the boundries and outside. Hence

$$\psi(0) = \psi(l) = 0 \qquad (11 \cdot 2\text{-}1)$$

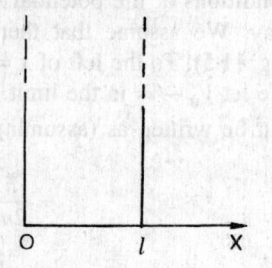

Fig. 11·2. One-dimensional potential box.

Since the potential energy is time-independent, we use the time-independent Schröndinger equation for the region $0 < x < l$.

$$\hat{H}\psi(x) = -\frac{\hbar^2}{2m} \frac{d^2\psi(x)}{dx^2} = E\psi(x)$$

Therefore $\qquad \dfrac{d^2\psi(x)}{dx^2} + \alpha^2 \psi(x) = 0 \qquad (11 \cdot 2\text{-}2)$

Solution of Schrödinger Equation in Some Simple Cases

where $\alpha^2 = 2mE/\hbar^2$ or, $\alpha = \sqrt{2mE/\hbar^2}$ (11·2-3)

The solution is of the form

$$\psi(x) = A \sin \alpha x + B \cos \alpha x$$

From the boundary condition $\psi = 0$ at $x = 0$, we get $B = 0$. Hence

$$\psi(x) = A \sin \alpha x \qquad (11·2\text{-}4)$$

The boundary condition $\psi = 0$ at $x = l$ gives

$$A \sin \alpha l = 0$$

which is satisfied if

$$\sin \alpha l = \sin n\pi$$

where n is an integer; $n = 1, 2, 3$ *etc.* $n = 0$ is excluded since this will make $\psi = 0$ everywhere. So

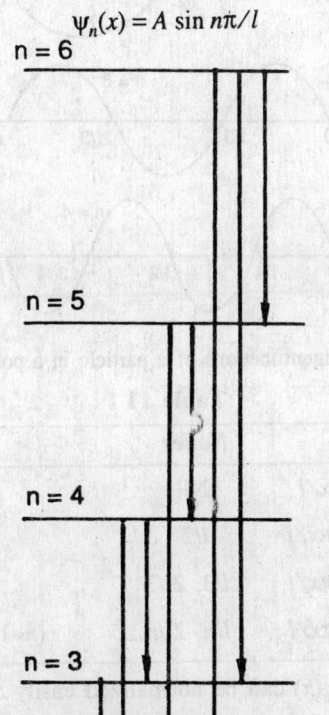

$$\alpha_n = n\pi/l \qquad (11·2\text{-}5)$$

and $\qquad \psi_n(x) = A \sin n\pi/l \qquad (11·2\text{-}6)$

Fig. 11·3. Energy spectrum of a particle in a potential box.

The eigenvalues thus form a discrete set as in the problem of a vibrating string with the two ends fixed. The energy of the particle will be

$$E_n = \frac{\hbar^2 \alpha_n^2}{2m} = \frac{n^2 \pi^2 \hbar^2}{2ml^2} \qquad (11 \cdot 2\text{-}7)$$

The energy spectrum is shown in Fig. 11·3. We list below some of the eigenfunctions ψ_n for different n which are shown in Fig. 11·4.

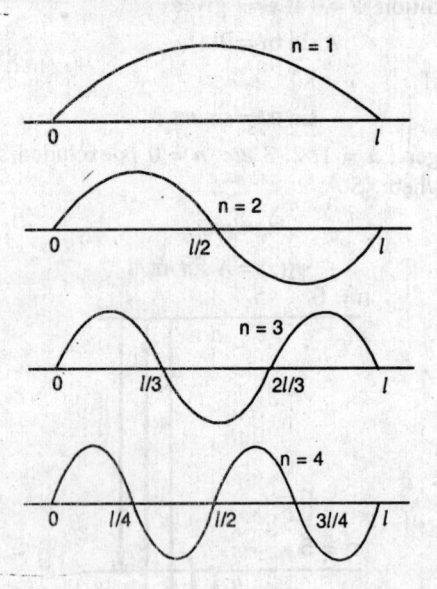

Fig. 11·4. Eigenfunctions of a particle in a potential box.

Table 11·1

n	$\psi_n(x)$	Nodes		No. of nodes
1	$A_1 \sin \pi x/l$	Nil		0
2	$A_2 \sin 2\pi x/l$	$l/2$		1
3	$A_3 \sin 3\pi x/l$	$l/3, 2l/3$		2
n	$A_n \sin n\pi x/l$	$l/n, 2l/n,\ldots$	$(n-1)l/n$	$n-1$

The eigenfunctions $\psi_n(x)$ can be normalized easily:

$$\int_0^l |\psi_n(x)|^2 \, dx = |A_n|^2 \int_0^l \sin^2 \frac{n\pi x}{l} \, dx = |A_n|^2 \frac{l}{2} = 1$$

$$A_n = \sqrt{\frac{2}{l}}$$

Solution of Schrödinger Equation in Some Simple Cases 285

and
$$\psi_n = \sqrt{\frac{2}{l}} \sin \frac{n\pi x}{l} \qquad (11\cdot2\text{-}8)$$

These functions are orthogonal and form a complete orthonormal set.

There is an infinite number of discrete energy levels corresponding to all positive integral values of the quantum number n. There is just one eigenfunction for each level and the number of nodes of the nth eigenfunction is $(n-1)$. Thus the energy spectrum is *discrete* and *non-degenerate*. It should be noted that the discreteness results from the application of the boundary conditions.

From Eq. (11·2-7) the spacing between the successive energy levels is given by

$$\Delta E = \frac{\pi^2 \hbar^2}{2ml^2}\left\{(n+1)^2 - n^2\right\} = \frac{\pi^2 \hbar^2}{2m}\frac{2n+1}{l^2} \qquad (11\cdot2\text{-}9)$$

As the dimension of the box increases, the levels come closer together. In the limit of $l \to \infty$, $\Delta E \to 0$ and the levels form a continuum as in the case of a free particle (see § 11·10).

11·3 Step potential

Consider a particle of mass m moving along the x-axis being acted upon by a constant potential V_0 at all points $x > 0$, while the potential is zero for all points $x < 0$. This type of potential has the appearance of a *step* (see Fig. 11·5) and hence it is known as the *step potential*.

We can divide the whole one-dimensional space into two regions, I for $x < 0$ and II for $x > 0$. Then we have

For region I $(x < 0)$, $V = 0$,

For region II $(x > 0)$, $V = V_0$

Denoting the wave functions in the two regions by $\psi_1(x)$ and $\psi_2(x)$, we can write down the time-independent Schrödinger equations for the two regions as follows:

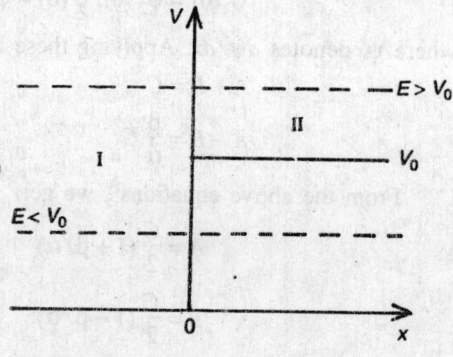

Fig. 11·5. Step potential.

$$-\frac{\hbar^2}{2m}\frac{d^2\psi_1}{dx^2} = E\psi_1 \qquad (11\cdot3\text{-}1)$$

$$-\frac{\hbar^2}{2m}\frac{d^2\psi_2}{dx^2} + V_0\psi_2 = E\psi_2 \qquad (11\cdot3\text{-}2)$$

The energy E of the particle may be greater or less than V_0. We shall consider the two cases separately.

(A) $\mathbf{E} > \mathbf{V_0}$

If we write $\alpha^2 = 2mE/\hbar^2$ and $\beta^2 = (2m/\hbar^2)(E - V_0)$ where α and β are real quantities, then from the equations (11·3-1) and (11·3-2), we get

$$\psi_1'' + \alpha^2 \psi_1 = 0 \qquad (11\cdot3\text{-}3)$$

$$\psi_2'' + \beta^2 \psi_2 = 0 \qquad (11\cdot3\text{-}4)$$

The solutions of the above two equations are

$$\psi_1 = A \exp(i\alpha x) + B \exp(-i\alpha x) \qquad (11\cdot3\text{-}5)$$

$$\psi_2 = C \exp(i\beta x) + D \exp(-i\beta x) \qquad (11\cdot3\text{-}6)$$

We assume that initially the particle moves from left to right. Then in Eq. (11·3-5), A denotes the amplitude of the incident wave ψ_{in} at the boundary. It is partly reflected and partly transmitted at the boundary $x = 0$. B is the amplitude of the reflected wave. The wave transmitted into region II has the amplitude C. Since there is no wave in this region moving from right to left, the amplitude of such a wave $D = 0$.

Thus Eq. (11·3-6) becomes

$$\psi_2 = C \exp(i\beta x) \qquad (11\cdot3\text{-}7)$$

The boundary conditions which must hold at $x = 0$ are

$$\psi_1(0) = \psi_2(0), \quad \psi'(0) = \psi_2'(0) \qquad (11\cdot3\text{-}8)$$

where ψ' denotes $d\psi/dx$. Applying these boundary conditions we get

$$A + B = C$$

$$A - B = \frac{\beta}{\alpha} C$$

From the above equations, we get

$$A = \frac{C}{2}(1 + \beta/\alpha)$$

$$B = \frac{C}{2}(1 - \beta/\alpha)$$

Hence we have
$$\frac{B}{A} = \frac{\alpha - \beta}{\alpha + \beta} \qquad (11\cdot3\text{-}9)$$

$$\frac{C}{A} = \frac{2\alpha}{\alpha + \beta} \qquad (11\cdot3\text{-}10)$$

Since $\psi_{in} = A \exp(i\alpha x)$, we have $\psi_{in}^* = A^* \exp(-i\alpha x)$.

So the probability current density for the incident wave is

Solution of Schrödinger Equation in Some Simple Cases

$$j_{in} = \frac{i\hbar}{2m}\left(\psi_{in}\frac{d\psi_{in}^*}{dx} - \psi_{in}^*\frac{d\psi_{in}}{dx}\right)$$

$$= \frac{i\hbar}{2m}|A|^2(-i\alpha - i\alpha) = \frac{\hbar\alpha}{m}|A|^2 \qquad (11\cdot3\text{-}11)$$

The wave reflected back into region I at the boundary is

$$\psi_r = B\exp(-i\alpha x)$$

The probability current density for the reflected wave is

$$j_r = \frac{i\hbar}{2m}\left(\psi_r\frac{d\psi_r^*}{dx} - \psi_r^*\frac{d\psi_r}{dx}\right)$$

$$= \frac{i\hbar}{2m}|B|^2(-i\alpha - i\alpha) = \frac{\hbar\alpha}{m}|B|^2 \qquad (11\cdot3\text{-}12)$$

The probability current density for the transmitted wave can be found using Eq. (11·3-7) for the transmitted wave $\psi_t = \psi_2$.

$$j_t = \frac{i\hbar}{2m}\left(\psi_t\frac{d\psi_t^*}{dx} - \psi_t^*\frac{d\psi_t}{dx}\right)$$

$$= \frac{i\hbar}{2m}|C|^2(-i\beta - i\beta) = \frac{\hbar\beta}{m}|C|^2 \qquad (11\cdot3\text{-}13)$$

Let us define the *transmission coefficient* as

$$T = \frac{\text{Probability current density for the transmitted wave}}{\text{Probability current density for the incident wave}}$$

Similarly the *reflection coefficient* is defined as

$$R = \frac{\text{Probability current density for the reflected wave}}{\text{Probability current density for the incident wave}}$$

Then using Eqs. (11·3-10), (11·3-11) and (11·3-13) we get

$$T = \frac{|j_t|}{|j_{in}|} = \frac{\beta}{\alpha}\cdot\frac{|C|^2}{|A|^2} = \frac{\beta}{\alpha}\cdot\frac{4\alpha^2}{(\alpha+\beta)^2} = \frac{4\alpha\beta}{(\alpha+\beta)^2} \qquad (11\cdot3\text{-}14)$$

Similarly using Eqs. (11·3-9), (11·3-11) and 11·3-12), we have

$$R = \frac{|j_r|}{|j_{in}|} = \frac{|B|^2}{|A|^2} = \frac{(\alpha-\beta)^2}{(\alpha+\beta)^2} \qquad (11\cdot3\text{-}15)$$

From Eqs. (11·3-14) and (11·3-15), we get

$$T + R = 1$$

Thus there is conservation of probability at the boundary between the two regions I and II.

According to Eq. (11·3-10), $|C| > |A|$, since $\alpha > \beta$. So the amplitude of the transmitted wave is greater than that of the incident wave. The nature of the wave functions in the two regions I and II is illustrated in

Fig. 11·6 a. Since the kinetic energy of the particle is greater in region I

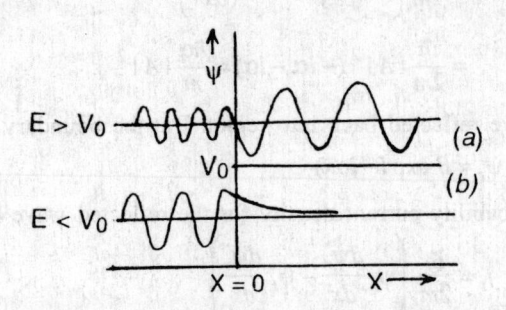

Fig. 11·6. Wave functions for a step potential.

than in region II, the de Broglie wavelength is shorter in region I which is also shown in the figure.

(B) $E < V_0$

If the energy E of the particle is less than the potential V_0 then we have

$$\beta^2 = \frac{2m}{\hbar^2}(E - V_0) = -\frac{2m}{\hbar^2}(V_0 - E) < 0$$

We write

$$\gamma^2 = -\beta^2 = \frac{2m}{\hbar^2}(V_0 - E) > 0$$

so that γ is real. Then from the Eqs. (11·3-3) and (11·3-4) we get

$$\psi_1'' + \alpha^2 \psi_1 = 0$$

$$\psi_2'' - \gamma^2 \psi_2 = 0$$

The solutions are $\quad \psi_1 = A \exp(i\alpha x) + B \exp(-i\alpha x) \quad$ (11·3-16)

$$\psi_2 = C \exp(-\gamma x) + D \exp(\gamma x) \quad (11\cdot3\text{-}17)$$

In Eq. (11·3-17), $\exp(\gamma x) \to \infty$ as $x \to \infty$. So we must put $D = 0$.

Hence $\quad \psi_2 = C \exp(-\gamma x) \quad$ (11·3-18)

From the boundary conditions (11·3-8), we have

$$A + B = C$$

and

$$i\alpha(A - B) = -\gamma C$$

or

$$A - B = \frac{i\gamma}{\alpha} C$$

These two equations give

$$A = \frac{C}{2}\left(1 + \frac{i\gamma}{\alpha}\right)$$

$$B = \frac{C}{2}\left(1 - \frac{i\gamma}{\alpha}\right)$$

So we get
$$\frac{B}{A} = \frac{\alpha - i\gamma}{\alpha + i\gamma}$$

and
$$\frac{C}{A} = \frac{2\alpha}{\alpha + i\gamma}$$

The reflection coefficient is

$$R = \left|\frac{B}{A}\right|^2 = \left|\frac{\alpha - i\gamma}{\alpha + i\gamma}\right|^2 = 1 \qquad (11\cdot3\text{-}19)$$

Since $T + R = 1$, the transmission coefficient $T = 0$. This can also be proved easily by calculating the probability current density for the transmitted wave the proof of which is left to the reader as a problem.

The nature of the wave function $\psi(x)$ in the two regions I and II is illustrated in Fig. 11·6b. It should be noted that there is no absorption of the wave in region II. If there were any such absorption then there would not be 100% reflection ($R = 1$) at the boundary Actually, the wave penetrating a small distance from the boundary into the region II is continually reflected till all the incident energy is turned back into region I. Due to such continual reflection, the amplitude of the wave penetrating into region II falls off exponentially.

The above situation is analogous to the total reflection of electromagnetic waves travelling from an optically denser to a rarer medium.

It should be noted that according to classical mechanics, a particle of energy $E < V_0$ can never penetrate into region II, since its kinetic energy would be negative. But in quantum mechanics, there is a finite probability of the particle wave penetrating a short distance into region II.

11·4 One dimensional rectangular potential barrier

Suppose a particle of mass m and energy E is acted upon by a constant potential $V = +V_0$ in the region $0 \leq x \leq a$ while the potential is zero everywhere else. The potential is shown in Fig. 11·7.

We divide the whole one dimensional space into three potential regions as follows :

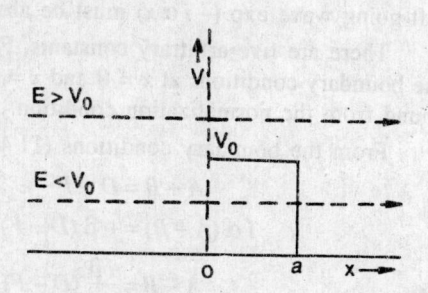

Fig. 11·7. One dimensional rectangular potential barrier.

Region I : For $x < 0$, $V = 0$
Region II : For $0 \leq x \leq a$, $V = V_0$
Region III : For $x > a$, $V = 0$

In the three regions let the wave functions be $\psi_1(x)$, $\psi_2(x)$ and $\psi_3(x)$. First derivatives with respect to x of the ψ's will be written as $\psi'(x)$ while the second derivatives as $\psi''(x)$.

Boundary conditions are :

At $x = 0$, $\psi_1(0) = \psi_2(0)$, $\psi_1'(0) = \psi_2'(0)$ (11·4-1)

At $x = a$, $\psi_2(a) = \psi_3(a)$, $\psi_2'(a) = \psi_3'(a)$ (11·4-2)

The initial condition is that the wave is incident in the region I from *left to right*.

The energy E can be either less than or greater than the barrier height V_0. The case of interest is when $E < V_0$.

Let us write $\alpha^2 = 2mE/\hbar^2$, $\beta^2 = (2m/\hbar^2)(V_0 - E)$.

From the time-independent Schrödinger wave equations in three regions we get

$$\psi_1'' + \alpha^2 \psi_1 = 0$$

$$\psi_2'' - \beta^2 \psi_2 = 0$$

$$\psi_3'' + \alpha^2 \psi_3 = 0$$

The solutions are :

$$\psi_1(x) = A \exp(i\alpha x) + B \exp(-i\alpha x) \quad (11\cdot4\text{-}3)$$

$$\psi_2(x) = D \exp(-\beta x) + F \exp(\beta x) \quad (11\cdot4\text{-}4)$$

$$\psi_3(x) = C \exp(i\alpha x) \quad (11\cdot4\text{-}5)$$

In region III, the wave propagates to the right only. Hence the left-going wave $\exp(-i\alpha x)$ must be absent.

There are five arbitrary constants. Four of them can be found from the boundary conditions at $x = 0$ and $x = a$ given above. The fifth can be found from the normalization condition.

From the boundary conditions (11·4-1) at $x = 0$, we get

$$A + B = D + F \quad (11\cdot4\text{-}6)$$

$$i\alpha(A - B) = -\beta(D - F)$$

or, $$A - B = \frac{i\beta}{\alpha}(D - F) \quad (11\cdot4\text{-}7)$$

From the boundary conditions (11·4-2) at $x = a$, we have

$$D \exp(-\beta a) + F \exp(\beta a) = C \exp(i\alpha a)$$

Solution of Schrödinger Equation in Some Simple Cases

$$-\beta \{D \exp(-\beta a) - F \exp(\beta a)\} = i\alpha C \exp(i\alpha a)$$

Solving for D and F we get

$$D = \frac{C}{2} \exp(i\alpha a) \cdot \left(1 - \frac{i\alpha}{\beta}\right) \cdot \exp(\beta a)$$

$$= \frac{C}{2\beta}(\beta - i\alpha) \cdot \exp(i\alpha a) \exp(\beta a)$$

$$F = \frac{C}{2} \exp(i\alpha a) \cdot \left(1 + \frac{i\alpha}{\beta}\right) \cdot \exp(-\beta a)$$

$$= \frac{C}{2\beta}(\beta + i\alpha) \exp(i\alpha a) \exp(-\beta a)$$

Using Eqs. (11·4-6) and (11·4-7) we then get

$$A = \frac{C}{4} \exp(i\alpha a) \left\{ -\frac{(\beta - i\alpha)^2}{i\alpha\beta} \exp(\beta a) + \frac{(\beta + i\alpha)^2}{i\alpha\beta} \cdot \exp(-\beta a) \right\} \tag{11·4-8}$$

$$B = \frac{C}{4} \exp(i\alpha a) \left\{ \frac{\beta^2 + \alpha^2}{i\alpha\beta} \exp(\beta a) - \frac{\beta^2 + \alpha^2}{i\alpha\beta} \exp(-\beta a) \right\} \tag{11·4-9}$$

The incident probability current density is

$$j_{in} = \frac{i\hbar}{2m}\left(\psi_{in}\frac{d\psi_{in}^*}{dx} - \psi_{in}^*\frac{d\psi_{in}}{dx}\right)$$

where $\psi_{in} = A \exp(i\alpha x), \quad \dfrac{d\psi_{in}}{dx} = i\alpha\psi_{in}$

$\psi_{in}^* = A^* \exp(-i\alpha x), \quad \dfrac{d\psi_{in}^*}{dx} = -i\alpha\psi_{in}^*$

Hence $\quad j_{in} = \dfrac{i\hbar}{2m}(-i\alpha|\psi_{in}|^2 - i\alpha|\psi_{in}|^2) = \dfrac{\hbar\alpha}{m}|A|^2 \tag{11·4-10}$

Similarly the transmitted probability current density is

$$j_t = \frac{i\hbar}{2m}\left(\psi_t\frac{d\psi_t^*}{dx} - \psi_t^*\frac{d\psi_t}{dx}\right)$$

where $\psi_t = \psi_3 = C \exp(i\alpha x), \quad \psi_t^* = C^* \exp(-i\alpha x)$

$\dfrac{d\psi_t}{dx} = i\alpha\psi_3, \quad \dfrac{d\psi_t^*}{dx} = -i\alpha\psi_3^*$

Hence $\quad j_t = \dfrac{i\hbar}{2m}(-i\alpha|\psi_3|^2 - i\alpha|\psi_3|^2) = \dfrac{\hbar\alpha}{m}|C|^2 \tag{11·4-11}$

So the transmission coefficient is

$$T = \frac{j_t}{j_{in}} = \frac{|C|^2}{|A|^2}$$

But from Eq. (11·4-8), we have

$$\frac{C}{A} = \frac{4i\alpha\beta \exp(-i\alpha a)}{(\beta + i\alpha)^2 \exp(-\beta a) - (\beta - i\alpha)^2 \exp(\beta a)}$$

$$= \frac{4i\alpha\beta \exp(-i\alpha a)}{-2(\alpha^2 - \beta^2) \sinh \beta a + 4i\alpha\beta \cosh \beta a}$$

Hence

$$\left|\frac{C}{A}\right|^2 = \frac{4\alpha^2\beta^2}{(\alpha^2 - \beta^2)^2 \sinh^2 \beta a + 4\alpha^2\beta^2 \cosh^2 \beta a}$$

$$= \frac{4\alpha^2\beta^2}{4\alpha^2\beta^2 + (\alpha^2 + \beta^2)^2 \sinh^2 \beta a}$$

So

$$T = \left\{1 + \frac{(\alpha^2 + \beta^2)^2}{4\alpha^2\beta^2} \sinh^2 \beta a\right\}^{-1} \tag{11·4-12}$$

But $(\alpha^2 + \beta^2)^2 = (2m V_0/\hbar^2)^2$, $\alpha^2\beta^2 = (2m/\hbar^2)^2 E(V_0 - E)$

So we get finally

$$T = \left\{1 + \frac{V_0^2}{4E(V_0 - E)} \sinh^2 \beta a\right\}^{-1} \tag{11·4-13}$$

Classically the particle with energy $E < V_0$ in region I can never go into Region III and T would be zero. This is because in Region II, the kinetic energy of the particle would be less than zero. However quantum mechanically there is a finite probability of the particle crossing into III. As E decreases below V_0, T monotonically decreases because of the factor $E(V_0 - E)$. Also as the thickness a of the barrier increases, $\sinh \beta a$ increases rapidly and hence T decreases. For $\beta a >> 1$, $\sinh^2 \beta a \sim \frac{1}{4} \exp(2\beta a) >> 1$.

We get in this case

$$T = \frac{16 E(V_0 - E)}{V_0^2} \exp(-2\beta a) \tag{11·4-14}$$

11·5 Explanation of α-decay

The quantum mechanical effect of a particle in Region I with energy E lower than the height V_0 of the potential barrier crossing into Region III with a finite probability is known as *tunnel effect*.

This quantum mechanical tunnel effect of penetration through a potential barrier can be applied to explain the α-decay of radioactive nuclei (Gamow, Gurney and Condon). The actual situation is much more complicated in the case of an α-particle escaping from the nucleus. Within the nucleus it is in an attractive potential well of depth about 40 MeV (for heavy nuclei). Just outside the nucleus it is acted on by the repulsive Coulomb potential $2Ze^2/r$ at a distance r from the centre, Ze being the nuclear charge (Fig. 11·8). The barrier in this case extends from the nuclear radius R upto a distance b where the Coulomb potential energy becomes equal to the α-energy E so that $b = 2Z\ e^2/E$. At the nuclear surface the barrier height is $2Ze^2/R$. The nuclear radius R is proportional to $A^{1/3}$ where A is the mass number of the nucleus : $R = r_0 A^{1/3}$ where $r_0 = 1.2 \times 10^{-15}$ m. It is then possible to calculate the average barrier height

$$<V> = \int_R^b \frac{2Ze^2/r}{b-R} dr$$

Calculations using the above average barrier do not usually give satisfactory results.

If we take arbitrarily an equivalent rectangular barrier of height $V_0 = 15$ MeV and the α-energy $E = 5$ MeV, then we get for $a = 2 \times 10^{-14}$ m

$$\beta a = \frac{a}{\hbar} \sqrt{2M(V_0 - E)} = \frac{2 \times 10^{-14}}{1 \cdot 05 \times 10^{-34}}$$
$$\times (2 \times 4 \times 1 \cdot 67 \times 10^{-27} \times 10$$
$$\times 1 \cdot 6 \times 10^{-13})^{1/2} = 27 \cdot 85$$

$$\therefore T = \frac{16 \times 5 \times 10}{15 \times 15} \exp(-55 \cdot 7) = 2 \cdot 3 \times 10^{-24}$$

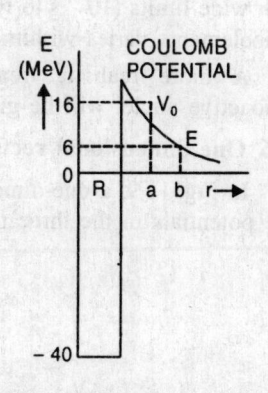

Fig. 11·8. Potential barrier for alpha decay.

Thus the transmission coefficient which gives the probability of penetration through the barrier for each collision against the barrier wall is quite small. The α-particle has about 2·3 chances in 10^{24} collisions against the barrier to escape from the nucleus. But in one second, the number of collisions made by it against the barrier (assuming the nuclear radius to be 10^{-14} m) is

$$n = \text{velocity/diameter of the nucleus} = \frac{1 \cdot 7 \times 10^7}{2 \times 10^{-14}} = 8 \cdot 5 \times 10^{20} \text{ per s}$$

Hence the probability of escape *per second* is

$$P = nT = 8.5 \times 10^{20} \times 2.3 \times 10^{-24} = 1.96 \times 10^{-3}$$

i.e., the mean life time against α-decay is

$$\tau = \frac{1}{P} = \frac{1}{nT} = \frac{10^3}{1.96} = 510 \text{ s} = 8 \text{ min } 30 \text{ s}$$

Since the transmission coefficient T and hence the decay probability P decreases exponentially with βa, small changes in the value of βa will produce very large changes in P with consequent large changes in the α-decay life time.

For barrier width 20% greater than that considered above, $2\beta a = 66.8$ which gives a mean life

$$\tau = 3.4 \times 10^7 \text{ s} = 1.08 \text{ yr}$$

which is 6.65×10^4 times as large as in the previous case. This explains why the observed α-emission half-lives of natural radioelements vary over such wide limits (10^{-7} s to 10^{10} yr) even through the α-energy for natural radioelements varies within a factor of only about 2.5.

A more realistic treatment of the problem of α-decay from radioactive nuclei will be given in Vol. II.

11·6 One dimensional rectangular potential well

In Fig. 11·9, a one-dimensional rectangular potential well is shown. The potentials in the three regions, I, II and III, are given by

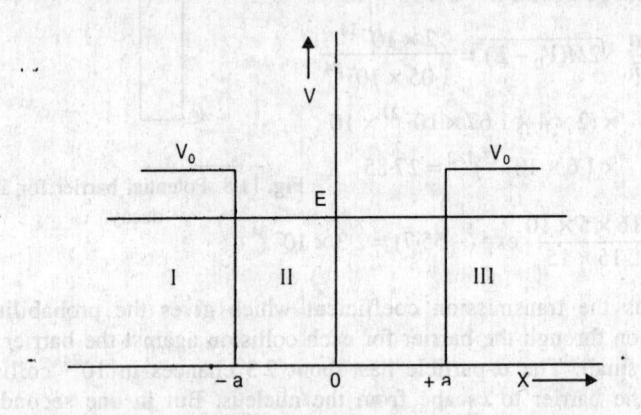

Fig. 11·9. One dimensional rectangular potential well.

Region I ($x < -a$) : $V = V_0$

Region II ($-a \leq x \leq a$) : $V = 0$

Region III ($x > a$) : $V = V_0$

Consider a particle of mass m and energy E to be initially in the

Solution of Schrödinger Equation in Some Simple Cases

potential well, *i.e.*, in region II. The Schrödinger equations for the three regions are :

$$-\frac{\hbar^2}{2m}\frac{d^2\psi_1}{dx^2} + V_0\psi_1 = E\psi_1 \qquad (11\cdot6\text{-}1)$$

$$-\frac{\hbar^2}{2m}\frac{d^2\psi_2}{dx^2} = E\psi_2 \qquad (11\cdot6\text{-}2)$$

$$-\frac{\hbar^2}{2m}\frac{d^2\psi_3}{dx^2} + V_0\psi_3 = E\psi_3 \qquad (11\cdot6\text{-}3)$$

The energy E of the particle may be greater or less than V_0. We first consider the case of $E < V_0$ which corresponds to bound state. We write

$$\alpha^2 = \frac{2mE}{\hbar^2} \quad \text{and} \quad \beta^2 = \frac{2m(V_0 - E)}{\hbar^2} \qquad (11\cdot6\text{-}4)$$

α and β are both real. Substituting in the Schrödinger equations, we get

$$\psi_1'' - \beta^2\psi_1 = 0 \qquad (11\cdot6\text{-}5)$$

$$\psi_2'' + \alpha^2\psi_2 = 0 \qquad (11\cdot6\text{-}6)$$

$$\psi_3'' - \beta^2\psi_3 = 0 \qquad (11\cdot6\text{-}7)$$

Remembering that the wave functions must vanish as $r \to \pm\infty$, the well-behaved solutions of the above equations can be written as

$$\psi_1 = A\exp(\beta x) \qquad (11\cdot6\text{-}8)$$

$$\psi_2 = C\exp(i\alpha x) + D\exp(-i\alpha x) \qquad (11\cdot6\text{-}9)$$

$$\psi_3 = F\exp(-\beta x) \qquad (11\cdot6\text{-}10)$$

The boundary conditions at $x = -a$ and $x = +a$ are :

$$\psi_1(-a) = \psi_2(-a),\ \psi_1'(-a) = \psi_2'(-a) \qquad (11\cdot6\text{-}11)$$

$$\psi_3(a) = \psi_2(a),\ \psi_3'(a) = \psi_2'(a)$$

where we have written ψ' for $d\psi/dx$. From the boundary conditions at $x = -a$ we get

$$A\exp(-\beta a) = C\exp(-i\alpha a) + D\exp(i\alpha a)$$

$$\beta A\exp(-\beta a) = i\alpha[C\exp(-i\alpha a) - D\exp(i\alpha a)]$$

These give $\quad \dfrac{C}{D} = \dfrac{\alpha - i\beta}{\alpha + i\beta}\exp(2i\alpha a) \qquad (11\cdot6\text{-}12)$

From the boundary condition at $x = +a$, we get

$$F\exp(-\beta a) = C\exp(i\alpha a) + D\exp(-i\alpha a)$$

$$-\beta F\exp(-\beta a) = i\alpha\{C\exp(i\alpha a) - D\exp(-i\alpha a)\}$$

These give
$$\frac{C}{D} = \frac{\alpha + i\beta}{\alpha - i\beta} \exp(-2i\alpha a) \qquad (11\cdot6\text{-}13)$$

From Eqs. (11·6-12) and (11·6-13) we have
$$\frac{C^2}{D^2} = 1 \quad \text{or,} \quad \frac{C}{D} = \pm 1$$

For $C = +D$, $\qquad \psi_2 = 2C \cos \alpha x \qquad (11\cdot6\text{-}14)$

and $\qquad A = +F \qquad (11\cdot6\text{-}15)$

The eigenfunction ψ_2 within the well is a *symmetric function* about the centre of the well in this case as at the energies E_0 and E_2 shown in Fig. 11·10. Obviously $\psi(-x) = \psi(x)$ in this case and the eigenfunctions are said to have *even parity* (see § 11·8).

For $C = -D$, $\psi_2 = 2i C \sin \alpha x \qquad (11\cdot6\text{-}16)$

and $\qquad A = -F \qquad (11\cdot6\text{-}17)$

ψ is an *antisymmetric function* about $x = 0$ in this case as at the energy E_1; $\psi(-x) = -\psi(x)$ and the eigenfunctions have *odd parity*.

The two equations (11·6-12) and (11·6-13) are usually incompatible except for some special values of the energy E of the particle. In such cases these equations can be combined to give

$(\alpha + i\beta)^2 \exp(-2i\alpha a)$
$\quad = (\alpha - i\beta)^2 \exp(2i\alpha a)$

or $(\alpha + i\beta) \exp(-i\alpha a)$
$\quad = \pm (\alpha - i\beta) \exp(i\alpha a)$

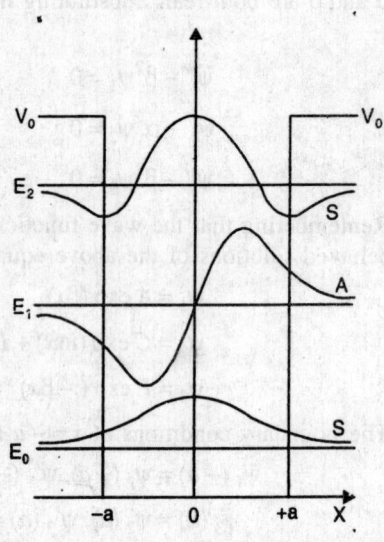

Fig. 11·10. Eigenfunctions for the rectangular well.

From the two cases (+) and (−) in the above equation, we get the following two relations

$$\tan \alpha a = \frac{\beta}{\alpha} \quad \text{or,} \quad \tan \frac{a}{\hbar} \sqrt{2mE} = \sqrt{\frac{V_0 - E}{E}} \qquad (11\cdot6\text{-}18)$$

and $\cot \alpha a = -\dfrac{\beta}{\alpha} \quad \text{or,} \quad \cot \dfrac{a}{\hbar} \sqrt{2mE} = -\sqrt{\dfrac{V_0 - E}{E}} \qquad (11\cdot6\text{-}19)$

We also have another relation between α and β, viz.,

$$(\alpha^2 + \beta^2) a^2 = \frac{2m V_0}{\hbar^2} a^2 \tag{11·6-20}$$

We thus have to solve the following two pairs of equations each of which can be solved graphically.

$$\beta a = \alpha a \tan \alpha a, \quad a^2(\alpha^2 + \beta^2) = 2mV_0 a^2/\hbar^2 \tag{11·6-21}$$

$$\beta a = -\alpha a \cot \alpha a, \quad a^2(\alpha^2 + \beta^2) = 2mV_0 a^2/\hbar^2 \tag{11·6-22}$$

The points of intersections of the graphs give the possible solutions in each case.

Eqs. (11·6-21) can be solved graphically to give the energies E_0, E_2 etc. for the symmetric solution (11·6-14).

Similarly Eqs. (11·6-22) can be solved graphically to yield the energies E_1, E_3 etc. for the antisymmetric solution (11·6-16).

These solutions are sketched graphically in Fig. 11·11 for the two cases $(\alpha^2 + \beta^2)^{1/2} a = 4$ and $(\alpha^2 + \beta^2)^{1/2} a = 2$ for an arbitrary a. These give

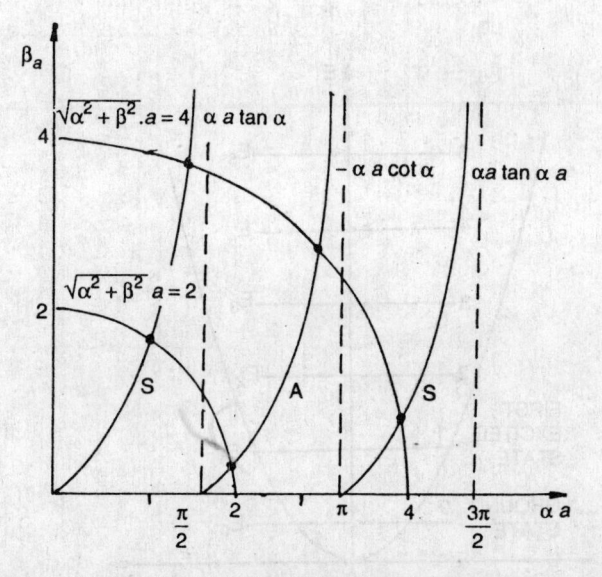

Fig. 11.11 Graphical solutions for a particle in rectangular well.

the two concentric quarter circles as shown. Their intersections with the $\alpha a \tan \alpha a$ and $-\alpha a \cot \alpha a$ curves give the permissible energy values. Note that for $(\alpha^2 + \beta^2)^{1/2} a = 2$ there is one symmetric solution and one antisymmetric solution while for $(\alpha^2 + \beta^2)^{1/2} a = 4$ there are two symmetric solutions and one antisymmetric solution. The number of possible energy levels increases with the increasing values of V_0 and a.

From the above discussions, it is clear that the existence of the discrete energy eigenvalues for a particle in a potential well is due to the requirements of finding well-behaved solutions of the Schrödinger equations in the different regions, consistent with the boundary conditions. This is a general result valid for any bound system.

For $E > 0$, well-behaved solutions can be obtained for all values of E so that the energy spectrum is a continuum in this case.*

11·7 Linear harmonic oscillator

The problem of the linear harmonic oscillator is of great importance both in classical physics as also in quantum mechanics. Many complex systems can be reduced to a collection of harmonic oscillators of different frequencies by Fourier analysis. The method is particularly suitable in many body systems. For example, in his theory of the specific heat of solids, Debye considered the collective motion of the atoms in the crystalline solid as made up of the harmonic vibrations of different frequencies upto a maximum cut off frequency.

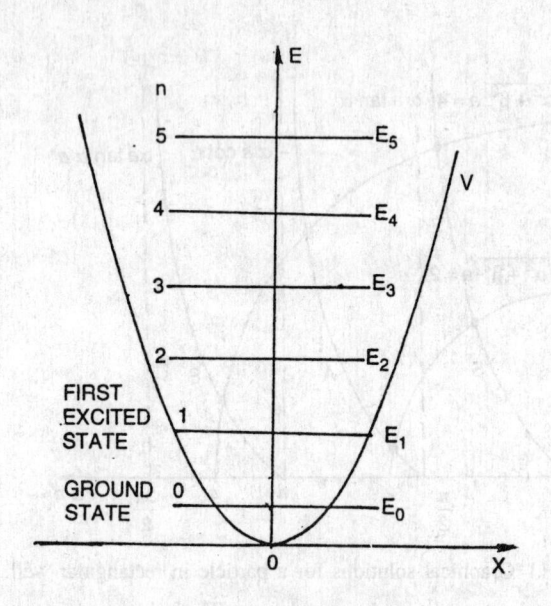

Fig. 11·12. Harmic oscillator potential.

---

* See Introductory Quantum Mechanics by the author.

Solution of Schrödinger Equation in Some Simple Cases

For a linear harmonic oscillator of angular frequency $\omega = 2\pi \nu$, the potential energy is given by

$$V(x) = \frac{1}{2} k x^2 = \frac{1}{2} m \omega^2 x^2 \qquad (11\cdot7\text{-}1)$$

where $k = m\omega^2$ (Fig. 11·12).

The Hamiltonian is

$$\hat{H} = \frac{\hat{p}_x^2}{2m} + V(x) = -\frac{\hbar^2}{2m} \frac{d^2}{dx^2} + \frac{1}{2} m \omega^2 x^2$$

The time-independent Schrödinger equation becomes

$$-\frac{\hbar^2}{2m} \frac{d^2 \psi}{dx^2} + V(x) \psi = E \psi$$

$$\frac{d^2 \psi}{dx^2} + \frac{2mE}{\hbar^2} \psi - \frac{2m}{\hbar^2} \frac{1}{2} m \omega^2 x^2 \psi = 0$$

or,

$$\frac{d^2 \psi}{dx^2} + \frac{2mE}{\hbar^2} \psi - \frac{m^2 \omega^2}{\hbar^2} x^2 \psi = 0$$

Let

$$\beta = 2mE/\hbar^2, \quad \alpha = m\omega/\hbar^2 \qquad (11\cdot7\text{-}2)$$

Then

$$\frac{d^2 \psi}{dx^2} + (\beta - \alpha^2 x^2) \psi = 0 \qquad (11\cdot7\text{-}3)$$

Introducing the variable $q = \sqrt{\alpha} x$, we have

$$\frac{d}{dx} = \frac{dq}{dx} \frac{d}{dq} = \sqrt{\alpha} \frac{d}{dq} \text{ and } \frac{d^2}{dx^2} = \alpha \frac{d^2}{dq^2}$$

Substituting in the wave equation we get

$$\alpha \frac{d^2 \psi}{dq^2} + (\beta - \alpha q^2) \psi = 0$$

Dividing by α we get

$$\frac{d^2 \psi}{dq^2} + (\epsilon - q^2) \psi = 0 \qquad (11\cdot7\text{-}4)$$

where

$$\epsilon = \beta/\alpha = \frac{2mE}{\hbar^2} \cdot \frac{\hbar^2}{m\omega} \sim \frac{2E}{\hbar \omega} \qquad (11\cdot7\text{-}5)$$

ϵ is a dimensionless quantity. Asymptotically (*i.e.*, for large q) the above equation reduces to

$$\frac{d^2 \psi_a}{dq^2} - q^2 \psi_a = 0$$

which has an approximate solution of the form
$$\psi_\alpha \sim A \exp(-q^2/2)$$
So we write the general solution in the form
$$\psi = \upsilon(q) \exp(-q^2/2) \tag{11.7-6}$$
This gives $\psi'' = (\upsilon'' - 2q\upsilon' + q^2\upsilon - \upsilon) \exp(-q^2/2)$
so that $\quad (\upsilon'' - 2q\upsilon' + q^2\upsilon - \upsilon + \epsilon\upsilon - q^2\upsilon) \exp(-q^2/2) = 0$
or, $\quad \upsilon'' - 2q\upsilon' + (\epsilon - 1)\upsilon = 0 \tag{11.7-7}$

This is the differential equation for $\upsilon(q)$. It can be solved by assuming a power series form:
$$\upsilon(q) = q^p \sum_i a_i q^i \tag{11.7-8}$$

i can take up integral values 0, 1, 2, 3, *etc.* p is a real positive number to be determined below. Differentiating Eq. (11.7-8), we get

$$\upsilon'(q) = q^p \sum_i (p+i) a_i q^{i-1}$$

$$\upsilon''(q) = q^p \sum_i (p+i)(p+i-1) a_i q^{i-2}$$

Substituting in Eq. (11.7-7) we get

$$\sum_i \left\{ (p+i)(p+i-1) a_i q^{p+i-2} - 2(p+i) a_i q^{p+i} + (\epsilon - 1) a_i q^{p+i} \right\} = 0$$

Since this must hold for all values of q the coefficient of each power of q must be zero separately. For $i = 0$ and $i = 1$, we get two *indicial equations* determining the index p:

For $i = 0$: coefficient of $q^{p-2} = p(p-1) a_0 = 0 \tag{11.7-9}$

For $i = 1$: coefficient of $q^{p-1} = p(p+1) a_1 = 0 \tag{11.7-10}$

For any general value of i, the coefficient of q^{p+i} can be evaluated from the relation

$$(p+i+2)(p+i+1) a_{i+2} - 2(p+i) a_i + (\epsilon - 1) a_i = 0 \tag{11.7-11}$$

This gives the *recursion relation* for the coefficients a_i:

$$\frac{a_{i+2}}{a_i} = \frac{2p + 2i + 1 - \epsilon}{(p+i+2)(p+i+1)} \tag{11.7-12}$$

Thus the alternate coefficients a_{i+2}, a_{i+4} etc. can be determined if a_i is known.

Solution of Schrödinger Equation in Some Simple Cases 301

Now if the series (11·7-8) is non-terminating, *i.e.*, it contains an infinite number of terms, then for large i, Eq. (11·7-12) gives

$$\frac{a_{i+2}}{a_i} \longrightarrow \frac{2}{i} \qquad (11\cdot7\text{-}13)$$

Thus the ratio of the coefficients a_{i+2} and a_i approaches the ratio of the coefficients of the terms q^{2i+2} and q^{2i} in the series expansion of exp (q^2) for large i. So if the series (11·7-8) contains an infinite number of terms, the solution (11·7-6) behaves like the function exp $(q^2/2)$ for large values of q. So the function is not well–behaved, because it is not square-integrable if the series is infinite. Hence the series $\upsilon(q)$ must terminate.

The necessary condition for the series to terminate is that the energy eigenvalues of the oscillator form a discrete set determined by the following equation* :

$$\epsilon_n = 4s + 2p + 1$$
$$= 2n + 1 \qquad (11\cdot7\text{-}14)$$

where it is assumed that the series terminates at $i = 2s$ and $n = 2s + p = 0, 1, 2, 3$ *etc.*, since the possible values of s are 0, 1, 2, 3 *etc.* and $p = 0$ or 1.

The energy eigenvalues are then

$$E_n = \epsilon_n \frac{\hbar\omega}{2} = \left(n + \frac{1}{2}\right)\hbar\omega \qquad (11\cdot7\text{-}15)$$

Thus we get a discrete series of equispaced energy eigenvalues, the interval being $\hbar\omega$. The ground state has the energy (*zero point energy*)

$$E_0 = \hbar\omega/2 = h\nu/2 \qquad (11\cdot7\text{-}16)$$

This is contrary to expectation from the old quantum theory which gives $E_n = nh\nu$ so that the ground state energy $E_0 = 0$. However the existence of the zero point energy in the quantum mechanical theory is consistent with the uncertainty relation. If the oscillator has zero energy (both kinetic and potential) in the ground state, then it is at $x = 0$ in this state, *i.e.*, at the bottom of the potential well. Obviously then at this position $\Delta x = 0$. At the same time, because its kinetic energy is also zero, the uncertainty in its momentum $\Delta p = 0$ so that both Δx and Δp are zero which contradicts the uncertainty relation.

11·8 Oscillator wave functions ; Parity

Using Eq. (11·7-14), we can transform Eq. (11·7-7) as follows :

---
* For a proof of this see *Introductory Quantum Mechanics* by the author.

$$v'' - 2qv' + 2nv = 0 \tag{11.8-1}$$

This is known as the Hermite differential equation. It can be proved that it has a solution of the form.*

$$v(q) = \exp(q^2)\left(\frac{d}{dq}\right)^n \exp(-q^2) \tag{11.8-2}$$

Eq. (11·8-2) multiplied by $(-1)^n$ gives the Hermite polynomial of order n (see Appendix A-I):

$$H_n(q) = (-1)^n \exp(q^2)\left(\frac{d}{dq}\right)^n \exp(-q^2) \tag{11.8-3}$$

That it is a polynomial of order n can be proved by term by term differentiation of the above expression.

We list below a few of the Hermite polynomials for different values of n:

$H_0(q) = 1$ $\qquad\qquad\qquad H_1(q) = 2q$

$H_2(q) = 4q^2 - 2$ $\qquad\qquad H_3(q) = 8q^3 - 12q$

$H_4(q) = 16q^4 - 48q^2 + 12 \qquad H_5(q) = 32q^5 - 160q^3 + 120q$

$$\tag{11.8-4}$$

The normalized wave function of a one dimensional linear harmonic oscillator is given by

$$\psi_n(q) = N_n \exp(-q^2/2) H_n(q) \tag{11.8-5}$$

The normalization constant is

$$N_n = \left(\frac{1}{\sqrt{\pi}\, n!\, 2^n}\right)^{1/2} \tag{11.8-6}$$

For the ground state ($n = 0$), we have

$$\psi_0 = N_0 \exp(-q^2/2) = \left(\frac{1}{\sqrt{\pi}}\right)^{1/2} \exp(-q^2/2)$$

For $n = 2$, we have

$$N_2 = \left(\frac{1}{\sqrt{\pi}\, 2!\, 2^2}\right)^{1/2} = \left(\frac{1}{8\sqrt{\pi}}\right)^{1/2}$$

So using the value of $H_2(q)$ from Eq. (11·8-4), we get

$$\psi_2 = \left(\frac{1}{8\sqrt{\pi}}\right)^{1/2} (4q^2 - 2) \exp(-q^2/2)$$

---

* See *ibid*.

Solution of Schrödinger Equation in Some Simple Cases

$$= \left(\frac{1}{2\sqrt{\pi}}\right)^{1/2} (2q^2 - 1) \exp(-q^2/2)$$

Similarly we can get the wave functions for different odd values of n.

Some useful relations involving the Hermite polynomials are given in Appendix A-I.

A few normalized oscillator wave functions for different *even* and *odd* n are listed below:

$$\psi_0(q) = \left(\frac{1}{\sqrt{\pi}}\right)^{1/2} \exp(-q^2/2)$$

$$\psi_1(q) = \left(\frac{2}{\sqrt{\pi}}\right)^{1/2} q \exp(-q^2/2)$$

$$\psi_2(q) = \left(\frac{1}{2\sqrt{\pi}}\right)^{1/2} (2q^2 - 1) \exp(-q^2/2)$$

$$\psi_3(q) = \left(\frac{1}{3\sqrt{\pi}}\right)^{1/2} (2q^3 - 3q) \exp(-q^2/2)$$

$$\psi_4(q) = \frac{1}{2}\left(\frac{1}{6\sqrt{\pi}}\right)^{1/2} (4q^4 - 12q^2 + 3) \exp(-q^2/2)$$

$$\psi_5(q) = \frac{1}{2}\left(\frac{1}{15\sqrt{\pi}}\right)^{1/2} (4q^5 - 20q^3 + 15q) \exp(-q^2/2)$$

(11·8-7)

It should be noted that ψ_0 has no node. For larger values of n, the number of nodes progressively increases with n. For example for ψ_1, there is a single node at $q = 0$. For ψ_2, there are two nodes at $q = \pm \frac{1}{\sqrt{2}}$; and so on.

In Fig. 11·13, a few normalized harmonic oscillator wave functions are shown.

The wave functions of the linear harmonic oscillators can be classified into two groups. They are either *even functions* of x or *odd functions* of x.

As x is changed to $-x$, $\psi(x)$ becomes $\psi(-x)$. For $n = 0$ or even, $\psi(-x) = +\psi(x)$ so that the wave function is an even function of x. On the other hand for odd n, $\psi(-x) = -\psi(x)$, so that $\psi(x)$ is an odd function of x. From these characteristics of the wave functions, we conclude that for $n = 0$ or even, $\psi(x)$ has *even-parity* while for odd n, $\psi(x)$ has *odd parity*.

The above classification of the wave functions according to *parity* is of fundamental importance in quantum mechanics. It is related to the

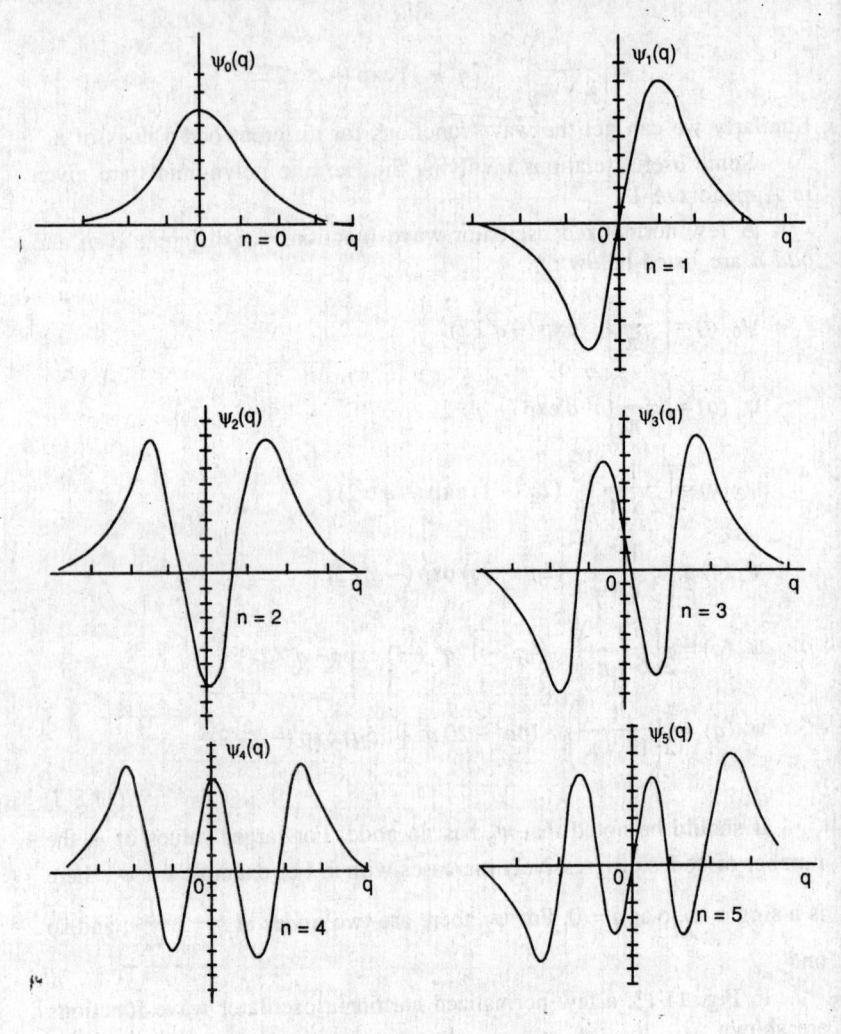

Fig. 11·13. Oscillator wave functions.

invariance of the Hamiltonian $\hat{H}$ under the inversion of the coordinate system at the origin. We shall discuss about it in more detail in §11·18.

In Fig. 11·14 the probability density $|\psi_0(x)|^2$ for the ground state (solid curve) has been compared with the classical probability density (dashed curve) for the same energy $\hbar\omega/2$ of the oscillator. As can be seen, there is no agreement between the two. Whereas the quantum mechanical probability density is the highest at $q = 0$ (*i.e.*, at $x = 0$) and falls off to $1/e$ th of its value at the classical limits of motion ($q = \pm 1$), the classical probability density is the lowest at $q = 0$ and approaches infinity as

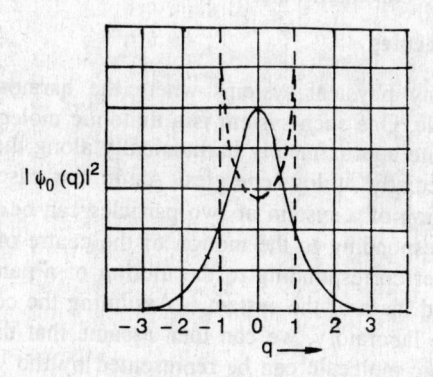

Solution of Schrödinger Equation in Some Simple Cases

[Figure: Plot of $|\psi_0(q)|^2$ vs q from -3 to 3, showing a bell-shaped curve with dashed classical probability curves rising at $q = \pm 1$.]

Fig. 11·14. Probability density for the harmonic oscillator in the ground state.

$q \longrightarrow \pm 1$. However the quantum mechanical probability density for larger values of n more nearly approximates te classical probability density for the same energy as can be seen from Fig. 11·15 for $n = 10$.

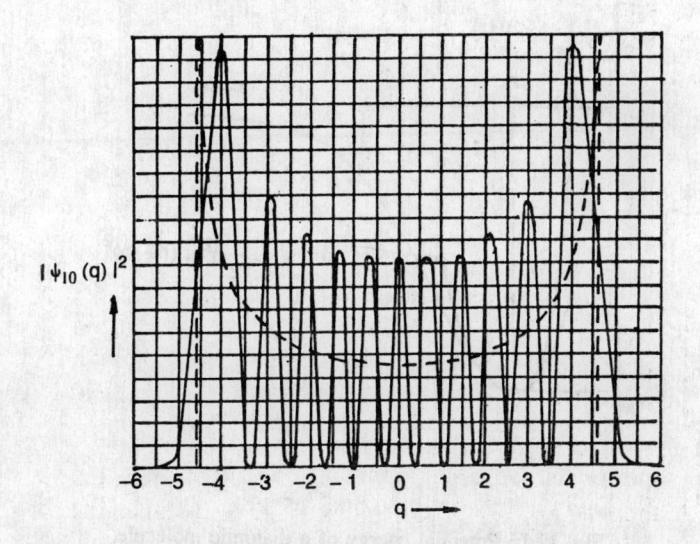

[Figure: Plot of $|\psi_{10}(q)|^2$ vs q from -6 to 6, showing ten oscillating peaks with dashed classical probability curve overlaid.]

Fig. 11·15. Probability density for the harmonic oscillator in the excited state with $n = 10$.

There is an alternative and more elegant method of solving the harmonic oscillator problem, known as the operator method which

11·9 Diatomic molecules

There are many physical systems where the harmonic oscillator solution is applicable. One such system is a diatomic molecule in which the two atoms vibrate approximately harmonically along the line joining the two atoms, specially at low energies. As in the case of classical mechanics, the motion of a system of two particles can be resolved into two parts, one corresponding to the motion of the centre of mass of the particles and another corresponding to the motion of a particle of mass equal to the reduced mass of the system.** Assuming the centre of mass to be at rest in the laboratory, we can then assume that the vibrational motion in a diatomic molecule can be represented by the vibration of a single particle of mass equal to the reduced mass μ of the two atoms.

The potential $V(r)$ has the shape as shown in Fig. 11·16. When the atoms come close together they are strongly repelled and move away. If they move too far away, the molecule will dissociate. Near about the

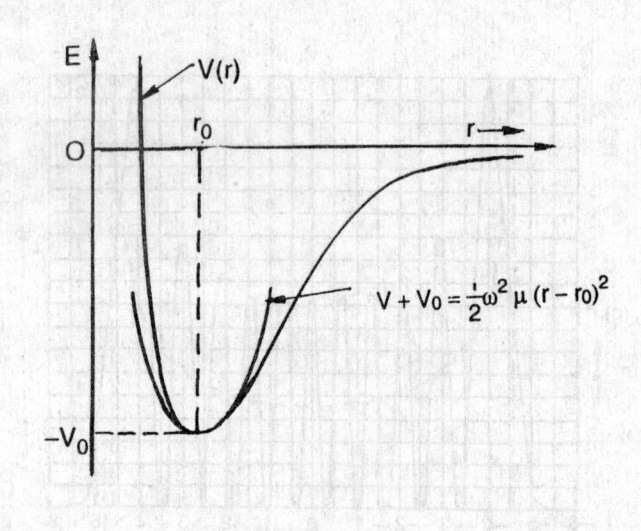

Fig. 11·16 Potential energy of a diatomic molecule.

equilibrium position r_0 the potential curve can be approximated to that for a linear harmonic oscillator.

---

* See *ibid*.
** See § 11.20

Solution of Schrödinger Equation in Some Simple Cases

$$V(r) = \frac{1}{2}\mu\omega^2(r-r_0)^2$$

Thus the vibrational energy levels near the bottom of the potential curve follow the behaviour of the harmonic oscillator levels :

$$E_\upsilon = \left(\upsilon + \frac{1}{2}\right)\hbar\omega$$

where υ is the vibrational quantum number which can have the values 0, 1, 2, etc. (See chapter XVI for further discussions).

11·10 Free particle with one dimensional motion ; Momentum eigenfunctions

We consider a free particle for which the potential $V = 0$ for all values of x. So the Hamiltonian is

$$\hat{H} = \frac{\hat{p}_x^2}{2m} + V(x) = -\frac{\hbar^2}{2m}\frac{d^2}{dx^2}$$

The time-independent Schrödinger equation is

$$-\frac{\hbar^2}{2m}\frac{d^2\psi}{dx^2} = E\psi$$

Writing $k^2 = 2mE/\hbar^2$, we get

$$\psi'' + \frac{2mE}{\hbar^2}\psi = \psi'' + k^2\psi = 0 \qquad (11\cdot10\text{-}1)$$

We assume $E > 0$, i.e., the energy of the particle is positive. Then $k^2 > 0$ and k is real. The solution of Eq. (11·10-1) is

$$\psi_k(x) = A_k \exp(ikx) + B_k \exp(-ikx) \qquad (11\cdot10\text{-}2)$$

The above solution $\psi(x)$ is finite at every point for all positive values of the energy E.

The wave function $\psi(x)$ is a linear combination of two linearly independent functions $\exp(ikx)$ and $\exp(-ikx)$ which means that for a definite value of the energy $E > 0$, there are two linearly independent eigenstates. Thus each energy $E > 0$ is a *doubly degenerate* eigenvalue of the Hamiltonian $\hat{H}$. So the energy spectrum of a free particle is *degenerate and continuous*.

It should be noted that the two linearly independent eigenfunctions $\exp(ikx)$ and $\exp(-ikx)$ of the Hamiltonian operator are simultaneous eigenfunctions of the momentum operator $\hat{p}_x$ of the free particle belonging to the respective eigenvalues $+\sqrt{2mE}$ and $-\sqrt{2mE}$ corresponding to the rightward and leftward motions along the x-axis.*

Since the wave functions of the free particle remain finite everywhere from $-\infty$ to $+\infty$, they are not square-integrable.

Taking the positive values of k only, we get

$$|\psi_k|^2 = |A_k|^2 = \text{constant}$$

so that

$$\int_0^\infty |\psi_k|^2 \, dx = |A_k|^2 \int_0^\infty dx \longrightarrow \infty$$

Similarly for the negative k solutions. Thus the wave functions of the free particle are not normalizable by the usual method. For this reason, the theory of the free particle is somewhat more involved than, say, the theory of the linear harmonic oscillator discussed earlier. The normalization of the free particle wave function can be carried out by the method of box normalization.*

Let us now consider the case $E < 0$ so that $k^2 < 0$. Let us write $k^2 = -\alpha^2$.

Here $\alpha^2 > 0$, so that α is real. From the Schrödinger equation we get

$$\frac{d^2\psi}{dx^2} - \alpha^2 \psi = 0$$

The solution is $\quad \psi(x) = A \exp(\alpha x) + B \exp(-\alpha x)$

As $x \to \infty$, the first term in the above solution containing $\exp(\alpha x)$ goes to ∞. On the other hand as $x \to -\infty$, the second term containing $\exp(-\alpha x)$ goes to infinity. Thus in this case $\psi(x)$ cannot be well-behaved everywhere for *any value* of energy $E < 0$. This means that the free particle cannot have negative energy.

(B) Solution of the three dimensional wave equation

In Chapter X, we deduced the three dimensional time-independent wave equation (10·7-6). The equation can be solved analytically in only a limited number of cases. Some of these are more conveniently treated by using cartesian coordinates as in the case of a particle in a three dimensional potential box with rigid boundaries or the three dimensional harmonic oscillator. These however will not be considered in the present book.*

---

* See *ibid.*

Solution of Schrödinger Equation in Some Simple Cases

In the following few sections we shall consider the central force problem, using spherical polar coordinates. For this it is necessary to be familiar with the properties of the angular momentum operator which will therefore be briefly considered first.

It may be noted that for most problems of physical interest, exact analytical solutions cannot be obtained. One has to use various approximation methods which have been developed. These will be considered in Ch. XII.*

11·11 Angular momentum operator

In classical mechanics, angular momentum is defined as

$$\mathbf{L} = \mathbf{r} \times \mathbf{p}$$

where $\mathbf{r}$ is the position vector and $\mathbf{p}$ is the linear momentum. The cartesian components of $\mathbf{L}$ are:

$$L_x = yp_z - zp_y, \quad L_y = zp_x - xp_z, \quad L_z = xp_y - yp_x$$

To obtain the quantum mechanical operators for the components of the angular momentum we follow the rules enunciated in § 10·3, viz., the cartesian components of the position vector $\mathbf{r}$ and the linear momentum $\mathbf{p}$ are to be replaced by the corresponding quantum mechanical operators in Schrödinger representation. Using Eqs. (10·3-9), (10·3-15) and 10·3-16), we get,

$$\begin{aligned}\hat{L}_x &= -i\hbar \left(y\frac{\partial}{\partial z} - z\frac{\partial}{\partial y} \right) \\ \hat{L}_y &= -i\hbar \left(z\frac{\partial}{\partial x} - x\frac{\partial}{\partial z} \right) \\ \hat{L}_z &= -i\hbar \left(x\frac{\partial}{\partial y} - y\frac{\partial}{\partial x} \right)\end{aligned} \quad (11\cdot11\text{-}1)$$

$\hat{L}_x, \hat{L}_y, \hat{L}_z$ do not commute with each other. The following commutation relations hold between them as can be easily verified:

$$[\hat{L}_x, \hat{L}_y] = i\hbar \hat{L}_z \qquad [\hat{L}_y, \hat{L}_z] = i\hbar \hat{L}_x \qquad [\hat{L}_z, \hat{L}_x] = i\hbar \hat{L}_y \qquad (11\cdot11\text{-}2)$$

The square of the total angular momentum is

$$L^2 = L_x^2 + L_y^2 + L_z^2$$

---

* See also *Quantum Mechanics* by L.I. Schiff.

The operator form for L^2 can be obtained by substituting the operator forms for L_x, L_y and L_z in the above equation. It is more convenient to express these in spherical polar coordinates in this case which are given below :*

$$\hat{L}_x = i\hbar \left(\sin\varphi \frac{\partial}{\partial\theta} + \cot\theta \cos\varphi \frac{\partial}{\partial\varphi} \right) \quad (11\cdot11\text{-}3)$$

$$\hat{L}_y = -i\hbar \left(\cos\varphi \frac{\partial}{\partial\theta} - \cot\theta \sin\varphi \frac{\partial}{\partial\varphi} \right) \quad (11\cdot11\text{-}4)$$

$$\hat{L}_z = -i\hbar \frac{\partial}{\partial\varphi} \quad (11\cdot11\text{-}5)$$

Substituting for $\hat{L}_x, \hat{L}_y, \hat{L}_z$ it can be proved that

$$\hat{L}^2 = \hat{L}_x^2 + \hat{L}_y^2 + \hat{L}_z^2$$

$$= -\hbar^2 \left(\frac{1}{\sin\theta} \frac{\partial}{\partial\theta} \sin\theta \frac{\partial}{\partial\theta} + \frac{1}{\sin^2\theta} \frac{\partial^2}{\partial\varphi^2} \right) \quad (11\cdot11\text{-}6)$$

It can be shown that $\hat{L}^2$ commutes with each of the three components $\hat{L}_x, \hat{L}_y, \hat{L}_z$:

$$[\hat{L}^2, \hat{L}_x] = 0 \quad [\hat{L}^2, \hat{L}_y] = 0, \quad [\hat{L}^2, \hat{L}_z] = 0 \quad (11\cdot11\text{-}7)$$

11·12 Three dimensional motion in a central field ; Problem of the hydrogen atom

The motion of an electron in a central field is one of the principal problems in the quantum mechanics of the atom. Though the field is not strictly Coulombian in character in most atoms because of the influence of the other electrons, for the hydrogen or hydrogen-like atoms the field may be regarded as Coulombian. The potential is given by

$$V(r) = -Ze^2/r \quad (11\cdot12\text{-}1)$$

where $+Ze$ is the nuclear charge.

The three-dimensional time-independent Schrödinger wave equation can be written as

$$\hat{H}\psi = -\frac{\hbar^2}{2m} \nabla^2 \psi + V(r)\psi = E\psi \quad (11\cdot12\text{-}2)$$

where $\psi = \psi(r) = \psi(r, \theta, \varphi)$. Here we assume the nucleus to be infinitely heavy, m being the mass of the electron. In the case of finite nuclear mass,

---

* See *Introductory Quantum Mechanics* by the author.

Solution of Schrödinger Equation in Some Simple Cases

m should be replaced by the reduced mass μ of the electron-nucleus system.

The Laplacian operator in spherical polar coordinates is given by

$$\nabla^2 = \frac{1}{r^2}\frac{\partial}{\partial r}r^2\frac{\partial}{\partial r} + \frac{1}{r^2\sin\theta}\frac{\partial}{\partial\theta}\sin\theta\frac{\partial}{\partial\theta} + \frac{1}{r^2\sin^2\theta}\frac{\partial^2}{\partial\varphi^2}$$

The Hamiltonian $\hat{H}$ then becomes

$$\hat{H} = -\frac{\hbar^2}{2m}\nabla^2 + V(r)$$

$$= -\frac{\hbar^2}{2m}\left(\frac{1}{r^2}\frac{\partial}{\partial r}r^2\frac{\partial}{\partial r} + \frac{1}{r^2\sin\theta}\frac{\partial}{\partial\theta}\sin\theta\frac{\partial}{\partial\theta} + \frac{1}{r^2\sin^2\theta}\frac{\partial^2}{\partial\varphi^2}\right)$$
$$+ V(r) \quad (11 \cdot 12\text{-}3)$$

Using Eq. (11·11-6) we then get

$$\hat{H} = -\frac{\hbar^2}{2m}\frac{1}{r^2}\frac{\partial}{\partial r}r^2\frac{\partial}{\partial r} + \frac{\hat{L}^2}{2mr^2} + V(r) \quad (11 \cdot 12\text{-}4)$$

The wave equation (11·12-2) then becomes

$$-\frac{\hbar^2}{2m}\frac{1}{r^2}\frac{\partial}{\partial r}r^2\frac{\partial\psi}{\partial r} + \frac{1}{2mr^2}\hat{L}^2\psi + V(r)\psi = E\psi \quad (11 \cdot 12\text{-}5)$$

The above equation can be solved by the method of separation of variables. We write

$$\psi(\mathbf{r}) = \psi(r, \theta, \varphi) = R(r)Y(\theta, \varphi) \quad (11 \cdot 12\text{-}6)$$

Substituting in the wave equation (11·12-5) we get, since $\hat{L}^2$ operates on the angle-dependent part $Y(\theta, \varphi)$ of the wave function only,

$$-\frac{\hbar^2}{2m}\frac{Y}{r^2}\frac{d}{dr}\left(r^2\frac{dR}{dr}\right) + \frac{R}{2mr^2}\hat{L}^2 Y + V(r)\psi = E\psi$$

Multiplying by $-(2m/\hbar^2)r^2/\psi$ we have

$$\frac{1}{R}\frac{d}{dr}\left(r^2\frac{dR}{dr}\right) + \frac{2mr^2}{\hbar^2}\{E - V(r)\} = \frac{1}{\hbar^2 Y}\hat{L}^2 Y = \lambda \text{ (say)} \quad (11 \cdot 12\text{-}7)$$

Since the l.h.s. is a function of r alone while the r.h.s. is a function of θ and φ, each side must be separately equal to a constant λ which is known as the *separation constant*. Then the radial equation becomes

$$\frac{1}{r^2}\frac{d}{dr}\left(r^2\frac{dR}{dr}\right) + \frac{2m}{\hbar^2}\{E - V(r)\}R - \frac{\lambda}{r^2}R = 0 \quad (11 \cdot 12\text{-}8)$$

From Eq. (11·12-7), we get the differential equation for $Y(\theta, \varphi)$ as

$$\frac{1}{\hbar^2} \hat{L}^2 Y(\theta, \varphi) = \lambda Y(\theta, \varphi)$$

With the help of Eq. (11·11-6) this becomes

$$\frac{1}{\sin\theta} \frac{\partial}{\partial\theta}\left(\sin\theta \frac{\partial Y}{\partial\theta}\right) + \frac{1}{\sin^2\theta} \frac{\partial^2 Y}{\partial\varphi^2} = -\lambda Y \qquad (11\cdot12\text{-}9)$$

This equation can be separated into two equations, one in θ and the other in φ. We write

$$Y(\theta, \varphi) = \Theta(\theta) \Phi(\varphi) \qquad (11\cdot12\text{-}10)$$

Then $\quad \dfrac{\Phi}{\sin\theta} \dfrac{d}{d\theta}\left(\sin\theta \dfrac{d\Theta}{d\theta}\right) + \dfrac{\Theta}{\sin^2\theta} \dfrac{d^2\Phi}{d\varphi^2} = -\lambda Y$

Multiplying by $\sin^2\theta / Y$ we get,

$$\frac{\sin\theta}{\Theta} \frac{d}{d\theta}\left(\sin\theta \frac{d\Theta}{d\theta}\right) + \lambda \sin^2\theta = -\frac{1}{\Phi} \frac{d^2\Phi}{d\varphi^2} = m^2 \text{ (say)} \quad (11\cdot12\text{-}11)$$

where m^2 is the *separation constant*. Hence we get two separate equations in θ and φ.

11·13 φ- equation

The φ-equation gives $\qquad \dfrac{d^2\Phi}{d\varphi^2} + m^2 \Phi = 0 \qquad (11\cdot13\text{-}1)$

The solution is $\qquad \Phi = A \exp(im\varphi) \qquad (11\cdot13\text{-}2)$

Since the wave function must be single-valued, a change of φ by 2π must give the same value of the function Φ *i.e.*, $\Phi(\varphi + 2\pi) = \Phi(\varphi)$

Hence $\qquad \exp im(\varphi + 2\pi) = \exp(im\varphi)$

or $\qquad \exp(2im\pi) = 1$

which gives $m = 0, \pm 1, \pm 2, \pm 3$ etc.

m is known as the *magnetic quantum number*.

11·14 θ-equation

From Eq. (11·12-11), we get the θ-equation as follows :

$$\sin\theta \frac{d}{d\theta}\left(\sin\theta \frac{d\Theta}{d\theta}\right) + (\lambda \sin^2\theta - m^2)\Theta = 0 \qquad (11\cdot14\text{-}1)$$

Substituting $\mu = \cos\theta$, Eq. (11·14-1) can be transformed as follows :

$$(1 - \mu^2) \frac{d^2\Theta}{d\mu^2} - 2\mu \frac{d\Theta}{d\mu} + \left(\lambda - \frac{m^2}{1-\mu^2}\right)\Theta = 0 \qquad (11\cdot14\text{-}2)$$

To solve this equation, we assume

$$\Theta(\mu) = (1-\mu^2)^{m/2} \upsilon(\mu) \quad (-1 \leq \mu \leq 1) \qquad (11\cdot14\text{-}3)$$

Solution of Schrödinger Equation in Some Simple Cases

with $m > 0$. Differentiating $\Theta(\mu)$ with respect to μ twice and substituting in Eq. (11·14-2) we get the differential equation for $\upsilon(\mu)$:

$$(1-\mu^2)\upsilon'' - 2(m+1)\mu\upsilon' + (\lambda - m - m^2)\upsilon = 0 \qquad (11\cdot14\text{-}4)$$

where $\upsilon' = d\upsilon/d\mu$ and $\upsilon'' = d^2\upsilon/d\mu^2$.

We try the series solution for $\upsilon(\mu)$:

$$\upsilon(\mu) = \sum_i a_i \mu^i \qquad (11\cdot14\text{-}5)$$

Differentiating $\upsilon(\mu)$ with respect to μ we get

$$\upsilon'(\mu) = \sum_i i a_i \mu^{i-1}, \quad \upsilon''(\mu) = \sum_i i(i-1) a_i \mu^{i-2}$$

Substituting is Eq. (11·14-4) we get

$$\sum_i i(i-1) a_i (1-\mu^2) \mu^{i-2} - 2(m+1)\mu \sum_i i a_i \mu^{i-1}$$
$$+ (\lambda - m - m^2) \sum_i a_i \mu_i = 0 \qquad (11\cdot14\text{-}6)$$

The above equation holds for all values of μ between -1 and $+1$. So the coefficient of μ^i for each i must be separately equated to zero. The coefficient of μ^i is

$$(i+2)(i+1)a_{i+2} - i(i-1)a_i - 2(m+1)ia_i + (\lambda - m - m^2)a_i = 0$$

So

$$\frac{a_{i+2}}{a_i} = \frac{(i+m)(i+m+1) - \lambda}{(i+1)(i+2)} \qquad (11\cdot14\text{-}7)$$

As $i \longrightarrow \infty$, $a_{i+2}/a_i \longrightarrow 1$. This means that the series diverges as $|\mu| \longrightarrow 1$. Hence the series must terminate. If it terminates at some value of $i = s$ then from Eq. (11·14-7) we see that a_{s+1}, a_{s+2} etc. must all be zero. So the numerator in Eq. (11·14-7) must be zero for $i = s$. i.e.,

$$(s+m)(s+m+1) - \lambda = 0$$
or, $$\lambda = (s+m)(s+m+1) \qquad (11\cdot14\text{-}8)$$

Now both s and m can take up the values 0, 1, 2, 3 *etc*. So we can write

$$l = s + m \qquad (11\cdot14\text{-}9)$$

where l is an integer or zero : $l = 0, 1, 2, 3$ *etc*. It is known as the *azimuthal quantum number*.

So we get $$\lambda = l(l+1) \qquad (11\cdot14\text{-}10)$$

Thus the eigenvalues of the square of the orbital angular momentum operator $\hat{L}^2$ are

$$\lambda \hbar^2 = l(l+1) \hbar^2 \qquad (11\cdot14\text{-}11)$$

The above value is different from the value of the square of the orbital angular momentum assumed in the old quantum theory according to which $L = k\hbar$ with $k = 1, 2, 3$ etc. So the quantum mechanical value of the orbital angular momentum quantum number is $l = (k - 1)$. If we operate the function $\Phi = \exp(im\varphi)$ by $\hat{L}_z = -i\hbar\,\partial/\partial\varphi$, we get

$$\hat{L}_z \Phi = -i\hbar \frac{\partial}{\partial \varphi} \exp(im\varphi) = m\hbar \exp(im\varphi) = m\hbar\,\Phi$$

So the solution of the φ-equation given by Eq. (11·13-2) is the eigenfunction of the operator $\hat{L}_z$ belonging to the eigenvalue $m\hbar$, where $m = 0, \pm 1, \pm 2 \ldots \pm l$. The upper limit $|m| = l$ follows from Eq. (11·14-9).

The solutions of the θ-equation are the *associated Legendre functions* $P_l^m(\mu)$ determined by the relation (A2·2) given in Appendix A-II:

$$\Theta = P_l^m(\mu) = \frac{1}{2^l\, l!} (1-\mu^2)^{m/2} \left(\frac{d}{d\mu}\right)^{l+m} (\mu^2 - 1)^l \qquad (11\cdot14\text{-}12)$$

The solutions of Eq. (11·12-9) are the spherical harmonics $Y_l^m(\theta, \varphi)$ defined in Appendix A-III by Eq. (A3·1). These functions are given by

$$Y_l^m(\theta, \varphi) = (-1)^m\, C_l^m\, P_l^m(\mu) \exp(im\varphi) \qquad (11\cdot14\text{-}13)$$

where
$$C_l^m = \sqrt{\frac{2l+1}{4\pi} \frac{(l-m)!}{(l+m)!}} \qquad (11\cdot14\text{-}14)$$

A few of the functions $Y_l^m(\theta, \varphi)$ for different l and m are given below:

$$Y_0^0 = 1/\sqrt{4\pi}$$

$$Y_1^0 = \sqrt{\frac{3}{4\pi}} \cos\theta,\quad Y_1^{\pm 1} = \mp \sqrt{\frac{3}{8\pi}} \sin\theta \exp(\pm i\varphi)$$

$$Y_2^0 = \sqrt{\frac{5}{16\pi}} (3\cos^2\theta - 1),\quad Y_2^{\pm 1} = \mp \sqrt{\frac{15}{8\pi}} \sin\theta \cos\theta \exp(\pm i\varphi)$$

$$Y_2^{\pm 2} = \sqrt{\frac{15}{32\pi}} \sin^2\theta \exp(\pm 2 i\varphi) \qquad (11\cdot14\text{-}15)$$

$Y_l^m(\theta, \varphi)$ are the simultaneous eigenfunctions of the operators $\hat{L}^2$ and $\hat{L}_z$ belonging to the eigenvalues $l(l+1)\hbar^2$ and $m\hbar$ respectively, where $l = 0, 1, 2, 3$ etc. and $m = 0, \pm 1, \pm 2, \ldots \pm l$.

11·15 Radial equation

The radial equation (11·12-8) for a given l is of the form

Solution of Schrödinger Equation in Some Simple Cases

$$\frac{1}{r^2}\frac{d}{dr}\left(r^2\frac{dR_l}{dr}\right) + \frac{2m}{\hbar^2}\{E - V(r)\}R_l - \frac{l(l+l)}{r^2}R_l = 0$$

Writing $V(r) = -Ze^2/4\pi\varepsilon_0 r$, we get

$$\frac{d^2 R_l}{dr^2} + \frac{2}{r}\frac{dR_l}{dr} + \frac{2m}{\hbar^2}\left(E + \frac{Ze^2}{4\pi\varepsilon_0 r}\right)R_l - \frac{l(l+1)}{r^2}R_l = 0 \quad (11\cdot15\text{-}1)$$

Let $R_l(r) = u_l(r)/r$. Then we get by substituting

$$\frac{d^2 u_l}{dr^2} + \frac{2m}{\hbar^2}\left\{E + \frac{Ze^2}{4\pi\varepsilon_0 r} - \frac{\hbar^2}{2m}\frac{l(l+1)}{r^2}\right\}u_l = 0 \quad (11\cdot15\text{-}2)$$

$\left(-\frac{1}{4\pi\varepsilon_0}Ze^2/r\right)$ is the Coulomb potential while $\frac{\hbar^2}{2m}\frac{l(l+1)}{r^2}$ is known as the *centrifugal potential*. The first is attractive while the second is repulsive.

The variation of these two potentials with r is shown in Fig. 11·17 in which the resultant of the two is also shown. We are interested in those solutions for which $E < 0$. They correspond to the bound states of the electron in the atom.

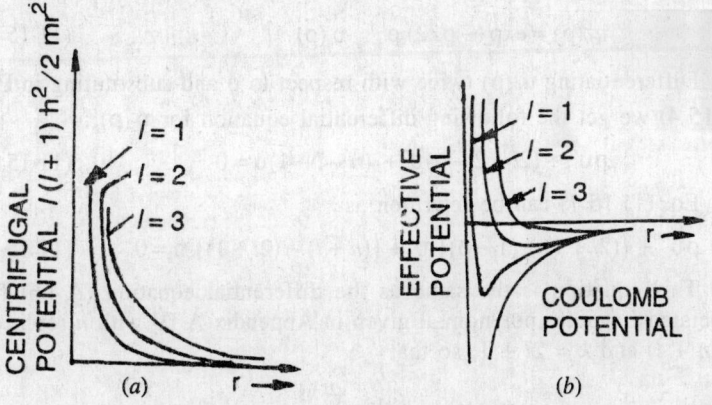

Fig. 11·17 Variation of Coulomb and centrifugal potentials.

Let $E = -E'$, $\beta = 2m/\hbar^2$; $E' > 0$. Then writing

$$\rho = 2r\sqrt{-\beta E} = 2r\sqrt{\beta E'} \quad (11\cdot15\text{-}3)$$

we can transform Eq. (11·15-2) as follows :*

---

* See *Introductory Quantum Mechanics* by the author.

$$u_l'' + \left\{ -\frac{1}{4} + \frac{n}{\rho} - \frac{l(l+1)}{\rho^2} \right\} u_l = 0 \qquad (11 \cdot 15\text{-}4)$$

where we have written u_l'' for $d^2 u_l/d\rho^2$ and

$$n = \frac{Ze^2}{8\pi\varepsilon_0} \sqrt{\frac{\beta}{E'}} \qquad (11 \cdot 15\text{-}5)$$

Eq. (11·15-4) is the dimensionless form of Eq. (11·15-2).

For $\rho \longrightarrow \infty$, this equation reduces to

$$u_l'' - \frac{u_l}{4} = 0$$

which has a solution $u_l \approx \exp(-\rho/2)$. Another solution $\exp(\rho/2)$ is not acceptable since it becomes infinite as $\rho \longrightarrow \infty$. On the other hand for $\rho \longrightarrow 0$, Eq. (11·15-4) reduces to

$$u_l'' - \frac{l(l+1)}{\rho^2} u_l = 0$$

which has a solution of the form $u_l \approx \rho^{l+1}$. The other solution ρ^{-l} is not acceptable, since it becomes infinite as $\rho \longrightarrow 0$. So for the general case for any value of ρ, let us try the solution

$$u_l(\rho) = \exp(-\rho/2) \rho^{l+1} \upsilon(\rho) \qquad (11 \cdot 15\text{-}6)$$

Differentiating $u_l(\rho)$ twice with respect to ρ and substituting in Eq. (11·15-4) we get the following differential equation for $\upsilon(\rho)$:

$$\rho \upsilon'' + (2l + 2 - \rho) \upsilon' + (n - l - 1) \upsilon = 0 \qquad (11 \cdot 15\text{-}7)$$

Eq. (11·15-7) can be rewritten as

$$\rho \upsilon'' + \{(2l+1) + (1-\rho)\} \upsilon' + \{(n+l) - (2l+1)\} \upsilon = 0 \qquad (11\text{-}15 \cdot 8)$$

This equation is the same as the differential equation (A 4·6) for associated Laguerre polynomial given in Appendix A-IV with n replaced by $(n + l)$ and $k = 2l + 1$, so that

$$\upsilon(\rho) = L_{n+l}^{2l+1}(\rho) = \left(\frac{d}{d\rho}\right)^{2l+1} L_{n+l}(\rho) \qquad (11 \cdot 15\text{-}9)$$

We write $\upsilon(\rho)$ in the power series

$$\upsilon(\rho) = \sum_i a_i \rho^{i+p} \qquad (11 \cdot 15\text{-}10)$$

where i has the successive integral values 0, 1, 2, 3 etc. p is to be determined. We have

Solution of Schrödinger Equation in Some Simple Cases

$$\upsilon'(\rho) = \rho^p \sum_i (i+p) \rho^{i-1}$$

$$\upsilon''(\rho) = \rho^p \sum_i (i+p)(i+p-1) \rho^{i-2}$$

Substituting in Eq. (11·15-8) we get

$$\sum_i \{(i+p)(i+p-1) + 2(i+p)(l+1)\} a_i \rho^{i+p-1}$$

$$+ \sum_i \{(n-l-1) - (i+p)\} a_i \rho^{i+p} = 0 \quad (11\cdot15\text{-}11)$$

Eq. (11·15-11) must hold for all values of ρ. Hence the co-efficient of each power of ρ must vanish separately. The lowest power of ρ is $(p-1)$ for $i = 0$.

Coefficient of $\rho^{p-1} = \{p(p-1) + 2p(l+1)\} a_0$

$$= p(p + 2l + 1) a_0 = 0$$

So the starting power of the series (11·15-10) is $p = 0$ or $p = -(2l+1)$. However the second solution for p is not acceptable, since it would make $\upsilon(\rho) \longrightarrow \infty$ as $\rho \longrightarrow 0$. So we must put $p = 0$. Coefficient of ρ^{i+p} is

$$\{(i+p)(i+p+1) + 2(i+p+1)(l+1)\} a_{i+1} + \{(n-l-1) - (i+p)\} a_i = 0$$

This gives

$$\frac{a_{i+1}}{a_i} = \frac{i+p+l+1-n}{(i+p+1)(i+p+2l+2)} \quad (11\cdot15\text{-}12)$$

Eq. (11·15-12) shows that for large i, $a_{i+1}/a_i = 1/i$.

This is also the ratio of the coefficients of the successive terms in the expansion of exp (ρ). So the radial function behaves like exp $(\rho/2)$ at large ρ which therefore does not remain square integrable. So the series expansion (11·15-10) for $\upsilon(\rho)$ must terminate (with $p = 0$) for the radial function to be well-behaved. If it terminates at the sth term, then the coefficients a_{s+1} and higher must all be zero. The necessary condition for this follows from Eq. (11·15-12):

$$(s+p+l+1) - n = (s+l+1) - n = 0$$

or, $$n = s + l + 1 \quad (11\cdot15\text{-}13)$$

Since both s and l can take up the values 0, 1, 2 *etc.* it is obvious that n is an integer which can take up the values

$$n = 1, 2, 3, 4, \ldots\ldots$$

n is called the *principal quantum number*.

For a given l, n can take up the values $(l+1), (l+2), (l+3)$, *etc.* On the other hand for a given n, l can take up the values $l = 0, 1, 2, \ldots\ldots, (n-1)$.

The polynomial $\upsilon(\rho)$ is related to the *Laguerre polynomial* of the ith degree defined as follows :

$$L_i(\rho) = \exp(\rho) \left(\frac{d}{d\rho}\right)^i \{\rho^i \exp(-\rho)\} \qquad (11\cdot15\text{-}14)$$

$\upsilon(\rho)$ is then given by (see Appendix A-IV)

$$\upsilon(\rho) = L_{n+l}^{2l+1}(\rho) = \left(\frac{d}{d\rho}\right)^{2l+1} L_{n+l}(\rho) \qquad (11\cdot15\text{-}15)$$

$L_{n+l}^{2l+1}(\rho)$ is known as the associated Laguerre polynomial of degree $(n+l) - (2l+1) = n-l-1$, which is defined in A-IV.

The energy of the electron is obtained from Eq. (11·15-5)

$$E = -E' = -\left(\frac{Ze^2}{8\pi\varepsilon_0 n}\sqrt{\beta}\right)^2 = -\frac{Z^2 e^4 \beta}{64\pi^2 \varepsilon_0^2 n^2} = -\frac{Z^2 e^4}{64\pi^2 \varepsilon_0^2 n^2} \frac{2m}{\hbar^2}$$

so that
$$E_n = -\frac{mZ^2 e^4}{32\pi^2 \varepsilon_0^2 n^2 \hbar^2} \qquad (11\cdot15\text{-}16)$$

This is the same expression as deduced from Bohr's theory (see § 4·4).

These energy eigenvalues form a discrete spectrum corresponding to different n values, with $n = 1, 2, 3, 4$ *etc.*

The corresponding energy eigenfunctions can be written as

$$\Psi(r, \theta, \varphi, t) = \psi_{nlm}(r, \theta, \varphi) \exp(-iE_n t/\hbar)$$

$$= R_{nl}(r) Y_l^m(\theta, \phi) \exp(-iE_n t/\hbar) \qquad (11\cdot15\text{-}17)$$

where the radial function $R_{nl}(r)$ is given by Eq. (11·16-1) while the angular part $Y_l^m(\theta, \varphi)$ is the spherical harmonic given by Eq. (11·14-13).

Eq. (11·15-16) shows that the energy of the hydrogen atom depends only on the principal quantum number n. Now for a given n there are a number of different linearly independent eigenfunctions for the hydrogen-like atom, depending on the values of l and m. Firstly, for each l, there are $2l + 1$ different possible integral values of m extending from $-l$ to $+l$ and varying by unit integral steps. Secondly for a given n, there are n possible integral values of l : $l = 0, 1, 2\ldots\ldots(n-1)$. Each pair of m and l values for a given n gives a separate eigenfunction. Their number is given by

$$\sum_{l=0}^{n-1} (2l+1) = 1 + 3 + 5 + \ldots + (2n-1) = n^2 \qquad (11 \cdot 15\text{-}18)$$

Thus the energy level of the hydrogen-like atom for a given n is n^2-fold degenerate. Obviously the ground state with $n=1$ is non-degenerate.

It is possible to *remove the degeneracy* of a state by the application of suitable external fields on the system. Thus for instance, if an atom is placed in a magnetic field, there is energy splitting of the $(2l+1)$ substates of different m for a given l so that they have all different energy eigenvalues now and the degeneracy is removed. Similarly due to the spin-orbit interaction (see Ch VI and XIII), the states of the same l but different j (total quantum number) split up and have different energies. In fact their energies now depend both on l and j so that the degeneracy of the n substates for the same n are removed.

It can be shown that for positive values of energy $(E > 0)$, the energy eigenvalues form a continuous spectrum. The discrete spectrum for $E < 0$ actually merges with this continuum as $n \rightarrow \infty$.

11·16 The wave functions of the hydrogen-like atoms

The normalized radial functions are given by

$$R_{nl}(r) = -2 \left(\frac{Z}{na_0} \right)^{3/2} \sqrt{\frac{(n-l-1)!}{n\{(n+l)!\}^3}} \left(\frac{2Zr}{na_0} \right)^l$$

$$\times \exp\left(-\frac{Zr}{na_0} \right) L_{n+l}^{2l+1}\left(\frac{2Zr}{na_0} \right) \qquad (11 \cdot 16\text{-}1)$$

were $a_0 = 4\pi \varepsilon_0 \hbar^2 / me^2 = 0.529$Å is the radius of the Bohr orbit of the hydrogen atom. n and l are the principal and azimuthal quantum numbers respectively. The normalization constant can be determined by using the integral relation (A 4·7).

As stated previously $L_{n+l}^{2l+1}(\rho)$ is the associated Laguerre polynomial, related to the Laguerre polynomial $L_{n+1}(\rho)$ defined by the Eqs. (11·15-14) and (11·15-15). It may be noted that (see Eq. 11·15-3).

$$\rho = 2r\sqrt{\beta E'} = 2r \left\{ \frac{2m}{\hbar^2} \frac{mZ^2 e^4}{32\pi^2 \varepsilon_0^2 n^2 \hbar^2} \right\}^{1/2} = \frac{mZe^2}{2\pi \varepsilon_0 n \hbar^2} r = \frac{2Zr}{na_0}$$

$$(11 \cdot 16\text{-}2)$$

A few normalized radial functions of the hydrogen-like atoms for different n and l are given below :

$$R_{10} = \left(\frac{Z}{a_0}\right)^{3/2} \cdot 2 \exp(-Zr/a_0)$$

$$R_{20} = \left(\frac{Z}{2a_0}\right)^{3/2} \left(2 - \frac{Zr}{a_0}\right) \exp(-Zr/2a_0).$$

$$R_{21} = \left(\frac{Z}{2a_0}\right)^{3/2} \left(\frac{Zr}{\sqrt{3}\,a_0}\right) \exp(-Zr/2a_0)$$

$$R_{30} = \left(\frac{Z}{3a_0}\right)^{3/2} \cdot 2\left\{1 - \frac{2}{3}\frac{Zr}{a_0} + \frac{2}{27}\left(\frac{Zr}{a_0}\right)^2\right\} \exp(-Zr/3a_0)$$

$$R_{31} = \left(\frac{Z}{3a_0}\right)^{3/2} \left(\frac{4\sqrt{2}\,Zr}{9a_0}\right)\left\{1 - \frac{1}{6}\left(\frac{Zr}{a_0}\right)\right\} \exp(-Zr/3a_0)$$

$$R_{32} = \left(\frac{Z}{3a_0}\right)^{3/2} \frac{2\sqrt{2}}{27\sqrt{5}} \left(\frac{Zr}{a_0}\right)^2 \exp(-Zr/3a) \qquad (11\cdot16\text{-}3)$$

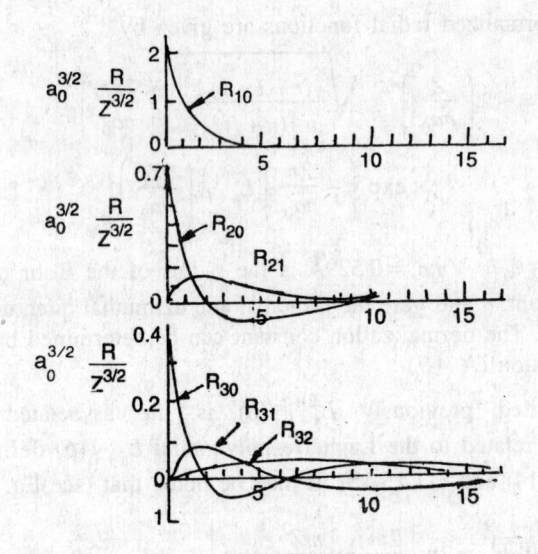

Fig.11·18. Radial wave functions for a hydrogen like atom for different n and l. The abscissa r is in units of a_0/Z.

The variation of the radial functions for different n and l is shown in Fig. 11·18. The time-independent part of the total wave function $\psi_{nlm}(r, \theta, \varphi)$ is given by

Solution of Schrödinger Equation in Some Simple Cases 321

$$\psi_{nlm}(r, \theta, \varphi) = R_{nl}(r) \, Y_l^m(\theta, \varphi) \qquad (11 \cdot 16\text{-}4)$$

A list of hydrogenic total wave functions is given in Appendix A-V. The probability of finding the electron in a volume element $d\tau$ is

$$|\psi_{nlm}|^2 \, d\tau = |\psi_{nlm}|^2 \, r^2 \, dr \, d\Omega$$

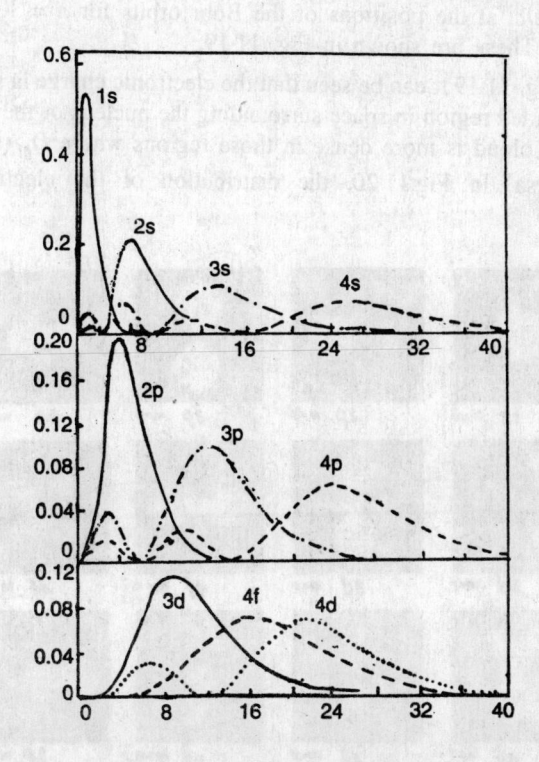

Fig. 11·19. Variation of the radial probability density $D_{nl}(r)$ with r (in units of a_o for different n and l. The positions of the maxima for the 1s, 2p and 3d curves correspond to the respective Bohr orbit radii.

Here $d\Omega$ is the element of solid angle in the direction θ, φ and the volume element $d\tau = r^2 dr \, d\Omega = r^2 \sin\theta \, dr \, d\theta \, d\varphi$. So the probability of finding the electron in a spherical shell of radius r and thickness dr is

$$\int_\Omega |\psi_{nlm}|^2 \, r^2 \, dr \, d\Omega = |R_{nl}(r)|^2 \, r^2 \, dr \int_\Omega |Y_l^m(\theta, \phi)|^2 \, d\Omega$$

$$= r^2 \, |R_{nl}(r)|^2 \, dr$$

$$= D_{n_l}(r)\, dr \tag{11.16-5}$$

where
$$D_{nl}(r) = r^2 |R_{nl}(r)|^2 \tag{11.16-6}$$

Notice that the spherical harmonics $Y_l^m(\theta, \varphi)$'s are normalized. $D_{nl}(r)$ is known as the *radial probability density*. By differentiating $D_{nl}(r)$ with respect to r and equating it to zero, it can be shown that D_{nl} is maximum for the states $1s$ ($n = 1$, $l = 0$), $2p$ ($n = 2$, $l = 1$), $3d$ ($n = 3$, $l = 2$) *etc.* at the positions of the Bohr orbits for $n = 1, 2, 3$ *etc.* respectively. These are shown in Fig. 11.19.

From Fig. 11.19 it can be seen that the electronic charge is spread out over an extended region in space surrounding the nucleus of the atom like a cloud. The cloud is more dense in those regions where $D_{nl}(r)$ is large and vice-versa. In Fig. 11.20, the distribution of the electron cloud

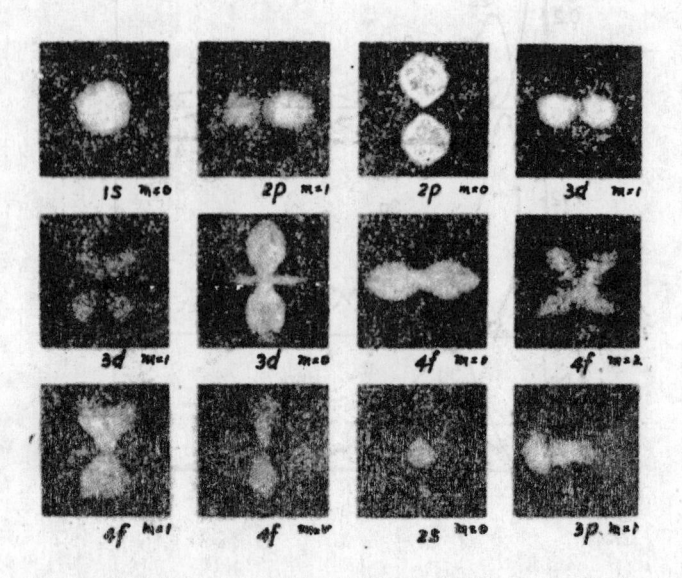

Fig. 11.20. Electron cloud. (Courtsey : Introduction to Atomic Physics by H.E. White, published by McGraw Hill Book Co. Inc., New York).

surrounding the nucleus for different n and l values are shown.

Atomic orbitals :

From the above discussions, it is evident that unlike in the Bohr-Sommerfeld theory, the electron wave in the hydrogen-like atom is spread over a region in which the probability density has an appreciable value determined by the wave function. This region of space is called an *orbital*. Within an orbital the electron can move according to the wave

Solution of Schrödinger Equation in Some Simple Cases

mechanical theory of Schrödinger. It corresponds approximately to the earlier idea of the Bohr-Sommerfeld theory that the electrons move in circular or elliptic orbits like the planets round the sun. The energy and size of the orbitals are such that smaller the orbital, closer is the electron to the nucleus and hence more firmly bound.

An orbital is actually represented by the wave function which is determined by the quantum numbers n, l, m_l and m_s. m_s is the magnetic quantum number for the spin angular momentum s of the electron where $s = 1/2$ and $m_s = \pm 1/2$ (see later). As has been seen, the wave function and hence the probability density for a given state is spread over a region of space extending up to infinity, though it is mainly confined in the neighburhood of the Bohr orbits for given n and l values. Thus it is not possible to give a precise definition of the shape and size of an orbital, unlike in the case of the orbits in the Bohr-Sommerfeld theory. However for convenient pictorial representation, an orbital is usually taken to be the three dimensional shape within which the electron spends a major fraction of its time (say 95%)

The shapes and sizes of the orbitals vary with n, l and m_l values. They are the pictorial representations of the wave functions. The (ns) orbitals with $l = 0$ for different n ($n = 1, 2, 3, \ldots\ldots$) are spherical in shape and are larger for higher values of n. For higher l values (p, d, etc. orbitals), a number of lobes appear whose number is $(2l+1)$ corresponding to different values of m_l.

The concept of orbitals is particularly useful in the elucidation of the binding of the molecules.

11.17 Constants of motion in a central field

The separation of the wave function into a radial and an angular part cannot be effected in general unless $V(r)$ is spherically symmetric. This corresponds to the classical result that the angular momentum is a constant of motion for a central field. However, the difference between the classical and the quantum mechanical cases lies in the fact that all the three components L_x, L_y, L_z of $\mathbf{L}$ can be precisely specified in classical mechanics, whereas only L_z and L^2 can be simultaneously specified in quantum mechanics.

This is due to the fact that though $Y_l^m(\theta, \varphi)$ is a simultaneous eigenfunction of $\hat{L}^2$ and $\hat{L}_z$ belonging to the eigenvalues $l(l+1)\hbar^2$ and $m\hbar$ respectively, it is not in general an eigenfunction of $\hat{L}_x$ and $\hat{L}_y$ at the same time (except for $l = 0$). This result is related to the uncertainty principle.

The eigenvalue $m\hbar$ of the operator $\hat{L}_z$ is always an integral multiple of $\hbar$ i.e., m = integer. The highest value of m is l, while the square of the orbital angular momentum $\hat{L}^2$ has the eigenvalue $l(l+l)\hbar^2 = (l^2+l)\hbar^2$

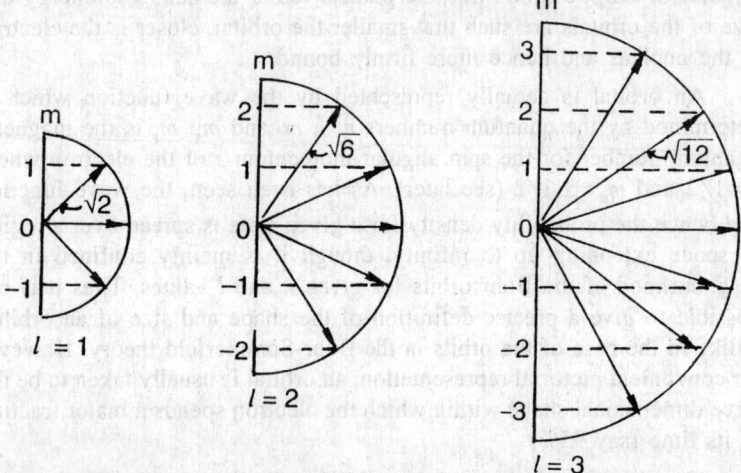

Fig. 11·21. Projections of the angular momentum vector.

which is greater than $(l\hbar)^2$. This shows that one can never measure a component of the angular momentum equal in magnitude to the total angular momentum. This is illustrated in Fig. 11·21.

The situation is different in classical mechanics, according to which the maximum value of the projection of any vector in a specified direction is equal to the length of the vector itself. The latter result also holds in old quantum theory according to which the components of the azimuthal angular momentum $k\hbar$ along a specified direction in space is $m\hbar$ where m can take up the integral values $m = 0, \pm 1, \pm 2, ... \pm k$, k being the azimuthal quantum number of the old quantum theory. Hence the maximum value of the component is equal to the length of the vector itself, viz., $k\hbar$.

The fact that $Y_l^m(\theta, \varphi)$ is the eigenfunction of both the operators $\hat{L}^2$ and $\hat{L}_z$ belonging to the eigenvalues $l(l+1)\hbar^2$ and $m\hbar$ respectively means that both the dynamical variables, the square of the orbital angular momentum and its z-component, have definite values in the same state described by the eigenfunction $Y_l^m(\theta, \varphi)$.

The simultaneous measurability of different physical quantities yielding definite values in a particular state is actually related to the commutability of the operators representing these quantities in quantum mechanics.

Let $\hat{\alpha}$ and $\hat{\beta}$ be two operators representing the two dynamical

Solution of Schrödinger Equation in Some Simple Cases

variables α and β which commute with each other so that

$$\hat{\alpha}\hat{\beta} = \hat{\beta}\hat{\alpha} \tag{11.17-1}$$

Let $\psi(x)$ be an eigenfunction of $\hat{\alpha}$ belonging to the eigenvalue a so that $\hat{\alpha}\psi = a\psi$.

This means that in the state $\psi(x)$, the dynamical variable α has the definite value a. We then have

$$\hat{\alpha}\hat{\beta}\psi = \hat{\beta}\hat{\alpha}\psi = \hat{\beta}a\psi = a\hat{\beta}\psi$$

Thus $\hat{\beta}\psi$ and ψ are both simultaneous eigenfunctions of the operator $\hat{\alpha}$ belonging to the same eigenvalue a. Unless the state is degenerate, $\hat{\beta}\psi$ must be a multiple of ψ:

$$\hat{\beta}\psi = b\psi \tag{11.17-2}$$

where b is a constant (complex).

However Eq. (11.17-2) is precisely the condition for $\psi(x)$ to be an eigenfunction of $\hat{\beta}$ belonging to the eigenvalue b. This means that the dynamical variable β has the definite value b in the state $\psi(x)$. Thus in the state represented by the wave function $\psi(x)$, both the dynamical variables α and β have simultaneously the definite values a and b respectively, if the operators $\hat{\alpha}$ and $\hat{\beta}$ commute.

The fact that the operators $\hat{L}^2$ and $\hat{L}_z$ commute tells us that the two dynamical vatiables L^2 and L_z can be measured precisely in the same state simultaneously.

However any two of the three components $\hat{L}_x, \hat{L}_y, \hat{L}_z$ are non-commuting (see Eq. 11.11-2). Hence, according to the above discussion L_x and L_y cannot be simultaneously determined with L_z for the same state. This means that the direction of L will remain indeterminate, though the magnitude of its expectation value has a definite value $\sqrt{l(l+1)}\,\hbar^2$. This fact can be shown to be connected with the uncertainty relation.

Since L_z has a definite value in the state $Y_l^m(\theta, \varphi)$ so its conjugate, the azimuthal angle φ, becomes completely indeterminate in this state. Hence the direction of the component of L in the $x - y$ plane is completely indeterminate which means that L_x and L_y are completely indeterminate. As a result the direction of L is also indeterminate.

The simultaneous measurability of L^2 and L_z along with energy in an energy eigenstate for a central potential shows that the operators L^2 and L_z must commute with the Hamiltonian operator in this case.

Thus both L^2 and L_z are constants of motion for a central potential. (see § 10.17).

11·18 Parity operator

We had introduced the concept of the parity of a wave function while discussing about the rectangular well and harmonic oscillator wave functions. The concept of parity as a dynamical variable is purely quantum mechanical. It is related to the invariance of the Hamiltonian operator $\hat{H}(\mathbf{r})$ under reflection at the origin of the co-ordinate system which results in transforming x, y, z into $-x, -y, -z$.

We introduce the parity operator $\hat{P}$ such that $\hat{P}$ operating on any function or any other operator changes the position vector $\mathbf{r}$ to $-\mathbf{r}$, *i.e.*, it changes the co-ordinates x, y, z to $-x, -y, -z$:

$$\hat{P} f(\mathbf{r}) = f(-\mathbf{r})$$
$$\hat{P} f(x, y, z) = f(-x, -y, -z)$$

Consider the three dimensional Hamiltonian

$$\hat{H} = -\frac{\hbar^2}{2m} \nabla^2 + V(\mathbf{r}) \tag{11·18-1}$$

Now $\nabla^2 = \partial^2/\partial x^2 + \partial^2/\partial y^2 + \partial^2/\partial z^2$ remains unchanged as x, y, z are changed to $-x, -y, -z$. Further, if we assume that $V(\mathbf{r})$ remains unchanged as $\mathbf{r}$ is changed to $-\mathbf{r}$, then the Hamiltonian operator $\hat{H}$ remains unchanged by parity operation.

We can write the time-independent Schrödinger equation as

$$\hat{H}(\mathbf{r}) \psi(\mathbf{r}) = E \psi(\mathbf{r}) \tag{11·18-2}$$

Operating by the parity operator $\hat{P}$ we get

$$\hat{P} \hat{H}(\mathbf{r}) \psi(\mathbf{r}) = \hat{P} E \psi(\mathbf{r}) \tag{11·18-3}$$

Since $\hat{H}$ remains unchanged by parity operation, we have

$$\hat{P} \hat{H}(\mathbf{r}) \psi(\mathbf{r}) = \hat{H}(\mathbf{r}) \psi(-\mathbf{r}) \tag{11·18-4}$$

Further since E is a constant, we get from Eqs. (11·18-3) and (11·18-4)

$$\hat{H}(\mathbf{r}) \psi(-\mathbf{r}) = E \hat{P} \psi(\mathbf{r}) = E \psi(-\mathbf{r}) \tag{11·18-5}$$

Comparing Eqs. (11·18-2) and (11·18-5), we see that $\psi(\mathbf{r})$ and $\psi(-\mathbf{r})$ are two simultaneous eigenfunctions of the Hamiltonian operator belonging to the same eigenvalue E. Since these two are non-degenerate eigenfunctions, they can differ only by a multiplying constant λ. So we can write

$$\hat{P} \psi(\mathbf{r}) = \psi(-\mathbf{r}) = \lambda \psi(\mathbf{r}) \tag{11·18-6}$$

Operating by $\hat{P}$ again we get

$$\hat{P}^2 \psi(\mathbf{r}) = \hat{P} \hat{P} \psi(\mathbf{r}) = \hat{P} \lambda \psi(\mathbf{r}) = \lambda \hat{P} \psi(\mathbf{r}) = \lambda^2 \psi(r) \tag{11·18-7}$$

But $\quad \hat{P}^2 \psi(\mathbf{r}) = \hat{P} \hat{P} \psi(\mathbf{r}) = \hat{P} \psi(-\mathbf{r}) = \psi(\mathbf{r}) \tag{11·18-8}$

Solution of Schrödinger Equation in Some Simple Cases

From Eqs. (11·18-7) and (11·18-8), we have

$$\lambda^2 \psi(\mathbf{r}) = \psi(\mathbf{r})$$

Hence $\quad\quad\quad\quad \lambda^2 = 1$

So that $\quad\quad\quad\quad \lambda = \pm 1 \quad\quad\quad\quad (11\cdot18\text{-}9)$

Hence we have the result

$$\psi(-\mathbf{r}) = \pm \psi(\mathbf{r}) \quad\quad\quad\quad (11\cdot18\text{-}10)$$

Thus the wave functions $\psi(\mathbf{r})$ can be grouped into two classes. If $\lambda = +1$, we have $\psi(-\mathbf{r}) = \psi(\mathbf{r})$. In this case the wave function $\psi(\mathbf{r})$ is said to have *even parity*.

If $\lambda = -1$, we have $\psi(-\mathbf{r}) = -\psi(\mathbf{r})$. In this case $\psi(\mathbf{r})$ is said to have *odd parity*.

The above classification of the wave functions according to parity is possible only if the Hamiltonian $\hat{H}$ remains invariant under parity operation, *i.e.*, if the interaction potential $V(\mathbf{r})$ remains unchanged by parity operation.

It can be shown that the eigenfunctions of the square of the orbital angular momentum operator $\hat{L}^2$ can be classified according to parity. We have seen before (§ 11·14) that these are the spherical harmonics $Y_l^m(\theta, \varphi)$. In spherical polar coordinates, reflection of the coordinate system at the origin has the effect of changing θ to $(\pi - \theta)$ and

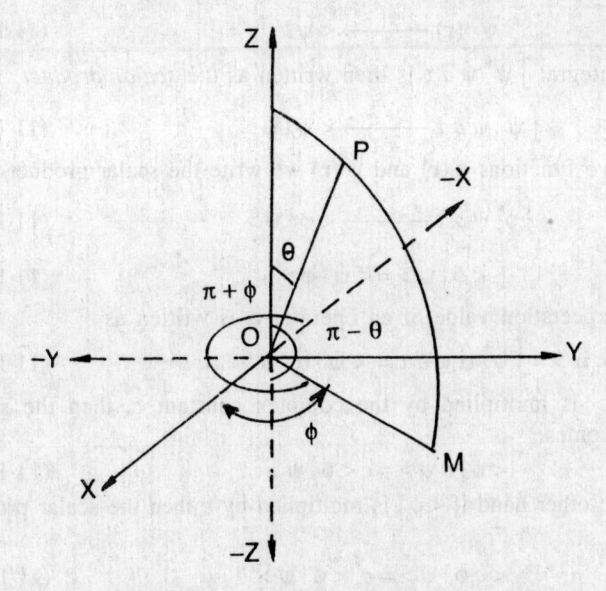

Fig. 11·22. Coordinate change on inversion.

ϕ to $(\pi + \phi)$ while r remains unchanged (see Fig. 11·22). It can be shown that§§

$$\hat{P} Y_l^m (\theta, \phi) = (-1)^l Y_l^m (\theta, \phi) \qquad (11\cdot18\text{--}11)$$

Hence if the azimuthal quantum number $l = 0$ or even, then the sign of the eigenfunctions of $\hat{L}^2$ remains unchanged by parity operation; *i.e.*, they have even parity. If l is odd then the parity of the eigenfunctions is odd.

11·19 Dirac notations

In developing the formal structure of quantum mechanics, based on a number of fundamental postulates (see § 10·18). Dirac introduced certain abbreviated notations for the expectation values and matrix elements. In the present section we shall give a brief introduction to these without explanation since they will be found useful in some of the succeeding sections.

The time-independent wave function $\psi(\mathbf{r})$ is treated as a vector in an n-dimensional space and is called the ket-vector or simply *ket*. It is written symbolically as

$$\psi(r) \longrightarrow |\psi\rangle \qquad (11\cdot19\text{--}1)$$

The complex conjugate $\psi^*(\mathbf{r})$ is written as the bra-vector (or *bra*) as follows :

$$\psi^*(\mathbf{r}) \longrightarrow \langle\psi| \qquad (11\cdot19\text{--}2)$$

The integral $\int \psi^* \psi \, d\tau$ is then written as the *scalar product*

$$\int \psi^* \psi \, d\tau \longrightarrow \langle\psi|\psi\rangle \qquad (11\cdot19\text{--}3)$$

For two functions $\phi(\mathbf{r})$ and $\psi(\mathbf{r})$ we write the scalar product as

$$\int \phi^* \psi \, d\tau \longrightarrow \langle\phi|\psi\rangle \qquad (11\cdot19\text{--}4)$$

Obviously $\qquad \langle\phi|\psi\rangle = \langle\psi|\phi\rangle^* \qquad (11\cdot19\text{--}5)$

The expectation value of an operator $\hat{\alpha}$ is written as

$$\langle\alpha\rangle = \int \psi^* \hat{\alpha} \psi \, d\tau = \langle\psi|\hat{\alpha}|\psi\rangle \qquad (11\cdot19\text{--}6)$$

If $|\psi\rangle$ is multiplied by the complex constant c, then the scalar product becomes

$$\langle\phi|c\psi\rangle = c \langle\phi|\psi\rangle \qquad (11\cdot19\text{--}7)$$

On the other hand if $\langle\phi|$ is multiplied by c then the scalar product becomes

$$\langle c\phi|\psi\rangle = c^* \langle\phi|\psi\rangle \qquad (11\cdot19\text{--}8)$$

---

§ See Introductory Quantum Mechanics by the author.

Solution of Schrödinger Equation in Some Simple Cases

Note that in the second case, since c appears to the left of the vertical bar within the angular brackets we have to multiply by the complex conjugate c^* outside the bracket (see below).

The matrix element of an operator $\hat{\alpha}$ between the states $\psi(\mathbf{r})$ and $\phi(\mathbf{r})$ is written as

$$\int \phi^* \hat{\alpha} \psi \, d\tau \longrightarrow <\phi|\hat{\alpha}|\psi> \qquad (11\cdot19\text{–}9)$$

If $\hat{\alpha}$ is Hermitian, then this is usually written as $<\phi|\hat{\alpha}|\psi>$

It should be noted that the initial state ψ appears on the r.h.s. of the operator while the final state ϕ appears on the l.h.s. of the operator. Also in this mode of writing the symbol for the complex conjugate is avoided. The function appearing on the left side of the bar (|) is to be taken as its complex conjugate while making the actual calculation.

The above notations are not introduced merely for the sake of convenience in writing. They have deep physical significance. As stated $|\psi>$ represents a vector in the n-dimensional linar vector space. The algebra of the linear vector space is very well-developed and forms the starting point of the formal structure of quantum mechanics. (see *Quantum Mechanics* by L.I. Schiff).

Some important relations in Dirac notation :

(a) The Hermitian property of an operator $\hat{\alpha}$ can be expressed in Dirac notation as follows (see Eq. 10·14–5). For two state vectors $|f>$ and $|g>$

$$<f|\hat{\alpha} g> = <\hat{\alpha} f|g> \qquad (11\cdot19\text{–}10)$$

(b) Hermitian conjugate $\hat{\alpha}^\dagger$ is defined by the following relation

$$<f|\hat{\alpha}^\dagger g> = <\hat{\alpha} f|g> \qquad (11\cdot19\text{–}11)$$

11·20 System of two particles ; Motion in the centre of mass frame

In the treatment of the hydrogen problem in § 11·12 it was assumed that the nucleus of the atom was infinitely heavy and its motion was neglected. Only the motion of the electron was considered so that the problem was reduced to the motion of a single particle. In reality, the nucleus of the hydrogen atom, viz. the proton which is very heavy compared to the electron ($\sim 1836\, m$), has nevertheless a finite mass. As such the electron-nucleus system as a whole moves relative to the centre of mass of the system. Classically the motion of such a two-particle system can be reduced to the motion of a free particle of mass equal to the sum of the masses of the two particles located at the centre of mass of the system and their relative motion which is equivalent to that of a particle of mass equal to the reduced mass of the system acted upon by a fixed potential. It can be shown that this is also true in quantum mechanics.

Consider a system of two particles of masses m_1 and m_2 having the position vector $\mathbf{r}_1$ and $\mathbf{r}_2$ in the laboratory system of coordinates, which have the cartesian components (x_1, y_1, z_1) and (x_2, y_2, z_2) respectively. Let $\mathbf{R}$ be the position vector of the C.M. of the system having the components X, Y and Z so that

$$\mathbf{R} = \frac{m_1 \mathbf{r}_1 + m_2 \mathbf{r}_2}{m_1 + m_2}$$

Here $X = (m_1 x_1 + m_2 x_2)/(m_2 + m_2)$ and similarly for Y and Z. The position vector of m_1 relative to m_2 is given by $\mathbf{r} = \mathbf{r}_1 - \mathbf{r}_2$ so that its components are $x = x_1 - x_2$ and so on.

The Hamiltonian of the two particle system is

$$\hat{H} = -\frac{\hbar^2}{2m_1} \nabla_1^2 - \frac{\hbar^2}{2m_2} \nabla_2^2 + V(\mathbf{r}) \qquad (11\cdot20\text{--}1)$$

where $V(\mathbf{r})$ is the potential of the interaction between the two particles which is assumed to be a function of the relative distance r.

$$\nabla_1^2 = \frac{\partial^2}{\partial x_1^2} + \frac{\partial^2}{\partial y_1^2} + \frac{\partial^2}{\partial z_1^2} \quad \text{and} \quad \nabla_2^2 = \frac{\partial^2}{\partial x_2^2} + \frac{\partial^2}{\partial y_2^2} + \frac{\partial^2}{\partial z_2^2}$$

are the Laplacian operators. Let $\psi(\mathbf{r}_1, \mathbf{r}_2) = \psi(\mathbf{R}, \mathbf{r})$ denote the wave function of the two particle system. The functional form on the r.h.s. may be different. Then the time-independent Schrödinger equation can be written as

$$\hat{H} \psi(\mathbf{r}_1, \mathbf{r}_2) = E \psi(\mathbf{r}_1, \mathbf{r}_2)$$

where E is the energy of the system.

Now

$$\frac{\partial}{\partial x_1} = \frac{\partial X}{\partial x_1} \frac{\partial}{\partial X} + \frac{\partial x}{\partial x_1} \frac{\partial}{\partial x} = \frac{m_1}{M} \frac{\partial}{\partial X} + \frac{\partial}{\partial x}$$

where $M = m_1 + m_2$. Then

$$\frac{\partial^2}{\partial x_1^2} = \frac{\partial}{\partial x_1} \frac{\partial}{\partial x_1} = \left(\frac{m_1}{M} \frac{\partial}{\partial X} + \frac{\partial}{\partial x}\right)\left(\frac{m_1}{M} \frac{\partial}{\partial X} + \frac{\partial}{\partial x}\right)$$

$$= \left(\frac{m_1}{M}\right)^2 \frac{\partial^2}{\partial X^2} + \frac{\partial^2}{\partial x^2} + \frac{2m_1}{M} \cdot \frac{\partial^2}{\partial X \, dx}$$

Similarly it can be shown that

$$\frac{\partial}{\partial x_2} = \frac{m_2}{M} \frac{\partial}{\partial X} - \frac{\partial}{\partial x}$$

Solution of Schrödinger Equation in Some Simple Cases

and
$$\frac{\partial^2}{\partial x_2^2} = \left(\frac{m_2}{M}\right)^2 \frac{\partial^2}{\partial X^2} + \frac{\partial^2}{\partial x^2} - \frac{2m_2}{M}\frac{\partial^2}{\partial X \partial x}$$

Hence
$$\frac{1}{m_1}\frac{\partial^2}{\partial x_1^2} + \frac{1}{m_2}\frac{\partial^2}{\partial x_2^2} = \frac{m_1 + m_2}{M^2}\frac{\partial^2}{\partial X^2} + \left(\frac{1}{m_1} + \frac{1}{m_2}\right)\frac{\partial^2}{\partial x^2}$$

$$= \frac{1}{M}\frac{\partial^2}{\partial X^2} + \frac{1}{\mu}\frac{\partial^2}{\partial x^2}$$

where $\mu = \dfrac{m_1 m_2}{m_1 + m_2}$ is the reduced mass of the system.

We can get similar expressions for the other components. Hence we finally get

$$\frac{1}{m_1}\nabla_1^2 + \frac{1}{m_2}\nabla_2^2 = \frac{1}{M}\nabla_R^2 + \frac{1}{\mu}\nabla_r^2$$

where $\nabla_R^2 = \dfrac{\partial^2}{\partial X^2} + \dfrac{\partial^2}{\partial Y^2} + \dfrac{\partial^2}{\partial Z^2}$ and $\nabla_r^2 = \dfrac{\partial^2}{\partial x^2} + \dfrac{\partial^2}{\partial y^2} + \dfrac{\partial^2}{\partial z^2}$

So the Schrödinger equation reduces to

$$\hat{H}\psi(\mathbf{R}, \mathbf{r}) = -\frac{\hbar^2}{2M}\nabla_R^2 \psi(\mathbf{R}, \mathbf{r}) - \frac{\hbar^2}{2\mu}\nabla_r^2 \psi(\mathbf{R}, \mathbf{r})$$
$$+ V(r)\psi(\mathbf{R}, \mathbf{r}) = E\psi(\mathbf{R}, \mathbf{r}) \qquad (11\cdot20\text{-}2)$$

Since the operators in the above equation depend either on the C.M. coordinates (**R**) or on the relative coordinates (**r**) and not on both, we can try to solve the equation by separation of the variables. We write

$$\psi(\mathbf{R}, \mathbf{r}) = \chi(\mathbf{R})\phi(\mathbf{r})$$

Substituting in Eq. (11.20–2) and dividing by $\psi(\mathbf{R}, \mathbf{r})$, we get

$$-\frac{1}{\phi(\mathbf{r})} \cdot \frac{\hbar^2}{2\mu}\nabla_r^2 \cdot \phi(\mathbf{r}) + V(r) - E = 1/\chi(\mathbf{R}) \cdot \frac{\hbar^2}{2M}\nabla_R^2 \chi(\mathbf{R}) = -E_c \text{ (say)}$$

The l.h.s. in the above equation is a function of the relative coordinates (**r**) alone while the r.h.s. is a function of the C.M. coordinates (**R**). Since these are independent of one another, each side must be put equal of some constant which we write as $(-E_0)$, so that we get the following two independent equations:

$$-\frac{\hbar^2}{2\mu}\nabla_r^2 \phi(\mathbf{r}) + V(r)\phi(\mathbf{r}) = (E - E_c)\phi(\mathbf{r}) = E_i \phi(\mathbf{r}) \qquad (11\cdot20\text{-}3)$$

$$-\frac{\hbar^2}{2M}\nabla_R^2 \chi(\mathbf{R}) = E_0 \chi(\mathbf{R}) \qquad (11\cdot20\text{-}4)$$

The second of these equations is the eigenvalue equation for the motion of a free particle of mass $M = m_1 + m_2$ having the energy E_0 which may vary continuously so that the C.M. of the system moves like a free particle, having a continuous energy spectrum.

The first is the eigenvalue equation for the relative motion of the two particles, which has been reduced to a form equivalent to the motion of a single particle of mass $\mu = m_1 m_2/(m_1 + m_2)$ under the influence of the potential $V(\mathbf{r})$, having an energy $E_i = E - E_c$.

The above analysis shows that if the motion of the nucleus is taken into account in the hydrogen problem, then the electron mass m in §11·12 is to be replaced by the reduced mass μ of the electron-nucleus system, everything else remaining unchanged.

Problems

1. Calculate the expectation values $<x>$, $<x^2>$, $<p_x>$ and $<p_x^2>$ in the ground state of a particle free to move within a potential box with rigid walls extending from $x = 0$ to $x = l$. $(l/2\,;\ l^2/3 - l^2/2\pi^2\,;\ 0\,;\ \pi^2 \hbar^2/l^2)$

2. Solve the problem of the one dimensional rectangular potential barrier of height V_0 and width a when the beam of particle has energy $E > V_0$. Show that the transmission co-efficient is given by

$$T = \left\{1 + \frac{V_0^2}{4E(E-V_0)} \sin^2 \beta a\right\}^{-1}$$

where $\beta^2 = 2m(E-V_0)/\hbar^2$. Sketch the variation of T with energy. What is the limiting value of T when $E \gg V_0$?

3. Calculate the transmission coefficient in Prob. 2 for a beam of electrons of energy 10 eV if the barrier has a height 5 eV and width 2 Å. (89%)

4. A one dimensional rectangular potential well has a height 20 eV and width 2 Å. Calculate graphically the energies of the first three energy levels for an electron confined within the well using the equations (11·6–18), (11·6–19) and (11·6–20).

5. Prove that the normalized wave function for the harmonic oscillator with x as the variable is given by

$$\psi_n(x) = \left\{\frac{\sqrt{\alpha}}{2^n n! \sqrt{\pi}}\right\}^{\frac{1}{2}} \exp\left(\frac{-\alpha x^2}{2}\right) H_n(\sqrt{\alpha}\, x)$$

where H_n is the Hermite function.

6. Show that the classical probability of finding the linear harmonic oscillator in dx at x is given by

$$P(x)\, dx = \frac{dx}{\pi \sqrt{a^2 - x^2}}$$

Compare this with the quantum mechanical probability for $n = 0$ graphically for the oscillator.

7. Prove that the quantum mechanical probability of finding the linear harmonic oscillator for $n = 0$ outside the classical limits is approximately 16%. Also

prove that the probability density at the classical limits is $1/e$th of that at the origin.

8. Using the one-dimensional harmonic oscillator wave functions for $n=0$ and $n=1$, prove that $\psi_0(q)$ and $\psi_1(q)$ given in Eqs. (11·8–7) are normalized wave functions and are orthogonal to each other.

9. Prove that the commutation relations between the components of the angular momentum operator can be written in the form
$$\hat{L} \times \hat{L} = i\hbar \hat{L}$$

10. Assuming a diatomic molecule to be a rigid rotator with the fixed inter-nuclear separation r_0, prove that the energy levels are given by
$$E_j = j(j+1)\hbar^2 / 2\mu r_0^2$$
where $j = 0, 1, 2, 3, \ldots$ and μ is the reduced mass.

11. Show that for $m=0$, the differential equation for the angular function $\Theta(\theta)$ reduces to the Legendre equation (A 2·5). Solve this equation by assuming a series solution of the form
$$\Theta(\theta) = \sum_i a_i \mu^i$$
where $\mu = \cos\theta$. Prove that well-behaved solutions exist only if the series terminates at some value $i = l$ such that $\lambda = l(l+1)$ in Eq. (11·14–2) (see Appendix A-II).

12. Calculate the expectation values $<L_z>$ and $<L_z^2>$ in the state of given l and m.
$$(m, m^2)$$

13. Prove that the radial probability densities for the states ψ_{100}, ψ_{210} and ψ_{320} of the hydrogen atom assume maximum values at $r = a_0, 4a_0$ and $9a_0$ respectively, where a_0 is the Bohr radius of the hydrogen atom.

14. Calculate the expectation values $<r>$ in the states ψ_{100} and ψ_{210} of the hydrogen atom.
$$(3a_0/2, 5a_0)$$

15. Calculate the expectation values $<1/r>$ and $<p^2>$ for the hydrogen atom in the ψ_{210} state.
$$(1/4a_0 \,;\, \hbar^2/4a_0)$$

16. From the results obtained in Prob. 15, calculate the expection values of the kinetic, potential and total energies in the state ψ_{210}. Compare the results with those found from Bohr's theory.
$$(me^4/128\pi^2 \varepsilon_0^2 \hbar^2 \,;\, -me^4/64\pi^2 \varepsilon_0^2 \hbar^2 \,;\, -me^4/128\pi^2 \varepsilon_0^2 \hbar^2)$$

17. Prove that the wave functions for the hydrogen atom in the states ψ_{100} and ψ_{200} are orthogonal.

18. Prove that ρ in Eq. (11·15–3) and n in Eq. (11·15–5) are dimensionless quantities.

19. Calculate the expectation values of the potential, kinetic and total energies in the state ψ_{100} of the hydrogen atom.

20. A particle of mass m is located in a two dimensional potential box extending from $x=0$ to $x=a$ and $y=0$ to $y=b$. Solve the Schrödinger equation by the separation of variables and find the eigenvalues and eigenfunctions. Hence find the energies of the first four levels.

12

APPROXIMATE METHODS IN QUANTUM MECHANICS

The number of problems involving physical systems which can be solved exactly by analytical methods in quantum mechanics is very limited. Most of the problems of physical interest require different types of approximate method for their solution. We shall now discuss some of these approximate methods.

12·1 Stationary perturbation method

We assume that the Hamiltonian for the system can be written as the sum of two parts.

$$H = H_0 + \lambda H' \qquad (12 \cdot 1\text{-}1)$$

where λ is a parameter with possible values between 0 and 1.

Here H_0 is the unperturbed Hamiltonian for which the wave equation is exactly solvable. $\lambda H'$ is a perturbation which is assumed to be small. Let the stationary state wave function and the energy of the system for the perturbed Hamiltonian for the state n be Ψ_n and W_n respectively so that

$$H \Psi_n = W_n \Psi_n \qquad (12 \cdot 1\text{-}2)$$

Let $u_n = \Psi_n^{(0)}$ and $E_n = W_n^{(0)}$ be the corresponding quantities for the unperturbed Hamiltonian H_0 for the same state, so that

$$H_0 u_n = E_n u_n \qquad (12 \cdot 1\text{-}3)$$

Non-degenerate case :

We assume that the stationary state under consideration is non-degenerate. Since the perturbation is small the perturbed energy level W_n is much closer to E_n than to any other unperturbed energy level so that we can expand Ψ_n and W_n in power series of λ. We assume the two series to be analytic and continuous functions of λ. In the final result we shall put λ equal to 1. We write

$$\Psi_n = \sum_i \lambda^i \Psi_n^{(i)} \qquad (3312 \cdot 1\text{-}4)$$

Approximate Methods in Quantum Mechanics

$$W_n = \sum_i \lambda^i W_n^{(i)} \tag{12.1-5}$$

where $i = 0, 1, 2, 3, \ldots$ refers to the order of perturbation. The wave equation (12·1-2) can be written as

$$(H_0 + \lambda H') \sum_i \lambda^i \Psi_n^{(i)} = \sum_i \lambda^i W_n^{(i)} \sum_i \lambda^i \Psi_n^{(i)} \tag{12.1-6}$$

Comparing the coefficients of different powers of λ, we get the series of equations :

For $i = 0$, $(H_0 - W_n^{(0)}) \Psi_n^{(0)} = 0$

$i = 1$, $(H_0 - W_n^{(0)}) \Psi_n^{(1)} = (W_n^{(1)} - H') \Psi_n^{(0)}$

$i = 2$, $(H_0 - W_n^{(0)}) \Psi_n^{(2)} = (W_n^{(1)} - H') \Psi_n^{(1)} + W_n^{(2)} \Psi_n^{(0)}$

$i = 3$, $(H_0 - W_n^{(0)}) \Psi_n^{(3)} = (W_n^{(1)} - H') \Psi_n^{(2)} + W_n^{(2)} \Psi_n^{(1)} + W_n^{(3)} \Psi_n^{(0)}$

and so on. (12·1-7)

The first of the above equations gives $\Psi_n^{(0)}$ as the unperturbed wave function for which the energy eigenvalue is $W_n^{(0)}$. Comparing with Eq. (12·1-3), we then have

$$\Psi_n^{(0)} = u_n \text{ and } W_n^{(0)} = E_n \tag{12.1-8}$$

The state u_n is assumed to be a discrete bound state which is non-degenerate. The other unperturbed states may not be non-degenerate.

12·2 First order perturbation

The second of the equations (12·1-7) for the first order perturbation is

$$(H_0 - W_n^{(0)}) \Psi_n^{(1)} = (W_n^{(1)} - H') \Psi_n^{(0)} \tag{12.2-1}$$

We expand the first order correction to the wave function in terms of the unperturbed eigenfunctions, i.e.

$$\Psi_n^{(1)} = \sum_j a_j^{(1)} \Psi_j^{(0)} = \sum_j a_j^{(1)} u_j \tag{12.2-2}$$

Substitution gives

$$\sum_j a_j^{(1)} (H_0 - E_n) u_j = (W_n^{(1)} - H') u_n \tag{12.2-3}$$

where we have put $W_n^{(0)} = E_n$.

Since $H_0 u_j = E_j u_j$, we get

$$\sum_j a_j^{(1)} (E_j - E_n) u_j = (W_n^{(1)} - H') u_n \tag{12.2-4}$$

Multiplying Eq. (12·2-4) by u_k^* on the left and integrating we get

$$\sum_j a_j^{(1)} (E_j - E_n) \int u_k^* u_j \, d\tau$$

$$= W_n^{(1)} \int u_k^* u_n \, d\tau - \int u_k^* H' u_n \, d\tau \qquad (12\cdot2\text{-}5)$$

In Dirac notation this becomes

$$\sum_j a_j^{(1)} (E_j - E_n) <u_k | u_j> = W_n^{(1)} <u_k | u_n> - <u_k | H' | u_n> \qquad (12\cdot2\text{-}6)$$

Since the eigenfunctions u_j form an orthonormal set, we get

$$\sum_j a_j^{(1)} (E_j - E_n) \delta_{kj} = W_n^{(1)} \delta_{kn} - H'_{kn} \qquad (12\cdot2\text{-}7)$$

where we have put $\qquad H'_{kn} = <u_k | H' | u_n> \qquad (12\cdot2\text{-}8)$

$$\delta_{kj} = 1 \text{ for } k = j \text{ and } \delta_{kj} = 0 \text{ for } k \neq j.$$

Similarly for δ_{kn}. All terms on the l.h.s. of Eq. (12·2-7) vanish except the term $j = k$ so that we get

$$a_k^{(1)} (E_k - E_n) = W_n^{(1)} \delta_{kn} - H'_{kn} \qquad (12\cdot2\text{-}9)$$

H'_{kn} given by Eq (12·2-8) is the *matrix element* of the perturbation H' with the unperturbed wave functions as the basis. Then for $k \neq n$, we get ($\because \delta_{kn} = 0$)

$$a_k^{(1)} = \frac{H'_{kn}}{E_n - E_k} \qquad (12\cdot2\text{-}10)$$

For $k = n$, we have from Eq (12·2-9),

$$0 = W_n^{(1)} - H'_{nn} \qquad (12\cdot2\text{-}11)$$

so that $\qquad W_n^{(1)} = H'_{nn} = <u_n | H' | u_n> \qquad (12\cdot2\text{-}12)$

Eq. (12·2-10) gives the first order correction to the eigenfunction for the nth state in the presence of the perturbation H'.

$$\Psi_n^{(1)} = \sum_j a_j^{(1)} u_j = \sum_j{'} \frac{H'_{jn}}{E_n - E_j} u_j \qquad (12\cdot2\text{-}13)$$

and $\qquad \Psi_n = \Psi_n^{(0)} + \lambda \Psi_n^{(1)} = u_n + \sum_j{'} a_j^{(1)} \Psi_j^{(0)}$

$$= u_n + \sum_j{'} \frac{H'_{jn}}{E_n - E_j} u_j \qquad (12\cdot2\text{-}13a)$$

Here we have put $\lambda = 1$. Σ' denotes sum over all terms except $j = n$.

Eq. (12·2-12) gives the energy for the nth state corrected to first order as

Approximate Methods in Quantum Mechanics 337

$$W_n = W_n^{(0)} + \lambda W_n^{(1)} = E_n + H'_{nn}$$
$$= E_n + \langle u_n | H' | u_n \rangle \qquad (12\cdot2\text{-}14)$$

Notice that the denominator of Eq. (12·2-10) can never be zero since we have excluded the possibility of degeneracy in the derivation. The theory developed above shows that the eigenstate of the first order perturbation is an admixture of the unperturbed states (basis states) with coefficients depending on the matrix elements of the perturbation between the unperturbed non-degenerate wave function u_n and the set of unperturbed states u_j with $j \neq n$. These coefficients also depend on the energy difference $(E_i - E_j)$. Larger this energy difference is, smaller is the admixture.

It should be noted that the sum in Eq. (12·2-13) for $\Psi_n^{(1)}$ does not contain the term $j = n$. It can be shown that if we add an arbitrary multiple of u_n to $\Psi_n^{(1)}$ then the first order perturbation equation remains unaltered. So without loss of generality we may assume $a_n^{(1)} = 0$ (see *Quantum Mechanics* by L.I. Schiff).

12·3 Second order perturbation

From Eq. (12·1-7) we get the second order perturbation equation as

$$(H_0 - E_n) \Psi_n^{(2)} = (W_n^{(1)} - H') \Psi_n^{(1)} + W_n^{(2)} u_n \qquad (12\cdot3\text{-}1)$$

As in the first order theory we expand $\Psi_n^{(2)}$ in terms of the unperturbed eigenfunctions:

$$\Psi_n^{(2)} = \sum_m a_m^{(2)} u_m \qquad (12\cdot3\text{-}2)$$

Using Eq. (12·2-2) we then get

$$\sum_m a_m^{(2)} (H_0 - E_n) u_m = \sum_j (W_n^{(1)} - H') a_j^{(1)} u_j + W_n^{(2)} u_n$$

or, $$(H_0 - E_n) \sum_m a_m^{(2)} u_m = \sum_j a_j^{(1)} (W_n^{(1)} - H') u_j + W_n^{(2)} u_n$$

Multiplying by u_k^* and integrating we get (since $H_0 u_m = E_m u_m$ and $W_n^{(1)} = H'_{nn}$).

$$\sum_m a_m^{(2)} (E_m - E_n) \langle u_k | u_m \rangle$$
$$= \sum_j a_j^{(1)} \{ H'_{nn} \langle u_k | u_j \rangle - \langle u_k | H' | u_j \rangle \} + W_n^{(2)} \langle u_k | u_n \rangle$$

$$\therefore \sum_m a_m^{(2)} (E_m - E_n) \delta_{km} = \sum_j a_j^{(1)} (H'_{nn} \delta_{kj} - H'_{kj}) + W_n^{(2)} \delta_{kn} \qquad (12\cdot3\text{-}3)$$

This gives after simplification (for $k \neq n$)

$$a_k^{(2)} = \sum_j' \frac{H'_{jn} H'_{kj}}{(E_n - E_j)(E_n - E_k)} - \frac{H'_{kn} H'_{nn}}{(E_n - E_k)^2} \quad (12\cdot3\text{-}4)$$

We then get

$$\Psi_n^{(2)} = \sum_k a_k^{(2)} u_k = \sum_k \sum_j' \frac{H'_{jn} H'_{kj} u_k}{(E_n - E_j)(E_n - E_k)} - \sum_k \frac{H'_{kn} H'_{nn} u_k}{(E_n - E_k)^2} \quad (12\cdot3\text{-}5)$$

For $k = n$ Eq. (12·3-3) gives for the l.h.s.

$$\sum_m a_m^{(2)} (E_m - E_n) \delta_{km} = \sum_m a_m^{(2)} (E_m - E_n) \delta_{nm} = 0 \quad (12\cdot3\text{-}6)$$

Hence we get $\sum_j a_j^{(1)} (H'_{nn} \delta_{nj} - H'_{nj}) + W_n^{(2)} \delta_{nn} = 0$

On simplification we then get

$$W_n^{(2)} = \sum_j a_j^{(1)} H'_{nj} - a_n^{(1)} H'_{nn} = \sum_j' a_j^{(1)} H'_{nj}$$

$$= \sum_j' \frac{H'_{jn} H'_{nj}}{E_n - E_j} \quad (12\cdot3\text{-}7)$$

But $(H'_{jn})^* = <u_n | H' | u_j >^* = \left(\int u_n^* H' u_j \, d\tau \right)^*$

$$= \int u_n (H')^* u_j^* \, d\tau = \int u_j^* H' u_n \, d\tau$$

$$= <u_j | H' | u_n > = H'_{nj}$$

because of the Hermitian character of H'. So

$$H'_{jn} H'_{nj} = (H'_{nj})^* H'_{nj} = |H'_{nj}|^2$$

Hence $$W_n^{(2)} = \sum_j' \frac{|H'_{nj}|^2}{E_n - E_j} \quad (12\cdot3\text{-}8)$$

Eqs. (12·3-5) and (12·3-6) give the second order corrections to the wave function Ψ_n and the energy W_n. The wave function corrected to second order is then

$$\Psi_n = \Psi_n^{(0)} + \Psi_n^{(1)} + \Psi_n^{(2)}$$

$$= u_n + \sum_j' \frac{H'_{jn} u_j}{E_n - E_j} + \sum_k \sum_j' \frac{H'_{kj} H'_{jn} u_k}{(E_n - E_k)(E_n - E_j)} - \sum_k \frac{H'_{kn} H'_{nn} u_k}{(E_n - E_k)^2} \quad (12\cdot3\text{-}9)$$

$$W_n = W_n^{(0)} + W_n^{(1)} + W_n^{(2)}$$

$$= E_n + H'_{nn} + \sum_j' \frac{|H'_{nj}|^2}{E_n - E_j} \quad (12\cdot3\text{-}10)$$

12·4 Degenerate case

In the preceding sections we have applied the perturbation method to non-degenerate systems. In the present section we shall consider a system with one or more degenerate levels. The methods developed so far are however not applicable to degenerate case. This can be easily seen by considering the expansion of the wave function of the perturbed system (Eq. 12·2-2) in the first order theory. The expansion coefficients $a_j^{(1)}$ are given by Eq. (12·2-10) :

$$a_j^{(1)} = \frac{H'_{jn}}{E_n - E_j}$$

where $H'_{jn} = <u_j|H'|u_n>$ is the matrix element of the perturbing potential H'. If there is a degenerate level so that $E_n = E_j$, then obviously the term $a_j^{(1)}$ in the expansion for Ψ_n becomes infinity and hence the method fails.

The reason for this failure lies in the fact that for the degenerate unperturbed level, the wave functions u_n and u_j are different though the energy is the same. The expansion (12·2-13a) for Ψ_n gives.

$$\Psi_n = u_n + \sum_j{}' \frac{H'_{jn}}{E_n - E_j} \cdot u_j$$

When there is no perturbation $\Psi_n \longrightarrow u_n$ which is the zero–order wave function $\Psi_n^{(0)}$ for the unperturbed level. However for the degenerate case, since there may be more than one such zero-order unperturbed functions, it is not possible for us to say to which function Ψ_n approaches in the limit of withdrawal of the perturbation. It is this lack of knowledge of the correct zero–order function which is responsible for the appearance of the infinity.

For simplicity we consider two-fold degeneracy for the level n such that $E_n = E_m$ for the unperturbed level. u_n and u_m are the two degenerate wave functions for the level.

To get over the difficulty of the expansion coefficient $a_m^{(1)}$ going to infinity in this case we have to choose u_n and u_m in such a way that the expansion coefficient $a_n^{(1)}$ or $a_m^{(1)}$ becomes indeterminate.

We know that if u_n and u_m are the eigenfunctions of the unperturbed Hamiltonian H_0, then any linear combination of them is also an eigenfunction of H_0. So we write the correct zero order eigenfunction to which Ψ_n reduces in the absence of the perturbation H' as

$$\Psi_n^{(0)} = \upsilon_n = C_n u_n + C_m u_m \qquad (12 \cdot 4\text{-}1)$$

We now substitute this in the first-order wave equation for the perturbed Hamiltonian $H = H_0 + \lambda H'$ (Eq. 12·2-1):

$$(H_0 - E_n) \Psi_n^{(1)} = (W_n^{(1)} - H') u_n$$

where $\Psi_n^{(1)}$ is the second term in the expansion for Eq. (12·1-4):

$$\Psi_n = \sum_i \lambda^i \Psi_n^{(i)}$$

Also W_n is given by Eq. (12·1-5):

$$W_n = \sum_i \lambda^i W_n^{(i)}$$

so that $W_n^{(1)}$ is the second term in the above expansion.

We now evaluate the coefficients C_n and C_m in Eq. (12·4-1). To do this we write down the first order wave equation with u_n replaced by the correct zero order wave function υ_n:

$$\begin{aligned}(H' - W_n^{(1)}) \upsilon_n &= (H' - W_n^{(1)})(C_n u_n + C_m u_m) \\ &= -(H_0 - E_n) \sum_j{}' a_j^{(1)} u_j \\ &= \sum_j{}' (E_n - E_j) a_j^{(1)} u_j \end{aligned} \quad (12\cdot4\text{-}2)$$

Multiplying the above equation by u_n^* and integrating, we get

$$\begin{aligned}\int u_n^* (H' - W_n^{(1)})(C_n u_n + C_m u_m) \, d\tau \\ = \sum{}' \int u_n^* (E_n - E_j) a_j^{(1)} u_j \, d\tau \end{aligned}$$

or, $C_n \langle u_n | H' | u_n \rangle - C_n W_n^{(1)} \langle u_n | u_n \rangle + C_m \langle u_n | H' | u_m \rangle$
$$- C_m W_n^{(1)} \langle u_n | u_m \rangle$$
$$= \sum_j{}' a_i^{(1)} (E_n - E_i) \langle u_n | u_j \rangle \quad (12\cdot4\text{-}3)$$

$$\therefore \quad C_n (H'_{nn} - W_n^{(1)}) + C_m H'_{nm} - C_m W_n^{(1)} \delta_{nm}$$
$$= \sum_j{}' a_j^{(1)} (E_n - E_j) \delta_{nj} \quad (12\cdot4\text{-}4)$$

Since $n \neq m$ and $j \neq n$, we get

$$C_n (H'_{nn} - W_n^{(1)}) + C_m H'_{nm} = 0 \quad (12\cdot4\text{-}5)$$

Similarly multiplying Eq. (12·4-2) by u_m^* and integrating we get

$$C_n H'_{mn} + C_m (H'_{mm} - W_n^{(1)}) = 0 \quad (12\cdot4\text{-}6)$$

The two algebraic equations (12·4-5) and (12·4-6) have non-trivial

Approximate Methods in Quantum Mechanics

solutions only if the determinant of the unknown coefficients C_n and C_m vanishes.

$$\begin{vmatrix} H'_{nn} - W_n^{(1)} & H'_{nm} \\ H'_{mn} & H'_{mm} - W_n^{(1)} \end{vmatrix} = 0 \qquad (12\cdot4\text{-}7)$$

The solutions of the equations (12·4-7) gives the first order energy correction $W_n^{(1)}$ for the degenerate level whose unperturbed energy is E_n. Since this is a quadratic equation, the level now splits into two in the presence of the perturbation H'. Substituting these energy values in the wave equation (12·4-2), we get two sets of values for the coefficients C_n and C_m. In doing this we must remember that the zero-order eigenfunction must be normalized.

Thus the two zero-order wave functions in the presence of the perturbation H' can be found which are linear combinations of u_n and u_m. So the effect of the perturbation has been to remove the degeneracy of the unperturbed level.

The above considerations can be generalised to the case of an unperturbed level which has g-fold degeneracy ($g > 2$). In this case the correct zero-order eigenfunction can be written as a linear combination of the g degenerate unperturbed functions u_{nk} with $k = 1, 2, 3, \ldots g$:

$$\upsilon_n = \sum_{k=1}^{g} C_{nk} u_{nk} \qquad (12\cdot4\text{-}8)$$

The first order correction to the energy of the degenerate level is then obtained by solving the determinantal equation :

$$\begin{vmatrix} H'_{11} - W_n^{(1)} & H'_{12} & \cdots & H'_{1g} \\ H'_{21} & H'_{22} - W_n^{(1)} & \cdots & H'_{2g} \\ \cdots & \cdots & & \\ \cdots & \cdots & & \\ H'_{g1} & H'_{g2} & \cdots & H'_{gg} - W_n^{(1)} \end{vmatrix} = 0 \qquad (12\cdot4\text{-}9)$$

H'_{11}, H'_{12} etc. are the matrix elements of the perturbation.

This gives g different values of $W_n^{(1)}$ and the unperturbed degenerate level splits into g different levels of different energy, so that the degeneracy is removed by the perturbation.

If all the roots of Eq. (12·4-9) are different the degeneracy is removed completely in the first order theory.

12·5 Time dependent perturbation

We assume a Hamiltonian of the form

$$H' = H_0 + \lambda H'(t) \qquad (12\cdot5\text{-}1)$$

where H_0 is the original time-independent Hamiltonian while $\lambda H'(t)$ is a small perturbation depending on time.

We assume that the solution for the unperturbed Hamiltonian H_0 is known:

$$H_0 u_n = E_n u_n \qquad (12 \cdot 5 \text{-} 2)$$

where E_n is the eigenvalue of the unperturbed Hamiltonian belonging to the stationary eigenfunction u_n.

Let $\Psi(t)$ be the solution for the time dependent wave equation

$$i\hbar \frac{\partial \Psi}{\partial t} = H\Psi \qquad (12 \cdot 5 \text{-} 3)$$

in the presence of the perturbation superimposed upon the unperturbed Hamiltonian.

We now expand $\Psi(t)$ in terms of the unperturbed stationary state wave functions.

$$\Psi(t) = \sum_n a_n(t) u_n \exp(-i E_n t/\hbar) \qquad (12 \cdot 5 \text{-} 4)$$

The expansion coefficients $a_n(t)$ are dependent on time. We interpret the quantity $|a_n(t)|^2$ as the probability of finding the system in the nth eigenstate at the time t. We now substitute the solution (12·5-4) in the wave equation (12·5-3):

$$\sum_n i\hbar \left\{ \dot{a}_n(t) - \frac{iE_n}{\hbar} a_n(t) \right\} u_n \exp(-iE_n t/\hbar)$$

$$= \{H_0 + \lambda H'(t)\} \sum_n a_n(t) u_n \exp(-iE_n t/\hbar) \qquad (12 \cdot 5 \text{-} 5)$$

Because of Eq. (12·5-2), this reduces to

$$\sum_n (i\hbar \dot{a}_n + a_n E_n - a_n E_n - \lambda H' a_n) u_n \exp(-i E_n t/\hbar) = 0$$

or, $\quad i\hbar \sum_n \dot{a}_n u_n \exp(-iE_n t/\hbar) = \sum_n \lambda H' a_n u_n \exp(-iE_n t/\hbar) \qquad (12 \cdot 5 \text{-} 6)$

Premultiplying both sides of Eq. (12·5-6) by u_k^* and integrating we get

$$i\hbar \sum_n \dot{a}_n \exp(-iE_n t/\hbar) <u_k | u_n>$$

$$= \lambda \sum_n a_n \exp(-iE_n t/\hbar) <u_k | H' | u_n> \qquad (12 \cdot 5 \text{-} 7)$$

Because of the orthonormality of the functions $u_n(t)$ we have

$$<u_k | u_n> = \delta_{kn} \qquad (12 \cdot 5 \text{-} 8)$$

Then writing $\quad H'_{kn} = <u_k | H' | u_n> \qquad (12 \cdot 5 \text{-} 9)$

Approximate Methods in Quantum Mechanics 343

we get the system of coupled equations

$$\dot{a}_k = \frac{\lambda}{i\hbar} \sum_n a_n(t) \, H'_{kn} \exp\{i(E_k - E_n)t/\hbar\}$$

$$= \frac{\lambda}{i\hbar} \sum_n a_n(t) \, H'_{kn} \exp(i\omega_{kn} t) \qquad (12 \cdot 5\text{-}10)$$

where $\qquad \omega_{kn} = (E_k - E_n)/\hbar$. $\qquad (12 \cdot 5\text{-}11)$

Since the perturbation $\lambda H'$ is weak, λ being small, we expand the coefficients $a_n(t)$ in powers of λ:

$$a_n = a_n^{(0)} + \lambda a_n^{(1)} + \lambda^2 a_n^{(2)} + \dots \qquad (12 \cdot 5\text{-}12)$$

We assume the series to be analytic in λ between 0 and 1. Substitution in Eq. (12·5-10) gives

$$\dot{a}_k^{(0)} + \lambda \dot{a}_k^{(1)} + \lambda^2 \dot{a}_k^{(2)} + \dots$$

$$= \frac{\lambda}{i\hbar} \sum_n \{a_n^{(0)} + \lambda a_n^{(1)} + \lambda^2 a_n^{(2)} + \dots\} H'_{kn}(t) \exp(i\omega_{kn} t)$$

Equating the coefficients of λ on either side of Eq. (12·5-12) we get the uncoupled equations

$$\dot{a}_k^{(0)} = 0 \qquad (12 \cdot 5\text{-}13a)$$

$$\dot{a}_k^{(1)} = \frac{1}{i\hbar} \sum_n a_n^{(0)} H'_{kn}(t) \exp(i\omega_{kn} t) \qquad (12 \cdot 5\text{-}13b)$$

$$\dot{a}_k^{(2)} = \frac{1}{i\hbar} \sum_n a_n^{(1)} H'_{kn}(t) \exp(i\omega_{kn} t) \qquad (12 \cdot 5\text{-}13c)$$

The first of these equations shows that the zero order coefficients $a_k^{(0)}$ are constant as already stated. The first order equation (12·5-13b) gives the coefficient $a_k^{(1)}$. We assume that initially (*i.e.* at $t = 0$), the system is definitely in the state l with energy E_l so that

$a_n^{(0)} = \delta_{nl}$ (for discrete states)

$a_n^{(0)} = \delta(n - l)$ (for continuum of states)

where δ_{nl} is the Kroenecker delta while $\delta(n - l)$ is the delta function.

Then from Eq. (12·5-13b) we have

$$\dot{a}_k^{(1)} = \frac{1}{i\hbar} \sum_n \delta_{nl} \exp(i\omega_{kn} t) H'_{kn}(t)$$

$$= -\frac{i}{\hbar} H'_{kl}(t) \exp(i\omega_{kl} t) \qquad (12 \cdot 5\text{-}14)$$

Integration of Eq 12·5-14) gives

$$a_k^{(1)}(t) = -\frac{i}{\hbar} \int_0^t H'_{kl}(t') \exp(i\omega_{kl}t') \, dt' \qquad (12\cdot5\text{-}15)$$

12·6. Constant perturbation lasting for a finite time

Suppose a constant perturbation is switched on at time $t' = 0$ and lasts for a time t. Then for $k \neq l$ we have

$$\begin{aligned}a_k^{(1)}(t) &= -\frac{i}{\hbar} H'_{kl} \int_0^t \exp(i\omega_{kl}t') \, dt' \\ &= -\frac{i}{\hbar} H'_{kl} \frac{\exp(i\omega_{kl}t) - 1}{i\omega_{kl}} \\ &= \frac{H'_{kl}}{\hbar\omega_{kl}} \{1 - \exp(i\omega_{kl}t)\}\end{aligned} \qquad (12\cdot6\text{-}1)$$

Then the first order probability of transition from the state l to the state k is given by

$$\begin{aligned}p_{kl}(t) &= |a_k^{(1)}(t)|^2 \\ &= |2H'_{kl}|^2 \frac{\sin^2(\omega_{kl}t/2)}{\hbar^2 \omega_{kl}^2}\end{aligned} \qquad (12\cdot6\text{-}2)$$

We write $x = \omega_{kl}t/2$. Then Eq. (12·6-2) can be rewritten as

$$p_{kl}(t) = \frac{|H'_{kl}|^2}{\hbar^2} t^2 \frac{\sin^2 x}{x^2} \qquad (12\cdot6\text{-}3)$$

For a given t the function $p_{kl}(t)$ given by Eq.(12·6-3) has the form shown in Fig. 12·1. It shows a sharp primary peak at $\omega_{kl} = 0$ with secondary peaks on either side. The primary peak has the height $|H'_{kl}|^2 t^2/\hbar^2$ which increases with increasing t. The half width of the peak has the value $\Delta\omega = 2\pi/t$ and decreases with increasing t. In the limit of $t \to \infty$, the peak Fig. 12·1. reduces to a δ-function.

The above discussion shows that the probability of transition from the initial state l to a final state has appreciable value only if the final level k has an energy E_k close to the initial energy E_l of the system. It is maximum for $E_l = E_k$.

The total probability of transition from the initial state l to any one of the final states k is obtained by integration over the final states, assuming the latter to form a continuum. Consider a range of energy ΔE for the final states in which the number of levels is Δn_k (see Fig. 12·2). The probability of transition to these Δn_k states is obtained by summing over all these states :

Approximate Methods in Quantum Mechanics

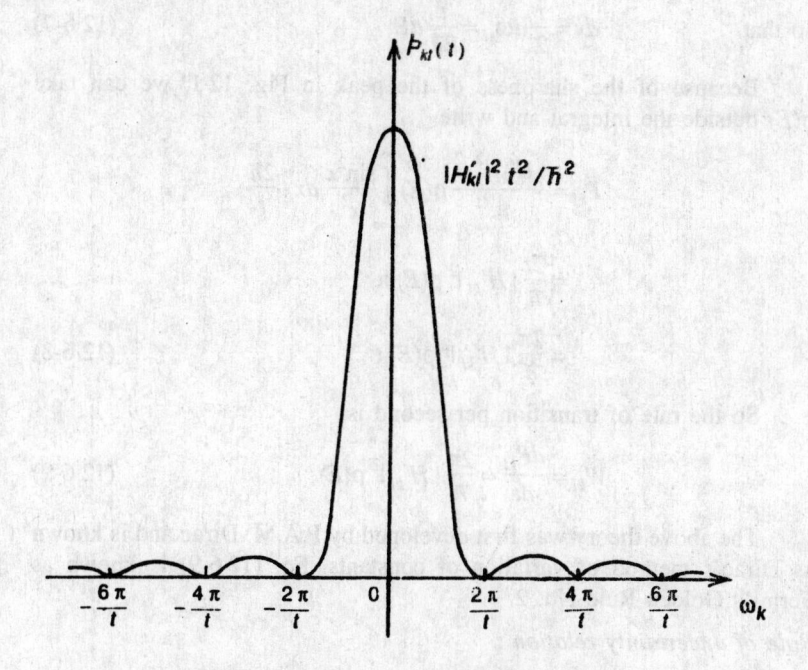

Fig. 12·1. Variation of 'transition amplitude'.

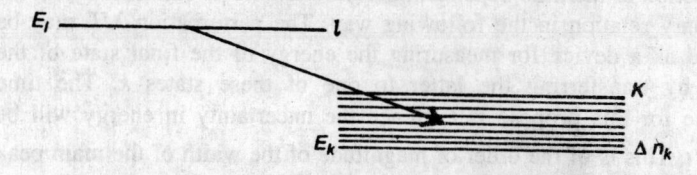

Fig. 12·2. Transition to final state.

$$p_{kl} \Delta n_k = p_{kl} \frac{\Delta n_k}{\Delta E} \Delta E \qquad (12\cdot 6\text{-}4)$$

If $\rho(E)$ denotes the density of the states we can write

$$\rho(E) = \lim_{\Delta E \to 0} \frac{\Delta n_k}{\Delta E} = \frac{dn_k}{dE} \qquad (12\cdot 6\text{-}5)$$

So the total probability of transition to any of the final states becomes

$$P_{kl} = \int_{-\infty}^{\infty} p_{kl} \rho(E) \, dE = \frac{|H'_{kl}|^2 t^2}{\hbar^2} \int \frac{\sin^2 x}{x^2} \rho(E) \, dE \qquad (12\cdot 6\text{-}6)$$

where $\qquad x = \omega_{kl} t/2$

so that
$$dx = \frac{t}{2} d\omega_{kl} = \frac{t}{2\hbar} dE \qquad (12\cdot 6\text{-}7)$$

Because of the sharpness of the peak in Fig. 12·1, we can take $\rho(E)$ outside the integral and write

$$P_{kl} = \frac{|H'_{kl}|^2 t^2}{\hbar^2} \rho(E) \int_{-\infty}^{\infty} \frac{\sin^2 x}{x^2} dx \cdot \frac{2\hbar}{t}$$

$$= \frac{2t}{\hbar} |H'_{kl}|^2 \rho(E) \pi$$

$$= \frac{2\pi}{\hbar} |H'_{kl}|^2 \rho(E) t \qquad (12\cdot 6\text{-}8)$$

So the rate of transition per second is

$$W_{kl} = \frac{dP_{kl}}{dt} = \frac{2\pi}{\hbar} |H'_{kl}|^2 \rho(E) \qquad (12\cdot 6\text{-}9)$$

The above theory was first developed by P.A.M. Dirac and is known as Dirac's method of variation of constants. Eq. (12·6-9) is known as Fermi's Golden Rule No. 2.

Role of uncertainty relation :

The energy spread in the final states k to which the transitions occur by the action of the time dependent perturbation $\lambda H'$ is connected with the uncertainty relation in the following way. The perturbation $\lambda H'$ may be regarded as a device for measuring the energy of the final state of the system by transferring the latter to one of these states k. The time available for this propose is t. Hence the uncertainty in energy will be $\Delta E \sim \hbar/t$. This is of the order of magnitude of the width of the main peak in Fig. 12·1 which is

$$\Delta w = \frac{4\pi}{t} = \frac{\Delta E}{\hbar}$$

This means that there is a whole lot of final states to which the transitions may occur consistent with the energy conservation principle with an uncertainty of magnitude $\Delta E \sim \hbar/t$. Larger t is, smaller is ΔE which becomes vanishingly small ($\Delta E \to 0$) for $t \to \infty$. Thus under reasonance condition ($\Delta \omega_{kl} = 0$), the transition will be of very long duration. In this case the radiation field ensures energy conservation by the creation and annihilation of photons. In the case discussed above conservation of energy is assured by the assumption that the transitions occur between states which are closely packed around $E_k = E_l$. (See *Quantum Mechanics* by L.I. Schiff).

12·7 Variational method

This method is applicable for the approximate calculation of the ground state energy of a system when no exact solution can be obtained for a problem closely related to the problem under consideration. Obviously the stationary perturbation method can not be applied in this case.

Let u_n be the eigenfunction of the Hamiltonian of the problem under consideration belonging to the eigenvalue E_n, so that we have

$$H u_n = E_n u_n \tag{12·7-1}$$

An arbitrary function Ψ can be expanded in terms of the complete set of orthonormal functions u_n:

$$\Psi = \sum_n a_n u_n \tag{12·7-2}$$

Then the expectation value of the energy for the state represented by the function Ψ is

$$\begin{aligned}
<H> &= <\Psi|H|\Psi> \\
&= <\sum_n a_n u_n | H | \sum_m a_m u_m > \\
&= \sum_n \sum_m a_n^* a_m <u_n | H | u_m> \\
&= \sum_n \sum_m a_n^* a_m E_m <u_n | u_m> \\
&= \sum_{nm} a_n^* a_m E_m \delta_{nm}
\end{aligned}$$

$$\therefore \quad <H> = \sum_n |a_n|^2 E_n \tag{12·7-3}$$

Since Ψ is normalized, we get

$$\begin{aligned}
<\Psi|\Psi> &= <\sum_n a_n u_n | \sum_m a_m u_m> \\
&= \sum_n \sum_m a_n^* a_m <u_n | u_m> \\
&= \sum_n \sum_m a_n^* a_m \delta_{nm} \\
&= \sum_n |a_n|^2 = 1
\end{aligned} \tag{12·7-4}$$

We now subtract the ground state energy E_0 from $<H>$:

$$<H> - E_0 = \sum_n |a_n|^2 E_n - E_0$$

$$= \sum_n |a_n|^2 (E_n - E_0) \geq 0 \tag{12·7-5}$$

since E_0 is the lowermost energy state so that $E_n > E_0$ for all states. The inequality (12·7-5) tells us that the upper bound to the ground state energy is $<H>$.

The variational method consists of using a suitable *trial function* Ψ to calculate $<H>$ which depends on a number of parameters $p_1, p_2, p_3, \ldots p_k$ and of minimising $<H>$ with the help of the set of k equations :

$$\frac{\partial <H>}{\partial p_1}=0, \frac{\partial <H>}{\partial p_2}=0, \frac{\partial <H>}{\partial p_3}=0, \ldots \frac{\partial <H>}{\partial p_k}=0 \qquad (12\cdot7\text{-}6)$$

$<H>_{min}$ so obtained gives an upper bound to the ground state energy E_0. With a judicious choice of the trial function Ψ, the upper bound $<H>_{min}$ may be brought quite close to the actual ground state energy E_0.

Application to excited states :

Suppose the trial function Ψ is orthogonal to the first N eigenfunctions $u_0, u_1, u_2, \ldots u_{N-1}$ belonging to the respective energy eigenvalues $E_0, E_1, E_2, \ldots E_{N-1}$ in ascending order. Then it is possible to calculate the upper bounds to the energies for the higher states $N, N+1, N+2, \ldots$ etc.

We choose the trial function as

$$\Psi = \phi - \sum_{n=0}^{N-1} u_n^* \phi \, d\tau \qquad (12\cdot7\text{-}7)$$

where ϕ is an arbitrary function. Then

$$<u_m|\Psi> = <u_m|\phi> - \sum_{n=0}^{N-1} <u_m|u_n><u_n|\phi>$$

$$= <u_m|\phi> - \sum_{n=0}^{N-1} \delta_{mn} <u_n|\phi>$$

$$= <u_m|\phi> - <u_m|\phi> = 0 \qquad (12\cdot7\text{-}8)$$

Thus Ψ is orthogonal to u_m for $m < N-1$. We write

$$\Psi = \sum_n a_n u_n \qquad (12\cdot7\text{-}9)$$

$$<u_m|\Psi> = <u_m|\sum_n a_n u_n> = \sum_n a_n <u_m|u_n>$$

$$= \sum_n a_n \delta_{mn} = a_m$$

i.e., for $\qquad m < N-1, \quad a_m = 0 \qquad (12\cdot7\text{-}10)$

So $a_1 = a_2 = a_3 = \ldots = a_{N-1} = 0$

$\therefore \quad <H> = <\Psi|H|\Psi> = <\sum_n a_n u_n|H|\sum_m a_m u_m>$

$$= \sum_n \sum_m a_n^* a_m <u_n|H|u_m> = \sum_n \sum_m a_n^* a_m E_m <u_n|u_m>$$

$$= \sum_n \sum_m a_n^* a_m E_m \delta_{nm} = \sum_n |a_n|^2 E_n \qquad (12\cdot 7\text{-}11)$$

Since Ψ is normalized

$$<\Psi|\Psi> = \sum_n |a_n|^2 = 1 \qquad (12\cdot 7\text{-}12)$$

So we can write

$$<H> - E_N = \sum_n |a_n|^2 E_n - E_N = \sum_n |a_n|^2 (E_n - E_N) \qquad (12\cdot 7\text{-}13)$$

Since E_N, E_{N+1}, E_{N+2} .... etc. are the energies of the levels $N, N+1, N+2,$ .... etc. in the ascending order, $E_n > E_N$ with $n = N+1, N+2,$ .... etc.

Hence $\qquad <H> - E_N \geq 0 \qquad (12\cdot 7\text{-}14)$

Thus $<H>$ gives the upper bound for E_N. As before by a judicious choice of the function ϕ, it is possible to minimise $<H>$ by adjustment of the parameters $p_1, p_2, p_3,$ .... etc. through the conditions (12·7-6) so that $<H>$ is quite close to the value of the excited state energy.

12·8. WKB approximation

(a) Criterion of applicability :

When the potential acting on a particle changes sharply there is usually reflection of the particle at the point of discontinuity similar to the reflection of light at the surface of separation between two media of different refractive indices. On the other hand, a slowly varying potential corresponds to a medium in which the refractive index varies slowly and continuously. There is no reflection in such a medium ; only a gradual bending of the ray occurs so that the ray is curved.

We first investigate the criterion for the non-occurrence of reflection. The necessary condition is that the wavelength λ changes by a small fraction of itself in a distance of the order of λ i.e.

$$\Delta \lambda = \left|\frac{\partial \lambda}{\partial x} \lambda\right| << \lambda \qquad (12\cdot 8\text{-}1)$$

or, $\qquad \left|\frac{\partial \lambda}{\partial x}\right| << 1 \qquad (12\cdot 8\text{-}2)$

Similar criterion applies in the case of the de Broglie wavelength $\lambda = h/p$ of a particle. The above condition then reduces to

$$\left|\frac{\partial \lambda}{\partial x}\right| = \left|\frac{h}{p^2} \frac{\partial p}{\partial x}\right| << 1 \qquad (12\cdot 8\text{-}3)$$

Eq. (12·8-3) is known as the *semi-classical approximation*. It is not applicable at points where the momentum p, has small values such as at the point where $E = V$ so that the momentum $p = \sqrt{2m(E-V)}$.

In this case the de Broglie wavelength $\lambda = h/p \to \infty$ so that the property of the particle is strongly manifested. Small changes in x from this point makes λ finite so that λ is very large even for a small value of Δx and the criterion $|\Delta \lambda|/\Delta x| << 1$ can not be satisfied.

Since $p^2 = 2m(E-V)$, we get $p\,dp = -m\,dV$

so that
$$\frac{\partial p}{\partial x} = -\frac{m}{p}\frac{\partial V}{\partial x} \tag{12·8-4}$$

Thus the condition (12·8-3) reduces to

$$\left|\frac{mh}{p^3}\frac{\partial V}{\partial x}\right| = \left|\frac{mh/p}{2m(E-V)}\frac{\partial V}{\partial x}\right| << 1$$

or,
$$\left|\frac{\lambda}{2(E-V)}\frac{\partial V}{\partial x}\right| << 1$$

$$\therefore \quad \frac{\lambda}{2}\left|\frac{\partial V}{\partial x}\right| << (E-V) \tag{12·8-5}$$

So the change in the potential V in a distance of the order of λ must be small compared to the kinetic energy $(E - V)$ of the particle. This means that for no reflection to occur, the potential V must change slowly. In three dimensions, this would correspond to a curved trajectory for the particle (or the wave packet). A classical description is adequate in this case.

It may be noted that such a classical description is possible because h is small compared to $|p^2 \partial p/\partial x|^{-1}$ (see Eq. 12·8-3). If however h were large compared to the above quantity then the quantum phenomena would be manifested even in the macroscopic world.

(b) Method of classical approximation :

The method of classical approximation has been developed, among others, by G. Wentzel, H.A. Kramers and L. Brillouin and is commonly known as the *WKB approximation*.

We have seen above that the method is applicable when the potential $V(x)$ is very slowly varying. Hence the wave function $\Psi(x)$ is not much different from the case in which V is constant. In the latter case, the wave function is

$$\Psi = \exp(ikx) = \exp(ipx/\hbar) \tag{12·8-6}$$

where $p = \pm\sqrt{2m(E-V)}$, V being constant.

We therefore write the wave function in the case of a slowly varying potential

$$\Psi(x) = \exp\{i\,S(x)/\hbar\} \tag{12·8-7}$$

Approximate Methods in Quantum Mechanics 351

where $S(x)$ is a slowly varying function and must reduce to the product px in the limit of a constant potential.

The Schrödinger equation for one-dimensional motion is

$$\frac{d^2 \Psi}{dx^2} + \frac{2m}{\hbar^2} \{E - V(x)\} \Psi = 0$$

We insert the values of Ψ and $d^2\Psi/dx^2$ into this from Eq. (12·8-7):

$$\frac{d\Psi}{dx} = \frac{i S'(x)}{\hbar} \exp\{iS(x)/\hbar\} = \frac{i}{\hbar} S'(x) \Psi \qquad (12·8-8)$$

$$\frac{d^2\Psi}{dx^2} = \left(\frac{i}{\hbar} S'' - \frac{S'^2}{\hbar^2}\right) \exp\{iS(x)/\hbar\}$$

$$= \left(\frac{i}{\hbar} S'' - \frac{S'^2}{\hbar^2}\right) \Psi \qquad (12·8-9)$$

Hence we get from the Schrödinger equation.

$$\left(\frac{i}{\hbar} S'' - \frac{S'^2}{\hbar^2}\right) \Psi + \frac{2m}{\hbar^2} \{E - V(x)\} \Psi = 0$$

or, $$\frac{1}{2m} S'^2 + (V - E) - \frac{i\hbar}{2m} S'' = 0 \qquad (12·8-10)$$

Eq. (12·8-10) cannot be solved exactly. However if the last term is small, which is true in the semiclassical approximation, we can solve this equation by an expansion of $S(x)$ in a power series of $\hbar$:

$$S(x) = S_0(x) + \hbar S_1(x) + \frac{\hbar^2}{2!} S_2(x) + ... \qquad (12·8-11)$$

For the method to yield a reasonably good approximation, the terms $\hbar S_1/S_0, \hbar S_2/2S_1$ etc. must be assumed to be small. Eq. (12·8-11) gives

$$S' = S'_0 + \hbar S'_1 + \frac{\hbar^2}{2!} S'_2 + ... \qquad (12·8-12a)$$

$$S'' = S''_0 + \hbar S''_1 + \frac{\hbar^2}{2!} S''_2 + ... \qquad (12·8-12b)$$

Inserting in Eq. (12·8-10), we have

$$\frac{1}{2m} (S'_0 + \hbar S'_1 + \frac{\hbar^2}{2!} S'_2 + ...)^2 + (V - E)$$

$$= \frac{i\hbar}{2m} (S''_0 + \hbar S''_1 + \frac{\hbar^2}{2!} S''_2 + ...)$$

or, $\dfrac{1}{2m} \{S_0'^2 + 2\hbar S_0' S_1' + \hbar^2 (S_1'^2 + S_0' S_2') + ...\}$

$$+ V - E = \dfrac{i}{2m} (\hbar S''_0 + \hbar^2 S''_1 + ...) \qquad (12\cdot 8\text{-}13)$$

Equating the coefficients of different powers of $\hbar$ on the two sides of Eq. (12·8-13), we get

$$\dfrac{S_0'^2}{2m} + V - E = 0 \qquad (12\cdot 8\text{-}14a)$$

$$\dfrac{S_0' S_1'}{m} = \dfrac{i}{2m} S''_0 \qquad (12\cdot 8\text{-}14b)$$

$$\dfrac{1}{2m}(S_1'^2 + S_0' S_2') = \dfrac{i}{2m} S''_1 \qquad (12\cdot 8\text{-}14c)$$

and so on.

The first of the above equations gives

$$S_0'^2 = 2m(E - V)$$
or, $$S_0' = \pm \sqrt{2m(E - V)} \qquad (12\cdot 8\text{-}15)$$

This gives

$$S_0 = \pm \int_{x_0}^{x} \sqrt{2m(E - V)}\, dx \qquad (12\cdot 8\text{-}16)$$

From the second of the above equations (12·8-14b) we get

$$S_1' = \dfrac{i\, S''_0}{2\, S_0'} = \dfrac{i}{2\, S_0'} \dfrac{\partial S_0'}{\partial x} \qquad (12\cdot 8\text{-}17)$$

This gives $\qquad S_1 = \dfrac{i}{2} \ln S_0' \qquad (12\cdot 8\text{-}18)$

Then $\qquad i S_1 = -\dfrac{\ln S_0'}{2} = \ln (S_0')^{-1/2}$

or, $\qquad \exp(i S_1) = (S_0')^{-1/2} = \dfrac{1}{\sqrt{S_0'}}$

$$= \dfrac{1}{\{2m(E-V)\}^{1/4}} \qquad (12\cdot 8\text{-}19)$$

Retaining only the first two terms in the expansion for $S(x)$ in Eq. (12·8-11), we then obtain from Eq. (12·8-7)

Approximate Methods in Quantum Mechanics

$$\Psi(x) = \exp\{iS(x)/\hbar\} = \exp\left\{\frac{i}{\hbar}(S_0 + \hbar S_1)\right\}$$

$$= \exp(iS_1)\exp\left(\frac{iS_0}{\hbar}\right) \qquad (12\cdot8\text{-}20)$$

Using Eqs. (12·8-16) and (12·8-19), we then get

$$\Psi(x) = \frac{1}{(E-V)^{1/4}}\left\{A\exp\left(\frac{iS_0}{\hbar}\right) + B\exp\left(-\frac{iS_0}{\hbar}\right)\right\}$$

$$= \frac{1}{(E-V)^{1/4}}\left[A\exp\left\{\frac{i}{\hbar}\int_{x_0}^{x}\sqrt{2m(E-V)}\,dx\right\}\right.$$

$$\left. + B\exp\left\{-\frac{i}{\hbar}\int_{x_0}^{x}\sqrt{2m(E-V)}\,dx\right\}\right] \qquad (12\cdot8\text{-}21)$$

Here we have provided for both positive and negative signs in Eq (12·8-16).

Eq. (12·8-21) is not applicable at the point where $E = V$, since the denominator becomes zero at this point. This poses a serious difficulty in applying the WKB method to the problem of penetration of a potential barrier of arbitrary shape which is one of the main fields of application of this method.

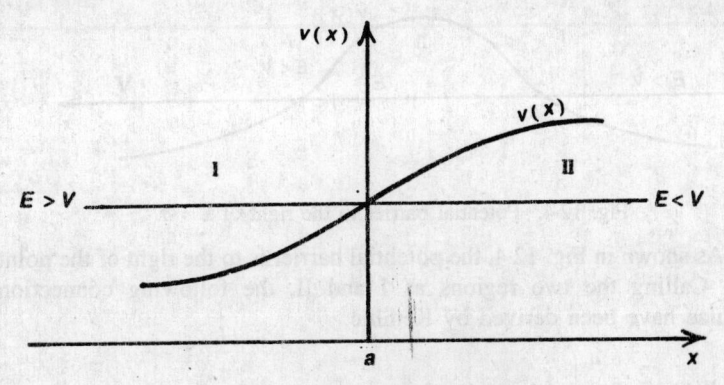

Fig. 12·3. Slowly varying potential.

Referring to Fig. (12·3) we note that in region I ($E > V$), the wave function should be oscillating while in region II ($E < V$), it should be a decreasing exponential. These conclusions follow from the facts that in region I, S_0 is real (see Eq. 12·8-16) so that the $\exp(iS_0/\hbar)$ in Eq. (12·8-21) should be an oscillating function of x. On the other hand in region II, S_0 is imaginary and $\exp(iS_0/\hbar)$ is a decreasing exponential. The two solutions must be matched at the boundary $x = a$ where $E = V$. But this is a point of singularity so that the WKB method fails here as we saw above.

The way out of the difficulty is to exclude the two solutions in the two regions in the immediate neighbourhood of the singularity $x = a$ and to derive the connection formula to join $\Psi (E > V)$ with $\Psi (E < V)$ at this point smoothly.

Assuming the region of inapplicability of the WKB method to be small, we assume a linear variation of the potential in the region so that we write for points near $x = a$

$$V - E = \left(\frac{\partial V}{\partial x}\right)_a (x - a) \qquad (12 \cdot 8\text{-}22)$$

The Schrödinger equation becomes

$$\frac{d^2 \Psi}{dx^2} + C(x - a) = 0 \qquad (12 \cdot 8\text{-}23)$$

where C is a constant. This equation can be solved by using Bessel functions of order 1/3.

(e) Connection formulae*
(i) Barrier to the right :

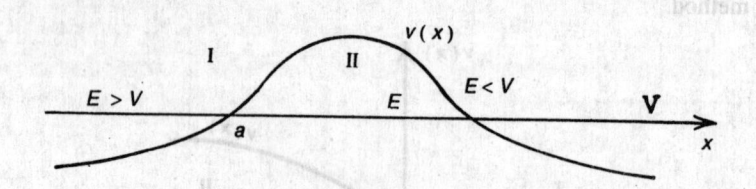

Fig. 12·4. Potential barrier to the right of $x = a$.

As shown in Fig. 12·4, the potential barrier is to the right of the point $x = a$. Calling the two regions as I and II, the following connection formulae have been derived by Kemble.

$$\Psi_I = \frac{2}{\sqrt{k_1}} \cos\left(\int_x^a k_1 \, dx - \pi/4\right)$$

$$\rightleftarrows \Psi_{II} = \frac{1}{\sqrt{k_2}} \exp\left(-\int_a^x k_2 \, dx\right) \qquad (12 \cdot 8\text{-}24a)$$

---

* See *Fundamental Principles of Quantum Mechanics* by E.C. Kemble.

$$\Psi_I = \frac{1}{\sqrt{k_1}} \sin\left(\int_x^a k_1\, dx - \frac{\pi}{4}\right)$$

$$\rightleftarrows \Psi_{II} = -\frac{1}{\sqrt{k_2}} \exp\left(\int_a^x k_2\, dx\right) \qquad (12\cdot 8\text{-}24b)$$

The symbol $\rightleftarrows$ signifies "goes over to". Also $\hbar k_1 = \sqrt{2m(E-V)}$ and $\hbar k_2 = \sqrt{2m(V-E)}$.

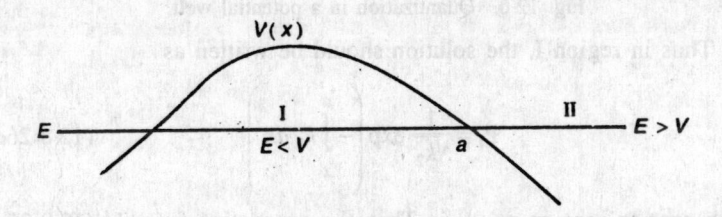

Fig. 12·5. Potential barrier to the left of $x = a$.

(ii) Barrier to the left:

In this case the barrier is in region I to the left of the point $x = a$ (Fig. 12·5). The connection formulae are:

$$\Psi_I = \frac{1}{\sqrt{k_2}} \exp\left(-\int_x^a k_2\, dx\right)$$

$$\rightleftarrows \Psi_{II} = \frac{2}{\sqrt{k_1}} \cos\left(\int_a^x k_1\, dx - \frac{\pi}{4}\right) \qquad (12\cdot 8\text{-}25a)$$

$$\Psi_I = \frac{1}{\sqrt{k_2}} \exp\left(\int_x^a k_2\, dx\right)$$

$$\rightleftarrows \Psi_{II} = -\frac{1}{\sqrt{k_1}} \sin\left(\int_a^x k_1\, dx - \frac{\pi}{4}\right) \qquad (12\cdot 8\text{-}25b)$$

(d) Quantization of the motion of a particle in a potential well of arbitrary shape:

As seen from Fig. 12·6, in regions I and III, $E < V(x)$ so that the wave functions are exponential in nature and must vanish as $x \to \infty$ in region III and as $x \to -\infty$ in region I. In region II, $E > V(x)$ so that the wave function is oscillatory.

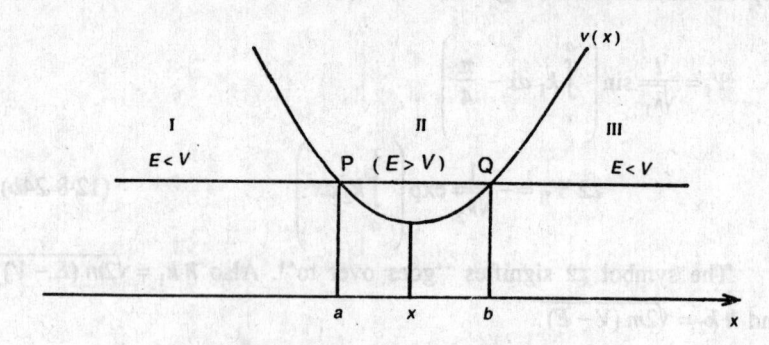

Fig. 12·6. Quantization in a potential well.

Thus in region I, the solution should be written as

$$\Psi_\text{I} = \frac{1}{\sqrt{k_2}} \exp\left(-\int_x^a k_2\, dx\right) \qquad (12\cdot8\text{-}26a)$$

which goes to zero as $x \to -\infty$. Then the connection formula (12·8-25a) gives for region II (barrier to the left).

$$\Psi_\text{II} = \frac{2A}{\sqrt{k_1}} \cos\left(\int_a^x k_1\, dx - \frac{\pi}{4}\right) \qquad (12\cdot8\text{-}26b)$$

On the other hand the solution in region III should be

$$\Psi_\text{III} = \frac{1}{\sqrt{k_2}} \exp\left(-\int_b^x k_2\, dx\right) \qquad (12\cdot8\text{-}27a)$$

which vanishes as $x \to \infty$. The connection formula (12·8-24a) gives for the region II (barrier to the right)

$$\Psi_\text{II} = \frac{2B}{\sqrt{k_1}} \cos\left(\int_x^b k_1\, dx - \frac{\pi}{4}\right) \qquad (12\cdot8\text{-}27b)$$

At any point in region II the expressions (12·8-26b) and (12·8-27b) for Ψ_II and its derivative Ψ'_II should give identical values which is possible by proper choice of the arbitrary constants A and B in Eqs. (12·8-26b) and (12·8-27b) respectively. Thus we get

$$\frac{2A}{\sqrt{k_1}} \cos\left(\int_a^x k_1\, dx - \frac{\pi}{4}\right) - \frac{2B}{\sqrt{k_1}} \cos\left(\int_x^b k_1\, dx - \frac{\pi}{4}\right) = 0 \qquad (12\cdot8\text{-}28a)$$

$$\frac{2Ak_1}{\sqrt{k_1}} \sin\left(\int_a^x k_1\, dx - \frac{\pi}{4}\right) + \frac{2Bk_1}{\sqrt{k_1}} \sin\left(\int_x^b k_1\, dx - \frac{\pi}{4}\right) = 0 \qquad (12\cdot 8\text{-}28b)$$

These are two homogeneous equations in the two unknowns A and B which can have non-trivial solutions if the determinant of the coefficients of A and B in Eqs $(12\cdot 8\text{-}28a)$ and $(12\cdot 8\text{-}28b)$ is zero:

$$\sin\left(\int_x^b k_1\, dx - \frac{\pi}{4}\right)\cos\left(\int_a^x k_1\, dx - \frac{\pi}{4}\right) + \cos\left(\int_x^b k_1\, dx - \frac{\pi}{4}\right)\sin\left(\int_a^x k_1\, dx - \frac{\pi}{4}\right) = 0$$

or, $\qquad \sin\left(\int_x^b k_1\, dx - \frac{\pi}{4} + \int_a^x k_1\, dx - \frac{\pi}{4}\right) = 0$

so that $\qquad \sin\left(\int_a^b k_1\, dx - \frac{\pi}{2}\right) = 0 \qquad (12\cdot 8\text{-}29)$

$$\therefore \qquad \int_a^b k_1\, dx - \frac{\pi}{2} = n\pi, \quad n = 0, 1, 2, \ldots \qquad (12\cdot 8\text{-}30)$$

If we write $\hbar k_1 = p_1$, we have

$$\int_a^b p_1\, dx = \left(n + \frac{1}{2}\right)\frac{h}{2} \qquad (12\cdot 8\text{-}31)$$

where h is Planck's constant.

Since the regions I and III are inaccessible to the particle in the classical limit, it can only oscillate back and forth between $x = a$ to $x = b$. A complete cycle of oscillation thus corresponds to the motion of the particle from $x = a$ to $x = b$ and back to $x = a$. Thus for a complete cycle.

$$\oint p\, dx = \left(n + \frac{1}{2}\right)h \qquad (12\cdot 8\text{-}32)$$

Eq. $(12\cdot 8\text{-}32)$ is similar to the well-known Bohr-Sommerfeld quantum condition except for the additional term $h/2$ on the r.h.s.

(e) *Potential Barrier Penetration* (by WKB method):

Consider a potential barrier of arbitrary shape, as shown in Fig. 12·7. We divide the whole one-dimensional space into three regions:

Region I : $\qquad -\infty < x < a, \quad E > V(x)$

Region II : $\qquad a \leq x \leq b, \quad E < V(x)$

Region III : $\quad b \leq x \leq \infty, \quad E > V(x)$

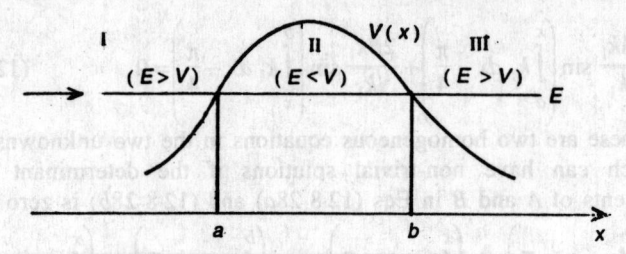

Fig. 12·7. Potential barrier of arbitrary shape.

The wave functions are oscillatory in regions I and III while it is exponential in region II.

We assume that a particle of mass m and energy E to be incident on the barrier from left to right in region I. Then the oscillatory wave function in region III consists only of a wave travelling to the right which is given by (see Eq. 12·8-21)

$$\Psi_{III} = \frac{A}{\sqrt{k_1}} \exp i\left(\int_b^x k_1 \, dx - \pi/4\right)$$

$$= \frac{A}{\sqrt{k_1}}\left[\cos\left(\int_b^x k_1 \, dx - \pi/4\right) + i \sin\left(\int_b^x k_1 \, dx - \frac{\pi}{4}\right)\right] \quad (12\cdot8\text{-}33)$$

The connection formulae (12·8-25a and b) give for region II

$$\Psi_{II} = \frac{A}{2\sqrt{k_2}}\left[\exp\left(-\int_x^b k_2 \, dx\right) - 2i \exp\left(\int_x^b k_2 \, dx\right)\right] \quad (12\cdot8\text{-}34)$$

To find Ψ_I, we use the connection formulae (12·8-24a and b). To do this we rewrite the expression for Ψ_{II} as below :

$$\Psi_{II} = \frac{A}{2\sqrt{k_2}}\left[\exp\left(-\int_a^b k_2 \, dx\right) \cdot \exp\left(\int_a^x k_2 \, dx\right)\right.$$

$$\left. - 2i \exp\left(\int_a^b k_2 \, dx\right) \cdot \exp\left(-\int_a^x k_2 \, dx\right)\right] \quad (12\cdot8\text{-}35)$$

Then we get

$$\Psi_I = -\frac{A}{2\sqrt{k_1}}\left[\exp\left(-\int_a^b k_2 \, dx\right) \sin\left(\int_x^a k_1 \, dx - \pi/4\right)\right.$$

$$+ 4i \exp\left(\int_a^b k_2\, dx\right) \cos\left(\int_x^a k_1\, dx - \pi/4\right)\right]\right] \quad (12\cdot8\text{-}36)$$

If we write

$$\alpha = \int_a^b k_2\, dx,\ \beta = \int_x^a k_1\, dx - \pi/4 \quad (12\cdot8\text{-}37)$$

we get $\quad \Psi_1 = -\dfrac{A}{2\sqrt{k_1}} [\exp(-\alpha)\sin\beta + 4i \exp\alpha \cdot \cos\beta]$

or, $\quad \Psi_1 = -\dfrac{A}{2\sqrt{k_1}} \left[\exp(-\alpha)\dfrac{\exp(i\beta) - \exp(-i\beta)}{2i}\right.$

$$\left. + 4i \exp\alpha\, \dfrac{\exp(i\beta) + \exp(-i\beta)}{2}\right]$$

$$= -\dfrac{iA}{2\sqrt{k_1}}[\{4\exp\alpha + \exp(-\alpha)\}\exp(-i\beta)$$

$$+ \{4\exp\alpha - \exp(-\alpha)\}\exp(i\beta)] \quad (12\cdot8\text{-}38)$$

The $\exp(+i\beta)$ term in Eq. (12·8-38) represents the incident wave in region I while the $\exp(-i\beta)$ term is the reflected wave from the barrier in region I. So we can write the incident wave in region I as

$$\Psi_{in} = -\dfrac{iA}{4\sqrt{k_1}}\{4\exp\alpha - \exp(-\alpha)\}\exp(i\beta) \quad (12\cdot8\text{-}39)$$

The transmission coefficient through the barrier is given by (see § 11·4)

$$T = |J_{tr}/J_{in}| \quad (12\cdot8\text{-}40)$$

where the probability current density

$$J = \dfrac{i\hbar}{2m}\left(\Psi\dfrac{\partial \Psi^*}{\partial x} - \Psi^*\dfrac{\partial \Psi}{\partial x}\right) \quad (12\cdot8\text{-}41)$$

Substituting for $\Psi, \Psi', \Psi^*, \Psi^{*'}$ for the incident wave and the transmitted wave (see Eqs. 12·8-39 and 12·8-33), we get

$$T = 16\{4\exp\alpha - \exp(-\alpha)\}^{-2}$$

i.e., $\quad T = 16\left\{4\exp\left(\int_a^b k_2\, dx\right) - \exp\left(-\int_a^b k_2\, dx\right)\right\}^{-2} \quad (12\cdot8\text{-}41)$

For a barrier of sufficient thickness, we can neglect the negative exponential term which is small. We then get

$$T = 16 \Big/ \left[4 \exp\left(\int_a^b k_2 \, dx\right) \right]^2$$

$$= \exp\left(-2 \int_a^b k_2 \, dx\right)$$

$$= \exp\left(-\frac{2}{\hbar} \int_a^b \sqrt{2m\,[V(x) - E]} \, dx\right) \qquad (12 \cdot 8\text{-}43)$$

For a rectangular barrier of height V_0 and width a, this reduces to Eq. (11·4-14) in Ch XI, except for the factor before the exponential.

Problems

1. The Hamintonian of a one-dimensional anharmonic oscillator has the form

$$H = \frac{p_x^2}{2\mu} + \frac{1}{2} kx^2 + bx^3$$

where μ is the mass; κ and b are constants. Taking the annarmonicity term bx^3 to be a small perturbation, show that the first order correction to the energy of the oscillator is zero while the second order correction to the nth eigenstate is

$$W_n^{(2)} = -\frac{b^2}{8\alpha^2 \mu \omega^2}(30n^2 + 30n + 11)$$

where $\alpha = \mu\omega/\hbar$ (ω is the circular frequency). Hence find the corrected energies of the first three states of the oscillator.

2. A charged linear harmonic oscillator is acted upon by a uniform electric field X. Show by the application of the stationary perturbation theory, that the first order correction to the energy of the oscillator vanishes while the second order correction to the energy of the nth eigenstate is

where e and μ are the charge and mass of the oscillator and ω is the circular frequency.

Note that this problem is solvable exactly. Show that the exact solution gives the same result as is obtained by the approximate method.

3. A plane rigid rotator of charge e and mass μ is acted on by a weak electric field X along the x-axis. Show that the first order correction to its energy is (for the nth state)

$$W_n^{(1)} = \frac{X^2 e\, a^4 \mu}{\hbar^2} \cdot \frac{1}{4n^2 - 1}$$

where a is the distance of the particle from the centre.

4. Using the variational method show that the ground state energy of a three-dimensional oscillator is given by

Approximate Methods in Quantum Mechanics

$$<H>_{min} = \frac{9}{7}\sqrt{\frac{3}{2}}\,\hbar\omega$$

Choose as trial function

$$\Psi = a(1+pr)\exp(-pr)$$

where p is the variational parameter and a is a constant.

5. A one-dimensional potential barrier of arbitrary shape can be divided into a large number of narrow rectangular potential barriers (width Δx) by drawing lines parallel to the ordinate in the V vs. x graph. Show that in the limit of $\Delta x \to 0$, the probability of peneration through the barrier of a particle of mass μ and energy E. can be obtained by multiplying the probabilities of peneration through the successive rectangular barriers of width Δx each and is given by

$$T \propto \exp\left\{-\frac{2}{\hbar}\int_{x_1}^{x_2}\sqrt{2\mu(V-E)}\,dx\right\}$$

where x_1 and x_2 are the limits of the barrier region. (Note that this expression is the same as that obtained by the WKB method).

6. In the free electron theory of the metal, the potential energy of a free electron in the metal acted upon by an electric field X perpendicular to the metal surface can be written as $V(x) = V_0 - Xex$ if we neglect the image force. V_0 is the height of the potential well, assumed rectangular, in which the electrons are distributed in the metal. Show that the probability of penetration of the potential barrier at the metal surface by an electron of energy E (cold emission) is given by

$$T = \exp\left\{-\frac{4}{3}\sqrt{2\mu}\,\frac{(V_0-E)^{3/2}}{Xe\hbar}\right\}$$

7. A one-dimensional potential barrier has the shape

$$V(x) = V_0(1 - x^2/l^2) \text{ for } -l \le x \le +l$$

and $\qquad V = 0$ for $x < -l$ and $x > +l$

Show that the probability of penetration through the barrier for a particle of mass μ and energy E is given by

$$T = \exp\left\{-\frac{\pi l}{\hbar}\sqrt{\frac{2\mu}{V_0}}(V_0 - E)\right\}$$

8. A one-dimensional potential barrier has the shape

$$V = 0 \text{ for } x < 0$$

$$V(x) = \frac{V_0}{1 + x/l} \text{ for } x \ge 0$$

Show by the substitution $\sin^2\phi = E/V(x)$ that the probability of penetration through the barrier for a particle of mass μ and energy E is

$$T \approx \exp\{-G(\pi - 2\phi_0 - \sin 2\phi_0)\}$$

where $\qquad \sin\phi_0 = \sqrt{E/V_0}$ and $G = \frac{V_0 l}{\hbar}\sqrt{\frac{2\mu}{E}}$

13

QUANTUM MECHANICAL THEORY OF ATOMIC STRUCTURE

13·1 Quantum mechanical theory of the fine structure of hydrogen-like atoms

The fine structure of the energy levels of the hydrogen-like atoms is due to relativistic effects. A complete theory of the fine structure must therefore the based on Dirac's theory of the relativistic electron (see Appendix A-XI) in the Coulomb field of the nucleus. However since the calculations are rather lengthy, it is more convenient to use the perturbation method using the relativistic Hamiltonian and retaining terms upto first order in v^2/c^2.

The relativistic Hamiltonian for the electron in an electromagnetic field can be written as

$$H = H_0 + H' \qquad (13\cdot1\text{-}1)$$

where H_0 is the unperturbed (non-relativistic) Hamiltonian :

$$H_0 = \frac{p^2}{2m} + V(r) \qquad (13\cdot1\text{-}2)$$

H' may be regarded as the perturbation given by

$$H' = H_1' + H_2' + H_3' \qquad (13\cdot1\text{-}3)$$

where

$$H_1' = -p^4/8m^3 c^2 \qquad (13\cdot1\text{-}4)$$

$$H_2' = \frac{1}{2m^2 c^2} \cdot \frac{1}{r}\frac{dV}{dr}(\mathbf{L}\cdot\mathbf{S}) \qquad (13\cdot1\text{-}5)$$

$$H_3' = \frac{\pi \hbar^2}{2m^2 c^2}\left(\frac{Ze^2}{4\pi(\epsilon)_0}\right)\delta(\mathbf{r}) \qquad (13\cdot1\text{-}6)$$

H_1' and H_2' are respectively the relativistic correction to the kinetic energy, and the spin orbit interaction. H_3' is known as the Darwin term. $\delta(\mathbf{r})$ is Dirac's δ-function. It applies to the case $l=0$.

$$V(r) = -\frac{1}{4\pi\epsilon_0}\frac{Ze^2}{r} \qquad (13\cdot1\text{-}7)$$

is the Coulomb potential due to the nucleus. Writing the zero order wave function (without perturbation) as

$$\Psi^{(0)} = \Psi_{nlm_l m_s}$$

we have $\qquad H_0 \Psi^{(0)} = E_n \Psi^{(0)} \qquad (13\cdot1\text{-}8)$

E_n is the unperturbed Schrödinger eigenvalue given by Eq. (11·15–16)

$$E_n = -\frac{mZ^2 e^4}{32 n^2 \pi^2 \hbar^2 \epsilon_0^2} \qquad (13\cdot1\text{-}9)$$

Here m_l is the magnetic quantum number for the orbital angular momentum $l\hbar$, its possible values being $m_l = 0, \pm 1, \pm 2, \ldots \pm l$.

The zero order wave function can be written as

$$\Psi^{(0)} = \Psi_{nlm_l m_s} = \Psi_{nlm_l}\chi_{sm_s} \qquad (13\cdot1\text{-}10)$$

where for spin up, $\qquad \chi_{sm_s} = \chi_{1/2\ 1/2} = \alpha = \begin{pmatrix} 1 \\ 0 \end{pmatrix} \qquad (13\cdot1\text{-}11)$

and for spin down, $\qquad \chi_{sm_s} = \chi_{1/2,\ -1/2} = \beta = \begin{pmatrix} 0 \\ 1 \end{pmatrix} \qquad (13\cdot1\text{-}12)$

(See Appendix A-X). m_s is the magnetic quantum number for the spin angular momentum and can have only two values $m_s = \pm 1/2$, the spin quantum number having the value $s = 1/2$.

Since H_0 does not operate on χ_{sms} it is possible to separate the space and spin functions as in Eq. (13·1-10). Thus the state with a given set of values of n, l, m_l is doubly degenerate and the eigenvalue E_n has a degeneracy of $2n^2$.

Fine structure splitting :

We now consider the effects of the three perturbing terms on the energy of the hydrogen-like atom.

(a) *Effect of H_1'* : The first term on the r.h.s. of Eq. (13·1-3) is

$$H_1' = -\frac{p^4}{8m^3 c^2}$$

The unperturbed level E_n is degenerate with a degeneracy $2n^2$ and hence we have to use the perturbation theory for degenerate levels discussed in § 12·4.

It may however be noted that p^4 commutes with the components of the orbital angular momentum :

$$[p^4, L_x] = [p^4, L_y] = [p^4, L_z] = 0 \qquad (13\cdot1\text{-}13)$$

Further H_1' does not act on the spin variables. The first order energy correction due to H_1' then becomes (relativity correction)

$$\Delta E_{nl} = \left\langle \Psi_{nlm_l} | H_1' | \Psi_{nlm_l} \right\rangle$$

$$= \left\langle \Psi_{nlm_l} \left| \left(-\frac{p^4}{8m^3 c^2} \right) \right| \Psi_{nlm_l} \right\rangle$$

$$= -\frac{1}{2mc^2} \left\langle \Psi_{nlm_l} | E_k^2 | \Psi_{nlm_l} \right\rangle \qquad (13\cdot1\text{-}14)$$

where E_k is the kinetic energy operator given by

$$E_k = \frac{p^2}{2m} = H_0 - V(r) = H_0 + \frac{1}{4\pi\epsilon_0} \frac{Ze^2}{r} \qquad (13\cdot1\text{-}15)$$

So we get

$$\Delta E_{\text{rel}} = -\frac{1}{2mc^2} \Bigg[\left\langle \Psi_{nlm_l} | H_0^2 | \Psi_{nlm_l} \right\rangle + \left(\frac{1}{4\pi\epsilon_0} \right)^2 \left\langle \Psi_{nlm_l} \left| \left(\frac{Ze^2}{r} \right)^2 \right| \Psi_{nlm_l} \right\rangle$$

$$+ \frac{1}{4\pi\epsilon_0} \left\{ \left\langle \Psi_{nlm_l} \left| H_0 \frac{Ze^2}{r} \right| \Psi_{nlm_l} \right\rangle + \left\langle \Psi_{nlm_l} \left| \frac{Ze^2}{r} H_0 \right| \Psi_{nlm_l} \right\rangle \right\} \Bigg]$$

$$= -\frac{1}{2mc^2} \Bigg[E_n^2 \left\langle \Psi_{nlm_l} | \Psi_{nlm_l} \right\rangle + \left(\frac{Ze^2}{4\pi\epsilon_0} \right)^2 \left\langle \Psi_{nlm_l} \left| \frac{1}{r^2} \right| \Psi_{nlm_l} \right\rangle$$

$$+ \frac{Ze^2}{4\pi\epsilon_0} 2E_n \left\langle \Psi_{nlm_l} \left| \frac{1}{r} \right| \Psi_{nlm_l} \right\rangle \Bigg]$$

$$= -\frac{1}{2mc^2} \Bigg[E_n^2 + 2E_n \frac{Ze^2}{4\pi\epsilon_0} \left\langle \frac{1}{r} \right\rangle_{nlm_l} + \left(\frac{Ze^2}{4\pi\epsilon_0} \right)^2 \left\langle \frac{1}{r^2} \right\rangle_{nlm_l} \Bigg] \qquad (13\cdot1\text{-}16)$$

where $\left\langle \frac{1}{r} \right\rangle$ and $\left\langle \frac{1}{r^2} \right\rangle$ are the expectation values which can easily be evaluated by substituting for the wave function Ψ_{nlm_l} (see Ch. XI) :

$$\Psi_{nlm_l} = R_{nl}(r) Y_{lm_l}(\theta, \phi)$$

Quantum Mechanical Theory of Atomic Structure

$$= (-1)^m C_{lm_l} P_l^{m_l}(\mu) \exp(im_l \phi) R_{nl}(r)$$

It is found that (see Physics of Atoms and Molecules by Bransden and Joachain) for an infinitely heavy nucleus :

$$\left\langle \frac{1}{r} \right\rangle_{nlm_l} = \frac{Z}{a_0 n^2} \tag{13$\cdot$1-17}$$

$$\left\langle \frac{1}{r^2} \right\rangle_{nlm_l} = \frac{Z^2}{a_0^2 n^3 (l+1/2)} \tag{13$\cdot$1-18}$$

$$\left\langle \frac{1}{r^3} \right\rangle_{nlm_l} = \frac{Z^3}{a_0^3 n^3 l(l+1/2)(l+1)} \tag{13$\cdot$1-19}$$

Here a_0 is the Bohr radius for an infinitely heavy nucleus (Eq. 4·4·6).

Since $\quad E_n = -\dfrac{mZ^2 e^4}{32 \pi^2 (\rho)_0^2 n^2 \hbar^2} = -\dfrac{mc^2 (Z\alpha)^2}{2n^2}$

we get with the help of Eqs. (13·1-17) and (13·1-18)

$$\Delta E_{\text{rel}} = -\frac{1}{2mc^2} \left[E_n^2 + 2E_n \frac{Ze^2}{4\pi(\rho)_0} \cdot \frac{Z}{a_0 n^2} + \left(\frac{Ze^2}{4\pi(\rho)_0} \right)^2 \frac{Z^2}{a_0^2 n^3 (l+1/2)} \right]$$

$$= -E_n \frac{(Z\alpha)^2}{n^2} \left(\frac{3}{4} - \frac{n}{l+1/2} \right) \tag{13$\cdot$1-20}$$

It should be noted that this expression is very similar to the relativistic correction term obtained by Sommerfeld with k replaced by $(l+1/2)$ (see Eq. 4·12-22). α is fine structure constant (see Eq. 4.12-20)

(b) Effect of the spin-orbit term:

Eq. (13·1-5) for the spin-orbit energy term can be written as

$$\Delta E_{LS} = f(r) \, \mathbf{L} \cdot \mathbf{S} \tag{13$\cdot$1-21}$$

where $\quad f(r) = \dfrac{Ze^2}{4\pi(\rho)_0 \, r^3} \cdot \dfrac{1}{2m^2 c^2} \tag{13$\cdot$1-22}$

From Fig 13·1 we see that

$$J^2 = L^2 + S^2 - 2\mathbf{L} \cdot \mathbf{S}$$

So that

$$\mathbf{L} \cdot \mathbf{S} = \frac{J^2 - L^2 - S^2}{2} \tag{13$\cdot$1-23}$$

We have to substitute Eq.

Fig. 13·1. Compounding the vectors **L** and **S** to get **J**.

(13·1-22) in Eq. (13·1-21) and take the expectation values of **L**·**S** and $1/r^3$:

$$\langle \mathbf{L}\cdot\mathbf{S}\rangle = \frac{j(j+1)-l(l+1)-3/4}{2} \quad (13\cdot1\text{-}24)$$

Then $\Delta E_{LS} = \frac{1}{2m^2 c^2} \cdot \frac{Ze^2}{4\pi(\rho)_0} \cdot \frac{Z^3}{a_0^3 n^3} \cdot \frac{j(j+1)-l(l+1)-3/4}{2l(l+\frac{1}{2})(l+1)} \quad (13\cdot1\text{-}25)$

We then get for the two possible values of $j = l \pm \frac{1}{2}$:

For $j = l + 1/2$, $j(j+1) - l(l+1) - 3/4$
$= (l+1/2)(l+3/2) - l(l+1) - 3/4 = l$

so that

$$\Delta E_{LS} = \frac{1}{2m^2 c^2} \frac{Ze^2}{4\pi(\rho)_0} \frac{Z^3}{a_0^3 n^3} \cdot \frac{l}{2l(l+1/2)(l+1)}$$

$$= -E_n \frac{(Z\alpha)^2}{2nl(l+1/2)(l+1)} \quad (13\cdot1\text{-}26)$$

For $j = l - 1/2$, $j(j+1) - l(l+1) - 3/4$
$= (l-1/2)(l+1/2) - l(l+1) - 3/4 = -(l+1)$

so that

$$\Delta E_{LS} = E_n \frac{(Z\alpha)^2}{2nl(l+1/2)} \quad (13\cdot1\text{-}27)$$

For $l = 0$, the spin-orbit interaction vanishes so that $\Delta E_{LS} = 0$.

(c) Effect of the Darwin term: (H_3'):

Using Eq. (13·1-6) we get for $l = 0$

$$\Delta E_{Dar} = \frac{\pi \hbar^2}{2m^2 c^2} \left(\frac{Ze^2}{4\pi(\rho)_0}\right) \langle \Psi_{n00} | \delta(r) | \Psi_{n00} \rangle$$

$$= -E_n \frac{(Z\alpha)^2}{n}, \quad (l=0) \quad (13\cdot1\text{-}28)$$

Combining the three terms (13·1-19), (13·1-26) or (13·1-27) and (13·1-28), we then have

$$\Delta E_{nj} = \Delta E_{rel} + \Delta E_{LS} + \Delta E_{Dar}$$

$$= E_n \frac{(Z\alpha)^2}{n^2}\left(\frac{n}{j+\frac{1}{2}} - \frac{3}{4}\right) \quad (13\cdot1\text{-}29)$$

The total energy including the fine-structure correction then becomes

$$E_{nj} = E_n + \Delta E_{nj}$$

$$= E_n \left[1 + \frac{(Z\alpha)^2}{n^2} \left(\frac{n}{j+1/2} - \frac{3}{4} \right) \right] \qquad (13\cdot1\text{-}30)$$

Again the similarity of this expression with Sommerfeld's expression (4·12–22) should be noted.

Eqs. (13·1-20), (13·1-26), (13·1-27) and (13·1-28) show that the fine structure splitting due to the three types of perturbation discussed above all depend on the value of the azimuthal quantum number l. However the total splitting given by Eq (13·1-29) does not depend on l, but on the total quantum number $j = l \pm 1/2$ and on n. Larger the values of j and n, smaller is the total splitting. Further it increases with increasing Z.

As mentioned at the beginning of this section, the total fine structure splitting can be calculated exactly by solving the Dirac equation for the Coulomb potential $V(r) = -Ze^2/4\pi\varepsilon_0 r$. Eq. (13·1-29) is found to agree with the more exact formula upto terms of the order of $(Z\alpha)^2$.

13·2. Fine structure of the hydrogen spectral terms

Since j can assume only two values ($j = l \pm 1/2$) for hydrogen-like atoms, it is evident from Eq. (13·1-29) that for two consecutive values of l for which j is the same, viz., $l_1 = j - 1/2$ and $l_2 = j + 1/2$, the fine structure splitting is the same. Thus $^2S_{1/2}$ and $^2P_{1/2}$ terms for the same n have the same energy according to this theory. Similarly $^2P_{3/2}$ and $^2D_{3/2}$ terms for same n have the same energy; and so forth.

In Table 13·1, we list the values of these splittings for hydrogen (Z = 1) in units of $E_n \alpha^2$ and compare them with the experimental values for n = 1, 2 and 3. The agreement is very good (see columns 5 and 7).

Fig. 13·2 shows the positions of the corresponding split levels with the unperturbed energy levels (E_n) shown on the left. The scales are different in the three cases. See also Fig. 6.13. in Fig. 6.13

Table 13·1

n	l	j	Level	ΔE_{nj} (calc)	ΔE_{nj} (obs) cm^{-1}	ΔE_{nj} (obs) $E_n \alpha^2$
1	0	1/2	$1^2S_{1/2}$	1/4	1·46	0·25
2	0	1/2	$2^2S_{1/2}$	5/16	0·445	0·305
	1	1/2	$2^2P_{1/2}$	5/16	0·445	0·305
	1	3/2	$2^2P_{3/2}$	1/16	0·091	0·0624

3	0	1/2	$3^2S_{1/2}$	1/4	0.162	0.25
	1	1/2	$3^2P_{1/2}$	1/4	0.162	0.25
	1	3/2	$3^2P_{3/2}$	1/12	0.054	0.0833
	2	3/2	$3^2D_{3/2}$	1/12	0.054	0.0833
	2	5/2	$3^2D_{5/2}$	1/36	0.018	0.0278

```
n = 3 ─────────────────────────────
                           5/2  0.018  ($3\,^2D_{5/2}$)
                           3/2  0.054  ($3\,^2P_{3/2}$; $3\,^2D_{3/2}$)

                           1/2  0.162  ($3\,^2S_{1/2}$; $3\,^2D_{3/2}$)

n = 2 ─────────────────────────────
                           3/2  0.091  ($2\,^2D_{3/2}$)

                           3/2  0.445  ($2\,^2S_{1/2}$; $2\,^2P_{1/2}$)

n = 1 ─────────────────────────────

                           1/2  1.46   ($1\,^2S_{1/2}$)
```

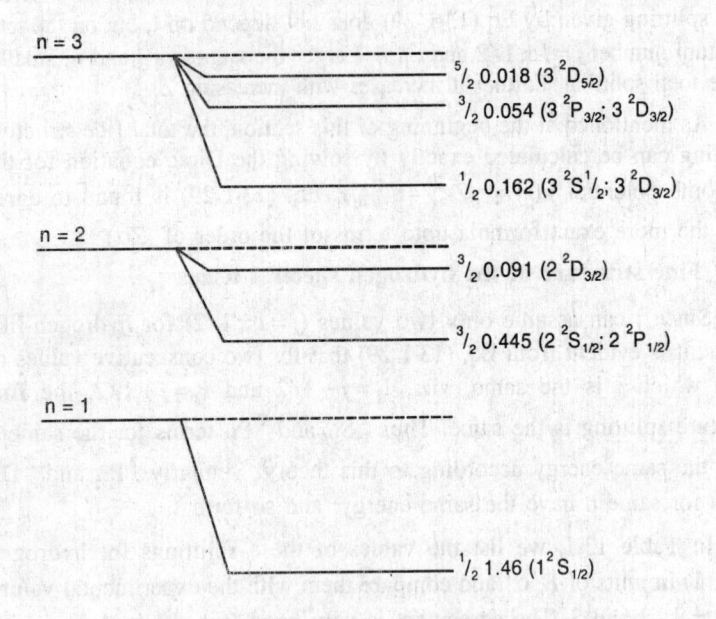

Fig. 13·2. Fine structure splitting of hydrogen levels.

In Fig. 6.13 we show the possible transitions between the upper and lower fine structure levels for the H_α line of hydrogen consistent with selection rules $\Delta l = \pm 1, \Delta j = 0, \pm 1$. As can be seen there should be five components into which the H_α line should split. However the observed pattern does not fully agree with the pattern expected on the basis of Dirac theory.

The problem was finally solved by considering the interaction between the electron and the radiation field.

The theory of fine structure splitting discussed in this section shows that the energy levels of a one electron atom with the same j but different l should coincide. However as early as 1938, there were some experimental evidences which pointed out to the contrary and it was believed that the $2^2 S_{1/2}$ and $2^2 P_{1/2}$ levels did not exactly coincide; the $2^2 S_{1/2}$ level had a slight upward shift (0.03 cm^{-1}) level.

The question was finally settled in 1947 by the very elegant experiment of two American physicists, W.E. Lamb and R.E. Retherford who showed conclusively that the $2^2 S_{1/2}$ level was shifted upwards w.r.t. the $2^2 P_{1/2}$ level by about 1000 MHz. Later more precise measurements gave the energy shift $(2^2 S_{1/2} - 2^2 P_{1/2})$ to be 1057.77 MHz. We shall describe the experiment of Lamb and Retherford in § 13.4.

Lamb shift is usually explained by considering the interaction of the electron with the zero-point vibrations of the electromagnetic field in vacuum. The fluctuations in the electromagnetic field are different from zero even when there are no real photons. As a result of the interaction with vacuum, the electron begins to "vibrate" in its orbit which causes it to spread out in space. So its interaction with the nucleus becomes weaker and the stationary energy levels are raised.

The theory based on second quantization of the electromagnetic field has been developed by H.A. Bethe, J. Tomonaga, J. Schwinger, R.P. Feynman and F.J. Dyson. The calculations are quite complicated. A simpler non-relativistic theory proposed by Welton gives the following expresson for the Lamb shift:

$$\delta E_{\text{vac}} = \frac{4}{3} Z e^2 \alpha \left(\frac{\hbar}{mc}\right)^2 |\Psi(0)|^2 \ln \frac{2n^2}{(Z\alpha)^2} \qquad (13\cdot2\text{-}1)$$

which shows that the shift takes place only for the S-states $(l = 0)$. For higher l values $|\Psi(0)|^2$ vanishes. For l = 0 we get

$$|\Psi(0)|^2 = \frac{Z^3}{\pi n^3 a_0^3} \qquad (13\cdot2\text{-}2)$$

where a_0 is the Bohr radius and n is the principal quantum number. So we get finally

$$\delta E_{\text{vac}} = \frac{8\alpha^3}{3\pi} \cdot \frac{Z^4}{n^3} \cdot R\hbar \ln \frac{2n^2}{(Z\alpha)^2} \qquad (13\cdot2\text{-}3)$$

This formula was first obtained by H.A. Bethe (1947). Here R is the Rydberg constant for an infinitely heavy nucleus.

For a hydrogen-like atom the Lamb shift w.r.t. to the unperturbed level is of the order

$$\frac{\delta E_{\text{vac}}}{|E|} \sim \frac{\alpha^3 Z^4 R \hbar}{R \hbar Z^2} = \alpha^3 Z^2 \qquad (13\cdot2\text{-}4)$$

Since the fine-structure splitting predicted by Dirac theory is $\sim R\hbar Z^4 \alpha^2$, the Lamb-shift is found to be α times smaller (see Eq. 13.2-3).

13.3. Hyperfine Structure of Spectral Lines

When the individual fine structure components are examined under a high resolution spectroscope, they are found to consist of a number of very close lying components known as the hyperfine structure of a spectral line. The splitting between the hyperfine components is of the order of 2 cm^{-1} or ~ 0.4 A° in the visible region of the spectrum.

W. Pauli was the first to point out that hyperfine structure of the spectral lines was due to the influence of the nucleus of the atom.

The influence of the nucleus may be either due to its mass (isotope shift) or due to the nuclear spin. These will be discussed in detail in Vol. II.

13.4. Lamb and Retherford's Experiment

The experiment is based on the microwave method of analysing the relative positions of the $2^2 S_{1/2}$ and $2^2 P_{1/2}$ levels of hydrogen. The method utilizes the metastable nature of the $2^2 S_{1/2}$ level.

Because of the metastability of the $2^2 S_{1/2}$ level, dipole transition from this level to the lower $1^2 S_{1/2}$ level is forbidden by the selection rules ($\Delta l = 0$ in this case). However a transition from the metastable level by two photon emission (life time $\sim 1/7$ s) or through the 2P level is possible. Lamb and Retherford decided to investigate the latter process.

The experimental set up is shown in Fig. 13.3. A beam of hydrogen atoms in the unexcited $1^2 S_{1/2}$ level is obtained through the dissociation of molecular hydrogen at high temperatures (in a tungsten oven O). About 1 in 10^8 atoms of hydrogen is excited to the metastable level by the impinging electron beam E. These metastable atoms readily give away their excitation energy to the electrons in a tungsten target (T), thereby tearing them out to produce a current through the galvanometer (G).

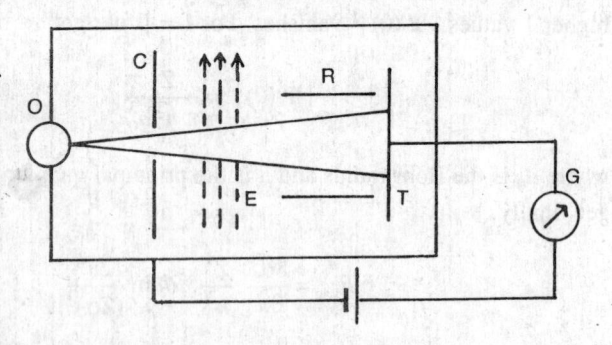

Fig. 13.3. Lamb – Retherford experiment.

If now the metastable atoms are subjected to a radio frequency field (in the region R), they undergo the transitions 2S→2P. These immediately go over to the $1^2 S_{1/2}$ state (in about 1.6×10^{-9} second) so that they move only through a distance of about 1.3×10^{-3} cm (moving with a velocity ~ 8×10^5 cm per second) and are unable to reach the target. As a result the current through the galvanometer decreases.

The transitions 2S→2P were induced by the r.f. field and a stronger absorption was observed at a resonance frequency ω. The energy $\hbar \omega$ thus corresponds to the energy difference between the 2S and 2P (both $2^2 P_{3/2}$ and $2^2 P_{3/2}$) levels. So it was possible to measure the energy difference between the levels with great accuracy and it was definitely established that the $2^2 S_{1/2}$ level was shifted upwards w.r.t. to the $2^2 P_{1/2}$ level by about one-tenth of the energy difference between the $2^2 P_{1/2}$ and $2^2 P_{3/2}$ levels which is equal to $\alpha^2 R/16$.

In the r.f. field region, the atomic beam passes through a variable magnetic field. This made possible the separation not only of the Zeeman components of the $2 S_{1/2}$ and $2 P_{1/2}$ levels but also reduced the possibility of fortuitous disintegration of the $2^2 S_{1/2}$ state due to Stark effect and mixing of the $2 S_{1/2}$ and the 2P levels caused by the perturbing electric fields. The presence of the magnetic field also helped produce the r.f. field of variable frequency but with constant r.f. power. Actually by this method it was possible to operate at a fixed frequency r.f. field and observe the resonance by varying the magnetic field.

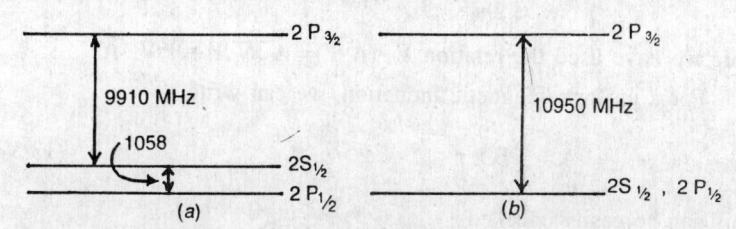

Fig. 13.4. Comparison of the actual splitting of the 2S and 2P levels of hydrogen with predictions from Dirac theory.

The arrangement of the levels of a hydrogen-like atom for $n = 2$ in the Lamb-Retherford experiment is shown in Fig. 13.4 (a). For the sake of comparison the arrangement of the same levels as predicted by Dirac theory is also shown in the figure on the right (b).

Lamb shift has been calculated and measured for other levels of atomic hydrogen as also for other hydrogen-like atoms (e.g., He^+, Li^{2+} etc.)

13·5 Quantum Mechanical Theory of Zeeman effect

The splitting of a spectral line in a magnetic field is known as the Zeeman effect. Two types of effect are known, the normal and the anomalous Zeeman effects. (see Ch. VI). In the first, a spectral line is split up into two or three components and can be explained on the basis of the classical electrodynamics. Quantum theoretical treatment of the problem is based on Sommerfeld's rules for space quantization. The explanation of the anomalous Zeeman effect requires the introduction of the concept of electron spin. The quantum theory of the normal Zeeman effect based on the vector model of the atom was discussed in § 6.2 while that of anomalous Zeman effect in § 6.9. We shall give a quantum mechanical treatment of the problem in the present section.

As shown in Appendix A-IX, the Hamiltonian of a charged particle in an electromagnetic field is (see Eq. A9-16).

$$H = \frac{1}{2m}\left(\frac{\hbar}{i}\nabla - e\mathbf{A}\right)^2 + e\phi \qquad (13\cdot5\text{-}1)$$

where $\mathbf{A}$ is the magnetic vector potential and ϕ is the scalar potential. e is the charge. The Schrödinger equation then becomes

$$\frac{1}{2m}\left(\frac{\hbar}{i}\nabla - e\mathbf{A}\right)^2 \Psi + e\phi\Psi = E\Psi \qquad (13\cdot5\text{-}2)$$

or, $\quad -\frac{\hbar^2}{2m}\nabla^2\Psi + \frac{ie\hbar}{2m}(\nabla\cdot\mathbf{A})\Psi + \frac{ie\hbar}{m}\mathbf{A}\cdot\nabla\Psi$

$$+ \frac{e^2}{2m}A^2\Psi + e\phi\Psi = E\Psi \qquad (13\cdot5\text{-}3)$$

Here we have used the relation $\nabla\cdot(\mathbf{A}\Psi) = \mathbf{A}\cdot\nabla\Psi + \Psi\nabla\cdot\mathbf{A}$.

For a uniform magnetic induction, we can write

$$\mathbf{A} = \frac{1}{2}\mathbf{B}\times\mathbf{r} \qquad (13\cdot5\text{-}4)$$

This can be easily checked :

$$\nabla\times\mathbf{A} = \frac{1}{2}\nabla\times(\mathbf{B}\times\mathbf{r})$$

$$= \frac{1}{2}\{\mathbf{B}(\nabla\cdot\mathbf{r}) - \mathbf{r}(\nabla\cdot\mathbf{B}) + (\mathbf{r}\cdot\nabla)\mathbf{B} - (\mathbf{B}\cdot\nabla)\mathbf{r}$$

The second and the third terms vanish.

Also $\nabla\cdot\mathbf{r} = 3$ and $(\mathbf{B}\cdot\nabla)\mathbf{r} = \mathbf{B}$. Hence $\nabla\times\mathbf{A} = \frac{1}{2}(3\mathbf{B} - \mathbf{B}) = \mathbf{B}$, which agrees with the usual expression for $\mathbf{B}$.

$$-\frac{ie\hbar}{m}\mathbf{A}\cdot\nabla\Psi = \frac{ie\hbar}{2m}(\mathbf{B}\times\mathbf{r})\cdot\nabla\Psi$$

$$= \frac{ie\hbar}{2m}\mathbf{B}\cdot(\mathbf{r}\times\nabla\Psi)$$

$$= -\frac{e}{2m}\mathbf{B}\cdot\left(\mathbf{r}\times\frac{\hbar}{i}\nabla\right)\Psi$$

$$= -\frac{e}{2m}(\mathbf{B}\cdot\mathbf{p_L})\Psi$$

where $\mathbf{p_L} = \mathbf{r}\times\frac{\hbar}{i}\nabla = \mathbf{r}\times\mathbf{p}$ is the angular momentum operator.

The Hamiltonian operator given by Eq. (13·5-1) for an electron in the hydrogen atom then reduces to (since e is negative)

$$\hat{H} = -\frac{\hbar^2}{2m}\nabla^2 + \frac{e}{2m}\mathbf{B}\cdot\mathbf{p_L} + \frac{e^2}{8m}(\mathbf{B}\times\mathbf{r})^2 - e\phi \qquad (13\cdot5\text{-}5)$$

(since $\nabla\cdot\mathbf{A} = 0$ because of gauge invariance)

The second term on the r.h.s. in the above equation is the well-known Zeeman term discussed in § 6·2. This can be easily seen by writing down the expression for the gyromagnetic ratio for the orbital motion of the electron:

$$\frac{\mu_L}{p_L} = \frac{e}{2m}$$

or, $\qquad \mu_L = \frac{e}{2m}p_L$ = magnetic moment for orbital motion.

The quadratic term in B2, known as the diamagnetic term, in Eq. (13·5-5) is usually small compared to the Zeeman term and can be neglected if the field B is not very high (weak field case). So we can write

$$\hat{H} = \hat{H}_0 + \frac{e}{2m}\mathbf{B}\cdot\mathbf{p_L} \qquad (13\cdot5\text{-}6)$$

where $\qquad \hat{H}_0 = -\frac{\hbar^2}{2m}\nabla^2 - e\phi \qquad (13\cdot5\text{-}7)$

Then we get $\qquad H\Psi = E\Psi \qquad (13\cdot5\text{-}8)$

Ψ is the wave function. Expressing μ_L in term of the Bohr magneton $\mu_B = e\hbar/2m$ we then have

$$\hat{H}\Psi = \left(H_0 + \frac{e}{2m}\mathbf{B}\cdot\mathbf{p_L}\right)\Psi$$

$$= \left(H_0 + \frac{e\hbar}{2m}BL_z\right)\Psi \qquad (13\cdot5\text{-}9)$$

$L_z\hbar$ is the component of the orbital angular momentum $\mathbf{p_L} = \mathbf{L}\hbar$ in the direction of the magnetic field $\mathbf{B}$. Writing the magnetic quantum number as m_L we then have

$$H\Psi = (H_0 + m_L\mu_B B)\Psi \qquad (13\cdot5\text{-}10)$$

If Ψ_0 is the unperturturbed wave function and E_0 is the energy of the state before the application of the field, then $\hat{H}_0\Psi_0 = E_0\Psi_0$.

The energy of the state in the presence of the field then becomes
$$E = <H> = <\Psi_0|H|\Psi_0> = <\Psi_0|H_0|\Psi_0> + <\Psi_0|m_L\mu_B B|\Psi_0>$$

$$= E_0 + m_L\mu_B B <\Psi_0|\Psi_0>$$

$$= E_0 + m_L\mu_B B \qquad (13\cdot5\text{-}11)$$

and $$\Delta E = +m_L\mu_B B \qquad (13\cdot5\text{-}12)$$

Since the possible values of m_L are $L, L-1, L-2, \ldots -L+1, -L$, the original state is split up into $(2L+1)$ states of different m_L values, L being the azimuthal quantum number

Eq. (13·5-11) for the energy of the state agrees with that given in § 6·2 obtained on the basis of the vector model for normal Zeeman effect and holds for the weak field case.

The above treatment can be generalized to include the electron spin. Eq. (13·5-6) can then be rewritten as

$$\hat{H} = \hat{H}_0 + \frac{e}{2m}\mathbf{B}\cdot(\mathbf{p_L} + 2\mathbf{p_s}) \qquad (13\cdot5\text{-}13)$$

In Eq. (13·5-13), we have provided for the gyromagnetic ratio for the spin motion being twice that for orbital notion.

Here $\mathbf{p_s} = \mathbf{S}\hbar$ is the spin angular momentum. We then have

$$\hat{H} = \hat{H}_0 + \frac{e\hbar}{2m}B(L_z + 2S_z)$$

$$= \hat{H}_0 + B\mu_B(m_L + 2m_S) \qquad (13\cdot5\text{-}14)$$

where m_S is the magnetic quantum number associated with spin angular momentum.

Writing $J_z = L_z + S_z$ we then have
$$\hat{H} = \hat{H}_0 + B\mu_B g\, m_J \qquad (13\cdot5\text{-}15)$$

where m_J is the magnetic quantum number for the total angular momentum $J\hbar$ where $\mathbf{J} = \mathbf{L} + \mathbf{S}$ and $m_J = m_L + m_S$. g is the Landé splitting factor (see Ch. VI).

The energy of the Zeeman split level then becomes, for the weak field case

$$E = E_0 + B\mu_B\, g\, m_J \qquad (13\cdot 5\text{-}16)$$

and

$$\Delta E = + B\mu_B\, g\, m_J \qquad (13\cdot 5\text{-}17)$$

This is in agreement with the expression obtained in § 6·9. in Ch. VI for the anomalous Zeeman effect.

g is given by Eq. (6·9–3) in Ch. VI.

In the strong field case (Paschen–Back affect), the splitting is given by the second term in the r.h.s. of Eq. (13·5-14) in agreement with Eq. (6·10–1) in Ch. VI. As we shall see in § 13·6, this expression is modified due the spin-orbit interaction.

13·6 Strong Field Case (Paschen–Back Effect)

In the preceding analysis, we did not take into account the spin–orbit interaction which is of the form $f(r)\,\mathbf{L}\cdot\mathbf{S}$ (see § 13·1). We can then write the Hamiltonian for the hydrogen atom as

$$H = H_0 + \frac{e}{2m}\mathbf{B}\cdot(\mathbf{L} + 2\mathbf{S}) + f(r)\,\mathbf{L}\cdot\mathbf{S} \qquad (13\cdot 6\text{-}1)$$

The spin–orbit interaction (the last term on the r.h.s. of Eq. (13·6-1)) may be regarded as a perturbation.

The unperturbed Hamiltonian is

$$H'_0 = H_0 + \frac{e}{2m}\mathbf{B}\cdot(\mathbf{L} + 2\mathbf{S})$$

The wave functions are $\psi(n, l, m_l, m_s)$; L^2, L_z and S_z are well defined for this state and Ψ is an eigenfunction of H'_0. The corresponding energy eigenvalues are (see Eq. 13·5-14)

$$E'_n = E_n + B\mu_B(m_l + 2m_s) \qquad (13\cdot 6\text{-}2)$$

Here m_l can have the $(2l+1)$ values $+l, l-1,\ldots\ldots -l+1, -l$ and $m_s = \pm 1/2$.

Since the expectation values of L_x and L_y are zero for a state with a definite value of L_z we get the expectation value of the perturbation as

$$\langle f(r)\,\mathbf{L}\cdot\mathbf{S}\rangle = \langle f(r)\rangle\langle L_z S_z\rangle$$

$$= \langle f(r)\rangle \langle l\,1/2\,m_l\,m_s|L_z S_z|l\,1/2\,m_l\,m'_s\rangle$$

where $|l\,1/2\,m_l\,m_x\rangle$ is the eigenket of the state of the single electron of orbital angular momentum l and spin angular momentum s and their projections m_l, m_s.

$$\therefore \quad \langle f(r)\, \mathbf{L}\cdot\mathbf{S}\rangle = \langle f(r)\rangle\, m_l\, m_s\, \hbar^2 \qquad (13\cdot 6\text{-}3)$$

Introducing the radial function $R_{nl}(r)$ for hydrogen to calculate $f(r)$ given by Eq. (13·1-22) we get the contribution of the perturbation to total energy as (see Eq. 13·1-19)

$$\Delta E = m_l\, m_s\, \hbar^2 \int_0^\infty [R_{nl}(r)]^2 f(r)\, r^2\, dr$$

$$= \lambda_{nl}\, m_l\, m_s \quad (l \neq 0) \qquad (13\cdot 6\text{-}4)$$

where
$$\lambda_{nl} = -\frac{Z^2 \alpha^2 E_n}{n} \cdot \frac{1}{l(l+1/2)(l+1)} \qquad (13\cdot 6\text{-}5)$$

E_n is the energy of nth state given by Eq. (13·1-9).

For $l = 0$, $\Delta E = 0$.

For a given l, the states of different m_l have different energies. The energy of the atom is given by

$$E'_n = E_n + B\, \mu_B (m_l + 1) + \frac{\lambda_{nl}\, m_l}{2} \quad \text{for } m_s = +1/2$$

$$= E_n + B\, \mu_B (m_l - 1) - \frac{\lambda_{nl}\, m_l}{2} \quad \text{for } m_s = -1/2 \qquad (13\cdot 6\text{-}6)$$

The perturbation due to L–S coupling in a strong magnetic field (Eq. 13·6-4) was neglected in the approximate treatment of the Paschen–Back effect given in § 13·5. For a complete description of the Paschen–Back effect this term must be taken into account in addition to the terms appearing in Eq. (6·10-1).

13·7 Two electron atoms

The study of the two-electron atoms provides an insight into the implications of Pauli's exclusion principle which is of great importance in atomic and molecular physics. Further the mathematical treatment of the problem illustrates the use of the different types of approximation methods in atomic structure calculation discussed in Ch. XII.

The simplest two electron structure consists of a nucleus of charge $+Ze$ and two electrons such as the H^- ion, He atom and Li^+ ion. In these the nuclear charges are $+e$, $+2e$ and $+3e$ respectively.

The potential energy of a system of two electrons in the Coulomb field of the nuclear charge $+Ze$ is

$$V = -\frac{Ze^2}{4\pi\epsilon_0 r_1} - \frac{Ze^2}{4\pi\epsilon_0 r_2} + \frac{e^2}{4\pi\epsilon_0 r_{12}} \qquad (13\cdot 7\text{-}1)$$

We assume the nucleus to be infinitely heavy.

The Schrödinger equation for the two-electron atom can then be written as

$$\left\{ -\frac{\hbar^2}{2m}\nabla_1^2 - \frac{\hbar^2}{2m}\nabla_2^2 - \frac{Ze^2}{4\pi\epsilon_0 r_1} - \frac{Ze^2}{4\pi\epsilon_0 r_2} + \frac{e^2}{4\pi\epsilon_0 r_{12}} \right\} \Psi(\mathbf{r}_1, \mathbf{r}_2)$$

$$= E\Psi(\mathbf{r}_1, \mathbf{r}_2) \quad (13 \cdot 7\text{-}2)$$

$\mathbf{r}_1$ and $\mathbf{r}_2$ being the position vectors of the two electrons. $r_{12} = |\mathbf{r}_1 - \mathbf{r}_2|$ is the relative distance between the two electrons. The last term on the r.h.s. of Eq. (13·7-1) represents the interaction energy between the two electrons.

If the two electrons 1 and 2 are interchanged, the Hamiltonian within the double bracket on the l.h.s. of Eq. (13·7-2) remains unchanged. Hence the wave function $\Psi(\mathbf{r}_2, \mathbf{r}_1)$ satisfies the same wave equation. This is known as *exchange degeneracy*.

We introduce the permutation operator P_{12} such that

$$P_{12}\Psi(\mathbf{r}_1, \mathbf{r}_2) = \Psi(\mathbf{r}_2, \mathbf{r}_1) \quad (13 \cdot 7\text{-}3)$$

If $\Psi(\mathbf{r}_1, \mathbf{r}_2)$ is non-degenerate, then the two functions $\Psi(\mathbf{r}_2, \mathbf{r}_1)$ and $\Psi(\mathbf{r}_1, \mathbf{r}_2)$ can only differ by a constant multiple K, so that we can write

$$\Psi(\mathbf{r}_2, \mathbf{r}_1) = P_{12}\Psi(\mathbf{r}_1, \mathbf{r}_2) = K\Psi(\mathbf{r}_1, \mathbf{r}_2) \quad (13 \cdot 7\text{-}4)$$

or, $$P_{12}^2 \Psi(\mathbf{r}_1, \mathbf{r}_2) = K P_{12} \Psi(\mathbf{r}_2, \mathbf{r}_1)$$

$$= K^2 \Psi(\mathbf{r}_1, \mathbf{r}_2) = \Psi(\mathbf{r}_1, \mathbf{r}_2)$$

So $K^2 = 1$ which gives $K = \pm 1$ and

$$\Psi(\mathbf{r}_2, \mathbf{r}_1) = \pm \Psi(\mathbf{r}_1, \mathbf{r}_2) \quad (13 \cdot 7\text{-}5)$$

The plus sign on the r.h.s. of Eq. (13·7-5) gives the symmetric function $\Psi_S(\mathbf{r}_1, \mathbf{r}_2)$ while the minus sign gives the antisymmetric function $\Psi_A(\mathbf{r}_1, \mathbf{r}_2)$.

We then have

$$\Psi_S(\mathbf{r}_2, \mathbf{r}_1) = +\Psi_S(\mathbf{r}_1, \mathbf{r}_2) \quad (13 \cdot 7\text{-}6a)$$

$$\Psi_A(\mathbf{r}_2, \mathbf{r}_1) = -\Psi_A(\mathbf{r}_1, \mathbf{r}_2) \quad (13 \cdot 7\text{-}6b)$$

Note that the symmetry mentioned above refers to spatial symmetry. Ψ_S and Ψ_A are symmetric and antisymmetric respectively when the spatial coordinates of the two electrons are interchanged. The electron spin is not considered in the above discussion.

Spin wave functions (see Appendix A–X) :

Since the electrons possess spin $s\hbar = \hbar/2$ each, the total wave function Ψ of the two-electron atom is a product of the spatial function

$\Psi(\mathbf{r}_1, \mathbf{r}_2)$ given above and the spin wave function $\chi(\sigma_1, \sigma_2)$ where σ_1 and σ_2 are the Pauli spin operators for the two electrons related to their spin vectors through the equations $\sigma_1 = 2\mathbf{s}_1$ and $\sigma_2 = 2\mathbf{s}_2$. The components of σ_1 and σ_2 are the 2×2 Pauli spin matrices given by §§

$$\sigma_x = \begin{pmatrix} 0 & 1 \\ 1 & 0 \end{pmatrix}, \quad \sigma_y = \begin{pmatrix} 0 & -i \\ i & 0 \end{pmatrix}, \quad \sigma_z = \begin{pmatrix} 1 & 0 \\ 0 & -1 \end{pmatrix} \tag{13.7-7}$$

for each electron.

The spins of the two electrons can be either parallel or antiparallel, giving the total spin $\mathbf{S} = \mathbf{s}_1 + \mathbf{s}_2$ which can have the possible values $S = 1$ or $S = 0$.

The first of these gives a spin triplet $(2S + 1 = 3)$ while the second gives spin singlet $(2S + 1 = 1)$.

Writing the spin functions as α and β for the spin 'up' and spin 'down' states for each electron, we then have the following possible spin functions with the spins aligned parallel to z:

$$\alpha_1 \alpha_2 (\uparrow \uparrow), \quad \alpha_1 \beta_2 = (\uparrow, \downarrow), \quad \beta_1 \alpha_2 = (\downarrow \uparrow), \quad \beta_1 \beta_2 (\downarrow \downarrow) \tag{13.7-8}$$

The possible spin alignments for the two electrons (1 and 2) are shown by the arrows within the brackets. The values of the z component of the resultant spin in the four cases are $S_z = +1, 0, 0$ and -1 respectively.

Of the four spin functions given above, the first $(\alpha_1 \alpha_2)$ and the last $(\beta_1 \beta_2)$ are symmetric in the interchange of the spin coordinates σ_1 and σ_2 of the two electrons. The other two $\alpha_1 \beta_2$ and $\beta_1 \alpha_2$ are neither symmetric nor antisymmetric. However it is possible to make suitable linear combinations of them to get one symmetric and one antisymmetric spin function. Thus we have the following four possible spin functions which are properly symmetrized and nomalized :

Symmetric triplet :
$(S = 1)$
$$\chi_{11} = \alpha_1 \alpha_2, \quad S_z = +1$$
$$\chi_{10} = \frac{\alpha_1 \beta_2 + \beta_1 \alpha_2}{\sqrt{2}}, \quad S_z = 0$$
$$\chi_{1,-1} = \beta_1 \beta_2, \quad S_z = -1$$

Antisymmetric singlet : $\chi_{00} = \dfrac{\alpha_1 \beta_2 - \beta_1 \alpha_2}{\sqrt{2}}, \quad S_z = 0 \tag{13.7-9}$

$(S = 0)$

The subscripts for the total spin functions χ_{S, S_z} refer to the values of S and $S_z = s_{1z} + s_{2z}$ for the two electrons.

---

§§ See Appendix A-X

It can be easily checked that the first three constitute a symmetric triplet with $S = 1$ while the last is an antisymmetric singlet with $S = 0$ when the spins σ_1 and σ_2 are interchanged.

The two spin functions α and β are the *Pauli spinors* given by the column matrices

$$\alpha = \begin{pmatrix} 1 \\ 0 \end{pmatrix}, \quad \beta = \begin{pmatrix} 0 \\ 1 \end{pmatrix} \tag{13.7-10}$$

Then operating by σ_Z given by Eq. (13.7-9), we get

$$\sigma_z \alpha = \alpha, \quad \sigma_z \beta = -\beta \tag{13.7-11}$$

So if we operate the three triplet and one singlet spin functions given by Eq. (13.7-9) by the resultant $S_z = s_{1z} + s_{2z} = \frac{1}{2}(\sigma_{1z} + \sigma_{2z})$ we get §§

$$S_z \alpha_1 \alpha_2 = \frac{1}{2}(\sigma_{1z} \alpha_1) \alpha_2 + \frac{1}{2} \alpha_1 (\sigma_{2z} \alpha_2)$$

$$= \frac{1}{2} \alpha_1 \alpha_2 + \frac{1}{2} \alpha_1 \alpha_2$$

$$= +1 \, (\alpha_1 \alpha_2)$$

$$S_z \frac{\alpha_1 \beta_2 + \beta_1 \alpha_2}{\sqrt{2}} = 0$$

$$S_z \beta_1 \beta_2 = -1 \, (\beta_1 \beta_2)$$

$$S_z \frac{\alpha_1 \beta_2 - \beta_1 \alpha_2}{\sqrt{2}} = 0 \tag{13.7-12}$$

These are in agreement with the results expressed by Eq. (13.7-9)

Effect of Pauli's exclusion principle :

In Ch. VI we saw that the periodic classification of elements is based on Pauli's exclusion principle according to which no two electrons can occupy a state for which all the quantum numbers are the same. This is equivalent to the statement that no two electrons can have all their spatial and spin coordinates ($r_1, r_2, \sigma_1, \sigma_2$) identical.

We shall see later (see Ch. XX) that all elementary particles can be classified into two groups, those obeying Fermi-Dirac (F-D) statistics (e.g. electrons, protons, neutrons etc.) and those obeying Bose-Einstein (B-E) statistics (e.g. photons, pi-mesons α-particles etc.) The first group of particles has half odd-integral spins ($s_e = 1/2, s_p = 1/2, s_n = 1/2$).

The second group has integral spins ($s_\gamma = 1, s_\pi = 0 \, s_\alpha = 0$). The above classification is based on the symmetry properties of the total wave

---

§§ See *ibid.*

function of the particles. For the particles obeying F-D statistics the total wave function is antisymmetric in all the coordinates (space plus spin). For the particles obeying B-E statistics, the total wave function is symmetric in all the coordinates. It will be seen that this behaviour of the wave functions lead to Pauli's exclusion principle for the particles obeying F-D statistics for which the total wave functions Ψ must be anti-symmetric in the interchange of the spatial and spin coordinates of all the identical particles comprising the system. (see Ch. XX).

Reverting to the case of the two-electron atoms, we then see that for $\Psi\ (r_1, r_2, \sigma_1, \sigma_2 = \Psi\ (r_1, r_2) \cdot \chi\ (\sigma_1, \sigma_2)$ to be antisymmetric, the symmetric space function $\Psi_S\ (r_1, r_2)$ must be combined with the single antisymmetric spin function $\chi_A\ (\sigma_1, \sigma_2) = \chi_{00}$ as given in Eq. (13.7.9). Similarly the antisymmetric space function $\Psi_A\ (r_1, r_2)$ has to be combined with any one of the three symmetric spin functions $\chi_S\ (\sigma_1, \sigma_2) = \chi_{11}, \chi_{10}$, or $\chi_{1,-1}$ in Eq. (13.7.9). Thus we get

$$\Psi\ (r_1, r_2, \sigma_1, \sigma_2) = \Psi_S\ (r_1, r_2)\ \chi_A\ (\sigma_1, \sigma_2) \qquad (13\cdot7\text{-}13a)$$

and $\qquad \Psi\ (r_1, r_2, \sigma_1, \sigma_2) = \Psi_A(r_1, r_2)\ \chi_s\ (\sigma_1, \sigma_2) \qquad (13\cdot7\text{-}13b)$

The states described by the symmetric space functions as in Eq. (13.7-13a) are known as *para-states*. For these, the spin functions are anti-symmetric. The states for which the space functions are antisymmetric as in Eq. (13.7-13b) are known as *ortho-states*. The corresponding spin functions are symmetric. Since for the two-electron system, the anti-symmetric spin functions are spin singlets (Eq. 13.7-9), *the para-states are singlets* for them. On the other hand, since the symmetric spin states are triplet, *the ortho-states are triplets*.

13.8 Term scheme for the helium atom

The spectroscopic terms (or the energy levels) of the helium atom form two groups, those corresponding to ortho-helium (triplets) and those corresponding to parhelium (singlets) as discussed above. These are shown in Fig. 13.5.

The spectroscopic terms are denoted, as usual, by the letters S, P, D, F... corresponding to $L = 0, 1, 2, 3, \ldots$ Thus for parhelium we have the terms $1^1S, 2^1S, 3^1S$ etc. for different principal quantum numbers $n = 1, 2, 3, \ldots$; the terms $2^1P, 3^1P, 4^1P$ etc.; the terms $3^1D, 4^1D$ etc. For orthohelium the terms are $2^3S, 3^3S, 4^3S$ etc.; $2^3P, 3^3P, 4^3P$ etc.; $3^3D, 4^3D$ etc. In Fig. 13.5, the different terms are shown at approximately the experimentally determined energy values. The different terms correspond to the electron configuration $1s\,nl$ in which one electron is in the $1s$ (with $n = 1, l = 0$) state while the other is in the nl state.

The ionization potential at $E = 0$ corresponds to a single electron being left in the $1s$ state while the other is just released from the neutral

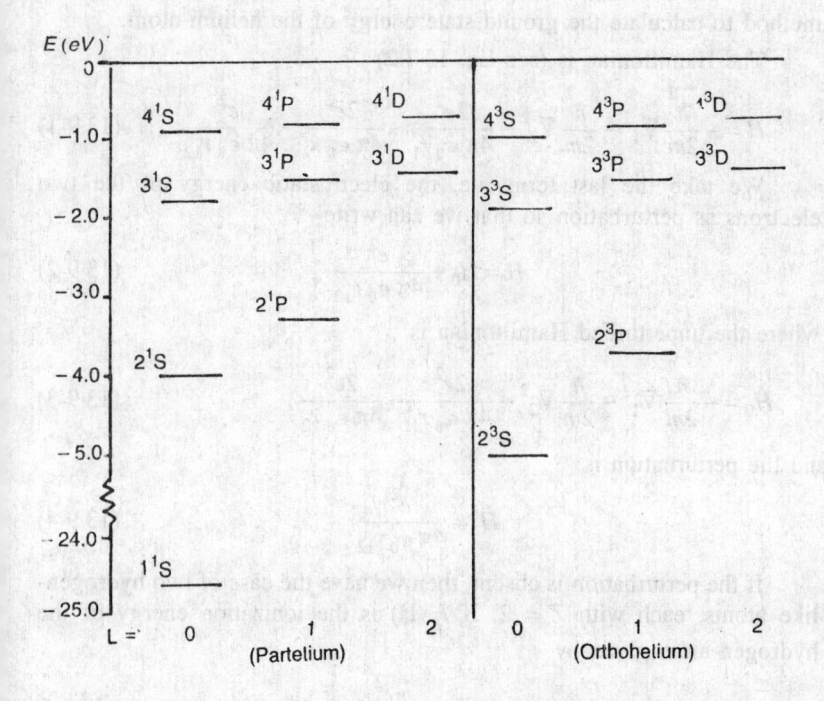

Quantum Mechanical Theory of Atomic Structure 381

Fig 13·5. Term scheme of Helium.

helium atom $(1s + e^-)$ forming the He^+ ion.

The experimentally determined energy value of the helium ground state (1^1S para-state) is -79.0 eV while the $1s$ state of He^+ ion has the energy of the hydrogen like atom : $Z^2 E_H = 4E_H$ which gives $E_{1s}(He^+) = -13.6 \times 4 = -54.4$ eV. The ionization energy then comes out to be $-54.4 + 79.0 = 24.6$ eV for the helium ground state (1^1S).

In the above discussion, the fine structure of the levels has been neglected.

13·9 Calculation of the ground state energy of the helium atom

The division of the spectroscopic terms of the helium (and other two-electron) atoms into para and ortho states results from the existence of the electron spin which have different energies. However the energy difference between them is not determined by the spin. It results from the electrostatic interaction between the two electrons (the last term on the r.h.s. of Eq. 13·7-1).

(a) Perturbation calculation :

We shall first use the first order time-independent perturbation method to calculate the ground state energy of the helium atom.

The Hamiltonian is (see Eq. 13·7-2)

$$H = -\frac{\hbar^2}{2m}\nabla_1^2 - \frac{\hbar^2}{2m}\nabla_2^2 - \frac{2e^2}{4\pi\epsilon_0 r_1} - \frac{2e^2}{4\pi\epsilon_0 r_2} + \frac{e^2}{4\pi\epsilon_0 r_{12}} \qquad (13\cdot9\text{-}1)$$

We take the last term, i.e. the electrostatic energy of the two electrons as perturbation so that we can write

$$H = H_0 + \frac{e^2}{4\pi\epsilon_0 r_{12}} \qquad (13\cdot9\text{-}2)$$

where the unperturbed Hamiltonian is

$$H_0 = -\frac{\hbar^2}{2m}\nabla_1^2 - \frac{\hbar^2}{2m}\nabla_2^2 - \frac{2e^2}{4\pi\epsilon_0 r_1} - \frac{2e^2}{4\pi\epsilon_0 r_2} \qquad (13\cdot9\text{-}3)$$

and the perturbation is·

$$H' = \frac{e^2}{4\pi\epsilon_0 r_{12}} \qquad (13\cdot9\text{-}4)$$

If the perturbation is absent, then we have the case of two hydrogen-like atoms, each with $Z = 2$. If $I_p(H)$ is the ionization energy of the hydrogen atom given by

$$I_p(H) = -E_H = \frac{me^4}{32\pi^2\epsilon_0^2\hbar^2} = 13\cdot6 \text{ eV} \qquad (13\cdot9\text{-}5)$$

then the unperturbed energy in the ground state would be

$$E_0 = +2Z^2 E_H = -2 \times 4 \times 13\cdot6 = -108\cdot8 \text{ eV} \qquad (13\cdot9\text{-}6)$$

E_H is the ground state energy of the hydrogen atom.

The unperturbed wave function would be the product of two hydrogen–like wave functions with $Z = 2$. For the ground state ($n = 1$, $l = 0$) the wave function for the hydrogen like atom is

$$\Psi_{1s} = \Psi_{100} = \sqrt{\frac{Z^3}{\pi a_0^3}} \exp(-\rho/2) \qquad (13\cdot9\text{-}7)$$

where
$$\rho = 2Zr/a_0 \qquad (13\cdot9\text{-}7a)$$

a_0 is the Bohr radius:

$$a_0 = \frac{4\pi\epsilon_0\hbar^2}{me^2} \qquad (13\cdot9\text{-}8)$$

So the unperturbed wave function for the two-electron atom in the 1s state would be

$$u_0 = \Psi_{100}\,\Psi_{100} = \frac{Z^3}{\pi a_0^3}\exp\{-(\rho_1+\rho_2)/2\} \qquad (13\cdot 9\text{-}9)$$

Here ρ_1 and ρ_2 are related to r_1 and r_2, the radial distances of the two electrons from the nucleus, through Eq. (13·9-7a).

Using the result discussed in § 12·2 of Ch. XII, we have from Eq. (13·9-9) the first order perturbation correction to energy for the helium ground state

$$W_0^{(1)} = H'_{00} = \langle u_0 | H' | u_0 \rangle$$

$$= \left(\frac{Z^3}{\pi a_0^3}\right)^2 \frac{e^2}{4\pi\epsilon_0} \left\langle u_0 \left| \exp\left(-\frac{\rho_1+\rho_2}{2}\right) \right| u_0 \right\rangle \qquad (13\cdot 9\text{-}10)$$

When the integrals in Eq. (13·9-10) are evaluated, we get (see Appendix A–VIII).

$$W_0^{(1)} = \frac{5Ze^2}{32\pi\epsilon_0 a_0} = -\frac{5}{4} Z E_H \qquad (13\cdot 9\text{-}11)$$

where the ground state energy of hydrogen is E_H given by (see Eq. 13·9-5)

$$E_H = -\frac{e^2}{8\pi\epsilon_0 a_0} = -13\cdot 6\,\text{eV} \qquad (13\cdot 9\text{-}12)$$

So the perturbation in the energy of the helium atom in the ground state is (with $Z = 2$)

$$W_0^{(1)} = \frac{5}{4} \times 13\cdot 6 = 34\,\text{eV} \qquad (13\cdot 9\text{-}13)$$

Hence the ground state energy of the helium atom after first order perturbation correction becomes

$$W_0 = E_0 + W_0^{(1)} = -108\cdot 8 + 34 = -74\cdot 8\,\text{eV} \qquad (13\cdot 9\text{-}14)$$

This may be compared with the experimental value of the ground state energy of the helium atom which is $W_0^{\text{exp}} = -79\,\text{eV}$

The agreement between the two values is actually not so bad if one considers the fact that the perturbation $e^2/4\pi\epsilon_0 r_{12}$ is rather large and is of the same order as the interaction energy between the nucleus and the electrons (see Eq. 13·9-1). Better agreement is obtained by considering the screening of the nuclear charge by the second electron and using variational method.

The above considerations can be applied to other two-electron systems, e.g. H^-, Li^+, Be^{++}, B^{3+} etc. ions to estimate the approximate values of their ground state energies. These estimated values are listed in Table 13·1

(b) Variational calculation to estimate the ground state energy of the helium atom :

The variational method was outlined in § 12.7. We choose the trial wave function for the ground state of the two electron system of nuclear charge Ze as

$$\Psi = \Psi_{100}(\mathbf{r}_1)\,\Psi_{100}(\mathbf{r}_2) \tag{13.9-15}$$

where Ψ_{100} is the ground state wave function for each of the electrons in a hydrogen-like atom of nuclear charge Ze :

$$\Psi_{100}(\mathbf{r}_1) = \frac{1}{\sqrt{\pi}}\left(\frac{Z}{a_0}\right)^{3/2}\exp(-Zr_1/a_0)$$

$$\Psi_{100}(\mathbf{r}_2) = \frac{1}{\sqrt{\pi}}\left(\frac{Z}{a_0}\right)^{3/2}\exp(-Zr_2/a_0) \tag{13.9-16}$$

$a_0 = 4\pi\hbar^2/me^2$ is the Bohr radius.

The Hamiltonian is (see Eq. 13.9-2)

$$\hat{H} = -\frac{\hbar^2}{2m}(\nabla_1^2 + \nabla_2^2) - \frac{Ze^2}{4\pi\epsilon_0 r_1} - \frac{Ze^2}{4\pi\epsilon_0 r_2} + \frac{e^2}{4\pi\epsilon_0 r_{12}} \tag{13.9-17}$$

We use the nuclear charge as the variatonal parameter and write $Z = Z'$ in the trial function (13.9-15) :

$$\Psi = \frac{1}{\pi}\left(\frac{Z'}{a_0}\right)^3 \exp-\left\{\frac{Z'(r_1+r_2)}{a_0}\right\} \tag{13.9-18}$$

We then get the expectation value for energy as

$$\langle H \rangle = \langle \Psi | H | \Psi \rangle$$

$$= \left\langle \Psi_{100}(\mathbf{r}_1, Z')\,\Psi_{100}(\mathbf{r}_2, Z') \left| -\frac{\hbar^2}{2m}(\nabla_1^2 + \nabla_2^2) - \frac{Ze^2}{4\pi\epsilon_0 r_1} \right.\right.$$

$$\left.\left. - \frac{Ze^2}{4\pi\epsilon_0 r_2} + \frac{e^2}{4\pi\epsilon_0 r_{12}} \right| \Psi_{100}(\mathbf{r}_1, Z')\,\Psi_{100}(\mathbf{r}_2, Z') \right\rangle \tag{13.9-19}$$

Now $\Psi_{100}(\mathbf{r}_1, Z')$ and $\Psi_{100}(\mathbf{r}_2, Z')$ obey the following wave equations :

$$\left(-\frac{\hbar^2}{2m}\nabla_1^2 - \frac{Z'e^2}{4\pi\epsilon_0 r_1}\right)\Psi_{100}(\mathbf{r}_1) = \varepsilon_1 \Psi_{100}(\mathbf{r}_1) \tag{13.9-20a}$$

$$\left(-\frac{\hbar^2}{2m}\nabla_2^2 - \frac{Z'e^2}{4\pi\epsilon_0 r_2}\right)\Psi_{100}(\mathbf{r}_2) = \varepsilon_2 \Psi_{100}(\mathbf{r}_2) \tag{13.9-20b}$$

Here $\varepsilon_1 = \varepsilon_2 = Z'^2 E_H$ for the two electrons in the ground state of a

Quantum Mechanical Theory of Atomic Structure

hydrogen–like atom of nuclear charge $Z'e$.

From Eq. (13·9-19) we get, writing $Z = Z' + (Z - Z')$

$$\langle H \rangle = \langle \Psi_{100}(\mathbf{r}_2) | \Psi_{100}(\mathbf{r}_2) \rangle$$

$$\times \left\langle \Psi_{100}(\mathbf{r}_1) \left| -\frac{\hbar^2}{2m}\nabla_1^2 - \frac{Z'e^2}{4\pi\epsilon_0 r_1} \right| \Psi_{100}(\mathbf{r}_1) \right\rangle$$

$$+ \langle \Psi_{100}(\mathbf{r}_1) | \Psi_{100}(\mathbf{r}_1) \rangle$$

$$\times \left\langle \Psi_{100}(\mathbf{r}_2) \left| -\frac{\hbar^2}{2m}\nabla_2^2 - \frac{Z'e^2}{4\pi\epsilon_0 r_2} \right| \Psi_{100}(\mathbf{r}_2) \right\rangle$$

$$+ (Z'-Z)\frac{e^2}{4\pi\epsilon_0}\left\langle \Psi \left| \frac{1}{r_1}+\frac{1}{r_2} \right| \Psi \right\rangle + \left\langle \Psi \left| \frac{e^2}{4\pi\epsilon_0 r_{12}} \right| \Psi \right\rangle$$

(13.9-21)

Evaluating the integrals in Eq. (13.9-21) (see Appendix A-VIII) we get finally

$$<H> = \left\{ 2Z'^2 - 4Z'(Z'-Z) - \frac{5}{4}Z' \right\} E_H \quad (13\cdot 9\text{-}22)$$

We now minimise $<H>$:

$$\frac{\partial <H>}{\partial Z'} = \left(4Z' - 8Z' + 4Z - \frac{5}{4} \right) E_H = 0$$

$$\therefore \quad 4Z' - 4Z + \frac{5}{4} = 0$$

or,
$$Z'_{min} = Z - \frac{5}{16} \quad (13\cdot 9\text{-}23)$$

Hence $<H>_{min} = \left\{ 2\left(Z - \frac{5}{16}\right)^2 + 4 \times \frac{5}{16}Z - \frac{5}{4}Z \right\} E_H$

$$= 2\left(Z - \frac{5}{16}\right)^2 E_H = 2Z'_{min} E_H \quad (13\cdot 9\text{-}24)$$

Putting $Z = 2$ (for He atom), we get $Z'_{min} = 1\cdot 6875$

and
$$<H>_{min} = -2 \times 1\cdot 6878^2 \times 13\cdot 6$$
$$= -77\cdot 46 \text{ eV} \quad (13\cdot 9\text{-}25)$$

Comparison with the experimental value for the ground state energy of helium (−79·0 eV) shows better agreement than obtained by perturbation calculation discussed previously.

In Table 13·1 are listed the ground state energies of some two electron systems calculated by variational method for comparison with the corresponding values estimated by perturbation calculation.

Table 13·1

Two-electron system	Unperturbed energy E_0 in eV	First order perturbation $W_0^{(1)}$ in eV	Perturbed energy W_0 in eV	Ground state energy by variational method (eV)
H^-	-27.2	17.0	-10.2	-12.9
He	-108.85	34.0	-74.85	-77.5
Li^+	-244.9	51.0	-193.9	-196.5
Be^{2+}	-435.4	68.0	-367.4	-370.1
B^{3+}	-680.3	84.9	-595.4	-597.8
C^{4+}	-979.6	102.1	-877.5	-880.3

It may be noted that the difference between the values estimated by the two methods (listed in columns 4 and 5) becomes less pronounced, percentage wise, for higher values of Z. This is expected, since the perturbation in the Hamiltonian becomes smaller compared to the interaction energy between the nucleus and the electrons as Z increases.

Comparison with experiments :

Both the approximation methods discussed above give values of the ground state energies of the two-electron systems which do not agree well with the experimental values. Later more accurate methods have been developed for the calculation of these energies, taking into account the reduced mass correction, the mass-polarization correction, relativistic corrections, and radiative corrections (Lamb shift). These yield the theoretical ionization potentials $I_p^{(th)}$ which are in very good agreement with the experimental values. For comparison, the values of I_p are usually expressed in wave numbers (cm^{-1}). For example in the case of He, the corresponding values are :

$$I_p^{(th)} = 198310.699 \text{ cm}^{-1}$$
$$I_p^{(exp)} = 198310.82 \text{ cm}^{-1}$$

This shows the degree of precision achieved in the theoretical calculations and vindicates quantum mechanical method in calculating the atomic energy states.

For further discussion, see *Physics of Atoms and Molecules* by B.H. Bransden and C.J. Joachain.

13·10 Excited states of the helium atom

In the excited state of the helium atom or of other two electron helium–like atoms, one of the electrons may be in the 1s state ($n = 1$, $l = 0$) while the other may be in a state of given n, l values. These are the 'genuinely discrete' excited states. For the second electron, there will be two excited states for which the electron configurations are 1s 2s ($l = 0$ for both) and 1s 2p ($l_1 = 0$, $l_2 = 1$).

Apart from the genuinely excited states in which one electron is in the ground state orbital while the other is in an excited orbital there may also be doubly excited states in which both the electrons may be in excited orbitals (e.g. 2s 2s, 2s 2p etc.) which lie above the ionization threshold and are discrete states within the region of the continuum.

If the electron spin is not considered, the wave functions for the two configurations 1s 2s and 1s 2p can be written as

$\Psi_{100} \Psi_{200}$, $\Psi_{100} \Psi_{211}$, $\Psi_{100} \Psi_{210}$, $\Psi_{100} \Psi_{21,-1}$

$\Psi = \Psi_{nlm_l}$, n, l, m_l being the principal, azimuthal and magnetic quantum numbers respectively.

These constitute a four-fold degenerate level if the electron spin is not taken into account. Actually, because of exchange degeneracy (see later) the level would be eight-fold degenerate.

When the electron spin is considered, the levels split up into *ortho* (spin-symmetric) and *para* (spin-antisymmetric) groups which split up due to the Coulomb and spin-orbit interactions. It can be seen that the 16-fold degenerate first excited state for the configurations 1s 2s and 1s 2p split up into sixteen substates due to above two effects. Thus Coulomb interaction alone would split up the 1s 2s and 1s 2p states into different states.

The remaining degeneracies are removed by the spin-orbit interaction. The corresponding levels are shown in Table 13·2.

Table 13·2

Electron Configuration	L	S	J	Level	J_z	No. of Levels	Total
1s 2s	0	0	0	$^1S_0$	0	1	
	0	1	1	$^3S_1$	1, 0, –1	3	4
1s 2p	1	0	1	$^1P_1$	1, 0, –1	3	
	1	1	2	$^3P_2$	2, 1, 0, –1, –2	5	12
	1	1	1	$^3P_1$	1, 0, –1	3	
	1	1	0	$^3P_0$	0	1	

The singlets (^{1}S and ^{1}P) are the para-states in which the spins of the two electrons are antiparallel and the spin state is a singlet. The triplets (^{3}S and ^{3}P) are the ortho-states in which the spins are parallel and the spin states are triplets (see §13·7)

The configuration $1s\ 2s$ gives rise of the two states $2\ ^3S_1$ and $2\ ^1S_0$ of which the $2\ ^3S_1$ (with 3 sublevels) is the lowest excited state of helium which is an ortho state (see Fig. 13·5). The $2\ ^1S_0$ which is a parastate comes above the $2\ ^3S_1$ state.

The configuration $1s\ 2p$ gives rise to the two states $2\ ^3P$ (with 9 sublevels) and the $2\ ^1P$ (with 3 sublevels) which appear above the S states.

The energies of the states can be calculated by the first order perturbation method. The unperturbed state for the $1s\ 2s$ configuration is

$$\Psi_2^{(0)} = u_2 = \frac{1}{\sqrt{2}} \{\Psi_{100}(1)\Psi_{200}(2) \pm \Psi_{200}(1)\Psi_{100}(2)\} \chi_{S\,m_S} \quad (13\cdot10\text{-}1)$$

where $\chi_{S\,m_S}$ is the spin wave function. Here exchange effect has been taken into account. The plus sign in Eq. (13·10-1) gives a space symmetric state which must be combined with the spin antisymmetric singlet state. On the other hand the minus sign in Eq. (13·10-1) gives a space antisymmetric state which must be combined with the spin symmetric triplet state.

Since the Coulomb perturbation is independent of the spin coordinates, the first order correction to the first excited state of helium will be

$$W_2^{(1)} = \left\langle u_2 \left| \frac{e^2}{4\pi\epsilon_0 r_{12}} \right| u_2 \right\rangle$$

$$= \frac{e^2}{8\pi\epsilon_0} \left\langle \Psi_{100}(1)\Psi_{200}(2) \pm \Psi_{200}(1)\Psi_{100}(2) \left| \frac{1}{r_{12}} \right| \right.$$

$$\left. \Psi_{100}(1)\Psi_{200}(2) \pm \Psi_{200}(1)\Psi_{100}(2) \right\rangle \left\langle \chi_{S\,m_S} \middle| \chi_{S\,m_S} \right\rangle \quad (13\cdot10\text{-}2)$$

This can be written as

$$W_2^{(1)} = J \pm K \quad (13\cdot10\text{-}3)$$

where

$$J = \frac{e^2}{4\pi\epsilon_0} \left\langle \Psi_{100}(1)\Psi_{200}(2) \left| \frac{1}{r_{12}} \right| \Psi_{100}(1)\Psi_{200}(2) \right\rangle \quad (13\cdot10\text{-}4a)$$

$$K = \frac{e^2}{4\pi\epsilon_0} \left\langle \Psi_{100}(1)\Psi_{200}(2) \left| \frac{1}{r_{12}} \right| \Psi_{100}(2)\Psi_{200}(1) \right\rangle \quad (13\cdot10\text{-}4b)$$

The integral J is normally known as the Coulomb (or direct) integral. It represents the Coulomb interaction energy of the charge distributions of the two electrons. The integral K is known as the *exchange integral*. It is the matrix element of $1/r_{12}$ between the two states such that the electrons 1 and 2 have exchanged their quantum numbers.

In evaluating the integrals J and K, the orthonormality of the spin functions have been taken into account.

The energy of the first excited state of helium comes out to be

$$W_2^{(1)} = W_2^{(0)} + J \pm K \tag{13.10-5}$$

In Eq. (13·10-5), the plus sign refers to the singlet para state ($^1S_1$) while the minus sign refers to the triplet (ortho state ^3S). The energies of both the ^1S and ^3S states are raised by the Coulomb interaction by the amount $J(1s\ 2s)$ over the unperturbed state. The state however lies above the ^3S state by an amount $2K\ (1s\ 2s)$.

Numerical calculations give the following values for helium (see *Modern Physics and Quantum Mechanics* by E.E. Anderson.

$$J\ (1s\ 2s) \approx 9.1\ \text{eV} \tag{13.10-6}$$

$$K\ (1s\ 2s) \approx 0.4\ \text{eV} \tag{13.10-7}$$

The integrals $J\ (1s\ 2p)$ and $K\ (1s\ 2p)$ for the configuration $1s\ 2p$ of the two electrons are :

$$J\ (1s\ 2p) = \frac{e^2}{4\pi\epsilon_0} \left\langle \Psi_{100}(1)\ \Psi_{2p}(2) \left| \frac{1}{r_{12}} \right| \Psi_{100}(1)\ \Psi_{2p}(2) \right\rangle \tag{13.10-8}$$

$$K\ (1s\ 2p) = \frac{e^2}{4\pi\epsilon_0} \left\langle \Psi_{100}(1)\ \Psi_{2p}(2) \left| \frac{1}{r_{12}} \right| \Psi_{100}(2)\ \Psi_{2p}(1) \right\rangle$$

$$\tag{13.10-9}$$

Numerical calculations give $J\ (1s\ 2p) \approx 10$ eV and $K\ (1s\ 2p) \approx 0.1$ eV.

The energies of the two P states (^1P) and (^3P) are raised by the amount $2J\ (1s\ 2p)$ above the unperturbed state.

Both the states ^1P and ^3P lie above the two S states. As before the ^1P para-state lies above the ^3P ortho-state by an amount $2K\ (1s\ 2p)$.

The above features of the splitting of the first excited state of helium are shown in Fig. 13·6.

Variational calculations :

Eckart has applied variational method to calculate the eneries of the ^3S, ^1S, ^3P ad ^1P states of helium using a simple trial function which is the antisymmetrised product of an inner ($1s$) orbital $u\ (1s)$ corresponding to the effective charge Z_i and an outer ($2s$) orbital $\upsilon\ (2s)$ corresponding to the effective charge Z_0. The minimum energy is obtained by varying the

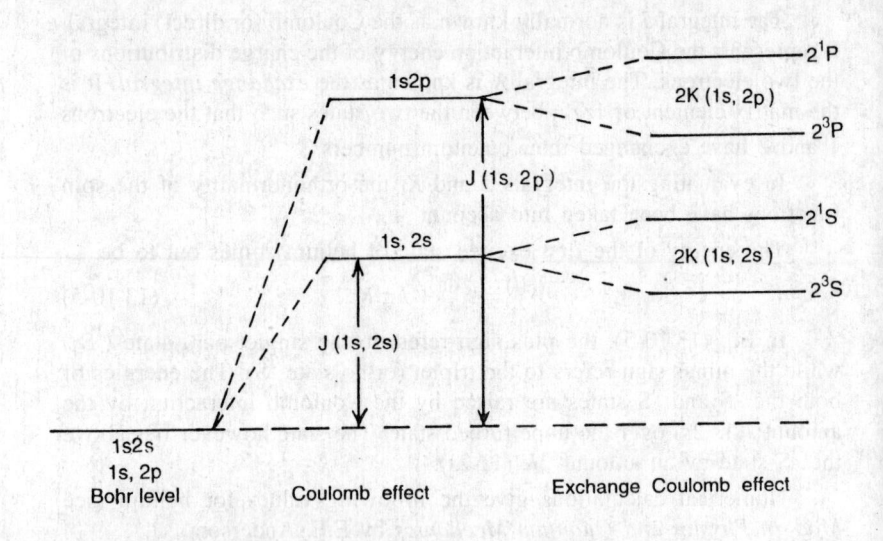

```
                                                    ────── 2¹P
                              1s2p ╱──── 2K(1s, 2p)
                             ╱           ────── 2³P
                            ↕ J(1s, 2p)
                           ╱
                          ╱                 ────── 2¹S
                    1s, 2s ────── 2K(1s, 2s)
                         ╲           ────── 2³S
                          ↕ J(1s, 2s)
```

Fig. 13·6. First excited state of Helium.

parameters Z_i and Z_0 which give $E(2\ ^3S) = 58.97$ eV. Better results are obtained by using elaborate Hylleraas type variational functions which give $E(2\ ^3S) = -59.17$ eV. For further details see *Physics of Atoms and Molecules* by Bransden and Joachain.

13·11 Identical particles §§

Identical particles cannot be distinguished from one another by means of any inherent characteristics. Otherwise they would not be identical in all respects. However the concept of identical particles in quantum mechanics differs from that in classical physics in several respects. Thus identical particles are said to be *indistinguishable* in quantum mechanics. In classical mechanics two or more identical particles can be distinguished by following their trajectories which are sharply defined. However because of uncertainty principle this is not possible in quantum mechanics, specially when they are in close proximity which makes it possible for them to interact with one another. Thus in experiments involving collisions between identical particles or observations on electrons within the same atom, the effect of indistinguishability must be taken into account. In quantum mechanical language we say that there is overlapping of the wave functions of the particles in this case. However when there is no such overlapping, the identical particles can be distinguished from one another. This is the case, for example, of the electrons in different atoms.

---

§§ See Quantum Mechanics by L.I. Schiff.

Another important feature of identical particles appears if they possess intrinsic spin, such as electrons, protons of neutrons which are all spin 1/2 particles. Then observations on the orientation of the spin in some specified direction in space (which remains unchanged in time) can make it possible to distinguish between the particles.

Symmetry properties of the wave functions :

We have seen that in the case of a system of two identical particles (such as in the case of a two-electron atom) it is possible to make a linear combination of the wave functions of the particles obtained by an exchange of the particles so as to produce a symmetric and an antisymmetric function. The first of these remains unchanged by an interchange of the particles while the second changes sign due to such interchange. We can similarly set up symmetric and antisymmetric wave functions for a system of n identical particles.

The time dependent Schrödinger equation for a system of n identical particles can be written as

$$\hat{H}(1, 2, 3, ...n, t)\,\Psi(1, 2, 3, ...n, t) = i\hbar \frac{\partial}{\partial t}\Psi(1, 2, 3, ...n, t) \quad (13\cdot 11\text{-}1)$$

where $\hat{H}$ is the Hamiltonian.

Since the particles are identical, an interchange of a pair of the particles does not change $\hat{H}$. Thus if the second and the third particles are interchanged, Eq. (13·11-1) transforms to

$$\hat{H}(1, 2, 3, ...n, t)\,\Psi(1, 3, 2, ...n, t) = i\hbar \frac{\partial}{\partial t}\Psi(1, 3, 2, ...n, t)$$

$$(13\cdot 11\text{-}2)$$

This shows that the wave function obtained by the interchange of any pair of particles of the system obeys the same wave equation. This is known as *exchange degeneracy* since the energy is the same in both cases. It is possible to make linear combinations of all such wave functions which may either be symmetric or antisymmetric.

It can be shown that the symmetry character of the wave function does not change with time. To prove this, suppose $\Psi = \Psi_S$ is symmetric at a given time t. Then since $\hat{H}$ is symmetric in its arguments in Eq. (13·11-1), $\hat{H}\Psi_S$ must be symmetric. Hence, according to Eq. (13·11-1) $\partial \Psi_S/\partial t$ is symmetric. Then since Ψ_S and $\partial \Psi_S/\partial t$ are symmetric at some instant t, Ψ_S at the instant $t + dt$ which is $\Psi_S + (\partial \Psi_S/\partial t)\,dt$ is also symmetric. We can repeat the argument at successive instants of time so that Ψ_S remains symmetric at all instants. This is also true for the antisymmetric combination Ψ_A.

The above arguments also apply in the case of groups of identical particles (e.g. α-particles or deuterons) which thus, must have definite symmetry characteristics.

Different types of identical particles have definite symmetry characters for their wave functions which remain unchanged. Experimental evidence shows that electrons, protons and neutrons have antisymmetric wave functions, while photons, pions and α-particles have symmetric wave functions.

Symmetric and antisymmetric wave functions :

For a system of n identical particles it is possible to obtain $n!$ solutions from any one solution, each of which is obtained by one of the $n!$ permutations of the n particles of the system. All these have exchange degeneracy. Any linear combination of these $n!$ solutions is also a solution of the wave equation (13·11-1).

The sum of all these $n!$ functions which are linearly independent is an unnormalized symmetric function Ψ_S. This can be easily checked in the case of two or three identical particles. For two particles the number of possible permutations is $2! = 2$, the corresponding wave functions being $\Psi(1, 2)$ and $\Psi(2, 1)$. The sum of these two $\Psi(1, 2) + \Psi(2, 1)$ becomes on interchange of the particles $\Psi(2, 1) + \Psi(1, 2)$ which remains the same as the original and is thus a symmetric function. On the other hand the function $\Psi(1, 2) - \Psi(2, 1)$ is antisymmetric.

In the case of three particles, permutation gives $3! = 6$ functions which are $\Psi(123)$, $\Psi(132)$, $\Psi(231)$, $\Psi(213)$, $\Psi(312)$ and $\Psi(321)$.

The symmetric and antisymmetric combinations of these six functions can be written as

$$\Psi = \Psi_{123} + \Psi_{231} + \Psi_{312} \pm (\Psi_{213} + \Psi_{321} + \Psi_{132}) \qquad (13 \cdot 11\text{-}3)$$

The plus sign gives the symmetric while the minus sign gives the antisymmetric function.

Pauli's exclusion principle :

Neglecting mutual interactions, the Hamiltonia of a system of n identical particles can be written as

$$H(1, 2, 3, \ldots n) = H_0(1) + H_0(2) + H_0(3) + \ldots + H_0(n) \qquad (13 \cdot 11\text{-}4)$$

The approximate energy eigenfunction can the be written as the product of n one particle eigenfunctions of H_0 :

$$\Psi_0(1, 2, 3, \ldots n) = u_\alpha(1) u_\beta(2) \ldots u_\nu(n) \qquad (13 \cdot 11\text{-}5)$$

where $u_\alpha(1), u_\beta(2), \ldots$ etc. obey the wave equations.

$$H_0 u_\alpha(1) = E_\alpha u_\alpha(1) \ ; \ H_0 u_\beta(2) = E_\beta u_\beta(2) \ ; \ \ldots \qquad (13 \cdot 11\text{-}6)$$

The total energy is

$$E = E_\alpha + E_\beta + \ldots + E_\nu \qquad (13 \cdot 11\text{-}7)$$

The total wave function $\Psi(1, 2, 3, \ldots n)$ is a solution of the wave equation

$$(H - E)\Psi = 0 \tag{13.11-8}$$

Ψ can be either symmetric or antisymmetric in the interchange of any pair of the particles. Particles for which Ψ is antisymmetric obey Pauli's exclusion principle so that no two of them can occupy the same quantum state.

The exclusion principle was first proposed by Wolfgang Pauli to explain the periodic classification of elements.

Particles for which the total wave function is symmetric do not obey Pauli's exclusion principle and any number of them can occupy one and the same state.

We shall later see (in Ch. XX) that the two groups of particles having antisymmetric and symmetric wave functions are governed by statistical laws which are different from those formulated by Maxwell and Boltzmann which is known as classical statistics. The statistics followed by the particles discussed above are known as quantum statistics. There are two types, Fermi-Dirac statistics obeyed by particles having antisymmetric wave functions and Bose-Einstein statistics obeyed by particles having symmetric wave functions. To the former group belong particles having half odd-integral spins while particles having integral or zero spin belong to the latter group.

13.12 Many-electron atoms

The starting point of calculations on many-electron atoms (except the lightest) is the *central field approximation*, according to which it is assumed that each electron in a many electron complex atom is acted upon by a central field produced by the nucleus (assumed small and infinitely heavy) and all the other electrons. This approximation holds because the deviation of the central potential $V(r)$ due to the close passage of all the other electrons is relatively small. This is so because the nuclear potential is Z times larger than the effect of the above fluctuating potential.

In the case of the two-electron atom, it is possible to treat the mutual interaction between the two electrons as a perturbation which is added to the central potential due to nucleus (see §13.7). To do the same in the case of the many electron–atom we have to write the unperturbed Hamiltonian so as to include the total effective central potential as stated above. The perturbation then contains the remaining spherical and all the non-spherical parts of the electronic interactions. If the perturbation is neglected then the central field Hamiltonian H_C can be written as

$$H_C = \sum \left\{ -\frac{\hbar^2}{2m} \nabla_j^2 + V(r_j) \right\} \tag{13.12-1}$$

where $\mathbf{r}_j$ denotes the relative coordinate of the jth electron w.r.t. the nucleus. If $\Psi_C(\mathbf{r}_1, \mathbf{r}_2, ... \mathbf{r}_N)$ denotes the spatial part of the N-electron central field wave function, we get the unperturbed wave equation as

$$H_C \Psi_C = E_C \Psi_C \qquad (13 \cdot 12\text{-}2)$$

Then Eq. (13·12-2) is separable into N equations, one for each electron so that we can write

$$\Psi_c = u_\alpha(\mathbf{r}_1) u_\beta(\mathbf{r}_2) .... u_\nu(\mathbf{r}_N) \qquad (13 \cdot 12\text{-}3)$$

The normalized individual electron orbitals obey equations of the form

$$\left\{ -\frac{\hbar^2}{2m} \nabla^2 + V(r) \right\} u_{n l m_l}(\mathbf{r}) = E u_{n l m_l}(\mathbf{r}) \qquad (13 \cdot 12\text{-}4)$$

$\alpha, \beta, ...$ etc. refer to the states of the individual electrons determined by the quantum numbers n, l, m_l. These have the forms

$$u_{nlm_l}(\mathbf{r}) = R_{nl}(r) Y_{lm_l}(\theta, \phi) \qquad (13 \cdot 12\text{-}5)$$

where R_{nl} is the radial part and Y_{lm_l} is the angular part ; n, l and m_l can assume the values :

$$n = 1, 2, 3, ...$$
$$l = 0, 1, 2, ... (n-1)$$
$$m_l = 0, \pm 1, \pm 2, ... \pm l \qquad (13 \cdot 12\text{-}6)$$

The functions u_{nlm_l} are different from the hydrogenic wave functions since the nature of the central potential is different in the two cases.

$R_{nl}(r)$ obeys the radial wave equation of the type (11.15-1) given in Ch. XI for the hydrogen-like atom with the Coulomb potential $Ze^2/4\pi\epsilon_0 r$ replaced by $V(r)$.

At a large distance r from the nucleus, the nuclear charge is screened by all the electrons except the one under observation so that the potential has the form $-e^2/4\pi\epsilon_0 r$. This has the form of the Coulomb potential acting on the electron in the hydrogen atom so that the system has an infinite number of bound energy levels as in hydrogen. In hydrogen, the levels of the same principal quantum number n but different values of l and m_l are degenerate. In the present case, however, degeneracy with respect to l is removed by a non-Coulomb central field, so that the levels of the same n but different l have different energies. This happens because the electrons with smaller angular momentum penetrate deeper into the atom so that the nuclear charge is less effectively screened for them. Thus $V(r)$ is stronger than $-e^2/4\pi\epsilon_0 r$ for them. This makes the state of the lowest l

Quantum Mechanical Theory of Atomic Structure

(smallest angular momentum) for a given n to have the lowest energy.

When the spin of the electron is taken into account, each state of the electron is characterized by the four quantum numbers n, l, m_l and m_s where m_s can have the values $\pm 1/2$.

Because of the degeneracy w.r.t. m_l and m_s there can be a maximum of $2(2l+1)$ electrons in a state of given l consistent with the exclusion principle. The maximum number of electrons that can be accommodated in a state of given n is $\sum_0^{n-1} 2(2l+1) = 2n^2$ (see §6.6).

The total wave function taking spin into account for any individual electron can be written as

$$u_{nlm_l m_s} = u_{nlm_l}(\mathbf{r}) \chi_{1/2\, m_s} \tag{13.12-7}$$

where
$$\chi_{1/2\, m_s} = \chi_{1/2\, 1/2} = \alpha = \begin{pmatrix} 1 \\ 0 \end{pmatrix} \quad \text{for } m_s = +1/2$$

$$\chi_{1/2\, m_s} = \chi_{1/2,\, -1/2} = \beta = \begin{pmatrix} 0 \\ 1 \end{pmatrix} \quad \text{for } m_s = -1/2$$

The two spin functions correspond to the spin 'up' and spin 'down' states. The energy remains the same, equal to E_{nl} independent of m_l and m_s.

Because of Pauli's exclusion principle, the total wave function Ψ_C must be antisymmetric in all coordinates, spatial and spin. This is given by the following determinant:

$$\Psi_C = \frac{1}{\sqrt{N!}} \begin{vmatrix} u_\alpha(q_1) & u_\beta(q_1) & \cdots & u_\nu(q_1) \\ u_\alpha(q_2) & u_\beta(q_2) & \cdots & u_\nu(q_2) \\ & \cdots\cdots\cdots & & \\ & \cdots\cdots\cdots & & \\ u_\alpha(q_N) & u_\beta(q_N) & \cdots & u_\nu(q_N) \end{vmatrix} \tag{13.12-8}$$

Here $q_1, q_2, \ldots q_N$ denote the spatial and the spin coordinates of the different electrons. The determinant in Eq. (13·12-8) is known as the *Slater determinant*. If we interchange the space and spin coordinates of any two electrons (say q_1 and q_2), which is equivalent to interchange of two rows of the determinant, then the determinant changes sign. This shows that Ψ_C given by Eq. (13·12-8) is antisymmetric in the interchange of any pair of coordinates q. The factor $1/\sqrt{N!}$ normalizes ψ_c. The energy $E_C = \sum_i E_{n_i l_i}$ is the eigenvalue of the central field Hamiltonian H_C corresponding to a given Slater determinant. It is the sum of the energies of the N individual electrons present in the determinant.

Since the determinant vanishes when two columns or rows are equal, we have the result that no two individual electrons can have all the four quantum numbers n, l, m_l, m_s equal. This is the statement of Pauli's exclusion principle.

The individual electron wave function given by Eq. (13·12-7) has definite parity $(-1)^l$ determined by the azimuthal quantum number l of the electron under consideration. Then the parity of the Slater determinant which is $(-1)^{\Sigma l_j}$ is either even or odd according as Σl_j is even or odd, l_j being the azimuthal quantum number of the jth electron and the summation is over all electrons in the atom.

It can be easily seen that H_C commutes with L and S where $\mathbf{L} = \Sigma \mathbf{l}_j$ is the total orbital angular momentum operator and $\mathbf{S} = \Sigma \mathbf{s}_j$ is the total spin operator. So it is possible to obtain eigenfunctions of H_C which are simultaneous eigenfunctions of L^2, S^2, L_z and S_z with the respective eigenvalues $L(L+1)\hbar^2$, $S(S+1)\hbar^2, m_L \hbar$. The Slater determinant is an eigenfunction of L_z and S_z but generally not of L^2 and S^2. However it is possible to form linear combinations of the Slater determinants which are simultaneous eigenfunctions of L^2, S^2, L_z and S_z.

13·13 Periodic classification of elements

The discussions in the previous sections provide a theoretical basis of the periodic classification of the elements first proposed by Dimitri Mendeleev which has been discussed in detail in §6.6 in Ch. VI. As seen above, because of the non-Coulomb central field due to the interelectronic repulsion, a state of given principal quantum number n splits up into $(n-1)$ sublevels of different l values ($l = 0, 1, 2, n-1$). So a state of given (n, l) values can accommodate a maximum of $2(2l+1)$ electrons consistent with Pauli's exclusion principle. The factor 2 arises due to the two possible values of $m_s (= \pm 1/2)$. For a state of given n there can thus be a maximum of $\sum_{0}^{n-1} 2(2l+1) = 2n^2$ electrons.

Electrons having the same (n, l) values belong to the same *subshell*. Such states are known as equivalent and the electrons occupying the same subshell of given (n, l) values are called *equivalent electrons*. The electronic states for a given value of n having different possible values of l (0 to $n-1$) constitute a shell. The different shells are also designated by the capital letters K, L, M, N etc. corresponding to $n = 1, 2, 3, 4$ etc. respectively. When a shell contains the maximum possible number $2n^2$ of electrons, it becomes closed. When a subshell of given l contains the maximum possible number $2(2l+1)$ electrons the subshell is closed.

We now consider the building up of the atoms of different atomic numbers (Z) by filling up of the electronic states of different n and l. The energy of the electron in a given subshell depends on the values of n and l such that for a given n, it is an increasing function of l. For a given l, the energy is an increasing function of n. The energy is in general, an increasing function of $(n + l)$. Furthermore, the gaps between the levels of different l within a shell are usually small compared to the gaps between the energies of different shells.

In the *ground state* of an atom of given Z, the Z electrons are so distributed that a certain number $(f - 1)$ of subshells are completely filled and the last subshell (i.e., the fth subshell) is in general, only partially filled. These have the highest energy so that they are least tightly bound. In the ground state of the atom, the electrons in the different subshells occupy the lowest energy levels consistent with the exclusion principle. For $Z = 2, 4, 10, 12, 18$ etc. the subshells are completely filled, (for the configurations $1s^2$, $1s^2\,2s^2$, $1s^2\,2s^2\,2p^6$, $1s^2\,2s^2\,2p^6\,3s^2$, $1s^2\,2s^2\,2p^6\,3s^2\,3p^6$ etc. respectively). Here the last subshells are $1s$, $2s$, $2p$, $3s$ and $3p$ respectively.

The ground state term of an atom of given Z is written as $^{2S+1}L_J$ where the superscript denoting the multiplicity $2S + 1$ depends on the spin quantum number S while J is the total quantum number. As seen in Ch. VI the values of the total azimuthal quantum number $L = 0, 1, 2, 3$ etc. correspond to the spectroscopic terms S, P, D, F etc. The above notation is in accordance with Russel-Saunders (or L–S) coupling.

13·14 The nature of the central field ; Thomas-Fermi model of the atom

The detailed calculation of the energy levels of complex atoms is possible only if we know the nature of the central field $V(r)$. There are a number of methods for finding it of which we shall consider two. The first of these is based on the Thomas-Fermi statistical model of the atom. The other method, to be considered in the next section, is known as the Hartree-Fock method of self-consistent fields.

In the Thomas-Fermi model, the potential is assumed to vary slowly over a distance equal to the de Broglie wavelength of the electron. Hence there will be enough electrons within a volume of such dimension in which $V(r)$ may be taken to be constant. We can apply Fermi gas model in which the electrons are treated by statistical mechanics obeying Fermi-Dirac statistics (see Ch. XX).

Fermi gas model :

It can be shown that the energy of a particle in a three dimensional potential box of sides a, b and c is given by (see *Introductory Quantum Mechanics* by S.N. Ghoshal ; see also §11·2) :

$$E = \frac{\pi^2 \hbar^2}{2m}\left(\frac{n_1^2}{a^2} + \frac{n_2^2}{b^2} + \frac{n_3^2}{c^2}\right) \qquad (13\cdot14\text{-}1)$$

where n_1, n_2 and n_3 are three positive integers excluding zero. The wave function of the particle is

$$\Psi = \sqrt{\frac{8}{abc}} \sin\frac{n_1 \pi x}{a} \sin\frac{n_2 \pi y}{b} \sin\frac{n_3 \pi z}{c} \qquad (13\cdot14\text{-}2)$$

The lowest level corresponds to $n_1 = n_2 = n_3 = 1$. The same energy can usually be obtained for a level with different sets of values of n_1, n_2 and n_3 and hence the level is degenerate.

When the spin of a particle is considered, we have to multiply the spatial part of the wave function (Eq. 13·14–2) by the spin function $\chi_{s\,m_s}$ to give the total wave function $\Psi n_1 n_2 n_3 (\mathbf{r}) \chi_{s\,m_s}$. For electrons $s = 1/2$ and m_s can have only two values $\pm 1/2$.

The energy gaps between consecutive levels are usually so small for a macroscopic box that it is possible to regard the energy levels as distributed almost continuously. It is therefore possible to apply statistical considerations for treating the problem of electron distribution within the atom.

We introduce the density function $D(E)$ for the states of the electron which is defined as the number of energy states in a unit energy interval at the energy E. Then $D(E)\,dE$ is the number of states in the energy interval E to $E + dE$.

Following the method outlined in Ch. III (§3·2) for the calculation of the number of classical oscillators with frequencies lying between ν and $\nu + d\nu$ in a spatial volume V we can calculate the number of states in the energy interval E to $E + dE$. According to Eq. (3·2-16) the number of classical oscillators in the frequency interval ν to $\nu + d\nu$ is

$$n_\nu\, d\nu = \frac{4\pi V}{c^3} \nu^2\, d\nu \qquad (13\cdot14\text{-}3)$$

Integration of Eqs. (3·14-3) gives the number of classical oscillators in the volume V with frequencies upto ν:

$$N_\nu = \frac{4\pi V}{3c^3} \nu^3 \qquad (13\cdot14\text{-}4)$$

The differential equation governing the behaviour of the classical oscillators (the equation of wave propagation of e.m. waves) $\nabla^2 \phi + (\omega^2/c^2)\phi = 0$ and the Schrödinger equation for the quantum states of the electron in the potential box $\nabla^2 \Psi + (2mE/\hbar^2)\Psi = 0$ are identical in form with (ω^2/c^2) replaced by $(2mE/\hbar^2)$. So we can substitute in Eq.

(13·14-4), $(2mE/4\pi^2\hbar^2)^{3/2}$ in place of v^3/c^3 to get the number of states at the energy E which is thus given by

$$N_E = 2 \times \frac{4\pi V}{3} \cdot \frac{1}{8\pi^3} \left(\frac{2m}{\hbar^2}\right)^{3/2} E^{3/2}$$

$$= \frac{V}{3\pi^2} \left(\frac{2m}{\hbar^2}\right)^{3/2} E^{3/2} \qquad (13\cdot14\text{-}5)$$

The factor 2 in Eq. (13·14-5) appears because of the spin of the electron.

The density of the electronic states then becomes

$$D(E) = \frac{dN_E}{dE} = \frac{V}{2\pi^2} \left(\frac{2m}{\hbar^2}\right)^{3/2} E^{1/2} \qquad (13\cdot14\text{-}6)$$

As the electrons obey Pauli's exclusion principle, the total wave functions, given by the Slater determinant (see §13·12) must be antisymmetric in all coordinates, spatial as well as spin. Since a particular state can contain only one electron, the total energy of a system of N electrons is equal to the sum of the energies of all the electrons filling up the different states upto a maximum E_f, known as the Fermi energy.

The electron distribution given by Eq. (13·14-6) is illustrated in Fig. 13·7 (see also Ch. XVIII) in which the density $D(E)$ is plotted as a function of the energy E. The occupied states are shown by the shaded area. The states of energy $E > E_f$ are all empty. We then get the total number of electrons filling up the states upto the Fermi energy E_f by integrating Eq. (13·14-6).

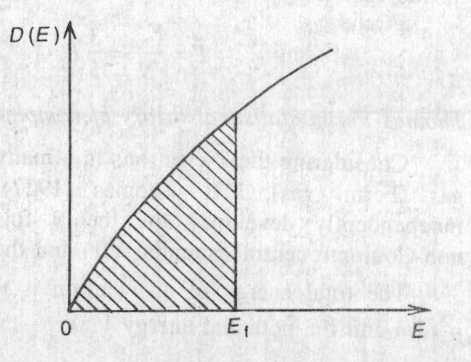

Fig. 13·7. Electron distribution according to F-D statistics.

$$N = \int_0^{E_f} D(E)\, dE = \frac{V}{2\pi^2} \left(\frac{2m}{\hbar^2}\right)^{3/2} \int_0^{E_f} E^{1/2}\, dE$$

$$= \frac{V}{3\pi^2} \left(\frac{2m}{\hbar^2}\right)^{3/2} E_f^{3/2} \qquad (13\cdot14\text{-}7)$$

This is the same as Eq. (18·7-1a) derived in Ch. XVIII using the Fermi-Dirac distribution law for an electron gas in a metal. From Eq.

13·14-7) we get the Fermi energy E_f (see Eq. 18·7-2):

$$E_f = \left(\frac{3\pi^2 N}{V}\right)^{2/3} \frac{\hbar^2}{2m} = (3\pi^2 \rho)^{2/3} \frac{\hbar^2}{2m} \qquad (13\cdot14\text{-}8)$$

where we have written $\rho = N/V$ as the density of the electrons:

$$\rho = \frac{N}{V} = \frac{1}{3\pi^2}\left(\frac{2m}{\hbar^2}\right)^{3/2} E_f^{3/2} \qquad (13\cdot14\text{-}7a)$$

The total energy of all the electrons, assuming a continuous distribution, is given by

$$E_{\text{tot}} = \int_0^{E_f} E \cdot D(E)\, dE = \frac{V}{2\pi^2}\left(\frac{2m}{\hbar^2}\right)^{3/2} \int_0^{E_f} E^{3/2}\, dE$$

$$= \frac{V}{2\pi^2}\left(\frac{2m}{\hbar^2}\right)^{3/2} \cdot \frac{2}{5} E_f^{5/2} = \frac{3}{5} N E_f \qquad (13\cdot14\text{-}9)$$

The above derivations correspond to the ground state which means that the electron gas is at $T = 0$ K. The average energy of the electrons in this case is

$$\bar{E} = \frac{E_{\text{tot}}}{N} = \frac{3}{5} E_f \qquad (13\cdot14\text{-}10)$$

Thomas-Fermi statistical theory of complex atoms :

Considering the Z electrons in a many-electron atom as an electron gas (Fermi gas), L.H. Thomas (1927) and Enrico Fermi (1928) independently developed the theory for finding the nature of the non-Coulomb central potential $V(r)$ and the electron density $\rho(r)$.

The total energy of an electron is the sum of its kinetic energy $p^2/2m$ and the potential energy $V(r)$:

$$E_{\text{tot}} = \frac{p^2}{2m} + V(r) \qquad (13\cdot14\text{-}11)$$

This energy E_{tot} is negative as otherwise the electrons will escape to infinity. Since the maximum kinetic energy of the electron is equal to the Fermi energy E_f, we get the maximum total energy E_m as

$$E_m = E_f + V(r) \qquad (13\cdot14\text{-}12)$$

Hence $\qquad E_f = E_m - V(r) \qquad (13\cdot14\text{-}12a)$

Then from Eq. (13·14-7a), we get

$$\rho(r) = \frac{1}{3\pi^2}\left(\frac{2m}{\hbar^2}\right)^{3/2} \{E_m - V(r)\}^{3/2}$$

Quantum Mechanical Theory of Atomic Structure

We write
$$e\phi(r) = E_m - V(r) \tag{13·14-13}$$

Then we get
$$\rho(r) = \frac{1}{3\pi^2}\left(\frac{2m}{\hbar^2}\right)^{3/2} \{e\phi(r)\}^{3/2} \tag{13·14-14}$$

Eq. (13·14-14) holds for $V(r) < E_m$ which means $\phi(r) > 0$. It shows that when $V(r) = E_m$, $\phi(r)$ and ρ vanish. This is also the case for $V(r) > E_m$ for which $\phi(r) < 0$. So $\rho = 0$ in this case.

Eq. (13·14-13) defines a limiting radius $r = r_0$ for which $V(r_0) = E_m$ so that $\phi(r_0) = 0$ and the electron density $\rho(r) = 0$. Thus we may say that the electrons in the Thomas-Fermi model are contained in a sphere of radius r_0. This is true for a positive ion. For neutral atoms, the electrostatic potential must vanish at the boundary since the net charge within the sphere of radius r_0 is zero (positive and negative charges cancel). So in this case we must put $E_m = 0$.

We now write down the Poisson's equation of electrostatics:

$$\nabla^2 \phi(r) = \frac{1}{r}\frac{d^2}{dr^2}\{r\phi(r)\} = \frac{e\rho(r)}{\epsilon_0} \tag{13·14-15}$$

for $\phi(r) \geq 0$; $e\rho(r)$ is the (negative) charge density of the electron distribution in the atom. $\epsilon_0 = 10^{-9}/36\pi$ F/m is the permittivity of vacuum. On the other hand $\nabla^2 \phi(r) = 0$ for $\phi(r) < 0$.

Since the potential $\phi(r)$ is equal to the Coulomb potential close to the nucleus $(r \to 0)$ and must fall off more rapidly than $1/r$ at large r (since there must not be any net charge inside the sphere of radius r), we have the following two boundary conditions

For
$$r \to 0, \quad \phi(r) = \frac{Ze}{4\pi\epsilon_0 r} \tag{13·14-16a}$$

For
$$r \to \infty, \quad r\phi(r) \to 0 \tag{13·14-16b}$$

From Eqs. (13·14-14) and (13·14-15) we get

$$\frac{1}{r}\frac{d^2}{dr^2}\{r\phi(r)\} = \frac{e}{3\pi^2\epsilon_0}\left(\frac{2m}{\hbar^2}\right)^{3/2} \{e\phi(r)\}^{3/2} \tag{13·14-17}$$

It is possible to transform Eq. (13·14-17) into a dimensionless form by the following substitutions:

$$r = bx, \quad r\phi(r) = \frac{Ze}{4\pi\epsilon_0}\chi(x) \tag{13·14-18}$$

where
$$b = \frac{1}{2}\left(\frac{3\pi}{4}\right)^{2/3}\frac{a_0}{Z^{1/3}} = \frac{0.8853\, a_0}{Z^{1/3}} \tag{13·14-19}$$

a_0 is the Bohr radius : $a_0 = 4\pi\epsilon_0 \hbar^2 / m_e^2$. Since b has the dimension of length, x is dimensionless. Since Ze/r has the same dimension as that of $\phi(r)$ i.e. of a potential, $\chi(x)$ is also dimensionless. Substitution in Eq.(13·14-17) finally gives the following dimensionless equation :

$$\frac{d^2\chi}{dx^2} = \frac{\chi^{3/2}}{\sqrt{x}} \qquad (13\cdot14\text{-}20)$$

for $\chi \geq 0$. This is known as Thomas-Fermi equation. For $\chi < 0$, we get

$$\frac{d^2\chi}{dx^2} = 0 \qquad (13\cdot14\text{-}21)$$

The electron density ρ can also be expressed in terms of χ and x, using Eq. (13·14-14).

$$\rho = \frac{Z}{4\pi b^3}\left(\frac{\chi}{x}\right)^{3/2} \text{ for } \chi \geq 0 \qquad (13\cdot14\text{-}22a)$$

$$= 0 \text{ for } \chi < 0 \qquad (13\cdot14\text{-}22b)$$

The boundary conditions (13·14-16a) and 16b) give

$$\chi(0) = 1 \qquad (13\cdot14\text{-}23a)$$

$$\chi(\infty) = 0 \qquad (13\cdot14\text{-}23b)$$

The solution of Thomas-Fermi Eq. (13·14-20) is obtained by numerical integration. For neutral atoms, $\chi(x)$ is a monotonically decreasing function of x. It can be shown that the asymptotic form of $\chi(x)$ for large x (i.e., large r) has the form

$$\chi(x) = \frac{144}{x^3} \text{ for } x \to \infty \qquad (13\cdot14\text{-}24)$$

On the other hand for small r, i.e. as $x \to 0$, we have

$$\chi(x) = 1 - 1\cdot588x + ... \qquad (13\cdot14\text{-}25)$$

Then substitution in Eq. (13·14-18) gives with help of Eq. (13·14-13) and (13·14-19) and putting $E_m = 0$

$$V(r) = -e\phi(r) = -\frac{Ze^2}{4\pi\epsilon_0 r}\chi(x)$$

$$= -\frac{Ze^2}{4\pi\epsilon_0 r}(1 - 1\cdot588x + ...)$$

or, $$V(r) = \frac{e^2}{4\pi\epsilon_0}\left(-\frac{Z}{r} + 1\cdot794\frac{Z^{4/3}}{a_0}\right) \qquad (13\cdot14\text{-}26)$$

for small r. Here we have used the value of x given by (13·14-18).

Eq. (13·14-26) shows that at small r, the dominant term in $V(r)$ is the

Coulomb potential energy due to nucleus ($\propto Z/r$) which is attractive (negative). The second term in Eq. (13·14-26) is due to the electrons and is repulsive (positive).

For large r, since $\chi \propto 1/r^5$ (see Eq. 13·14-24), $V(r)$ falls off more rapidly than $1/r$ as stated earlier.

Thomas-Fermi statistical model gives an electron distribution in atoms which is in good agreement with observed values when average values are considered inspite of the relatively small number of electrons involved. These results have been used widely in calculating the scattering powers of atoms for x-rays. However the finer details of the electron distribution such as the appearance of the atomic shells are not reproduced in this model.

The nature of electron distribution function calculated by using the Thomas-Fermi model when compared with the results obtained by more accurate Hartree-Fock calculations shows broad agreement of the average values, (see Introduction to *Quantum Mechanics* by Pauling and Wilson).

Thomas-Fermi model gives better results for intermediate distances from the centre of the atomic sphere. It does not work very well either close to the nucleus ($r \to 0$) or far way from the centre ($r \to \infty$).

For further details see *Physics of Atoms and Molecules* by Bransden and Joachain.

13·15 Hartree-Fock method of self-consistent field

In this method originally suggested by D.R. Hartree (1928) the non-Coulomb central potential $V(r)$ is derived from the average charge density of all the electrons except the one under consideration. Thus this last electron is assumed to be in the field produced by all the other electrons as well as by the nucleus. The electronic part of the assumed central potential is calculated by taking the space average of each individual unperturbed wave function $\Psi_i^{(0)}$. Thus the potential on the jth electron becomes

$$V_j = \sum_i{}' \frac{e\rho_i}{4\pi\epsilon_0 r_{ij}} = \sum_j{}' \left\langle \Psi_i^{(0)} \left| \frac{e}{4\pi\epsilon_0 r_{ij}} \right| \Psi_i^{(0)} \right\rangle \qquad (13\cdot15\text{-}1)$$

where $\sum_j{}'$ denotes summation over all electrons i except $i = j$.

Since the unperturbed wave functions $\Psi_i^{(0)}$ of the individual electrons are assumed to be known, it is possible to calculate V_j from Eq. (13·15-1).

$\Psi_i^{(0)}$ for any individual electron is found by solving the Schrödinger equation, taking into account only the Coulomb potential due to the nucleus.

We then substitute V_j in the Schrödinger equation along with the Coulomb potential due to the nucleus to solve for the new (first-order) wave function $\Psi_j^{(1)}$ for the electron under consideration *self-consistently* :

$$\nabla_j^2 \Psi_j^{(1)} + \frac{2m}{\hbar^2}\left(E_j + \frac{Ze^2}{4\pi\epsilon_0 r_j} - V_j\right)\Psi_j^{(1)} = 0 \qquad (13\cdot15\text{-}2)$$

These new wave functions $\Psi_j^{(1)}$ for the different individual electrons are substituted in Eq. (13·15-1) to obtain a better approximation value V'_j of the potential. This iteration process is continued until the calculated potential does not change any further in the next approximation.

The chief drawback of Hartree's method is that the wave function is not antisymmetric in the electron coordinates as required by Pauli's exclusion principle. This drawback was removed by V. Fock (1930) who used properly symmetrized initial wave functions for the electrons §§

The Hartree-Fock method has become an important tool for calculations not only for many-electron atoms but also for other many body systems e.g. electrons in a molecule or in a solid as also for nuclei because it can be readily programmed for digital computation (see *Quantum Theory of Matter* by J.C. Stater, 2nd Edition, 1969).

Fock has shown by variational method that the complete wave equation for the *j*th electron, taking into account the *quantum exchange effect*, can be written as

$$\{H(r) + V_j(r)\}\Psi_j(r) - \frac{e^2}{4\pi\epsilon_0}\int\frac{\rho_j(r',r)}{|\mathbf{r}-\mathbf{r}'|}\Psi_j(r')\,d\tau'$$
$$= E_j \Psi_j(r) \qquad (13\cdot15\text{-}3)$$

If the integral term in Eq. (13·15-3) is omitted then we get the Schrödinger equation for an electron in a field with the potential energy $U(r) + V_j(r)$ where $U(r)$ is the potential energy of the external field produced by all electrons except the *j*th. $V_j(r)$ corresponds to a potential for the charge density $\rho - |\Psi_j|^2$ where

$$\rho = \rho(r, r') = \sum_{k=1}^{n/2} \Psi_k(r')\Psi_k(r) \qquad (13\cdot15\text{-}4)$$

n is the number of electrons in the system.

The integral term has no classical analogue. It represents the correction for *quantum exchange*.

Without the integral term in Eq. (13·15-3), the equation becomes the same as had been proposed by Hartree. The complete equation (13·15-3) is called the *Hartree-Fock equation*.

---

§§ See Fundamentals of Quantum Mechanics by V.A. Fock.

13·16 Application of Hartree-Fock method to many electron atoms

As was seen in §6·6, Ch. VI, the periodic classification of the elements was explained theoretically, even before the advent of quantum mechanics, on the basis of Bohr's theory. Bohr based his interpretation on the experimentally determined spectra and the chemical properties of the atoms. Bohr's explanation found a more rigorous theoretical confirmation when quantum mechanics was invented and the self-consistent field method was developed. According to the latter theory, the total wave function of a complex atom is expressed in terms of one-electron wave functions which are determined by the quantum numbers of the different electrons in an atom.

The wave function of an electron with the quantum numbers n, l, m_l can be written as

$$\Psi_{nlm_l} = \frac{1}{\sqrt{4\pi}} R_{nl}(r) Y_{lm_l}(\theta, \phi) \tag{13·16-1}$$

where Y_{lm_l} is the normalized spherical harmonic.

If we consider the electrons in a closed subshell of equivalent electrons (see §13·19), then the *mixed charge density* of the electronic cloud of this subshell of given n and l is written as

$$\rho_{nl}(r, r') = \sum_{m_l} \Psi^*_{nlm_l}(r, \theta, \phi) \Psi_{nlm_l}(r', \theta', \phi') \tag{13·16-2}$$

Substituting for Ψ_{nlm_l} from Eq. (13·16-1), we then get

$$\rho_{nl}(r, r') = \frac{1}{4} \sum_{m_l} R_{nl}(r) R_{nl}(r') Y^*_{lm_l}(\theta, \phi) Y_{lm_l}(\theta', \phi')$$

$$= \frac{2l+1}{4\pi} R_{nl}(r) R_{nl}(r') P_l(\cos \gamma) \tag{13·16-3}$$

where we have made use of the addition theorem of the spherical harmonics according to which

$$P_l(\cos \gamma) = \frac{4\pi}{2l+1} \sum_{m_l} Y^*_{lm_l}(\theta, \phi) Y_{lm_l}(\theta', \phi') \tag{13·16-4}$$

Here γ is the angle between the two vectors $\mathbf{r}$ and $\mathbf{r}'$ and $P_l(\cos \gamma)$ is the Legendre polynomial of lth order.

$$\cos \gamma = \cos \theta \cos \theta' + \sin \theta \sin \theta' \cos(\phi - \phi') \tag{13·16-5}$$

γ does not depend on the choice of the polar axis. Eq. (13·16-5) tells us that the closed subshell (n, l) is spherically symmetric.

According to Bohr, most of the electrons in an atom make up closed subshells (see §13·13). The few remaining outer electrons are placed in a subshell outside the closed subshells which is usually unfilled (except in the case of the rare earth and a few other elements which have two such unfilled subshells).

We now consider the case of an alkali atom (e.g. sodium with $Z = 11$) which has a single valence electron outside the closed subshells. In this case the inner closed subshells have n, l values 1, 0 (1s), 2, 0 (2s) and 2,1 (2p). The corresponding wave functions are $\Psi_{100}, \Psi_{200}, \Psi_{21,-1}$ Ψ_{210}, Ψ_{211}.

So the mixed charge density for the subshells will be (see Eq. 13·16-3)

$$\rho(r, r') = \frac{1}{4\pi} \{R_{10}(r) R_{10}(r') + R_{20}(r) R_{20}(r') + 3.R_{21}(r) R_{21}(r') \cos \gamma\} \quad (13·16-6)$$

Notice that for the s states $l = 0$ and $P_l (\cos \gamma) = P_0 (\cos \gamma) = 1$; for the p-states, $l = 1$ and $P_1 (\cos \gamma) = \cos \gamma$.

The energy of the inner shell electrons is given by §§

$$W_0 = \langle \Psi(r) | H(r) | \Psi(r) \rangle$$
$$+ \frac{e^2}{4\pi\epsilon_0} \int\int \frac{2\rho(r, r') \rho(r', r) - |\rho(r, r')|^2}{|\mathbf{r} - \mathbf{r}'|^2} d\tau \, d\tau' \quad (13·16-7)$$

where $H(r)$ is the Hamiltonian.

Substituting for ρ from Eq. (13·16-6), W_0 can be found by integrating over the angles in Eq. (13·16-7). It will obviously depend upon the radial functions R_{10}, R_{20} and R_{21}.

Fock has solved the problem by choosing some reasonable initial trial functions with a small number of variational parameters. The following trial functions were chosen;

$$R_{10} = a \exp(-\alpha r), \quad R_{20} = b \left\{1 - \frac{1}{3}(\alpha + \beta)r\right\} \exp(-\beta r)$$
$$R_{21} = cr \exp(-\gamma r) \quad (13·16-8)$$

a, b and c are found from the normalization condition:

$$|R_{nl}(r)|^2 = 1 \quad (13·16-9)$$

These are thus expressed in terms of α, β and γ so that W_0 is found as a linear fractional function of α, β and γ. By equating the derivatives of W_0 w.r.t. α, β and γ to zero these parameters are the found:

$$\alpha = 10·68, \quad \beta = 4·22, \quad \gamma = 3·49 \quad (13·16-10)$$

---

§§ See *ibid*.

Thus the trial functions reduce to

$$R_{10}(r) = 69 \cdot 80 \exp(-10 \cdot 68\, r)$$
$$R_{20}(r) = 13 \cdot 602\, (1 - 4 \cdot 967\, r) \exp(-4 \cdot 22\, r)$$
$$R_{30}(r) = 26 \cdot 276 \exp(-3 \cdot 49\, r) \tag{13.6-11}$$

These trial functions give better approximation value for W_0 than from the variational equations.

The functions (13·16-11) give the value of the energy of the ionized sodium atom which is in good agreement with the value obtained by using more exact functions found by numerical solution method.

To find the energies of the valence electrons in different states of the sodium atom, it is necessary to solve an integro–differential equation obeyed by the valence electron which is of the type

$$\{H(r) + V(r)\}\, \Psi(r) - \frac{e^2}{4\pi\epsilon_0} \int \frac{\rho(r, r')}{|\mathbf{r} - \mathbf{r}'|}\, d\tau' = E\, \Psi(r) \tag{13.16-12}$$

where $\rho(r, r')$ is the mixed charge density introduced earlier. It is given by

$$\rho(r, r') = \sum_{j=1}^{p} \Psi_j^*(r')\, \Psi_j(r) \tag{13.16-13}$$

when there are $2p$ inner shell electrons with compensated spins, i.e. two oppositely aligned spins in each state. Assuming that the wave function Ψ_{nlm_l} for the valence electron to be of the type (13·16-1), $R_{nl}(r)$ can be found by solving certain equations numerically. The energies of the lower states (optical terms) are found to agree fairly well with the observed values.

13·17 Results of Hartree-Fock calculations

Results of some Hartree-Fock calculations for the neon atoms are shown graphically in Fig. 13·8 (see *Physics of Atoms and Molecules* by Bransden and Joachain) in which the radial functions $R_{nl}(r)$ for the states 1s, 2s and 2p are plotted as functions of r/a_0, a_0 being the Bohr radius Fig. 13·9 shows the radial density function $D(r)$ as a function of r/a_0 for singly ionized rubidium atom which brings out the shell structure.

13·18 Corrections to the central field approximation

Hartree-Fock method is a variational method and hence the energies and the wave functions calculated by the method cannot be expected to agree with the "exact" values of these quantities. The difference between the exact values of the energies and wave functions of an atomic system from their Hartree-Fock values is known as *correlation effect*. The correlation effect arises due to the difference between the actual electrostatic

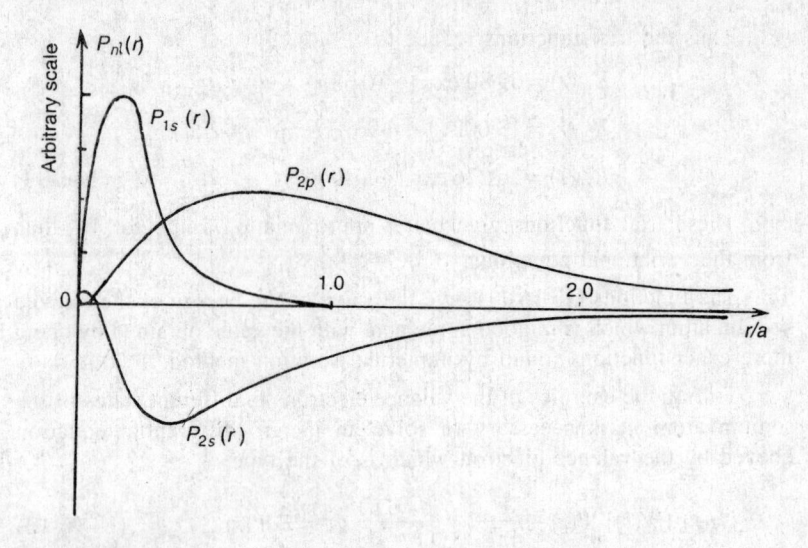

Fig 13·8. Radial function $R_{np}(v)$ for neon for the states 1s, 2s and 2p.

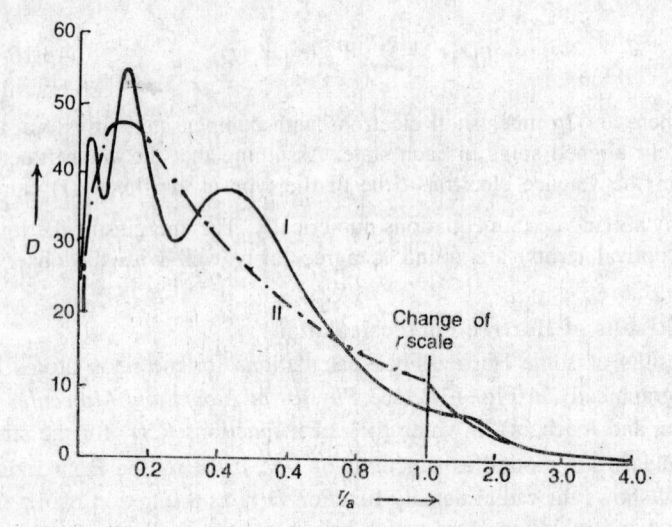

Fig 13·9. Radial density function $D(r)$ for normal rubidium atom (I) by Hartree's method and (II) by Thomas – Fermi method.

interaction between the electrons and the average interaction that is included in the central field. These have been studied by perturbation theory, starting with the H-F energies and wave functions as zero-order approximations. The *correlation energies* are usually small, being only about one percent or less of the "exact" values. However the Hartree-Fock wave functions may be quite inaccurate in some cases and hence serious errors may be made in calculating the matrix elements, transition rates or hyperfine structure effects, using these wave functions.

Spin-orbit coupling :

Apart from the correlation correction mentioned above, a number of other corrections must be introduced in the central field approximation calculation. The most important of these is the spin-orbit coupling correction. As in the case of the hydrogen atom, the spin-orbit correction term for a many electron atom can be written as (see § 13·1)

$$H_{LS} = \sum_i \xi(r_i)\, \mathbf{l}_i \cdot \mathbf{s}_i \qquad (13\cdot18\text{-}1)$$

The function $\xi(r_i)$ is given by

$$\xi(r_i) = \frac{1}{2m^2 c^2}\, \frac{1}{r_i}\, \frac{dV(r_i)}{dr_i} \qquad (13\cdot18\text{-}2)$$

The contribution of the spin-orbit term (13·18-1) can be shown to vanish for the electrons in the closed subshells so that the sum in Eq. (13·18-1) has to be taken only for the electrons outside the closed subshells.

The total parity and the total angular momentum $\mathbf{J}$ are constants of motion where

$$\mathbf{J} = \mathbf{L} + \mathbf{S} = \sum_{i=1}^{N} \mathbf{j}_i \qquad (13\cdot18\text{-}3)$$

$$\mathbf{j}_i = \mathbf{l}_i + \mathbf{s}_i \qquad (13\cdot18\text{-}4)$$

for the individual electrons. Here $\mathbf{L}$ stands for the total orbital angular momentum operator and $\mathbf{S}$ is the spin angular momentum of the atom.

It may be noted that the correlation correction to the central field approximation is usually large compared to that for H_{LS} for atoms (or ions) with small or intermediate values of Z. This is known as the *Russel-Saunders* or $L-S$ coupling. The case in which the correction due to H_{cor} is small compared to that due to H_{LS} occurs for atomic systems with large Z and is known as $j-j$ coupling.

To calculate the effect of the spin-orbit interaction (13·18-1), we consider the sum $H = H_c + H_{cor}$ to be the unperturbed Hamiltonian where H_c is the central force Hamiltonian. Then the approximate eigenfunctions and energy eigenvalues of this unperturbed Hamiltonian are found.

The unperturbed Hamiltonian H does not contain the spin-orbit term ; so it commutes with the total angular momentum **J** as also with the total orbital angular momentum **L** and the total spin angular momentum **S**. So the eigenvalues of H are dependent on L and S. But they are independent of the quantum numbers m_L and m_S. Thus these energy levels are $(2L+1)(2S+1)$ fold degenerate.

For a completed subshell of given n, l containing $2(2l+1)$ equivalent electrons, the values of L and S are obtained by the vector addition of the individual electron orbital angular momenta (l_i) and of the individual spin angular momenta (s_i) respectively consistent with the exclusion principle. For the completed subshell the values of m_L and m_S are

$$m_L = \Sigma\, m_{li} = 0 \qquad (13\cdot18\text{-}5a)$$
$$m_S = \Sigma\, m_{si} = 0 \qquad (13\cdot18\text{-}5b)$$

The spin of the equivalent electrons are aligned oppositely in pairs. This means $L=0$ and $S=0$ for a completed subshell. Thus for $2(2l+1)$ equivalent electrons in a completed subshell, the only possible term is $^1S_0$.

We now consider the case of atoms with incomplete subshells. Since $L=0$ and $S=0$ for the filled subshells only the electrons outside the completed subshells determine the L and S values for the atom. The following possibilities arise (see § 6·8 in Ch. VI) :

(a) Non-equivalent electrons outside the completed subshells :

Suppose two non-equivalent electrons are in subshells with different n so the there is no possibility of the exclusion principle being violated. If l_1 and l_2 are the orbital angular momenta of the two electrons in the subshells (n, l_1) and (n', l_2), the possible total orbital angular momentum values are

$$L = |l_1 - l_2|, |l_1 - l_2| + 1, \ldots (l_1 + l_2) \qquad (13\cdot18\text{-}6)$$

Similarly the possible S values are

$$S = |s_1 - s_2|, |s_1 - s_2| + 1, \ldots (s_1 + s_2) \qquad (13\cdot18\text{-}7)$$

Since s can be only $1/2$ for each electron, the possible S values are $S=0$ and 1 (a singlet and a triplet respectively) for two electrons.

As an example for the configuration $(np^1), (n'p^1)$, $l_1=1$ and $l_2=1$ for the two non-equivalent electrons with different n values. The possible L values are $L=0, 1, 2$ so that the possible terms are $^1S, ^1P, ^1D, ^3S, ^3P$ and ^3D.

In the case of three non-equivalent electrons in the subshells $nl_1, n'l_2, n''l_3$, the l value of the third electron has to be added to the possible L values for the first two. Similarly for S. Thus if there is a third non-equivalent electron $n''d^1$ (with $l_3 = 2$), the possible S values for the three are obtained by adding $s_3 = 1/2$ for the third electron to the S values for the others two in the previous case. So the possible S values for the three electrons are $S = 1/2$ (doublet) and $S = 3/2$ (quartet). So the possible terms for the three non-equivalent electrons considered above are

($\because L = 0, 1, 2, 3$ and 4)

^2S,	^2P,	^2D,	^2F,	^2G,	^4S,	^4P,	^4D,	^4F,	^4G
2	4	6	4	2		2	3	2	

A particular term in the above table may be formed by different possible combinations of the l and s values for the three electrons. These possible numbers of identical terms are indicated by the numbers below the different terms.

(b) Equivalent electrons outside the completed subshell:

In this case certain L and S values are not possible due to the exclusion principle.

First we consider the case of two equivalent electrons in the $l = 0$ subshells, i.e. the case of the configuration ns^2 As seen above $m_L = 0, m_s = 0$ for them since the ns^2 is a completed subshell. Obviously $L = l_1 + l_2 = 0$; S can have the values 0 or 1. However $S = 1$ is excluded by Pauli's exclusion principle. So there is only one possible term $^1S_0$.

Now consider the case of two equivalent p electrons (np^2). As seen before for two non-equivalent p electrons, the possible terms are $^1S, ^1P, ^1D, ^3S, ^3P$ and ^3D. However all these terms are not possible for two equivalent p electrons.

In this case since n and l are the same for both electrons, either n_l or m_s must be different for the two. If for example, the two p electrons have both l in the same direction, we get a D term and m_l is the same for both ($+1, 0$ or -1). So according to Pauli principle, the two electrons cannot both have $m_s = +1/2$ or both $m_s = -1/2$. Hence their spins can only be antiparallel for a D term giving ^1D only and ^3D is not possible. Similar argument shows that only ^1S term can occur. ^3S would violate the exclusion principle since $l = 0$ and $m_l = 0, m_s = 0$ for both electrons in this case.

We shall now show that only ^3P and not ^1P can occur. To do this we first write the possible m_l and m_s values of each p electron:

$m_l =$	1	0	-1	1	0	-1
$m_s =$	1/2	1/2	1/2	$-1/2$	$-1/2$	$-1/2$
States =	a	b	c	d	e	f

For the two equivalent p electrons, the possible combinations with m_L and m_S values are :

ab	ac	ad	ae	af
(1, 1)	(0, 1)	(2, 0)	(1, 0)	(0, 0)

bc	bd	be	bf
(−1, 1)	(1, 0)	(0, 0)	(−1, 0)

cd	ce	cf
(0, 0)	(−1, 0)	(−2, 0)

de	df
(1, −1)	(0, −1)

ef
(−1, −1)

In the above table, Pauli's exclusion principle has been taken into account so that either m_l or m_s for the two electrons are different in all cases.

The largest value of $m_L = 2$ must belong to a D term ($L = 2$) for which $m_S = 0$ (the combination ad). Thus it belongs to the ^1D term. The above table gives the following five combinations for this state.

^1D:(2,0) (1,0) (0,0) (−1,0) (−2,0)

The remaining ten combinations can be grouped as follows to give a ^3P and a ^1S term

^3P : (1, 1) (0,1) (−1,1) ; (1,0) (0,0) (−1,0) ; (1,−1) (0−1) (−1,−1)

^1S ; (0,0)

Thus the two equivalent p electrons give rise to ^1S, ^3P and ^1D terms. Pauli's exclusion principle excludes the ^1P, ^3S and ^3D terms which are also present in the case of two non-equivalent p-electrons in addition to the other three.

Hund's rule : The angular momenta **l** and **s** of the individual electrons have well defined meanings only if there is negligible interaction between them. In this case all terms of a given electron configuration would have the same energy (degenerate). However the spin-orbit interaction removes the degeneracy and the energy difference that appears between the different terms becomes larger for stronger interaction. F. Hund formulated an empirical rule according to which the term with the largest multiplicity (largest S) for a given configuration of *equivalent electrons* has the lowest energy and lies deepest in the energy level diagram. For a given S, the term with the maximum L has the lowest energy. This is known as *Hund's rule*.

As an example the possible terms for two equivalent d electrons

(d^2) are ^1S, ^1D, ^1G, ^3P and ^3F. Of these, the term ^3F with $S = 1$ and $L = 3$ lies deepest in accordance with Hund's rule.

Table 13·3
Terms of equivalent electrons

Electron Configuration	Terms
s^2	^1S
p^2	^1S, ^1D, ^3P
p^3	^2P, ^2D, ^4S
p^4	^1S, ^1D, ^3P
p^5	^2P
p^6	^1S
d^2, d^8	^1S, ^1D, ^1G, ^3P, ^3F
d^3, d^7	^2P, $^2D(2)$, ^2F, ^2G, ^2H, ^4P, ^4F
d^4, d^6	$^1S(2)$, $^1D(2)$, ^1F, $^1G(2)$, ^1I, $^3P(2)$, ^3D, $^3F(2)$, ^3G, ^3H, ^5D
d^5	^2S, ^2P, $^2D(3)$, $^2F(2)$, $^2G(2)$, ^2H, ^2I, ^4P, ^4D, ^4F, ^4G, ^6S

Another rule in this connection states that the multiplets formed from equivalent electrons are *regular* when less than half the shell is occupied, but *inverted* when more than half the shell is occupied.

From Hund's rule, it follows that pairs of electronic configurations in the unfilled subshells in which the numbers of equivalent electrons add up to produce a completed subshell have identical sets of allowed terms, e.g., np^1 and np^5, np^2 and np^4, nd^3 and nd^7, etc. (see Table 13·3).

Lande' interval rule :

In § 13·12 it was seen that for a many electron atom, the levels with the same n but different l are non-degenerate due to the penetration of the electrons with smaller angular momentum deeper into the atom so that the nuclear charge is less effectively screened for them.

A state can be specified by the quantum numbers J, L, S, m_J, m_L and m_S which are connected with the eigenvalues of the angular momentum operators through the following relations :

$$\mathbf{J}^2 = J(J+1)\hbar^2, \quad J_z = m_J \hbar$$
$$\mathbf{L}^2 = L(L+1)\hbar^2, \quad L_z = m_L \hbar$$
$$\mathbf{S}^2 = S(S+1)\hbar^2, \quad S_z = m_S \hbar \tag{13·18-8}$$

When the H_{LS} term is neglected, only two of the four quantum numbers are independent so that either the $<LS\,m_L\,m_S|$ representation or the $<LSJ\,m_J|$ representation will specify a state. The energy is, in this case, independent of m_L and m_S because of the spherical symmetry of the Hamiltonian w.r.t. the space and spin parts. This makes the levels $(2L+1)(2S+1)$ fold degenerate as stated above.

If the spin-orbit term H_{LS} given by Eq. (13·18-1) is included, and $H_{LS} < H_{cor}$ the total Hamiltonian $H = H_c + H_{cor} + H_{LS}$ commutes with the total angular momentum $\mathbf{J} = \mathbf{L} + \mathbf{S}$, though it does not commute with L and S so that L and S are not constants of motion. In this case we get Russel-Saunders or $L-S$ coupling and the level with given n and L splits into a number of fine structure components (multiplets) determined by the possible values of J which are

$$J = |L-S|, |L-S|+1, \ldots (L+S) \qquad (13\cdot18\text{-}9)$$

If $L > S$ the number of multiplets is $2S+1$ while if $S > L$ the numbers of multiplets is $2L+1$. In both cases the multiplicity is given by $2S+1$, so that the various multiplets are designated by the values of the multiplicity $2S+1$, the values of L and J : $^{2S+1}L_J$. (see Ch. VI). The energy difference between the multiplets is relatively small in this type (LS) of coupling and is given by the expectation value of H_{LS} for the wave function which depends on J, L, S, m_L and m_S.

The energy is independent of m_J so that the levels are $(2J+1)$–fold degenerate.

The difference in energy between states of different J is due to the $\mathbf{L}\cdot\mathbf{S}$ term and can be found from its expectation value or the diagonal matrix elements H'_{nm} in the first order perturbation theory (see Eq. 12·2-12). Since

$$J^2 = (\mathbf{L}+\mathbf{S})^2 = L^2 + S^2 + 2\mathbf{L}\cdot\mathbf{S}$$

we get

$$2\,\mathbf{L}\cdot\mathbf{S} = J^2 - L^2 - S^2$$

Written in terms of the expectation values this gives

$$E_J = \frac{A}{2}\{J(J+1) - L(L+1) - S(S+1)\} \qquad (13\cdot18\text{-}10)$$

where A is a constant. Thus the unperturbed level of given L and S splits into $(2S+1)$ sublevels if $L > S$ or $(2L+1)$ sublevels if $L < S$. The separation between two adjacent fine-structure levels E_J and E_{J-1} is given by

$$\begin{aligned}E_J - E_{J-1} &= \frac{A}{2}\{J(J+1) - L(L+1) - S(S+1) - (J-1)J \\ &\quad + L(L+1) + S(S+1)\} \\ &= AJ \end{aligned} \qquad (13\cdot18\text{-}11)$$

Quantum Mechanical Theory of Atomic Structure

Thus the splitting is linearly proportional to J. This is known as *Lande' interval rule*. Experimental results confirm this rule for the case of $L-S$ coupling.

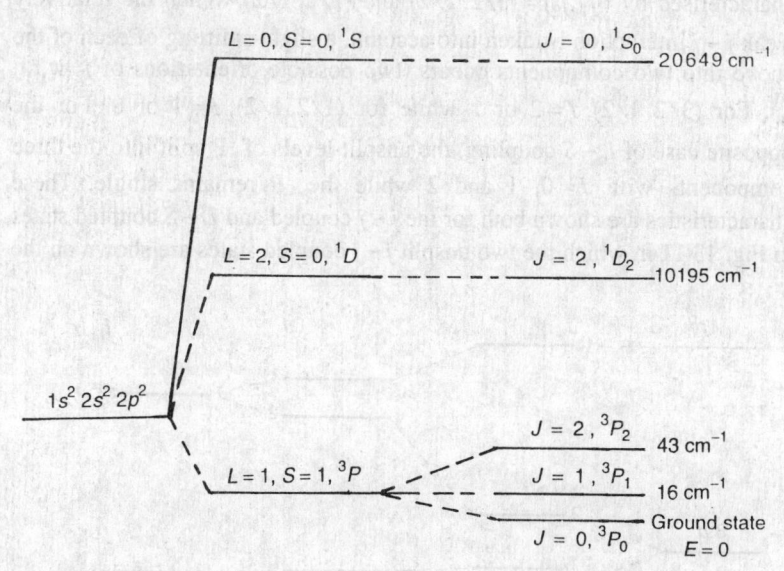

Fig. 13·10. Splitting of the ground state of Carbon ($Z=6$) for which $L-S$ coupling applies.

Fig. 13·10 shows the splitting of the ground state of the carbon atom ($Z=6$) for which $L-S$ coupling scheme applies.

$j-j$ *coupling*: If the spin-orbit energy H_{LS} is large compared to the correlation energy H_{cor}, each electron can be characterized by the quantum numbers n, l, j, m_j rather than by n, l, m_l, m_s where $(\mathbf{l}_i + \mathbf{s}_i)^2 = j_i (j_i + 1)\hbar^2$ and $l_{iz} + s_{iz} = j_{iz} = m_{ji}\hbar$. The electrostatic energy (correlation energy) then splits the states of different J where $\mathbf{J} = \sum_i \mathbf{j}_i$. This is known as the $j-j$ coupling scheme and is of interest for heavy atoms for which large $V(r)$ makes the spin-orbit energy for the individual electrons the dominant perturbation. This is due to the fact that while the spin-orbit energy for the individual electrons is proportional to Z^4 (see Eq. 13·1-26) the first order electrostatic energy correction is proportional to Z (see Eq. 13·9-11).

To illustrate $j-j$ coupling, we may consider the configuration $p^1 s^1$ which gives a $^3P_{0,1,2}$ and a $^1P_1$ term in $L-S$ coupling scheme. If, however, we assume (j,j) coupling the two electrons have the following (l, s) values : $l_1 = 1$ and $s_1 = 1/2$ for the p electron and $l_2 = 0$ and

$s_2 = 1/2$ for the s electron. Thus the possible values of j_1 are 3/2 and 1/2 while the only value of $j_2 = 1/2$. Thus the two possible states may be characterised by $(j_1, j_2) = (3/2, 1/2)$ and $(1/2, 1/2)$. When the relatively weak $j - j$ interaction is taken into account, a slight splitting of each of the above into two components occurs (two possible orientations of j_2 w.r.t. j_1). For $(3/2, 1/2), J = 2$ or 1 while for $(1/2, 1/2), J = 1$ or 0. For the opposite case of $L - S$ coupling, the unsplit levels of ^3P split into the three components with $J = 0$, 1 and 2 while the ^1P remains single. These characteristics are shown both for the $j - j$ coupled and $L - S$ coupled states in Fig. 13·11 in which the two unsplit $j - j$ coupled states are shown on the

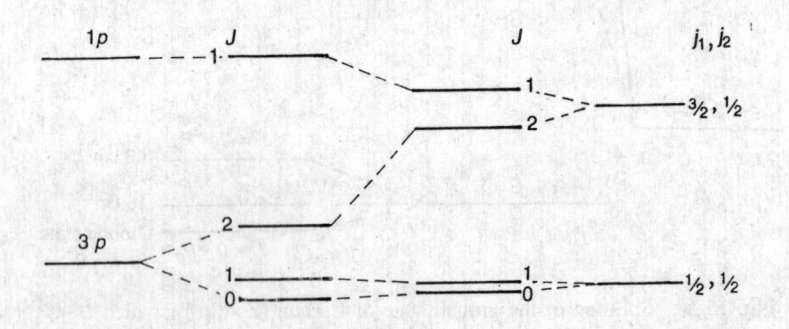

Fig. 13·11. Relative positions of the terms of a ps electronic configuration.

extreme right while the two unsplit $L - S$ coupled states are shown on the extreme left. As is evident the number of states in both cases eventually remains the same.

$j - j$ coupling is usually not found in pure form, though for some heavy atoms the spectra exhibit structures close to those predicted by $j - j$ coupling scheme..

Fig. 13·12 shows the transition from the $L - S$ coupling scheme to the $j - j$ coupling scheme for the first excited ^3P and ^1P terms of the carbon groups of elements which have

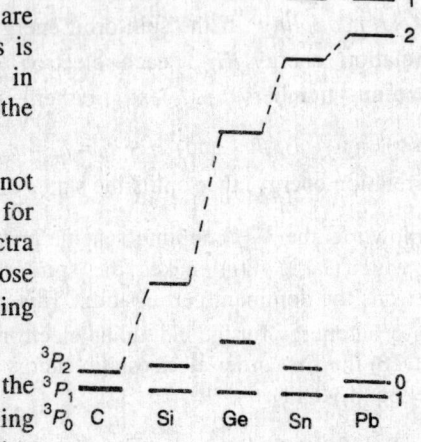

Fig. 13·12. Gradual transition from $L - S$ to $j - j$ coupling for the first excited $3p$ and $1s$ terms of the carbon group of elements

the configuration $(p^1 s^1)$. While carbon and silicon show almost pure $L-S$ coupling, Ge, Sn and Pb approach more and more to $j-j$ coupling. The effect is most prominent for the $J=2$ term which moves farther and farther away from the neighbourhood of the lowest $J=0$ term to that of the uppermost $J=1$ ($^1P_1$) terms.

For further details see *Atomic Spectra* and *Atomic Structure* by G. Hertzberg.

Spin-orbit interaction in alkali atoms :

In alkali atoms, there is a single valence electron in the outermost S-state. The inner electronic configuration constitutes completed shells. The valence electron moves in a non-Coulomb central field so that the Hamiltonian can be written as

$$H = -\frac{\hbar^2}{2m}\nabla^2 + V(r) + \xi(r)\mathbf{L}\cdot\mathbf{S} \qquad (13\cdot18\text{-}12)$$

$$= H_0 + \xi(r)\mathbf{L}\cdot\mathbf{S} \qquad (13\cdot18\text{-}13)$$

where $H_0 = -(\hbar^2/2m)\nabla^2 + V(r)$ is the unperturbed Hamiltonian. The last term on the r.h.s. of Eq. (13·18-13) is due the spin-orbit interaction which may be regarded as a perturbation.

Now the total angular momentum $\mathbf{J}$ is the vector sum of $\mathbf{L}$ and $\mathbf{S}$ so that we have $\mathbf{J} = \mathbf{L} + \mathbf{S}$ and

$$J^2 = L^2 + S^2 + 2\,\mathbf{L}\cdot\mathbf{S} \qquad (13\cdot18\text{-}14)$$

Further $\qquad J_z = L_z + S_z \qquad (13\cdot18\text{-}15)$

The scalar product $\mathbf{L}\cdot\mathbf{S}$ is defined in the usual way as

$$\mathbf{L}\cdot\mathbf{S} = L_x S_x + L_y S_y + L_z S_z \qquad (13\cdot18\text{-}16)$$

J^2 commutes with L^2, S^2 and J_z. But J^2 does not commute with L_z or S_z, since the scalar product $\mathbf{L}\cdot\mathbf{S}$ given by Eq. (13·18-16) does not commute with L_z or S_z. We thus conclude that the perturbed Hamiltonian H (Eq. 13·18-13) does not commute with L_z or S_z. So m_L and m_S are not good quantum numbers. Since J^2 commutes with J_z, it follows that $\mathbf{L}\cdot\mathbf{S}$ and hence H commutes with J_z so that m_J is a good quantum number. Hence the states can be represented by the set of quantum numbers n, L, J, m_J and $<n\,L\,J\,m_J|$ is a valid representation in the presence of the spin-orbit interaction.

To calculate the splitting of the different alkali levels for a given L, we have to find the expectation value or the diagonal matrix element of the $\mathbf{L}\cdot\mathbf{S}$ term (perturbation) for the wave functions $|\gamma L S m_L m_S>$ which corresponds to given L and S for a particular energy level. γ is a possible non-angular momentum quantum number. The matrix elements of the spin-orbit interaction term are the same as those of the operator $A\,(\mathbf{L}\cdot\mathbf{S})$ where A is a constant characteristic of unperturbed level. So the matrix elements are

$$<\gamma L S m_L m_S | H_{LS} | \gamma L S m'_L m'_S>$$
$$= A <\gamma L S m_L m_S | \mathbf{L}\cdot\mathbf{S} | \gamma L S m'_L m'_S> \qquad (13\cdot18\text{-}17)$$

In general H_{LS} is not diagonal. However linear combinations of these can be made using Clebach-Gordan coefficients (see Vol. II) to obtain the new unperturbed wave functions $|\gamma L S J m_J>$ which are the eigenfunctions of L^2, S^2, J^2 and J_z. We thus have the diagonal elements:

$$<\gamma L S J m_J | H_{LS} | \gamma L S J m_J>$$
$$= \frac{A}{2} \{j(j+1) - l(l+1) - s(s+1)\} \qquad (13\cdot18\text{-}18)$$

which is the same as Eq. (13·18-11) as already obtained (in units of $\hbar^2$). This gives the positions of split levels which in the case of a single electron alkali atom with $s = 1/2$ is a doublet with $j = l \pm 1/2$. The factor $A/2$ can be calculated by finding the expectation value of $\xi(r)$. The shift in energy of the level (γLS) thus becomes

$$\delta E_{nlsj} = <\xi(r)> \{j(j+1) - l(l+1) - s(s+1)\} \qquad (13\cdot18\text{-}19)$$

The expectation value $<\xi(r)>$ is obtained by using the radial part of the unperturbed wave functions.

It may be noted that the corrected level thus obtained is $(2j+1)$-fold degenerate with different values of $m_j = -j, -j+1, \ldots j-1, j$. The degeneracy w.r.t. m_j can be removed by the application of a magnetic field.

The doublet separation between the two levels j and $j-1$ is, as before, given by

$$E_j - E_{j-1} = \frac{A}{2} [j(j+1) - l(l+1) - s(s+1)$$
$$- (j-1)j + l(l+1) + s(s+1)]$$
$$= Aj \qquad (13\cdot18\text{-}20)$$

This is the empirically obtained Lande' interval rule (see Eq. 13·18-11).

If an approximation is made by assuming that the non-Coulomb central potential $V(r)$ can be replaced by the hydrogenic Coulomb potential ($-Ze^2/4\pi\epsilon_0 r$) for a one-electron atom (alkali atom), we get

$$<\xi(r)> = \frac{Ze^2\hbar^2}{8\pi\epsilon_0 m^2 c^2} \cdot \frac{Z^3}{a_0^3 n^3 l(l+1/2)(l+1)} \tag{13.18-21}$$

where a_0 is the Bohr radius (see § 13·1). This is valid for $l > 0$.

Doublet intensity :

To find the relative intensities of the two lines in an alkali doublet (e.g., the D_1 and D_2 lines of sodium), it is necessary to calculate the angular integrals in the expressions for the matrix element for the transitions in the dipole approximation (see § 14·1) assuming the radial integrals to be the same for such transitions as $np_{3/2} \longrightarrow n's_{1/2}$ and $np_{1/2} \longrightarrow n's_{1/2}$. When this is done we get the ratio 2 : 1 for the two lines of the doublet in the above case (*sp*-transition).

For further details see *Physics of Atoms and Molecules* by Bransden and Joachain ; *Quantum Mechanics* by L.I. Schiff.

Problems

1. Using the expression for the fine structure splitting of the levels of a one-electron atom, show that the doublet separation between the split levels decreases with increasing l. Hence show that the doublet splitting of the alkali spectral lines of the principal series decreases with increasing l while that of the sharp series is constant.

2. A free electron is acted upon by a uniform magnetic field along the z-axis. Write down the Hamiltonian and show that the Schrödinger equation can be reduced to a form similar to the reduced differential equation for the linear harmonic oscillator. Solve the equation and show that the energy spectrum is continuous for motion along z while it is discrete for motion in a plane perpendicular to z.

3. Find the magnetic induction B for which the lowest Zeeman sublevel of the $^2P_{3/2}$ state of sodium conicides with the upper Zeeman sublevel of the $^2P_{1/2}$ state. (14·28 T)

4. Using Hund's rule show that the ground states of the electronic configurations nd^3 and np^5 are $^4F_{3/2}$ and $^2P_{1/2}$ respectively. Calculate the magnetic moments of the atoms having the above outermost electronic configurations in their ground state (in Bohr magnetons). ($\sqrt{3/5}$; $\sqrt{20/3}$)

5. Show that the variables x and $\chi(x)$ in Eq. (13·14-18) are dimensionless.

6. Transform the differential equation (13·14-17) into the reduced equation (13·14-20) by the substitutions given in Eqs. (13·14-18) and (13·14-19).

7. Show that the electron density ρ can be expressed in terms of the variables x and χ as given by Eqs. (13·14-22a and b).

8. Show that the solution for the potential in Thomas-Fermi model for small r is given by Eq. (13·14-26).

9. If we define the radius of a neutral atom as the radius of a sphere $R(\alpha)$ within which a fixed fraction $(1-\alpha)$ of the total electronic charge of the atom is located then we can write.

$$\int_0^R 4\pi r^2 \rho(r)\, dr = (1-\alpha) Z$$

Apply Thomas-Fermi model to prove that $R(\alpha) \propto Z^{-1/3}$.

10. Prove that the Thomas-Fermi model improves with increasing Z.

14

ATOMS IN RADIATION FIELDS

14·1. Interaction of electromagnetic field with atomic systems

The study of the interaction of electromagnetic field with atomic systems is based on the theory of time-dependent perturbation discussed in § 12·5. The perturbation in this case varies harmonically with time and may be written in the form

$$2H' \cos \omega t' = H'\{\exp(i\omega t') + \exp(-i\omega t')\} \qquad (14\cdot1\text{-}1)$$

ω is the circular frequency of the e.m. field which may either be the electric or the magnetic vector.

Eq. (12·5-14) gives the time rate of variation of the transition amplitude in the first order theory,

$$\dot{a}_k^{(1)} = -\frac{i}{\hbar} H'_{kl}(t') \exp(i\omega_{kl} t') \qquad (14\cdot1\text{-}2)$$

H'_{kl} is the matrix element of the transition given by $H'_{kl} = \int \Psi_k^* H' \Psi_l d\tau = \langle \Psi_k | H' | \Psi_l \rangle$. It is assumed that the system is definitely in the state l of energy E_l initially. $a_k^{(1)}(t)$ obtained by the integration of Eq. (14·1-2) over time gives the amplitude of transition to the state k of energy E_k after a time t :

$$a_k = -\frac{i}{\hbar} \int_0^t H'_{kl} \exp(i\omega_{kl} t') dt', \quad (k \neq l) \qquad (14\cdot1\text{-}3)$$

Here $\omega_{kl} = (E_k - E_l)/\hbar$. We have dropped the superfix (1) in Eq. (14·1-3). We get, using Eq. (14·1-1),

$$a_k = -\frac{i}{\hbar} H'_{kl} \int_0^t \left[\exp\{i(\omega_{kl}+\omega)t'\} + \exp\{i(\omega_{kl}-\omega)t'\} \right] dt'$$

$$= -\frac{i}{\hbar} H'_{kl} \left[\frac{\exp\{i(\omega_{kl}+\omega)t'\}}{i(\omega_{kl}+\omega)} + \frac{\exp\{i(\omega_{kl}-\omega)t'\}}{i(\omega_{kl}-\omega)} \right]_0^t$$

$$= -H'_{kl} \left[\frac{\exp\{i(\omega_{kl}+\omega)t\}-1}{\hbar(\omega_{kl}+\omega)} + \frac{\exp\{i(\omega_{kl}-\omega)t\}-1}{\hbar(\omega_{kl}-\omega)} \right]$$
(14·1-4)

Two cases may arise :

Case I : If $\omega_{kl}+\omega \longrightarrow 0$ so that $\omega_{kl}=-\omega$, the first term is much larger than the second. This corresponds to the *stimulated emission* of radiation. The incident radiation of energy $\hbar\omega$ stimulates transition of the system from an initial state l of energy E_l to a final state k of lower energy E_k. The maximum probability of transition occurs when $\hbar\omega = E_l - E_k$.

Case II : If $\omega_{kl}-\omega \longrightarrow 0$ so that $\omega_{kl}=\omega$, the second term is much greater than the first. This corresponds to *resonance absorption* of radiation of energy $\hbar\omega$ which can cause the transition of the system to a state of higher energy $E_k > E_l$. The maximum probability of the transition occurs when $\hbar\omega = E_k - E_l$.

For ω far removed from ω_{kl}, the probability of transition is negligibly small.

Since only one term is important in the above expression at a time, we can write

$$a_k(t) = -H'_{kl} \frac{\exp\{i(\omega_{kl}\pm\omega)t\}-1}{\hbar(\omega_{kl}\pm\omega)}$$

$$= -H'_{kl} \frac{\exp\{i(\omega_{kl}\pm\omega)t/2\}}{\hbar(\omega_{kl}\pm\omega)}$$

$$\times \left[\exp\{i(\omega_{kl}\pm\omega)t/2\} - \exp\{-i(\omega_{kl}\pm\omega)t/2\} \right]$$

$$= -2iH'_{kl} \frac{\exp\{i(\omega_{kl}\pm\omega)t/2\}}{\hbar(\omega_{kl}\pm\omega)} \sin(\omega_{kl}\pm\omega)t/2 \quad (14·1\text{-}5)$$

The probability of transition is given by

$$|a_k(t)|^2 = \left| \frac{2H'_{kl}}{\hbar(\omega_{kl}\pm\omega)} \right|^2 \sin^2(\omega_{kl}\pm\omega)t/2 \quad (14·1\text{-}6)$$

The (+) sign corresponds to stimulated emission and the (−) sign to resonance absorption.

Eq. (14·1-6) can be used to calculate Einstein's B coefficient (see Ch. VII).

Charged particles in electromagnetic field :

The following treatment of the interaction of radiation with atomic systems is semi-classical. The e.m. field is treated classically by

Maxwell's equations while the particles interacting with the field by quantum mechanics.

The Hamiltonian of a particle of charge e in an electromagnetic field is given by (see Eq. (13·5-1)

$$H = \frac{1}{2m}\left(\frac{\hbar}{i}\nabla - e\mathbf{A}\right)^2 + e\phi \qquad (14\cdot1\text{-}7)$$

Here ϕ is the scalar potential while $\mathbf{A}$ is the magnetic vector potential. Assuming gauge invariance $\nabla \cdot \mathbf{A} = 0$, we then write the Schrödinger equation for an electron of mass m as (see Eq. (13·5-3)

$$i\hbar\frac{\partial \Psi}{\partial t} = \left\{-\frac{\hbar^2}{2m}\nabla^2 - \frac{Ze^2}{4\pi\epsilon_0 r} - \frac{i\hbar e}{m}\mathbf{A}\cdot\nabla + \frac{e^2}{2m}A^2\right\}\Psi(\mathbf{r},t) \quad (14\cdot1\text{-}8)$$

For a weak field, the term in $\mathbf{A}^2$ may be neglected. Treating the term linear in $\mathbf{A}$ as a small perturbation, we write,

$$H'(t) = -\frac{i\hbar e}{m}\mathbf{A}\cdot\nabla \qquad (14\cdot1\text{-}9)$$

The explicit time-dependance of H' is through the vector potential $\mathbf{A}$. If Ψ_κ is the unperturbed wave function for the state k, we can write

$$H_0\Psi_\kappa = E_k\Psi_\kappa \qquad (14\cdot1\text{-}10)$$

where
$$H_0 = -\frac{\hbar^2}{2m}\nabla^2 - \frac{Ze^2}{4\pi\epsilon_0 r} \qquad (14\cdot1\text{-}11)$$

For transition from an initial state l of energy E_l to a final state k of energy E_k, the first order transition amplitude at time t is given by an expression similar to Eq. (14·1-4):

$$\begin{aligned}a_k(t) &= -\left\{\frac{H'_{kl}}{\hbar}\frac{\exp[i(\omega_{kl}+\omega)t]-1}{\omega_{kl}+\omega} + \frac{H''_{kl}}{\hbar}\frac{\exp[i(\omega_{kl}-\omega)t]-1}{\omega_{kl}-\omega}\right\}\\ &= -\frac{2iH'_{kl}}{\hbar}\frac{\exp i(\omega_{kl}+\omega)t/2}{(\omega_{kl}+\omega)/2}\sin(\omega_{kl}+\omega)t/2\\ &= -\frac{2iH''_{kl}}{\hbar}\frac{\exp i(\omega_{kl}-\omega)t/2}{(\omega_{kl}-\omega)/2}\sin[\omega_{kl}-\omega]t/2 \quad (14\cdot1\text{-}12)\end{aligned}$$

As before the first term in Eq. (14·1-12) corresponds to the case of stimulated emission while the second term corresponds to resonance absorption.

The electric and the magnetic fields associated with the vector protential $\mathbf{A}$ can be written as (taking $\phi = 0$)

$$\mathbf{E} = -\frac{\partial \mathbf{A}}{\partial t} \quad\text{and}\quad \mathbf{B} = \nabla\times\mathbf{A} \qquad (14\cdot1\text{-}13)$$

We write
$$\mathbf{A} = 2\mathbf{A_0}\cos(\mathbf{k}\cdot\mathbf{r} - \omega t + \alpha)$$
$$= \mathbf{A_0}\exp\{i(\mathbf{k}\cdot\mathbf{r} - \omega t + \alpha)\} + c.c. \quad (14\cdot1\text{-}14)$$

where c.c. means the complex conjugate. The wave number k and ω are related through the equation $\omega/k = c$.

We then have

$$a_k(t) = -\frac{i}{\hbar}\frac{e}{m}\int_0^t dt'\left[\left\langle\Psi_k\middle|\mathbf{A_0}\cdot\exp i(\mathbf{k}\cdot\mathbf{r} - \omega t' + \alpha_\omega)\cdot\mathbf{p}\middle|\Psi_l\right\rangle\right.$$
$$\left. + \left\langle\Psi_k\middle|\mathbf{A_0}\exp -i(\mathbf{k}\cdot\mathbf{r} + \omega t' - \alpha_\omega)\cdot\mathbf{p}\middle|\Psi_l\right\rangle\right]\times\exp(i\omega_{kl}t')$$

$$= -\frac{e}{m}\int_0^t dt'\left[\left\langle\Psi_k\middle|\mathbf{A_0}\exp i(\mathbf{k}\cdot\mathbf{r} + \alpha_\omega)\cdot\nabla\middle|\Psi_l\right\rangle\right.$$
$$\times\exp i(\omega_{kl} - \omega)t'$$
$$\left. + \left\langle\Psi_k\middle|\mathbf{A_0}\exp\left\{-i\left((\mathbf{k}\cdot\mathbf{r} + \alpha_\omega)\right)\right\}\cdot\nabla\middle|\Psi_l\right\rangle\times\exp i(\omega_{kl} + \omega)t'\right]$$

$$= -\frac{e}{m}\left\langle\Psi_k\middle|\mathbf{A_0}\exp\left\{i(\mathbf{k}\cdot\mathbf{r} + \alpha_\omega)\right\}\cdot\nabla\middle|\Psi_l\right\rangle$$
$$\times\frac{\exp i(\omega_{kl} - \omega)t/2}{(\omega_{kl} - \omega)/2}\sin(\omega_{kl} - \omega)t/2$$

$$+ \left\langle\Psi_k\middle|\mathbf{A_0} - \frac{e}{m}\left\{-i(\mathbf{k}\cdot\mathbf{r} + \alpha_\omega)\right\}\cdot\nabla\middle|\Psi_l\right\rangle$$
$$\times\frac{\exp i(\omega_{kl} + \omega)t/2}{(\omega_{kl} + \omega)/2}\sin(\omega_k + \omega)t/2 \quad (14\cdot1\text{-}15)$$

Comparing with Eq. (14.1-12), we get

$$H''_{kl} = -i\frac{e\hbar}{m}\left\langle\Psi_k\middle|\mathbf{A_0}\exp i(\mathbf{k}\cdot\mathbf{r} + \alpha_\omega)\cdot\nabla\middle|\Psi_l\right\rangle$$

$$H'_{kl} = -i\frac{e\hbar}{m}\left\langle\Psi_k\middle|\mathbf{A_0}\exp\left\{-i(\mathbf{k}\cdot\mathbf{r} + \alpha_\omega)\right\}\cdot\nabla\middle|\Psi_l\right\rangle \quad (14.1\text{-}16)$$

The second term in Eq. (14.1-13) corresponds to stimulated emission while the first term corresponds to absorption.

Then from Eq. (14·1·14) we get
$$\mathbf{E} = -2\omega\mathbf{A_0}\sin(\mathbf{k}\cdot\mathbf{r} - \omega t + \alpha) \quad (14\cdot1\text{-}17)$$
$$\mathbf{B} = -2(\mathbf{k}\times\mathbf{A_0})\sin(\mathbf{k}\cdot\mathbf{r} - \omega t + \alpha) \quad (14\cdot1\text{-}18)$$

Atoms in Radiation Fields

The condition $\nabla \cdot \mathbf{A} = 0$ gives $\mathbf{k} \cdot \mathbf{A_0} = 0$ which shows that $\mathbf{A_0}$ is perpendicular to $\mathbf{k}$ so that the wave is transverse.

Further $\mathbf{E}$ and $\mathbf{B}$ are mutually perpendicular as Eqs. (14·1-17) and (14·1-18) show.

The energy carried by the wave is given by the Poynting vector

$$\mathbf{N} = \mathbf{E} \times (\mathbf{B}/\mu_0)$$

$$= \frac{4\omega}{\mu_0} \mathbf{A_0} \times (\mathbf{k} \times \mathbf{A_0}) \sin^2 (\mathbf{k} \cdot \mathbf{r} - \omega t + \alpha) \qquad (14\cdot1\text{-}19)$$

where $\mathbf{B}/\mu_0$ is the magnetic field intensity. $\mathbf{N}$ is the energy flow per unit area per second normally. When averaged over a complete period $2\pi/\omega$, we get

$$<\mathbf{N}>_t = \frac{4\omega}{\mu_0} \{\mathbf{k} A_0^2 - \mathbf{A_0}(\mathbf{k} \cdot \mathbf{A_0})\} <\sin^2 (\mathbf{k} \cdot \mathbf{r} - \omega t + \alpha)>$$

which gives ($\because \mathbf{k} \cdot \mathbf{A_0} = 0$)

$$<N>_t = 2\epsilon_0 c \omega^2 A_0^2 \qquad (14\cdot1\text{-}20)$$

where we have used the relation $c^2 = 1/\epsilon_0 \mu_0$.

The energy density of the e.m. field in vacuum is given by

$$\frac{1}{2}(\epsilon_0 E^2 + B^2/\mu_0) = \frac{|A_0|^2}{2}\left(4\epsilon_0 \omega^2 + \frac{4k^2}{\mu_0}\right)\sin^2(\mathbf{k} \cdot \mathbf{r} - \omega t + \alpha)$$

$$= 4\epsilon_0 \omega^2 |\mathbf{A_0}|^2 \sin^2(\mathbf{k} \cdot \mathbf{r} - \omega t + \alpha) \qquad (14\cdot1\text{-}21)$$

The average energy density over the period $2\pi/\omega$ is then

$$<\rho(\omega)>_t = 2\epsilon_0 \omega^2 |A_0|^2 \qquad (14\cdot1\text{-}22)$$

We can correlate this classical result derived from Maxwell's equations with the quantum description of the e.m. field. If $n(\omega)$ denotes the number of photons of angular frequency ω in a unit volume then

$$\rho(\omega) = n(\omega) \hbar \omega \qquad (14\cdot1\text{-}23)$$

Since $<N(\omega)>_t$ is a measure of the intensity $I(\omega)$ of the radiation for the angular frequency ω, we get

$$I(\omega) = \rho(\omega) c \qquad (14\cdot1\text{-}24)$$

The vector potential amplitude $\mathbf{A_0}$ can be related to the intensity of the radiation with the help of Eq. (14·1-21) for the range $\Delta \omega$:

$$|A_0|^2 = \frac{<N(\omega)>_t \Delta \omega}{2\epsilon_0 c \omega^2}$$

$$= \{I(\omega)/2\epsilon_0 c \omega^2\} \Delta \omega \qquad (14\cdot1\text{-}25)$$

The probability of transition to *a higher energy state* $E_k = E_l + \hbar\omega$ at t is, for the frequency range $\Delta\omega$ (see Eq. 14·1-12)

$$|a_k(t)|^2 = \sum_{\omega} \frac{4|H''_{kl}|^2 \sin^2(\omega_{kl}-\omega)t/2}{\hbar^2(\omega_{kl}-\omega)^2} \qquad (14\cdot1\text{-}26)$$

Here we have summed over all possible angular frequencies in the neighbourhood of $\omega_{kl} = (E_k - E_l)/\hbar$.

$$\therefore \quad |a_k(t)|^2 = \sum_{\omega} \frac{2e^2 t}{m^2 \epsilon_0 c \omega^2} I(\omega)\,\Delta\omega$$

$$\times \left| <\Psi_k \left| \exp(i\mathbf{k}\cdot\mathbf{r})\nabla \right| \Psi_l> \right|^2 \frac{\sin^2(\omega_{kl}-\omega)t/2}{(\omega_{kl}-\omega)^2 t} \qquad (14\cdot1\text{-}27)$$

The gradient is to be evaluated for the polarization direction along **A**. The contributions to the probability for the different frequencies are additive. Each frequency range $\Delta\omega$ can be made infinitesimally small and the summation in Eq. (14·1-28) can be replaced by an integral. Since the time factor has a sharp maximum at $\omega = \omega_{kl}$ we can evaluate the terms involving ω at $\omega = \omega_{kl}$ while performing the integration over ω and write

$$\int_{-\infty}^{\infty} \frac{\sin^2(\omega_{kl}-\omega)t/2}{(\omega_{kl}-\omega)t}\,d\omega = \frac{\pi}{2} \qquad (14\cdot1\text{-}28)$$

So we get finally for transition to a state of higher energy (resonance absorption).

$$|a''_k(t)|^2 = \frac{\pi e^2 t}{\epsilon_0 m^2 c \omega_{kl}^2} I(\omega_{kl})$$

$$\times \left| <\Psi_k | \exp(i\mathbf{k}\cdot\mathbf{r}) \nabla_A | \Psi_l> \right|^2 \qquad (14\cdot1\text{-}29)$$

Similarly the probability of transition to a state of lower energy (stimulated emission) is given by

$$|a'_k(t)|^2 = \frac{\pi e^2 t}{\epsilon_0 m^2 c \omega_{kl}^2} I(\omega_{kl})$$

$$\times \left| <\Psi_k | \exp(-i\mathbf{k}\cdot\mathbf{r}) \nabla_A | \Psi_l> \right|^2 \qquad (14\cdot1\text{-}30)$$

Electric dipole transition :

In many cases, the spatial variation of the electric vector over the dimensions r of the atomic system can be neglected. For example, in the case of visible radiation the wavelength $\lambda \sim 10^{-7}$ m, while the size of the atomic system is $r \sim 10^{-10}$ m so that $kr = r/\lambda \sim 10^{-3} << 1$. Hence the exponential in Eq. (14·1-29) or (14·1-30) can be approximated as $|\exp(i\mathbf{k}\cdot\mathbf{r})| \sim 1$ and $|\exp(-i\mathbf{k}\cdot\mathbf{r})| \sim 1$.

The integrals in these equations then reduce to

$$\int \Psi_k^* \exp(\pm i\mathbf{k}\cdot\mathbf{r}) \nabla_A \Psi_l \, d\tau \approx \int \Psi_k^* \nabla_A \Psi_l \, d\tau$$

$$= \frac{i}{\hbar} \int \Psi_k^* \hat{p}_A \Psi_l \, d\tau = \frac{i}{\hbar} (p_A)_{kl} \qquad (14\cdot1\text{-}31)$$

Here p_A is the component of the momentum operator along the direction of polarization of the vector potential $\mathbf{A}$ for the incident radiation. $(p_A)_{kl}$ is the matrix element of $\hat{p}_A$ between the states k and l.

We can apply Heisenberg's equation of motion (10·17-5) to the dynamical variable $\mathbf{r}$ to get

$$\mathbf{p} = m\dot{\mathbf{r}} = -\frac{m}{i\hbar}[\mathbf{r}, H_0] = -\frac{m}{i\hbar}(\mathbf{r}H_0 - H_0\mathbf{r}) \qquad (14\cdot1\text{-}32)$$

so that we get

$$(p_A)_{kl} = <\Psi_k|p_A|\Psi_l> = -\frac{m}{i\hbar} <\Psi_k|r_A H_0 - H_0 r_A|\Psi_l|$$

$$= -\frac{m}{i\hbar}\left\{<\Psi_k|r_A H_0|\Psi_l> - <\Psi_k|H_0 r_A|\Psi_l>\right\}$$

$$= -\frac{m}{i\hbar}(E_l - E_k)<\Psi_k|r_A|\Psi_l>$$

$$= -\frac{m\omega_{kl}}{i}(r_A)_{kl} \qquad (14\cdot1\text{-}34)$$

Using the above result, we get the transition probability per unit time from Eq. (14·1-30):

$$W_{kl} = \frac{1}{t}|a_k(t)|^2 = \frac{\pi e^2}{\epsilon_0 m^2 c \omega_{kl}^2} I(\omega_{kl}) \frac{m^2 \omega_{kl}^2}{\hbar^2} |(r_A)_{kl}|^2$$

$$= \frac{\pi e^2}{\epsilon_0 c \hbar^2} I(\omega_{kl}) |(r_A)_{kl}|^2 \qquad (14\cdot1\text{-}34)$$

Writing the electric dipole moment as $\mathbf{D} = e\mathbf{r}$ we get the rate of transition at time t as

$$W_{kl} = \frac{1}{t}|a_k(t)|^2 = \frac{\pi}{\epsilon_0 c \hbar^2} I(\omega_{kl}) |(D_A)_{kl}|^2 \qquad (14\cdot1\text{-}35)$$

$(D_A)_{kl}$ is the matrix element of the *electric dipole moment* in the direction of polarization of the incident wave between the states k and l. If $(D_A)_{kl}$ is non-vanishing, the transition is allowed. If $(D_A)_{kl} = 0$, the transition is forbidden. For forbidden transitions, higher terms in the series expansion of $\exp(\pm i\mathbf{k}\cdot\mathbf{r})$ must be taken into account. These will give magnetic dipole, electric quadrupole etc. transitions.

If θ is the angle between **r** and the direction of polarization **A,** then a factor $\cos^2\theta$ will appear in the expression for the transition probability (Eq. 14·1-34 or 14·1-35). For unpolarized incident radiation, the average value of $\cos^2\theta$ has to be taken which is 1/3. The rate of absorptive transition from l to k which is also equal to the rate of stimulated emission then becomes (see Eq. 14·1-34).

$$W_{kl} = \frac{\pi e^2}{3\epsilon_0 c \hbar^2} I(\omega_{kl}) |(r_A)_{kl}|^2 \qquad (14\cdot1\text{-}36)$$

This gives the following expression for the probability of spontaneous emission of radiation in the dipole approximation (see *Physics of Atoms and Molecules* by Bransden and Joachain) :

$$W_{kl}^{sp} = \frac{\omega_{kl}^3}{3\epsilon_0 \pi \hbar c^3} |(D_A)_{kl}|^2 \qquad (14\cdot1\text{-}37)$$

Relation with Einstein's A and B coefficients :

We saw in Ch. VII, that the rate of stimulated emission of radiation of frequency ν between two states 1 and 2 is given by $B_{21} n_2 u_\nu$ where n_2 is the number of atoms in the upper energy state E_2 and u_ν is the energy density of the radiation. B_{21} is called Einstein's *B*-coefficient. If A_{21} is a measure of the rate of spontaneous transition from the upper energy state (2) of energy E_2 to the lower energy state (1) of energy E_1, then we have the following relations :

$$\frac{A_{21}}{B_{21}} = \frac{\hbar\omega^3}{\pi^2 c^3} \qquad (14\cdot1\text{-}38)$$

$$B_{12} = B_{21} \qquad (14\cdot1\text{-}39)$$

(See Eq. 7·1-5 in Ch. VII).

These lead to the well-known radiation formula due to Planck :

$$\rho_\omega = \frac{\hbar\omega^3}{\pi^2 c^3} \cdot \frac{1}{\exp(\hbar\omega/kT) - 1} \qquad (14\cdot1\text{-}40)$$

Since B_{12} is the rate of absorption of radiation per unit energy density ρ_ω where $\rho_\omega = I_\omega/c$, I_ω being the intensity, we get, using Eq. (14·1-36), for transition between the states l and k

$$B_{kl} = \frac{W_{kl}}{\rho} = \frac{\pi e^2}{3\epsilon_0 \hbar^2} (r_{kl})^2 \qquad (14\cdot1\text{-}41)$$

The coefficient A_{21} for the rate of spontaneous transition is then given by (for the transition $l \to k$)

$$A_{kl} = \frac{\hbar \omega_{kl}^3}{\pi^2 c^3} B_{kl} = \frac{e^2 \omega_{kl}^3}{3 \epsilon_0 \pi c^3 \hbar} (r_{kl})^2$$

$$= \frac{\omega_{kl}^3}{3 \epsilon_0 \pi c^3 \hbar} (D_{kl})^2 \tag{14.1-42}$$

which is the same as the expression for the spontaneous transition probability given by Eq. (14.1-37).

14.2 Dipole selection rules for one electron atom

The probability of transition in the dipole approximation given by Eq. (14.1-36) depends on the matrix element D_{kl} of the electric dipole moment of the atom. In order that the atom may radiate $D_{kl} \neq 0$. We shall now investigate the conditions under which the matrix element of the electric dipole moment of the single-electron atom (e.g., H or alkalis) may be non-vanishing.

Assuming the potential $V(r)$ in the unperturbed Hamiltonian to be spherically symmetric, the energy eignfunctions of the one-electron atom can be written as a product of the radial function and the spherical harmonies $Y_{lm}(\theta, \varphi)$ where

$$Y_{lm}(\theta, \varphi) = (-1)^m C_{lm} P_l^m (\mu) \exp(im\varphi) \tag{14.2-1}$$

Here $P_{lm}(\mu)$ is the associated Legindre function and $\mu = \cos\theta$. The magnetic quantum number m can take up the values : $m = 0, \pm 1, \pm 2, \ldots \pm l$ while the azimuthal quantum number can be $l = 0, 1, 2, \ldots n-1$; n is the principal quantum number.

$\mathbf{r}$ has the component $r \cos\theta$ along z so that the component of the matrix element along z depends on the integral

$$\int \cos\theta \, P_l^m (\cos\theta) P_{l'}^{m'} (\cos\theta) \, d(\cos\theta)$$

$$= \int_{-1}^{1} \mu \, P_l^m (\mu) P_{l'}^{m'} (\mu) \, d\mu \tag{14.2-2}$$

To evaluate the above integral we make use of the recurrence relations :

$$\cos\theta \, Y_{lm} = \left\{ \frac{(l+1+m)(l+1-m)}{(2l+1)(2l+3)} \right\}^{1/2} \cdot Y_{l+1, m}$$

$$+ \left\{ \frac{(l+m)(l-m)}{(2l+1)(2l-1)} \right\}^{1/2} Y_{l-1, m} \tag{14.2-3}$$

$$\sin\theta\, Y_{lm} = \left[-\left\{ \frac{(l+1-m)(l+2-m)}{(2l+1)(2l+3)} \right\}^{1/2} \cdot Y_{l+1,m-1} \right.$$
$$\left. + \left\{ \frac{(l+m)(l+m-1)}{(2l+1)(2l+3)} \right\}^{1/2} \cdot Y_{l-1,m-1} \right] \exp(i\varphi) \quad (14\cdot2\text{-}4)$$

Orthonormality relation of the spherical harmonics give

$$\int Y_{lm}\, Y_{l'm'}\, d\Omega = \int_0^{2\pi} d\varphi \int_0^{\pi} \sin\theta\, d\theta \cdot Y_{lm}^{*}\, Y_{l'm'} = \delta_{ll'}\, \delta_{mm'} \quad (14\cdot2\text{-}5)$$

The above equation shows that the integral vanishes unless $l = l'$ and $m = m'$. Thus the selection rule for the magnetic quantum number gives $\Delta m = m - m' = 0$ where m and m' refer to the two states between which the transition takes place.

With the help of the recurrence relations the integration in the case of the z component of the matrix element of the dipole moment can be split up into two integrals of the type

$$\int_{-1}^{1} P_{l+1}^{m}(\mu)\, P_l^{m}(\mu)\, d\mu \quad (14\cdot2\text{-}6a)$$

and

$$\int_{-1}^{1} P_{l-1}^{m}(\mu)\, P_l^{m}(\mu)\, d\mu \quad (14\cdot2\text{-}6b)$$

These will be zero unless $l' = l+1$ or $l' = l-1$ respectively. These therefore give the selection rules $\Delta l = \pm 1$.

To calculate the components of the matrix elements along x and y it is convenient to calculate the matrix elements of $x + iy$ and $x - iy$. A similar treatment as above shows that the matrix element for $x + iy$ which involves $\sin\theta \exp(i\varphi)$ vanishes unless $m' = m - 1$ and $l' = l \pm 1$; the matrix element for $x - iy$ which involves $\sin\theta \exp(-i\varphi)$ vanishes unless $m' = m + 1$ and $l' = l \pm 1$. Finally we get the selection rules for the one-electron atom

$$\Delta l = \pm 1, \quad \Delta m = 0 \pm 1 \quad (14\cdot2\text{-}7)$$

Parity selection rules :

The parity of a state is determined by the value of the azimuthal quantum number l and is even for $l = 0$, even, while it is odd for l odd.

Since l must change by one unit in dipole transition as seen above, the parity of the sate must change (Yes) in dipole transition.

Polarization of the emitted radiation :

The polarization of the emitted radiation is determined by the dipole matrix element $\mathbf{D}_{kl}$. When $\Delta l = 0$ and $\Delta m = 0$, only the z-component of the matrix element is non-vanishing. The emitted radiation is then polarized along the z-axis if viewed from a direction in the $x - y$ plane. There is no radiation along the z axis in this case.

If, on the other hand, $\Delta l = 0$ and $\Delta m = \pm 1$, the x and y components of the matrix element are non-vanishing and are $90°$ out of phase. The z-component vanishes. When viewed along the z-axis, one observes left and right circularly polarized light. This can be seen from the fact that $x + iy$ represents a left circularly polarized light while $x - iy$, represents a right circularly polarized light.

If viewed in the $x - y$ plane, the radiation appears linearly polarized perpendicular to the z-axis. These results are of interest in connection with the study of Zeeman effect.

Dependence on spin :

In the above discussion the spin of the electron was not taken into account. In fact the electric dipole operator $\hat{\mathbf{D}} = e\mathbf{r}$ and hence the matrix element of the electric dipole operator does not depend on the spin of the electron.

Spin of the photon :

The selection rules for the electric dipole transitions are intimately connected with the spin of the photon. Since the e.m. waves are transverse, it is obvious that the photon cannot have a component of the orbital angular momentum in the direction of proporgation of the radiation. On the other hand the orbital angular momentum selection rule for dipole transition requires that l must change by one unit ($\Delta l = \pm 1$). To conserve angular momentum, the photon should then carry away one unit of angular momentum which must necessarily be due to the intrinsic spin of the photon. We therefore conclude that the photon has a spin of unit magnitude ($\hbar$).

14.3 Forbidden transitions

If the matrix element of the electric dipole vanishes, the transition may occur through higher order terms in the expansion of $\exp(i\mathbf{k}\cdot\mathbf{r})$ in Eq. (14.1-29) or (14.1-30). We write the matrix element

$$M_{kl} = \langle \Psi_k | \exp(i\mathbf{k}\cdot\mathbf{r}) \nabla_A | \Psi_l \rangle \simeq \langle \Psi_k | ik_z z \frac{i}{\hbar} p_A | \Psi_l \rangle$$

$$= \langle \Psi_k | ik_z z \frac{i}{\hbar} m\dot{x} | \Psi_l \rangle$$

$$= -\frac{m\omega_{kl}}{c\hbar} \langle \Psi_k | z\dot{x} | \Psi_l \rangle \qquad (14.3\text{-}1)$$

assuming the propagation to be along the z-direction and the polarization along the x-axis ($A = A_x$) and neglecting the first term in the expansion of $\exp(i\mathbf{k}\cdot\mathbf{r})$.

Writing the y-component of the orbital angular momentum as
$$L_y = zp_x - xp_z = m(z\dot{x} - x\dot{z})$$
we have
$$mz\dot{x} = L_y + mx\dot{z}$$
We then get since $\dot{z} = i\omega z$

$$M_{kl} = -\frac{m}{\hbar}\left\langle \Psi_k \left| \omega\frac{z}{c}\dot{x} \right| \Psi_l \right\rangle$$

$$= -\frac{\omega_{kl}}{2\hbar c}\left\langle \Psi_k | L_y | \Psi_l \right\rangle - \frac{im\,\omega_{kl}^2}{2\hbar c}\left\langle \Psi_k | zx | \Psi_l \right\rangle \quad (14\cdot 3\text{-}2)$$

Both terms are of the same order of magnitude. The first term represents the matrix element of the orbital magnetic moment $\mu_y = eL_y/2m$. The second represents the matrix element of the electric quadruple moment of the atom. Both terms are smaller than the electric dipole moment contribution by a factor of the order of the fine structure constant α.

In addition to the contribution from the orbital magnetic moment, we must also take into account the contribution due to the spin magnetic moment (intrinsic magnetic moment).

The selection rules for magnetic dipole radiation are

$$\Delta l = 0, \Delta j = 0, \pm 1 \quad (j = 0 \text{ to } j' = 0 \text{ forbidden})$$

$$\Delta m_j = 0, \pm 1$$

Here $\mathbf{J} = \mathbf{L} + \mathbf{S}$ is the total angular momentum, j being the total angular momentum quantum number. Obviously there is no change of parity (No), since l does not change.

For the electric quadruple transition, parity does not change (No) since the matrix elements of the products, such as zx, xy, yz etc. vanish unless the two states k and l have the same parity. The other selection rules are

$$\Delta l = 0, \pm 2$$

$$\Delta m = 0, \pm 1, \pm 2$$

$$(l = 0 \text{ to } l' = 0 \text{ forbidden}).$$

$\Delta l = \pm 2$ is due to electric quadruple transition only.

14·4 Width of a spectral line

The radiation emitted or absorbed due to transition between two atomic levels should be strictly monochromatic according to the theory developed in the foregoing sections. The angular frequency should be $\omega = \omega_{kl} = (E_k - E_l)/\hbar$. This means that the spectral line should be absolutely sharp. However the observed lines are not absolutely sharp, but have small but finite widths showing that they have small frequency spreads. The broadening of a spectral line arises due to a number of reasons, viz. broadening due to the finite gas pressure within the material (pressure broadening), broadening due to the relative velocity between the emitting (or absorbing) atoms and the observer (Doppler broadening) and broadening due to radiation damping (instrinsic boraending). We shall consider the last of these effects here.

A classical oscillator emitting e.m. radiation loses energy. Hence the amptitude of its oscillations decreases with time. The amplitudes of the electric or magnetic vectors associated with the emitted e.m. wave will thus be damped so that the radiated field may be represented by exp $(-\gamma t/2) \exp(i\omega_0 t)$ where the first exponential represents the damping factor. By Fourier analysis, we then get the frequency spectrum of the radiation from the classical oscillator. The radiated intensity per unit angular frequency range is found to be proportional to §§

$$I_\omega \propto \frac{1}{(\omega - \omega_0)^2 + \frac{1}{4}\gamma^2} \qquad (14\cdot4\text{-}1)$$

The above expression shows that the frequency variation of the intensity has a Gaussian shape with a maximum at $\omega = \omega_0$. For $\omega = \omega_0 \pm \gamma/2$ the intensity falls to half the maximum value. The quantity γ is called the *natural line width* of a spectral line which is usually small in compasion with ω_0.

Since the energy W of the oscillator is proportional to the square of the amplitude, its decrease with time is determined by the exponential $\exp(-\gamma t)$ and we can write

$$W \propto \exp(-\gamma t)$$

So the rate of decrease of energy with time $dW/dt \propto \gamma \exp(-\gamma t)$. Thus the fractional rate of decrease of energy of the oscillator with time at $t = 0$ is equal to the line-width γ.

It is possible to correlate the rate of change of energy of the classical oscillator with the rate of change of the probability of the corresponding quantum system being in the initial upper energy state. Then γ can be interpreted as being equal to the probability per unit time of the quantum system making spontaneous transition from the initial state.

The above interpretation is consistent with Heisenberg's uncertainty principle. The transition probability is of the order of the reciprocal of the mean life τ of the initial state. According to the uncertainty relation $\Delta E \cdot \Delta t = \Delta E \cdot \tau \sim \hbar$. A quantum system having a mean life time of τ has

---

§§ See *Classical Electricity and Magnetism by* Panofsky and Phillips.

an uncertainty in energy of the order of $\Delta E \sim \hbar/\tau$ which determines the energy width of the level. If the transition is to a lower state k having an infinite mean life ($\tau \to \infty$), then this state has an absolutely sharply defined energy ($\Delta E_k = 0$). Hence the spectral line emitted due to the transition from the initial upper state of energy E_l with a width ΔE_l to the state of energy E_k has a line width $\Delta W = \Delta E_l/\hbar = \gamma$.

The ratio of the natural line width to the angular frequency ω of the spectral line $\Delta\omega/\omega_0$ is expected to be quite small ($\sim 10^{-6}$) for typical dipole transitions.

14·5 Selection rules for many electron atoms

In the case of many electron atoms the perturbation can be written as (see Eq. 14·1-8)

$$H'(t) = -\frac{i\hbar}{e} \sum_j A_j \cdot \nabla \tag{14·5-1}$$

where we sum over all the electrons in the atom (or ion).

In the dipole approximation, the matrix element of the perturbation becomes

$$H'_{kl} = -im\,\omega_{kl} \sum_j (D_{Aj})_{kl} \tag{14·5-2}$$

We get the transition probability per unit time for spontaneous emission of radiation (see Eq. 14·1-30) into the solid angle $d\Omega$ as

$$W_{kl}\,d\Omega = \frac{1}{t}\,\left|a'_k(t)\right|^2 d\Omega$$

$$= \frac{\omega_{kl}^3}{8\pi^2 \epsilon_0 \hbar c^3}\,\left|D_{kl}\right|^2 d\Omega \tag{14·5-3}$$

where $\qquad \mathbf{D} = \sum_j e\mathbf{r_j}$

is the total dipole moment operator of the atom.

The angular momentum radiated by an oscillating dipole cannot exceed $\hbar$. This classical argument holds even when several electrons contribute to the dipole moment. Quantum electrodynamical calculations show that the total angular momentum quantum number of the system can remain unchanged or can change by one unit. However no zero to zero transition can take place.

We summarize the electric dipole selection rules as below :

$$\Delta J = 0, \pm 1 \qquad \text{(No } J = 0 \text{ to } J' = 0)$$

$$\Delta M_j = 0, \pm 1$$

Here $\mathbf{J} = \mathbf{L} + \mathbf{S}$ is the total angular momentum of the atom.

Parity selection rule, as before, states that parity must change in the case of electric dipole transition (Laporte rule). In case of L–S coupling, we have the additional selection rule for the orbital angular momentum quantum number L:

$$\Delta L = 0, \pm 1$$
$$\Delta S = 0$$

(No $L = 0$ to $L' = 0$ transition is possible).

Problems

1. A particle of mass μ and charge e is confined within a one-dimentional potential box extending from $x = 0$ to $x = l$. Apply the time dependent perturbation theory to show that the matrix elements for the transition between two states j and k due to a plane monochromatic e.m. wave is

$$M_{kj} = \langle \Psi_k | x | \Psi_j \rangle$$
$$= 0 \text{ if } j \pm k \text{ is even}$$
$$= -\frac{8l}{k^2} \exp\{i(E_j - E_k)t/\hbar\} \cdot \frac{jk}{(j^2 - k^2)^2} \text{ if } j \pm k \text{ is odd.}$$

2. Using the expressions given in § 14·2, prove that the matrix elements of the components of the dipole moment for the transition $2P \rightarrow 1S$ of the hydrogen atom are given by

$$(D_x)_{2P \rightarrow 1S} = \left(\frac{128}{243}\right) a_0 e \left(\delta_{m,-1} - \delta_{m,1}\right)$$

$$(D_y)_{2P \rightarrow 1S} = -\left(\frac{128i}{243}\right) a_0 e \left(\delta_{m,-1} - \delta_{m,1}\right)$$

$$(D_z)_{2p \rightarrow 1S} = \sqrt{2} \left(\frac{128}{243}\right) a_0 e \, \delta_{m,0}$$

Hence show that the probability of spontaneous transition for all possible values of m is given by $W_{2P \rightarrow 1S}^{(S)} = \frac{32}{3} \left(\frac{2}{3}\right)^{10} \frac{\omega^3 a_0^2 e^2}{\pi \hbar c^3}$

where a_0 is the Bohr radius and $\hbar \omega = 3e^2/8a_0$ is the transition energy.

3. Prove that the probability of transition from the $2p_{3/2}$ state of hydrogen to the $1s_{1/2}$ state is twice that for the transition $2p_{1/2} \rightarrow 1s_{1/2}$.

4. Prove that the selection rules for the allowed transition of a linear harmonic oscillator under the action of an incident plane monochromatic radiation is $\Delta n = \pm 1$.

5. Using Eq. (14·1-2), show by integrating by parts that in the limit of the perturbing field of the radiation changing slowly (adiabatic approximation), the transition amplitude can be written as

$$a_l = \sum_k c_k \langle \Psi_l | u_k \rangle \exp\left[-i(E_k - E_l) t/\hbar\right]$$

where the unperturbed and perturbed wave functions can be expressed as

$$\Psi_{un} = \sum_k c_k u_k \exp(-i E_k t/\hbar) \text{ for } t < 0$$

and $\Psi_{per} = \sum_l a_l \psi_l \exp(-i E_l t/\hbar)$ for $t > 0$

The continuity condition gives

$$\sum_k c_k u_k = \sum_l a_l \psi_l$$

Assume the perturbation to act for a time $t = \Delta t << 1$.

6. Using the result of Prob. 5, show that if the perturbation acts suddenly ($\Delta t \rightarrow 0$) on a linear harmonic oscillator to change its force constant to half the original value, the probability of the oscillator to remain in the ground state is proportional to $|a_o|^2 \sim 98\%$.

7. Show that the probability of the oscillator of the previous problem making a transition to the second excited state is proportional to $|a_2|^2 \sim 1\%$.

8. Photoionization is the process of ionization of an atom by the action of e.m. radiation (photo-electric effect). Defining the photo-ionization cross section σ_{ph} as the rate of transition of an atom from a given bound state to a final state f in the continum per atom per unit flux of the incident radiation, prove that the photo-ionization cross section of the hydrogen atom per unit solid angle (differential cross section) from the ground state (1s) is given by

$$\left(\frac{d\sigma}{d\Omega}\right)_{ph} = 32\, \alpha \left(\frac{\hbar}{m}\right) \frac{Z^5}{\omega (k_f a_o)^5} \cdot \frac{\sin^2\theta \cos^2\phi}{(1 - v_f \cos\theta/c)^4}$$

where $k_f = p_f/\hbar$; v_f and p_f are the velocity and momentum of the electron emitted into the solid angle $d\Omega$ at the angles (θ, ϕ). α is Sommerfeld's fine structure constant and a_o is the Bohr radius.

15

SPECIAL THEORY OF RELATIVITY

15·1 Frames of reference ; Newtonian relativity

It is well-known that the position of a body at any instant of time is determined by measuring its distances from a set of *coordinate axes* at that particular instant of time. For instance, the position of an object on the floor of a room is determined if its distances from the lines of intersection of two mutually perpendicular walls with the floor of the room are measured. Again for determining the position of an overhead electric bulb hanging from the ceiling of a room, it is necessary to measure the distance of some particular point on the bulb from the floor and two adjacent walls of the room. We have assumed here that the object in question is at rest in the room. If however the object is moving, then we must also specify the time at which the position of the object is determined. The time has to be measured on a definite scale which may be any subdivision of the period of rotation of the earth about its axis. The set of coordinate axes with reference to which the position of a particle is measured and time scale with reference to which the time of these measurements is expressed are jointly known as a *frame of reference*. It is possible for an observer at rest relative to a given frame of reference to determine the position of an object or of a physical system at every instant of time with reference to that frame.

Frames of reference can be of different types. We are familiar with the rectangular, spherical or cylindrical systems of coordinates. The numerical values of the coordinates giving the position of an object at a given instant of time are different for the different systems. It is however possible to determine simple mathematical relationships between the coordinates in the different systems.

In the above discussion, we have assumed that the different frames of reference are at rest relative to one another. However there may be cases in which one frame of reference may be moving relative to another. If we consider one of the frames to be at rest (*rest frame*) and if the nature of motion of the other relative to it is known, then it is possible to determine the mathematical relationships between the coordinates of an object measured in the two frames. Our classical notion is of course to regard the measurement of time to be the same in both the frames.

As an example consider a car C moving with a uniform velocity in a straight line along a road. An observer sitting in the car can determine the position of an object inside the car by choosing a frame of reference fixed in the car. If the object is moving relative to the car, then the observer can also determine its velocity relative to the car. However this velocity will be different from the velocity of the same object as measured by an observer standing on the road. This second observer will have to use a frame of reference fixed on the road. For instance, suppose our object is a toy car A moving with a velocity of 1 m/s relative to the car C in the above example which has a velocity of 10 m/s along the road. Suppose the toy car A is moving in the same direction as the car C. If the observer sitting in the car measures the velocity of the toy car C, he will get a value of 1 m/s for the velocity. However the observer standing on the road will measure the velocity of A to be 11 m/s.

Consider another simple experiment. If the observer sitting in the car C throws a ball vertically downward upon the floor of the car, assumed to be perfectly smooth, then to him the ball will appear to rebound vertically upward. The movement of the ball is determined by Newton's laws of motion. If however, an observer standing on the road side watches the ball, to him it will appear to follow a parabolic path going downward and then a similar parabolic path after it rebounds upward. If the velocity of the car is known, then the observer on the road can determine the path of the ball with the help of Newton's laws of motion. If the opposite experiment is performed in which the observer on the road throws a ball vertically downward, to him the ball will appear to rebound vertically upward, while to the observer in the moving car C, the ball will appear to follow parabolic paths both downward and upward.

From the above discussions, it will be evident that though according to our usual notions, we regard the observer on the road as the observer at rest while that in the moving car as the moving observer, to the latter, it will be just the other way around *i.e.*, to him the observer on the road will be the moving observer. Thus the notions of the rest frame of reference and moving frame of reference are purely *relative*. It is a common experience that when an observer sitting in the compartment of a railway train standing at rest on the track watches another train passing by with uniform speed parallel to the first train, it is usually difficult for him to decide which train is moving. Similar is the situation if the observer's train is moving with uniform speed and the other train is at rest on the track. On the other hand if the first train begins to move starting from rest, *i.e.* if its motion is accelerated, while the other train is at rest on the track it is possible for the observer in the first train to decide which train is in motion.

Thus we may say that two frames of reference which are in uniform

Special Theory of Relativity

rectilinear motion with respect to each other are *equivalent*. In both cases, Newton's laws of motion are equally applicable. According to classical mechanics, these laws are applicable to the motion of all physical systems. If a force F acts on a body of mass m along the x-axis, then the acceleration $\ddot{x}$ of the body in that direction is given by Newton's second law of motion :

$$m\ddot{x} = F \qquad (15\cdot1\text{-}1)$$

The coordinate x and the time t are measured in a frame of reference (x, y, z, t) in which Newton's laws of motion are applicable. Now consider another observer in a frame of reference (x', y', z', t') which is moving with a uniform velocity v relative to the first frame (x, y, z, t) in the x-direction, the corresponding axes of the two frames being parallel to one another. If the origins of the two frames coincide at $t = 0$, then the following transformation relations hold between the cordinates and times in the two frames of reference (see Fig. 15·1.).

$$x' = x - vt \qquad (15\cdot1\text{-}2a)$$
$$y' = y \qquad (15\cdot1\text{-}2b)$$
$$z' = z \qquad (15\cdot1\text{-}2c)$$
$$t' = t \qquad (15\cdot1\text{-}2d)$$

With the help of the above transformation relations, it is possible to transform the coordinates and time from the (x, y, z, t) frame to the (x', y', z', t') frame. Similarly the following transformation relations hold in transforming from (x', y', z', t') frame to the (x, y, z, t) frame :

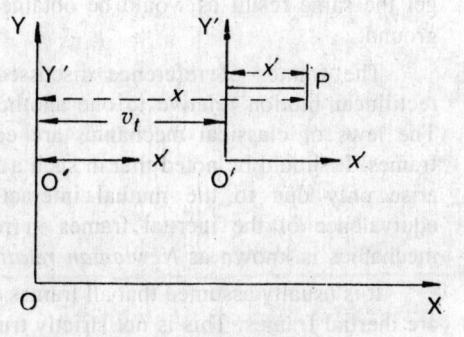

Fig. 15.1. Galilean transformation

$$x = x' + vt' \qquad (15\cdot1\text{-}3a)$$
$$y = y' \qquad (15\cdot1\text{-}3b)$$
$$z = z' \qquad (15\cdot1\text{-}3c)$$
$$t = t' \qquad (15\cdot1\text{-}3d)$$

The transformation relations (15·1-2) and (15·1-3) are known as Galilean transformation equations.

Eq. (15·1-2a) gives

$$\dot{x}' = \dot{x} - v \qquad (15\cdot1\text{-}4)$$

Eq. (15·1-4) can be used to transform velocity from one reference frame to another. Differenting the above equation, we get

$$\ddot{x}' = \ddot{x} \qquad (15\cdot1\text{-}5)$$

So Newton's second law of motion (Eq. 15·1-1) can be written in the (x', y', z', t') frame of reference as

$$m\ddot{x}' = F \tag{15.1-6}$$

Thus the second law of motion remains unchanged in the frame (x', y', z', t') moving with uniform velocity along a straight line relative to the frame (x, y, z, t). So it may be stated that Newton's laws of motion remain the same in frames of reference which are either at rest or are in uniform rectilinear motion relative to one another, *i.e.*, they are *covariant*. This means that an observer at rest in a particular frame of reference will not be able to decide whether his frame is at rest or is in uniform rectilinear motion by performing some mechanical experiment within the frame. For example if the experiment on ball throwing discussed above is performed within the closed chamber of an aeroplane flying with a uniform velocity, the observer in the plane will not be able to tell whether the plane in flying or not by noting the results of the experiment. He will get the same result as would be obtained by an observer at rest on the ground.*

The frames of reference discussed above which are in uniform rectilinear motion relative to one another are known as inertial *frames*. The laws of classical mechanics are equally applicable in all inertial frames. It should be noted that in such a frame accelerations of all bodies arise only due to the mutual interactions between the bodies. The equivalence of the inertial frames in respect to the laws of classical mechanics is known as *Newtonian relativity*.

It is usually assumed that all frames of reference attached to the earth are inertial frames. This is not strictly true, because of the rotation of the earth about its own axis and also due to its revolution around the sun. Thus, all bodies on the earth are subjected to a centripetal force and hence have accelerated motion. Thus no frame of reference attached to the earth can be regarded as uniformly moving and hence these cannot be inertial frames. However in practice these may still be regarded approximately as inertial frames since the velocity of translation of the earth in space which has a value of about 30 km/sec is large compared to the velocity of rotation of the earth about its axis (0·46 km/sec). Newton himself was aware of the fact that no frame of reference on the earth can be regarded strictly as an inertial frame. However he assumed that there is an *absolute space*, "in its own nature, and without reference to anything external remains always immovable". He further contended that relative space is "some movable dimension or measure of the absolute space which our senses determine by its position to other bodies and which is commonly taken for absolute space". So according to him, the second law of motion

---

* It is assumed here that the "ground", *i.e.*, the earth is at absolute rest which is not strictly true.

Special Theory of Relativity

strictly holds good in frames of reference at rest in this absolute frame or as seen above in frames of reference in uniform rectilinear motion with respect to it.

Though it was proved later that absolute space does not exist and is a fiction of imagination, Newton believed in its existence, According to him, the fixed stars constituted the absolute frame of reference which were at absolute rest in space. But we know that even these so called fixed stars are not at absolute rest. Because of their vast distance from the solar system, they appeared to be at fixed positions in the sky and hence were taken to be at absolute rest in space. However their motions have been detected by the modern techniques of measurement.

So the question remains as to which frame or frames can be regarded as inertial frames. Albert Einstein examined this question in depth which finally resulted in the formulation of his famous special theory of relativity at the beginning of the present century.

15·2 Galilean transformation and electromagnetic theory

We have seen that Newton's laws of motion remain the same (*i.e.*, covariant) in all inertial frames. However difficulty arises in trying to transform the Maxwell's equations, which are the fundamental equations of the electromagnetic theory of light, using the Galilean transformation equations (15·1-2). They no longer remain the same when transformed from one inertial frame to another. Thus the Galilean transformation equations are not universally applicable in all fields of physics.

Another problem arose in applying the law of addition of velocities in the case of the propagation of electromagnetic waves using the Galilean transformation equations (Eq. 15·1-4). Before the advent of the electromagnetic theory, it was believed that light waves were of the same type as the waves in an elastic medium, e.g., the sound waves. Such waves have velocities determined by the elasticity and density of the medium. Since light waves are transverse in nature as the polarisation experiments show, the medium needed for their propagation must have the properties of a solid. Further, the medium must be infinitely rigid in order that there might be no longitudinal light waves. Since the velocity of light is very high, the density of the medium must be very low while its elasticity must be very high. The medium must pervade all space including the interstices of matter in order to account for the propagation of light through matter. However it must not hinder the movement of celestial bodies which are known to move in non-viscous medium. This hypothetical solid medium, endowed with such contradictory properties as very high elasticity, very low density and infinite rigidity was called the *ether*. Many scientists believed that the absolute frame of reference about which Newton had talked and in which his laws of motion were supposed to be strictly valid, was nothing but ether.

The formulation of the electromagnetic theory of light by Maxwell in 1864 and its experimental confirmation through the series of experiments by Heinrich Hertz, led to the realization that light was a kind of electromagnetic wave which is a combination of very rapidly oscillating and mutually perpendicular electric and magnetic fields propagated through space with the velocity $c = 3 \times 10^8$ m/s. Later researches have established that not only light waves, but x-rays and γ-rays on the short wavelength side and radiowaves, radar, microwaves *etc.* on the long wavelength side are examples of electromagnetic waves. After the formulation of the tromagnetic theory of light, the view that light waves are a kind of elastic waves in the hypothetical medium ether was given up.

Maxwell, Hertz and others still held the view that the electromagnetic waves required a medium for propagation and ether was supposed to be this medium. However the expected properties of ether were so strange that serious doubts were expressed about its existence. So it became necessary to investigate very carefully the reality of the existence of ether. The most important of these experimental investigations was carried out by the two American physicists A.A. Michelson and E.W. Morley in 1887. This along with some other important experiments having a bearing upon the question of the existence of ether are discussed below.

(A) Aberration of light

It is well-known that the observed positions of the stars vary slightly throughout the year. There are two reasons for such variation, viz., *parallax* and *aberration.* Parallax is important for the nearer stars and is due to the finite distances of these stars from the solar system. On the other hand, the aberration of star light is due to the fact that the apparent direction of a star in the sky is different from its true direction. This causes the apparent position of the star in the sky to describe a circle or ellipse during the whole year and was first studied by J. Bradley in 1728. The shift in the observed direction of a star due to aberration which is much greater than that caused by parallax depends on the relative velocity of the light source (*i.e.*, the star) and the observer (*i.e.*, the earth). Since this is the same for all stars, the shift due to aberration of light has the same magnitude for all stars.

The earth moves in its orbit with a velocity $\upsilon \approx 30$ km/s. As will be evident from Fig. 15·5., to see a star the telescope should be tilted towards the direction of the earth's motion from the true direction of the star by an angle α given by.

$$\alpha \approx \tan\alpha = \frac{\upsilon}{c} = \frac{30 \times 10^3}{3 \times 10^8} = 10^{-4} \text{ radian} \approx 20\cdot5'' \text{ of arc}$$

Special Theory of Relativity

Six months later, the telescope has to be tilted by the same angle in the opposite direction.

The conclusions drawn from the observations of stellar aberration are that if there is such a medium as ether through which light waves are propagated, then the stars must be at rest with respect to it while *the earth moves through ether with its observed velocity* (see § 15·10).

(B) Trouton and Noble's experiment

If the earth moves through ether, there should be an *ether wind* like the wind felt by a passenger in a railway compartment blowing in a direction opposite to the motion of the train. A number of experiments were performed to detect this ether wind. The most famous of these experiments was the Michelson-Morley experiment to be described in the next section. Another such experiment was performed by Trouton and Noble (1903) who tried to detect the torque on a charged capacitor due to the motion of the earth through ether.

The charge on any one plate of the capacitor moving with the earth behaves like an electric current which produces a magnetic field at the position of the opposite plate. Since the charge on this second plate also moves with the same velocity it is equivalent to an electric current flowing in the opposite direction so that the magnetic field exerts a force on it, determined by Biot-Savart's law. An equal and opposite force acts on the first plate for the same reason. The two together give rise to a torque on the suspended condenser which should thus be turned parallel to the *ether wind*. By measuring the angle through which the condenser is turned, the turning moment can be measured.*

Very careful measurements by Trouton and Noble and later by others failed to detect any such turning moment. Thus the expected ether wind could not be detected.

15·3 Michelson-Morley experiment

It is well-known that the velocity of elastic waves depends upon the velocity of the medium through which it travels. Thus, the velocity of sound waves in air is different if a wind is blowing, from its value in still air. Michelson and Morley's experiment was designed to detect whether there was any such effect in the case of light waves.

A Michelson's interferometer was used in this experiment. As shown in Fig. 15·2, the interferometer consists of two plane mirrors A and B whose planes are mutually perpendicular. Monochromatic light from the sources fall on the parallel plate glass plate C one face of which is half-silvered. Light ray SC falling on this face is partly reflected and partly refracted. The reflected beam CA is incident normally on the mirror

* For further details see *Theoretical Physics* by Georg Joos

A and is reflected back in the opposite direction along AC. The refracted beam CB is transmitted to the other side of the plate C and after being incident on the mirror B normally is reflected back along BC to be reunited with the beam AC reflected back from A at a point on the silvered face of C. A compensating plane-parallel plate (not shown) of the same thickness as C is interposed between C and A to compensate for the additional path traversed by the beam CB in going through C and back after reflection by B.

The two beams AC and BC after being reunited produce interference fringes which can be observed by the microscope D.

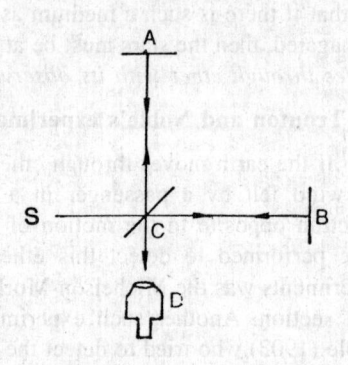

Fig. 15.2. Michelson-Morley experiment.

Since the experiment is performed on the surface of the earth, the interferometer travels through space with the velocity of the earth. Suppose that the arm CB of the interferometer is parallel to the direction of motion of the earth. If ether exists, then the earth along with all objects on it moves through ether with a velocity v relative to it. So an ether wind should be produced in the opposite direction having the velocity $-v$ similar to the wind blowing relative to a running train in the direction opposite to the motion of the train. If ether acts as the medium for the propagation of light waves, then it may be expected that the relative motion between the earth and ether will change the velocity of the light wave. This velocity should be $(c - v)$ for the light travelling from C to B while it should be $(c + v)$ for the return journey of the light travelling from B to C, where c is the velocity of light in vacuum.

If L be the length of the arm CB, then the time taken by light to travel from C to B and back is (see Fig. 15.3).

$$t_1 = \frac{L}{c-v} + \frac{L}{c+v} = \frac{2Lc}{c^2 - v^2}$$

So the length of the path traversed by light in going from C to B and back C is

$$L_1 = ct_1 = \frac{2Lc^2}{c^2 - v^2} = \frac{2L}{1 - v^2/c^2} \approx 2L(1 + v^2/c^2) \qquad (15 \cdot 3 \cdot 1)$$

since $v \ll c$.

On the other hand, due to the motion of the earth the path of the light in going from C to the mirror A is CA' while the return path from the mirror A to C' is A'C'. Suppose the length of the arm CA is equal to that of CB, *i.e.*, L. The time taken the light to travel from C to A is CA'/c. During this time the mirror A moves through AA' = v. A'C/c. Hence

$$CA'^2 = CA^2 + AA'^2 = L^2 + CA'^2 \frac{v^2}{c^2}$$

or, $CA'^2 (1 - v^2/c^2) = L^2$

i.e., $CA' = L/\sqrt{1 - v^2/c^2}$

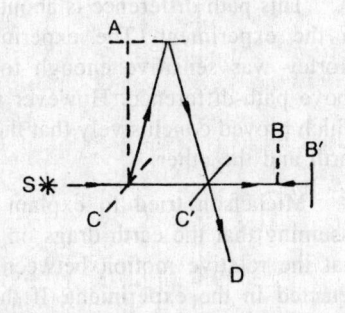

Fig. 15.3. Path of rays in Michelson-Morley experiment.

Since CA' = A'C', so the length of the path traversed by light in going from C to the mirror A and back to C at the position C' is

$$L_2 = CA' + AC' = \frac{2L}{\sqrt{1 - v^2/c^2}} \approx 2L(1 + v^2/2c^2) \qquad (15\cdot3\text{-}2)$$

So the path difference between the two rays in going from C to the two mirrors B and A and back to C after reflections from the respective mirrors is

$$L_1 - L_2 = Lv^2/c^2 \qquad (15\cdot3\text{-}3)$$

If the earth were not in motion, there would not be any such path difference. So if it is imagined that the earth had suddenly come to a stop during the course of the experiment, then the path difference between the two rays would suffer a sudden change and there would be some displacement of the interference fringes observed through the microscope. This of course cannot happen in practice. However if the whole apparatus is turned through 90°, then the arm CA of the interferometer would occupy the position of the arm CB while the arm CB would be placed in a position which is parallel to the original orientation of CA. So the light ray along CA would now traverse a path longer by Lv^2/c^2 compared to that traversed along CB. Thus due to the rotation of the apparatus through 90°, the path difference between the two rays would change by $2Lv^2/c^2$ which would cause a displacement of the fringe system.

Since the velocity of the earth is $v \approx 30$ km/s, $v^2/c^2 \sim 10^{-8}$. In the original experiment of Michelson and Morley, the effective length of the interferometer arms were 11 m long so that the expected change in the path difference due to 90° rotation would be

$$\frac{2Lv^2}{c^2} = 2 \times 11 \times 10^{-8} \text{ m} = 2200 \text{ Å}$$

This path difference is about 2/5 of the wavelength of the light used in the experiment. The experimental arrangement of Michelson and Morley was sensitive enough to detect the fringe-shift caused by the above path-difference. However no such fringe-shift could be detected, which proved conclusively that there was no relative velocity between the earth and the ether.

Michelson tried to explain the null result of the experiment by assuming that the earth drags on the ether in contact with its surface so that the relative motion between the earth and the ether could not be detected in the experiment. If this were true, then the relative motion between the earth and the ether could be detected if the experiment were performed high above the earth's surface, say on the top of a high mountain. However even when the experiment was repeated at such a high altitude, no ether-wind could be detected.

It may be mentioned that the result of the experiment on the aberration of starlight discussed previously is in direct contradiction to the ether-drag hypothesis of Michelson.

Thus the null results of Michelson-Morley experiment forced the physicists to abandon the ether hypothesis and to conclude that the absolute frame of reference proposed by Newton did not exist in reality. The results of the experiment also proved that the velocity of light is not influenced by the motion of the observer's frame of reference. It is the same in all frames of reference which are in uniform rectilinear motion with respect to one another, *i.e.*, in all inertial frames. As is well-known, this velocity is

$$c = 2 \cdot 997925 \times 10^8 \text{ m/s}$$

Many other workers repeated Michelson-Morley type experiment in later years, most of which corroborated the null result proving the absence of any ether wind. In 1958, the American physicist C.H. Townes, performed an experiment using *maser*, discovered by him, to lend a very convincing support to the results of Michelson-Morley experiment. Many other recent experiments based on improved techniques, have conclusively shown that the absolute velocity of the earth, if any, cannot be more than 5 m/s.

Michelson-Morley experiment does not contradict the relativity principle, *i.e.*, it does not contradict the notion of the equivalence of different inertial frames. On the contrary, it lends support to this principle. However it is in conflict with the *law of addition* of velocities (Eq. 15·1-4) deduced from the Galilean transformation relations, since the experiment shows that the velocity of light c is the same in all inertial frames. This result points to the necessity of revising the Galilean transformation equations in favour of a new set of transformation equations which would make both the laws of the electrodynamics and the laws of mechanics covariant in all inertial systems.

15·4 Lorentz-Fitzgerald contraction hypothesis

The null result of Michelson-Morley experiment was sought to be explained by H.A. Lorentz and G.F. Fitzgerald by proposing the hypothesis that the length of a body contracts in the direction of its motion.

If the length of a body is l_0 when it is at rest, then its length l when it is moving with a velocity υ parallel to its length is given by

$$l = l_0 \sqrt{1 - \beta^2}$$

where $\beta = \upsilon/c$. This is known as Lorentz-Fitzgerald contraction hypothesis. According to this hypothesis, the length L of the arm of the interferometer which is parallel to the velocity of the earth in Michelson-Morley experiment (the arm CB in Fig. 15.3) contracts and becomes $L\sqrt{1-\beta^2}$. Hence the length of the path followed by light along CB, as given by Eq. (15·3-1) becomes

$$L_1' = \frac{2L\sqrt{1-\beta^2}}{1-\beta^2} = \frac{2L}{\sqrt{1-\beta^2}}$$

Comparing with Eq. (15·3-2) we see that this is equal to the length L_2 of the path of the light beam along the perpendicular direction CA, so that there is no path-difference between the two rays when they reunite. Thus there is no fringe-shift, even when the interferometer is turned through 90°.

If the contraction hypothesis is accepted then the null result of Michelson-Morley-experiment can be explained without abandoning the idea of an absolute frame of reference (*i.e.*, ether). The contraction assumed by Lorentz and Fitzgerald occurs when the object moves relative to the absolute frame. However as we shall see later, the contraction of the length of an object along the direction of its motion also follows from the special theory of relativity proposed by Einstein in which there is no place of any absolute frame of reference.

15·5 Einstein's special theory of relativity

We have seen that Maxwell's electromagnetic equations do not obey Newtonian principle of relativity. They are not covariant under Galilean transformation. On the other hand the laws of classical mechanics obey Newtonian relativity. There are two alternative ways in which the crisis created by this contradictory behaviour of the laws of electromagnetism and of classical mechanics can be resolved.

(*a*) Relativity principle is applicable both for the laws of electromagnetism and for the laws of mechanics. However, Maxwell's equations are not valid representation of electromagnetic phenomena.

(b) Relativity principle is applicable both for the laws of electromagnetism and for the laws of mechanics. However Newton's equations of motion are not valid representation of the laws of mechanics and should be revised.

Einstein, after a thorough analysis of the various relevant experimental facts, decided in favour of the latter course. However it was clear to him that a new set of transformation equations was required to retain the covariance of Maxwell's electromagnetic equations since the null result of Michelson-Morley experiment had pointed out the inadequacy of the Galilean transformation equations (see at the end of § 15.3). Obviously therefore the laws of mechanics must be framed in such a way as to retain their forms, *i.e.*, make them covariant under the new transformation. Since the Newtonian equations of motion are covariant under Galilean transformation which must be abandoned in favour of the new transformation relations, the Newtonian equations themselves require revision.

In 1905, Einstein proposed his new principle of relativity, known as the special theory of relativity which was based on the following two fundamental postulates.

(a) All physical laws, electromagnetic and mechanical, must be covariant in all frames of reference which are in uniform rectilinear motion relative to one another, *i.e.* in all inertial frames.

(b) The velocity of light is the same in all inertial frames. It does not depend on the velocity of the observer or on the velocity of the source of light.

At first sight, the first postulate may appear to be similar to Newtonian relativity. However on closer scrutiny, it will be seen that they are fundamentally different. Newton had postulated that there existed an absolute frame of reference in which his laws of motion were strictly valid and hence the laws held in all frames of reference in uniform rectilinear motion with respect to that frame. Michelson-Morley experiment proved conclusively that no absolute frame of reference existed and hence Einstein formulated his relativity principle without reference to any absolute frame. According to him the relativity principle was valid in all frames of reference in uniform rectilinear motion relative to one another. Further the relativity principle must be applicable both to the laws of mechanics as well as to the laws of electromagnetism (Maxwell's equations).

The second postulate of Einstein requires replacement of the Galilean transformation relations by a new set of transformation relations between two inertial frames of reference S (x, y, z, t) and S' (x', y', z', t'). For these, he chose the new transformation relations discovered earlier by H.A. Lorentz, which he himself rediscovered who had proved that the Maxwell's equations remained covariant under this new transformation.

Lorentz had deduced his transformation equations by assuming that the velocity of light remained invariant under the new transformation. He had also assumed that the *time coordinate* (t) might not be the same in different inertial systems. This was a radical departure from the basic assumption underlying the Galilean transformation relations, in which time is taken to be the same in all frames of reference.*

Let us examine this point in further detail. Suppose an observer measures times at regular intervals with the help of a clock which is at rest with respect to him. The difference of time in different systems implies that if he measures the same time-intervals with the help of a clock which is moving relative to him, then the measured intervals will appear to be different from the previous measurements. According to the same argument, the notion of simultaneity of two events also depends on the choice of the coordinate system. Two events, which appear simultaneous to an observer in a particular frame of reference, may not appear so to another observer in a different frame of reference.

As an example, consider a railway train of length $2l$ moving with a uniform velocity v along the railway track. Suppose two light-signals are simultaneously flashed and sent out from the centre towards the two ends of the train. To an observer sitting in the train, the signals will appear to reach the two ends exactly at the same moment, *i.e.*, simultaneously. The time for the signals to reach the two ends of the train from the centre will be $t = l/c$ as measured by the observer sitting in the train On the other hand to an observer located by the trackside, the two signals will appear to reach the two ends of the train at different instants of time. According to Einstein, the velocity of light is the same for both the observers. As a result, to the second observer, the light-signal will appear to reach the front-end a little later. If the times taken by the light signal to reach the front and back ends as measured by the second observer are t_1 and t_2 respectively, then we get (see Fig. 15.4.).

$$c\,t_1 = l + v\,t_1$$

$$c\,t_2 = l - v\,t_2$$

So that $\qquad t_1 = \dfrac{l}{c-v} \quad$ and $\quad t_2 = \dfrac{l}{c+v}$

---

* Lorentz had called t' the local time in the inertial frame S', but had not realized its true significance. To him the time t in the so called absolute frame was the real time in all frames and t' was simply a mathematical contrivance introduced to make the Maxwell's equations covariant in different inertial frames. It was, Einstein who first realized that t' played the same role in the frame S' as t in the frame S.

Thus the two events, which appear to occur simultaneously to the observer sitting in the train, appear to take place at different instants of time to the observer standing by the trackside.

This shows that the notion of simultaneity is purely relative and times, as measured by observers in different frames of reference in relative motion, are not the same.

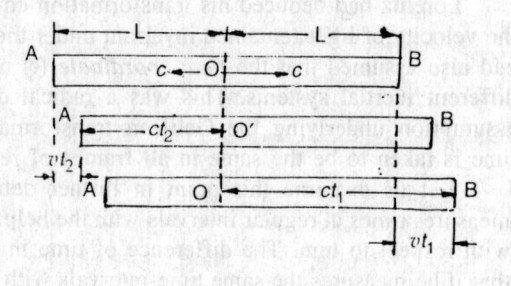

Fig. 15·4. Relativity of simultaneity

If the velocity of light were infinitely great ($c \to \infty$), then the time would be the same in all inertial systems. This implies instantaneous action at a distance which was precisely what was assumed by Newton.

15·6 Lorentz transformation equations

These transformation equations may be derived by different methods. They should have the following characteristics :

(a) In order that two inertial frames (x, y, z, t) and (x', y', z', t') may be completely equivalent, the transformation equations must be symmetrical with respect to both. In other words, the mathematical expressions for transformation from one system to the other must be the same as those for the reverse transformation, except for the sign of the relative velocity between the two systems, which should be *positive* in one case and *negative* in the reverse case.

(b) If all the quantities (x, y, z, t) are finite then the quantities (x', y', z', t') obtained by the transformation must also be finite.

(c) When the relative velocity of the two frames is zero, the transformation relations must give identical values of the coordinates and time for the two systems, i.e., $x = x'$, $y = y'$, $z = z'$ and $t = t'$.

(d) The law of addition of velocities obtained by using the transformation relations must be such that velocity of light is the same (i.e., invariant) in all inertial frames.

The first two conditions require that the transformation relations must be *linear*. Consider two frames of reference S and S' whose corresponding axes of coordinates are parallel. Suppose S' frame moves along x-direction with a uniform velocity v with respect to the S frame. If to an observer in the S frame an event appears to take place at the point (x, y, z) at the instant t, then to an observer in S' frame, the same event will appear to take place at the point (x', y', z') at the instant t'. As stated previously t and t' may not be the same.

Special Theory of Relativity

Suppose during the course of its motion, the origin O' of the S' frame at some instant becomes coincident with the origin O of the S frame. The observers in the two frames regulate their respective clocks at this instant to indicate zero time, i.e., $t = t' = 0$.

Exactly at this instant a light-signal is sent out from a source located at O. Since the velocity of light is the same in both the frames S and S', to the observers in both frames, it will appear that a spherical light wave is propogated outwards from the respective origins of the coordinate systems with the velocity c. To the observer in the S frame, the position of this spherical wave at the time t will be determined by the equation.

$$x^2 + y^2 + z^2 = c^2 t^2 \qquad (15 \cdot 6\text{-}1)$$

On the other hand, to the observer in the S' frame the equation determining the position of the spherical wave will be

$$x'^2 + y'^2 + z'^2 = c^2 t'^2 \qquad (15 \cdot 6\text{-}2)$$

Suppose the transformation relations between the two systems of coordinates are

$$x' = k(x - vt) \qquad (15 \cdot 6\text{-}3a)$$
$$y' = y \qquad (15 \cdot 6\text{-}3b)$$
$$z' = z \qquad (15 \cdot 6\text{-}3c)$$
$$t' = \alpha t + \gamma x \qquad (15 \cdot 6\text{-}3d)$$

k, α and γ are three constants which have to be determined.

We get from Eqs. (15·6-2) and (15·6-3)

$$k^2(x - vt)^2 + y^2 + z^2 = c^2(\alpha t + \gamma x)^2$$

or

$$x^2(k^2 - \gamma^2 c^2) + y^2 + z^2 - 2xt(k^2 v + \alpha \gamma c^2)$$
$$= t^2(\alpha^2 c^2 - k^2 v^2)$$

Comparing coefficients of the different terms in the above equation with the corresponding terms in Eq. (15·6-1), we get

$$k^2 - \gamma^2 c^2 = 1 \qquad (15 \cdot 6\text{-}4a)$$
$$k^2 v + \alpha \gamma c^2 = 0 \qquad (15 \cdot 6\text{-}4b)$$
$$\alpha^2 c^2 - k^2 v^2 = c^2 \qquad (15 \cdot 6\text{-}4c)$$

From Eqs. (15·6-4a) and (15·6-4b), we get

$$\gamma^2 c^2 = k^2 - 1$$
$$\alpha^2 \gamma^2 c^4 = k^4 v^2$$

Taking the ratio of the above two equations, we get by using Eq. (15·6-4c)

$$\alpha^2 c^2 = \frac{k^4 v^2}{k^2 - 1} = k^2 v^2 + c^2$$

or $\quad k^4 v^2 = k^4 v^2 + k^2 c^2 - k^2 v^2 - c^2$

Again from Eq. (15·6-4c), we have

$$(\alpha^2 - 1) c^2 = k^2 v^2 = \frac{v^2}{1 - \beta^2} = \frac{c^2 v^2}{c^2 - v^2}$$

so that $\quad \alpha^2 = 1 + \dfrac{v^2}{c^2 - v^2} = \dfrac{c^2}{c^2 - v^2} = \dfrac{1}{1 - \beta^2}$

Thus we have $\quad k = \alpha = \dfrac{1}{\sqrt{1 - \beta^2}}$

where $\beta = \dfrac{v}{c}$.

$\hfill (15\cdot6\text{-}5)$

Again from Eq. (15·6-4b), we have

$$\gamma = -\frac{k^2 v}{\alpha c^2} = -\frac{kv}{c^2} = -\frac{v/c^2}{\sqrt{1 - \beta^2}} \qquad (15\cdot6\text{-}6)$$

Thus the transformation equations (15·6-3) become

$$x' = \frac{x - vt}{\sqrt{1 - \beta^2}} \qquad (15\cdot6\text{-}7a)$$

$$y' = y \qquad (15\cdot6\text{-}7b)$$

$$z' = z \qquad (15\cdot6\text{-}7c)$$

$$t' = \frac{t - vx/c^2}{\sqrt{1 - \beta^2}} \qquad (15\cdot6\text{-}7d)$$

These are known as Lorentz transformation equations. The reverse transformation relations for transforming from S' to S are

$$x = \frac{x' + vt'}{\sqrt{1 - \beta^2}} \qquad (15\cdot6\text{-}8a)$$

$$y = y' \qquad (15\cdot6\text{-}8b)$$

$$z = z' \qquad (15\cdot6\text{-}8c)$$

$$t = \frac{t' + vx'/c^2}{\sqrt{1 - \beta^2}} \qquad (15\cdot6\text{-}8d)$$

It should be noted that the transformation relations (15·6-7) and (15·6-8) have the same forms. Only the sign of v is opposite which is in conformity with the requirement (a) given at the outset. If $v = 0$, then x, y, z, t are equal to the corresponding quantities x', y', z', t' in the S' frame which is in agreement with the condition (c) above. That the condition (b) also holds good for these transformations can easily be verified.

If the relative velocity v between S and S' frames is small, *i.e.*, $v << c$, then the above transformation equations reduce to the Galilean transformation equations (15·1–2) and (15·1–3). This is left to the reader as an exercise.

15·7 Relativity of length measurement

Suppose we want to measure the length of a rod made up of some very rigid material, *e.g.* steel. The rod is at rest parallel to the x'-axis in the frame of reference S'. An observer at rest in this frame measures the x-coordinates of the two ends of the rod as x'_1 and x'_2. Then the length of the rod as measured by this observed *at rest* is

$$L_0 = x'_2 - x'_1$$

Let S be another frame of reference with respect to which the frame S' is moving with a uniform velocity v along the x-axis which is parallel to the x' axis in the S' frame. If the observer in the S frame measures the coordinates x_1 and x_2 of the two ends of the rod in his frame at the instant t, then he will find the length of the rod as

$$L = x_2 - x_1$$

From the transformation equation (15·6–7a), we get

$$x'_1 = \frac{x_1 - vt}{\sqrt{1 - \beta^2}} \text{ and } x'_2 = \frac{x_2 - vt}{\sqrt{1 - \beta^2}}$$

So we get $\quad L_0 = x'_2 - x'_1 = \dfrac{x_2 - x_1}{\sqrt{1 - \beta^2}} = \dfrac{L}{\sqrt{1 - \beta^2}}$

i.e.,
$$L = L_0 \sqrt{1 - \beta^2} \qquad (15 \cdot 7\text{-}1)$$

Thus the length of the rod (L) as measured by the observer moving parallel to its length will be shorter than the length (L_0) of the rod as measured by the observer at rest. This contraction of length is the same as the Lorentz-Fitzgerald contraction discussed in § 15·4. However as stated there, the two are fundamentally different. The contraction of length predicted by the special theory of relativity is a two way process. The contraction in the length observed by the moving observer in the S frame of the rod lying at rest in the S' frame is exactly equal to the contraction that would be observed by the observer in the S' frame if the rod were lying at rest in the S frame. On the other hand, the contraction hypothesis postulated by Lorentz to explain the null result of Michelson-Morley experiment occurs when the rod moves relative to an absolute frame of reference which, as we have seen, does not exist in reality.

15·8 Relativity of time measurement

Suppose an observer in a frame of reference S measures times at different instants with the help of a clock or a system of clocks, properly synchronized, situated at rest in his frame. It t_1 and t_2 are two successive instants of time measured by him, the time-interval between the two will be

$$\Delta t = t_2 - t_1$$

It should be noted that the clock used for time-measurement is at the same position in the S frame at different instants of time, i.e., $x_1 = x_2$.

If now an observer in another frame S′ moving relative to S with the velocity v measures the same instants of time with the clock situated in the S frame, he will get (see Eq. 15·6–7d)

$$t_1' = \frac{t_1 - vx_1/c^2}{\sqrt{1 - \beta^2}} \text{ and } t_2' = \frac{t_2 - vx_2/c^2}{\sqrt{1 - \beta^2}}$$

Since $x_1 = x_2$ when the observer in S′ frame measures the same time interval with the help of a clock moving relative to him, he will find

$$\Delta t' = t_2' - t_1' = \frac{t_2 - t_1}{\sqrt{1 - \beta^2}} = \frac{\Delta t}{\sqrt{1 - \beta^2}} \qquad (15\cdot 8\text{-}1)$$

Since $\sqrt{1 - \beta^2} < 1$, we have $\Delta t' > \Delta t$. This means that a clock moving with uniform velocity relative to an observer runs more slowly than a clock which is at rest in the observer's frame of reference.

In the reverse case, if the clock is at rest in the S′ frame, it will appear to run at a slower rate to an observer in the S frame. Thus the scale of time measurement is relative.

The lengthening of the time intervals measured by a moving observer is known as *time-dilatation*. This *time dilatation* as predicted by the special theory of relativity, has been verified experimentally by measuring the mean life of disintegration of the *muons* which are a type of unstable elementary particle first discovered in the cosmic rays (see Vol. II). These have a mean life of disintegration $T = 1\cdot 5 \times 10^{-6}$ s when they are at rest. But when they are moving with a velocity close to the velocity of light, then an experimenter in the laboratory will find the mean life to be much longer. For instance, if the velocity of the muon is $v = 0\cdot 95c$, then the experimenter in the laboratory will measure the mean life to be 3·2 times as long :

$$T' = \frac{T}{\sqrt{1 - \beta^2}} = \frac{1\cdot 5 \times 10^{-6}}{\sqrt{1 - 0\cdot 9025}} = 1\cdot 5 \times 3\cdot 2 \times 10^{-6} = 4\cdot 8 \times 10^{-6} \text{ s}$$

The differences in time measurements in different inertial frames prompt us also to re-examine the question of simultaneity of two events in different systems.

Special Theory of Relativity

Suppose two events occur at the times t_1 and t_2 at two points x_1 and x_2 respectively in a frame S. If $t_1 = t_2$, then to an observer in the S frame, the events will appear simultaneous. Obviously x_1 and x_2 must be different in order that the two events may appear as different. If now to an observer in another frame S' moving with uniform velocity relative to S the events appear to occur at the times t_1' and t_2' then from Eq. (15·6–7d), we have

$$t_1' = \frac{t_1 - vx_1/c^2}{\sqrt{1-\beta^2}} \text{ and } t_2' = \frac{t_2 - vx_2/c^2}{\sqrt{1-\beta^2}}$$

Since $t_1 = t_2$, we get

$$t_2' - t_1' = \frac{-v(x_2 - x_1)/c^2}{\sqrt{1-\beta^2}}$$

Since x_1 and x_2 are different, t_1' and t_2' will also be different so that two events which appear simultaneous in the S frame will not appear to be simultaneous to an observer in the S' frame *i.e.*, simultaneity of events is relative.

In this connection we may also consider the question of the ordering of events. Suppose an event occurs in the S frame at x_1 at the instant t_1 and another event occurs in the same frame at x_2 at a *later instant* t_2 so that $t_2 > t_1$. The question is whether this ordering of the events in the S frame is also maintained in another frame S' which is in uniform relative motion with respect to S. Let the times of occurrence of the same events be t_1' and t_2' as measured by the observer in the S' frame. Then Eq. (15·6–7d) gives us

$$t_2' - t_1' = \frac{(t_2 - t_1) - v(x_2 - x_1)/c^2}{\sqrt{1-\beta^2}}$$

In order that the ordering of the events may remain unchanged to the observer in the S' frame, we must have $t_2' - t_1' > 0$. So from the above equation, the necessary condition becomes

$$t_2 - t_1 > \frac{v(x_2 - x_1)}{c^2}$$

or

$$\frac{x_2 - x_1}{t_2 - t_1} < \frac{c^2}{v}$$

In order that the interval of time $(t_2' - t_1')$ may be real, we must have $v < c$ so that $c^2/v > c$. So the above inequality will certainly be satisfied if

$$\frac{x_2 - x_1}{t_2 - t_1} < c$$

This means that the ordering of events will remain unaltered in two inertial frames moving with uniform velocity relative to each other provided that it is not possible to send any signal with a veloity greater than the velocity of light. One of the consequences of the special theory of relativity is that no velocity greater, than the velocity of light can be produced (see below). So the ordering of events can never be changed. It should be noted that the above conclusion applies only for two events which are *causally connected*. If two events are not causally connected, their ordering may not remain unchanged.

If an event at x_1 occurs at the time t_1, then an event at x_2 occurring at a later instant t_2 cannot be causally connected to the first event unless some signal could have been propagated from x_1 to x_2 in the interval (t_2-t_1). Since according to the special theory of relatively the highest attainable velocity of any signal is the velocity of light, two events may be regarded as causally connected only if the greatest velocity with which an agency can connect cause and effect is the velocity of light.

15·9 Einstein's velocity addition theorem

Consider a particle moving with a velocity $v_x' = w$ in the S' frame along x'-axis, We want to know what the velocity of the particle v_x will be to an observer in a different frame S if S' is moving relative to S with a uniform velocity u along the x-axis which is assumed parallel to x'. In Newtonian relativity, the answer is simple; the velocity as measured by the observer in the S frame will be $v_x = u + w$. However, according to the special theory of relativity, the answer in different, specially when u and w are close to the velocity of light c.

By defiition, the velocity components in the two systems are

$$v_x = \frac{dx}{dt} \qquad\qquad v'_x = \frac{dx'}{dt'}$$

$$v_y = \frac{dy}{dt} \qquad\qquad v'_y = \frac{dy'}{dt'}$$

$$v_z = \frac{dz}{dt} \qquad\qquad v'_z + \frac{dz'}{dt'}$$

From the Lorentz transformation equations (15·6–8), we have

$$dx = \frac{dx' + udt'}{\sqrt{1-\beta^2}} = \gamma(dx' + udt'),\ dy = dy',\ dz = dz'$$

$$dt = \frac{dt' + \frac{u}{c^2}dx'}{\sqrt{1-\beta^2}} = \gamma\left(dt' + \frac{u}{c^2}dx'\right) \qquad (15·9\text{-}1)$$

where $\gamma = 1/\sqrt{1-\beta^2}$ and $\beta = u/c$

Then the velocity components of the particle in the S frame are

$$v_x = \frac{dx}{dt} = \frac{dx' + udt'}{dt' + \frac{u}{c^2}dx'} = \frac{dx'/dt' + u}{1 + \frac{u}{c^2}\frac{dx'}{dt'}} = \frac{v_x' + u}{1 + uv_x'/c^2}$$

$$v_y = \frac{dy}{dt} = \frac{dy'}{\gamma(dt' + \frac{u}{c^2}dx')} = \frac{\frac{dy'}{dt'}\sqrt{1-\beta^2}}{1 + \frac{u}{c^2}\frac{dx'}{dt'}} = \frac{v_y'\sqrt{1-\beta^2}}{1 + uv_x'/c^2}$$

$$= \frac{v_y'}{\gamma(1 + uv_x'/c^2)}$$

Similar equation can be obtained for v_z. Thus the transformation equations from S' to S frame for the velocity components are

$$v_x = \frac{v_x' + u}{1 + uv_x'/c^2}, v_y = \frac{v_y'}{\gamma(1 + uv_x'/c^2)}, v_z = \frac{v_z'}{\gamma(1 + uv_x'/c^2)} \quad (15\cdot9\text{-}2)$$

The transformation equations for the velocity components in the reverse case, *i.e.* from S to S' can similarly be obtained with the help of Eqs. (15·6-7) :

$$v_x' = \frac{v_x - u}{1 - uv_x/c^2}, v_y' = \frac{v_y}{\gamma(1 - uv_x/c^2)}, v_z' = \frac{v_z}{\gamma(1 - uv_x/c^2)} \quad (15\cdot9\text{-}3)$$

In the example of the particle moving with the velocity w along x' in the S' frame as mentioned in the first para above, we get the velocity v_x of the particle in the S frame by using Eq. (15·9-2) as

$$v_x = \frac{dx}{dt} = \frac{v_x' + u}{1 + uv_x'/c^2} = \frac{w + u}{1 + uw/c^2} \quad (15\cdot9\text{-}4)$$

Eq. (15·9-4) thus gives the resultant of the two velocities w and u along the same direction. If both w and u are small compared to c, then $uw/c^2 \ll 1$ and we get $v_x = w + u$ in agreement with the classical result.

On the other hand if $w = c$, then

$$v_x = \frac{c + u}{1 + uc/c^2} = c \cdot \frac{c + u}{c + u} = c \quad (15\cdot9\text{-}5)$$

Similarly if $u = c$, the resultant is $v_x = c$. Even when both the velocities u and w are equal to c, the resultant is $v_x = c$. This proves that it is not possible to have any velocity greater than the velocity of light c, even by compounding two (or more) velocities, each of which is equal to c. *Further the velocity of light is the same in all systems.* This can be proved easily by using the velocity transformation equations deduced above.

Let the velocity in the frame S′ be

$$v' = \sqrt{v_x'^2 + v_y'^2 + v_z'^2} = c$$

Then in the S frame we have, using Eqs. (15·9-2)

$$v^2 = v_x^2 + v_y^2 + v_z^2$$

$$= \frac{(v_x' + u)^2 + (v_y'^2 + v_z'^2)(1 - u^2/c^2)}{(1 + uv_x'/c^2)^2}$$

$$= \frac{v_x'^2 + v_y'^2 + v_z'^2 + 2v_x'u + u^2 - (v_y'^2 + v_z'^2)u^2/c^2}{(1 + uv_x'/c^2)^2}$$

$$= \frac{c^2\{1 + 2v_x'u/c^2 + u^2/c^2 - (c^2 - v_x'^2)u^2/c^4\}}{(1 + uv_x'/c^2)^2}$$

$$= \frac{c^2\{1 + 2v_x'u/c^2 + u^2/c^2 - u^2/c^2 + v_x'^2 u^2/c^4\}}{(1 + uv_x'/c^2)^2}$$

$$= \frac{c^2(1 + uv_x'/c^2)^2}{(1 + uv_x'/c^2)^2} = c^2$$

∴ $v = c$

Thus the absolute value of the velocity of light does not change in passing from one inertial frame to another. The components of the velocity may however be different in different frames which means that the direction of the light ray will generally appear to be different for the observers in different frames.

15·10 Explanation of stellar aberration from the special theory of relativity

The phenomenon of stellar aberration discussed in § 15·2 finds an easy explanation from the special theory of relativity.

Consider a star from which light arrives at the earth in a direction at right angles to the plane of the earth's orbit round the sun (*ecliptic*). In a frame of reference S in which the observer is at rest relative to the sun, the direction of the light ray is PS along y so that its velocity is $u_y = -c$ while $u_x = 0$ where x is the direction of the earth's velocity in its orbit (Fig. 15·5 *a*).

From the point of view of observer on the earth (S′ frame), the ray appears to come from the direction P′S and the angle P′SP = α is the angle of aberration (Fig. 15·5*b*). In this frame the velocity of light has two components u_x' and u_y' along x' in the plane of the earth's orbit and y' perpendicular to it. From Eqs. (15·9-3) we have

$$u_x' = \frac{u_x - v}{1 - vu_x/c^2}, \quad u_y' = \frac{u_y}{\gamma(1 - vu_x/c^2)}$$

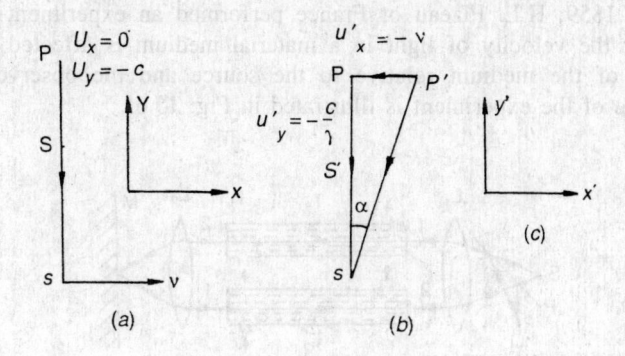

Fig. 15·5. Stellar aberration.

where v is the velocity of the earth relative to the sun which is along x. Substituting the values of u_x and u_y, we get

$$u_x' = -v, \quad u_y' = \frac{u_y}{\gamma} = -\frac{c}{\gamma}$$

Squaring and adding we get

$$u_x'^2 + u_y'^2 = v^2 + c^2(1 - \beta^2) = c^2$$

Hence the resultant velocity of light is c as it should be. The angle of aberration is given by

$$\sin \alpha \cong \alpha = \frac{u_x'}{c} = \frac{v}{c} = \beta$$

Since $v \sim 30$ km/s and $c = 3 \times 10^5$ km/s, we get

$$\alpha \sim 10^{-4} \text{ radian} = 20.63''$$

This is in agreement with the observed value. We have seen (§15·2A) the explanation of stellar aberration on the stationary ether theory gives $\tan \alpha = \beta$ which in practice is indistinguishable from the above result.

However the special theory of relativity removes the apparent contradiction between the results of the aberration experiment and Michelson-Morley experiment. The null result of the latter experiment showed that there was no relative motion between the earth and the ether. On the other hand the aberration experiment of Bradley can be explained on the ether hypothesis only if it is assumed that there is a relative motion between the earth and the ether. Special theory of relativity explains the result of both and experiments without requiring the hypothesis of the existence of ether as the medium for the propagation of light wave.

15·11 Fizeau's experiment

In 1859, H.L. Fizeau of France performed an experiment to test whether the velocity of light in a material medium is affected by the motion of the medium relative to the source and the observer. The principle of the experiment is illustrated in Fig. 15·6.

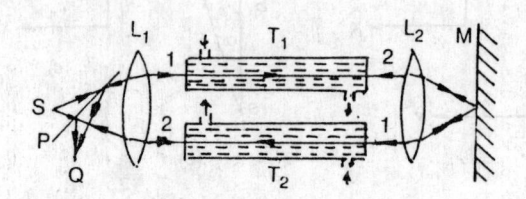

Fig. 15·6. Fizeau's ether drag experiment.

Light from the source S made parallel by means of the lens L_1 is split into two beams which travel through the two tubes T_1 and T_2 through which water flows in opposite directions. On reflection from the mirror M, the beams interchange their paths and are reunited at Q after reflection from the half-silvered glass plate P to produce a system of dark and bright fringes as in a Rayleigh refractometer. The beam 1 which travels in the direction of water flow in the tube T_1 during its onward course again travels in the direction of water flow in the tube T_2 during its return path. Similarly, the beam 2 travels opposite to the direction of water flow in the two tubes T_2 and T_1 before being reunited with the other beam at Q. If the velocity of light is affected by the velocity of the medium as would be expected from classical mechanics, then there would be a shift in the fringe system relative to its position when water in both the tubes remains stationary.

Using tubes 1·5 m long and a water velocity of 7 m/s, Fizeau observed a shift of 0·46 of a fringe when the direction of water flow was reversed. This would correspond to an increase in the velocity of light in one of the tubes and a decrease in the other by about half the velocity of flow of the water.

The above result is in apparent contradiction to the result of Michelson-Morley experiment, according to which the velocity of light remains unaffected by the velocity of the source or of the observer. However a close analysis shows that the result does not in any way contradict the general idea of the relativity of motion and can actually be explained with the help of the velocity addition theorem.

If n be the refractive index of water, then the velocity of light in

Special Theory of Relativity

water is $v'_x = c/n$. This would be the velocity of light as observed by an observer moving with the water in the tube. To get the velocity as observed by the observer at rest relative to the tube, we use Eq. (15.9–2). If u be the velocity of water, we get

$$v_x = \frac{v'_x \pm u}{1 \pm u v'_x / c^2} = \frac{c/n \pm u}{1 \pm u c/nc^2} = \frac{c/n \pm u}{1 \pm u/nc}$$

Since $u << c$. we can write

$$v_x = \left(\frac{c}{n} \pm u\right)\left(1 \pm \frac{u}{nc}\right)^{-1} \approx \left(\frac{c}{n} \pm u\right)\left(1 \mp \frac{u}{nc}\right)$$

$$= \frac{c}{n} \pm u\left(1 - \frac{1}{n^2}\right)$$

neglecting $\frac{u^2}{nc}$.

Hence the change in the velocity of light due to the flow of the water is given by

$$\Delta v_x = u\left(1 - \frac{1}{n^2}\right)$$

For water $n = 1.33$, which gives

$$\Delta v_x = u\left(1 - \frac{9}{16}\right) = \frac{7}{16} u \approx 0.44\, u$$

i.e., the change in the velocity of light is about half the velocity of flow of water in the tubes, which is in agreement with Fizeau's observations.

15.12 Change of the mass of a body with velocity

In classical mechanics, the mass of a body is assumed to be an invariant, independent of its velocity. But according to the special theory of relativity, the mass of a body increases with increasing velocity. To prove this we consider the following *thought experiment* proposed by R.C. Tolman.

Let S' be a frame of reference relative to which another frame S moves with a velocity $-V$ along the x-axis. Two bodies a and b moving along the x-axis with the velocities $+u'$ and $-u'$ in the S' frame collide with each other elastically. Assume the bodies to be perfectly elastic and identical in the S' frame. Immediately on collision they come to momentary rest and then fly apart in opposite directions (see Fig. 15.7a). Since the collision is elastic, there is conservation of mass and momentum in the S' frame after the collision.

Let us now consider the above collision from the point of view of an observer in the S frame. Let u_a and u_b be the velocities of the two bodies in this frame before collision and m_a and m_b their masses. Here we have considered the possibility of the dependence of the mass of a body on its

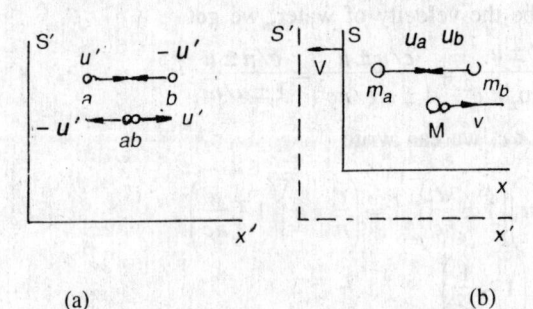

Fig. 15·7. Collision between two bodies observed by two observes.

velocity so that $m_a = m_a(u_a)$ and $m_b = m_b(u_b)$ may not be equal to one another.

Suppose that the total mass of the two bodies immediately on collision (when they are momentarily at rest) is M. Obviously at the time of collision, the above combination of the two bodies has a velocity $+V$ relative to the S system. According to the laws of conservation of mass and momentum, the sum of the masses and momenta of the two bodies before collision must be equal to their combined mass and momentum at the time of collision when they are at relative rest to each other. So we can write

$$m_a + m_b = M \tag{15.12-1}$$

$$m_a u_a + m_b u_b = MV \tag{15.12-2}$$

From the velocity addition theorem (15·9–2), we have

$$u_a = \frac{u' + V}{1 + u'V/c^2}, \quad u_b = \frac{-u' + V}{1 - u'V/c^2} \tag{15.12-3}$$

So we get from Eqs. (15·12-1) and (15·12-2)

$$\frac{m_a(u' + V)}{1 + u'V/c^2} - \frac{m_b(u' - V)}{1 - u'V/c^2} = (m_a + m_b)V$$

or, $$\frac{m_a}{1 + u'V/c^2}(u' + V - V - u'V^2/c^2)$$

$$= \frac{m_b}{1 - u'V/c^2}(u' - V + V - u'V^2/c^2)$$

$$\frac{m_a}{m_b} = \frac{1 + u'V/c^2}{1 - u'V/c^2} \tag{15.12-4}$$

Special Theory of Relativity

From Eq. (15·12-3), we have

$$1 - \frac{u_a^2}{c^2} = 1 - \frac{(u'+V)^2}{c^2(1+u'V/c^2)^2} = \frac{c^2(1+u'^2V^2/c^4) - (u'^2+V^2)}{c^2(1+u'V/c^2)^2}$$

$$= \frac{c^2(1-V^2/c^2) - u'^2(1-V^2/c^2)}{c^2(1+u'V/c^2)^2}$$

$$= \frac{(1-u'^2/c^2)(1-V^2/c^2)}{(1+u'V/c^2)^2}$$

$$\therefore \quad \sqrt{1-u_a^2/c^2} = \frac{\sqrt{1-u'^2/c^2} \cdot \sqrt{1-V^2/c^2}}{1+u'V/c^2}$$

Hence
$$1 + u'V/c^2 = \frac{\sqrt{1-u'^2/c^2} \cdot \sqrt{1-V^2/c^2}}{\sqrt{1-u_a^2/c^2}} \tag{15·12-5}$$

Similarly it can be shown that

$$\sqrt{1-u_b^2/c^2} = \frac{\sqrt{1-u'^2/c^2} \cdot \sqrt{1-V^2/c^2}}{1 - u'V/c^2}$$

so that
$$1 - u'V/c^2 = \frac{\sqrt{1-u'^2/c^2} \cdot \sqrt{1-V^2/c^2}}{\sqrt{1-u_b^2/c^2}} \tag{15·12-6}$$

Hence we get from Eq. (15·12-4).

$$\frac{m_a}{m_b} = \frac{\sqrt{1-u_b^2/c^2}}{\sqrt{1-u_a^2/c^2}} \tag{15·12-7}$$

The two particles have been assumed to be identical when they are at rest in the same frame of reference. Let m_0 be their mass under this condition. Then according to Eq. (15·12-7), the mass of the bodies must be inversely proportional to the quantity $\sqrt{1-u^2/c^2}$ when they are moving with the velocity u. So we can write the mass $m(u)$ of a particle moving with the velocity u as

$$m = m(u) = \frac{m_0}{\sqrt{1-u^2/c^2}} = m_0 \gamma \tag{15·12-8}$$

where $\gamma = (1-\beta^2)^{-1/2}$, $\beta = \frac{u}{c}$. The mass m_0 of the body at rest is usually called its *rest mass*. Eq. (15·12-8) then gives the relativistic increase of the mass of the body when it is in motion. This was first experimentally verified for electrons by the German scientist Bucherer experimenting with the β-particles emitted by radioactive substances (see Vol II).

15·13 Mass-energy equivalence

From Eq. (15·12–8), we arrive at another important conclusion. Let m be the mass of a body moving with the velocity v. Suppose a force F

acts on the body which produces a displacement dx of body. During this displacement the energy E of the body increases by an amount dE. Then we can write

$$dE = F\,dx$$

Since F is equal to the time rate of change of the momentum of the body, we can write

$$F = \frac{d}{dt}(mv)$$

Then $\quad dE = \frac{d}{dt}(mv)\,dx = \frac{d}{dt}(mv)\frac{dx}{dt}dt$

$$= v\,d(mv) = v^2\,dm + mv\,dv$$

$\therefore \qquad\qquad m v\,dv = dE - v^2\,dm \qquad\qquad (15\cdot13\text{-}1)$

If m_0 be the rest mass of the body, then from Eq. (15·12–8). we have

$$m^2(1 - v^2/c^2) = m_0^2$$

Differentiating we get

$$2m\,dm(1 - v^2/c^2) - 2m^2 v\,dv/c^2 \quad 0$$

or, $\qquad\qquad mv\,dv = (c^2 - v^2)\,dm \qquad\qquad (15\cdot13\text{-}2)$

Using Eqs. (15·13-1) and (15·13-2) we then get

$$dE - v^2 dm = c^2 dm - v^2\,dm$$

or, $\qquad\qquad dE = c^2\,dm$

which on integration gives

$$E = mc^2 = \frac{m_0 c^2}{\sqrt{1-\beta^2}} = m_0 c^2\,\gamma \qquad\qquad (15\cdot13\text{-}3)$$

From Eq. (15·13-3) we conclude that according to the special theory of relatively, if m be the mass of a body moving with the velocity v, then the body possesses an amount of energy $mc^2 = m_0 c^2\,\gamma$. This is known as the principle of *mass-energy equivalence*. If the body is at rest having the rest mass m_0, then according to the above principle, the body possesses an amount of energy $m_0 c^2$ which is known as its *rest energy*. If the entire mass m_0 of the body is converted into energy, then the above amount of energy will be produced. The principle of mass-energy equivalence is of great importance in the release of energy during differnt types of nuclear transformations including nuclear fission and nuclear fusion (see Vol. II). The quantity of energy released due to the transformation of mass into energy is very great. If a mass of 10^{-3} kg (*i.e.* 1 g) is completely converted into energy, then we get about 9×10^{13} joules or $2\cdot5 \times 10^7$ kilowatt-hours of energy. Practical utilization of this energy has been made possible due to the discovery of nuclear fission (see Vol. II).

15·14 Some important mathematical relationships

We have seen above that the special theory of relatively predicts equivalence of mass and energy. If a body moving with a velocity υ has a mass m, then its total energy is mc^2 which is made up partly of its rest energy $m_0 c^2$ and partly by virtue of its motion. The latter, which is nothing but the kinetic energy of the body, is thus given by

$$E_k = mc^2 - m_0 c^2 = m_0 c^2 \left\{ \frac{1}{\sqrt{1-\beta^2}} - 1 \right\} \qquad (15\cdot14\text{-}1)$$

$$= m_0 c^2 (\gamma - 1)$$

where $\beta = \upsilon/c$. This is different from the kinetic energy $m_0 \upsilon^2/2$ obtained from classical mechanics. The difference becomes more prominent when the velocity υ of the body becomes comparable to the velocity of light c. If $\upsilon << c$. then $\beta << 1$, and we get

$$E_k = m_0 c^2 \{ (1-\beta^2)^{-1/2} - 1 \}$$

$$= m_0 c^2 \left\{ 1 + \frac{1}{2} \beta^2 \ldots - 1 \right\}$$

$$= m_0 c^2 \cdot \frac{1}{2} \frac{\upsilon^2}{c^2} = \frac{1}{2} m_0 \upsilon^2$$

which is the same as that obtained from classical mechanics.

Again according to the special theory of relativity, the momentum p of a body is given by

$$p = m\upsilon$$

where m is the mass of a body moving with a velocity υ. Here m is a function of the velocity υ. Using Eq. (15·12–8), we get

$$p = \frac{m_0 \upsilon}{\sqrt{1-\beta^2}} = \frac{m_0 \beta c}{\sqrt{1-\beta^2}} \qquad (15\cdot14\text{-}2)$$

$$\therefore \quad p^2 c^2 + m_0^2 c^4 = \frac{m_0^2 \beta^2 c^4}{1-\beta^2} + m_0^2 c^4$$

$$= m_0^2 c^4 \left\{ \frac{\beta^2}{1-\beta^2} + 1 \right\}$$

$$= \frac{m_0^2 c^4}{1-\beta^2} = m^2 c^4$$

Since the total energy of the body is $E = mc^2$, we get

$$E^2 = m^2 c^4 = p^2 c^2 + m_0^2 c^4 \qquad (15\cdot14\text{-}3)$$

or,
$$E = \sqrt{p^2c^2 + m_0^2 c^4}$$

Then from Eq. (15·14-1), the kinetic energy of the body will be

$$E_k = E - m_0 c^2 = \sqrt{p^2 c^2 + m_0^2 c^4} - m_0 c^2 \qquad (15 \cdot 14\text{-}4)$$

15·15 Graphical representation of Lorentz transformation equations

The Lorentz transformation equations can be represented graphically as was originally suggested by H. Minkowski.

To see how this can be done, we first consider the graphical representation of the Galilean transformation equations (15·1–2). Since y and z coordinates remain unaffected by the transformation from one inertial frame S to another frame S' we draw a two dimensional diagram with x along and abscissa and t along the ordinate for the S frame. It should be noted that though x and t are drawn mutually perpendicular in Fig. 15·8 this need not be so. The choice of the angle between them is arbitrary since they do not represent a pair of orthogonal coordinates e.g. x-y, y-z or z-x axes.

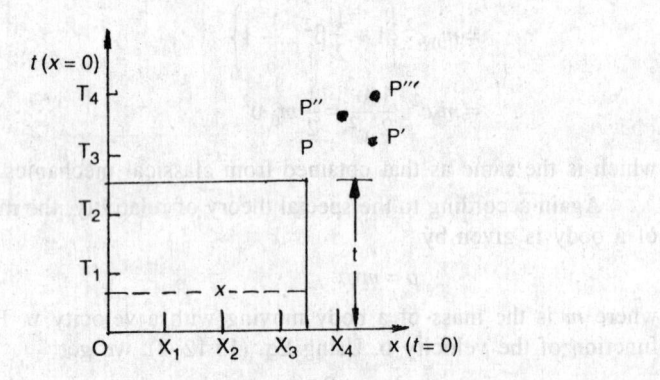

Fig. 15·8. World points in space-time diagram.

Along the x-axis in Fig. 15·8, $t = 0$ at all points. Similarly $x = 0$ at all points along the t-axis. If a series of *events* takes place at different instants of time at the point $x = 0$, then these will be represented by a series of points T_1, T_2, T_3,...along the t-axis for all of which $x = 0$. The *events* could be, for instance the firing of a gun successively at the instants $0, t_1, t_2, t_3,...$by an observer sitting stationary at the origin O in the S frame where $x = 0$.

On the other hand, if a number of *different events* occur at different points along the x-axis, all at the *same* instant of time $t = 0$, these can be represented by a series of points X_1, X_2, X_3...

Special Theory of Relativity

In general if an event takes place at x at the instant t, it will be represented by a point P known as *world-point*, with the coordinates (x, t) in the space-time diagram. A series of events occurring at the different points x_1, x_2, x_3...at different instants t_1, t_2, t_3...will be represented by a series of such world points P', P'' P''' .....

Suppose now we transform to a new inertial frame S'; then the space-time coordinates of a world-point, such as P, will be different. Let these be $(x' \, t')$. We shall see how we can find the new axes x' and t'. Obviously $t' = 0$ at all points along the x'-axis while $x' = 0$ at all points along the t'-axis. Eq. (15·1–2d) gives $t = 0$ for $t' = 0$, so that $x' = x$ which shows that the x'-axis is along the x-axis. On the other hand for $x' = 0$ which is the t'-axis, we have $x = \upsilon t$ from Eq. (15·1–2a). In Fig. (15·9) we

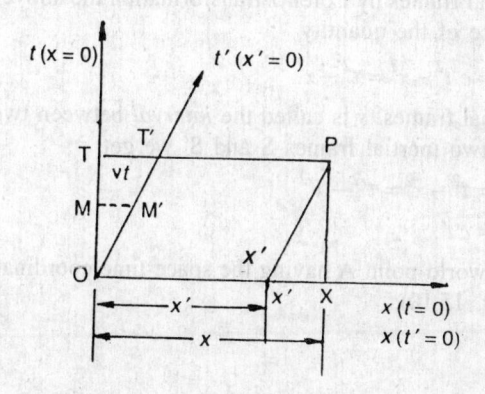

Fig. 15·9. Graphical representation of Galilean transformation.

have for the world-point P, $x = OX = TP$ and $t = OT = XP$. Now we measure off a length $TT' = \upsilon t$ along TP. Then the straight line OT' defines the t'-axis along which $x' = 0$ at all instants of time. The distance T'P of P from the t'-axis (OT') measured parallel to the x'-axis (OX') gives x':

$$T'P = TP - TT' = x - \upsilon t = x'$$

Now if we draw PX' parallel to OT' then $OX' = T'P = x'$. Obviously $t' = 0$ along the x'-axis. The coordinates in S' frame are thus obtained by drawing the projections on the x' and t'-axes parallel to these axes. This gives $t' = OT'$. As seen above $x' = OX'$.

As seen from Fig. 15·9 the t'-axis has a direction different from the t-axis. To satisfy the transformation relation $t' = t$ (Eq. 15·1–2d) it is necessary to choose different scales along t' and t-axes. If OM represents the unit of the time-scale along the t-axis in the S frame, OM' will represent the unit of the time-scale along the t'-axis in the S' frame. The number of units OM contained in OT is equal to the number of units

OM' contained in OT', so that OT'/OM' = OT/OM which ensures $t' = t$.

The motion of an object in space-time would be represented by a continuous series of world-points in such a diagram. The line through these world-points is known as a *world-line*.

Lorentz transformation and Minkowski diagram

We are now in a position to consider graphical representation of the Lorentz transformation, which is known as the *Minkowsky diagram* in the special theory of relativity.

The basic property of Lorentz transformation is the invariance of the quantity

$$c^2 t^2 - x^2 - y^2 - z^2 = \tau^2 - x^2 - y^2 - z^2$$

in different inertial frames, where $\tau = ct$. Since y and z remain unchanged in different inertial frames by Lorentz transformation the above invariance implies invariance of the quantity

$$s^2 = c^2 t^2 - x^2 = \tau^2 - x^2$$

in different inertial frames. s is called the *interval* between two events in space-time. For two inertial frames S and S' we get

$$s^2 = \tau^2 - x^2 = \tau'^2 - x'^2$$

where $\tau' = c t'$.

Consider a world-point A having the space-time coordinates (x, τ) in S frame (see Fig. 15·10).

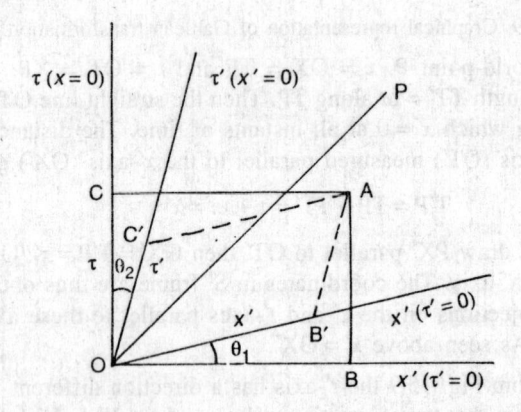

Fig. 15·10. Minkowski diagram for Lorentz transformation.

From Lorentz transformation equations (15·6–7) the coordinates

$$x' = \frac{x - \beta\tau}{\sqrt{1-\beta^2}}, \quad \tau' = \frac{\tau - \beta x}{\sqrt{1-\beta^2}}$$

where $\beta = \upsilon/c$. It can be readily seen that in this case, the x'-axis is not along the x-axis. Similarly the τ'-axis is different from the τ-axis. These are shown in Fig. 15·10 in which the coordinates have been chosen as x and $\tau = ct$ for the S frame and x' and $\tau' = ct'$ for the S' frame.

The direction of the x'-axis is determined by the condition $\tau' = 0$ along this axis, so that from Eq. (15·6–7d) we get

$$\tau' = \frac{\tau - \beta x}{\sqrt{1-\beta^2}} = 0$$

or, $\qquad \tau - \beta x = 0$

which gives $\qquad \tan\theta_1 = \dfrac{\tau}{x} = \beta$

$\therefore \qquad\qquad\qquad \theta_1 = \tan^{-1}\beta$

Here θ_1 is the inclination of the x'-axis to the x-axis.

Similarly, the equation for the τ'-axis for which $x' = 0$ is given by

$$x' = \frac{x - \beta\tau}{\sqrt{1-\beta^2}} = 0$$

or, $\qquad x - \beta\tau = 0$

which gives $\qquad \tan\theta_2 = \dfrac{x}{\tau} = \beta$

$\therefore \qquad\qquad\qquad \theta_2 = \tan^{-1}\beta$

Here θ_2 is the inclination of the τ'-axis to the τ-axis. Obviously $\theta_1 = \theta_2$. Thus the axes x', τ' in the S' frame are equally inclined to axes x, τ respectively and are symmetrical about the line OP which is equally inclined to the axes x, τ.

The above discussion shows that Lorentz transformation corresponds to equal and symmetrical rotation of the space-time axes.

Now consider an event such as the firing of a bullet by an observer at $x = 0$ at the time $t = 0$ (*i.e.* $\tau = 0$) in the S frame. This is represented by the world-point O in the Minkowski diagram (Fig. 15·11). The bullet moving with a velocity υ hits a target at x at the time t in the same frame (S). This second event is represented by the world-point O' (say). The space-time interval s between the two events is given by

$$s^2 = \tau^2 - x^2$$

Since this remains invariant under Lorentz transformation, we have for the S' frame

$$s^2 = \tau'^2 - x'^2$$

If we plot the values of τ and x for the interval between two definite events measured in different inertial frames, we get the hyperbola BAC (or B′A′C′) as shown in Fig. 15·11 for which s^2 has a definite value.

For all positive values of s^2 (i.e., $s^2 > 0$), we have $x^2 = \tau^2 - s^2 < \tau^2$ so that $|x| < \tau$. So in this case, the hyperbolas all lie either above or below the two asymptotic lines P′OP and Q′OQ in Fig. 15·11 for which $x = \tau = ct$ and $x = -\tau = -ct$ respectively.

If an object travels with a velocity v, then it will cover a distance $x = vt$ in time t. For all world points lying above the line P′OP or below the line Q′OQ, $v < c$ (since $x = vt < ct$). Since, according to the special theory of relativity no material object can travel with a velocity v greater than c, the events represented by the world-points in these regions, shown shaded in the diagram, can be *causally connected*, as in the case of the firing of a bullet at $x = 0$, $t = 0$ and its hitting a target at the point x at a later instant t.

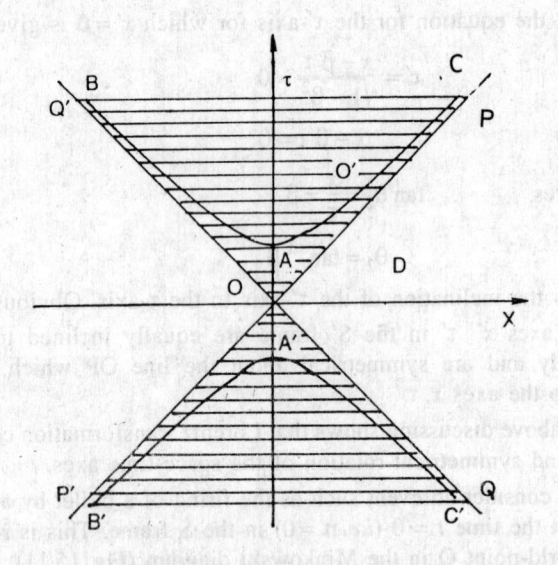

Fig. 15·11. Space-time intervals between events.

In Fig. 15·11 the world-points in the shaded region between the asymptotes OQ′ and OP represent all *future events* (for which t > 0) which have been causally conditioned by the present (t = 0) represented by the world-point O. Similarly, all events represented by the world-points in the shaded region between OP′ and OQ belong to the *past* (t < 0) which have conditioned the present event O.

Special Theory of Relativity

Events represented by the world-points (*e.g.*, D) in the unshaded regions of the diagram cannot have any causal connection with the present event O. For in this case, space-time coordinate systems may be found in which the second event occurs *after* the first while there are systems in which the second event occurs *before* the first.

As an example, consider an event occurring on the sun and a second event which is observed by an observer on the earth about 4 min later. These two events cannot be causally connected. We know that the minimum time in which the first event on the sun can make its effect felt on the earth is about 8 min, which is the time required by a light-signal to reach the earth from the sun. So the second event as observed on the earth cannot be causally connected to the first event on the sun.

Now consider the two events from the point of view of an observer who is moving from the earth towards the sun with a velocity $v = c/2$. The time of occurrence of the second event for this observer can be obtained from the Lorentz transformation equation :

$$t' = \frac{t - vx/c^2}{\sqrt{1-\beta^2}}$$

Here $t = 4$ min; x is the sun to earth distance given by $x = c\, t_0$ where t_0 is the time for the light signal to reach the earth from the sun ($t_0 = 8$ min). Then

$$t' = \frac{4 - \left(\frac{c}{2} \times \frac{8c}{c^2}\right)}{\sqrt{1-\beta^2}}$$
$$= \frac{4 - 4}{\sqrt{1-\beta^2}}$$
$$= 0$$

If $v > c/2$, then $t' < 0$. So there are frames of reference in which the second event will appear to occur *earlier* than the first event, though in the frame considered at the beginning, the second event occurs *after* the first event.

15·16 Relativistic Doppler effect

Doppler effect is a well-known phenomenon in classical wave theory. Everybody is familiar with the sudden change of pitch of the whistle of a railway engine as it passes by an observer standing by the track-side which occurs due to the change in the apparent frequency of the whistle as the velocity of the engine changes sign, *i.e.*, as the approaching velocity suddenly changes into receding velocity.

Doppler effect occurs both due to the motion of the source and that of the observer relative to the medium. The effect is not symmetric for the two cases. If ν is the apparent frequency as recorded by the observer and v_0 the frequency of the sound wave as emitted by the source, we have for motions along the line joining the source and the observer (see any text book on *Acoustics*):

(a) For the source in motion with the velocity u_1, the observer being stationary

$$v = \frac{v_0}{1 \pm u_1/V} \qquad (15\cdot 16\text{-}1)$$

where the plus sign in the denominator applies when the source moves away from the observer while the *minus* sign applies when they approach one another.

(b) For the observer in motion with the velocity u_2, the source being stationary

$$v = v_0 (1 \mp u_2/V) \qquad (15\cdot 16\text{-}2)$$

where the *minus* and *plus* signs refer to receding and approaching motions respectively between the source and the observer. V is the velocity of sound for the source at rest relative to the medium.

Doppler shift of frequency also takes place in the case of light wave. However in this case the effect must be symmetric in the two cases discussed above, since light waves do not require any medium for propagation. So the change of frequency can depend only on the relative motion of the source and the observer. Since the velocity of light remains the same in all inertial frames, we have to use the Lorentz transformations equations in this case.

We note that the phase of the light wave must remain *invariant* in transforming from one inertial frame to another. This can be easily understood by referring to a point x where the phase is such as to produce a minimum disturbance (zero intensity) at some instant t in the S frame. Then in another frame S′ moving with uniform velocity v relative to S, the disturbance must be minimum at x' at time t' where x' and t' are related to x and t by the Lorentz transformation equations.

Now the phase of the light wave can be written as (in S frame)

$$\varphi = \mathbf{k}\cdot\mathbf{r} - \omega t = k_x x + k_y y + k_z z - \omega t$$

In S′ frame, we have

$$\varphi' = \mathbf{k}'\cdot\mathbf{r}' - \omega' t' = k_x' x' + k_y' y' + k_z' z' - \omega' t'$$

Then because of the invariance of the phase ($\varphi = \varphi'$), we get by using the Lorentz transformation equations :

$$k_x x + k_y y + k_z z - \omega t = k_x' x' + k_y' y' + k_z' z' - \omega' t'$$
$$= \gamma k_x'(x - vt) + k_y' y + k_z' z - \omega' \gamma(t - vx/c^2)$$

where $\qquad \gamma = 1/\sqrt{1-\beta^2},\ \beta = v/c$

Equating the coefficients of x, y, z and t separately on the two sides, we get

Special Theory of Relativity

$$k_x = \gamma(k_x' + \omega' v/c^2), \, k_y = k_y', \, k_z = k_z' \qquad (15 \cdot 16\text{-}3)$$

$$\omega = \gamma(\omega' + k_x' v) \qquad (15 \cdot 16\text{-}4)$$

These are the transformation equations for the components of the wave vector **k** and the angular frequency ω between the two inertial frames.

Now consider a source of light Q moving with the velocity v relative to the observer O along x (see Fig. 15.12). If $\omega' = \omega_0$ be the angular frequency of the light waves emitted by Q in its own frame S' and θ' the angle between the line of sight QO and the direction of motion in the same frame, then we have $k_x' = k' \cos \theta' = (\omega_0/c) \cos \theta'$. The frequency as recorded by the observer in its own frame S is then given by

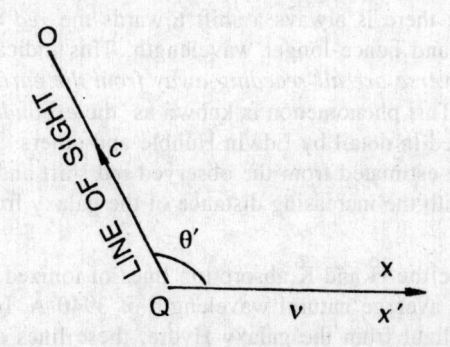

Fig. 15.12. Path of ray in relativistic Dopper effect

$$\omega = \gamma(\omega' + k_x' v) = \gamma \left(\omega_0 + \frac{\omega_0 v}{c} \cos \theta' \right)$$

$$= \gamma \omega_0 (1 + \beta \cos \theta')$$

If the source moves along the line of sight *away from the observer*, then $\theta' = \pi$ and we have

$$\omega = \gamma \omega_0 (1 - \beta) = \omega_0 \sqrt{\frac{1-\beta}{1+\beta}} \qquad (15 \cdot 16\text{-}5a)$$

Written in terms of frequency $\nu = \dfrac{\omega}{2\pi}$ we get

$$\nu = \nu_0 \sqrt{\frac{1-\beta}{1+\beta}} \qquad (15 \cdot 16\text{-}5b)$$

If λ be the wavelength, we have

$$\lambda = \lambda_0 \sqrt{\frac{1+\beta}{1-\beta}} \qquad (15 \cdot 16\text{-}6)$$

where λ is the wavelength measured by the observer and λ_0 is the *natural* wavelength as measured in the frame of the emitter.

Eq. (15·16-6) shows that $\lambda > \lambda_0$ so that there is a shift towards the longer wavelength when the source is *moving away* from the observer. For *the source and the observer approaching one another*, the signs before β in Eq. (15·16-6) are to be interchanged so that the wavelength is decreased in this case :

$$\lambda = \lambda_0 \sqrt{\frac{1-\beta}{1+\beta}} \qquad (15\cdot16\text{-}7)$$

Experimental evidence in support of the Doppler shift of the wavelength of light is provided by the well-known phenomenon of *galactic red shift*. The natural frequencies of the spectral lines emitted by the galaxies which are in motion relative to the earth are found to change in such a way that there is always a shift towards the *red* which has a smaller frequency and hence longer wavelength. This indicates that the *galaxies in the universe are all receding away from the earth as well as from one another*. This phenomenon is known as the *expanding universe* and has been studied in detail by Edwin Hubble and others. The velocity of recession can be estimated from the observed red shift and is found to increase linearly with the increasing distance of the galaxy from the earth (*Hubble's law*).

As an example the H and K absorption lines of ionized calcium are known to have the average natural wavelength of 3940 Å. In a study of the spectra of the light from the galaxy Hydra, these lines are found to have the wavelength 4750 Å. Hence we have

$$\Delta\lambda = \lambda - \lambda_0$$

$$= \lambda_0 \left(\sqrt{\frac{1+\beta}{1-\beta}} - 1 \right)$$

$$\therefore \quad \sqrt{\frac{1+\beta}{1-\beta}} = 1 + \frac{\Delta\lambda}{\lambda_0} = \frac{4750}{3940} = 1\cdot205$$

This gives $\beta = 0\cdot184$ so that the velocity of recession is $\upsilon = 0\cdot184 c$ which is nearly the one-fifth the velocity of light.

Eq. (15·16-5b) gives a shift of frequency of the light wave due to the relative motion of the source and the observer which, to a first order of approximation, is the same as that obtained from the classical expressions of the Doppler shift if we write $V = c$ and $u_1 = u_2 = \upsilon$. Expanding Eq. (15·16-5b) in a power series of $\beta = \upsilon/c$ we get

$$\nu = \nu_0 (1-\beta)^{1/2} (1+\beta)^{-1/2}$$

$$\nu = \nu_0 \left(1 - \frac{1}{2}\beta - \frac{1}{8}\beta^2 \ldots \right) \left(1 - \frac{1}{2}\beta + \frac{3}{8}\beta^2 \right) \ldots$$

$$= v_0 \left(1 - \beta + \frac{1}{2}\beta^2 \ldots \right)$$

$$\approx v_0(1-\beta)$$

if we retain terms upto the first power β.

Alternatively, we can write

$$v = \frac{v_0}{(1+\beta)^{1/2}(1-\beta)^{-1/2}}$$

$$= \frac{v_0}{\left(1 + \frac{1}{2}\beta - \frac{1}{8}\beta^2 \ldots\right)\left(1 + \frac{1}{2}\beta + \frac{3}{8}\beta^2 \ldots\right)}$$

$$= \frac{v_0}{1 + \beta + \frac{1}{2}\beta^2 \ldots} \approx \frac{v_0}{1+\beta}$$

if only terms upto β are retained. These first order expressions are the same as the classical expressions (15·16-1) and (15·16-2) respectively.

If $\beta << 1$, the second order terms in β^2 in the above expansions are small compared to the first order terms in β and hence no distinction can be made between the classical and relativistic expressions in this case.

To distinguish between the classical and relativistic cases, it is thus necessary to measure the second order effect depending on β^2. Einstein, as early as 1907 had pointed out that this could be done by looking for the so called *transverse Doppler effect*. If observations are made perpendicular to the direction of motion of the source in the observer's frame of reference S then $\theta = \frac{\pi}{2}$ and $k_x = k \cos \theta = 0$, so that we get from Eq. (15·16-3)

$$k_x = \gamma(k_x' + \omega' v/c^2) = 0$$

or, $$k_x' = -\frac{\omega' v}{c^2}$$

Hence $\omega = \gamma(\omega' + k_x' v)$

$$= \gamma \omega'(1 - v^2/c^2) = \frac{\omega'(1-\beta^2)}{\sqrt{(1-\beta^2)}}$$

or, $$\omega = \omega' \sqrt{1-\beta^2}$$

This gives $v = v'\sqrt{1-\beta^2}$ and $\lambda = \lambda'/\sqrt{1-\beta^2}$

Expansion in powers of β^2 gives (writing $v' = v_0$)

$$\nu = \nu_0 \left(1 - \frac{1}{2}\beta^2 \ldots\right) \approx \nu_0 \left(1 - \frac{1}{2}\beta^2\right)$$

Thus the frequency shift is a second order effect proportional to β^2. Obviously the effect will be very small. It is relativistic in origin. Experimental measurement of transverse Doppler effect is extremely difficult since even a small departure from $\theta = \pi/2$ will produce a first order effect which will be much larger than the second order relativistic effect. However the difficulty can be surmounted by making observations in the longitudinal direction both for the approaching and receding motions. Then from Eqs. (15·16-6) and (15·16-7) we get:

For motion of recession

$$\lambda_r = \lambda_0 \sqrt{\frac{1+\beta}{1-\beta}}$$

For motion of approach

$$\lambda_a = \lambda_0 \sqrt{\frac{1-\beta}{1+\beta}}$$

$$\therefore \quad \lambda_r + \lambda_a = \lambda_0 \left(\sqrt{\frac{1+\beta}{1-\beta}} + \sqrt{\frac{1-\beta}{1+\beta}}\right)$$

$$= \frac{2\lambda_0}{\sqrt{1-\beta^2}} = 2\gamma\lambda_0$$

The mean wavelength of the two lines is then

$$\lambda_m = \frac{\lambda_r + \lambda_a}{2} = \gamma \lambda_0$$

Then the shift in the mean wavelength is

$$\lambda_m - \lambda_0 = (\gamma - 1)\lambda_0 = \left(\frac{1}{\sqrt{1-\beta^2}} - 1\right)\lambda_0 \approx \frac{\lambda_0 \beta^2}{2}$$

which is a second order effect and is purely relativistic in origin. For the non-relativistic case (first order effect only) there would be no such shift of the mean of the two wavelengths which would be the same as the original wavelength λ_0.

In 1938, H.E. Ives and G.R. Stilwell made measurements on the radiation in the forward and backward directions relative to the motion of the atoms in a discharge tube containing hydrogen. Observations were made on the H_β line of the Balmer series of hydrogen. The hydrogen ions in the discharge tube were accelerated through an accurately controlled voltage and the wavelengths of the H_β line for the two directions of emission were measured which confirmed the relativistic Doppler shift.

15·17. Four Vectors

One of the fundamental postulates of the special theory of relativity is that the velocity of light is the same in all inertial frames Consider two inertial frames S (x, y, z, t) and S' (x', y', z', t') having the same origin at $t = 0$. Let spherical light waves start out front the origin in the S frame at $t = 0$. Then after a time t these will spread out to a spherical surface in the S frame determined by the relation

$$x^2 + y^2 + z^2 = c^2 t^2$$

or,
$$c^2 t^2 - x^2 - y^2 - z^2 = 0 \qquad (15 \cdot 17\text{-}1)$$

Since the velocity c is the same in S' frame, we can write similar equation for this frame also :

$$x'^2 + y'^2 + z'^2 = c^2 t'^2$$

or,
$$c^2 t'^2 - x'^2 - y'^2 - z'^2 = 0 \qquad (15 \cdot 17\text{-}2)$$

If we introduce the coordinates

$x_1 = x$ $\qquad\qquad$ $x_1' = x'$

$x_2 = y$ $\qquad\qquad$ $x_2' = y'$

$x_3 = z$ $\qquad\qquad$ $x_3' = z'$

$x_4 = ict$ $\qquad\qquad$ $x_4' = ict'$

we get from Eqs. (15·17-1) and (15·17-2)

$$x_1^2 + x_2^2 + x_3^2 + x_4^2 = x_1'^2 + x_2'^2 + x_3'^2 + x_4'^2$$

The above result can also be proved by using the Lorentz transformation equations (15-6-7). Writing $\gamma = 1/\sqrt{1 - \beta^2}$ where $\beta = \upsilon/c$, we have

$$\sum_{i=1}^{4} x_i'^2 = x'^2 + y'^2 + z'^2 - c^2 t'^2$$

$$= \gamma^2 (x - \upsilon t)^2 + y^2 + z^2 - c^2 \gamma^2 (t - \upsilon x/c^2)^2$$

$$= \gamma^2 [x^2 - 2x\upsilon t + \upsilon^2 t^2 - c^2 t^2 + 2x\upsilon t - \upsilon^2 x^2/c^2] + y^2 + z^2$$

$$= \gamma^2 [x^2 (1 - \upsilon^2/c^2) - c^2 t^2 (1 - \upsilon^2/c^2)] + y^2 + z^2$$

$$= x^2 + y^2 + z^2 - c^2 t^2 = \sum_{i=1}^{4} x_i^2$$

Thus the quantity $\sum_{i=1}^{4} x_i^2$ is an invariant under Lorentz transformation. This result is very similar to the property of a vector in ordinary (3-dimensional) space. It is well-known that if x, y, z represent the position coordinates of a point in some reference frame Σ, then the position vector **r** of the point w.r.t. the origin has a mangnitude which is an invariant under the rotation of the coordinate system about the origin, *i.e.*, if x', y', z' are the position coordinates of the same point in another frame of reference Σ', obtained by a simple rotation of the coordinates of the Σ frame, then

$$r^2 = r'^2$$

i.e., $$x^2 + y^2 + z^2 = x'^2 + y'^2 + z'^2 = \text{Invariant}$$

As we have seen in § 15·15, Lorentz transformation represents a relative rotation of the coordinate system between two inertial frames in relativistic dynamics. This makes it possible for us to look upon the four quantities x_1, x_2, x_3 and x_4 ($= ict$) as the coordinates in a four dimensional space-time continuum similar to the coordinates x, y, z in three dimensional space. The quantity

$$|s| = \sqrt{\sum_{i=1}^{4} x_i^2} = \sqrt{x_1^2 + x_2^2 + x_3^2 + x_4^2}$$

is known as an "interval" and is an invariant under Lorentz transformation. Thus we may regard x_1, x_2, x_3, x_4 as the components of four-dimensional vector (or simply a 4-*vector*).

Now in three-dimensional space, the components of any vector transform like the components of the position vector **r** under a linear orthogonal transformation. Similarly in the four-demensional case, we take a quantity to be a 4-vector if its four components (three spatial components and one time component) transform like the components x_1, x_2, x_3 and x_4 of the "interval" under Lorentz transformation. From this it follows that if $|A|$ is the magnitude of a 4-vector having the components A_1, A_2, A_3, A_4, then $|A|$ must be an invariant under Lorentz transformation.

$$|A| = \sqrt{\sum_{i=1}^{4} A_i^2} = \sqrt{A_1^2 + A_2^2 + A_3^2 + A_4^2}$$

$$= \text{Invariant}$$

The components A_i transform like the components x_i of the "interval" in 4-dimensional space-time continuum.

We have seen above that the three spatial components x_1, x_2, x_3 of s are real while the fourth component x_4, which we may call the time component, is pure imaginary ($x_4 = ict$). This is also true for the transformed coordinates x_1', x_2', x_3', x_4', the last one being pure imaginary.

In the case of the "interval" 4-vector, the sum of the three space compnents $x_1^2 + x_2^2 + x_3^2 \leq x_4^2$, since no material particle can have velocity greater than the velocity of light c, which means that $x^2 + y^2 + z^2$ must either the less than or equal to $c^2 t^2$. Hence s^2 is either negative or may be even be zero.

In general if a 4-vector with components A_i is such that $|A|^2 > 0$, it is called *space-like*, while if $|A|^2 \leq 0$, it is called *time-like*.

Four-velocity : As an example, we first consider the case of 4-velocity.

We first introduce the concept of proper time τ which is the time measured in a coordinate system fixed to a particle. Since by definition, the particle has no spatial displacement in this system, the four-dimensional incremental interval is

$$|\Delta^s| = \sqrt{\Delta x_1^2 + \Delta x_2^2 + \Delta x_3^2 + \Delta x_4^2} = \sqrt{-c^2 \Delta \tau^2}$$

where $\Delta x_1 = \Delta x_2 = \Delta x_3 = 0$ and $\Delta x_4 = ic \Delta \tau$ in this "proper frame".

Hence $|\Delta^s| = ic \Delta \tau$

In another coordinate system moving with a velocity u relative to the partile, we have

$$|\Delta^s| = \sqrt{\Delta x_1'^2 + \Delta x_2'^2 + \Delta x_3'^2 + \Delta x_4'^2}$$
$$= \sqrt{\Delta x^2 + \Delta y^2 + \Delta z^2 - c^2 \Delta t^2}$$

where $\Delta x_1' = \Delta x$, $\Delta x_2' = \Delta y$, $\Delta x_3' = \Delta z$ and $\Delta x_4' = ic \Delta t$

From the invariance of the "interval" we then have

$$-c^2 \Delta \tau^2 = \Delta x^2 + \Delta y^2 + \Delta z^2 - c^2 \Delta t^2$$

which gives $\Delta \tau^2 = \Delta t^2 - \dfrac{\Delta x^2 + \Delta y^2 + \Delta^2}{c^2}$

or, $\Delta \tau = \Delta t \sqrt{1 - u^2/c^2} = \Delta t \sqrt{1 - \beta^2}$ \hfill (15·17-3)

where $\beta = u/c$

The motion of a material particle may be described by the specification of its four coordinates as functions of $\tau : x_i = x_i(\tau)$ where $i = 1, 2, 3, 4$. Then differentiating w.r.t. τ we get the components of the 4-velocity as

$$v_i = \lim_{\Delta\tau \to 0} \frac{\Delta x_i}{\Delta\tau} = \frac{dx_i}{d\tau}$$

Since Δx_i represents the component of the incremental interval in 4-space, it is the component of a 4-vector. Also the proper time interval $\Delta\tau$ is an invariant. Hence v_i represents the component of a 4-vector. It is equal the ratio of the change of the interval component of the particle in 4-dimensional space-time continuum to the change of proper time τ as measured by the particle in its own frame.

We now deduce the components of the 4-velocity. Eq. (15·17-3) gives

$$\frac{d\tau}{dt} = \sqrt{1-\beta^2} \qquad (15\cdot17\text{-}4)$$

Hence $\qquad v_1 = \dfrac{dx_1}{d\tau} = \dfrac{dx}{dt} \bigg/ \dfrac{d\tau}{dt} = \dfrac{u_x}{\sqrt{1-\beta^2}} \qquad (15\cdot17\text{-}5)$

where $u_x = dx/dt$ is the x-component of the ordinary (3-dimensional) velocity.

Similarly, we have

$$v_2 = \frac{u_y}{\sqrt{1-\beta^2}}, \quad v_3 = \frac{u_z}{\sqrt{1-\beta^2}} \qquad (15\cdot17\text{-}6)$$

where $\qquad u_y = dy/dt$ and $u_z = dz/dt$

The time component is

$$v_4 = \frac{dx_4}{d\tau} = \frac{dx_4}{dt} \bigg/ \frac{d\tau}{dt} = \frac{ic}{\sqrt{1-\beta^2}} \qquad (15\cdot17\text{-}7)$$

We thus have the components of the 4-velocity :

$$v_1 = \frac{u_x}{\sqrt{1-\beta^2}}, \, v_2 = \frac{u_y}{\sqrt{1-\beta^2}}, \, v_3 = \frac{u_z}{\sqrt{1-\beta^2}},$$

$$v_4 = \frac{ic}{\sqrt{1-\beta^2}} \qquad (15\cdot17\text{-}8)$$

In the coordinate system in which the particle is at rest ($u = 0$), the 4-velocity components are :

$$v_1^\circ = v_2^\circ = v_3^\circ = 0, \, v_4^\circ = ic \qquad (15\cdot17\text{-}9)$$

Since the magnitude of the 4-velocity must be an invariant, we can find its value from the special case given above :

$$\sum_{i=1}^{4} v_i^2 = \sum_{i=1}^{4} (v_i^\circ)^2 = -c^2 \qquad (15\cdot17\text{-}10)$$

Special Theory of Relativity

Energy-momentum invariance:

We next seek an example of a 4-vector amongst the dynamical quantities connected with the motion of a particle.

In § 15·14 we deduced the relativistic relation between the energy (E) and momentum (p) of a particle (Eq. 15·14-3) of rest mass m_0:

$$E^2 = p^2 c^2 + m_0^2 c^4$$

For a particle at rest, $\beta = 0$ so that its energy is $m_0 c^2$ which is an invariant. We can rewrite the above equation as

$$E^2 - p^2 c^2 = m_0^2 c^4 = \text{Invariant} \qquad (15\cdot17\text{-}11)$$

The above invariance relation suggests that it should be possible to express the total energy and the momentum as the components of a 4-vector.

Consider two inertial frames S and S' such that S' moves with a uniform velocity **v** in 3-dimensional space in the S frame and v_x', v_y', v_z' are the components of the velocity **v'** in the S' frame. We have from the velocity addition theorem (see Eqs. 15·9-3)

$$v_x' = \frac{v_x - u}{1 - v_x u / c^2} \qquad (15\cdot17\text{-}12)$$

$$v_y' = \frac{v_y \sqrt{1 - u^2/c^2}}{1 - v_x u / c^2} \qquad (15\cdot17\text{-}13)$$

$$v_z' = \frac{v_z \sqrt{1 - u^2/c^2}}{1 - v_x u / c^2} \qquad (15\cdot17\text{-}14)$$

Now in S frame :

$$p_x = \frac{m_0 v_x}{\sqrt{1 - v^2/c^2}}, \; p_y = \frac{m_0 v_y}{\sqrt{1 - v^2/c^2}}, \; p_z = \frac{m_0 v_z}{\sqrt{1 - v^2/c^2}} \qquad (15\cdot17\text{-}15)$$

$$E = \frac{m_0 c^2}{\sqrt{1 - v^2/c^2}}$$

In S' frame :

$$p_x' = \frac{m_0 v_x'}{\sqrt{1 - v'^2/c^2}}, \; p_y' = \frac{m_0 v_y'}{\sqrt{1 - v'^2/c^2}},$$

$$p_z' = \frac{m_0 v_z'}{\sqrt{1 - v'^2/c^2}} \qquad (15\cdot17\text{-}16)$$

$$E' = \frac{m_0 c^2}{\sqrt{1 - v'^2/c^2}}$$

With the help of Eqs. (15·17-12), (15·17-13), (15·17-14), it is possible to show that

$$1 - \frac{v'^2}{c^2} = \frac{(1 - v^2/c^2)(1 - u^2/c^2)}{(1 - v_x u/c^2)^2} \qquad (15\cdot 17\text{-}17)$$

We then have

$$E' = \frac{m_0 c^2}{\sqrt{1 - v'^2/c^2}} = \frac{m_0 c^2 (1 - v_x u/c^2)}{\sqrt{1 - v^2/c^2}\sqrt{1 - u^2/c^2}}$$

$$\therefore \qquad E' = \gamma_u (E - u p_x)$$

where $\qquad \gamma_u = \dfrac{1}{\sqrt{1 - u^2/c^2}}$

Similarly it can be shown that

$$p_x' = \gamma_u (p_x - uE/c^2),\ p_y' = p_y,\ p_z' = p_z \qquad (15\cdot 17\text{-}18)$$

Also $\qquad E' = \gamma_u (E - u p_x) \qquad (15\cdot 17\text{-}19)$

Eqs. (15·17-18) and (15·17-19) constitute the set of transformation equations for the momentum components and the energy between the two inertial frames S and S' in relativistic mechanics.

We can also prove the converse relations as follows:

$$p_x = \gamma_u (p_x' + uE'/c^2),\ p_y = p_y',\ p_z = p_z' \qquad (15\cdot 17\text{-}20)$$

$$E = \gamma_u (E' + u p_x') \qquad (15\cdot 17\text{-}21)$$

The set of transformation equations (15·17-18) and (15·17-19) or the reverse relations (15·17-20) and (15·17-21) are very similar to the Lorentz transformation equations (15·17-7) or (15·17-8). We can therefore regard the quantities p_x, p_y, p_z as the three spatial components and iE/c as the time component of the 4-*momentum* which is a 4-vector. The invariance of the magnitude of the 4-momentum can easily be verified:

$$p_x'^2 + p_y'^2 + p_z'^2 + (iE'/c)^2$$

$$= \gamma_u^2 (p_x - uE/c^2)^2 + p_y^2 + p_z^2 - \frac{\gamma_u^2 (E - u p_x)^2}{c^2}$$

$$= \gamma_u^2 [p_x^2 - 2u p_x E/c^2 + u^2 E^2/c^4 - E^2/c^2 + 2u p_x E/c^2$$
$$\qquad\qquad - u^2 p_x^2/c^2] + p_y^2 + p_z^2$$

$$= \gamma_u^2 \left[p_x^2 (1 - u^2/c^2) - \frac{E^2}{c^2}(1 - u^2/c^2) \right] + p_y^2 + p_z^2$$

$$= p_x^2 + p_y^2 + p_z^2 + (iE/c)^2$$

This shows that the quantity $p_x^2 + p_y^2 + p_z^2 - E^2/c^2$ is an invariant

Special Theory of Relativity

which is nothing but Eqs. (15·17-11).

15.18. Twin paradox

Some of the predictions of the special theory of relativity, such as the variation of the mass of a body with velocity, equivalence of mass and energy, time-dilatation, to quote only a few, appear paradoxical and pose conceptual difficulties, because they seem to be at variance with our everyday experience. However all these have been confirmed experimentally over the years and the special theory of relativity rests on solid experimental foundation.

One of the strangest and most controversial predictions of the relativity theory is the so called *twin paradox*. In his very first paper on relativity, Einstein pointed out (1905) that if there are two clocks which are set to agree at a certain instant of time at some point on a particular inertial frame and one of these is taken off by an observer setting out for a travel to some distant point and is ultimately brought back to the position of the first clock 'at rest', then the traveller's clock will be found to have *lost time* in comparison with the other.

Take for example the case of the identical twins A and B at rest at some point on the earth (assumed to be an inertial frame). They set their respective clocks to indicate zero time and then one of them B sets out on a travel in a space-vehicle to some star outside the solar system. If the space-vechicle moves with a uniform rectilinear velocity comparable to the velocity of light, then he will complete the trip in a few years' time.* He then turns the space-vehicle back and returns to the earth with the same velocity which takes the same time to return to the earth as it took for the outward journey.

From what was stated above regarding the slowing down of the traveller's clock, it is evident that B on his return to the earth, would find his twin brother A *much older than himself.*

We can explain this bizarre phenomena in the following way. If $\upsilon = \beta c$ be the velocity of the space-vehicle relative to the earth, the observer A on the earth measures the time taken by B to reach the star at the distance d as $t = d/\upsilon$. Obviously according to his clock, the time for the return journey of B to the earth is the same, so that the total time of the travel of B to the star and back, as measured by A with his own clock, is $2t$.

However according to A, the clock of B runs slow because of time-dilatation (see § 15·8) so that the total time of travel $2t'$ of B as measured by B's clock will appear to A to be less by the time-dilatation factor $\gamma^{-1} = \sqrt{1 - \beta^2}$. Hence we get

---

* The star nearest to the solar system is alpha-Centauri which is about 4.3 light-years away.

$$2t' = \frac{2t}{\gamma} = 2t\sqrt{1-\beta^2}$$

Thus when B returns to the earth, he finds that the total time of his travel, as recorded in his clock ($2t'$), is less than that recorded by A on the earth ($2t$). This means that to B, A appears to have become much older than himself. This is also the conclusion that A draws.

The paradox lies in the one-sidedness of the event which is against the basic postulate of relativity, viz., the equivalence of all inertial frames. The one-sidedness in the present case is due to the sudden reversal of the motion of B which means that he changes from one inertial frame to another while no such thing happens to A. The sudden reversal of the velocity of B implies an acceleration which actually can be detected by him by various methods. It is the presence of this momentary acceleration which goes against our assumption of the complete equivalence of the frames of reference of A and B. Hence the one-sidedness of the result is real.

15.19. Einstein's general theory of relativity

The special theory of relativity is applicable to the case of inertial frames, *i.e.*, frames of reference which are in uniform rectilinear motion relative to each other. Later Einstein extended the theory to the case of accelerated frames of reference. A new theory of gravitation follows from this *general theory of relativity*. According to this all inertial and gravitational forces are manifestations of the same phenomena which is known as Einstein's *principle of equivalence*. It follows from the exact equality of the *inertial mass* (defined by Newton's laws of motion) and the *gravitational mass*, which has been established by a series of very accurate measurements by R.V. Eotvos between 1890 and 1922 and later by R.H. Dicke (1961). The theory is very complicated and is beyond the scope of the present book.

Some of the important predictions of the general theory of relativity and attempts at their experimental verification are briefly discussed below.

(A) *Precession of the perihelion of the planet Mercury*:

There is an observed precession of the orbit of the planet mercury which is nearest to the sun in the solar system amounting to 5599·7" of arc per century. The precession has been calculated taking into account the perturbation of its motion due to other planets, especially Venus, which comes out to be 5557·2" of arc per century. The difference 42·5" of arc per century can only be explained with the help of the general theory of relativity which predicts an advance of the perihelion of mercury by 43·03" of arc per century which is in good agreement with the above mentioned discrepancy.

(B) Gravitational red shift :

The general theory of relativity predicts that the frequency of a spectral line due to an atomic transition will appear to be less if the light travels against the gravitational field before being measured by an observer at some distance from the source emitting the light. Hence the measured wavelength will shift towards the red (see § 15·16). Astronomical measurements of this gravitational red shift are difficult because of the much larger Doppler shifts. The effect has however been established with great accuracy (within a few per cent) by the experiment of R.V. Pound and G.A. Rebka (1960) who were able to measure the wavelength shift of light falling through a distance of 60 ft in the earth's gravitational field, by utilizing Mössbauer effect (see Vol. II) the effect being only about 2·5 parts in 10^{15}.

(C) Deflection of light in the gravitational field :

One of the most well-known predictions of the general theory of relativity is the deflection of a ray of light in the gravitational field of a very massive body like the sun and other stars. Since the sun is the star nearest to the earth, astronomers have concentrated their efforts to observe such a shift in the gravitational field of the sun.

The experiment involves the photographing of a star field around the sun during a total solar eclipse when the stars are visible during day time. Six months later when the same stars are visible at night, the star field is photographed again. The displacements in the apparent positions of the stars can be measured by comparing the photographs.

Starting from 1919, expeditions have been organized over the years to different parts of the world wherever total solar eclipse was visible to carry out these measurements. Results of the measurements are mostly in support of the prediction of the general theory of relativity, though some deviations have also been observed.

(D) Change in the period of pulsars :

Pulsars are pulsing radio stars. They have been found to emit pulsed radio waves at regular intervals ranging from several seconds down to milliseconds. The pulses flashed out by these magnetized neutron stars are so regular that a pulsar could be used as a clock that is accurate to one part in a hundred million. The pulsing of the intensity of the radio signals from the pulsars is believed to be due to their very rapid rotation. In a few cases the pulsar is believed to be a component of a *binary system*, a system of two stars which orbit around their common centre of mass. There are evidences to indicate a change in the binary period with time (pulsing period), the cause of which is the emission of *gravitational waves* predicted by the general theory of relativity. Such emission results in the reduction of the separation between the two stars which produces a change of the binary period. It may be noted that uptill now, there is no other explanation of the observed change of the binary period.

15.20 Acceleration and Force

(a) *Acceleration in special theory of relativity* :

In § 15.6, we derived the transformation equations for the components of the velocity of a particle from one inertial frame to another. Here we rewrite these :

$$v_x = \frac{v_x' + u}{1 + u v_x'/c^2} \tag{15·20-1}$$

$$v_y = \frac{v_y'}{\gamma(1 + u v_x'/c^2)} \tag{15·20-2}$$

$$v_z = \frac{v_z'}{\gamma(1 + u v_x'/c^2)} \tag{15·20-3}$$

u is along x and $\gamma = (1 - \beta^2)^{-1/2}$.

v_x' is the longitudinal component of the velocity while v_y', v_z' are the transverse components of the velocity in S'-frame.

The transformation equation for the time is (see Eq. (15·6-d)

$$t = \gamma (t' + u x'/c^2) \tag{15·20-4}$$

Differentiating Eqs. (15·20-4) we have, since u is taken to be constant

$$dv_x = \frac{dv_x'}{1 + u v_x'/c^2} - \frac{v_x' + u}{(1 + u v_x'/c^2)^2} \cdot \frac{u dv_x'}{c^2}$$

$$= \frac{(1 - u^2/c^2) dv_x'}{(1 + u v_x'/c^2)^2} = \frac{dv_x'}{\gamma^2 (1 + u v_x'/c^2)^2} \tag{15·20-5}$$

$$dt = \gamma (dt' + u dx'/c^2)$$

$$= \gamma (1 + u v_x'/c^2) dt' \tag{15·20-6}$$

where $\quad v_x' = \dfrac{dx'}{dt'}$

Dividing Eq. (15·20-5) by Eq. (15·20-6) we get the longitudinal component of the acceleration

$$a_x = \frac{dv_x}{dt} = \frac{dv_x'/dt'}{\gamma^3 (1 + u v_x'/c^2)^3}$$

Special Theory of Relativity

or,
$$a_x = \frac{a_x'}{\gamma^3 (1 + u v_x'/c^2)^3} \qquad (15 \cdot 20\text{-}7)$$

Eq. (15·20-7) gives the transformation equation of the longitudinal component of the acceleration from the inertial frame S' to S.

Similarly if we differentiate Eq. (15·20-2) we get
$$dv_y = dv_y'/\gamma (1 + u v_x'/c^2) - \frac{v_y'}{\gamma (1 + u v_x'/c^2)^2} \cdot \frac{u dv_n'}{c^2}$$

Dividing by dt we then get
$$a_y = \frac{dv_y}{dt} = \frac{dv_y'/dt'}{\gamma^2 (1 + u v_x'/c^2)^2}$$
$$- \frac{v_y'}{\gamma^2 (1 + u v_x'/c^2)^3} \cdot \frac{u(dv_x'/dt')}{c^2}$$
$$\therefore a_y = \frac{a_y'}{\gamma^2 (1 + u v_x'/c^2)^2} - \frac{(u v_y'/c^2) a_x'}{\gamma^2 (1 + u v_x'/c^2)^3} \qquad (15 \cdot 20\text{-}8)$$

Eq. (15·20-8) gives the transformation equation of the transverse component of the acceleration a_y' in the S'-fram into a_y in the S-frame.

Similarly for the other transvere component of the acceleration a_z' we get, starting from Eq. (15·20-3)
$$a_z = \frac{a_z'}{\gamma^2 (1 + u v_x'/c^2)^2} - \frac{(u v_z'/c^2) a_x'}{\gamma^2 (1 + u v_x'/c^2)^2} \qquad (15 \cdot 20\text{-}9)$$

For a special case $v_y' = 0, v_z' = 0$ and $a_x' = 0$, we have
$$a_x = 0, \ a_y = \frac{a_y'}{\gamma^2 (1 + u v_x'/c^2)^2}, \ a_z = \frac{a_z'}{\gamma^2 (1 + u v_x'/c^2)^3}$$
$$(15 \cdot 20\text{-}10)$$

If a body is instantaneously at rest in the S'-frame ($v_x' = v_y' = v_z' = 0$) we get
$$a_x = \frac{a_x'}{\gamma^3}, \ a_y = \frac{a_y'}{\gamma^2}, \ a_z = \frac{a_z'}{\gamma^2} \qquad (15 \cdot 20\text{-}11)$$

In other words, the longitudinal component is diminished by the factor $1/\gamma^3$ while the transverse components are diminished by the factor $1/\gamma^2$.

It may be mentioned that the expressions for the transformation of the components of the acceleration in the special theory of relativity are of limited usefulness, unlike in the case of Newtonian dynamics.

(d) Force in the special theory of relativity :

As in Newtonian dynamics, we define the force in relativistic dynamics by the equation

$$\mathbf{F} = \frac{d}{dt}(m\mathbf{u}) = d\mathbf{p}/dt \qquad (15\cdot20\text{-}12)$$

where $\mathbf{p} = m\mathbf{u}$ is the relativistic momentum, m being the relativistic mass

$$m = \frac{m_0}{\sqrt{1-\beta^2}} = m_0\gamma \qquad (15\cdot20\text{-}13)$$

If the nature of the force is known, then equating the r.h.s. of Eq. (15·20-12) to the analytical expression for the force, we should expect the two sides of the equation to transform similarly under Lorentz transformation.

Consider a force F_{ox} to act on the particle in a frame of reference in which the particle is instantaneously at rest so that it receives an acceleration a_{ox} in this frame. We than have

$$F_{ox} = m_0 a_{ox} \qquad (15\cdot20\text{-}14)$$

In another frame (the laboratory frame) moving with the velocity u w.r.t. to the rest frame, the particle has a momentum

$$p_x = \gamma m_0 u \qquad (15\cdot20\text{-}15)$$

So using Eq. (15·20-13), we get the x-component of the force as

$$F_x = \frac{dp_x}{dt} = \gamma m_0 \frac{du}{dt} + m_0 u \frac{d\gamma}{dt} \qquad (15\cdot20\text{-}16)$$

where $\gamma = 1/\sqrt{1-\beta^2}$ so that, writing $a_x = du/dt$ we have

$$\frac{d\gamma}{dt} = \frac{d}{dt}\frac{1}{\sqrt{1-u^2/c^2}} = \frac{u/c^2}{(1-u^2/c^2)^{3/2}}\frac{du}{dt} = \frac{u}{c^2}\gamma^3 a_x$$

Thus $F_x = \gamma m_0 a_x + \gamma^3 m_0 (u^2/c^2) a_x$

$$= m_0 a_x \gamma^3 (1 - u^2/c^2 + u^2/c^2)$$

or, $\qquad F_x = m_0 a_x \gamma^3 \qquad (15\cdot20\text{-}17)$

But from Eq. (15·20-11), we have the following relation between the longitudinal components of the acceleration in the two frames: $a_x = a_x'/\gamma^3$. So we get the following transformation relation between the longitudinal components of the force ($\because a_x' = a_{ox}$)

$$F_x = \gamma^3 m_0 a_x'/\gamma^3 = m_0 a_{ox}$$

or, $\qquad F_x = F_{ox} \qquad (15\cdot20\text{-}18)$

Special Theory of Relativity

Thus we get the rather surprising result that the longitudinal component of the force remains the same in the two inertial frames, *i.e.* it is invariant.

We now consider the transverse component of the force F_y perpendicular to the momentum vector $m\mathbf{u}$. For a very small interval of time Δt, this force changes only the direction of the velocity $\mathbf{u}$, but not its magnitude (which approximation holds strictly in the limit of $\Delta t \to 0$). So the mass $m = \gamma m_0$ may be regarded as remaining constant during this interval and the transverse impulse may be written as

$$F_y \Delta t = m\Delta u_y = \gamma m_0 \Delta u_y \tag{15·20-19}$$

or,
$$F_y = \gamma m_0 a_y \tag{15·20-20}$$

Again since $a_y = a_y'/\gamma^2$ (Eq. 15.20-11) we get the transverse component of the force as

$$F_y = \gamma m_0 (a_{oy}/\gamma^2) = \gamma F_{oy}/\gamma^2$$

$$F_y = F_{oy}/\gamma \tag{15·20-21}$$

We can get a similar relation between F_z and F_{oz}:

$$F_z = F_{oz}/\gamma \tag{15·20-22}$$

From Eqs. (15·20-18) and (15·20-21) we get

$$\frac{F_y}{F_x} = \frac{F_{oy}/\gamma}{F_{ox}} = \frac{m_0 a_{oy}/\gamma}{m_0 a_{ox}}$$

or,
$$\frac{F_y}{F_x} = \frac{a_{oy}}{\gamma a_{ox}} \tag{15·20-23}$$

This equation shows that in general the force in the special theory of relativity as defined here (rate of change of relativistic momentum) is not in the same direction as the acceleration.

Eq. (15·20-22) shows that the transverse component of the force is not an invariant.

It should be noted that the above analysis applies to the special case of transformation in which one of the two frames is the instantaneous rest frame of the particle.

In the more general case of transformation between two arbitrary inertial frames, the transformation equations for the components of force can be shown to be

$$F_x' = \frac{F_x - (u/c^2)(\mathbf{F} \cdot \mathbf{v})}{1 - u v_x/c^2} \tag{15·20-24}$$

$$F_y' = \frac{F_y}{\gamma(1 - u v_x/c^2)} \qquad (15\cdot 20\text{-}25)$$

$$F_z' = \frac{F_z}{\gamma(1 - u v_x/c^2)} \qquad (15\cdot 20\text{-}26)$$

It may be mentioned that the three components of the force F_x', F_y', F_z' constitute the three space components of a four-vector of which, the fourth or the time-component is $(\mathbf{F}\cdot\mathbf{v}/c)$, the corresponding transformation equation being

$$\mathbf{F}\cdot\mathbf{v} = \frac{\mathbf{F}\cdot\mathbf{v} - u F_x}{1 - u v_x/c^2} \qquad (15\cdot 20\text{-}27)$$

(See Eqs. 15.17-18 and 15.17-19).

For further details, see Special Relativity by A.P. French.

Problems

1. The velocity of a particle increases by 1%. What is the percentage increase of its momentum if (a) $v/c = 0.7$, (b) $v/c = 0.99$ (2.96%; 51.25%)

2. Prove that if a unidirectional force F acts on a particle of rest-mass m_0 which is at rest initially it acquires a velocity $v = cFt/\sqrt{m_0^2 c^2 + F^2 t^2}$ after a time t. Show that if t is small, then the above expression reduces to the classical result. What is the velocity after a long time?

3. If the total energy of a particle increases by 10%, what is the percentage increase of its velocity if the total energy is (a) twice, (b) 10 times, (c) 100 times the rest-energy? (3.33%; 0.10%; 0.001%)

4. If the kinetic energy of a particle is expressed by the nonrelativistic formula, there is an error of 1.5%. What is its kinetic energy if it is (a) an electron, (b) a proton? What is its velocity? (2.55 keV; 4.69 Mev; $c/10$)

5. Show that the kinetic energy and the total energy of a very high energy particle are both equal to pc approximately.

6. If the kinetic energy of a particle is written as $E_k = pc$, there is an error of 2%. What is its velocity? What is its kinetic energy if it is (a) an electron, (b) a proton? (0.9998 c; 24.99 Mev; 45962 MeV)

7. If the momentum of a particle is 0.7 MeV/c, what are its velocity and energy if it is (a) an electron, (b) a proton? (0.866 MeV : 938 MeV)

8. Two particles move in opposite directions with the same velocity 0.8 c. What is their relative velocity (0.9756 c)

9. Calculate the radius of curvature of a 100 MeV electron in a magnetic field of 1 T. (0.33 m)

10. A charged particle of velocity $v = c/2$ has a radius of curvature 0.46 m in a magnetic field of 1 T. Estimate its mass in the unit of electronic mass.

    (270 m_e)

11. Using the transformation equations for the components of the four momentum (Eqs. 15.17-18 and 15.17-19), prove the relations (15.20-24) to (15.20-26).

12. If a particle of rest mass m and relativistic velocity v_L in the laboratory frame of reference collides with an identical particle at rest in the laboratory (L), show that the kinetic energy T_C of the two particles in the centre of mass frame (C) is related to T_L as given below :

$$T_L = 2T_C + T_C^2/2W_0$$

where $W_0 = m_0 c^2$. T_C is equal to the energy available for the production of a nuclear reaction by collision.

13. What should be the energy of a proton colliding with another proton at rest in the laboratory if the total energy available for reaction is 10 $M_0 c^2$?

    (65.66 GeV)

14. Assuming the work done by the x-component of the force $F_x \Delta x$ to be equal to the energy $c^2 \Delta m$ of a particle, show that $F_x = \gamma^3 m_0 a_x$.

15. Prove that the relationship $F = Bqv$ between the force acting on a particle of charge q moving with a velocity $\mathbf{v}$ perpendicular to the magnetic induction vector $\mathbf{B}$ also holds in the special theory of relativity.

16

MOLECULAR SPECTRA

16·1 Introduction

In Chapters IV and VI, we discussed about the nature of the spectra of light emitted from gas discharge tubes which is generally found to consist of a number of discrete spectral lines. For this reason, this type of spectrum is known as the line spectrum. The origin of the line spectrum lies in the atoms of different elements. Their origin can be adequately explained with the help of Bohr-Sommerfeld theory and by quantum mechanics which have been discussed in detail in the earlier chapters.

In some special cases, the spectrum of the light emitted from discharge tubes has the appearance of a *band* of light, instead of discrete lines. Such a spectrum, known as the *band spectrum*, has its origin in the molecules of different elements or compounds.

Study of the band spectrum is usually done by experiments with the *absorption spectrum*. However the spectra of many diatomic molecules are more conveniently studied in emission. These often supplement the absorption data. For absorption measurements, the substance under investigation is enclosed within a cell through which monochromatic light of known wavelength is sent and its absorption (*i.e.*, change of intensity) measured by some suitable arrangement. This is done for different wavelengths. When the energy of the incident light $h\nu$ becomes equal to the energy difference between two quantum states of the given substance, there is a sudden decrease in the intensity of the transmitted light.

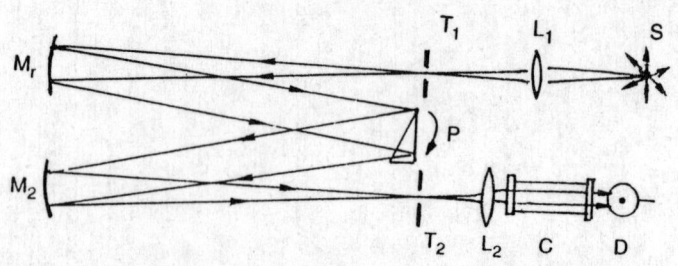

Fig. 16.1 Experimental set up for the measurement of absorption sprectra.

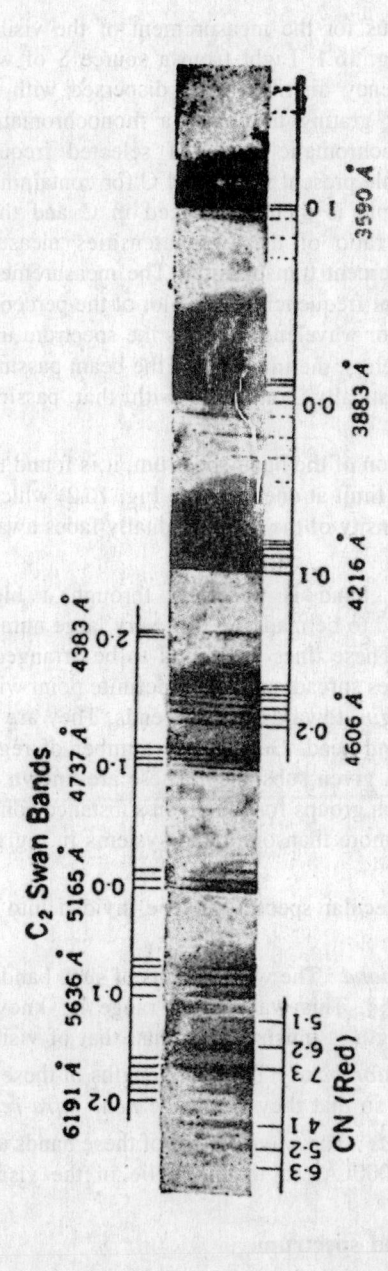

Fig. 16.2. Photograph of a band spectrum. Band heads on the left of the bands are to be noted. (From Herzberg : Spectra of Diatomic Molecules, Vol I, published by Van Nostrand Reinhold Co; Copyright 1950 by Litton Publishing Inc.)

A simple apparatus for the measurement of the visible absorption spectra is shown in Fig. 16.1. Light from a source S of white radiation with continuous frequency distribution is dispersed with the help of a prism P or diffraction grating to obtain a monochromatic beam. The intensity of the monochromatic beam of selected frequency is then measured with no sample present in the cell C for containing the material under study. The sample is then introduced in C and the intensity is measured again. The ratio of the two intensities measured with the detector D gives the percent transmission. The measurement is repeated for a number of different frequencies. The plot of the percent transmission against the frequency (or wavelength) gives the spectrum under study. In most modern spectrometers, the intensity of the beam passing through the sample cell is automatically compared with that passing through a reference cell.

On closer inspection of the band spectrum, it is found that each band extends upto a definite limit at one end (see Fig. 16.2) which is known as the *band-head* The intensity of the band gradually fades away towards the other end.

When a spectral band is observed through a high resolution spectroscope, it is found to be made up of a very large number of closely spaced spectral lines. These lines are found to be arranged in a regular manner. Usually the lines spread out from a definite point within the band, known as the *band origin*, towards the two ends. They are more densely packed towards the band-head. Generally a number of regularly spaced bands is observed for a given substance, These are known as a *group of bands*. A number of such groups for the given substance constitues a *band system*. There may be more than one band systems in a given molecular spectrum.

The bands in molecular spectra may be divided into three distinct classes :

(*a*) *Pure rotation band* : The wavelengths of such bands extend from about 150 μ to 30 μ §§. This wavelength range is known as the *far infra-red*. The wavelength is much longer than that of visible light.

(*b*) *Rotation-vibration band* : The wavelengths of these bands extend from about 5 μ to 1 μ so that they lie in the *near infra-red*.

(*c*) *Electronic bands* : The wavelengths of these bands usually extend from about 1000 to 7000 Å, so that they lie in the visible or in the ultraviolet region.

16·2 Origin of the band spectrum

As stated above, the origin of the band spectra lies, in the molecules of substance. If a substance giving rise to band spectrum is heated then the

---

§§ 1 μ = 1 micron= 10^{-4} cm = 10^3 nanometres (nm) = 10^4 Å.

nature of the spectrum is changed. The bands disappear at high temperature and discrete spectral lines characteristic of the constituent elements of the molecules of the substance are observed. As an example, when an asbestos fibre impregnated with the solution of common salt is held in the flame of a bunsen burner, then yellow D-lines of sodium appear in the spectrum. At the high temperature of the flame, NaCl molecules dissociate into sodium and chlorine atoms. As a result the characteristic lines of sodium are observed.

Molecular spectra are usually much more complex than the atomic spectra. We shall discuss below the origin of the spectra of the simplest of the molecules, *viz*., the diatomic molecules.

Diatomic molecules *e.g.*, H_2, N_2, O_2, HCl, CN, etc. have the structure of a dumb-bell. A strong attractive force acts between the two atoms of the molecule due to which they remain bound. The electrons in the two atoms exist in their respective energy states. These states are however different from the energy states of the individual unbound atoms. When bound in the molecule, each atom influences the motion of the electrons in the other atom which results in a change of their energy states. As in the case of the atomic energy levels, the electronic states of the molecules have energies (E_e) of the order of a few electron-volts. Transitions between such states result in the emission of light in the visible or ultraviolet regions.

Actually the energies of the molecular energy levels depend not only on the motion of the orbital electrons, but also on the motion of the atoms as a whole inside the molecule. In a diatomic molecule, the two atoms usually vibrate along the line joining them, somewhat like a linear harmonic oscillator, due to which the molecule possesses some *vibrational energy* E_v in addition to the electronic energy E_e, which affects the molecular energy levels.

In addition, the molecule as a whole may rotate due to which it will have some *rotational energy* E_r. This also affects the energy of the molecular levels. The rotation may be about an axis joining the two atoms or about an axis perpendicular to the former. The rotational energy depends on the *moment of inertia* of the molecule. We know that the moment of inetia of a body depends upon the distances of the different parts of the body from the axis of rotation and the mass of the body. Since the atomic mass is almost totally confined within its nucleus having dimensions of the order of 10^{-15} m which is small compared to atomic dimensions ($\sim 10^{-10}$ m), the distances of the different parts of the nucleus from the line joining them is negligibly small so that the moment of interia about this line is also negilible. Thus the contribution of the rotation about this axis to the total rotational energy may be neglected and we need consider the rotation only about the axes perpendicular to the line joining the two atoms.

In a quantum mechanical theory of the molecular structure, it is necessary not only to consider the above three types of motion, but also the mutual coupling of the different motions. It is difficult to obtain an exact solution of the Schrödinger wave equation when all these are taken into account. It is however possible to obtain an approximate solution by neglecting the coupling between the different types of motion as suggested by Max Born and J.R. Oppenheimer.* If this is done then to a first approximation, the energy of the molecular energy level may be written as

$$E = E_e + E_\upsilon + E_r \qquad (16 \cdot 2\text{-}1)$$

As seen above, the electronic energy E_e is of the order of a few electron-volts. The vibrational energy E_υ is much smaller than E_e and the rotational energy E_r is small compared to E_υ.

To get an idea of the order of magnitude of the electronic, vibrational and rotational energies of a diatomic molecule, we note that for a separation a between the nuclei of the two atoms in the molecule, the uncertainty principle gives the momentum of the valence electrons to be $p_e \approx \Delta p = \hbar/a$ so that the kinetic energies of the electrons should be of the order of

$$E_e = p_e^2/2m_e = \hbar^2/2m_e a^2 \qquad (16 \cdot 2\text{-}2)$$

Since $a \sim 1$ Å, we get $E_e \sim 3.80$ eV. Thus the electronic energy in the molecule is typically of the same order of magnitude as that for the valence electrons in the atoms.

To get an estimate of the vibrational energy of the nuclei, we use the following classical argument. If F be the force which binds the electrons to the molecule, the nuclei must be bound by an equal and opposite force. If the force is assumed simple harmonic with a force constant k we can write $F = -kx$ so that the angular frequency of electronic motion is $\omega_e \sim (k/m_e)^{1/2}$. Correspondingly the angular frequency of the vibrational motion of the nuclei will be $\omega_N \sim (k/M)^{1/2}$ where M is the nuclear mass. So the ratio of the energies of the vibrational motion of the nuclei to that of the electronic motion is typically

$$E_\upsilon/E_e = \hbar \omega_N/\hbar \omega_e = (m_e/M)^{1/2} \qquad (16 \cdot 2\text{-}3)$$

Since $m_e/M \sim 10^{-3}$ to 10^{-5}, the above ratio is of the order of 10^{-2}. As the electronic energy is ~ 1 eV, producing lines in the visible or ultra-violet regions, the vibrational energy should be in the infra-red region.

We now estimate the rotational energy E_r. If we take a diatomic

---

* See *Introduction to Quantum Mechanics* By L. Pauling and E.B. Wilson.

molecule with two nuclei of mass M each separated by a distance a, the rotational energy is

$$E_r \sim \hbar/2I = \hbar /Ma^2 \qquad (16 \cdot 2\text{-}4)$$

where $I = Ma^2/2$ is the moment of inertia. Then using Eq. (16·2-2) we get

$$E_r = (m_e/M) E_e \qquad (16 \cdot 2\text{-}5)$$

Thus the rotational energy of nuclear motion is lower than the electronic energy by a factor $\sim (m_e/M)$ and by a factor $\sim (m_e/M)^{1/2}$ compared to the vibrational energy. They are thus typically $\sim 10^{-3}$ eV or less while the vibrational energy is typically 10^{-1} to 10^{-2} eV.

The above estimates justify the Born-Oppenheimer approximation regarding the uncoupling of the electronic, vibrational and rotational motions mentioned above.

If E_1 and E_2 be the energies of two levels in a molecule, then the energy of the radiation resulting from a transition between the levels is

$$h_\nu = E_2 - E_1 \qquad (16 \cdot 2\text{-}6)$$

It is possible to explain the origin of the *pure-rotational* or of the *rotation-vibration* spectra of molecules with the help of Eqs. (16·2-1) and (16·2-2).

16·3 Pure rotational spectrum of a diatomic molecule

Suppose the electronic energy E_e of a molecule remains unchanged. In addition, we assume that the vibrational energy E_v of the molecule also remains unchanged. This is possible if we assume that the two atoms of the diatomic molecule are as if joined together by a *rigid rod* so that the distance between them remains unchanged as the molecule rotates as a whole. Such a model of the molecule is known as a *rigid rotator model*. As discussed, only the rotation about an axis perpendicular to the line joining the two atoms has to be considered.

If the masses of the two atoms separated by a distance r be m_1 and m_2 and their distances from the centre of mass be r_1 and r_2 then we can write

$$m_1 r_1 = m_2 r_2 \quad \text{and} \quad r = r_1 + r_2$$

$$\therefore \quad r_1 = \frac{m_2 r}{m_1 + m_2}, \quad r_2 = \frac{m_1 r}{m_1 + m_2}$$

So the moment of intertia of the molecule is

$$I = m_1 r_1^2 + m_2 r_2^2 = \frac{m_1 m_2}{m_1 + m_3} r^2 = \mu r^2 \qquad (16 \cdot 3\text{-}1)$$

where $\mu = \dfrac{m_1 m_2}{m_1 + m_2}$ is the reduced mass.

If ω be the angular velocity of rotation, then the angular momentum L and the rotational kinetic energy E_r of the molecule are

$$L = I\omega, \quad E_r = \tfrac{1}{2} I \omega^2$$

This gives $\qquad E_r = L^2/2I \qquad (16\cdot3\text{-}2)$

For a rigid rotator, there is no variation of r so that the Hamiltonian becomes (see Eq. 11.11-6)

$$\hat{H} = -\frac{\hbar^2}{2I} \left\{ \frac{1}{\sin\theta} \frac{\partial}{\partial\theta} \sin\theta \frac{\partial}{\partial\theta} + \frac{1}{\sin^2\theta} \frac{\partial^2}{\partial\phi^2} \right\} \qquad (16\cdot3\text{-}3)$$

The potential energy term is absent since there is no vibration. The Schrödinger wave equation $\hat{H}\psi = E\psi$ can be solved by the seperation of variables. The two equations in θ and ϕ (see § 11·13 and § 11·14) that result can be solved to yield the eigenvalues of the square of the angular momentum operator $\hat{L}^2$ as $j(j+1)\hbar$. So the energy eigenvalues are given by

$$Er = \frac{j(j+1)\hbar^2}{2I} \qquad (16\cdot3\text{-}4)$$

where j is the rotational quantum number : $j = 0, 1, 2, 3$ etc. The molecule can thus have a discrete set of rotational energy levels with different values of j. These constitute the *pure rotational* levels and are illustrated in Fig.16·3.

As in the case of the atomic spectra, the energies of the different molecular levels are expressed by their *term-values* obtained by dividing the energy by $2\pi\hbar c$. The unit of the term-value is m^{-1}. The difference between the term values of two levels gives the *wave number* of the emitted (or absorbed) spectral line due to transition between the two levels. The reciprocal of this gives the wavelength.

For the rotational energy levels, the term values can be written as (see Eq. 16·3-4)

$$F(j) = Bj(j+1) \qquad (16\cdot3\text{-}5)$$

where $\qquad B = \hbar/4\pi Ic \qquad (16\cdot3\text{-}6)$

B is known as the *rotational constant*

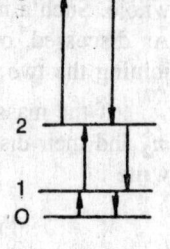

Fig. 16·3 Discrete set of pure rotational energy levels of a molecule.

According to the electromagnetic theory of light, rotational motion of only the *polar molecules e.g.*, HF, HCL *etc.* lead to the emission of radiation. In such molecules, the centres of positive and negative charges are permanently separated so that they possess permanent electric dipole moments. In *non-polar* molecules like N_2, O_2 *etc.* the centres of positive and negative changes are coincident so that they do not posses any permanent dipole moments. Since according to the e.m. theory, the rotation of a dipole can lead to the emission of radiation, there will be emission of radiation due to pure rotation of only polar molecules. Applying Bohr's correspondence principle, we then conclude that even in the quantum mechanical case, pure rotational spectrum can arise only in the case of polar molecules. The transitions between the pure rotational states are governed by the *setection rule* (see § 14·2)

$$\Delta j = \pm 1 \qquad (16\cdot3\text{-}7)$$

Applying the above selection rule, we see that the transition between two rotational states differing in j by one unit gives rise to a rotational spectral line having the wave number

$$\bar{v} = B\{(j+1)(j+2) - j(j+1)\}$$

$$= 2B(j+1) = \frac{\hbar}{2\pi I c}(j+1) \qquad (16\cdot3\text{-}8)$$

The spacing between the consecutive rotational lines is thus independent of j, given by $\Delta \bar{v} = 2B = \frac{\hbar}{2\pi I_C}$. The pure rational spectral lines are thus equispaced. This is illustrated in 16·4 which shows the pure rotational absorption spectrum of the HF molecule. The wave number difference between the consecutive rotational lines is found to be $\Delta \bar{v} = 4050$ m^{-1} in his case.

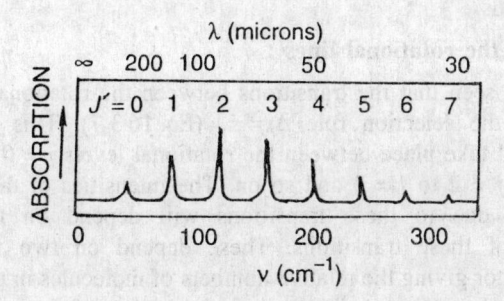

Fig. 16.4. Equispaced pure rotational spectral lines in the absorption spectrum of the HF molecule.

As seen from Fig. 16.4, the wave number $\bar{v}$ of the spectral lines in this case are of the order of a few hundred cm^{-1}. Writing $\bar{v} = 200$ cm$^{-1} = 20,000$ m^{-1} for a typical line, we get its wavelength as $\lambda = 1/\bar{v} = 50 \times 10^{-6}$ m $= 50\,\mu$ which is in the far infra-red.

As we estimate the moment of ineria and hence the distance between the two atoms (bond-length) of a diatomic molecule from the measured value of the rotational constant B. For the HF molecule we get

$$B = \frac{\Delta \bar{\nu}}{2} = \frac{4050}{2} = 2025 \text{ m}^{-1}$$

$$\therefore \quad B = \frac{\hbar}{4\pi I c} = 2025$$

or, $\quad I = \dfrac{\hbar}{4\pi B c} = \dfrac{1.05 \times 10^{-34}}{4\pi \times 2025 \times 3 \times 10^8}$

$$= 1.38 \times 10^{-47} \text{ kg m}^2$$

The reduced mass of the HF molecule is

$$\mu = \frac{m_1 m_2}{m_1 + m_2} = \frac{1 \times 19}{1 + 19} \times 1.66 \times 10^{-27} = 1.58 \times 10^{-27}$$

Since $I = \mu r^2$, we get the interatomic distance

$$r = \sqrt{\frac{I}{\mu}} = \left(\frac{1.38 \times 10^{-47}}{1.58 \times 10^{-27}}\right)^{1/2}$$

$$= 0.935 \times 10^{-10} \text{m} = 0.935 \text{ Å}$$

As stated above, the pure rotational lines lie in the far infra-red. These lines are also observed in the microwave region ($\lambda \sim 10^{-3}$ to 10^{-1}m). Discrete pure rotational lines observed in the case of gaseous substances can yield valuable information about their molecular structure. However it is not easy to obtain such information in the case of solids or liquids.

Intensities of the rotational lines :

We have seen that the transitions between the rotational levels are governed by the selection rule $\Delta j = \pm 1$ (Eq. 16·3-7). This means that transitions will take place between the rotational levels $j = 0$ to $j = 1$, $j = 1$ to $j = 2$, $j = 2$ to $j = 3$ and so on. The intensities of the rotational spectral lines due to these transitions will depend on the relative probabilities of these transitions. These depend on two factors, the Boltzmann factor giving the relative numbers of molecules in the different initial rotational states and the statistical weights of these states. The Boltzmann factor is given by

$$N_j = N_0 \exp(-E_j / k_B T) \qquad (16\cdot3\text{-}9)$$

where E_j is given by Eq. (16·3-4).

$$E_j = Bhcj(j+1) \qquad (16\cdot3\text{-}10)$$

B is the rotational constant (Eq. 13·3-6). k_B is Boltzmann constant : $k_B = 1.38 \times 10^{-23}$ J/K.

Typically for $T = 300$ K and $B = 200$ m^{-1}, and $j = 1$, we get

$$N_1 = N_0 \exp\left(-\frac{2 \times 10^2 \times 6.63 \times 10^{-34} \times 3 \times 10^8 \times 2}{1.38 \times 10^{-23} \times 300}\right)$$

$$= N_0 \exp(-0.019) \approx 0.98 N_0 \quad (16\cdot3\text{-}11)$$

The above estimate shows that at room temperature, the number of molecules in the $j = 1$ state is almost equal to the number of molecules in the lowest ($j = 0$) state.

The statistical factor of a level of given j is $(2j+1)$ so that the total population of the level j is determined by the factor $(2j+1) \exp(-E_i/k_BT)$. When plotted against j, this shows a maximum at a value of j determined by the rotational constant B and the temperature T.

So the intensities of the rotational lines j to $j+1$ will differ, depending on the value of j for the initial state, being maximum for

$$j = \sqrt{\frac{k_BT}{2Bhc}} - \frac{1}{2}$$

16·4 Rotation-vibration spectra

The rigid rotator model of a diatomic molecule is an idealized model. Actually the two atoms vibrate along the line joining them, as if they are connected by a massless spring. Thus the distance r between the atoms change in a periodic manner about a mean value r_0. If the amplitude of vibration Δr is small compared to r_0, then the vibration may be regarded as simple harmonic. Otherwise it will not be so.

The force F between the two atoms in the molecule is a function of the distance r between them. Corresponding to this force, the molecule has a potential energy $V = V(r)$ such that $F = -\partial V/\partial r$. In Fig. 16.5 we show a plot of V against r. We assume $V = 0$ for very large r. By evaluating the slope of the potential curve at every point we can get the force F as a function of r. The nature of the variation of $F(r)$ with r is also shown in Fig. 16.5. It is seen from the figure that at $r = r_0$, V has its minimum value $-V_o$ and the force $F = 0$ at this point. For $r > r_0$, the force is

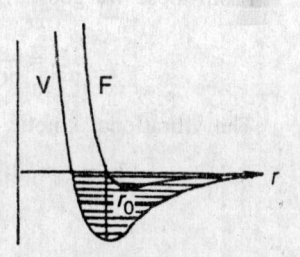

Fig. 16.5. Plot of the potential energy V and the distance r between two atoms of a molecule.

negative, i.e., attractive, while for $r < r_o$ it is positive, i.e., repulsive. Under the influence of such a force the two atoms would be at rest at the equilibrium distance r_0. If however they are disturbed even slightly, they will vibrate about this point. If the system is given a kinetic energy greater than $-V_0$ then the atoms will fly apart and the molecule will dissociate.

For small vibrations, the potential energy curve can be approximated to a parabolic potential about r_0 corresponding to the case of a linear harmonic oscillator (see § 11.9) where r_0 is the bond-length.

Writing $\Delta r = r - r_0 = \xi$, the vibrational potential energy in this case can be written as

$$V = \frac{1}{2} k \xi^2 = \frac{1}{2} \mu \omega_0^2 \xi^2 \qquad (16 \cdot 4 \text{-} 1)$$

where $k = \mu \omega_0^2$ is the *force constant*, μ being the reduce mass. If ξ_1 and ξ_2 represent the instantaneous distances of the two atoms (masses m_1 and m_2) from their equilibrium positions A and B, then we can write (see Fig. 16.6)

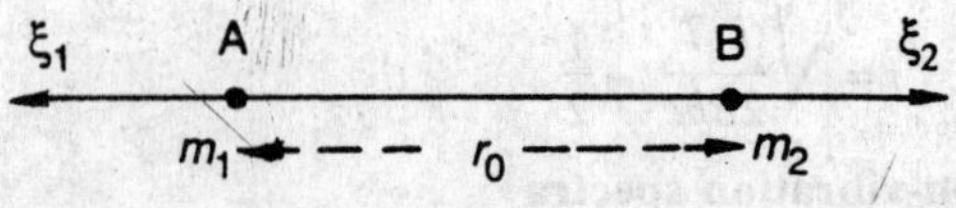

Fig. 16.6. Calculation of the total vibrational energy for small vibrations of two atoms (masses m_1, m_2) of a molecule from their equilibrium positions A and B.

$$\xi = \xi_1 + \xi_2 \qquad (16 \cdot 4 \text{-} 2)$$

$$m_1 \xi_1 = m_2 \xi_2 \qquad (16 \cdot 4 \text{-} 3)$$

From these we get

$$\xi_1 = \frac{m_2 \xi}{m_1 + m_2}, \; \xi_2 = \frac{m_1 \xi}{m_1 + m_2} \qquad (16 \cdot 4 \text{-} 4)$$

The vibrational kinetic energy of the atoms is

$$E_k = \frac{1}{2} m_1 \dot{\xi}_1^2 + \frac{1}{2} m_2 \dot{\xi}_2^2 = \frac{1}{2} \mu \dot{\xi}^2 \qquad (16 \cdot 4 \text{-} 5)$$

$$\mu = \frac{m_1 m_2}{m_1 + m_2}$$

The total vibrational energy is

$$E_v = E_k + V = \frac{1}{2} \mu \dot{\xi}^2 + \frac{1}{2} k \xi^2 = p^2 / 2\mu \qquad (16 \cdot 4 \text{-} 6)$$

The Hamiltonian for the vibrational motion is (see § 11.7)

$$\hat{H} = -\frac{p^2}{2\mu} + V = -\frac{\hbar}{2\mu} \frac{\partial^2}{\partial \xi^2} + \frac{1}{2} \mu \omega_0^2 \xi^2 \qquad (16 \cdot 4 \text{-} 7)$$

Molecular Spectra

so that the Schödinger equation can be written as

$$\hat{H}\psi = -\frac{\hbar^2}{2\mu}\frac{\partial^2\psi}{\partial\xi^2} + \frac{1}{2}\mu\omega_0^2\xi^2\psi = E_\nu\psi$$

The solution of the above equation is (see § 11.8)

$$\psi = \left(\frac{\sqrt{\alpha}}{2^\upsilon \upsilon! \sqrt{\pi}}\right)^{1/2} \exp(-\alpha\xi^2/2) H_\nu(\sqrt{\alpha}\xi) \qquad (16\cdot4\text{-}8)$$

where $\alpha = m\omega_0/\hbar$. $H_\upsilon(\sqrt{\alpha}\,\xi)$ is the Hermite polynominal of order υ. Writing $\sqrt{\alpha}\,\xi = q$, we have

$$H_\upsilon(q) = (-1)^\upsilon \exp(q^2)\left(\frac{d}{dq}\right)^\upsilon \exp(-q^2) \qquad (16\cdot4\text{-}9)$$

The energy eigenvalues are given by

$$E_\upsilon = \left(\upsilon + \frac{1}{2}\right)\hbar\omega_0 = \left(\upsilon + \frac{1}{2}\right)h\nu_0 \qquad (16\cdot4\text{-}10)$$

where, as before, the vibrational quantum number $\upsilon = 0, 1, 2, 3$, etc. For $\upsilon = 0$, we get the *zero-point energy* $E_0 = \hbar\omega/2 = h\nu_0/2$. ν_0 is called the proper frequency.

The *term-values* of the vibrational levels are given by

$$G_\upsilon = \frac{E_\upsilon}{\hbar c} = \left(\upsilon + \frac{1}{2}\right)\bar{\nu}_0 \qquad (16\cdot4\text{-}11)$$

where $\bar{\nu}_0 = \nu_0/c = \omega_0/2\pi c$. If the rotational motion is not taken into account, then the transitions between the pure vibrational levels are governed by the selection rule (see § 14·2)

$$\Delta\upsilon = \pm 1 \qquad (16\cdot4\text{-}12)$$

The spacing between such idealized *pure vibrational levels* are given by

$$\Delta G(\upsilon) = \bar{\nu}_0$$

When e.m. radiation is incident on a diatomic molecule, its electric vector many cause a transition if the electric dipole moment of the molecule varies with the vibrational coordinate ξ. Since *homonuclear* diatomic molecules *e.g.*, O_2, N_2, etc. have no permanent electric dipole moments for any value of ξ, there will be no such transition for them. Such molecules do not have any vibrational infrared spectrum. As we saw earlier, such molecules do not have any pure rotational infra-red spectrum either.

For a *heteronuclear* diatomic molecule, the dipole moment is a function of the internuclear seperation. Assuming linear variation of the dipole moment with the distance between the nuclei, it can be shown from the time-dependent pertubation theory (see § 12.6) that transitions between the different vibrational levels in such molecules are governed by the selection rule given above (Eq. 16.4-12). The energy level spectrum corresponds to that of a linear harmonic oscillator so that the levels are equispaced as shown in Fig. 16·7. The selection rules (Eq. 16.4-12) show that the allowed transitions between the vibrational levels give rise to a *single absorption band* with the wave number $\bar{\nu}_0$:

$$\bar{\nu}_\nu = G(\nu + 1) - G(\nu) = \bar{\nu}_0 \qquad (16\cdot4\text{-}13)$$

This agrees approximately with the experimental observation. The infrared spectra of heteronuclear diatomic molecules such as CO or HCl show only one strong absorption band. From a measurement of the wave number $\bar{\nu}_\nu$ of this band, it is possible to determine the force constant for the molecule concerned :

$$\bar{\nu}_\nu = \bar{\nu}_0 = \frac{\nu_0}{c} = \frac{\omega_0}{2\pi c} = \frac{1}{2\pi c}\sqrt{\frac{k}{\mu}}$$

So the force constant is given by

$$k = 4\pi^2 c^2 \mu \bar{\nu}_0^2 \qquad (16\cdot4\text{-}14)$$

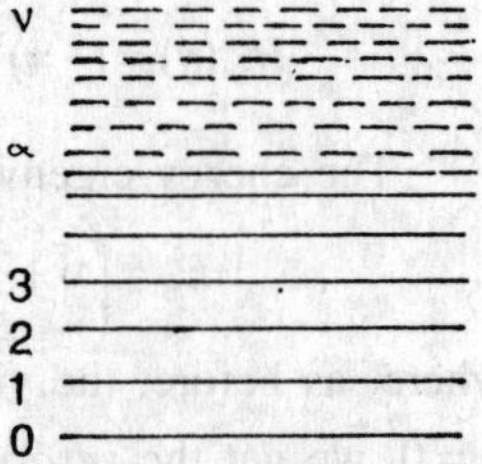

Fig. 16·7. Converging energy levels of a diatomic molecule.

Experimental values of $\bar{\nu}_0$ for HCl and CO molecules are found to be $2.886 \times 10^5 \text{m}^{-1}$ and $2.143 \times 10^5 \text{m}^{-1}$ respectively. These give the force constants as

$$k(\text{HCl}) = 480 \cdot 7 \text{ N/m}$$
$$k(\text{CO}) = 1855 \text{ N/m}$$

This shows that the CO bond strength is much stronger than the HCl bond strength. The amplitude ξ_o of the classical turning point for the ground state with $\nu = 0$ is given by

$$E_0 = \frac{1}{2}\mu\omega_0^2 \xi_o^2 = \frac{\hbar\omega_0}{2}$$

$$\therefore \quad \xi_o = \sqrt{\frac{\hbar}{\mu\omega_0^2}} = \sqrt{\frac{h}{4\pi^2\mu\,\bar{\nu}_0 c}} \qquad (16\cdot4\text{-}15)$$

For the values of $\bar{\nu}_\nu = \bar{\nu}_o$ given above we get

$$\xi_o(\text{HCl}) = 0\cdot11 \text{Å}$$
$$\xi_o(\text{CO}) = 0\cdot05 \text{Å}$$

The corresponding equilibrium bond-lengths of these molecules are 1.275Å and 1.128Å respectively. Thus even in the ground state, the amplitudes of vibration of the atoms in the diatomic molecules are 5% to 10% of the bond-lengths.

According to Eq. (16·4-10), the pure vibrational levels should be equispaced. However for large vibrational quantum numbers, the vibrational amplitude ξ becomes progressively larger so that the vibration is no longer simple harmonic. As a result, for large υ the levels are not equispaced but converge slowly as shown in Fig.16·7. As $\upsilon \to \infty$, the internuclear seperation r becomes very large and the energy levels extend upto a *convergence limit*. At this point the molecule dissociates into the constituent atoms.

Anharmonic oscillator :

The selection rules (16·4-12) are not strictly obeyed even for lower values of υ. This is due to the fact that as seen above, even for $\upsilon = 0$, ξ is a significant fraction of the bond length. The permanent dipole moments of heteronuclear molecules are strictly linear in ξ only for small values of ξ in which case only the selection rules (16·4-12) are strictly obeyed. Because of the relatively large values of ξ, anharmonic vibrational bands appear with $\Delta \upsilon > 1$. For example in the case of HCl, anharmonic band for 2–0 ($\upsilon' = 0$ to $\upsilon'' = 2$) and 3–0 transitions appear at the wave numbers 5.668×10^5 m^{-1} and 8.347×10^5 m^{-1}. These are slightly smaller than twice and three times the wave number 2.886×10^5 m^{-1} of the fundamental 1–0 band given above. This shows that even for such low values of υ, these levels are not strictly equispaced.

The anharmonicity of the vibrational levels can be represented by a formula of the type

$$G(\upsilon) = \bar{\nu}_e \left(\upsilon + \frac{1}{2}\right) - \alpha \bar{\nu}_e \left(\upsilon + \frac{1}{2}\right)^2 \qquad (16\cdot4\text{-}16)$$

α is the anharmonicity parameter which is small and positive so that the levels crowd together with increasing υ.

This formula replaces Eq. (16·4-11) for the harmonic oscillator. $\bar{\nu}_e$ depends on the force constant k for infinitesimal vibrations. For HCl, the values of $\bar{\nu}_e$ and $\alpha \bar{\nu}_e$ come out to be 2.990×10^5 m^{-1} and 5.2×10^3 m^{-1} respectively. Thus $\bar{\nu}_e$ differs slightly from $\bar{\nu}_0$ in Eq. (16·4-11). The latter is 2.886×10^5 m^{-1} for HCl which determines the mean force constant for the amplitude of the $\upsilon = 1$ level.

For more accurate representation of the anharmonicity of the levels, sometimes a small cubical term is to be added on the r. h. s. of Eq. (16·4-16) for some molecules.

To explain the gradual convergence of the vibrational levels for higher values of υ, P.M. Morse has suggested an empirical potential function of the form

$$V(r) = D \left[1 - \exp\{-\beta(r - r_0)\}\right]^2 \qquad (16\cdot4\text{-}17)$$

Morse potential is a fairly correct representation even for larger r. D and β are constants. It should noted that the Morse potential does not go to infinity as $r \to \infty$, but remains constant as shown in Fig. 16.5. If it is substituted in the Schrödinger equation in place of the harmonic oscillator potential, then the equation can be solved exactly. The term values are then given exactly by Eq. (16.4-16). The eigenfunctions no longer can be classified according to parity as in the case of the linear harmonic oscillator (see § 11.8). They have greater amplitudes for $\xi > 0$ than for $\xi < 0$ (anharmonic about $\xi = 0$) which is responsible for the violation of the selection rules (16.4-12).

The selection rules for the anharmonic oscillators are $\Delta \upsilon = \pm 1, \pm 2, \pm 3, \ldots$ In practice, the transitions with $|\Delta \upsilon| > 3$ have negligible probability. Further since the vibrational level spacings are of the order of 10^5 m^{-1}, the ratio of the populations between the $\upsilon = 1$ and $\upsilon = 0$ levels is (at room temperature)

$$N_1/N_0 = \exp(-E_1/k_B T)$$
$$= \exp(-\Delta G_\upsilon \, h\, c/k_B T)$$
$$= \exp\left(-\frac{10^5 \times 6\cdot 63 \times 10^{-34} \times 3 \times 10^8}{1\cdot 38 \times 10^{-23} \times 300}\right)$$
$$= \exp(-4\cdot 8) \approx 0\cdot 008 \qquad (16\cdot 4\text{-}18)$$

Thus the population of the $\upsilon = 1$ state is less than 1% of the population in the $\upsilon = 0$ state. So we can take all absorptive type of vibrational transitions to originate from the $\upsilon = 0$ level. The three transitions of importance are thus :

$$\upsilon = 0 \text{ to } \upsilon' = 1 \ (\Delta \upsilon = 1) :$$
$$\Delta G_\upsilon = G_1 - G_0 = \bar{\nu}_e (1 - 2\alpha) \qquad (16\cdot 4\text{-}19)$$

This has a large intensity.

$$\upsilon = 0 \text{ to } \upsilon' = 2 \ (\Delta \upsilon = 2) :$$
$$\Delta G_\upsilon = G_2 - G_0 = 2\bar{\nu}_e (1 - 3\alpha) \qquad (16\cdot 4\text{-}20)$$

$$\upsilon = 0 \text{ to } \upsilon' = 3 \ (\Delta \upsilon = 3) :$$
$$\Delta G_\upsilon = G_3 - G_0 = 3\bar{\nu}_e (1 - 4\alpha) \qquad (16\cdot 4\text{-}21)$$

The last two bands are of diminishing intensities, much weaker than the first.

Since α is small ($\sim 0\cdot 001$), the three lines lie close to $\bar{\nu}_e, 2\bar{\nu}_e$ and $3\bar{\nu}_e$. The first of these is called the fundamental absorption band while the other two are the first and second overtones. As stated above these conclusions agree with the observed wave numbers in the case of the HCl bands.

At higher temperatures, the population of the $\upsilon = 1$ level may become appreciable so that the transition $\upsilon = 1$ to $\upsilon' = 2$ may have appreciable probability. Thus at $T = 600$ K, the population of the $\upsilon = 1$ level is about 9% of that of the $\upsilon = 0$ level and the 1–2 band has an intensity of about 10% of that of the 0–1 band.

16·5 Rotational structure of the infra-red bands for diatomic molecules

In the above analysis, we have considered independent rotational and vibrational motions. In the actual diatomic molecule, both are present simulatneously and they influence each other.

Due to the vibration of the two atoms, the internuclear distance change continuously so that the moment of interia I does not remain constant. Even if an average value of I is taken for a given vibrational level, it will obviously be different for different vibrational levels. As a result, the rotational constant B given by Eq. (16·3-6) will be different for different vibrational states.

Again, the rotation of the molecule as a whole gives rise to a centrifugal acceleration which influences the vibrational potential energy. As a result the vibrational levels will not be represented correctly by Eq. (16·4-11) even if the potential were strictly harmonic.

If the mutual influences of the rotational and vibrational motions are not taken into account, then to a first approximation, the rotation-vibration energy of the molecule can be written as

$$E_{\upsilon r} = E_\upsilon + E_r = h\nu_0 \left(\upsilon + \frac{1}{2}\right) + \frac{h^2}{8\pi^2 I} j(j+1) \qquad (16\cdot5\text{-}1)$$

The wave numbers of the rotation-vibration spectral lines can then be given by §§

$$\bar{\nu}_{\upsilon r} = \{(E'_\upsilon + E'_r) - (E_\upsilon + E_r)\}/hc$$

$$= \bar{\nu}_0 (\upsilon' - \upsilon) + \frac{h}{8\pi^2 c} \left\{ \frac{j'(j'+1)}{I'} - \frac{j''(j''+1)}{I''} \right\} \qquad (16\cdot5\text{-}2)$$

So the transition takes place between a state of vibrational quantum number υ, rotational quantum number j'' and a state of vibrational quantum number υ', rotational quantum number j'. The selection rules governing the transitions between such states are :

$$\Delta\upsilon = \pm 1, \pm 2, \pm 3 \text{ etc.} \qquad (16\cdot5\text{-}3)$$

$$\Delta j = \pm 1$$

For some molecules $\Delta j = 0$ transition is also observed which however does not occur for diatomic molecules.

---

§§ It is usual to denote by j' and j'' the rotational quantum numbers for the upper

Since the moment of interia is different for different vibrational levels, we have represented these by I' and I'' for the two levels of vibrational quantum numbers υ *and* υ' *respectively*.

For a given υ, *the rotational quantum number j* can have different values so that for each vibrational state there are many close lying rotational levels as shown in Fig. 16·8. As stated previously, the vibrational energy is large compared to the rotational energy ($E_\upsilon >> E_r$). The gaps between the consecutive vibrational levels are also large compared to the gaps between the consecutive rotational levels.

In Fig. 16·8, the rotational levels corresponding to two different vibrational levels are shown. The spacings between the consecutive rotational levels have been shown highly exaggerated in the figure. Some of the possible transitions between the lower and upper states consistent with the selection rules (16·5-3) are also shown. It is evident that a large number of closely spaced vibration-rotation lines result due to the transitions between the two vibrational states. These lines are so close lying that they cannot be resolved by an ordinary spectroscope and appear as a *band of light*. This is the explanation of the appearance of the rotation-vibration bands.

In the case of the HCl molecule, the proper vibrational wave-number $\bar{\nu}_0 = \nu_0/c$ $= 2·866 \times 10^5$ m^{-1}. The corresponding wavelength is 34,650 Å or 3·465 μ. The rotation-vibration spectral lines have wavelengths of this order of magnitude so that they lie in the near infra-red. For this reason it is more convenient to study these spectra than the pure rotational spectra which lie in the far infra-red. From the wave number differences between the rotation-vibration lines, it is possible to obtain many relevant information regarding the rotational structure of the molecules.

It should be noted that as in the case of rotational spectra, rotation-vibration spectra are observed only for polar molecules.

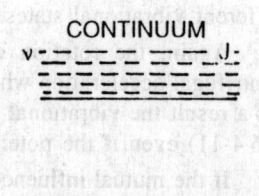

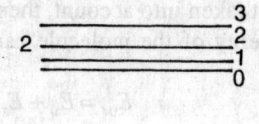

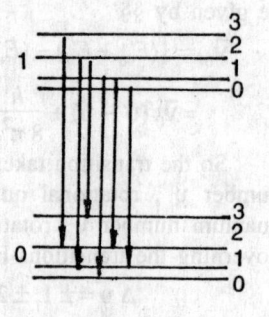

Fig. 16·8. Rotational and vibrational levels of a diatomic molecule.

Out of the different permissible values of $\Delta \upsilon$ in Eq. (16·5-3), the transition $\Delta \upsilon = \pm 1$ has the highest probability. Experiments show that the 2-0 absorption band has only a few percent intensity compared to the 1-0 band. Higher order

Molecular Spectra

transitions have still lower probabilities of occurrence.

As in the case of pure rotational spectrum, the rotational structure of the rotation-vibration spectra are observed in the case of gaseous diatomic substances. These are not observed in the case of solids or liquids, Because of the selection rules $\Delta j = \pm 1$, discrete lines are observed on both sides of the *band-origin* which is at the proper frequency v_0. The spectrum on the higher frequency side of v_0 is known as the *R-branch* while that on the lower frequency side is known as the *P-branch*. The wave numbers for the two branches are given by:

For R-branch $(\Delta j = +1)$:
$$R(j) = \bar{v}_0 + B'(j+1)(j+2) - B''j(j+1)$$
$$= \bar{v}_0 + (B' + B'')(j+1) + (B' - B'')(j+1)^2 \quad (16 \cdot 5 \text{-} 4)$$

For P-branch $(\Delta j = -1)$:
$$P(j) = \bar{v}_0 + B'(j-1)j - B''j(j+1)$$
$$= \bar{v}_0 - (B' + B'')j + (B' - B'')j^2 \quad (16 \cdot 5 \text{-} 5)$$

$j = 0, 1, 2, 3$ *etc*. Since the rotational constant B is almost the same for the different vibrational states, we can write $B' \approx B''$. Then the quadratic terms in Eqs. (16·5-4) and (16·5-5) can be neglected so that the lines are almost equispaced. The presence of the quadratic terms makes the lines of the R-branch slightly converging while those of the P-branch slightly diverging, since B'' is slighthy less than B'. No line is observed at the band-origin v_0 (see Fig. (16.9).

The rotational constants B' and B'' can be determined by measuring the wave number differences of the lines of R and P branches originating out of the transitions either from a lower rotational level of given j value *or* to a higher rotational level of given j value.

Thus the transitions from a given rotational level j in the lower vibrational state to the higher or lower rotational levels in the higher vibrational state give the R and P lines for which the wave number difference is given by

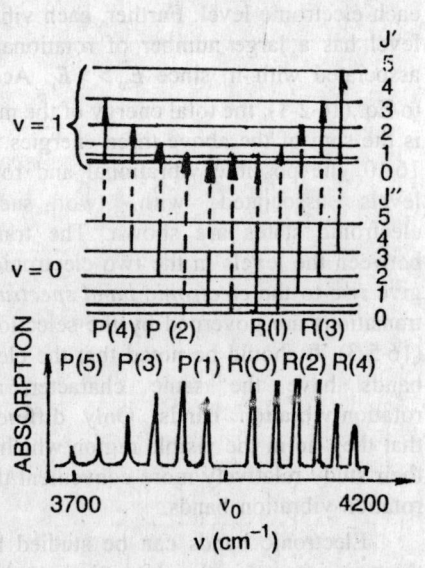

Fig. 16.9. Rotational structure of vibrational bands.

$$R(j) - P(j) = (B' + B'')(2j+1) + (B' - B'')(2j+1)$$
$$= 2B'(2j+1) \qquad (16\cdot5\text{-}6)$$

Knowing j, B' can be calcutated for the lower level.

Similarly the transitions from the rotational levels $(j-1)$ and $(j+1)$ in the lower vibrational state to a level of given j value in the upper vibrational state give the R and P lines whose wave number difference is

$$R(j-1) - P(j+1) = (B' + B'')(2j+1) - (B' - B'')(2j+1)$$
$$= 2B''(2j+1) \qquad (16\cdot5\text{-}7)$$

Knowing j, B'' can be found for the lower state.

16·6 Electronic bands and their vibrational structure

In the preceeding discussions, it was assumed that the electronic configuration of the atoms constituting the molecule remains unchanged while the molecule undergoes vibrational and rotational transitions. This means that the electronic energy of the molecule remains the same. The energy is of the same order of magnitude as the energies of the atomic levels. So the transitions between the electronic energy levels of the molecules give rise to the spectral lines which usually lie in the visible region. Their wave numbers are of the order of 10^6 m^{-1}. This shows that the electronic energy of the molecule is large compared to its vibrational energy ($E_e >> E_v$) so that there may be a large number of vibrational levels associated with each electronic level. Further, each vibrational level has a large number of rotational levels associated with it, since $E_v >> E_r$. According to Eq. (16·2-1), the total energy of the molecule is the sum of the above three energies. In Fig. 16·10, the possible vibrational and rotational levels associated with two successive electronic states are shown. The transitions between the levels in the two electronic states give rise to the *electronic band spectrum*. The transitions are governed by the selection rules (16·5-3). It should be noted that the electronic bands have the same character as the rotation-vibration bands. Only difference is that they lie in the visible region which makes their study relatively more convenient than the rotation-vibration bands.

Electronic bands can be studied both in the emission and in the absorption. At the room temperatures, most molecules are in the

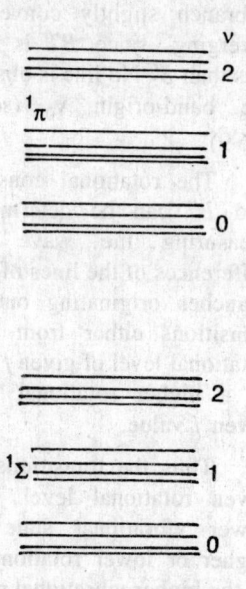

Fig. 16·10. Electronic states with vibrational and rotational levels.

$v'' = 0$ vibrational level of the lowest electronic state. When irradiated with visible radiation, they are raised to the different vibrational levels v' of the upper electronic state. These absorption bands constitute a *progression*. The absorption bands $0-0, 1-0, 2-0$ *etc.*, illustrated in Fig. 16·11, show the progression of the vibrational levels of the upper electronic state. It is possible to assign the correct vibrational quantum numbers to these levels from a study of these bands.

The emission spectra, on the other hand, are usually much more complex. At the higher temperatures necessary to excite these spectra, a large number of molecules are raised to the different vibrational levels of the upper electronic states. The transitions from these to the different vibrational levels of the lower electronic states give rise to a large number of emission bands. Hence it is much more difficult to assign correct vibrational quantum numbers to the associated levels from a study of the emission bands. However, one important advantage of the emission spectra is that it is possible to study the high vibrational levels of the ground state of the molecule which cannot usually be done by any other method.

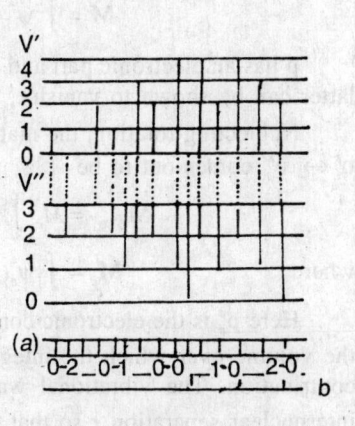

Fig. 16·11 Vibrational structure of electronic bands for (*a*) emission ; (*b*) absorption.

16·7 Frank-Condon principle

The vibrational intensity distribution in an electronic band varies in a complicated manner from one molecule to another. The intensity distribution gives important information regarding the vibrational structure, specially when no rotational analysis is possible. The intensity distribution can be explained with the help of *Frank-Condon principle* which may be stated as follows : *The electronic transition in a molecule takes place so rapidly in comparison with the vibrational motion of the atoms that immediately after the transition, the nuclei of the atoms are at the same relative positions and have the same relative velocity as before the transition.*

Classically the vibrational velocity of an atom in a diatomic molecule is zero at the two turning points. Consequently, it has the highest probability of being in a given element of length in its path near these turning points (see Prob. 6 in Ch. XI). So the probability of transition is the highest if one of the two turning points of a level in one electronic state is at the same internuclear distance as one of the two turning points of a level in the other electronic state. Such transitions are known as vertical *transitions* for obvious reasons.

According to the quantum mechanical time-dependent perturbation theory, the probability of transitions between two states is dependent on the square of the matrix element **M** of the electronic dipole moment **p** where

$$\mathbf{M} = \int \psi^* \mathbf{p} \psi \, d\tau \qquad (16\cdot7\text{-}1)$$

p has an electronic part and a nuclear part. The contribution from the latter can be shown to vanish.*

Neglecting rotation, the matrix element for the vibrational transition $\upsilon' \leftrightarrow \upsilon''$ comes out to be

$$\mathbf{M}_{\upsilon'\upsilon''} = M_e \int \psi_{\upsilon'} \psi_{\upsilon''} \, dr \qquad (16\cdot7\text{-}2)$$

where
$$\mathbf{M}_e = \int \psi_{e'}^* \mathbf{p}_e \psi_{e''} \, d\tau_e \qquad (16\cdot7\text{-}3)$$

Here $\mathbf{p}_e$ is the electronic contribution to the dipole moment and τ_e is the volume over which the integral is to be evaluated for the electronic contribution. The vibrational wave functions are real functions of the internuclear separation r so that $\psi_{\upsilon'}^* = \psi_{\upsilon'}$ (see Eq. 11·8-5 in Ch. XI).

The average electronic transition moment $\mathbf{M}_e$ determines the intensity of a whole band system and is the analogue of the transition moment for an atomic spectral line. The relative intensities of the different vibrational transitions making up the whole band system therefore depends on the vibrational *overlap integral* :

$$q_{\upsilon'\upsilon''} = \left| \int \psi_{\upsilon'} \psi_{\upsilon''} \, dr \right|^2 \qquad (16\cdot7\text{-}4)$$

Obviously $q_{\upsilon'\upsilon''}$ is appreciable for those values of υ' and υ'' for which both $\psi_{\upsilon'}$ and $\psi_{\upsilon''}$ have large amplitudes.

For $\upsilon = 0$ the vibrational wave function has one central maximum while for other values of υ, it has a number of maxima with one prominent maximum near each turning point on the inner side (see Fig. 11·13). Hence the vibrational transition will have an appreciable probability only if one of the turning points of a vibrational level of one electronic state is at the same internuclear distance as one of the turning points of a level of the other electronic state except when $\upsilon = 0$ in the upper state.

The above conclusions are illustrated graphically in Fig. 16·12 for the case in which vibrational potential energy curves have no relative displacement

---

* See *Introduction to Quantum Mechanics* by L. Pauling and E.B. Wilson.

Molecular Spectra

along r for the two electronic states. As a result the vibrational wave functions of one electronic state are identical with those having the same value of v for the other electronic state. The selection rule for the allowed transitions in this case is $\Delta v = 0$, since the vibrational wave functions for different v are orthogonal while those for the same v in these two electronic states have the maximum overlap.

For the case where the potential energy curves in the two electronic states have appreciable relative displacements along r differ appreciably, as shown in Fig. 16·13 the molecule in the lowest vibrational state $v' = 0$ of the upper electronic state A has a large probability of being at $r = r_{eA}$. So the transition will take place to the lower electronic state B in such a manner as to leave the molecule at a point P_1 on the lower potential energy curve which corresponds to vibrational levels $v'' = 7$ or 8 in the lower state. The vibrational wave functions have large amplitudes for these values of v'' and hence the bands 0–7 or 0–8 will be strong.

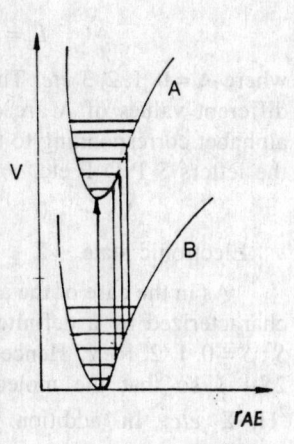

Fig. 16·12 Vertical transitions without relative displacement along r for the potential energy curves in two states.

16·8 Electronic states

The atomic energy states are characterized by definite values of the orbital angular momentum quantum number L. The molecular electrons do not have definite orbital angular momentum. On the other hand, the component of the orbital angular momentum L_z along the vibrational axis z of the atoms in the diatomic molecule

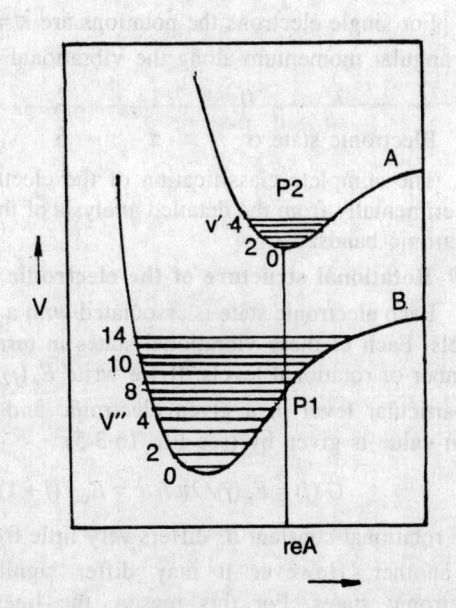

Fig. 16·13. Vertical transitions with relative displacements along r of the potential energy curves in two states.

has a definite value. We can then write

$$L_z = \Lambda \hbar$$

where $\Lambda = 0, 1, 2, 3$ *etc*. The molecular electronic states characterized by different values of Λ are designated by a set of letters from the Greek alphabet corresponding to the designations of the atomic orbital states by the letters S P D F etc.

Λ	0	1	2	3	4
Electronic state	Σ	Π	Δ	Φ	Γ

As in the case of the atomic states, the molecular electronic states are characterized by a definite value of the resultant spin quantum number $S: S = 0, 1/2, 1$ *etc*. Hence each electronic state has definite multiplicity $2S + 1$ so that the molecular electronic states can be designated as $^2\Pi, ^3\Sigma$ *etc*. In addition the Σ states may be symmetric (Σ^+) or antisymmetric (Σ^-) with respect to reflection at a plane containing the molecular axis. The other states are doubly degenerate with both a (+) and a (−) component. For homouclear diatomic molecules, all the terms are either *even* (*g*) or *odd* (*u*) with respect to inversion. As examples of the above classifications, the ground states of NO and of O_2 (homonuclear) are $^2\Pi$ and $^3\Sigma_g^-$ respectively.

For single electrons the notations are $\lambda = |m_l|$ for the component of the angular momentum along the vibrational axis and the states are

λ	0	1	2	3 ...
Electronic state	σ	π	δ	φ ...

The complete classification of the electronic states are determined experimentally from the detailed analysis of the rotational structure of the electronic bands.

16·9 Rotational structure of the electronic bands

Each electronic state is associated with a large number of vibrational levels. Each of these vibrational states in turn is associated with a large number of rotational levels. If we write $E_r(j)$ as the rotational energy of a particular level in a given electronic and vibrational state, then the term-value is given by (see Eq. 16·3-5).

$$G(j) = E_r(j)/2\pi \hbar \, c = B_v j(j+1)$$

The rotational constant B_v differs very little from the one vibrational state to another. However it may differ significantly between different electronic states. For this reason, the lines of the P ($\Delta j = -1$) and R ($\Delta j = +1$) branches either rapidly converge or rapidly diverge. Since B' and B'' are now different in the two electronic states, the quadratic term on the r.h.s. of Eqs. (16·5-4) or (16·5-5) no longer vanishes. Hence there

Molecular Spectra

arises the existence of clear cut *bandheads* in this case.

In addition, there is another branch, known as the Q-branch in the electronic bands for which the selection rule is $\Delta j = 0$. The lines of the Q-branch are also converging. The variation of the wave numbers of the lines of the P, Q and R branches are shown graphically in Fig. 16·14. These graphs are known as *Fortrat diagrams*.

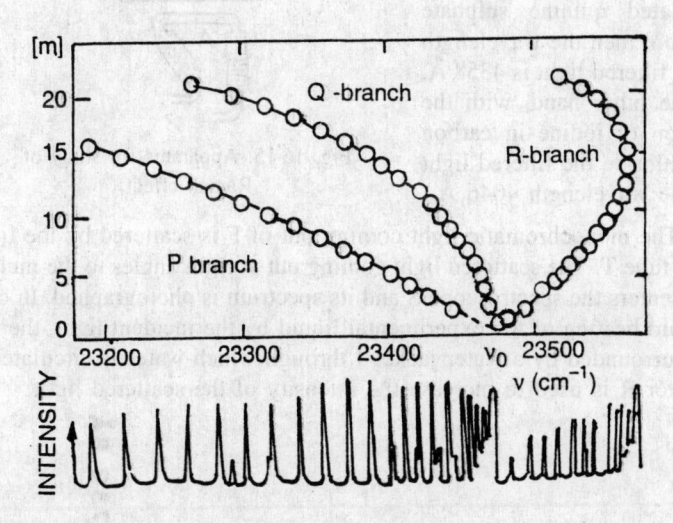

Fig. 16·14 Fortrat diagram for 4291 Å AlH band.

16·10 Raman effect

In 1928, the famous Indian physicist, C.V. Raman observed that when monochromatic visible light is passed through a transparent liquid *e.g.*, benzene or toluene, the light scattered at right angles to the direction of incidence had some components in it having wavelengths longer as well as shorter than the wavelength of the incident light in addition to the line of the original wavelength (parent line). This phenomenon is known as Raman effect. Later Raman effect has also been observed in solid and gaseous media. Raman was awarded Nobel Prize in Physics for this very important discovery in 1930.

Various improved methods of studying Raman effect have been developed since Raman's pioneering experiment. We describe below an apparatus developed by R.W. Wood for studying Raman effect.

In Fig. 16·15, T is a hornshaped glass tube, which contains the experimental liquid. The outer side of the curved portion of the tube is blackened. The other end of the tube, is covered by a plane glass plate through which the light scattered by the liquid enters the spectroscope S.

The light source A is usually a mercury arc lamp which gives out light of different wavelengths. The light from the source is passed through a *light filter* F so that one gets monochromatic light coming out of the filter at the other side. The filter is usually some liquid kept in a glass tube. If the liquid is acidulated quinine sulphate solution, then the wavelength of the filtered light is 4358 Å. On the other hand, with the solution of iodine in carbon tetrachloride the filtered light has the wavelength 4046 Å.

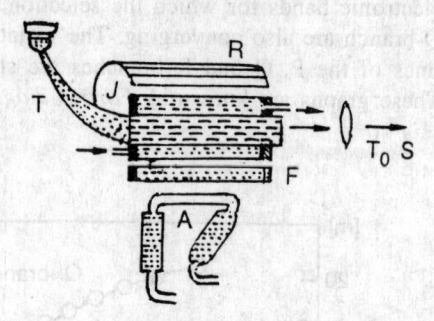

Fig. 16·15. Apparatus for study of Raman effect.

The monochromatic light coming out of F is scattered by the liquid in the tube T. The scattered light coming out at right angles to the incident beam enters the spectroscope S and its spectrum is photographed. In order to avoid heating of the experimental liquid by the incident light, the tube T is surrounded by a water jacket J through which water is circulated. A reflector R is used to increase the intensity of the scattered light.

2536.5 Å — exciting line

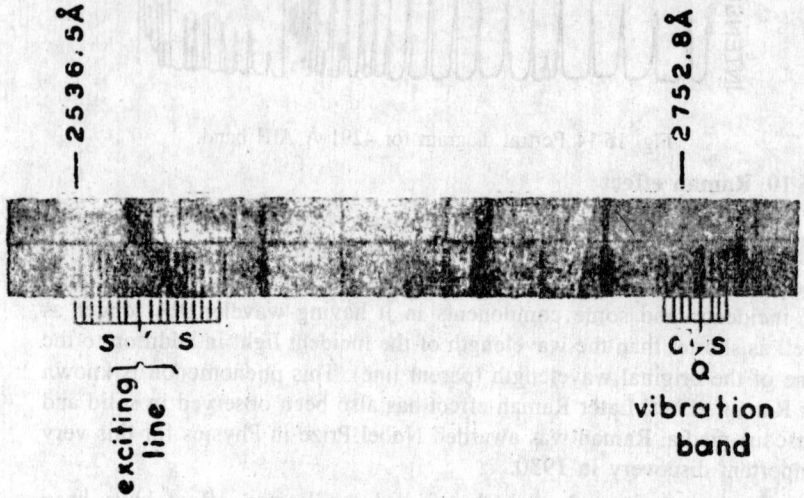

2752.8 Å — vibration band

Fig. 16·16. Photograph of Raman sepectrum. Exciting line is the Hg 2536·5 Å line. Mercury arc spectrum is shown in the upper portion. In the lower portion is shown the mercury arc radiation scattered by HCl gas (5 atm).
(From Herzberg : Spectra of Diatomic Molecules, Vol. I, published by Van Nostrand Reinhold Co. Copyright 1950 by Litton Publishing Inc.)

In the photographs of Raman spectra shown in Fig. 16·16, weak spectral lines (Raman lines) can be seen on either side of the exciting line. To improve the intensities of these lines it is necessary to increase the intensity of the incident light as much as possible,. Besides, the exposure time of the photographic plate should be suitably adjusted to get distinct lines. In modern arrangement for the study of Raman effect, high power laser beams are used. (see § 16·15)

16·11 Characteristics of Raman spectrum

The main characteristics of Raman spectrum are enumerated below :

1. In the Raman spectrum lines having wavelengths both longer and shorter than that of the exciting line are observed in addition to the parent line. The longer wavelength lines are known as *Stokes lines*. Their frequencies are lower than that of the parent line. The shorter wavelength lines are known as *anti-Stoke lines* and have frequencies higher than that of the parent line. It should be noted, that in the fluorescence spectrum (see Ch. IV) only the Stokes lines are observed. Anti-Stokes lines are observed only in the Raman spectrum.

2. The Stokes and anti-Stokes lines are symmetrically placed on either side of the exciting line so that the frequencies or wave-numbers of these lines depend on the wave number of the parent line. It should be noted that this is just the opposite of what is observed in the fluorescence spectra. In the latter case, the wave number of the Stokes line *does not depend* on the wave number of the exciting radiation. It depends on the nature of the fluorescent material only.

3. The wave number differences of the Stokes and anti-Stokes lines depend on the nature of the experimental substance. It is independent of the wave number $\bar{v}$ of the exciting radiation. The wave numbers of the Stokes and anti-Stokes lines are given by

$$\bar{v}' = \bar{v}_0 - \Delta \bar{v} \qquad \text{(Stokes line)} \qquad (16 \cdot 11 \text{-} 1)$$

$$\bar{v}'' = \bar{v}_0 + \Delta \bar{v} \qquad \text{(Anti-Stokes line)} \qquad (16 \cdot 11 \text{-} 2)$$

If $\bar{v}_0$ is changed, then the wave numbers of the Stokes and anti-Stokes lines also change, with their wave number difference $\Delta \bar{v}$ from the parent line remaining unchanged. For visible exciting radiation $\Delta \bar{v} << \bar{v}_0$. *i.e.*, the wave number difference is small compared to the wave number of the exciting radiation. $\Delta \bar{v}$ is usually found to be equal to the wave number of some rotational or rotation-vibration line of the scatterer lying in the infra-red region.

4. The intensities of the Raman lines are usually small compared to the intensity of the parent line. Of these the anti-Stokes lines are much weaker than the Stokes lines. If the temperature of the scatterer is raised, the intensities of the anti-Stokes lines increase to some extent.

5. The Raman lines exhibit different degrees of polarization which differ from one line to another.

6. We have mentioned earlier that Raman effect arises from the scattering of light. However this scattering is quite different from ordinary Rayleigh scattering. In Rayleigh scattering, there is no change in the wavelength of the exciting radiation. In Raman scattering there are both increase and decrease in the wavelength. Also in Rayleigh scattering, the intensity of the scattered radiation I depends on the wavelength $\lambda : I \propto 1/\lambda^4$. It is this variation of I with λ which is responsible for the blue of the sky which is due to the Rayleigh scattering of the sun-light from the air molecules. There is no such simple relationship between I and λ in the case of Raman scattering.

7. As seen above, Raman effect is basically different in nature from fluorescent scattering.

16·12 Classical theory of Raman effect

In Raman effect, the Stokes and anti-Stokes lines originate due to rotational or vibrational transitions of the molecules. The origin of the vibrational Raman effect can be explained on the basis of the e.m. theory of light in the following way.

Consider an e.m. wave of frequency v and wave number $\bar{v}$ incident upon a molecule. The electric vector $\mathbf{E} = \mathbf{E}_0 \cos 2\pi \bar{v} ct$ of this wave induces an electric dipole moment in the molecule, which can be expressed as

$$\mathbf{p} = \alpha \, \mathbf{E}_0 \cos 2\pi \bar{v} ct \qquad (16\cdot12\text{-}1)$$

α is known as the *polarizability* of the molecule. α depends on the relative orientation between the applied electric field and the dipole axis. Usually α is a constant. However in the present case, it is not constant. Due to the rotational and vibrational motion of the molecule, α changes with time. The dipole moment depends on the charges at the poles and their linear separation. Due to the vibrational motion of the molecule, this separation changes periodically with time. This causes the dipole moment $\mathbf{p}$ and hence the polarizability α to change periodically with time. If the frequency of vibration of the charges in the molecular dipole be v_1, then we can write

$$\alpha = \alpha_0 + \alpha_1 \cos(2\pi \bar{v}_1 ct + \delta) \qquad (16\cdot12\text{-}2)$$

where $\bar{v}_1 = v_1/c$ is the wave-number. δ is the phase of the vibration. α_0 is the static part of the polarizability. The induced dipole-moment is

$$\begin{aligned}\mathbf{p} &= \{\alpha_0 + \alpha_1 \cos(2\pi \bar{v}_1 ct + \delta)\} \, \mathbf{E}_0 \cos 2\pi \bar{v} ct \\ &= \alpha_0 \mathbf{E}_0 \cos 2\pi \bar{v} ct + \alpha_1 \mathbf{E}_0 [\cos\{2\pi(\bar{v} + \bar{v}_1) ct + \delta\} \\ &\quad + \cos\{2\pi(\bar{v} - \bar{v}_1) ct - \delta\}]\end{aligned} \qquad (16\cdot12\text{-}3)$$

From Eq. (16·12-3) we see that the change in the dipole moment may be regarded as being due to the superposition of three harmonically varying terms with the three different wave numbers $\bar{v}, \bar{v}+\bar{v}_1$ and $\bar{v}-\bar{v}_1$. The first of these represents the oscillation of the electric dipole with the frequency of the incident radiation. The other two represent two different linear harmonic oscillations of the dipole having these other frequencies. According to the e.m. theory of light, an oscillating electric dipole emits e.m. radiation of frequency equal to its frequency of oscillation. So in the present case, the molecule emits e.m. radiation of three different frequencies $v, v+v_1$ and $v-v_1$ or wave numbers $\bar{v}, \bar{v}+\bar{v}_1$ and $\bar{v}-\bar{v}_1$ The first of these is equal to the wave number of the incident radiation. The other two are the wave numbers of the of the Stokes and anti-Stokes lines respectively as observed in the Raman spectrum. The wave number difference $\Delta \bar{v} = \bar{v}_1$ of these lines from the parent line is equal to the rotational or vibrational wave number of the molecule. As was discussed in §16·1, spectral lines of these wave numbers are observed in the spectra of the molecules. They lie in the infrared region.

Though the classical e.m. theory can explain the origin of the Raman spectrum in a simple manner, it cannot explain the intensity or polarization of the Raman lines for which quantum mechanical theory is needed.

16·13 Quantum theory of Raman effect

According to the quantum theory of Raman effect when a photon of energy hv is incident upon a molecule, the following three processes may occur :

(a) The photon may be scattered without causing any change in the quantum state of the molecule. As a result there will be no change in the frequency or wave number of the photon. (b) A small fraction of the energy of the photon may be transferred to the molecule during the photon-molecule collision. As a result, the photon loses some energy so that its frequency and wave number are reduced and we get the Stokes line. (c) If the photon encounters a molecule in an excited state, then it may gain some energy from the molecule. As a result, the frequency and the wave number of the photon are increased which gives rise to the anti-Stokes line. In the process the molecule goes down to a lower energy state.

In case (a), the collision between the photon and the molecule may be regarded as elastic, while in the other two cases the collisions may be regarded as inelastic.

If the initial *intrinsic energy* of the molecule be W_1, that after the scattering be W_2 and the frequency of the scattered photon be ν' then from the law of conservation of energy, we get

$$W_1 + h\nu = W_2 + h\nu' \qquad (16\cdot13\text{-}1)$$

Eq. (16·13-1) holds if it is assumed that there is no change in the kinetic energy of the molecule during its encounter with the photon. We then get

$$\nu' = \nu + \frac{W_1 - W_2}{h}$$

In terms of wave numbers, we then have

$$\bar{\nu}' = \bar{\nu} + \frac{W_1 - W_2}{ch} \qquad (16\cdot13\text{-}2)$$

If $W_1 = W_2$, the intrinsic energy of the molecule remains unchanged and $\bar{\nu}' = \bar{\nu}$ so that the wave number of the scattered photon is the same as that of the incident photon. If $W_1 < W_2$ then the molecule undergoes transition from a lower energy state to a higher energy state and $\bar{\nu}' < \bar{\nu}$ so that the wave number of the scattered photon is less than that of the incident photon. This gives rise to the Stokes line.

If $W_1 > W_2$, then the molecule makes a transition from a higher to a lower energy state and $\bar{\nu}' > \bar{\nu}$ so that the scattered photon has a higher wave number than the incident photon. We get the anti-Stokes line in this case.

The difference in wave number of the Stokers or anti-Stokers line from the parent line is

$$\Delta\bar{\nu} = \bar{\nu}' - \bar{\nu} = \frac{W_1 - W_2}{ch}$$

This wave number difference depends upon the energy states of the molecule and not upon the wave number of the parent line.

In the vibrational Raman effect, assuming linear harmonic oscillation, the selection rule for the transition as obtained from quantum mechanics is given by

$$\Delta \upsilon = \pm 1$$

Here υ is the vibrational quantum number.

According to Maxwell-Boltzmann statistics, the number of molecules n per unit volume in an energy level W at the temperature T K is given by

$$n = n_0 \exp(-W/kT)$$

Obviously as W increases n decreases. The number of molecules n_0 in the ground state for which $W = 0$ is the highest.

The ratio of the number of molecules in an excited state to that in the ground state is

$$\frac{n}{n_0} = \exp(-W/kT)$$

At $T = 300$ K (room temperature), we then have for a rotational state of energy W_r just above the ground state

$$\frac{n_r}{n_0} = \exp(-W_r/kT)$$

Since the wave number difference between successive rotational states is of the order of $\Delta \bar{v}_r \sim 4000 \text{ m}^{-1}$ (see § 16·3), energy of the low-lying rotational state is

$$W_r \sim hc\, \Delta \bar{v}_r \sim 5 \times 10^{-3} \text{ eV}$$

At room temperature

$$kT = \frac{1 \cdot 38 \times 10^{-23} \times 300}{1 \cdot 6 \times 10^{-19}} = 2 \cdot 6 \times 10^{-2} \text{ eV}$$

Hence $W_r << kT$ and we get

$$\frac{n_r}{n_0} = \exp(-W_r/kT) = \exp(-5 \times 10^{-3}/2 \cdot 6 \times 10^{-2})$$

$$\simeq \exp(-0.2) = 0.82$$

So at room temperature quite a substantial fraction of the molecules are in the low lying excited rotational states.

On the other hand, the wave number difference between the successive vibrational levels is of the order of 10^5 m^{-1}. For HCl molecule, $\bar{v} = 2 \cdot 886 \times 10^5$ m^{-1}. So the energy of the low lying vibrational level is

$$W_v = hc\bar{v}_v = 0.36 \text{ eV}$$

Hence $W_v >> kT$ at the room temperature and we get

$$\frac{n_v}{n_0} = \exp(-W_v/kT)$$

$$= \exp(-36 \times 10^{-2}/2 \cdot 6 \times 10^{-2})$$
$$= \exp(-13 \cdot 85) = 9 \cdot 7 \times 10^{-7}$$

Thus the number of molecules in the low lying vibrational levels is very small.

Since the number of molecules is the highest in the ground state, the probability of transition from this state to higher excited states is greater than the probability of the reverse transition. So the Stokes lines have greater intensity than the anti-Stokes lines. The above calculations show that the probability of anti-Stokes transitions may be fairly high in the case of rotational Raman spectrum, specially at higher temperatures. The nature of the distribution of molecules in different energy levels is shown schematically in Fig. 16.17 at different temperatures.

If $\Delta \bar{v}$ be the wave number difference between the Stokes and the

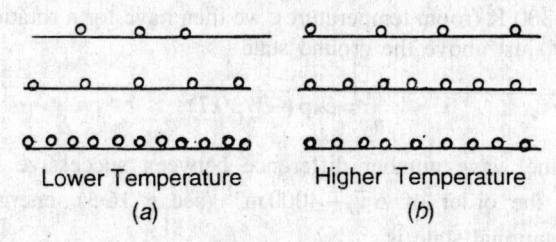

Fig. 16.17. The number of molecules in different levels. The increase in the number in the upper levels in (b) should be noted.

anti-Stokes lines from the parent line, then the ratio of the intensities of these lines I_s/I_a is determined by the ratio of the number of molecules in the excited state and the ground state, so that we have

$$\frac{I_a}{I_s} = \exp\left(-\frac{hc\Delta \bar{v}}{kT}\right)$$

Study of the Raman spectrum yields valuable information regarding the structure of molecules. Since the Raman lines lie in the visible region, their experimental study can be done more conveniently than the study of the pure rotation or the rotation-vibration spectra which lie in the infrared. Much of the information about molecular structure obtainable from the study of the latter can be more conveniently obtained from the study of the Raman spectra.

16·14. Rotational structure of the Raman spectrum

The Stokes and anti-Stokes lines in the Raman spectrum originate due to transitions between the vibrational or rotational states of the molecules. If the vibrational state of a molecule changes due to Raman transition, then there may be transition between the different rotational levels of the two vibrational states involved. Due to this, the vibrational Raman bands usually have rotational structure.

In the case of diatomic molecules, the following selection rules hold for the rotational Raman transitions :

$$\Delta j = 0, \pm 2 \qquad (16 \cdot 14\text{-}1)$$

If $\Delta j = 0$ there is no change of the wave number of the scattered radiation. In fact there is no difference between Raman and Rayleigh scattering in this case.

The rotational energy of a molecule is given by (see Eq. 16·3-5)

$$E_r(j) = Aj(j+1)$$

Here A is a constant. If there is no change in the vibrational state of the molecule due to the Raman scattering, then we have usually pure rotational Raman lines for which the change in wave number from the exciting radiation is given by

$$\left|\Delta \bar{v}_r\right| = B(j+2)(j+3) - Bj(j+1) = 4B\left(j + \frac{3}{2}\right) \qquad (16\cdot14\text{-}2)$$

$B = A/ch$ is a constant. In Fig. 16·18 is shown the origin of the Raman lines due to such pure rotational transitions. For gaseous diatomic substance, these lines are equispaced with the spacing $4B$. The smallest change in the wave number takes place due to the transition $j = 0$ to $j = 2$. In this case, the change in the wave number as obtained from (Eq. 16·14-2) is $6B$. These are shown in the inset of Fig. 16·18.

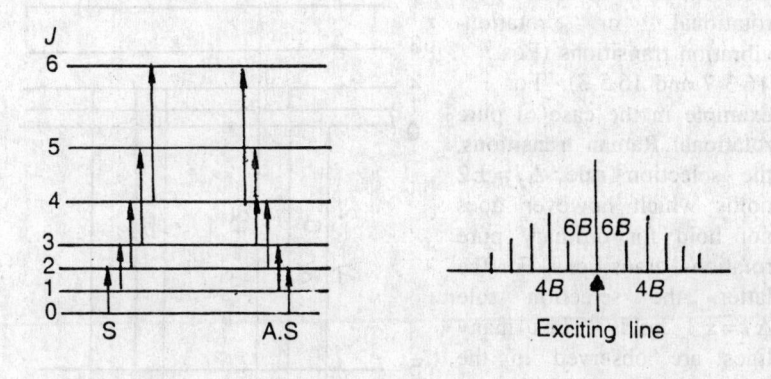

Fig. 16·18 Origin of Raman lines due to pure rotational transitions.

As seen before, due to the small energy difference between the rotational levels, there may be a substantial number of molecules in the exicited rotational states adjoining the ground state even at the room temperature. For this reason, the Stokes and anti-Stokes lines have comparable intensities at the room temperature of the pure rotational transitions. The study of these pure rotational Raman lines yield valuable information about the rotational structure of nonpolar molecules, which as we have seen before, do not give rise to rotational emission or absorption bands (see § 16·3).

If the vibrational state of the molecule changes due to Raman scattering, the Raman spectrum has rotational structure. The selection rules for the transition between the rotational levels associated with the two vibrational states are the same as those for the pure rotational transitions, viz., $\Delta j = 0, \pm 2$. In the case of the $\Delta j = 0$ transitions giving rise to the lines of the Q-branch, the change in the wave number is given by (see Eq. 16·5-1)

$$|\Delta \bar{v}| = \Delta G(v) + (B' - B'')j(j+1)$$

Here $\Delta G(v)$ is the change in the wave number due to change in the vibrational energy of the molecule. Since the rotational constant B is nearly the same in different vibrational states, $B' - B'' \approx 0$ so that the wave numbers and hence the wavelengths of all the lines of the Q-branch are nearly the same. Thus only one strong line is produced at the band-origin. One either side of this line, there are a few weaker line due to the transitions $\Delta j = \pm 2$. These transitions are illustrated in Fig. 16·19. The photograph of a Raman spectrum is shown in Fig.16·16.

From the above discussions, it is clear that the selection rules for the origin of the Raman lines are different from those for pure rotational or rotation-vibration transitions (Eqs. 16·3-7 and 16·5-3). For example in the case of pure rotational Raman transitions, the selection rule $\Delta j = \pm 2$ holds which however does not hold for ordinary pure rotational transitions. For the latter, the selection rule $\Delta j = \pm 1$ holds. Thus many lines are observed in the Raman spectra which are not found in the ordinary infrared absorption spectra. Conversely many lines observed in the infrared absorption spectra may not be present in the Raman spectra. Combined study of both types of spectra thus yield much valuable information about the rotational levels of the molecules.

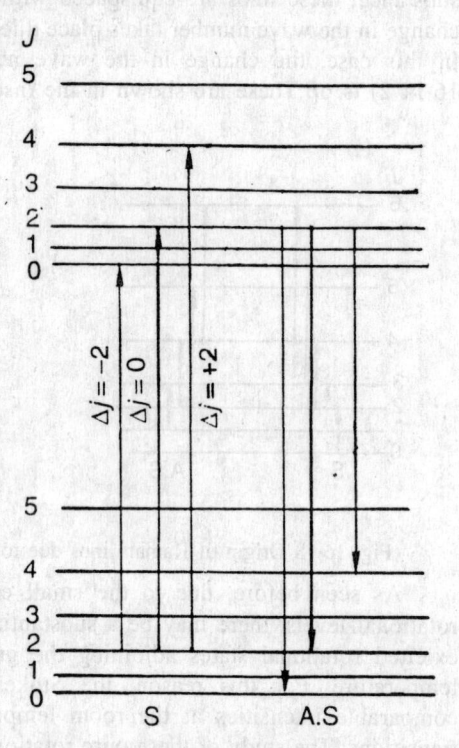

Fig. 16·19. Raman spectrum of a gas with diatomic molecule.

In the quantum mechanical theory of Raman effect, it is assumed that under the influence of the incident radiation, the molecule first makes a *virtual transition* from the initial state A to an intermediate state C from which it makes the transition to the final state B as shown in Fig. 16·20. Both these transitions are governed by the selection rules $\Delta j = \pm 1$ for allowed transition. It is possible to estimate the intensity, polarization *etc*.

Molecular Spectra

of the Raman lines wih the help of the above picture.

From Fig. 16·20, it is seen that a virtual transition takes place first from the initial state A of quantum number j to the intermediate state C of quantum number $(j+1)$ or $(j-1)$. In the second stage, the transition is

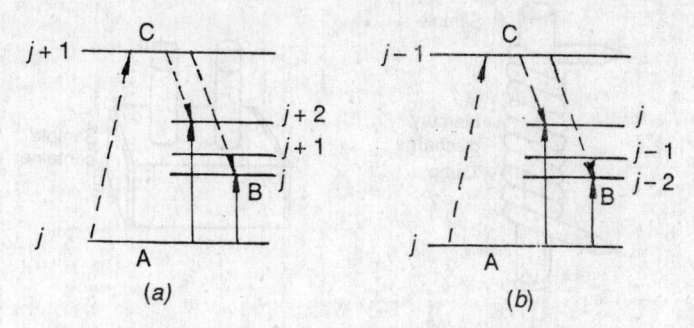

Fig. 16·20. Virtual transitions showing origin of Raman lines.

from the intermediate state C to the final state B which may have the quantum numbers j and $j+2$ (or j and $j-2$). The net result is that a transition takes place from the initial state of quantum number j to a final state of quantum number j or $j \pm 2$. Thus we get the selection rules for the transition $\Delta j = 0, \pm 2$ which are in agreement with observations.

16·15 Experimental techniques for the study of Raman effect using laser source

Raman effect is usually studied in emission using visible region spectroscopes. The most important component of the instrument in the study of Raman effect is the exciting source. The ready availabability of inexpensive laser sources has revolutionized the techniques of Raman effect studies which has practically eliminated the use of traditional mercury arc sources in many cases. Further, instead of using larger samples with 10-20 ml of the liquid, only about 1 ml of the samples can be used in the modern methods of study.

Laser sources are most suitable for the study of Raman effect since they give highly monochromatic beam of light which may be focused very finely into a small sample. concentrating relatively large power (few mW to few W) on to the latter.

Fig. 16·21 shows two different techniques for excitation. In Fig. 16·21a, a mercury discharge lamp in the form of a spiral surrounds the sample cell which is about 20-30 cm long and 1-2 cm in diameter. The main disadvantages of this type of excitation are : (i) The source being extended feeds a considerable amount of the exciting radiation directly into the spectrometer which makes it difficult to observe close lying Raman lines (less than ~ 100 cm^{-1}); (ii) Relatively large amount of the sample is required ; (iii) Relatively high frequency of the mercury radiation (4356Å and 2536Å) causes the sample to fluoresce in some cases so that the fluorescent spectrum swamps the very weak Raman spectrum. Mercury lamp is now a days used mostly with gaseous samples.

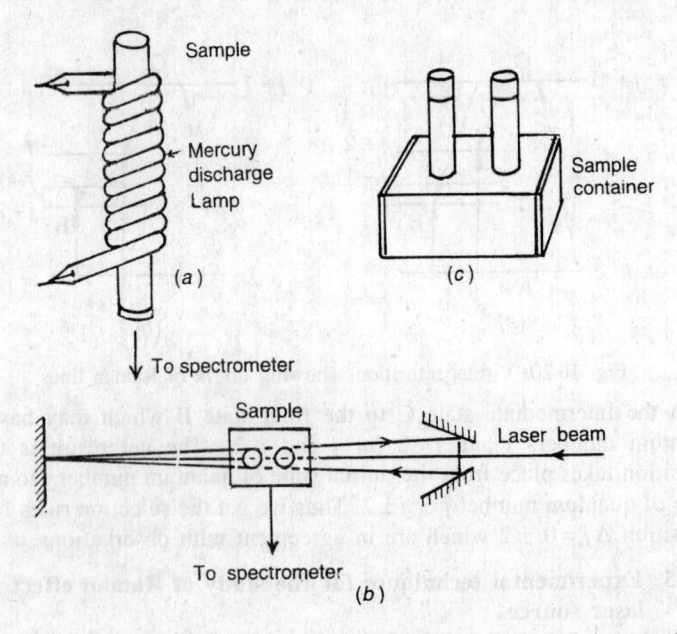

Fig. 16·21 (a) Mercury discharge lamp for the excitation of the sample in the study of Raman Effect (b) Multipass operation for excitation by laser beam (c) Sample container for excitation by laser.

Fig. 16·21b shows the method of multi-pass arrangement in which the laser beam is made to pass through the sample a number of times with the help of reflectors. The standard sample container is shown in Fig. 16·21c which consists of a quartz box in which the sample cells of about 1 ml capacity are mounted. If capillary tubes are used, the volume of the liquid required can be still further reduced. The multi-pass system cannot of course be used in this case. Even so, fairly high intensity spectrum can be obtained in this case using only 1 µl of the sample liquid. Fluorescence of the sample is avoided since the wavelength of the laser beam are usually much larger (λ 6328Å for He - Ne source). Rayleigh scattering is also much reduced with laser source due to very fine focusing of the beam.

Monochromators using a quartz prism or grating are used to select radiations of different wavelengths. Detection is usually made with the help of photoelectric detectors.

16·16 Applications of Raman effect

Study of Raman effect is particularly important is elucidating molecular structure. The diatomic molecules have characteristic vibrational frequencies which are usally higher for molecules made up of lighter atoms. Besides the characteristic frequency also increases with increasing binding force between the atoms, *i.e.*, with the increasing value of k in Eq. (16·4-14). Thus molecules with double bonds have higher characteristic frequencies than molecules with single bonds.

We have seen that in the case of nonpolar homonuclear molecules, no infrared rotation-vibration spectra are observed. Since they do not have any permanent electric dipole moments, they are not influenced by the electric vector of the incident e.m. radiation. So the vibrational motion of these molecules can only be analysed by the study of their Raman spectra. From the differences of their Raman lines, the vibrational frequency, bond strength *etc.*, of such molecules have been determined. As an example Rasetti gives the values of the vibrational wave numbers of H_2, N_2 and O_2 as 4162, 2331 and 1551 cm^{-1} respectively. These values show that the vibrational frequencies are higher for molecules with lighter atoms.

Since the heteronuclear molecules like HCl, HBr and HI have permanent electric dipole moments, they are influenced by the electric vector of the incident e.m. radiation. Due to this, infrared absorption spectra are observed for them. As an example, in the rotation vibration spectra of the HCl molecule, the infrared absorption bands for the transitions $\Delta j = \pm 1$ are observed (R and P branches). But the bands corresponding to the forbidden transition $\Delta j = 0$ (Q branch) are not observed. But in the Raman spectra of the HCl molecule, the lines due to the transition $\Delta j = 0$ are observed. In addition, the lines due to the transitions $\Delta j = \pm 2$ are also observed in the Raman spectra. In case of some of these lines the wave number differences between the Raman lines and the parent lines are found to be equal to the wave numbers of some rotation-vibration lines in their infra-red absorption spectra. As examples, the wave number differences $\Delta \bar{v}$ for the 0-2 1-3 and 2-4 Raman lines are found to be equal to the wave numbers of the 2-3, 4-5 and 6-7 lines in the infrared absorption spectra of HCl.

Raman shift does not depend on the wavelength λ of the incident radiation. But the intensities of the Raman lines depend on λ. If λ for the exciting line is close to an absorption line of the molecule, then a new type of Raman spectrum, known as the resonance Raman spectrum is observed. In such spectrum, a few lines of very high intensities appear which can be used to determine the structure of large organic molecules. Raman effect is a powerful tool in determining the characteristics of semi-conductors and superconductors, the origin of catalytic action, the nature of crystal vibration and the change of some properties of materials under high pressure.

16·17 Electronic structure of diatomic molecules

In Ch. XI we briefly introduced the idea of atomic orbitals (§ 11.16). The orbitals represent the space within which an electron can move in an atom according to wave mechanics and are determined by the wave functions obtained as solution of Schrödinger wave equation.

Different theories have been proposed for the formation of molecules from atoms. One of these is the molecular orbital theory which gives a pictorial representation of molecule formation and represents the electronic transitions in the molecules in a manner analogous to the case of atomic structure calculations.

In the molecular orbital theory, the orbitals enclose two or more nuclei and their shapes and sizes can be calculated by solving the Schroödinger wave equation. The calculations are relatively simpler for diatomic molecules since only two nuclei are involved.

16.18 Hydrogen molecular ion : LCAO approximation

We first consider the case of the hydrogen molecular ion (H_2^+) which is the simplest possible molecular system with two nuclei (protons) and one electron. The approximate shapes and sizes of the molecular orbitals can be calculated by assuming these to be made up by the superposition of the orbitals of the constituent atoms — which is known as the *linear combination of the atomic orbitals* (LCAO) approximation.

Using the notations discussed in §16.8 we write the H_2^+ molecular orbitals as $\sigma_g, \sigma_u, \pi_g, \pi_u$ etc., g and u denoting the *even* (gerade) and *odd* (ungerade) states respectively on inversion at the midpoint between the nuclei. The molecular states σ, π etc. depend on the component of the orbital angular momentum along the molecular axis. This has a value $\lambda \hbar$ where λ can be $0, \pm 1, \pm 2$ etc., Thus for the σ states $\lambda = 0$; for the π states $\lambda = 1$; for the δ states $\lambda = 2$; for the φ states $\lambda = 3$; and so on. Because both positive and negative $\lambda (=\pm m)$ are possible, the states are doubly degenerate.

The Hamiltonian of the H_2^+ ionic system comprising the two protons a and b separated by a distance R can be written as

$$\hat{H} = \frac{-\hbar^2}{2m} \nabla_r^2 - \frac{e^2}{4\pi\epsilon_0}\left(\frac{1}{r_a} + \frac{1}{r_b} - \frac{1}{R}\right) \quad (16\cdot17\text{-}1)$$

If r is the distance of the electron measured from the mid-point between a and b along the molecular axis, we have $\mathbf{r}_a = \mathbf{r} + \mathbf{R}/2$ and $\mathbf{r}_b = \mathbf{r} - \mathbf{R}/2$. We have to solve the Schrödinger equation $\hat{H}\psi_{mol} = E\psi_{mol}$.

According to the LCAO approximation, the wave functions of the two different molecular orbitals formed out of the atomic orbitals

Molecular Spectra

ψ_1 and ψ_2 of the two atoms comprising the molecule can be written as

$$\Psi_{mol} = \psi_1 + \psi_2 \qquad (16.18\text{-}2a)$$

or,
$$\Psi_{mol} = \psi_1 - \psi_2 \qquad (16.18\text{-}2b)$$

If the atomic orbitals correspond to the state of a hydrogen atom, we have the two molecular orbitals

$$\psi_g(\mathbf{R}, \mathbf{r}) = \frac{1}{\sqrt{2}}\left[\psi_{1s}(a) + \psi_{1s}(b)\right] \qquad (16\cdot18\text{-}3a)$$

and
$$\psi_u(\mathbf{R}, \mathbf{r}) = \frac{1}{\sqrt{2}}\left[\psi_{1s}(a) - \psi_{1s}(b)\right] \qquad (16\cdot18\text{-}3b)$$

$\psi_{1s}(a)$ and $\psi_{1s}(b)$ are the atomic orbitals with the electron in the nuclear electric fields of a and b respectively.

The first of these functions corresponds to the σ_g state while the second to the σ_u state.

Using the wave functions (16.18-3a) and (16.18-3b) as trial functions we can then carry out a variational calculation to find the energy E_{mol} of the systems as a function of the internuclear separation R:

$$E_{mol}(R) = \frac{<\psi(\mathbf{R},\mathbf{r})|H|\psi(\mathbf{R},\mathbf{r})>}{<\psi(\mathbf{R},\mathbf{r})|\psi(\mathbf{R},\mathbf{r})>} \qquad (16\cdot18\text{-}4)$$

where ψ can be ψ_g or ψ_u. For the two cases we have the energies E_g and E_u which can be calculated by solving the determinantal equation

$$\begin{vmatrix} J-E & K-IE \\ K-IE & J-E \end{vmatrix} = 0 \qquad (16\cdot18\text{-}5)$$

where

$$J = <\psi_{1s}(a)|H|\psi_{1s}(a)>$$
$$= <\psi_{1s}(b)|H|\psi_{1s}(b)> \qquad (16\cdot18\text{-}6)$$
$$K = <\psi_{1s}(b)|H|\psi_{1s}(a)>$$
$$= <\psi_{1s}(a)|H|\psi_{1s}(b)> \qquad (16\cdot18\text{-}7)$$

and
$$I = <\psi_{1s}(a)|\psi_{1s}(b)> \qquad (16\cdot18\text{-}8)$$

Detailed calculation gives the following expressions for the energies in the two cases (see *Quantum Theory of Molecules and Solids* Vol. I by J.G. Slater :

$$E_g(R) = E_{1s} + \frac{(1+R)\exp(-2R) + (1-2R^2/3)\exp(-R)}{R\{1 + (1+R+R^2/3)\exp(-R)\}} \qquad (16\cdot18\text{-}9)$$

$$E_u(R) = E_{1s} + \frac{(1+R)\exp(-2R) - (1 - 2R^2/3)\exp(-R)}{R\{1 - (1 + R + R^2/3)\exp(-R)\}} \quad (16\cdot18\text{-}10)$$

Fig. (16·22) shows the variation of $E_g - E_{1s}$ calculated from Eq. (16·17-9) as function of R (dashed curve) and is compared with more exact calculation (solid line) by numerical solution method (see below). There is qualitative agreement on the nature of variation with R.

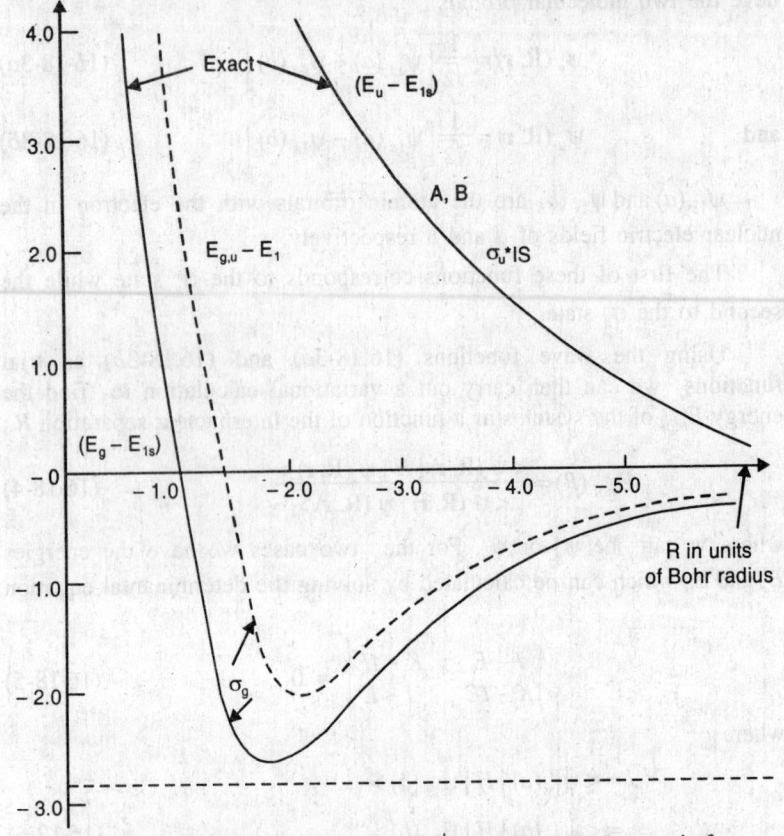

Fig. 16.22. Potential energy curves of H_2^+ ion. The dashed curve is for σ_{g1s} state, obtained by LCAO approximation; the the soild curves are obtained by more exact method.

The curve shows a minimum at the internuclear separation $R_0 = 1.32$ Å. Detailed comparison with the value obtained from more exact calculations shows the lack of agreement between the two cases. The molecular orbital corresponding to this state σ_g is called a *bonding orbital* and represents approximately the ground state σ_{g1s}. σ_{u1s} is antibonding state.

The variation of $E_u - E_{1s}$ on the other hand does not show any minimum and is positive all the way as R decreases from ∞. See Fig. 16·22.

It corresponds to the *antibonding orbital*.

Exact solution :

It may be noted that the Schrödinger equation for the hydrogen molecular ion H_2^+ admits of accurate numerical solutions by using hyperbolic coordinates

$$\xi = \frac{r_a + r_b}{R}, \eta = \frac{r_a - r_b}{R}, \varphi$$

φ is the azimuthal angle. It is possible to separate the variables and write the solution as

$$\psi = X(\xi) Y(\eta) \exp(im\varphi) \qquad (16\cdot18\text{-}11)$$

In terms of the hyperbolic coordinates, the wave equation becomes

$$\frac{\partial}{\partial \xi}(\xi^2 - 1)\frac{\partial \psi}{\partial \xi} + \frac{\partial}{\partial \eta}(1 - \eta^2)\frac{\partial \psi}{\partial \eta} + \left(\frac{1}{\xi^2 - 1} + \frac{1}{1 - \eta^2}\right)\frac{\partial^2 \psi}{\partial \varphi^2}$$
$$+ \frac{2mR^2}{\hbar^2}\left[\frac{E}{4}(\xi^2 - \eta^2) + \frac{1}{4\pi\epsilon_0}\frac{e^2}{R}\left(\xi - \frac{\xi^2 - \eta^2}{4}\right)\right]\psi = 0$$

$$(16\cdot18\text{-}12)$$

Separation of the variables yield the following equations (see *Introduction to Quantum Mechanics* by Pauling and Wilson)

$$\frac{d}{d\xi}(\xi^2 - 1)\frac{dX}{d\xi} + \left(-\lambda \xi^2 + 2D\xi - \frac{m^2}{\xi^2 - 1} + \mu\right)X = 0 \qquad (16\cdot18\text{-}13)$$

$$\frac{d}{d\eta}(1 - \eta^2)\frac{dY}{d\eta} + \left(\lambda \eta^2 - \frac{m^2}{1 - \eta^2} - \mu\right)Y = 0 \qquad (16\cdot18\text{-}14)$$

$$\lambda = -\frac{mR^2}{2\hbar^2}(E - e^2/4\pi\epsilon_0 R)$$

and $$D = R/a_0$$

where a_0 is the Bohr radius.

The solutions, as we have seen, yield the two functions σ_{g1s} and σ_{u1s} which have been calculated numerically by Burrau and later more accurately by Hylleraas and by Jaffe'. These are plotted in Fig. 16.22 (solid curves). The equilibrium values of the internuclear separation in the σ_{g1u} state comes out to be 1.06 Å and the dissociation energy is 2.79 eV.

Similar calculations can be made for any one electron diatomic molecule ion, both homonuclear and heteronuclear.

16·19 Hydrogen molecule

(i) Heitler-London method

The potential energy in this case is (see Fig. 16·23)

$$V = \frac{e^2}{4\pi\epsilon_0}\left[\frac{1}{r_{12}} - \frac{1}{r_{a1}} - \frac{1}{r_{b1}} - \frac{1}{r_{a2}} - \frac{1}{r_{b2}} + \frac{1}{R}\right] \qquad (16\cdot19\text{-}1)$$

The wave equation $\hat{H}\psi = E\psi$ then becomes

$$\nabla^2\psi + \frac{2m}{\hbar^2}(E - V)\psi = 0 \qquad (16\cdot19\text{-}2)$$

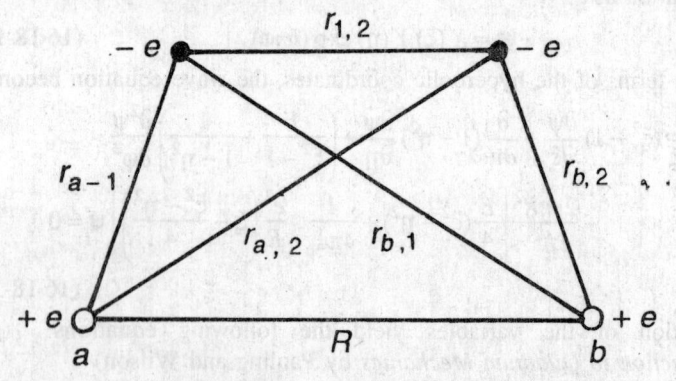

Fig 16·23. Coordinates used for hydrogen molecule.

Heitler and London assumed that the wave function of the H_2 molecule can be written as a linear combination of the wave funciton of the two separated H atoms in the 1s state properly antisymmetrized by multiplying by the spin functions χ_{sm_s}.

We thus have for the lowest state of H_2

$$\psi_{mol} = \frac{1}{2}\{\psi_{1s}(a1)\psi_{1s}(b2) + \psi_{1s}(a2)\psi_{1s}(b1)\}\chi_{s00}(1,2) \qquad (16\cdot19\text{-}3)$$

for the spin singlet state $\chi_{00}(1, 2)$ of the two electrons given by Eq. (13.7-9). $\psi_{1s}(a1)$ etc. are the atomic orbitals for the 1s state when the electron 1s in the nuclear electric field of the atom a etc.

The other possible antisymmetric wave function would be

$$\psi_{mol} = \frac{1}{2}\{\psi_{1s}(a1)\psi_{1s}(b2) - \psi_{1s}(a2)\psi_{1s}(b1)\}\chi_{1m_{s(1,2)}} \qquad (16\cdot19\text{-}4)$$

where χ_{1m_s} gives the spin triplet wavefunction with $m_s = 0, \pm 1$ (see Eq. 13·7-9).

Molecular Spectra

Using the functions ψ_{mol} given by Eqs. (16·19-3) or (16·19-4) as trial function we can then apply the variational method to calculate the possible energies of the molecule.

The two states given by Eqs. (16·19-3) and (16·19-4) are respectively the $^1\Sigma_g^+$ and $^3\Sigma_u^+$ states. The first is the *gerade* state while the second is the *ungerade* state. The orbital angular momentum component along the molecular axis is $\Lambda = m = 0$. The first is a singlet ($S = 0$, $m_S = 0$) while is the second is a triplet with $S = 1$, $m_S = 1, 0, -1$. Both are even (+) states.

The engeries for the two states are found by solving the determinantal equation:

$$\begin{vmatrix} J - E & K - I^2 E \\ K - I^2 E & J - E \end{vmatrix} = 0 \qquad (16\cdot19\text{-}5)$$

where I, J, K are given by

$$I = <\psi_{1s}(a1) | \psi_{1s}(b1)> = <\psi_{1s}(a2) | \psi_{1s}(b2)> \qquad (16\cdot19\text{-}6)$$

$$J = <\psi_{1s}(a1)\,\psi_{1s}(b2) | H | \psi_{1s}(a1)\,\psi_{1s}(b2)>$$

$$= <\psi_{1s}(b1)\,\psi_{1s}(a2) | H | \psi_{1s}(b1)\,\psi_{1s}(a2)> \qquad (16\cdot19\text{-}7)$$

$$K = <\psi_{1s}(a1)\,\psi_{1s}(b2) | H | \psi_{1s}(b1)\,\psi_{1s}(a2)> \qquad (16\cdot19\text{-}8)$$

Solving Eq. (16·19-5), the following energy values are obtained:

$$E_g = 2E_{1s} + \frac{J}{1 + I^2} + \frac{K}{1 + I^2} + \frac{e^2}{4\pi\epsilon_0 R} \qquad (16\cdot19\text{-}9)$$

$$E_u = 2E_{1s} + \frac{J}{1 - I^2} + \frac{K}{1 - I^2} + \frac{e^2}{4\pi\epsilon_0 R} \qquad (16\cdot19\text{-}10)$$

The first term in both equations is the total energy of the two separated hydrogen atoms.

The last terms are due to the Coulomb interaction between the two nuclei.

The term J (Eq. 16·19-7) is called the *Coulomb integral* while K (Eq. 16·19-8) is called the *exchange intergral*. K is found to be negative so that $^1\Sigma_g^+$ state with energy E_g is lower than the $^3\Sigma_u^+$ state with energy E_u.

The equilibrium distance for the $^1\Sigma_u^+$ state comes out to be $R_0 = 0.87$ Å while the dissociation energy is $D = 3.14$ eV. These are to be compared with the experimental values $R_0 = 0.74$ Å and $D = 4.75$ eV. $^3\Sigma_u^+$ state is not a bound state.

The energies calulated from Hetler-London theory for both the states are shown by the dashed curves in Fig. 16·24.

The H-L theory when applied to other homonuclear diatomic molecules is also known as the valence-bond method.

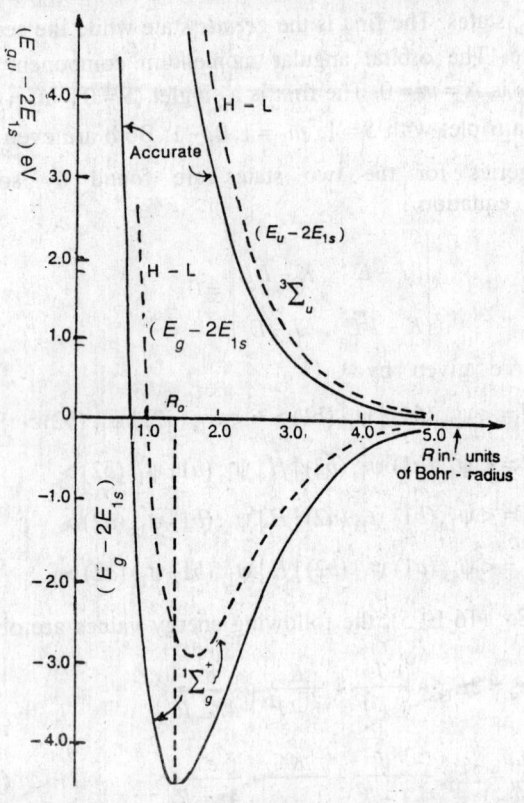

Fig. 16.24. Potential Energy curves for H_2 molecule for the lowest $^1\Sigma_g^+$ and $^3\Sigma_u^+$ state. The dashed curve is obtained by Heitler-London method. The solid curves are obtained by more accurate method.

Improvements of Hetler-London Theory :

In the original Heitler-London theory, the atomic orbitals used were those for the hydrogen atom. If however the atomic orbitals include a variational parameter α and are written as $\sqrt{\alpha^3/\pi}\ \exp(-\alpha r)$ in atomic units, then by varying the parameter α, the energy can be minimised. This was first done by S.C. Wang and later extended by N. Rosen. These result in considerable improvements over the original H-L method.

Another modification suggested by Rosen was that the polariztion of the atoms towards their neighbours should be taken into account. This would change the spherical atomic orbitals. A somewhat similar modification suggested by Gurnee and others takes into account the movement of the atomic orbital of each atom bodily towards the neighbouring atom so that the nucleus would no longer be the centre of symmetry (*method of floating wave function*). Both these modifications have improved the calculated energy value considerably.

(*ii*) **Molecular orbital method :**

In this method the lowest state of H_2 is written as a linear combination of the two H_2^+ orbitals ψ_g (H_2^+) and ψ_u (H_2^+) obtained by the LCAO method and given by Eqs. (16·18-3a) and (16·18-3b) respectively. These have to be properly antisymmetrized as in the Heitler-London method by combining with the spin functions χ_{Sm_S}.

Writing the H_2^+ orbitals for the two H_2^+ ions as $\psi_g(1)$, $\psi_u(1)$ and $\psi_g(2)$, $\psi_u(2)$ we then have the following four possible wave functions for H_2:

$$\psi_I(1,2) = \psi_g(1)\,\psi_g(2)\,\chi_{00}(1,2) \tag{16·9-11}$$

$$\psi_{II}(1,2) = \psi_u(1)\,\psi_u(2)\,\chi_{00}(1,2) \tag{16·9-12}$$

$$\psi_{III}(1,2) = \frac{1}{\sqrt{2}}\Big[\chi_g(1)\,\psi_u(2) + \psi_g(2)\,\psi_u(1)\Big]\chi_{00}(1,2) \tag{16·9-13}$$

$$\psi_{IV}(1,2) = \frac{1}{\sqrt{2}}\Big[\psi_g(1)\,\psi_u(2) - \psi_g(2)\,\psi_u(1)\Big]\chi_{1m_S}(1,2) \tag{16·9-14}$$

where m_S can be +1, 0, −1 in the last case.

Of these ψ_I and ψ_{II} both represent $^1\Sigma_g^+$ states while ψ_{III} gives a $^1\Sigma_u^+$ state and ψ_{IV} gives a $^3\Sigma_u^+$ state.

Using the Hamiltonian

$$\hat{H} = -\frac{\hbar^2}{2m}\nabla_1^2 - \frac{\hbar^2}{2m}\nabla_2^2 + V \tag{16·9-15}$$

where V is given by Eq. (16·9-7), the energy can be found by the variational method. For the $^1\Sigma_g^+$ state, using the wave function ψ_I, we then get

$$E_I = <\psi_I|H|\psi_I> \tag{16·9-16}$$

When written explicitly we have

$$\psi_I = \frac{1}{2}\{\psi_{1s}(a1) + \psi_{1s}(b1)\}\{\psi_{1s}(a2)\,\psi_{1s}(b2)\}\chi_{00}(1,2)$$

$$= \frac{1}{2}\{\psi_{1s}(a1)\psi_{1s}(a2) + \psi_{1s}(b1)\psi_{1s}(b2)\}\chi_{00}(1,2)$$

$$+ \frac{1}{2}\{\psi_{1s}(a1)\psi_{1s}(b2) + \psi_{1s}(a2)\psi_{1s}(b1)\}\chi_{00}(1,2) \qquad (16\cdot19\text{-}17)$$

Thus ψ_I can be split into two parts :

$$\psi_I^{(ion)} = \frac{1}{2}\{\psi_{1s}(a1)\psi_{1s}(a2) + \psi_{1s}(b1)\psi_{1s}(b2)\}\chi_{00}(1,2) \qquad (16\cdot9\text{-}18)$$

$$\psi_I^{(cov)} = \frac{1}{2}\{\psi_{1s}(a1)\psi_{1s}(b2) + \psi_{1s}(a2)\psi_{1s}(b1)\}\chi_{00}(1,2) \qquad (16\cdot9\text{-}19)$$

In $\psi_I^{(ion)}$ both the electrons 1 and 2 are attached to the same nucleus (either a or b) which means an unequal distribution of the electronic charge between the two atoms. This gives rise to the so called ionic bonding. On the other hand in $\psi_I^{(cov)}$ the two electrons 1 and 2 are associated with the two different nuclei a and b (or b and a) respectively and produces covalent bonding. When separated by a large distance it yields two isolated H atoms.

If either $\psi_I^{(ion)}$ or $\psi_I^{(cov)}$ is used for the calulation of the ground state energy of the molecule, the agreement with experimental value is very poor. However a linearc ombination of the two functions ψ_I and ψ_{II} for the ground state $^1\Sigma_g^+$ in the form $\psi_I + \lambda\,\psi_{II}$ with λ as the variational parameter gives much better result when λ is determined by minimising the energy by the usual variational method. Thus we get $R = 0.749$ Å and $D = 4.00$ eV which are in better agreement with the experimental values given earlier (See *Physics of Atoms and Molecules by* Bransden and Joachain).

Problems

1. The interatomic distance of the CO molecule is 1·13 Å. What is the energy difference between the $j = 0$ and $j = 1$ rotaional states of this molecule ?

   (0·000479 eV)

2. Show that the Morse potential (Eq. 16·4-17) is minimum at $r = r_0$. If the nature of the potential is the same as that for a harmonic oscillator near this point, then express the proper frequency ν_0 of the oscillator in terms of the constants D and β of the Morse potential.

3. Show that for pure rotational spectrum of a diatomic molecule the wave number difference between the consecutive lines $\Delta\bar{\nu}_r = \hbar/2\pi Ic$.

   If $\Delta\bar{\nu}_r = 4050$ m^{-1} for the HF molecule what is the iteratomic distance of this molecule ? (0·937 Å)

4. The most intense vibrational bands of CO and HCl molecules have the wave numbers 2.143×10^5 m^{-1} and 2.886×10^5 m^{-1} respectively. Calculate the force constants of these molecules. (1855 N/m ; 480.7 N/m)

5. The difference in the term values between the ground state and the first excited vibrational state of HCl molecule is 2.886×10^5 m^{-1}. Calculate the ratio of the number of molecules in these two levels at 300 K. (9.76×10^{-6})

6. The term value of the first rotational level of the molecule CO is 384 m^{-1}. Calculate the ratio of the number of molecules in this level to that in the ground state at 300 K. (0.9816)

7. Show that the lines of the Q branch of the electronic band of a diatomic molecule can be represented by the formula

$$Q(j) = \bar{v}_0 + (B' - B'')j(j+1)$$

Here B' and B'' are the rotational constants for the two electronic states and $\bar{v}_0$ is the wave number of the band-origin.

8. Prove that for diatomic molecules the wave number difference between the consecutive pure rotational Raman lines is $4B$ while the wave number shift for the line originating from the 2-0 transition is $6B$ where B is the rotational constant.

9. The fundamuntal vibrational wave number of HCl molecule is 2990 cm^{-1}. Calculate the energies of the lowest and the first excited vibrational levels of this molecule. Also find the force constant.

10. The dissociation energy of H_2 molecule is 4.74 eV while its fundamental vibrational wave number is 4395 cm^{-1}. Assuming the depth of the potential well and the force constant for the deuterium molecule D_2 to be equal to those for H_2 molecule, find the dissociation energy of the D_2 molecule.

(4.56 eV)

11. Show that the interatomic distance in a diatomic molecule increases by ΔR due to the rotation of the molecule given by

$$\Delta R = \frac{\hbar^2}{\mu k} \cdot \frac{j(j+1)}{R_0^3}$$

μ is the reduced mass, k is the force constant, j is the rotational quantum number and R_0 in the equilibrium distance between the nuclei when there is no rotation.

17

STRUCTURE AND PROPERTIES OF SOLIDS

17·1 Introduction

It is well-known that matter can exist in three states of aggregation, solid, liquid and gaseous. Substances in the solid state have the properties of *rigidity* and *elasticity*. Application of external tensile or shearing force generates elastic stresses in them. As a result, the solids have definite shape and size.

Solids can be grouped into two classes, the amorphous and the crystalline. In amorphous substances, even though the atoms or the molecules are strongly bound to one another, there is no regularity in their arrangement. As a result the molecules in them are arranged somewhat like the molecules in a liquid. For this reason, they are more like a super-cooled liquid (example *glass*).

On the other hand, in crystalline solids, a definite regularity is observed in the arrangement of the atoms or molecules. This was briefly mentioned in § 8·17 where the method of measurement of the x-ray wavelengths by diffraction from the regularly spaced atoms or groups of atoms in the crystal lattice was discussed. Because of this regularity in the arrangement of the atoms or molecules in crystals, it is easier to explain the various properties of crystalline solids than those of amorphous solids.

17·2 Interatomic force and the classification of solids

Two opposing forces act between the neighbouring atoms in a solid—attrative and repulsive. Since all solids (and liquids) occupy finite volume, the distances between their atoms cannot exceed a certain limiting value. This can happen only if an attractive force acts between them. Further if a high pressure is applied upon a solid or a liquid, there is very little decrease in their volume. This shows that when the atoms in them come close together, there must be a strong repulsion between them.

The general nature of the variation of interatomic attractive and repulsive force with the distance between the atoms are shown graphically in Fig. 17·1. In the diagram, the attractive force is shown negative while the repulsive force is shown positive. When the interatomic distance r is relatively large, the attractive force F_a predominates over the repulsive. With the decrease of r this force increases rapidly. On the other hand, when the interatomic distance is very short, the repulsive force F_r becomes predominant. With further decrease of r, the repulsive force increases more sharply than the attractive force. At a definite interatomic distance $r = r_0$, F_a and F_r cancel one another and the resultant force becomes zero. The nature of the variation of the resultant force is also shown Fig. 17·1 by the dashed line.

If instead of the interatomic force, we plot graphically the variation of the interatomic potential energy U with r, we obtain the graphs shown in Fig. 17·2. The negative attractive potential energy decreases with increasing r. This means that its derivative is positive which makes the attractive fore negative since force is the negative of the derivative of the potential energy : $F = -\partial U/\partial r$. On the other hand, the repulsive potential energy which is positive decreases very rapidly with increasing r which means that its derivative has large negative value. Hence the repulsive force is positive as shown in Fig. 17·1. At $r = r_0$, where the resultant force is zero, the resultant interatomic potential energy (shown by the dashed curve in Fig. 17·2) attains a minimum and rises on either side of this point. Thus the resultant potential energy has the appearance of a potential well. $r = r_0$ represents the position of stable equilibrium of the two atoms.

Fig. 17·1 Variation of interatomic forces in a solid.

It may be noted that there can be stable equilibrium in the positions of the two atoms if the range of the repulsive force is shorter than that of the attractive force.

In a crystalline solid, the atoms are arranged in a regular pattern, each atom (or a group of atoms) occupying the positions of stable quilibrium as mentioned above. When the atoms are displaced from these positions, forces come into play which

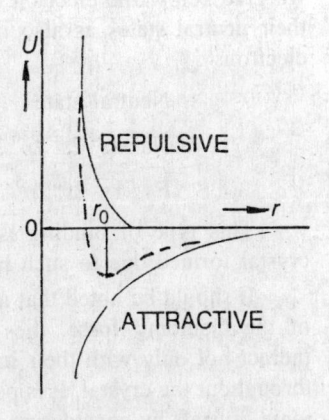

Fig. 17·2. Variation of interatomic potential energy in a solid.

tend to restore their equilibrium positions. As a result atoms vibrate about these positions.

In solids, the force acting upon the atoms is electrostatic in nature. The nature of the force depends upon the electron distribution in the atoms. On this basis, the crystalline solids are generally gouped into five classes : (a) Ionic crystals, e.g., NaF, KI etc.; (b) Covalent crystals, e.g.; diamond, SiC etc.; (c) Metallic crystals, e.g.; Cu, Ag, Mg, etc.; (d) Molecular crystals, e.g., solidified argon and different types of organic crystals; (e) Hydrogen-bonded crystals, e.g., the crystals of ice and HF.

We shall discuss briefly some important properties, nature of binding etc. of these different types of crystals below.

(a) *Ionic crystals* : In the periodic table, the atoms of the elements situated just before or after the inert gases have the tendency either to acquire or to lose an electron, so that their electron configuration becomes similar to that of the inert gas atoms. In the atoms of the alkaline elements just after the inert gas (e.g., Na or K), the outermost electron is very loosely bound in an s-shell. On the other hand, in the atoms of the halogens just before the inert gases (e.g., F or Cl) there is a deficit of one electron in the outermost p orbit.

When two such atoms (e.g., Na and F) come close together, the alkali atom loses the surplus valence electron to the electron-deficient halogen atom and becomes transformed into a positive ion (e.g., an Na^+ ion), thereby achieving the inert gas electron configuration. The halogen atom on acquiring the extra electron in its outermost p-orbit becomes transformed to a negative ion (e.g., F^- ion) and also achieves the inert gas electron configuration. These two oppositely charged ions attract each other according to Coulomb's law and become bound. As an illustration we give below the electron configurations of the atoms in NaF, both in their neutral states as also in their ionized states after the exchange of electrons :

Neutral state	Ionized state
Na : $1s^2 2s^2 2p^6 3s^1$	Na^+ : $1s^2 2s^2 2p^6$
F : $1s^2 2s^2 2p^5$	F^- : $1s^2 2s^2 2p^6$

This type of binding is known as the *ionic bond* and the type of crystal formed due to such bonds is known as an *ionic crystal*.

It should be noted that in ionic crystals, due to the long range nature of the Coulomb force, the ions located at the different lattice points ineract not only with their immediate neighbours but with all other ions throughout the crystal. It is possible to calculate the binding energy of the ionic crystals by considering such overall interaction between all the ions and the repulsion at very short distances (see § 17·4).

Different inorganic compounds e.g., NaF, KCl etc. form ionic crystals. In these crystals, the positive and negative ions are arranged in a regular array at the lattice points.

Ionic crystals have usually low electrical conductivity at low temperatures but exhibit ionic conductivity at higher temperatures. Their absorption coefficients are high for infra-red radiation. They can be cleaved easily. The melting points are usually high.

(b) *Covalent crystal* : In these crystals, the neighbouring atoms share their electrons which results in producing *homopolar* or *covalent bonds* in them.

As examples of covalent bonds we may consider the binding of the atoms in the hompolar molecules e.g., H_2, N_2, O_2 etc. When two hydrogen atoms come very lose together, the electrons in them are no longer bound

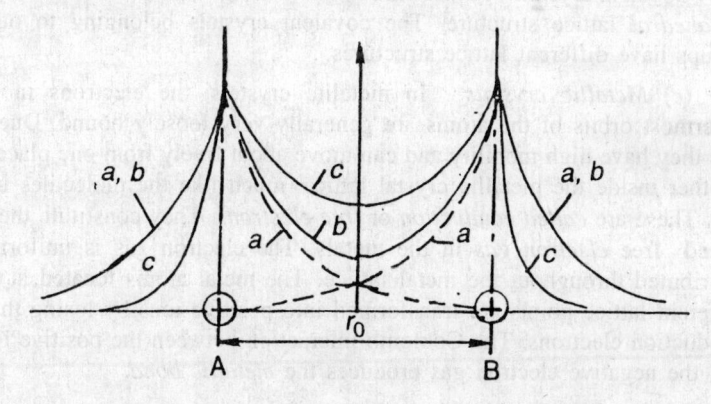

Fig. 17·3. Probability of locating two electrons in H_2 (*c*). Curve *a* is for two isolated H atoms. Curve *b* is for the two H atoms brought nearer.

to their respective nuclei. Instead, both of them are continually exchanged between the two atoms. As a result the probability of the electrons being found in a region between the atoms A and B is usually much greater than the probability of their location outside this region (see Fig. 17·3) as shown by the curve *c*.

Under this condition, the potential energy of the composite system made up of the two hydrogen atoms is reduced which gives rise to an attractive force between the two atoms. This force is known as the *exchange force*. It can be shown on the basis of the quantum mechanical theory, that the exchange interaction between the two hydrogen atoms becomes attractive when the spins of the two electrons are *antiparallel*. Such a state is known as the *symmetric state*. On the other hand, if the spins of the two electrons are parallel, then the exchange force becomes repulsive. The corresponding state is called *antisymmetric* (see § 17·10 and 19·6).

Similar exchange of the valence electrons takes place between the neighboring atoms in covalent crystals which gives rise to the strong attractive interaction between them. As an example, in the diamand crystal, the carbon atoms have four electrons ($2s^2\ 2p^2$) each in their outermost orbits. But the $n=2$ or L shell requires eight electrons for complete filling up. Each carbon atom shares one electron with four of its neighbours to fulfil this requirement. This produces the strong covalent binding between the neighbouring atoms.

Diamond, carborandum, silicon-carbide (SiC) are examples of covalent crystals. They have great hardness and cannot be cleaved easily. Their eletrical and heat conductivities are low. Their melting points are usually quite high.

Covalent crystals of diamond (C) and other group four elements have *tetrahedral* lattice structure. The covalent crystals belonging to other groups have different lattice structures.

(*c*) *Metallic crystals* : In metallic crystals, the electrons in the outermost orbits of the atoms are generally very loosely bound. Due to this they have high mobility and can move about freely from one place to another inside the metallic crystal lattice, much like the molecules in a gas. These are *called conduction* or *free electrons*. They constitute the so called free *electron gas* in the metals. The electron gas is uniformly distributed throughout the metal lattice. The metal atoms located at the different lattice points are transformed into positive ions by losing these conduction electrons. The Coulomb interaction between the positive ions and the negative electron gas produces the *metallic bond*.

In metallic crystals, three types of lattice structure are usually observed : face-centred cubic, hexagonal close-packed and body centred cubic. Metals are characterised by high electrical and thermal conductivities.

(*d*) *Molecular crystals* : The bond between the neighbouring atoms or molecules in these crystals is due to weak Van der Waal force between the dipole moments associated with them. Due to the spherically symmetric charge distribution, the inert gas atoms do not possess any permanent electric dipole moment. But the positively charged nuclei of these atoms and negatively charged orbital electrons in them produce rapidly time-varying dipole moments. The electrostatic interaction between these dipoles in the condensed states produce weak binding between the neighbouring atoms in these crystals. This type of force is known as the *dispersion* force.

Molecular bonds are also observed in many organic crystals. The molecules of some of them possess permanent dipole moments. The dipole-dipole interaction between the neighbouring molecules produce attraction between them. In addition, these permanent electric dipole moments induce dipole moments in the neighbouring molecules. The interaction between the permanent and induced moments also produces an attractive force. All these interactions along with the dispersion force mentioned above produce the binding in such crystals. The combination of these interactions is known as Van der Waal force. The interaction energy due to the Van der Waal force is proportional to r^{-6} where r is the distance between the molecules, and hence the force is quite weak.

These crystals have low melting and boiling points. They are very compressible. They have usually face-centred cubic structure.

(e) *Hydrogen bonded crystals* : In addition to the above four types, a fifth type, known as the *hydrogen bond* is found to be present in some crystals. An example is the crystal of ice. Such bonds are usually produced between a strongly electro-negative atom, *e.g.*, oxygen and the hydrogen atom. In the ice crystal, the oxygen atom of an ice molecule draws an electron from a hydrogen atom of a neighbouring molecule and

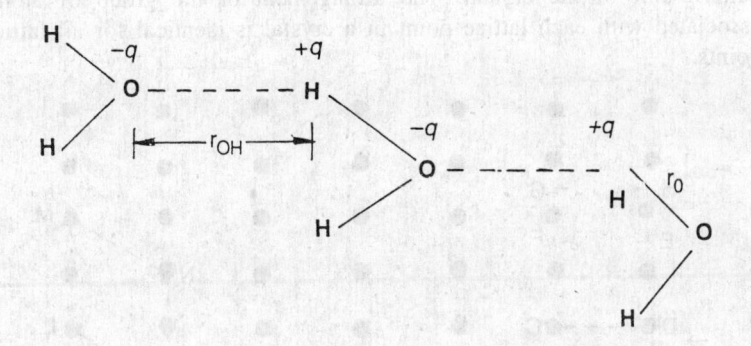

Fig. 17·4 Hydrogen bond.

becomes negatively charged. In the process, the hydrogen atom becomes positively charged. Together they constitute an electric dipole. The electrostatic interaction between these two oppositely charged ions produces the hydrogen bond. This is illustrated in Fig. 17·4 in which the hydrogen bonds between the neighbouring molecules is represented by the horizontal dashed lines. The charges on the ions are shown as $\pm q$.

In the ice crystal, each H_2O molecule is surrounded by four neighbouring molecules. They are located at the vertices of a regular tetrahedron and are bound together by hydrogen bond.

Hydrogen-bonded molecules have a tendency to polymerize (i.e., to form groups of many molecules). It may be noted that the strength of the O-H bond between the hydrogen-bonded neighbouring molecules in the ice crystal (distance $r_{OH} = 2.76$ Å) is much weaker than the strength of covalent O-H bond in an H_2O molecule itself (distance $r_0 = 0.96$ Å).

It may be mentioned that in an actual crystal none of the above bonds is found to exist exclusively. In general a mixture of two or more bonds is found to be present in a crystal. However usually one type of bond is found to dominate over the others in a particular type of crystal.

17·3 Lattice structure of crystals

The external structure of a crystalline solid shows a definite symmetry in its shape. From this, one can infer that the atoms or groups of atoms in a crystal must be arranged in a regular pattern in the form of a lattice. A lattice is an arrangement of an infinite number of points in the form of a two dimensional or a three dimensional network. In an ideal case the points are arranged along parallel lines. The appearance of the lattice as seen from any *lattice point* is identical with that seen from any other point in the lattice. In an actual crystal, an atom or a group of atoms is to be associated with each lattice point. These constitute the basis or the pattern unit of the crystal. The arrangement of the group of atoms associated with each lattice point in a crystal is identical for all lattice points.

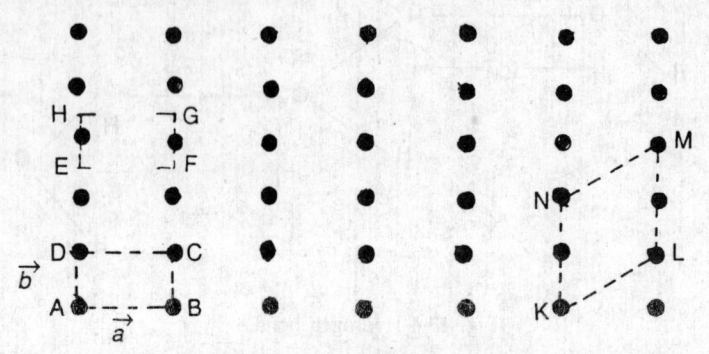

Fig. 17·5 Two dimensional lattice.

In order to understand the basic characteristics of the crystal structure, we show a two dimensional arrangement of the lattice points ("two dimensional crystal lattice") in Fig. 17·5, Each of the lattice points is associated with a *pattern unit* in the correspoding hypothetical two dimensional crystal which is shown by two small circles in Fig. 17·6. It should be noted that the nature and arrangement of the atoms in the pattern units are identical for all the lattice points.

In Fig. 17·7, we illustrate how the entire two dimensional lattice can be built up by repeatedly translating the parallelogram ABCD through the vector distances **a** and **b** represented by the two sides AB and AD respectively. If the translation is through the vector distance **a**, then we get the parallelogram BEFC as shown in Fig, 17·7a. On the other hand, translation through the vector distance **b** results in generating the

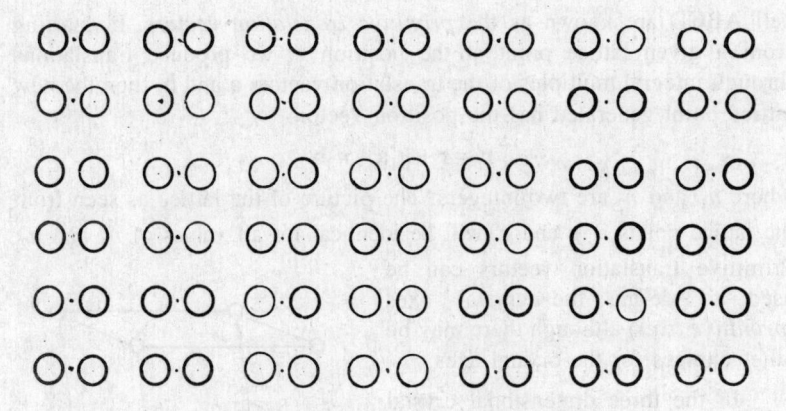

Fig. 17·6 Pattern unit at the lattice points.

parallelogram DCGH (Fig. 17·7b). If ABCD is first translated through **a** and then through **b**, we get the parallelogram CFIG as shown in Fig. 17·7c. In each of the above diagrams, the arrangement of the lattice points in the parallelogram produced by the translations is identical with the arrangement in the initial position ABCD. This is knows as the *translational symmetry* which is responsible for the periodic structure of the lattice.

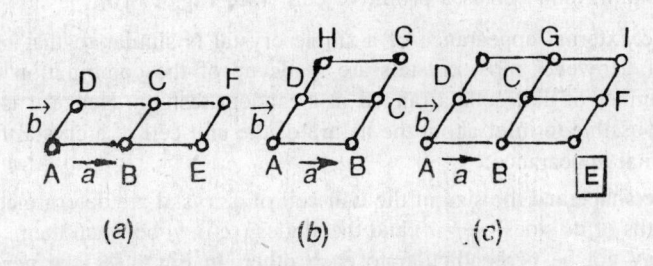

Fig. 17·7 Translational symmetry in two dimensional lattice.

The parallelogram ABCD is called the unit cell. The unit cell in a crystal lattice may be chosen in different manners. For example, in Fig. 17·5, instead of ABCD, we could have chosen the parallelograms EFGH or KLMN as the unit cell. Repeated translations of these would also result in reproducing the entire lattice structure shown in the figure. However it is the unit cell ABCD which has a lattice point at each of its four corners only. Such a unit cell is called the *primitive cell*. Obviously the unit cells EFGH or KLMN are not primitive cells. In EFGH, the lattice points are located at the centres of the sides EH and FG and not at the corner points. In KLMN, in addition to the lattice points at the corners there are lattice points also at the centres of the sides KN and LM.

In Fig. 17·7, the vectors **a** and **b** represented by the sides of the unit cell ABCD are known as the *primitive translation vectors*. If, starting from a given lattice point at the position **r**, we produce translations through integral multiples of the translation vectors **a** and **b**, then the new lattice point generated has the position vector

$$\mathbf{r'} = \mathbf{r} + n_1\mathbf{a} + n_2\mathbf{b}$$

where n_1 and n_2 are two integers. The picture of the lattice as seen from the lattice points at **r** and **r'** will be identical for all values of n_1 and n_2. Primitive translation vectors can be used to define the crystal axes (*primitive axes*) although there may be other choices for the orystal axes.

In the three dimensional crystal lattice, (space lattice) there are three primitive translation vectors **a**, **b** and **c**. The parallelopiped with the sides a, b, c along these three vectors constitute the unit cell in this case (see Fig. 17·8). The entire three dimensional lattice can be built up by repeated translations of the unit cell through integral multiples of the vectors **a**, **b** and **c**. If there is a lattice point only at the eight corner points of the above parallelopiped, the corresponding unit cell is a primitive cell (see Fig. 17·10).

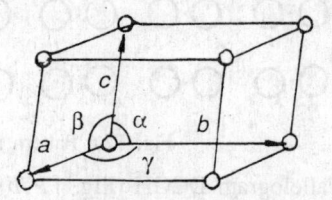

Fig. 17·8. Unit cell.

The external appearance of a single crystal is similar to that of its unit cell. However most crystals are made up of the combination of a large number of unit cells arranged in a random fashion. Hence it is not always possible to infer about the nature of the unit cell of a crystal from its external appearance.

The shape and the size of the unit cell of a crystal are determined by the lengths of its sides (a, b, c) and the angles (α, β, γ) between them. The sides may not be perpendicular to each other. In Fig. 17·8, the general appearance of the unit cell of a three dimensional crystal lattice is shown.

In 1848, the French crystallographer Bravais was the first to introduce the concept of the three dimensional lattice while explaining the structure of crystals. The Bravais lattice has an infinite number of lattice points in it. As explained above, in an actual crystal an atom or a group of atoms is to be associated with each lattice point. The arrangement of the atoms as seen from any lattice point is identical with that seen from any other point in the lattice. The truth of this statement can be understood with reference to the hypothetical two dimensional crystal shown in Fig. 17·6.

All in all, there are fourteen types of space lattices, differing in their space symmetry. These fourteen types belong to seven different crystal systems as grouped below :

1. Triclinic—one type.
2. Monoclinic—four types (simple, base-centred)
3. Orthorhombic—four types (simple, base-centred, body-centred and face-centred).
4. Hexagonal—one type.
5. Trigonal—one type.
6. Tetragonal—two types (simple and body-centred).
7. Cubic—three types (simple, body-centred and face-entred).

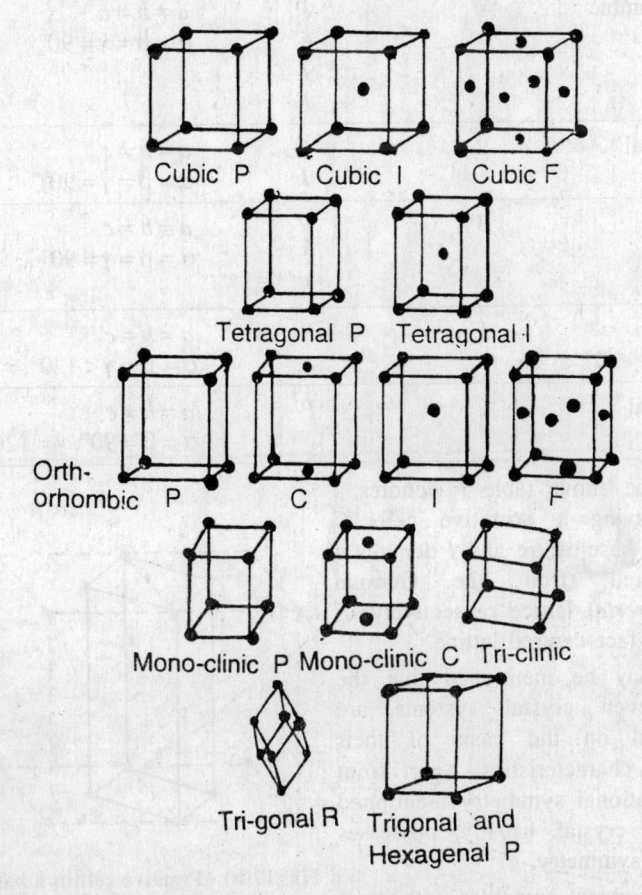

Fig. 17·9. Bravais lattices. P, I, F, C denote respectively a primitive cell, a body-centred, a face-centred and a base-centred lattice.

The fourteen types of Bravais lattices are illustrated in Fig. 17·9. Their important characteristics are listed in Table 17·1. Fig. 17·10 shows the primitive cell of a base-centred monoclinic crystal.

Table 17.1

Crystal system	Number of lattices in the system	Lattice symbol	Nature of the unit cell
Triclinic	1	P	$a \neq b \neq c$ $\alpha \neq \beta \neq \gamma$
Monoclinic	2	P C	$a \neq b \neq c$ $\alpha = \gamma = 90° \neq \beta$
Orthorhombic	4	P C I F	$a \neq b \neq c$ $\alpha = \beta = \gamma = 90°$
Tetragonal	2	P I	$a = b \neq c$ $\alpha = \beta = \gamma = 90°$
Cubic	3	P I F	$a = b = c$ $\alpha = \beta = \gamma = 90°$
Trigonal	1	R	$a = b = c$ $\alpha = \beta = \gamma < 120°, \neq 90°$
Hexagonal	1	P	$a = b \neq c$ $\alpha = \beta = 90°, \gamma = 120°$

In the above table P denotes a lattice having a primitive cell. C denotes a base-centre and I denotes a body-centred (from the German *Innenzentrerte*) lattice repsectively. F denotes a face-centred lattice.

It may be mentioned that the above seven crystal systems are determined on the basis of their symmetry characteristics. Apart from the translational symmetry mentioned above, a crystal usually possesses rotational symmetry.

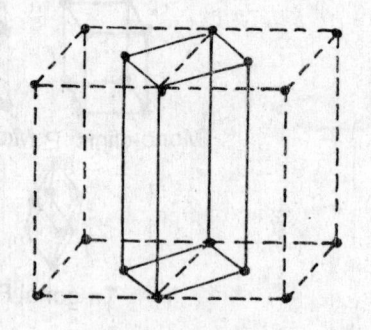

Fig. 17.10. Prmitive cell in a base centred crystal.

If an object coincides with itself exactly when turned about an axis through an angle $2\pi/n$ with $n = 2, 3, 4$ etc., we say that it has *rotational symmetry*. The axis is called the axis of *n-fold rotation*. As an example we notice, that the letter N when truned through an angle 180° about an axis through its centre perpendicular to its plane coincides with the original

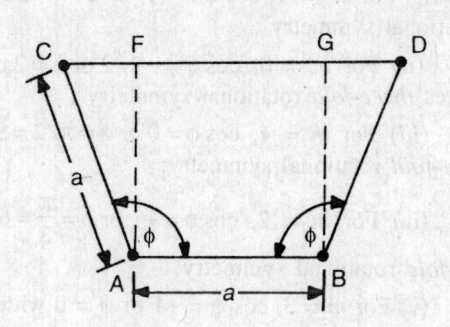

(a) An example of rotational symmetry.

(b) Effect of translational symmetry on rotational symmetry.

Fig. 17.11

(see Fig. 17·11 a). Thus n = 2 in this case so that it possesses *two-fold* rotational symmetry. Again a two dimensional square lattice can be brought into exact coincidence with itself by turning it through the angle 90° about an axis through its centre perpendicular to the plane of the square. So n = 4 in this case and it has *four-fold* rotational symmetry.

It can be proved quite generally, that in the presence of traslational symmetry, the only possible values of n are n = 2, 3, 4 and 6; *i.e.*, only two, three, four and six-fold rotational symmetries are possible.

To prove this, suppose A and B are two lattice points in a two dimensional lattice such that the lattice constant AB = a. Let an n-fold rotation axis pass through the lattice point perpendicular to the plane of the lattice. If the lattice is turned through the angle $\phi = 2\pi/n$ about A, then the lattice point B comes to the position C as shown in Fig. 17·11 (b). On the other hand, if we turn the lattice about B through ϕ then A would occupy the position D. The presence of translational symmetry requires that CD should be an integral multiple of a. i.e.,

$$CD = m \cdot AB = ma$$

Here m is an integer. From the geometry of the figure, since AC = BD = a we have

$$CD = AB + (CF + DG)$$
$$= a + 2a \cos \phi = ma$$

This gives $\qquad \cos \phi = (m-1)/2$

Since $\cos \phi < 1$, the limiting values of $(m-1)$ are :

$$-2 \leq (m - 1) \leq 2$$

Thus m can have the following five values : $m = -1, 0, 1, 2$ and 3. Corresponding to these we get :

(i) For $m = -1$, $\cos \phi = -1$. or $\phi = \pi$. So $n = 2$ which gives *two fold* rotational symmetry ;

(ii) For $m = 0$, $\cos \phi = -1/2$ or $\phi = 2\pi/3 = 120°$. So $n = 3$ which gives *three-fold* rotational symmetry ;

(iii) For $m = 1$, $\cos \phi = 0$ or $\phi = \pi/2 = 90°$. So $n = 4$ which gives *four-fold* rotational symmetry;

(iv) For $m = 2$, $\cos \phi = +\frac{1}{2}$ or $\phi = \frac{\pi}{3} = 60°$. So $n = 6$ which gives *six-fold* rotational symmetry;

(v) For $m = 3$, $\cos \phi = +1$ or $\phi = 0$ which gives $n = 1$ (trivial case.)

Thus *five-fold* rotation axis is impossible with translational symmetry. Also symmetries of higher order than six are not possible.

It may be noted that the system to which a crystal belongs may be determined by noting its external appearance. However the arrangements of the atoms at the lattice points cannot be determined from such observations. It is necessary to perform x-ray diffraction experiment for this purpose.

In addition to the translational and rotational symmetries discussed above, other types of symmetries have to be considered, e.g., mirror symmetry.

17.4 Lattice energy

The minimum energy required to separate all the ions from one another at the different lattice points in a crystal is known as the *lattice energy*.

The lattice energy can be calculated in a fairly simple manner for the ionic crystals by considering both the attractive and repulsive forces between the ions. Since the Coulomb force is of the long range type, the ions located at the different lattice points interact not only with their immediate neighbours but with all other ions throughout the crystal. As there are both positive and negative ions in the crystal, the interaction energy due to the Coulomb force between the ions i and j separated by a distance r_{ij} can be written as $\pm \dfrac{1}{4\pi\varepsilon_0} (e^2/r_{ij})$ where e is the charge on each ion. The + sign is to be taken for the like charges and the − sign for the opposite charges. In addition, one has to consider the positive interaction energy due to the very short range repulsive force which may be written as β/r_{ij}^n where β and n are constants to be determined from the observed values of the compressibility of the crystal and the attice constant which is the equilibrium distance r_0 between the

Structure and Properties of Solids

ions of the crystal. Thus the total potential energy of interaction between the pair of ions i and j is

$$u_{ij} = \frac{\beta}{r_{ij}^n} \pm \frac{e^2}{4\pi\varepsilon_0 r_{ij}} \qquad (17\cdot 4\text{-}1)$$

Total energy of the ith ion due to its interaction with all the other ions is

$$u_i = \sum_j{}' u_{ij} = \sum_j{}' \left(\frac{\beta}{r_{ij}^n} \pm \frac{e^2}{4\pi\varepsilon_0 r_{ij}} \right) \qquad (17\cdot 4\text{-}2)$$

where $\sum'$ represents the sum over all ions except $j = i$.

The nature of variation of u_{ij} with r_{ij} for two oppositely charged ions has the general form shown in Fig. 17.2. From the condition $\partial u_{ij}/\partial r_{ij} = 0$ at $r_{ij} = r_0$ where the potential energy minimum occurs, it is possible to express β in terms of r_0. Writing $r_{ij} = p_{ij}\rho$ and $r_0 = p_{ij}\rho_0$, we get*.

$$\beta = \frac{\alpha e^2}{4\pi\varepsilon_0 n A_n} \rho_0^{n-1} \qquad (17\cdot 4\text{-}3)$$

where
$$A_n = \sum_j{}' 1/p_{ij}^n$$

and
$$\alpha = \sum_j{}' \mp(1/p_{ij}) \qquad (17\cdot 4\text{-}4)$$

Substitution in Eq. (17·4-2) gives

$$(u_i)_{\rho_0} = -\frac{\alpha e^2}{4\pi\varepsilon_0 \rho_0}\left(1 - \frac{1}{n}\right) \qquad (17\cdot 4\text{-}5)$$

Hence the total lattice energy is

$$U_0 = N(u_i)_{\rho_0} = -\frac{N\alpha e^2}{4\pi\varepsilon_0 \rho_0}\left(1 - \frac{1}{n}\right) \qquad (17\cdot 4\text{-}6)$$

α is known as *Madelung constant*. It can be calculated on the basis of simple assumptions. More general methods have also been evolved.

The repulsive potential exponent n in β/r_{ij}^n can be correlated with the compressiblity of the crystal. For NaCl it is found to be of the order of 8 to 10 which shows the very strong short-range character of the repulsive force.

The experimental and theoretical values of the molar lattice energies for a number of ionic crystals are listed in Table 17.2. The agreement between the two sets of values is fairly good.

---

* See *Introduction to Solid State Physics* by Charles Kittel.

Table 17.2

Crystal	U_0 (obs) (J/mole)	U_0 (calc) (J/mole)
NaCl	7.52×10^5	7.54×10^5
KI	6.50×10^5	6.30×10^5
RbBr	6.35×10^5	6.45×10^5
CsI	5.95×10^5	5.85×10^5

In Table 17.3, we list the observed values of the molar lattice energies of some crystals belonging to the different classes discussed in the previous section.

Table 17.3

Type of crystal	Example	Lattice energy (J/mole)
Ionic	NaCl	7.52×10^5
	KI	6.50×10^5
Covalent	Diamond (C)	7.11×10^5
	Silicon	4.46×10^5
	Germanium	3.5×10^5
Metallic	Na	1.07×10^5
	Fe	4.13×10^5
Molecular	Ar	7.74×10^3
	CH_4	10.8×10^3
Hydrogen-bonded	Ice (H_2O)	5×10^4
	HF	2.89×10^4

From the above table, it is evident that the molecular bonds are much weaker than the others. The hydrogen bonds are also relatively weak.

17.5 Miller indices

In a crystal lattice, there are sets of planes, which have larger concentration of atoms (or group of atoms) at the lattice sites. The orientations of such parallel planes are usually expressed in terms of a set of three numbers known as *Miller indices*.

To understand the meaning of the Miller indices we first consider a two dimensional lattice as shown in Fig. 17.12. The lattice is built up by the repeated translations of the unit cell, e.g., OACB. In place of the crystal planes in an actual three dimensional crystal, we have the *crystal lines* e.g., KL, PQ etc. along which the atoms are more densely concentrated. Let a and b be the sides of the unit cell along the crystal axes OX and OY. From the figure it can be seen that the intercepts of the crystal lines KL, PQ etc. along OX and OY are equal to integral multiples of a and b. As for example the x and y intercepts of the line KL are OK = $3a$ and OL = $2b$ respectively. In units of a and b, these intercepts are 3 and 2 respectively. The ratio of the reciprocals of these numbers is

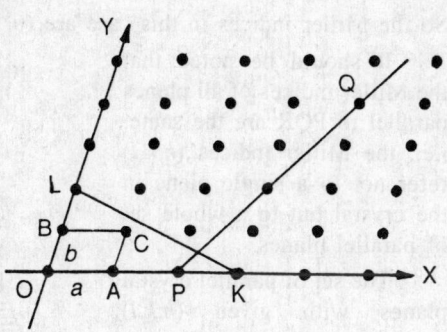

Fig. 17.12, Diagram to final Miller indices for a two dimensional lattice.

$$\frac{1}{3}:\frac{1}{2}=\frac{2}{6}:\frac{3}{6}=2:3$$

The two integers 2 and 3 on the r.h.s. are the Miller indices of the line KL.

Thus to get the Miller indices of a crystal line we have to first determine the intercepts of the line between two lattice points along the two crystal axes in units of the sides of the unit cell, take the ratio of their reciprocals by multiplying each of them by the lowest common multiple (*l.c.m.*) of their denominators. The numbers thus obtained are the Miller indices of the given crystal line.

We can similarly define the Miller indices in the case of a crystal plane in a three dimensional crystal. Let PQR be such a plane as shown in Fig. 17.13. The intercepts of the plane along the three crystal axes will be integral multiples of the sides a, b and c of the unit cell so that we can write these as

$$OP = Ha, \; OQ = Kb \text{ and } OR = Lc$$

where H, K and L are three integers. Expressed in terms of a, b and c, the intercepts have the lengths H, K and L. The ratios of their reciprocals can be written as

$$\frac{1}{H}:\frac{1}{K}:\frac{1}{L}=h:k:l$$

where h, k and l are three integers which are obtained by multiplying the reciprocals of H, K and L by their *l.c.m.* Then h,k,l are the Miller indices of the plane PQR.

As for example, in Fig. 17.13, the intercepts of the plane PQR along the crystal axes are $2a$, $3b$ and $4c$ respectively so that $H = 2$, $K = 3$ and

$L = 4$. The ratio of their reciprocals is
$$\frac{1}{2} : \frac{1}{3} : \frac{1}{4} = \frac{12}{2} : \frac{12}{3} : \frac{12}{4} = 6 : 4 : 3$$
So the Miller indices in this case are (6, 4, 3).

It should be noted that the Miller indices of all planes parallel to PQR are the same. i.e., the Miller indices (h,k,l) refer not to a single plane in the crystal but to a whole set of parallel planes.

The set of parallel crystal planes with given (h,k,l) values must be drawn so as to embrace *all the lattice points* in the crystal. It is then possible to find the distances between the points of intersection of the adjacent crystal planes with the crystal axes. The procedure for

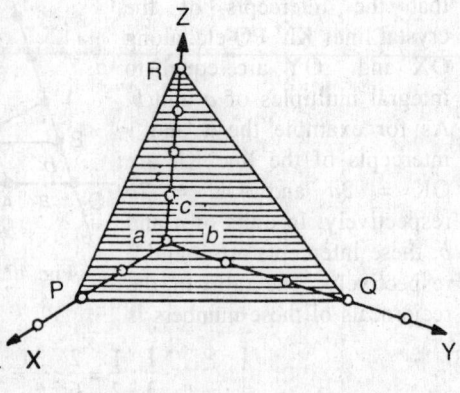

Fig. 17.13 Miller indices for a three dimensional lattice.

finding these can be more easily understood by referring to Fig. 17.14 for the two dimensional case.

Consider the set of parallel crystal lines with $h = 1$ and $k = 2$ passing through the lattice points. These (1,2) crystal lines are represented by the solid lines in the figure. The consecutive lines have intercepts separated by $2a$ and b along the crystal axes OX and OY respectively. However, in between these solid

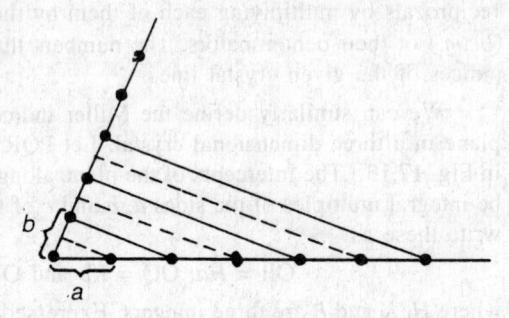

Fig. 17.14 Lattice points in a two dimensional lattice.

crystal lines, we can draw another set of lines, parallel to the former, shown by the dashed lines in the figure. These pass through the lattice points which do not fall on the previous set of lines. Since the crystal lines with given h and k values must cover *all the lattice points*, both the above sets of parallel lines constitute the full family of (1,2) crystal lines Obviously the distances between the points of intersection of any two adjacent lines with the crystal axes are $a/h = a$ and $b/k = b/2$ respectively.

The above arguments can be easily extended to the case of the actual three dimensional crystal. It can be proved that in this case the distances between the points of intersection of the adjacent crystal planes with a given set of Miller indices h, k, l with the crystal axes are equal to a/h, b/k and c/l respectively. Miller indices can be both positive and negative. In the latter case, a bar is placed above the integer expressing the index (*e.g.*, $\bar{2}$).

The advantage of denoting the crystal planes by the Miller indices (h,k,l) instead of the intercepts (H,K,L) is that the numbers h/a, k/b, and l/c are proportional to the direction cosines of the normal to the crystal plane. This can again be most easily understood by referring the two dimensional case (Fig. 17.15).

Consider a lattice point at O through which a crystal line (h,k) passes. We draw the perpendicular $OP = \vec{d}$ from O on the adjacent crystal line BA which makes the angles α and β with the crystal axes OX and OY respectively. From the previous discussion, it is clear that the intercepts OA and OB have the lengths a/h and b/k respectively. Then we have

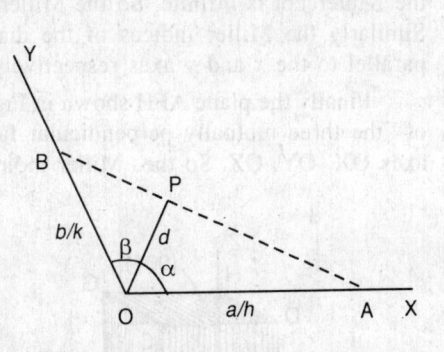

Fig. 17.15 A Calculation of the direction works

$$\cos \alpha = \frac{OP}{OA} = \frac{d}{a/h} = \frac{hd}{a}$$

$$\cos \beta = \frac{OP}{OB} = \frac{d}{b/k} = \frac{kd}{b}$$

Hence we get $\cos \alpha : \cos \beta = \dfrac{h}{a} : \dfrac{k}{b}$

Extension to the three dimensional case follows easily.

The distance d between the adjacent planes with the Miller indices (h,k,l) can also be determined. It can be shown that for a cubic lattice of sides a, this is given by

$$d_{hkl} = \frac{a}{\sqrt{h^2 + k^2 + l^2}}$$

Cubic lattice :

We now consider the case of a cubic lattice of sides a. The crystal axes OX, OY and OZ are mutually perpendicular. Referring to Fig. 17.16a we note that the face ABCD of the unit cell perpendicular to OX is parallel to the y and z axes so that the intercepts of this plane along these axes are infinite while that along the x-axis has a finite value, equal to an integral multiple of a. To get the Miller indices of the plane we have to take the reciprocals of the intercepts. Since the reciprocals of the infinite intercepts are zero, the Miller indices of these planes perpendicular to OX can be written as (1,0,0).

Similarly the Miller indices of the planes BCGF and DHGC perpendicular to OY and OZ are (0,1,0) and (0,0,1) respectively.

Again, referring to Fig. 17.16b, we note that the diagonal plane AFGD has equal intercepts along OX and OY and is parallel to OZ so that the z-intercept is infinite. So the Miller indices of this plane are (1,1,0). Similarly the Miller indices of the diagonal planes BDHF and ABGH parallel to the x and y axes respectively are (0,1,1) and (1,0,1).

Finally the plane AFH shown in Fig. 17.16c defined by the diagonals of the three mutually perpendicular faces have equal intercepts on the axes OX, OY, OZ. So the Miller indices of this plane are (1,1,1).

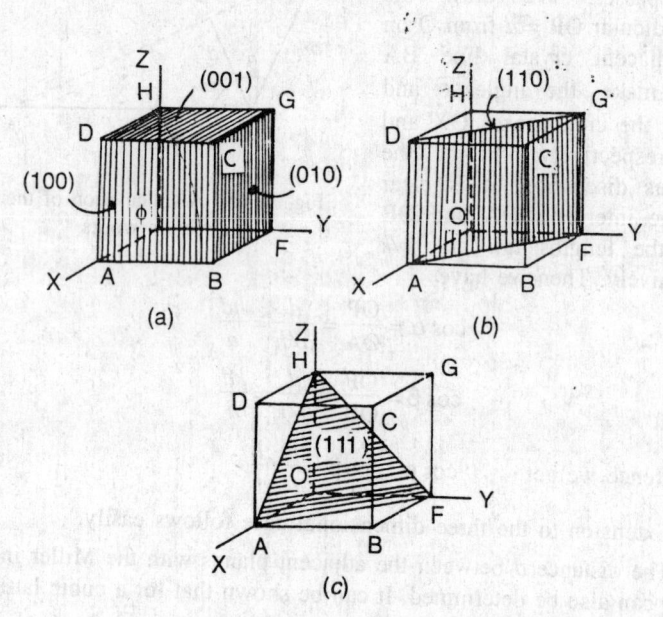

17.16. Crystal planes in a cubic lattice

17.6 Laue diffraction equations

The Bragg equation for x-ray diffraction was deduced in Chapter VIII by considering the reflection of a parallel beam of x-rays from two adjacent crystal planes. This is equivalent to the determination of the diffraction relation for a two dimensional diffraction grating. A crystal on the other hand behaves like a three dimensional grating. Von Laue

Structure and Properties of Solids

developed a theory of X-ray diffraction by crystals to explain the occurrence of the Laue diffraction by crystals to explain the occurrence of the Laue spots discussed in § 8·14. Bragg equation (8·14-1) can be deduced from Laue's theory.

Laue's theory is based on the simple static atomic model of the crystal according to which each lattice site has a simple static point atom which scatters the incident x-ray beam in different directions. The interference between the scattered radiation from the different atoms at the different lattice sites produces the diffraction maxima and minima.

Let us consider the scattering of x-rays from *any* two atoms in a crystal.

As shown in Fig. 17.17, P_1 and P_2 are any two lattice points, the vector distance between them being **r**. Suppose a parallel beam of monochromatic x-rays incident on the crystal is scattered from the atoms

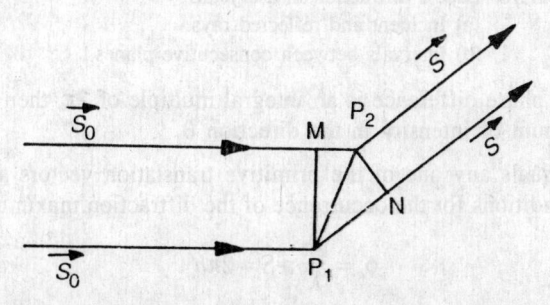

Fig. 17.17 Diffraction of x-rays from two lattice points.

located at the different lattice points. Let s_0 be the unit vector along the wave normal of the incident wave and **s** that along the wave normal of the scattered wave. P_1M and P_2N are the projections of the vector **r** perpendicular to the incident and scattered directions. Then from the figure, it is evident that the path difference between the waves scattered from P_2 and P_1 is given by

$$P_2M - P_1N = \mathbf{r} \cdot \mathbf{s}_0 - \mathbf{r} \cdot \mathbf{s} = \mathbf{r} \cdot \mathbf{S} \qquad (17 \cdot 6\text{-}1)$$

where $\mathbf{S} = \mathbf{s}_0 - \mathbf{s}$ is the vector difference between the two unit vectors $\mathbf{s}_0$ and **s**, known as the *scattering normal*.

Consider a plane perpendicular to the line CD shown in Fig. 17.18a such that the vectors $\mathbf{s}_0$ and **s** are equally inclined to this plane. Let the angle of inclination be θ for both.

Since the unit vectors $\mathbf{s}_0$ and **s** are each of the same (unit) magnitude, the vector difference **S** must be perpendicular CD as shown in the figure and its magnitude is

$$S = AB = 2 |\mathbf{s}_0| \sin \theta = 2 \sin \theta \qquad (17 \cdot 6\text{-}2)$$

The phase difference between the incident and scattered radiation is

$$\phi = \frac{2\pi}{\lambda} \times \text{path difference} = \frac{2\pi}{\lambda}(\mathbf{r}\cdot\mathbf{S}) \qquad (17\cdot6\text{-}3)$$

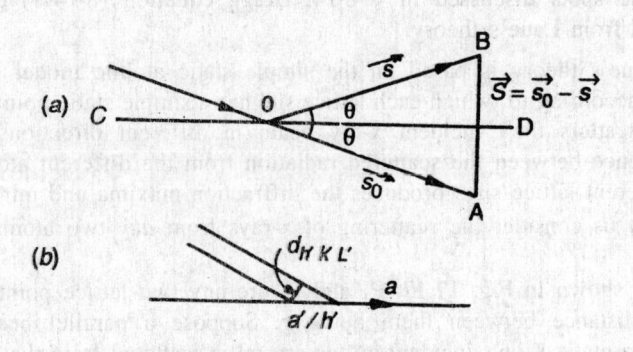

Fig. 17.18 Laue's diffraction in a crystal.
(a) Incident and reflected rays.
(b) Intervals between consecutive planes.

If this phase difference is an integral multiple of 2π, then there will be a maximum of intensity in the direction θ.

If $\mathbf{r}$ equals any one of the primitive translation vectors $\mathbf{a}$, $\mathbf{b}$ and $\mathbf{c}$ then the conditions for the occurrence of the diffraction maximum will be

$$\phi_a = \frac{2\pi}{\lambda}\mathbf{a}\cdot\mathbf{S} = 2\pi h' \qquad (17\cdot6\text{-}4a)$$

$$\phi_b = \frac{2\pi}{\lambda}\mathbf{b}\cdot\mathbf{S} = 2\pi k' \qquad (17\cdot6\text{-}4b)$$

$$\phi_c = \frac{2\pi}{\lambda}\mathbf{c}\cdot\mathbf{S} = 2\pi l' \qquad (17\cdot6\text{-}4c)$$

where h', k', l' are three integers. If $\mathbf{a}, \mathbf{b}, \mathbf{c}$ are inclined at the angles α, β and γ with respect to $\mathbf{S}$, then we have

$$\mathbf{a}\cdot\mathbf{S} = aS\cos\alpha = 2a\cos\alpha\sin\theta \qquad (17\cdot6\text{-}5a)$$

$$\mathbf{b}\cdot\mathbf{S} = bS\cos\beta = 2b\cos\beta\sin\theta \qquad (17\cdot6\text{-}5b)$$

$$\mathbf{c}\cdot\mathbf{S} = cS\cos\gamma = 2c\cos\gamma\sin\theta \qquad (17\cdot6\text{-}5c)$$

Hence we get by substitution

$$2a\cos\alpha\sin\theta = h'\lambda = nh\lambda \qquad (17\cdot6\text{-}6a)$$

$$2b\cos\beta\sin\theta = k'\lambda = nk\lambda \qquad (17\cdot6\text{-}6b)$$

$$2c\cos\gamma\sin\theta = l'\lambda = nl\lambda \qquad (14\cdot6\text{-}6c)$$

Here it is assumed that the integers h', k', l' have a highest common integral factor $n > 1$ so that we can write $h' = nh, k' = nk, l' = nl$. If h', k', l' have no such highest common integral factor, then $n = 1$.

The above equations are known as *Laue equations*. For a given λ and given values of h', k', l' the equations determine θ and two of the three direction cosines $\cos\alpha$, $\cos\beta$, $\cos\gamma$. However only two of these direction cosines are independent, the third being determined from the relationship connecting them. For instance, in the case of axes orthogonal to one another, this relation gives $\cos^2\alpha + \cos^2\beta + \cos^2\gamma = 1$. For non-orthogonal axes, the relationship will be different. So the solutions of Laue equations determine θ and $\mathbf{S}$ uniquely, thus defining a scattering direction. Eqs. (17·6-6) also show that for given values of λ and θ, the direction cosines of the scattering normal $\mathbf{S}$ are proportional to h/a, k/b, and l/c i.e.,

$$\cos\alpha : \cos\beta : \cos\gamma = \frac{h}{a} : \frac{k}{b} : \frac{l}{c} \qquad (17\cdot6\text{-}7)$$

But we have seen before that the direction cosines of the normal to the family of lattice planes with the Miller indices (h,k,l) are also proportional to h/a, k/b and l/c. The scattering normal $\mathbf{S}$ is thus identical with the normal to the family of (h,k,l) planes. So the (h,k,l) planes may be regarded as the planes from which Bragg reflection takes place.

Again since the adjacent planes with the Miller indices (h,k,l) interesect the crystal axes $\mathbf{a}$, $\mathbf{b}$, $\mathbf{c}$ at intervals of a/h, b/k, c/l we have (see Fig. 17.18b)

$$\cos\alpha = \frac{d}{a/h}, \ \cos\beta = \frac{d}{b/k}, \ \cos\gamma = \frac{d}{c/l} \qquad (17\cdot6\text{-}8)$$

where the distance between the adjacent (h,k,l) plane has been taken as d. So we have

$$a\cos\alpha = hd, \ b\cos\beta = kd, \ c\cos\gamma = ld, \qquad (17\cdot6\text{-}9)$$

Thus we get finally, using any one of the Laue equations

$$2\,hd\sin\theta = nh\,\lambda \qquad (17\cdot6\text{-}10)$$

or, $$2d\sin\theta = n\lambda, \qquad (17\cdot6\text{-}11)$$

which is Bragg equation (8·17–1).

17.7 Atomic scattering factor and geometrical structure factor

Laue equations were derived on the assumption of point atoms located at the lattice sites acting as scattering centres. Actually the atoms have an electronic structure and the scattering takes place due to the interaction between the atomic electrons and the incident x-ray beam.

When a monochromatic beam of x-rays is incident on the atom at the lattice site in a crystal, the atomic electrons are set into forced vibration by the electric vector of the incident radiation. These accelerated electrons emit radiation of the same frequency in different directions, as expected from the e.m. theory of light. This results in the scattering of the incident beam in different directions for which all the electrons in the scattering atom are responsible.

In the case of visible radiation, the wavelength (4000 to 8000Å) is large compared to the mean distance d (~1Å) between the electrons in the atoms. For this reason, the scattering from the different electrons in an atom is incoherent in this case. On the other hand, since the x-ray wavelength is of the order of an angstrom, we have $\lambda \sim d$ in this case. Hence the radiation scattered from the different electrons in an atom is coherent for X-ray scattering. So the scattered intensity is determined by both the amplitudes and the relative phases of the the waves scattered from the different electrons in the atom. Due to the interference between the waves scattered from the different electrons, the total scattered amplitude from an atom is different from the sum of the amplitudes scattered from each of the Z electrons in an atom where Z is the atomic number of the scatterer. The ratio of the scattered amplitude f_s from the whole atom to that from the individual electrons in the atom is usually less than Z (*i.e.*, $f_s < Z$). This ratio f_s is known as the atomic scattering factor. The intensity of the diffracted beam in a given direction depends on the value of f_s.

f_s can be calculated theoretically from a knowledge of the electronic charge distribution in the atom. Assuming a spherically symmetric continuous charge distribution with a density $\rho(r)$, it can be shown that f_s depends on the wavelength λ and the angle of scattering θ, being given by*

$$f_s(\theta) = 4\pi \int_0^\infty \rho(r) \frac{\sin \mu r}{\mu r} dr \qquad (17\cdot 7\text{-}1)$$

where $\mu = 4\pi \sin \theta / \lambda$. Since $\mu = 0$ for $\theta = 0$ so that $\lim_{\theta \to 0} (\sin \mu r / \mu r) = 1$ we get for forward scattering

$$\lim_{\theta \to 0} f_s(\theta) = Z \qquad (17\cdot 7\text{-}2)$$

As θ increases, f_s decreases. It is necessary to know $\rho(r)$ to calculate f_s which is usually obtained from a knowledge of the quantum mechanical wave functions of the atoms in the crystal. A simple approximation would be to use the wave functions of the free atoms which are actually different from those when the atoms are bound in the crystal lattice.

In deducing the Laue equations, it is assumed that there is just one scattering atom per unit cell which is possible if there are atoms only at the corners of the unit cell which is true for primitive unit cells. In many cases, the unit cells are more complex, having more than one atom per cell. For instance, in the case of a body-centred cubic (*b.c.c*) cell, if all the atoms in the crystal are identical, the number of atoms per unit cell

---

* See Introduction to *Solid State Physics* by Charles Kittel.

should be taken to be *two*, one corner atom and the other body-centred atom. Here we have to consider the interference between the two atoms to get the resultant scattered amplitude per unit cell.

In general for (h,k,l) reflection the ratio of the amplitude of the radiation scattered by all the atoms in a unit cell to that scattered by a point electron at the origin can be written as

$$F(h,k,l) = \sum_j f_j \cdot \exp \varphi_j \qquad (17 \cdot 7\text{-}3)$$

where f_j is the atomic scattering factor as deduced above for the jth atom in the unit cell and φ_j is the phase difference between the radiation scattered at the origin and that by the jth atom in the unit cell. The sum is evaluated for all atoms in the unit cell. For this, a knowledge of the position vectors of the different atoms in the unit cell must be known. If all the atoms in the crystal are identical, then $f_j = f$ is the same for all atoms and we can write

$$f(h, k, l) = f S \qquad (17 \cdot 7\text{-}4)$$

where S is known as the *geometrical structure factor* and depends on the geometrical arrangement of the atoms within the unit cell. It is given by*

$$S = \sum_j \exp 2\pi i (h x_j + k y_j + l z_j) \qquad (17 \cdot 7\text{-}5)$$

where x_j, y_j, z_j determine the position coordinates of the jth atom in terms of the basis vectors.

The intensity of the diffracted beam is proportional to $|F|^2$

As an example consider the case of a b.c.c. crystal with two atoms per unit cell as mentioned above, all the atoms in the crystal being identical. We assign arbitrarily the coordinates (0,0,0) to the corner atom of the unit cell so that the body-centred atom has the coordinates $\left(\frac{1}{2}, \frac{1}{2}, \frac{1}{2}\right)$ in terms of the sides of the unit cell. Then we get

$$F = f[1 + \exp i\pi (h + k + l)] \qquad (17 \cdot 7\text{-}6)$$

If $(h+k+l)$ is an odd integer, then F vanishes since the exponential term is -1. Reflections for these planes will be missing, through they will be present for the simple cubic structure. Thus (1, 0, 0) or (1, 1, 1) reflections will be missing in this case. On the other hand, if $(h+k+l)$ is zero or an even integer, we get maximum reflection. This occurs in the cases of (1, 1, 0), (2, 2, 2) etc. reflections.

17.8 Bragg's experiment on the structure of rock salt crystal

As stated in § 8.17, W.L Bragg and his son W.H. Bragg developed a method for the determination of the wavelength of X-rays, using an

---

* See *ibid*.

ionization chamber. W.H. Bragg (1913) was the first to use the method to investigate the structure of a crystal. He did this for the rock salt (sodium chloride) crystal.

In Fig. 17.19a, the peak of the ionization current obtained due to reflections from some important planes of an NaCl crystal are shown which should be compared with the similar peaks for the KCl crystal shown in Fig. 17.19b. Palladium ($Z = 46$) x-rays were used for obtaining these peaks. The glancing angles for the three peaks in the case of the KCl crystal (Fig. 17.19b) for reflection from the (1,0,0) plane in different orders are found to be $\theta_1 = 5° 23'$, $\theta_2 = 10°49'$ and $\theta_3 = 16° 20'$.

These give us

$$\sin \theta_1 : \sin \theta_2 : \sin \theta_3$$
$$= 0.09382 : 0.18767 : 0.28123$$
$$= 1:2:3 \text{ (appx.)} \qquad (17\cdot 8\text{-}1)$$

These values are in agreement with those found from the Bragg equation $2d \sin \theta = n\lambda$ with $n = 1, 2, 3$. Here $d = d_{100}$.

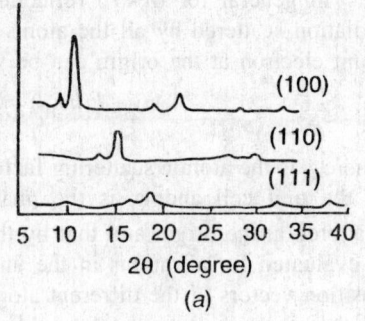

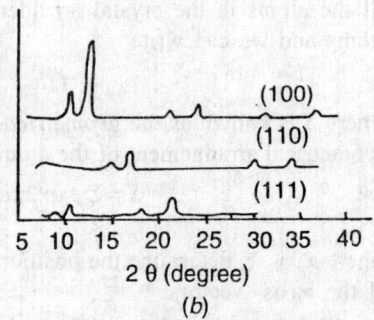

Fig. 17.19. Variation of ionization current (I) with angle (θ) of (a) NaCl (b) KCl crystal.

Again for reflections from the (1,1,0) and (1,1,1) planes of the KCl crystal, the glancing angles are

$$\theta = 7°37'$$
and
$$\theta = 9°25'$$

respectively for the first order ($n = 1$). So from the Bragg equation, we get

$$d_{100} : d_{110} : d_{111} = \frac{1}{\sin \theta_{100}} : \frac{1}{\sin \theta_{110}} : \frac{1}{\sin \theta_{111}}$$

$$= \frac{1}{\sin 5°23'} : \frac{1}{\sin 7°37'} : \frac{1}{\sin 9°25'}$$

$$= 1 : \frac{1}{\sqrt{2}} : \frac{1}{\sqrt{3}} \quad \text{(appx)} \qquad (17\cdot 8\text{-}2)$$

The interplanar distances d_{100}, d_{110} and d_{111} for a cubic crystal which has unit cells of sides a can easily be determined with the help of Fig. 17.20.

From Fig. 17.20a, it can be seen that the separation between the

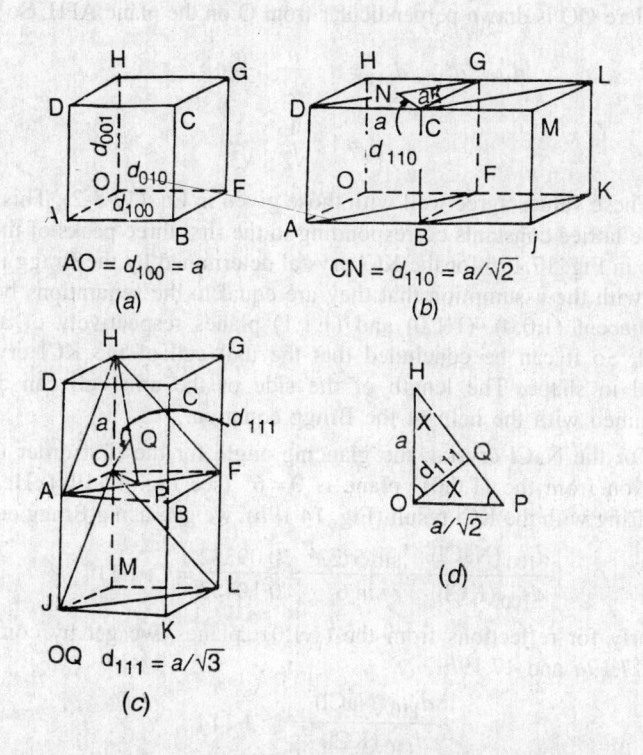

Fig. 17.20. Inter-planar distances for a cubic crystal

adjacent (1,0,0) planes *e.g.*, ABCD and OFGH is $d_{100} = a$. From Fig. 17.20b the separation between the adjacent (1,1,0) planes *e.g.*, AFGD and BKLC is found to be

$$d_{110} = CN = CD \sin 45° = a/\sqrt{2}$$

Finally, from Fig. 14.20c the separation between the adjacent (1,1,1) planes *e.g.*, AFH and JLO is found to be

$$d_{111} = OQ = a/\sqrt{3}*$$

---

* From the inset of Fig. 17.20c, we have

$$PH^2 = OH^2 + OP^2 = a^2 + \frac{a^2}{2} = \frac{3a^2}{2}$$

$$\therefore \quad PH = a\sqrt{\frac{3}{2}}$$

Further, $\quad \dfrac{OQ}{OP} = \dfrac{OH}{PH} = \dfrac{a}{a\sqrt{3/2}} = \sqrt{\dfrac{2}{3}}$

$$\therefore \quad OQ = d_{111} = \sqrt{\frac{2}{3}} \; OP = \sqrt{\frac{2}{3}} \cdot \frac{a}{\sqrt{2}} = \frac{a}{\sqrt{3}}$$

Here OQ is drawn perpendicular from O on the plane AFH. So we get

$$d_{100} : d_{110} : d_{111} = a : \frac{a}{\sqrt{2}} : \frac{a}{\sqrt{3}}$$

$$= 1 : \frac{1}{\sqrt{2}} : \frac{1}{\sqrt{3}} \qquad (17 \cdot 8 \text{-} 3)$$

These values agree well with those given in Eq. (17·8-2). This shows that the lattice constants corresponding to the first three peaks of the three curves in Fig. 17.19b for the KCl crystal determined by the Bragg method agree with the assumption that they are equal to the separations between the adjacent (1,0,0), (1,1,0) and (1,1,1) planes respectively of a cubic crystal. So it can be concluded that the unit cell of the KCl crystal is cubical in shape. The length of the side of the unit cell can also be determined with the help of the Bragg equation.

For the NaCl crystal, the glancing angle for the first order ($n = 1$) reflection from the (1,0,0) plane is $\theta = 6°$ (see Fig. 17·19a). Hence by comparing with the KCl result (Fig. 14.19b), we get, using Bragg equation

$$\frac{d_{100} (\text{NaCl})}{d_{100} (\text{KCl})} = \frac{\sin 5°23'}{\sin 6°} = \frac{0 \cdot 09382}{0 \cdot 10453} = 1 : 1 \cdot 114$$

Similarly for reflections from the (1,1,0) planes, we get by comparing Figs. 17.19a and 17.19b

$$\frac{d_{110} (\text{NaCl})}{d_{110} (\text{KCl})} = 1 : 1 \cdot 114$$

However, when similar comparison is made for the reflections from the (1,1,1) planes, we get

$$\frac{d_{111} (\text{NaCl})}{d_{111} (\text{KCl})} = 2 : 1 \cdot 114$$

Hence using the results given in Eq. (17.8-3) for the KCl crystal, we have for NaCl crystal

$$d_{100} : d_{110} : d_{111} = 1 : \frac{1}{\sqrt{2}} : \frac{2}{\sqrt{3}} \qquad (17 \cdot 8 \text{-} 4)$$

This shows that the interplanar separation for (1,1,1) planes of the NaCl crystal is $d_{111} = 2a/\sqrt{3}$ if it is assumed to have a cubic unit cell of sides a. This is twice that for the KCl crystal as found above.

The above difference in the results for the NaCl and KCl crystals was explained by Bragg by assuming that it is the ionized atoms forming the NaCl or KCl molecules which are located at the lattice sites of these crystals and not the molecules as a whole. It is known that the intensity of the x-rays scattered from an atom depends upon the number of electrons in the atom which is equal to the atomic number Z of the atom and is approximately proportional to Z^2. In the case of KCl crystal, the atomic numbers of K and Cl are 19 and 17 respectively. If we assume that these atoms exist in the ionized states as K^+ and Cl^- at the alternate lattice sites, then the number of electrons responsible for the scattering from each lattice site is the same, viz., 18. So the intensity of the x-rays scattered from each lattice site is the same.

On the other hand, in the case of the NaCl crystal, the atomic numbers of the constituent atoms Na and Cl are 11 and 17 respectively. Hence the numbers of electrons at the alternate lattice sites with Na^+ and Cl^- ions are 10 and 18 repectively. The scattering power of the Cl^-. ions is obviously much greater (~3 times) than that of the Na^+ ions. In Fig. 17.21, the arrangements of the two types of ions at the different lattice sites of the NaCl crystal is shown. From Fig. 17.21a, it can be seen that the numbers of both types of ions in all (1,0,0) planes are the same. This is also true for the (1,1,0) planes. But in the case of the (1,1,1) planes, a particular plane contains only one type of ion, either Na^+ or Cl^-, as can be readily seen from Fig. 17.21b. It will also be seen from the figure that the alternate planes contain oppositely charged ions. Thus the plane ABC in Fig. 17.21b contains only Cl^- ions while the next parallel plane DEF contains only Na^+ ions. Since the scattering power of the Na^+ ions is much smaller than that of the Cl^- ions, we can assume in the first approximation that there is no scattering from the (1,1,1) planes containing the Na^+ ions.

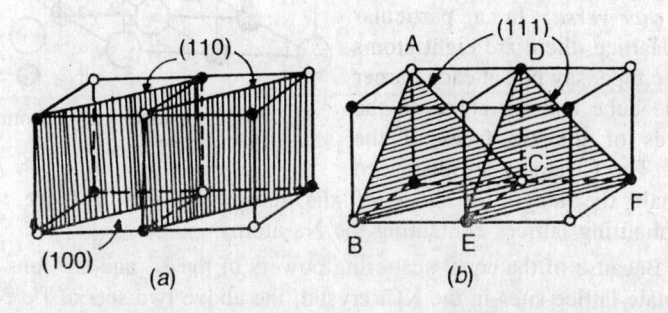

Fig. 17.21 Arrangement of Na^+ (●) and Cl^- (O) ions in NaCl crystal.

There is scattering only from those (1,1,1) planes which contain the Cl^- ions. The interplanar spacing between these latter planes is obviously $2a/\sqrt{3}$ and not $a/\sqrt{3}$ as in the case of the KCl crystal. Thus for reflections from the (1,1,1) planes of the NaCl crystal, the Bragg angle is half of what would be expected if the scattering were the same from all the planes.

In the above discussion, scattering from the Na^+ ions has been completely ignored. Actually, these ions also scatter the x-rays, through to a much lesser extent than the Cl^- ions. Since a plane containing the Na^+ ions is exactly midway between two successive planes containing the Cl^- ions in the case of the (1, 1, 1) planes, the path difference between the rays reflected in the first order from the (1,1,1) plane containing the Na^+ ions and those reflected in the same order from the next (1, 1, 1) plane containing the Cl^- ions is exactly equal to $\lambda/2$. This produces destructive interference between these rays. As a result, the intensity of the scattered radiation in the first order is considerably reduced. However for the second order reflection, the corresponding path difference is λ so that there is *constructive interference* between the rays reflected from the alternate planes in this case. Hence the intensites are relatively higher in the second order as can be seen from Fig. 17.19*b*.

From the experimental results obtained by Bragg as discussed above, it is clear that in the NaCl and KCl crystal, it is the ions of the constituent atoms of the corresponding molecules which are located at the alternate lattice sites and not the molecules as a whole. In Fig. 17.22, this arrangement of the atoms is shown for the NaCl crystal. The atoms of any one type (in the ionized state) are arranged at the lattice points of a *face-centred cubic crystal*. There are two such lattices for the two types of atoms which are so arranged that each Na atom has six Cl atoms as its nearest neighbours and *vice-versa*. In a particular f.c.c. lattice, there are eight atoms of one type, say Na, at each corner of the cube and six atoms at the centres of the six faces of the cube. The Cl atoms are displaced by half the length of sides of the unit cube with respect to the neighbouring lattices containing the Na atoms.

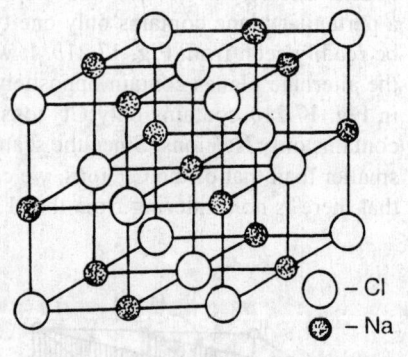

Fig. 17.22 Arrangement of atoms in NaCl crystal.

Because of the equal scattering powers of the K^+ and Cl^- ions at the alternate lattice sites in the KCl crystal, the above two sets of f.c.c cubic lattices, one for the K atoms and the other for the Cl atoms, may be regarded as forming the simple cubic lattice in this case.

17.9 Crystal structure determination

(A) Laue's method :

As stated in § 8.17, Von Laue in Germany was the first to point out that since the x-ray wavelengths were of the same order of magnitude as the spacing between the atoms in the crystal, the crystal could be used for diffracting an x-ray beam. Observations of the diffraction maxima and minima would therefore give an idea about the structure of the crystal.

In Laue's method a single crystal is exposed to a beam of continuous x-rays (white radiation) usually in the range 0.2 to 2 Å. The beam is collimated by a pinhole (see Fig. 17.23) and the diffracted beam is received on a flat film placed on the other side of the crystal. A relatively small crystal (dimension ~ 1 mm) may be used. As stated previously, the diffraction pattern consists of a series of spots as shown in Fig. 8.22.

Though continuous x-ray are used in this case, a given set of reflecting planes selects a wavelength satisfying the Bragg condition of reflection $2d \sin \theta = n\lambda$.

Laue pattern is specially suitable for checking the orientation of the crystals used in different types of experiments. This is due to the fact that the pattern reproduces the symmetry of the crystal in the orientation used for the experiment. Thus for a cubic crystal with its body diagonal having 3-fold symetry parallel to the direction of the collimated incident beam, the pattern will also show 3-fold symmetry.

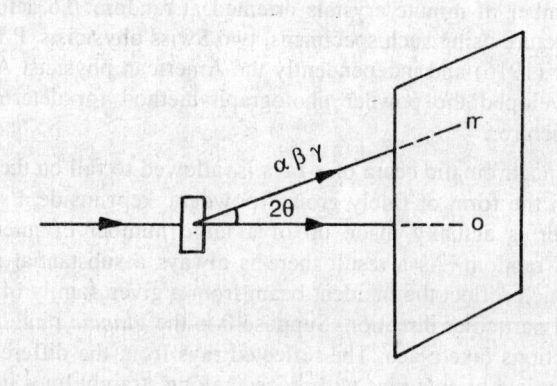

Fig. 17.23 Laue method.

The advantages of this method are that it is an easy method for obtaining qualitative data on the reflections from a large number of planes simultaneously and hence is much less time-consuming than the Bragg method.

There are however some disadvantages of the method also. Since continuous x-rays are used, a number of different wavelengths may be reflected from the same plane in different orders so that all these may be superposed on a single spot. This makes the determination of the reflected intensity in a given direction quite difficult.

(B) Bragg method :

In Bragg method, monochromatic x-rays are reflected from a single crystal and maxima of the reflected beam are observed with the help an ionization chamber (see § 8.14). As the crystal plane is rotated in small steps, the ionization chamber is turned through twice the angle at each step and the ionization current is measured as a function of the angle of reflection.

The method developed by the Braggs is of importance mainly for the accuracy of intensity measurements. It is also possible to obtain very accurate value of $\sin\theta$ by this method and hence of the dimensions of the unit cell. Its main disadvantage is that it is extremely time-consuming to take enough readings for reflections from sufficient number of planes to yield meaningful information about the structure of the crystal.

(C) Powder photograph method :

In both the above two methods, single crystals with greater than microscopic dimensions have to be used. It is usually difficult to prepare absolutely flawless single crystals in relatively larger sizes. Most crystalline substances are available in forms in which they are made up of a large number of minute crystals oriented at random. To determine the crystal structure using such specimens, two Swiss physicists, P.Debye and P. Scherrer (1916) and independently the American physicist A. W. Hull (1917) developed the powder photograph method for determining the crystal structure.

A monochromatic beam of x-rays is allowed to fall on the specimen which is in the form of finely ground powder, kept inside a small tube. The powder is actually made up of a large number of micro-crystals oriented at random. As a result there is always a substantial number of crystals which reflect the incident beam from a given family of important planes in a particular direction. Suppose θ is the *glancing angle* at which such reflections take place. The reflected rays from the different crystals with appropriate orientations will proceed along straight lines lying on the curved surface of a cone of semivertical angle 2θ (see Fig. 17.24). Due to reflections from a given set of planes, there will be a number of such cones corresponding to different orders of reflection. These reflected rays give rise to a series of concentric circular lines on a photographic plate kept at right angles to the incident beam. Usually a photographic film cut into a rectangular strip and bent in the form of a cylinder is used so that one gets a series of concentric circular arcs. With this arrangement, even those reflected rays for which $2\theta > 90°$ can also be received on the film. An example of a Debye-Scherrer powder photograph is shown in Fig. 17.25. It can be seen from the picture that further away from the central spot where direct beam hits the film, the curvatures of the lines become less. For $\theta = 90°$, the lines become straight. For $\theta > 90°$ the curvatures of the lines are reversed.

For crystals with high symmetry and with smaller number of atoms per unit cell, the diffraction patterns are relatively simple. Powder photograph method is thus specially suitable for the analysis of the structures in the case of many metals and alloys where the above

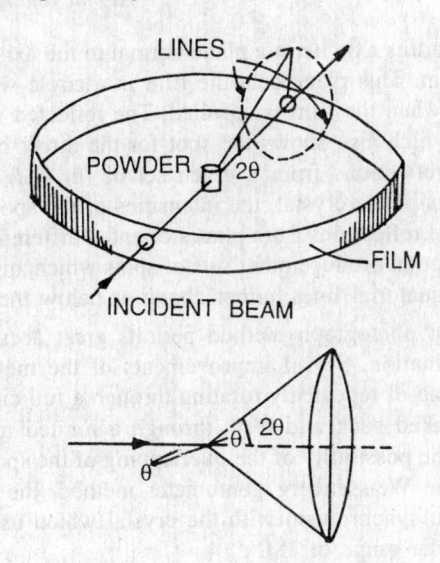

Fig. 17.24. Debye-Scherrer method.

conditions are fulfilled. In these cases powder photographs are often sufficient to establish the structure.

Fig. 17.25 Debye-Scherrer photograph.

(D) Rotation photograph method :

In this method a single crystal with sides about 1 mm in length is rotated about some crystal axis. A photographic film bent in the shape of a cylinder surrounds the crystal with its axis coinciding with the axis of rotation. A beam of monochromatic x-rays is allowed to fall on the crystal in a direction normal to the rotation axis (see Fig. 17.26). As the crystal is rotated slowly, the reflections from all

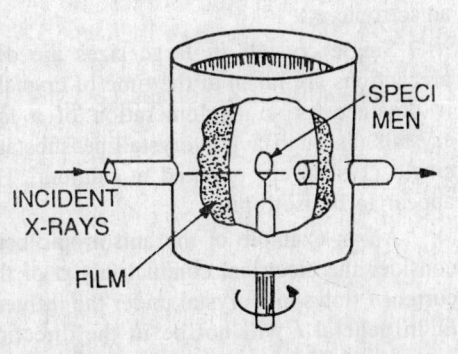

Fig.17.26. Rotation photograph method.

planes parallel to this axis lie in a plane normal to the axis and containing the incident beam. This plane cuts the film in a circle which appears as a horizntal line when the film is unrolled. The reflected spots lie on this equatorial line which also shows the spot for the direct beam. Each spot corresponds to reflections from a given set of (h, k, l) planes. Due to repeated rotations of the crystal, the intensities of the spots are enhanced considerably. The reflections from planes oriented differently with respect to the axis of rotation produce other sets of spots which appear along lines parallel to the equatorial lines, either above or below the latter.

The rotation photograph method permits great accuracy in crystal structure determination. Serval improvements of the method were later introduced. Instead of repeatedly rotating through a full circle, the crystal is sometimes rocked back and forth through a limited range of angles. This eliminates the possibility of the overlapping of the spots produced by reflections. In the Weissenberg goniometer method, the film is rocked back and forth in synchronism with the crystal which usually oscillates through the angular range of $180°$.

17.10 Anisotropy of the physical properties of single crystals

The different physical properties of single crystals e.g., elastic deformation, electrical and thermal conductivities, electric and magnetic susceptibilities, refractive index etc., are in general different in different directions within the crystal As an example, the Young's modulus of elasticity of a single crystal of zinc has the maximum and minimum values 126×10^9 Pa and 65×10^9 Pa along two different directions within the crystal, showing high degree of anisotropy.

Some crystals exhibit anisotropy in some of the above properties while they may be isotropic with respect to certain other properties. For instance, in all single crystals of the cubic type, the velocity of light is the same in all directions showing that the refractive index in isotropic in them. On the other hand, the elastic properties of such crystals are usually anisotropic.

Single crystals in large sizes are difficult to grow unless special precautions are taken at the time of crystallization. In general, the crystal is obtained as an agglomeration of a large number of minute single crystals. In such a polycrystalline substance, the small granules of the single crystals are oriented at random,. Hence their physical properties appear to be isotropic.

An an example of the anisotropic behaviour of a single crystal, we consider the electrical conductivity σ of the crystal. Suppose an electric current I flows in a crystal under the influence of an external electric field E. In general I will not be in the direction of E. It is found that each component of $\mathbf{I}$ is a linear function of all the three components of $\mathbf{E}$ in rectangular coordinate system :

$$I_x = \sigma_{xx}E_x + \sigma_{xy}E_y + \sigma_{xz}E_z$$
$$I_y = \sigma_{yx}E_x + \sigma_{yy}E_y + \sigma_{yz}E_z \qquad (17\cdot10\text{-}1)$$
$$I_z = \sigma_{zx}E_x + \sigma_{zy}E_y + \sigma_{zz}E_z$$

The nine coefficients σ_{ik} represent the components of the conductivity tensor. Each of the indices i and k can be x, y or z. The conductivity tensor is an example of a tensor of the second rank. It is actually a symmetric tensor so that $\sigma_{ik} = \sigma_{ki}$. Such a tensor has six components instead of nine.

In each crystalline substance, there are three special directions, known as the principal axes, with reference to which the Eqs. (17·10-1) can be transformed as follows*

$$I_x = \sigma_1 E_x, \; I_y = \sigma_2 E_y, \; I_z = \sigma_3 E_z \qquad (17\cdot10\text{-}2)$$

The three coefficients σ_1, σ_2 and σ_3 in Eq. (17·10-2) are known as the *principal conductivities*. The conductivity of a crystal is usually expressed in terms of the three principal conductivities. Their reciprocals are known as the principal resistivities:

$$\rho_1 = \frac{1}{\sigma_1}, \; \rho_2 = \frac{1}{\sigma_2}, \; \rho_3 = \frac{1}{\sigma_3}$$

If the electric field **E** acts along any one of the principal axes, then the current **I** has the direction of **E**. As an example if **E** is along x, so that $E = E_x$, then $I = I_x = \sigma_1 E_x$ while $I_y = I_z = 0$. So the current is along the x-axis in this case.

For a cubic crystal, $\sigma_1 = \sigma_2 = \sigma_3$ so that $\rho_1 = \rho_2 = \rho_3$. In this case ρ has the same value in all directions. In some crystals, e.g., in a hexagonal crystal two of the principal resistivities ρ_1, ρ_2 and ρ_3 are equal. In this case, the resistivity along a direction making an angle ϕ with the hexagonal axis depends on ϕ only and is given by

$$\rho(\phi) = \rho_s \sin^2\phi + \rho_p \cos^2\phi$$

If $\rho_p > \rho_s$ as in the zinc crystal, $\rho = \rho_p$ has the highest value along the hexagonal axis ($\phi = 0$) and lowest value $\rho = \rho_s$ at right angles to it ($\phi = \pi/2$)

17·11 Specific heat of solids

We have seen that in a crystalline solid, the atoms or groups of atoms are arranged in a regular three dimensional lattice. However the atoms are not stationary, but vibrate about their mean positions of rest at the lattice sites. Thus each of them behaves like a harmonic oscillator.

---

* See *Theoretical Physics* by George Joos.

Considering a one-dimensional array of N atoms vibrating harmonically about their respective mean positions of rest, it can be shown that their total vibrational energy is equal to the total energy of N linear harmonic oscillators*.

In a three dimensional arrangement of N atoms, the total vibrational energy is equal to that of $3N$ linear harmonic oscillators. When heat energy is applied to the substance, this total energy is changed which can be calculated theoretically, assuming suitable models. From such calculations, it is possible to develop the theory of the specific heat of solids.

17·12 Classical theory and its limitations

The total vibrational energy of a linear harmonic oscillator of mass m vibrating with the angular frequency ω is

$$\varepsilon = \varepsilon_k + V = \frac{p^2}{2m} + \frac{1}{2} m \omega^2 q^2 \qquad (17\cdot12\text{-}1)$$

where ε_k and V are the kinetic and potential energies respectively. q and p are the displacement and the linear momentum of the oscillator. It can be shown from classical statistical mechanics that such an oscillator has the mean energy at the absolute temperature T given by

$$<\varepsilon> = \int_0^\infty \varepsilon \exp(-\varepsilon/kT)\, d\varepsilon \bigg/ \int_0^\infty \exp(-\varepsilon/kT)\, d\varepsilon = kT \qquad (17\cdot12\text{-}2)$$

For three-dimensional motion, this becomes

$$<\varepsilon> = 3kT \qquad (17\cdot12\text{-}3)$$

where $k = 1\cdot38 \times 10^{-23}$ joule per K, k is the Boltzmann constant. From Eq. (17·12-3), it can be seen that the mean energy $<\varepsilon>$ is independent of the frequency of vibration of the oscillator. So for N oscillators, the total energy is

$$U = 3NkT \qquad (17\cdot12\text{-}4)$$

For one mole of the substance, $N = N_0 = $ Avogadro number. If the volume of the crystal is assumed to remain unchanged, then the *molar specific heat* of the substance comes out to be

$$C_v = \left(\frac{\partial U}{\partial T}\right)_v = 3N_0 k = 3R = \text{constant} \qquad (17\cdot12\text{-}5)$$

Here $\qquad R = N_0 k = 8\cdot314$ J/mole/K

So we get $\qquad C_v = 5\cdot96$ cal/deg/mole

---

* See *Solid State Physics* by A.J. Dekker

Eq. (17·12-5) is known as Dulong-Petit's law, which had been discovered as an empirical law by these two scientists in 1819. Though the law is obeyed at higher temperatures, it fails completely at very low temperatures. Experimental observations show that C_v decreases with decreasing temperature at very low temperatures.

17·13 Einstein's theory of specific heat

Albert Einstein was the first to explain the decrease of the specific heat of solids at low temperature on the basis of the then newly discovered quantum theory (1905).

The underlying assumption of Einstein's theory is that the crystal can be represented by $3N$ linear harmonic oscillators where N is the number of atoms in the crystal. According to this model the atoms are assumed to vibrate independently of each other. Further, instead of using Eq. (17·12-2) for the mean energy of the oscillator deduced on the basis of classical statistics, he used the expression for the mean energy (Eq. 3·3-3) deduced by Planck on the basis of quantum theory :

$$<\varepsilon> = \frac{h\nu}{\exp(h\nu/kT) - 1} = \frac{\hbar\omega}{\exp(\hbar\omega/kT) - 1}$$

As we saw in Ch. III, the above expression for $<\varepsilon>$ was deduced by assuming the energy of the linear harmonic oscillator to be given by $E_n = n\hbar\omega$. This was Planck's original formulation of the quantum theory. Later the expression for E_n was modified by quantum mechanics (Ch. XI) according to which

$$E_n = \left(n + \frac{1}{2}\right)\hbar\omega$$

n can take up the values 0, 1, 2, 3, ... The mean energy $<\varepsilon>$ of the oscillators comes out to be slightly different from that given above. As in § 3·3, we get

$$<\varepsilon> = \frac{\sum_n \left(n + \frac{1}{2}\right)\hbar\omega \exp\left\{-\left(n + \frac{1}{2}\right)\frac{\hbar\omega}{kT}\right\}}{\sum_n \exp\left\{-\left(n + \frac{1}{2}\right)\frac{\hbar\omega}{kT}\right\}}$$

Writing $x = -\dfrac{\hbar\omega}{kT}$, we get

$$<\varepsilon> = \frac{\hbar\omega \sum_n \left(n + \frac{1}{2}\right)\exp\left(n + \frac{1}{2}\right)x}{\sum_n \exp\left(n + \frac{1}{2}\right)x}$$

$$= \frac{\hbar\omega\left\{\frac{1}{2}\exp\frac{x}{2} + \frac{3}{2}\exp\frac{3x}{2} + \frac{5}{2}\exp\frac{5x}{2} + \ldots\ldots\right\}}{\exp\frac{x}{2} + \exp\frac{3x}{2} + \exp\frac{5x}{2} + \ldots\ldots}$$

$$= \hbar\omega \frac{d}{dx}\ln\left[\exp\left(\frac{x}{2}\right)(1 + \exp x + \exp 2x + \ldots)\right]$$

$$= \hbar\omega\left[\frac{d}{dx}\frac{x}{2} + \frac{d}{dx}\ln(1-\exp x)^{-1}\right]$$

$$= \hbar\omega\left[\frac{1}{2} - \frac{d}{dx}\ln(1-\exp x)\right]$$

$$= \hbar\omega\left[\frac{1}{2} + \frac{\exp x}{1-\exp x}\right] = \hbar\omega\left[\frac{1}{2} + \frac{1}{\exp x - 1}\right]$$

$$\therefore \quad <\varepsilon> = \hbar\omega\left[\frac{1}{2} + \frac{1}{\exp(\hbar\omega/kT - 1)}\right] \quad (17\cdot13\text{-}1)$$

This includes the zero point energy of the oscillator $\hbar\omega/2$. The second term in the expression for $<\varepsilon>$ depends on the frequency $\nu = \frac{\omega}{2\pi}$ of the oscillator. For one mole of the solid containidg N_0 atoms, the total energy becomes

$$U = 3N_0 <\varepsilon> = \frac{3N_0\hbar\omega}{2} + \frac{3N_0\hbar\omega}{\exp(\hbar\omega/kT) - 1} \quad (17\cdot13\text{-}2)$$

Then the molar specific heat is

$$C_v = \left(\frac{\partial U}{\partial T}\right)_v = 3N_0 \frac{\hbar\omega}{[\exp(\hbar\omega/kT) - 1]^2} \cdot \frac{\hbar\omega}{kT^2}\exp(\hbar\omega/kT)$$

$$= 3R\left(\frac{\hbar\omega}{kT}\right)^2 \frac{\exp(\hbar\omega/kT)}{[\exp(\hbar\omega/kT) - 1]^2} \quad (17\cdot13\text{-}3)$$

Since the first term in Eq. (17·13-2) is independent of temperature, it does not contribute to the specific heat C_v. So Eq. (17·13-3) for C_v is the same as would be obtained by using Planck's expression for $<\varepsilon>$.

Eq. (17·13-3) shows that C_v decreases with the decrease of temperature T.

At higher temperatures when $kT >> \hbar\omega$ so that $\hbar\omega/kT << 1$, we get

$$\exp(\hbar\omega/kT) \approx 1 + \frac{\hbar\omega}{kT}$$

Hence we have

$$C_v \approx 3R\left(\frac{\hbar\omega}{kT}\right)^2 \frac{\exp(\hbar\omega/kT)}{(\hbar\omega/kT)^2}$$

$$= 3R\left[1 + \frac{\hbar\omega}{kT} + \ldots\ldots\right] \longrightarrow 3R$$

Thus at higher temperatures, Einstein's theory gives results in agreement with Dulong-Petit's law.

If we write $\hbar\omega = kT_E$ then the *Einstein temperature* T_E is given by

$$T_E = \hbar\omega/k \qquad (17\cdot13\text{-}4)$$

We can then rewrite Eq. (17·13-3) as

$$\frac{C_v}{3R} = \left(\frac{T_E}{T}\right)^2 f_E(T_E/T) \qquad (17\cdot13\text{-}5)$$

Here $F_E(T_E/T)$, known as the Einstein function, is given by

$$F_E(T_E/T) = \frac{\exp(T_E/T)}{[\exp(T_E/T) - 1]^2} \qquad (17\cdot13\text{-}6)$$

As T decreases, F_E decreases and goes to zero as $T \to 0$. At very low temperatures, $T_E/T \gg 1$ so that we have

$$F_E(T_E/T) \approx \frac{\exp(T_E/T)}{\{\exp(T_E/T)\}^2} = \exp(-T_E/T) \to 0$$

as $T \to 0$. Thus at very low temperatures, F_E decreases exponentially with decreasing T and hence C_v given by Eq. (17·13-5) also falls off almost exponentially with decreasing T. However, the experimental values of C_v are found to decrease much more slowly with decreasing T (small circles in Fig. 17.27).

Thus though the decrease of C_v with decrease of T can be explained qualitatively by Einstein's theory, there is no quantitative agreement between theory and experiment at low temperatures.

T_E can be calculated if the natural vibrational frequency ω is known. ω can be found if the atomic mass and the elastic constants are known. $T_E \sim 100$ to 200 K for metallic elements. Transition from low to high temperature behaviour should occur in this temperature range.

17·14 Debye's theory of specific heat

As seen above, the primary assumption of Einstein's theory was that all the atoms in the crystal vibrate with the same frequency, independently of each other. Paul Debye in 1912 pointed out that this assumption cannot be correct, because there are strong interatomic forces between the atoms at the different lattice sites so that the vibrations of any one of the atoms strongly influence the vibrations of its neighbouring atoms. So Debye

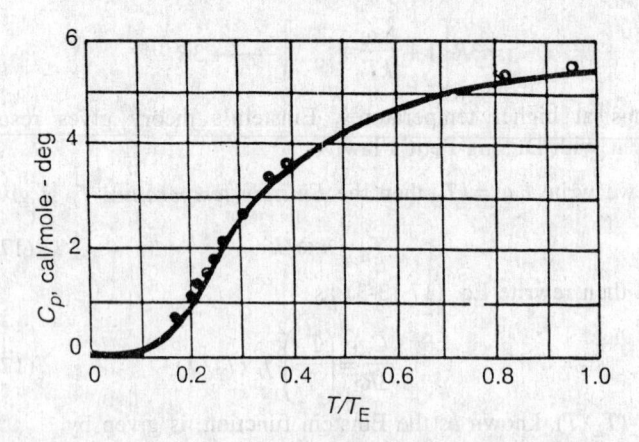

Fig. 17·27. C_v vs. T for diamond after Einstein.

proposed that instead of considering the independent vibrations of the individual atoms, it was necessary to consider the vibrational modes of the entire crystal.

The strong interactions between the neighbouring atoms in the crystal transmit the vibrations of any one atom to the others throughout the crystal and a collective motion in the nature of an elastic wave involving all the atoms is excited within the crystal. Such collective motion is known as the *normal mode* of a lattice. The number of normal modes is equal to the number of degrees of freedom $3N$ where N is the number of atoms.

Though in an actual crystal, the atoms are arranged in a regular pattern as *discrete mass points* with finite distances in between, in Debye's theory, the crystal is assumed to be a *continuous medium* through which stationary waves of different frequencies which constitute the normal modes of the crystal are propagated with the velocity of elastic waves (sound waves in the medium).

As an analogy we may consider the one-dimensional wave propagation in an elastic string of length l with rigidly fixed ends. Considering the string to be a *continuous medium*, we write down the differential equation for the wave propagation as

$$\frac{\partial^2 y}{\partial x^2} = \frac{1}{v^2}\frac{\partial^2 y}{dt^2}$$

where y is the displacement of the string at some point x at the instant t.

υ is the velocity of propagation of the elastic wave in the string. The solution consistent with the boundary conditions $y = 0$ at $x = 0$ and $x = l$ is

$$y_n = \sin \frac{n\pi x}{l} (a_n \cos \omega_n t + b_n \sin \omega_n t)$$

where n is a positive integer : $n > 0$. The wavelengths and the frequencies of the vibration for the nth mode of vibration are given by

$$\lambda_n = \frac{2l}{n}, \quad v_n = \frac{\omega_n}{2\pi} = \frac{\upsilon}{\lambda_n} = \frac{n\upsilon}{2l}$$

Thus the string can vibrate in different numbers of segments corresponding to the different modes n. Since $n = 2l v/\upsilon$ we have the number of modes of vibration in the frequency range v to $v + dv$

$$dn = \frac{2l}{\upsilon} dv \tag{17.14-1}$$

Similarly if we consider the propagation of elastic waves in a three dimensional continuous medium, assumed to be in the shape of a cube of sides l, the possible modes of vibration are determined by a set of three *positive integers* n_1, n_2, n_3 which are related by the equation

$$(n_1^2 + n_2^2 + n_3^2) \frac{\pi^2}{l^2} = \frac{4\pi^2 v^2}{\upsilon^2} \tag{17.14-2}$$

A similar equation was deduced in connection with the derivation of Rayleigh-Jeans law for black-body radiation (see § 3·2). It may be noted that in the above equation the velocity υ of the elastic wave propagation in the crystalline medium has replaced the velocity c of electromagnetic wave propagation within a cavity in Eq. (3·2-15). In analogy with Eq. (3·2-16) the number of normal modes of vibration within the angular frequency range ω to $\omega + d\omega$ in the present case is given by

$$z(\omega) \, d\omega = \frac{V}{2\pi^2 \upsilon^3} \omega^2 \, d\omega \tag{17.14-3}$$

where $V = l^3$ is the volume of the cube.

In the crystal, two types of elastic waves, longitudinal and transverse, can be propagated with the respective velocities υ_l and υ_t. Besides, there are two mutually perpendicular directions of vibration for the transverse waves. Hence for a crystalline medium, Eq. (17·14-3) has to be replaced by the following equation :

$$z(\omega) \, d\omega = \frac{V}{2\pi^2} \left(\frac{1}{\upsilon_l^3} + \frac{2}{\upsilon_t^3} \right) \omega^2 \, d\omega \tag{17.14-4}$$

In a continuous medium, the number of possible modes of vibration is infinite. Debye assumed that in a crystal with N_0 atoms, the number of normal modes of vibration is finite, equal to $3N_0$. Hence the frequency spectrum corresponding to a perfect continuum has to be cut off at some upper limit $\omega = \omega_m$ so as to be consistent with the finite number $3N_0$ of the modes of vibration. This maximum *Debye angular frequency* ω_m can be calculated from the relation

$$\int_0^{\omega_m} z(\omega)\, d\omega = \frac{V}{2\pi^2}\left(\frac{1}{v_l^3} + \frac{2}{v_t^3}\right) \int_0^{\omega_m} \omega^2\, d\omega = 3N_0$$

or, $$\frac{V}{2\pi^2}\left(\frac{1}{v_l^3} + \frac{2}{v_t^3}\right)\frac{\omega_m^3}{3} = 3N_0$$

This gives

$$\omega_m^3 = \frac{18\pi^2 N_0}{V}\left(\frac{1}{v_l^3} + \frac{2}{v_t^3}\right)^{-1} \qquad (17\cdot14\text{-}5)$$

It is assumed here that v_l and v_t are independent of the wavelength.

Using the quantum mechanical expression for the mean energy of the oscillators (Eq. 17·13-1), we then get the total energy of vibration for the whole crystal as

$$U = \int_0^{\omega_m} z(\omega) <\varepsilon> d\omega$$

$$= \frac{V}{2\pi^2}\left(\frac{1}{v_l^3} + \frac{2}{v_t^3}\right)\int_0^{\omega_m} \omega^2\, d\omega \left[\frac{\hbar\omega}{2} + \frac{\hbar\omega}{\exp(\hbar\omega/kT) - 1}\right]$$

$$= \frac{9N_0}{\omega_m^3}\left[\frac{\hbar\omega_m^4}{8} + \int_0^{\omega_m}\frac{\hbar\omega^3\, d\omega}{\exp(\hbar\omega/kT) - 1}\right]$$

$$= 9N_0\left[\frac{\hbar\omega_m}{8} + \frac{1}{\omega_m^3}\int_0^{\omega_m}\frac{\hbar\omega^3\, d\omega}{\exp(\hbar\omega/kT) - 1}\right] \qquad (17\cdot14\text{-}6)$$

The first term on the r.h.s. in Eq. (17·14-6) arises out of the contributions from the zero point energy of the oscillators. Since it is independent of T is does not contribute anything to C_v.

Writing $x = \hbar\omega/kT$ and $x_m = \hbar\omega_m/kT$, we then get from Eq. (17·14-6) by differentiation

$$C_v = \left(\frac{\partial U}{\partial T}\right)_v = \frac{9N_0}{\omega_m^3}\int_0^{\omega_m}\frac{\hbar\omega^3\,(\hbar\omega/kT^2)\exp(\hbar\omega/kT)\, d\omega}{[\exp(\hbar\omega/kT) - 1]^2}$$

$$= \frac{9 N_0}{\omega_m^3} \int_0^{x_m} \frac{\hbar^2}{kT^2} \left(\frac{kT}{\hbar}\right)^5 \frac{x^4 (\exp x)\, dx}{(\exp x - 1)^2}$$

$$= 3R \cdot 3\left(\frac{T}{T_D}\right)^3 \int_0^{x_m} \frac{x^4 \exp x\, dx}{(\exp x - 1)^2} \qquad (17\cdot14\text{-}7)$$

Here we have written the *Debye temperature* T_D as

$$T_D = \hbar \omega_m / k \qquad (17\cdot14\text{-}8)$$

The integral in Eq. (17·14-7) is a function of $x_m = \hbar \omega_m / kT = T_D/T$. Hence we get

$$C_v = 3R\, F_D(T_D/T) \qquad (17\cdot14\text{-}9)$$

Here the *Debye function* F_D is given by

$$F_D(T_D/T) = 3\left(\frac{T}{T_D}\right)^3 \int_0^{T_D/T} \frac{x^4 \exp x}{(\exp x - 1)^2}\, dx \qquad (17\cdot14\text{-}10)$$

Eq. (17·14-10) cannot be integrated in finite terms. T_D is determined by numerical integration for different values of x_m from which the values of C_v for different temperatures are obtained. These are shown graphically in Fig. 17·28.

From Eq. (17·14-10), we get after integrating by parts

$$\int_0^{x_m} \frac{x^4 (\exp x)\, dx}{(\exp x - 1)^2} = -\left.\frac{x^4}{\exp x - 1}\right]_0^{x_m} + 4 \int_0^{x_m} \frac{x^3\, dx}{\exp x - 1}$$

$$= 4 \int_0^{x_m} \frac{x^3\, dx}{\exp x - 1} - \frac{x_m^4}{\exp x_m - 1} \qquad (17\cdot14\text{-}11)$$

Hence we get from Eq. (17·14-7)

$$C_v = 3R \cdot \frac{3}{x_m^3}\left[4 \int_0^{x_m} \frac{x^3\, dx}{\exp x - 1} - \frac{x_m^4}{\exp x_m - 1}\right]$$

$$= 3R\left[\frac{12}{x_m^3} \int_0^{x_m} \frac{x^3\, dx}{\exp x - 1} - \frac{3 x_m}{\exp x_m - 1}\right] \qquad (17\cdot14\text{-}12)$$

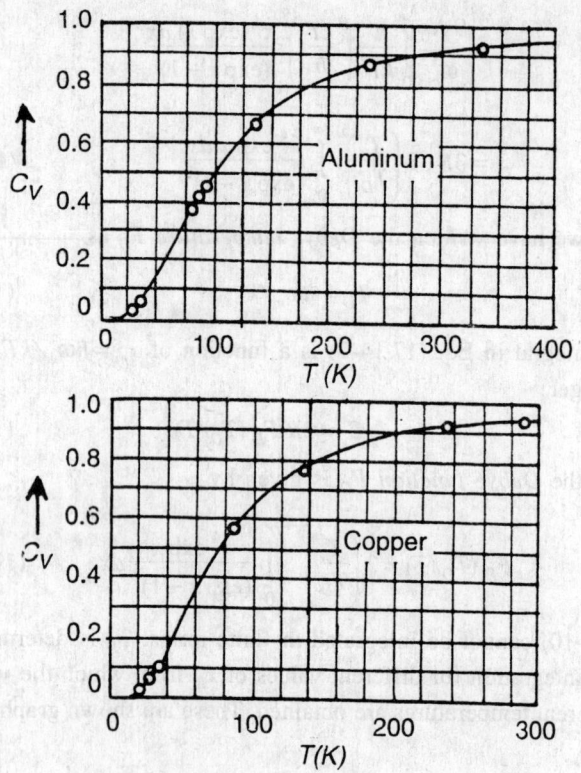

Fig. 17·28. Cv vs. T for Al ($T_D = 396$ K) and Cu ($T_D = 309$ K) in units of 3R after Debye

We now consider Eq. (17·14-12) for the two limiting cases of high and low temperatures.

(a) For $T >> T_D$: If ω_m is such that $\hbar \omega_m << kT$, then $x_m = \hbar \omega_m / kT << 1$. Obviously in this case $x = \hbar \omega / kT << 1$, since $\omega < \omega_m$. So we have

$$\exp x - 1 \approx x \quad \text{and} \quad \exp x_m - 1 \approx x_m$$

Hence
$$C_v = 3R \left[\frac{12}{x_m^3} \int_0^{x_m} \frac{x^3 \, dx}{x} - \frac{3x_m}{x_m} \right]$$

$$= 3R \left[\frac{12}{x_m^3} \cdot \frac{x_m^3}{3} - 3 \right] = 3R$$

This result is in agreement with Dulong-Petit's law.

(b) For $T << T_D$: In this case $x_m >> 1$. So we have for $x_m \to \infty$.

Structure and Properties of Solids

$$\frac{3x_m}{\exp x_m - 1} \approx 3x_m \exp(-x_m) \to 0$$

Again
$$\int_0^{x_m} \frac{x^3\, dx}{\exp x - 1} = \int_0^{x_m} x^3 \exp(-x) [1 - \exp(-x)]^{-1}\, dx$$

$$= \int_0^{x_m} x^3 \exp(-x) [1 + \exp(-x) + \exp(-2x) + \ldots]\, dx$$

$$\therefore \lim_{x_m \to \infty} \int_0^{x_m} \frac{x^3\, dx}{\exp x - 1}$$

$$= \int_0^{\infty} x^3 [\exp(-x) + \exp(-2x) + \ldots]\, dx$$

$$= \sum_{n=1}^{\infty} \int_0^{\infty} x^3 \exp(-nx)\, dx$$

But if r is an integer then

$$\int_0^{\infty} x^r \exp(-nx)\, dx = \frac{r!}{n^{r+1}}$$

Hence
$$\int_0^{\infty} x^3 \exp(-nx)\, dx = \frac{6}{n^4}$$

Also
$$\sum_{n=1}^{\infty} \frac{1}{n^4} = \frac{\pi^4}{90}$$

Hence we get

$$\sum_{n=1}^{\infty} \int_0^{\infty} x^3 \exp(-x)\, dx = \frac{6\pi^4}{90} = \frac{\pi^4}{15}$$

So we have at low temperatures

$$C_v = 3R \cdot \frac{12}{x_m^3} \int_0^{\infty} \frac{x^3\, dx}{\exp x - 1}$$

$$= 3R \cdot \frac{12}{x_m^3} \cdot \frac{\pi^4}{15}$$

$$= 3R \cdot \frac{4\pi^4}{5} \left(\frac{T}{T_D}\right)^3 \qquad (17\cdot14\text{-}13)$$

Thus at very low temperatures $C_v \propto T^3$. This is known as Debye's T^3 law. According to this, $C_v \to 0$ as $T \to 0$.

Debye's T^3 law has been verified experimentally for various materials. From Fig. 17·28, it will be seen that the agreement between the theoretical and experimental value of C_v for copper is fairly good. This is also true for other metals. According to Eq. (17·14-13), $\sqrt[3]{C_v}/T$ should be a constant. This is found to be approximately true in the case of calcium fluoride as can be checked from column 3 of Table 17·4.

Though Debye's theory has been generally successful in explaining the observed variation of C_v at low temperatures, deviations are observed in some cases when comparisons are made with very accurately measured values. According to Debye theory, the T^3 law should be valid for temperatures $T < 0 \cdot 1 \, T_D$. However this is not always the case, as can be seen from Table 17·5. The values of T_D determined from the measured values of C_v using Debye formula (Eq. 17·14-13) and listed in the table are found to differ somewhat at different temperatures. According to Debye's theory T_D should be the same for all temperatures.

Table 17·4

T (K)	C_v (J/K – mole)	$\sqrt[3]{(C_v/T)}$
17·5	0·0670	0·0232
19·9	0·1028	0·0236
21·5	0·1316	0·0237
23·5	0·1680	0·0235
25·6	0·2180	0·0235
27·6	0·2760	0·0236
29·1	0·3310	0·0238
34·0	0·5360	0·0239
36·8	0·6630	0·0237
37·8	0·7130	0·0238
39·8	0·8360	0·0239

Table 17.5

NaCl		KCl	
T(K)	T_D (K)	T(K)	T_D (K)
20	288	14	213
15	297	8	222
10	308	4	236
		3	227

The main limitation of Debye's theory is the assumption that a solid behaves like a continuous medium in which the elastic waves are set up. This cannot be a valid assumption, specially at high frequencies, since the solid is composed of atoms separated in space.

Using Eq. (17·14-5) we can estimate v_m, the cut off frequency used in Debye's theory. Since $N_0/V \sim 10^{28}$ atoms per m³ and the velocity of the elastic waves, $\upsilon \sim 10^3$ m/s, we get

$$v_m^3 \approx 10^{38}$$

so that $\quad v_m = \dfrac{\omega_m}{2\pi} = 10^{12}$ to 10^{13} hertz

This gives a wavelength

$$\lambda_m = \frac{\upsilon}{v_m} \sim \frac{10^3}{10^{13}} = 10^{-10}\,\text{m} = 1\,\text{Å}$$

Thus λ_m is of the same order as the atomic spacings within a crystal. So the assumption of a continuous medium cannot be valid at such frequencies. The relationship between the frequency and the wave number (dispersion relation) shows considerable dispersion at higher frequencies because of the above reason, whereas in Debye's theory no such dispersion is assumed upto the cut-off frequency v_m. Thus one has to take into account the dependence of the velocity of the elastic waves on the wavelength which would modify Debye's formula.

A modification in Debye's theory was introduced by Max Born who proposed a different cut-off procedure. According to Born, this should be based on a common minimum cut-off wavelength for the longitudinal and transverse waves. This would make v_m different for these two types of waves so that the Debye temperature and thence the Debye functions for these would be different. The expression for C_v would then involve the sum of these two functions. Such a procedure is in line with the theory of lattice vibrations developed by Born and von Karman.

Again in Debye's theory, a solid is regarded as isotropic. However as discussed in § 17·10, a crystalline solid shows marked anisotropy in many of its physical properties. Thus the velocity of the elastic waves is usually different in different directions. Slater has discussed this and has shown how the theory should be modified because of such variations.

17.15 Phonon

While discussing about Debye's theory of specific heat, we have seen that the vibrations of the N atoms in a crystal lattice are equivalent to $3N$ normal modes of vibration. Each normal mode has an energy equal to the energy of a linear hormonic oscillator which oscillates independently of the others. According to the theory of vibrations, such an oscillator may be regarded as having mass equal to that of the atoms in the crystal while its frequency is equal to that of the normal mode. Since the total vibrational energy of the N atoms within the crystal is equal to the total energy of the $3N$ normal modes excited within the crystal, it must be equal to the total vibrational energy of the $3N$ linear harmonic oscillators mentioned above.

According to quantum mechanics, a linear harmonic oscillator can exist in a discrete energy level of energy

$$E_n = \left(n + \frac{1}{2}\right)\hbar\omega$$

n can be 0, 1, 2, 3 *etc*. These energy levels are equispaced, their spacings being equal to $\hbar\omega = h\nu$.

When a crystal is heated, each of the linear harmonic oscillators can absorb a minimum amount of energy h_ν. This minimum amount of energy by which the vibrational energy of an oscillator can change at a time may be regarded as the quantum of energy associated with the elastic vibrations of the oscillators. They are called *phonons*. They are analogous to the quantum of energy associated with the electromagnetic oscillations which are known as photons. Like the photons, the phonons also carry an energy $h\nu$ each. If λ be the wavelength of the elastic waves and υ is their velocity in the crystal, then the momentum of the phonons is $p = h\nu/\upsilon = h/\lambda$.

In Ch. III, Planck's formula for black-body radiation was deduced considering the total number of standing waves set up inside a cavity within a given frequency range and multiplying it by the mean energy of the oscillators associated with these e.m. waves. An alternative way of deducing Planck's radiation formula was proposed by the famous Indian scientist S.N. Bose by regarding the cavity to be full of photons of energy $nh\nu$ (n = integer) and taking the *indistinguishability* of the photons into account (see Ch. XX). According to Bose, the resulting *statistics* obeyed by the *photon gas* is different from the classical statistics of Maxwell and Boltzmann. The new statistics proposed by Bose and later elaborated by

Structure and Properties of Solids

Einstein is known as *Bose-Einstein statistics*.

Like the photons, the phonons obey Bose-Einstein statistics. The number of phonons in the angular frequency range ω to $\omega + d\omega$ is then given by (see Eq. 20·8·7)

$$N(\omega)\, d\omega = \frac{z(\omega)\, d\omega}{\exp(\alpha + \hbar\omega/kT) - 1} \tag{17·15-1}$$

where $z(\omega)\, d\omega$ is the number of possible normal modes of vibration between ω and $\omega + d\omega$. As in the case of photons, the number of phonons is not constant with respect to temperature. In this case $\alpha = 0$ and we get

$$N(\omega)\, d\omega = \frac{z(\omega)\, d\omega}{\exp(\hbar\omega/kT) - 1}$$

$z(\omega)\, d\omega$ which gives the density of states for the *phonon gas* at the angular frequency ω is given by Eq. (17·14-4):

$$z(\omega)\, d\omega = \frac{V \omega^2\, d\omega}{2\pi^2 v^3}$$

where
$$\frac{1}{v^3} = \frac{1}{v_l^3} + \frac{2}{v_t^3}$$

We then get

$$N(\omega)\, d\omega = \frac{V}{2\pi^2 v^3} \cdot \frac{\omega^2\, d\omega}{\exp(\hbar\omega/kT) - 1} \tag{17·15-2}$$

This is the same as Eq. (20·9-4) for the frequency distribution of the photon gas first deduced by S. N. Bose, if we write $\omega = 2\pi\nu$ and replace c by v.

Since each phonon carries an energy $\hbar\omega$, the total energy of all the phonons in the frequency range ω and $\omega + d\omega$ is

$$dU = N(\omega)\, \hbar\omega\, d\omega$$
$$= \frac{V\hbar}{2\pi^2 v^3} \cdot \frac{\omega^3\, d\omega}{\exp(\hbar\omega/kT) - 1} \tag{17·15-3}$$

Integrating over all angular frequencies from 0 to ω_m, we get the total internal energy U of the crystal which comes out to be the same as given by Eq. (17·14-6) except for the contribution from the zero-point energy. The difference is due to the fact that the zero-point energy is not connected with the phonon distribution in any way. Since the zero-point energy term is independent of temperature, it does not contribute to the specific heat so that we get the same expression for C_v as given by Eq. (17·14-7) by carrying out calculations from the phonon point of view.

The above discussion shows that the crystal may be treated as an ensemble of distinguishable harmonic oscillators obeying Maxwell-

Boltzmann statistics or as a gas of phonons (*i.e.*, vibrational quanta), which are indistinguishable and which obey Bose-Einstein statistics.

The phonon concept has been found useful in explaining many other properties of solids apart from the specific heat. For example, the thermal conductivity of solids can be understood by considering the motion of the phonons in the solid much like the motion of the molecules in a gas. As is well-known the thermal conductivity of a gas can be explained on the basis of the kinetic theory of the transport of heat energy due to the motion of the gas molecules. Similar considerations can be applied to the motion of the phonons. It is necessary to know the mean free path against phonon-phonon interaction to carry out such calculations which are quite difficult to do as was shown by R. Peierls (1929).

Another field in which the concept of the phonon has been successfully applied is super-conductivity (see Ch. XVIII). The formation of the so called Cooper pairs which constitutes the basis of the microscopic theory of super-conductivity is believed to be due to the electron-phonon-electron interaction.

Problems

1. Prove that in a three dimensional lattice, the separation between the lattice planes (*hkl*) along the three crystal axes **a**, **b**, **c** are a/h, b/k and c/l respectively.

2. X-rays of wavelength 1.2 Å suffer first order reflection from the (100) plane of a cubic crystal at the glancing angle of 13.5°. Calculate the separation between the (100) planes of this crystal. (2.57 Å)

3. The atomic weight, density (ρ), bulk modulus (K) and shear modulus (n) of copper are respectively 63.5, 8.96×10^3 kg/m³, 10.2×10^{10} Pa and 4.6×10^{10} Pa.

   Calculate the velocities of transverse and longitudinal waves in copper using the formulae $v_t = \sqrt{n/\rho}$ and $v_l = \sqrt{(K + 4n/3)/\rho}$. Hence find the maximum cut off frequency and the Debye temperature of copper.

   (2.26×10^3 m/s ; 4.27×10^3 m/s ; 6.9×10^{12} ; 331 K)

4. If the Debye temperature of platinum is 240 K what are the heat energies required to heat 1 g of platinum from 5 K to 10 K and from 500 K to 600 K ?

5. The Debye temperature of silver is 210 K. Calculate the specific heat of silver at 20 K using T^3 law and compare it with the experimental value of 1.672 J/K-mole. What will be the value of the specific heat if it is calculated using Einstein's formula. Assume the Einstein temperature to be equal to the Debye temperature for silver. (1.67 J/K-mole ; 0.076 J/K-mole)

6. If T_{D1} and T_{D2} are the Debye temperatures of two crystals such that $T_{D1} > T_{D2}$, which of them has a higher low temperature specific heat ?

7. Prove that for an infinite one-dimensional chain of equidistant ions, the Madelung constant is 2 ln 2.

8. One thousand small cubes of sides 1 mm each are arranged face to face to construct a larger cube of sides 1 cm. Calculate the ratio of the number of common corners of groups of eight adjacent small cubes within the large cube to the total number of small cubes. Calculate the same ratio when the number of small cubes making up a larger cube is $10^6, 10^{12}$ and 10^{21} respectively. (0·729 ; 0·9703 ; 0·9997 ; 0·9999997)

9. Calculate by Debye method the specific heat C_l of a one-dimensional chain of N identical atoms and length l. Show that in the limit of $T \to 0$, $C_l = \pi^2 RT/3T_D$. What is the high-temperature limit of C_l?

10. Using Debye method, calculate the specific heat C_S of a two-dimensional periodic crystal lattice consisting of N atoms and an area S. Show that in the limit of $T \to 0$, $C_S \propto T^2$. What is the high-temperature limit of C_S?

11. In an electron diffraction experiment from a polycrystalline film the diffraction rings are formed on a flat screen at a distance 25 cm from the atom. If the diameter of the first ring is 12·5 mm, calculate the lattice constant if the lattice is a body centered cubic. The energy of the electron beam is 20 keV.

18

METALS AND SEMI-CONDUCTORS

18·1 Specific heat of metals

Metals, like other crystalline solids, have atoms in the ionized states at the lattice sites which vibrate about their equilibrium positions. In addition there are free electrons which move about randomly inside the metals like the molecules in a gas. Free electron theory of the metals was first proposed by H.A. Lorentz, P. Drude and others at the beginning of the twentieth century to explain the high electrical conductivity of the metals. This has later found justification on the basis of the band theory of solids to be discussed in § 18·8. There are evidences to indicate that the density of the free electrons in the metals is of the same order as the density of the atoms at the lattice sites.

When a metal is heated, the heat energy supplied to it can be distributed in three different ways. A part goes to increase the kinetic energy of the vibrating atoms while another part goes to increase the vibrational potential energy of the atoms due to the increase of their amplitudes of vibration. Finally, a part of the supplied heat energy is shared by the free conduction electrons whose random velocities are thereby increased. According to the law of equipartition of energy, the mean kinetic and potential energies of the metallic atoms at the lattice sites at the absolute temperature T are $3kT/2$ each where k is Boltzmann constant. In addition, the mean kinetic energy of the free electrons is also $3kT/2$. Assuming the number density of the free electrons in the metal to be equal to the number density of the atoms at the lattice sites, the total internal energy of the metal per mole at the temperature T will be

$$U = N_0 \left(\frac{3}{2} kT + \frac{3}{2} kT + \frac{3}{2} kT \right) = \frac{9}{2} RT \qquad (18\cdot1\text{-}1)$$

where N_0 is the Avogadro number and R is the universal gas constant. Hence the molar specific heat of the metal should be

$$C_v = \left(\frac{\partial U}{\partial T} \right)_v = \frac{9}{2} R = 9 \text{ cal/mole} \qquad (18\cdot1\text{-}2)$$

However the experimental value of C_v at relatively higher temperatures is found to be 6 cal/mole in agreement with Dulong-Petit's law. In other words, the metallic specific heat has the same value as would be obtained by neglecting the contribution from the motion of the free electrons altogether and is equal to the lattice specific heat $C_l = 3N_0 k$.

To explain the above anomaly it was pointed out by W. Pauli and A. Sommerfeld that we cannot apply the classical statistics of Maxwell and Boltzmann to the free electron gas in metals on the basis of which the equipartition law mentioned above is deduced. According to them, the electron gas in metals is degenerate and is governed by the Fermi-Dirac quantum statistics (see Ch. XX) which gives the number of electrons in the energy interval $d\varepsilon$ at ε per unit volume at the temperature T as (Eq. 20·7-9)

$$n(\varepsilon)\, d\varepsilon = \frac{\sqrt{2m^3}}{\pi^2 \hbar^3} \cdot \frac{\sqrt{\varepsilon}\, d\varepsilon}{\exp(\varepsilon - \varepsilon_f)/kT + 1} \qquad (18\cdot1\text{-}3)$$

Here m is the electron mass, $\hbar = h/2\pi$ where h is Planck's constant. ε_f is a constant known as the *Fermi energy*.

In Fig. 18·1. the Fermi-Dirac energy distribution law given by Eq. (18·1-3) at $T = 0$ K is shown graphically. For $\varepsilon < \varepsilon_f$, the exponential in the

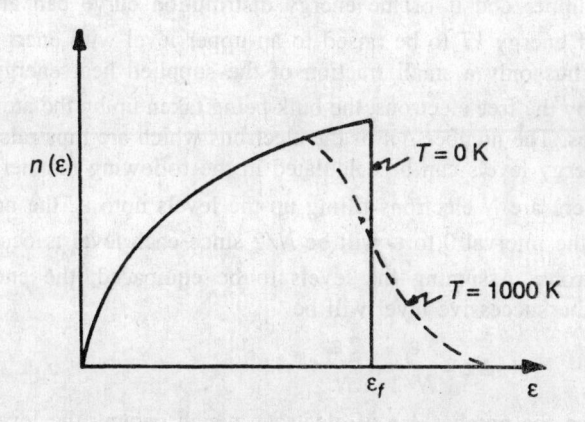

Fig. 18·1. Fermi-Dirac energy distribution curve $n(\varepsilon)$ $v\varepsilon$.

denominator is zero at $T = 0$ K which means that $n(\varepsilon) \propto \sqrt{\varepsilon}$. On the other hand for $\varepsilon > \varepsilon_f$, the exponential term is infinite at $T = 0$ K so that $n(\varepsilon) = 0$ for $\varepsilon > \varepsilon_f$. This means that at absolute zero, the maximum permissible energy of the electrons in the metal is equal to the Fermi energy ε_f.

Since the electrons obey Pauli's exclusion principle, each energy level can be occupied by only two electrons with spins aligned oppositely. Thus the electrons in the metal fill up all the energy levels starting from the bottom upto the highest at $\varepsilon = \varepsilon_f$, as shown in Fig. 18·2. As we shall see later ε_f is of the order of a few electron volts for most metals. On the

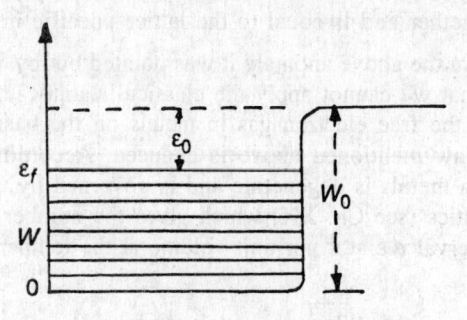

Fig. 18·2. Electronic levels in a metal.

other hand, when the metal is heated from 0 K to a few hundred kelvins, the energy gained by an electron is on the average kT which is of the order of one tenth of an electron volt or less for $T \sim 1000$ K. Since all the lower energy levels upto ε_f are filled up, the electrons in these levels have no available empty levels to which they can go by absorbing such a small amount of energy. Only the electrons near the top of the filled levels, *i.e.*, near the upper end ε_f of the energy distribution curve can absorb the amount of energy kT to be raised to an upper level with energy greater than ε_f. Thus only a small fraction of the supplied heat energy can be taken up by the free electrons, the bulk being taken up by the atoms at the lattice sites. The number Δn of the electrons which are thus raised to the higher energy levels can be calculated in the following manner.

If there are N electrons filling up the levels upto ε_f, the number of levels in the interval 0 to ε_f will be $N/2$ since each level is occupied by two electrons. Assuming the levels to be equipaced, the energy gap between the successive level will be

$$\Delta \varepsilon = \frac{\varepsilon_f}{N/2} = \frac{2\varepsilon_f}{N}$$

Hence the number Δn of electrons which occupy the levels in the energy interval kT just below the Fermi level ε_f is

$$\Delta n = \frac{kT}{\Delta \varepsilon} = \frac{NkT}{2\varepsilon_f}$$

For $T = 300$ K, $kT \sim 0.026$ eV so that for a typical metal e.g., silver for which $\varepsilon_f = 5.46$ eV, we get

$$\frac{\Delta n}{N} = \frac{0.026}{2 \times 5.46} \approx 0.00237$$

Thus the number of electrons thermally excited is only about

0·24% of the total number of free electrons present in the metal. The energy absorbed by these electrons from the supplied heat energy is

$$\Delta U_e = \Delta n \cdot kT = \frac{Nk^2 T^2}{2\varepsilon_f}$$

The contribution to the specific heat of the metal due to the free electrons then comes out to be

$$C_e = \frac{\partial U_e}{\partial T} = \frac{Nk^2 T}{\varepsilon_f} \qquad (18 \cdot 1\text{-}4)$$

A more precise treatment gives

$$C_e = \frac{\pi^2 Nk^2 T}{2\varepsilon_f} \qquad (18 \cdot 1\text{-}5)$$

Since the lattice contribution to the specific heat is $C_l = 3Nk$ we get

$$\frac{C_e}{C_l} = \frac{\pi^2 Nk^2 T}{2\varepsilon_f \times 3Nk} = \frac{\pi^2 kT}{6\varepsilon_f} = \frac{0 \cdot 04 \pi^2}{6 \times 5 \cdot 46} = 0 \cdot 012$$

Thus $C_e \ll C_l$ which means that there is very little contribution to the metallic specific heat from the free electrons inside the metal. The specific heat of the metal is almost entirely equal to the lattice specific heat and assumes the value $3N_0 k = 3R \sim 6$ cal per mole at higher temperatures in agreement with Dulong-Petit law.

At very low temperatures, however, since Debye's T^3-law holds for the lattice specific heat, the contribution from the free electrons may become significant, because C_e decreases much more slowly than C_l in this region.

18·2 Metallic conductivity ; Free electron theory

Amongst the most important properties of metals are their high thermal and electrical conductivities. It was Paul Drude in Germany who first tried to explain these properties by assuming that there were large concentrations of mobile free electrons in metals (1900). Drude's suggestions were latter elaborated by H.A. Lorentz to take into account the velocity distribution of the electrons which were assumed to obey the classical Maxwell-Boltzmann distribution law (1909).

According to Drude-Lorentz model, the free electrons move through the lattice structure of the metal and suffer repeated random collisions with the lattice vibrations, defects and impurities, much like the collisions between the gas molecules in a given volume of a gas. For this reason, the free electrons in the metals are termed *electron gas*. The mutual repulsion between the electrons is neglected in this model.

Because of the randomness of the motion of the electrons in a metal, there is no net electric current flowing in the metal unless an electric field is applied. When an electric field is applied across a metal, the free electrons acquire a *drift motion* in the direction of the applied field in addition to their random thermal motion. This gives rise to the flow of an electric current through the metal.

If a potential difference V is applied between the two ends of a metal rod of length L and cross-sectional area A, then the force acting on each electron carrying a charge e is

$$F = -\frac{Ve}{L} = -Xe$$

where $X = V/L$ is the electric field. The resulting acceleration of the electron in the field direction is

$$f = -\frac{Xe}{m}$$

The negative sign is due to the electron charge being negative.

If the electron motion under the action of the applied electric field were unopposed, the drift velocity of the electron due to the above acceleration would go on increasing indefinitely. Actually this does not happen, since according to Drude-Lorentz theory the electrons suffer collisions with the lattice vibrations, defects and impurities. The drift velocity gained by an electron due to acceleration by the applied electric field which is superimposed upon its random thermal velocity is lost completely as a result of the collision and only the random velocity remains. After this, the electron again begins to get accelerated by the applied field and loses the drift velocity thus gained at the next collision. The process goes on repeating so that we can talk of a constant average drift velocity v_d by the action of the field when a large number of collisions is considered. We can look upon the whole process by supposing that a resistive frictional force F_r acts upon the electron, opposing its gain of drift velocity due to the accelerating field.

The probability of the electron suffering a collision in time dt may be written as dt/τ where τ is taken to be a constant in the first approximation having the dimension of time. The rate of change of the drift velocity due to the action of the field may be written as

$$\left(\frac{dv_d}{dt}\right)_F = f = -\frac{Xe}{m} \qquad (18\cdot2\text{-}1)$$

The rate of change of v_d due to collisions at the lattice sites will be

$$\left(\frac{dv_d}{dt}\right)_C = -\frac{v_d}{\tau} \qquad (18\cdot2\text{-}2)$$

since $1/\tau$ is the probability of collision per second. Hence in the steady state we have

$$\left(\frac{dv_d}{dt}\right)_F + \left(\frac{dv_d}{dt}\right)_C = -\frac{Xe}{m} - \frac{v_d}{\tau} = 0 \qquad (18\cdot2\cdot3)$$

$$\therefore \qquad v_d = -\frac{Xe\tau}{m} = -\frac{Ve}{mL}\cdot\tau = -\frac{Ve}{L}\cdot\frac{\tau}{m} \qquad (18\cdot2\text{-}4)$$

If the electric field X is suddenly withdrawn at some instant $t=0$ when the electrons have acquired a drift velocity v_d, then due to repeated collisions at the lattice sites, the drift velocity gradually falls to zero and the electron velocity becomes completely random again. How soon this will happen is determined by the time τ. In this case, since $(dv_d/dt)_F = 0$, we can write

$$\frac{dv_d}{dt} = \left(\frac{dv_d}{dt}\right)_C = -\frac{v_d}{\tau} \qquad (18\cdot2\text{-}5)$$

which gives on integration

$$v_d = v_{do} \exp(-t/\tau) \qquad (18\cdot2\text{-}6)$$

where v_{do} is the average drift velocity at $t=0$. Because of the exponential fall of v_d with time, τ has been given the name *relaxation time*. For good conductors, e.g., copper, silver etc., $\tau \sim 10^{-14}$ s.

It can be shown that τ is equal to the mean time interval between successive collisions of the electrons at the lattice sites.

Let $P(t)$ be the probability that an electron does not suffer any collision upto the time t after suffering a collision at $t=0$. Since the probability of suffering a collision in an interval of time dt is dt/τ (see above), the probability that the electron does not suffer any further collision between t and $t+dt$ is $(1-dt/\tau)$. So the probability of the electron not suffering any collision upto the time $t+dt$ is given by

$$P(t+dt) = P(t)(1-dt/\tau)$$

Expanding the left hand side by Taylor series we get, after neglecting the higher order terms

$$P(t) + \frac{dP}{dt} dt = P(t) - \frac{P(t)}{\tau} dt$$

$$\therefore \qquad \frac{dP}{dt} = -\frac{P(t)}{\tau}$$

Integrating we get $\qquad P(t) = \exp(-t/\tau)$.

This gives $P(0) = 1$ as it should be. The mean free time between collisions will then be

$$\langle t \rangle = \int_0^\infty t\, P(t)\, dt = \int_0^\infty t \exp(-t/\tau)\, dt = \tau \qquad (18\cdot2\text{-}7)$$

The above result is true if it is assumed that there is no persistence of velocities after the collisions.

If the mean free path of the electrons between successive collisions is λ and the mean velocity of the electrons due to thermal motion is $\langle c \rangle$, then assuming $\upsilon_d \ll \langle c \rangle$ we get

$$\tau = \lambda/\langle c \rangle \qquad (18\cdot2\text{-}8)$$

which gives
$$\upsilon_d = -\frac{Ve}{mL}\tau = -\frac{Ve}{mL}\frac{\lambda}{\langle c \rangle} \qquad (18\cdot2\text{-}9)$$

Since the electric field $X = -V/L$, the mobility of the electron, defined as the drift velocity gained in a unit electric field, will be

$$u = \frac{\upsilon_d}{X} = \frac{e\tau}{m} = \frac{e\lambda}{m\langle c \rangle} \qquad (18\cdot2\text{-}10)$$

If there are n free electrons per unit volume of the metal, the electric current flowing through the rod due to the applied electric field will be

$$I = -ne\, A\upsilon_d = \frac{ne^2 A\lambda}{mL\langle c \rangle}\cdot V$$

So the current density is

$$j = \frac{I}{A} = \frac{ne^2 \lambda\, V}{mL\langle c \rangle}$$

From Ohm's law, we have $I = V/R$ where R is the resistance of the metal rod which is thus given by

$$R = \frac{V}{I} = \frac{mL\langle c \rangle}{ne^2 \lambda A} = \rho\frac{L}{A}$$

where ρ is the resistivity of the metal. So we get

$$\rho = \frac{m\langle c \rangle}{ne^2 \lambda} = \frac{m}{ne^2 \tau} \qquad (18\cdot2\text{-}11)$$

The conductivity of the metal is

$$\sigma = \frac{1}{\rho} = \frac{ne^2 \lambda}{m\langle c \rangle} = \frac{ne^2 \tau}{m} \qquad (18\cdot2\text{-}12)$$

In the above treatment τ has been assumed constant. Actually τ is a function of the velocity of the electrons. A more rigorous treatment using *Boltzmann transport equation* gives in place of Eq. (18·2-4) the following expressions for the mean drift velocity υ_d

$$\upsilon_d = -\frac{Xe}{m}\bar{\tau} \qquad (18\cdot2\text{-}13)$$

where the mean relaxation time $\bar{\tau}$ is defined as

$$\bar{\tau} = \frac{<v^2\tau>}{<v^2>} \quad (18\cdot 2\text{-}14)$$

Here it is assumed that the electron gas obeys Maxwell-Boltzmann statistics.

The electrical conductivity then comes out to be (see Eq. 18·2-12)

$$\sigma = \frac{ne^2}{m}\bar{\tau} \quad (18\cdot 2\text{-}15)$$

If we assume to a first approximation that τ on the r.h.s. of Eq. (18·2-14) is independent of the velocity, then we get by writing $\tau = \lambda/v$ where λ is also assumed to be independent of v,

$$\bar{\tau} = \frac{<v^2\tau>}{<v^2>} = \frac{<v^2\lambda/v>}{<v^2>} = \frac{\lambda<v>}{<v^2>}$$

But $<v> = \sqrt{8k\bar{T}/\pi m}$

according to Maxwell–Boltzmann statistics. Also since $m<v^2>/2 = 3kT/2$ we get

$$\bar{\tau} = \frac{\lambda<v>}{(3\,kT/m)}$$

$$= \frac{\lambda m}{3kT}\sqrt{\frac{8kT}{\lambda m}} \quad (18\cdot 2\text{-}16)$$

We then get from Eq. (18·2-15)

$$\sigma = \frac{ne^2\lambda}{3}\sqrt{\frac{8}{\lambda mkT}} \quad (18\cdot 2\text{-}17)$$

Assuming n and λ to be independent of T we then get $\sigma \propto 1/\sqrt{T}$. This does not agree with experimental results which is the main drawback of Drude's theory.

We have seen in § 18·1 that in order to explain the observed specific heat of metals, it is necessary to assume that the electron gas in the metals is degenerate and obeys Fermi-Dirac statistics. It is therefore necessary to treat the problem of the electrical conductivity of the metals on the basis of F – D statistics. This will modify the theory as developed above. The difference between the two cases can be understood as follows.

If the Maxwell-Boltzmann statistics is considered, the velocity distribution curve as a whole is shifted towards higher velocities due to the effect of the applied electric field X. This is illustrated in Fig. 18·3, considering only the x-component of the velocity for the sake of simplicity. The shift in the velocity distribution curve along x is through the drift velocity v_d. The nature of the distribution function in this case is of the form

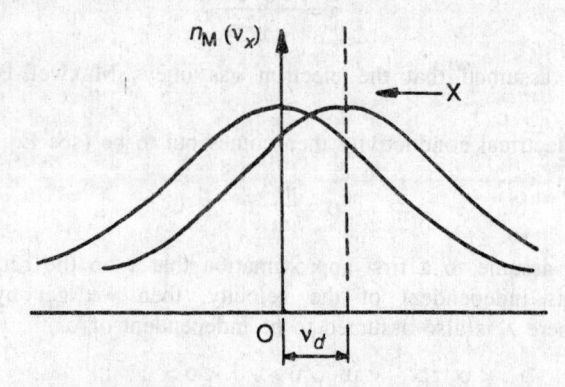

Fig. 18·3. Shift in the M—B velocity distribution curve at higher temperature.

$$n_M(v_x) = A \exp(-mv_x^2/2kT)$$

In the presence of the field this is changed to

$$n'_M(v_x) = A' \exp\{-m(v_x - v_d)^2/2kT\}$$

Thus the maximum of the distribution is shifted from $v_x = 0$ to $v_x = v_d$ where v_d is given by Eq. (18·2-4).

On the other hand if the Fermi-Dirac distribution is considered the one-dimensional distribution function has the form shown in Fig. 18·4. at the temperature T K. At 0 K, the distribution function has a sharp cut off at $v = v_f$, where v_f is determined by the Fermi energy $\varepsilon_f = mv_f^2/2 = 3k\, T_f/2$.

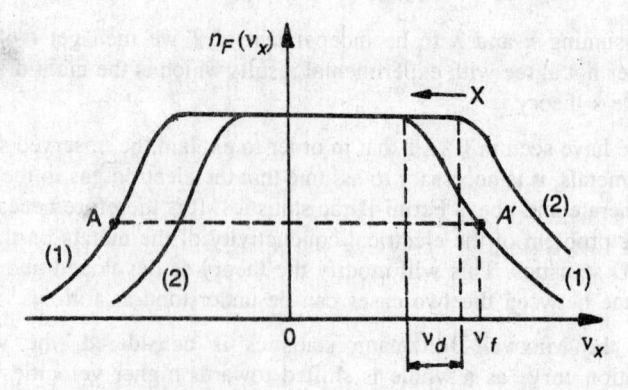

Fig. 18·4. Shift in the F-D velocity distribution curve at higher temperature.

As the temperature rises, the distribution function is rounded off near the edge as shown in the figure, all the lower energy levels remaining unaffected because of Pauli's exclusion principle. The external field can act effectively on the electrons close to the Fermi level, lifting them to the higher vacant levels from the left of the distribution function to the right. Thus for the degenerate gas in Eq. (18·2-15), we should take the value of the relaxation time corresponding the velocity v_f as defined above :

$$\sigma = \frac{ne^2}{m} \tau_f$$

where $\tau_f = \lambda/v_f$

is the relaxation time.

Thus the effect of using F-D statistics makes a difference only in the details of averaging and does not affect the conductivity by orders of magnitude.

18·3 Wiedemann-Franz law

It is well-known that good conductors of electricity are also good conductors of heat. G. Wiedemann and P. Franz (1853) noted that the ratio of the thermal conductivity K to the electrical conductivity σ of a metal is nearly the same for all metals. This is known as *Wiedemann-Franz law*. L. Lorenz (1881) later found that this correlation between K and σ could be written as

$$L = K/\sigma T = \text{constant} \qquad (18\cdot3\text{-}1)$$

where T is the absolute temperature of the metal. L is called the *Lorenz number*. Lorenz was the first to explain the Wiedemann-Franz law on the basis of the free electron theory of metals.

It is well-known that the thermal conductivity of gases can be explained on the basis of the kinetic theory. When one layer of the gas is heated, the mean thermal energy of the gas molecules in this heated layer is increased. Due to the collisions of the gas molecules in the heated layer with those of the neighbouring unheated layer, the mean thermal energy of these latter molecules is increased. Thus the heat energy is transferred from the heated layer to the neighbouring unheated layer. Subsequently, due to collisions of the molecules of this second layer with those in another layer, the mean energy of the molecules in the third layer is similarly increased. In this way, the heat energy is transferred from the heated region of the gas to the cooler regions.

Assuming the above mechanism and taking into account Maxwell-Boltzmann velocity distribution law of the gas molecules, the following expression for the thermal conductivity is obtained.[*]

---

[*] See *A Treatise on Heat* by Saha and Srivastava.

$$K = \frac{1}{3} \lambda <c> nk \qquad (18\cdot3\text{-}2)$$

where λ is the mean free path and $<c>$ is the mean thermal velocity of the gas molecules. n is the number of gas molecules per unit volume and k is Boltzmann constant so that nk is actually the specific heat of the gas per unit volume.

The above *transport theory* of gaseous thermal conductivity can be applied to the case of a dielectric solid by assuming that the transport of heat energy takes place by the so-called *phonon-phonon* scattering. Actually the process is quite complicated since one has to consider the polarization of the phonons and the frequency distribution of the normal modes, besides different types of scattering processes involving the phonons, impurities *etc*.

In the case of metals, the conduction of heat takes place primarily due to the free electrons. The lattice conduction is relatively much less important in this case. The electronic thermal conductivity, in analogy with the case of the gaseous thermal conductivity as given above, can be written as

$$K_e = \frac{1}{3} n <c> \lambda \frac{d}{dt} <E> = \frac{1}{2} nk <c> \lambda \qquad (18\cdot3\text{-}3)$$

where the mean kinetic energy $<E> = 3kT/2$.

Then from Eqs. (18·2-12) and (18·3-3), w get

$$\frac{K_e}{\sigma} = \frac{nk<c>\lambda}{2} \cdot \frac{m<c>}{ne^2 \lambda} = \frac{km<c>^2}{2e^2} \qquad (18\cdot3\text{-}4)$$

Assuming Maxwell-Boltzmann velocity distribution law to be valid in the case of the free electron gas in the metals, we have

$$<c> = \sqrt{\frac{8kT}{\pi m}}$$

Hence the Lorenz number is given by

$$L = \frac{K_e}{\sigma T} = \frac{4k^2}{\pi e^2} \qquad (18\cdot3\text{-}5)$$

Eq. (18·3-5) is in agreement with the empirical relationship (18·3-1) and gives a theoretical justification of Wiedemann-Franz law.

Using the free electron model and the Fermi-Dirac statistics, the thermal conductivity of the free electron gas is given by[*]

$$K_e = \frac{\pi^2}{3m} k^2 T n \tau \qquad (18\cdot3\text{-}6)$$

where $\tau = \lambda/v_f$ is the relaxation time, λ being the mean free path and

---

[*] See *Introduction to Solid State Physics* by Charles Kittel

v_f the Fermi velocity Then using Eq. (18·2-15) for σ we get

$$L = \frac{K_e}{\sigma T} = \frac{\pi^2}{3}\left(\frac{k}{e}\right)^2 \tag{18·3-7}$$

Substituting the values of k and e we get

$$L = 2\cdot 45 \times 10^{-8} \text{ watt} - \text{ohm/deg}^4$$

The observed values are in good agreement with the theoretical value calculated above at room temperature.

18·4 Thermionic emission ; Work function

We saw in Ch. V that an electron in a metal needs a minimum amount of energy for emission from the metal surface either by photo-electric process or by the heating of the metal. This minimum of energy ε_0 is known as the *work function* of the metal. If e is the electronic charge and ϕ is the potential difference through which an electron must be accelerated in order to gain an energy ε_0, then we may write $\varepsilon_0 = e\phi$. ϕ is also sometimes called the work function. It is measured in volts. ε_0 is usually expressed in electron-volts.

To understand the origin of the work function we note that though according to the free electron theory of metals the electrons are free to move about inside a metal, they cannot normally come out of the metal surface since they find themselves in an attractive electric field due to the positive ions at the lattice sites in the metals. The positive ions generate within the metal a positive potential field which varies periodically from point to point. However for the consideration of the motion of an electron in the metal such a periodically varying potential may, to a first approximation, be replaced by constant potential V_0 throughout the body of the metal due to which the electron will have a negative potential energy $U_0 = -eV_0$.

Thus the electrons in the metal are in a *potential well* of depth U_0 with respect to the potential $U \doteq 0$ in vacuum just outside the metal. So an electron requires a certain minimum amount of energy to be emitted from the metal surface which is the work function referred to above. If the electrons do not have any kinetic energy, then the minimum amount of energy needed for its emission is equal to the depth U_0 of the potential well. This would be true at the absolute zero of temperature if the electrons were to obey the classical Maxwell-Boltzmann statistics, since all the electrons would then be in the lowermost level at the bottom of the potential well.

However as we have seen, the free electron gas in the metal obeys Fermi-Dirac statistics. Because of Pauli's exclusion principle, since no two electrons can occupy the same level, the electrons in the metal fill up the successive energy levels, starting from the lowermost level at the bottom of the potential well upto the Fermi level at the energy ε_f at the absolute zero of temperature. At $T > 0$ K, there will be some spill-over of the electrons from the levels near the top (*i.e.*, at $\varepsilon = \varepsilon_f$) to the higher energy levels, though their number is not very large at temperatures upto a few thousand kelvins. So the minimum energy required for an electron to be lifted to the top of the potential well at $U = 0$ is nearly equal to the difference between the depth U_0 of the well and the energy of the Fermi level :

$$\varepsilon_0 = U_0 - \varepsilon_f$$

The free electron model can be used to deduce the Richardson-Dushman equation for the variation of saturation electron current density evaporated from a heated metallic body with temperature (see Ch. V). In his original deduction, O. W. Richardson (1903) had used the classical Maxwell-Boltzmann velocity distribution law for the free electron gas in a metal and had deduced the following expression for the saturation current at the temperature T K :

$$i_s = AT^{1/2} \exp(-\varepsilon_0/kT) \qquad (18 \cdot 4 \text{-} 1)$$

where A is a constant and ε_0 is the thermonic work function.

However, as seen above this is not a correct approach and we have to use Fermi-Dirac statistics for the velocity distribution of the free electrons in a metal. According to Fermi-Dirac distribution law, the number of electrons per unit volume of the electron gas having energies in the interval ε to $\varepsilon + d\varepsilon$ is given by (see Eq. 20·7-9)

$$n(\varepsilon) d\varepsilon = \frac{1}{2\pi^2} \left(\frac{2m}{\hbar^2}\right)^{3/2} \cdot \frac{\sqrt{\varepsilon}\, d\varepsilon}{\exp(\varepsilon - \varepsilon_f)/kT + 1} \qquad (18 \cdot 4 \text{-} 2)$$

Here m is the electronic mass, k is Boltzmann constant, $\hbar = h/2\pi$ where h is Planck's constant. ε_f is the Fermi energy mentioned before (see § 18·1).

For the emission of electrons ε must be greater than ε_f. In fact $\varepsilon - \varepsilon_f$ should be of the order of a few electron-volts for thermionic emission to take place. On the other hand at temperatures required for appreciable thermonic emission ($T \sim 2000$ K), the mean kinetic energy of the electrons kT is less than about 0·2 ev, *i.e.*, $\varepsilon - \varepsilon_f >> kT$ so that $\exp(\varepsilon - \varepsilon_f)/kT >> 1$. Hence we can neglect 1 in the denominator of Eq. (18·4-2) and write

$$n(\varepsilon) d\varepsilon = \frac{1}{2\pi^2} \left(\frac{2m}{\hbar^2}\right)^{3/2} \exp(\varepsilon_f/kT) \exp(-\varepsilon/kT) \sqrt{\varepsilon}\, d\varepsilon$$

Since $\varepsilon = mv^2/2$, we get the number of electrons in the velocity

Metals and Semi-conductors

interval v to $v + dv$ in a unit spatial volume as

$$n(v)\,dv = \frac{m^3}{\pi^2 \hbar^3} \exp(\varepsilon_f/kT) \times \exp(-mv^2/2kT)\,v^2\,dv \qquad (18\cdot4\text{-}3)$$

If v_x, v_y, v_z be the velocity components along x, y, z then the element of volume in the velocity space in the interval v_x to $v_x + dv_x$, v_y to $v_y + dv_y$ and v_z to $v_z + dv_z$ can be written as $dv_x\,dv_y\,dv_z$ in place of $4\pi v^2\,dv$.

Hence we get in terms of the velocity components

$$dn = \frac{m^3}{4\pi^3 \hbar^3} \exp(\varepsilon_f/kT) \exp\{-m(v_x^2 + v_y^2 + v_z^2)/kT\}\,dv_x\,dv_y\,dv_z$$

$$(18\cdot4\text{-}4)$$

Now in order that an electron may be emitted from a metal it has to overcome the attractive *image force* acting a right angles to the metal surface (see § 18·5). If the metal surface coincides with the y-z plane, then the image force will act along the x-direction. Hence the x-component of the electron velocity must be such that $mv_x^2/2$ exceeds the minimum energy W_0 needed for the emission of the electron, *i.e*, $mv_x^2/2 > W_0$ (see Fig. 18·2.).

The components v_y and v_z of the electron velocity may however extend from $-\infty$ to $+\infty$. The number of such electrons with the x-component of the velocity lying in the interval v_x to $v_x + dv_x$ is then given by

$$n(v_x)\,dv_x = \frac{m^3}{4\pi^3 \hbar^3} \exp(\varepsilon_f/kT) \exp(-mv_x^2/2kT)\,dv_x$$

$$\times \int_{-\infty}^{\infty} \exp(-mv_y^2/2kT)\,dv_y \int_{-\infty}^{\infty} \exp(-mv_z^2/2kT)\,dv_z \qquad (18\cdot4\text{-}5)$$

The values of the two integrals on the r.h.s. are well-known :

$$\int_{-\infty}^{\infty} \exp(-mv_y^2/2kT)\,dv_y = \exp(-mv_z^2/2kT)\,dv_z = \sqrt{\frac{2\pi kT}{m}}$$

Hence $n(v_x)\,dv_x = \dfrac{m^2 kT}{2\pi^2 \hbar^3} \exp(\varepsilon_f/kT) \exp(-mv_x^2/2kT)\,dv_x \qquad (18\cdot4\text{-}6)$

So the number of electrons crossing a unit area per second in the x-direction with $v_x > v_0$ is given by

$$n = \int_{v_0}^{\infty} v_x \, n(v_x) \, dv_x$$

$$\therefore \quad n = \frac{m^2 kT}{2\pi^2 \hbar^3} \exp(\varepsilon_f/kT) \int_{v_0}^{\infty} v_x \exp(-mv_x^2/2kT) \, dv_x \quad (18\cdot 4\text{-}7)$$

In the above integral the lower limit v_0 is determined by the minimum energy $U_0 = W_0$ required for the emission of an electron at rest from the metal and is given by

$$U_0 = \frac{1}{2} m v_0^2, \quad v_0 = \sqrt{2U_0/m}$$

Hence we get, by substituting ξ for $mv_x^2/2kT$

$$n = \frac{m^2 kT}{2\pi^2 \hbar^3} \cdot \frac{kT}{m} \exp(\varepsilon_f/kT) \int_{U_0/kT}^{\infty} \exp(-\xi) \, d\xi$$

$$= \frac{m(kT)^2}{2\pi^2 \hbar^3} \exp(\varepsilon_f - U_0)/kT$$

$$= \frac{m(kT)^2}{2\pi^2 \hbar^3} \exp(-\varepsilon_0/kT) \quad (18\cdot 4\text{-}8)$$

where $\varepsilon_0 = U_0 - \varepsilon_f$ is the thermionic work function. If e be the electronic charge, then the saturation thermionic current density will be

$$i_s = \frac{me(kT)^2}{2\pi^2 \hbar^3} \exp(-\varepsilon_0/kT) \quad (18\cdot 4\text{-}9)$$

Eq. (18·4-9) can be written as

$$i_s = AT^2 \exp(-\varepsilon_0/kT) \quad (18\cdot 4\text{-}9a)$$

where $\quad A = \frac{mek^2}{2\pi^2 \hbar^3} = 1\cdot 2 \times 10^6 \text{ amp/m}^2/\text{deg}^2 \quad (18\cdot 4\text{-}9b)$

Eq. (18·4-9), known as Richardson-Dushman equation, does not take into consideration the reflection of the electron wave at the potential barrier which it must cross at the time of emission from the metal surface (see Ch. XI). This is a quantum mechanical effect and can be taken into account by introducing a quantum mechanical reflection coefficient $R(v_x)$ which depends on the x-component of the electron velocity v_x. Hence Eq. (18·4-7) should be multiplied by $(1-R)$ within the integral sign. It is usually assumed that $R(v_x)$ does not vary appreciably over the small velocity range from which the major contribution to the integral in Eq. (18·4-7) comes. So we can take an average value of the reflection coefficient R and take the factor $(1-R)$ outside the integral sign so that we get the following expression for the saturation current density i_s in place of Eq. (18·4-9) :

$$i_s = \frac{me(kT)^2}{2\pi^2 \hbar^3}(1-R)\exp(-\varepsilon_0/kT)$$
$$= AT^2(1-R)\exp(-\varepsilon_0/kT) \qquad (18\cdot4\text{-}10)$$

This is the Richardson-Dushman equation corrected for the reflection of the electron wave at the potential barrier.

From Eq. (18·4-10) we get

$$\ln(i_s/T^2) = \ln A(1-R) - \varepsilon_0/kT$$

If the experimental values of $\ln i_s/T^2$ are plotted against $1/T$ we should get a straight line, which is known as the *Richardson plot*. Such a plot shown in Fig. 18·5 is in agreement with the theoretically expected nature of the graph. The values of ε_0 and $A(1-R)$ can be determined from the slope and the intercept along the ordinate of the straight line graph. The order of magnitude of A comes out to be in agreement with the theoretically expected value for pure (uncoated) metal surfaces.

In Table 18·1 are listed the values of the thermionic work functions ϕ and the *Richardson constant* $A(1-R)$ for a few typical metals as determined from Richardson plots for them.

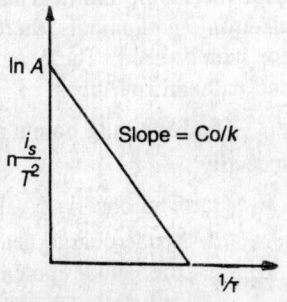

Fig. 18·5. Graph of $\left(\ln \dfrac{i_s}{T^2}\right)$ and $1/T$.

Table 18·1

Metal	Work function ϕ (electron volt)	$A(1-R)$ (ampere/m^2/deg^2)
Platinum	5·3	$0\cdot32 \times 10^6$
Tungsten	4·5	$0\cdot72 \times 10^6$
Molybdenum	4·4	$1\cdot15 \times 10^6$
Tantalum	4·1	$0\cdot37 \times 10^6$
Calcium	3·2	$0\cdot60 \times 10^6$
Thoriated tungsten	2·6	$0\cdot03 \times 10^6$
Barium	2·5	$0\cdot25 \times 10^6$
Caesium	1·8	$1\cdot60 \times 10^6$
Caesiated tungsten	1·4	$0\cdot03 \times 10^6$

From the table it is seen that for pure metal $A(1-R)$ has values of the same order of magnitude as the calculated value of A from Eq. (18·4-9b). The strong dependence of ϕ on the presence of impurities on the surface is evident from a comparison between the value of ϕ for pure tungsten and those for tungsten coated with thorium and caesium. For the coated surfaces, ϕ is considerably reduced which helps emission at much lower temperatures. Thus a reduction of ϕ from a typical value of 5 volts (say) by 1% causes the emission current to increase by 80% at $T = 1000$ K. Usually atoms of a low work function material present on a surface of a metal of larger ϕ cause an enormous increase in the emission current.

The reduction of work function by surface coating is of great importance in the operation of vacuum tubes without the necessity of making the filament *white-hot*. It may be noted that though vacuum tubes with thermionic emitters have been largely supplanted by transistors in electronic equipments, there are still some important fields in which they are widely used. These include high-power tubes for radio transmitters and cathode ray tubes.

The following points regarding the work function ε_0 are to be noted specially.

(1) The observed value of ε_0 depends on the presence of negative space charge near the emitting surface. During measurements, the anode potential should be kept sufficiently positive to prevent such space charge build up.

(2) The observed value of ε_0 is found to decrease with increasing field strength (see § 18·5).

(3) Due to thermal expansion, the observed values of ε_0 are found to depend somewhat on the temperature

(4) The work function ε_0 varies from one crystal plane to another and hence will be the same over the whole emitting area only for a single crystal emitter.

(5) Small amounts of adsorbed gases may influence ε_0 very strongly. Hence the emitting surface must be atomically clean.

It may be noted that Eq. (18·4-1) deduced by Richardson on the basis of classical statistics also gives a straight line plot if $\ln(i_s/\sqrt{T})$ is plotted against $1/T$. Indeed from the experimental results, it is difficult to decide between the two expressions for i_s, viz., Eqs. (18·4-1) and (18·4-9a). However as we have seen before, all evidences point to the fact that the behaviour of the electron gas in metals is determined by the Fermi-Dirac statistics and hence Eq. (18·4-9a) is to be taken as the correct representation of the behaviour of the saturation thermionic current.

18·5 Schottky effect

According to Richardson-Dushman equation (18·4-9), the saturation thermionic current i_s is independent of the anode potential. However this is not found to be strictly correct. Increase of the anode potential causes i_s to increase slightly (see Fig. 18·6).

This is known as Schottky effect and was first observed by W. Schottky in 1914.

We have seen before that for the emission from a metal surface, the electron potential energy has to change from $U_0 = -(\varepsilon_0 + \varepsilon_f)$ inside the metal to $U = 0$ outside in vacuum. Here $\varepsilon_0 = e\phi$ is the thermionic work function. This causes an abrupt change of the potential energy. However in the actual case, the change in potential energy is more gradual. This is due to the fact that at the time of emission, an attractive image force acts on the electron tending to hinder the emission. According to the electrostatic image theory, if a point charge e is at a distance x from a plane metal surface, the total effect of the opposite charges induced by it on the metal surface is equivalent to that of an equal and opposite point charge $-e$ situated at an equal distance x behind the metal surface such that the line joining the two charges is perpendicular to the surface. The force acting on the charge e due to this image charge is known as the *image force*. The potential energy of the electron due to the image force is

$$V_e(x) = -\frac{e^2}{16\pi\varepsilon_0 x}$$

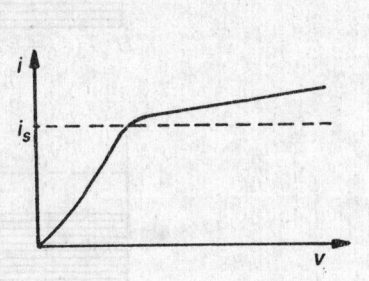

Fig. 18·6. Schottky effect.

This gives rise to a more gradual change in the potential energy of the electron in coming out from the metal as shown in Fig. 18·7a (curve B). Quantum mechanical theory shows that reflection effect at the wall of the potential well which introduces the factor $(1 - R)$ in Eq. (18·4-9) is considerably less in this case than in the case of an abrupt change of potential at the metal surface if the image force is not considered (curve A in Fig. 18·7a.).

Normally the attractive image force, which begins to become prominent at a distance of a few angstroms beyond the surface, goes to zero at infinity. However if an anode at a positive potential is placed facing the heated emitter surface, the electron experiences an outward force due to the anode potential which acts opposite to the image force. The two forces cancel one another at some distance from the surface of

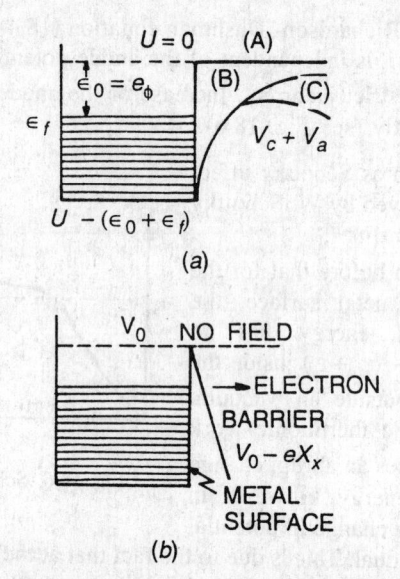

Fig. 18·7.(a) Effect of image force on the potential energy (A) of an electron in a metal (curve B) and with an anode potential (C). (b) Effect of applied field (10^8 V/m) on the shape of the barrier wall.

the emitter. So if an electron is able to come out from the emitter upto this point, it will be emitted. Since this point is nearer to the emitting surface than in the absence of the anode potential, an electron with lesser initial energy can be emitted from the heated surface in this case. This means that the work function is effectively reduced due to the action of the positive anode potential. Higher the positive potential on the anode, greater is the reduction in the effective work function.

If an e.s. field X acts due to the presence of the anode at a positive potential, $V(x)$ is changed to

$$V(x) = V_c(x) + V_a(x)$$

where $$V_a(x) = -Xex$$

Hence we can write for points at some distance from the emitter surface

$$V(x) = -\frac{e^2}{16\pi \varepsilon_0 x} - Xex$$

Differentiating $V(x)$ with respect to x and putting it equal to zero, we can easily see that $V(x)$ attains a maximum at $x = x_m$ where the force on the electron becomes zero :

$$x_m = \left(\frac{e}{16\pi\varepsilon_0 X}\right)^{1/2}$$

The maximum value of $V(x)$ is then

$$V_m(x) = -\left(\frac{Xe^3}{4\pi\varepsilon_0}\right)^{1/2} \qquad (18\text{·}5\text{-}1)$$

The variation of $V(x)$ with x in the presence of the anode potential is shown in Fig. 18·7a by the curve (C). As will be seen, the action of the anode potential reduces the barrier height so that the work function is also reduced effectively. Hence the probability of electron emission is considerably increased.

Using Eq. (18·5-1), we can write the reduced work function, after dividing by e, as

$$\phi' = \phi - \left(\frac{eX}{4\pi\varepsilon_0}\right)^{1/2} \qquad (18\text{·}5\text{-}2)$$

The saturation thermionic current i_s in the presence of the anode is then given by (see Eq. 18·4-9)

$$i_s = i_{s0}\exp(c\sqrt{X}/T) \qquad (18\text{·}5\text{-}3)$$

where c is a canstant and i_{s0} is the saturation thermionic current in the limit of $X = 0$. As the field X increases (*i.e.*, as the anode potential V is increased), i_s increases. This is in agreement with observation. The theoretically expected increase is about 10% for a change in the field by about 2×10^5 V/m (*i.e.*, 2000 volts/cm).

18·6 Field emission

The presence of an electric field of moderate strength causes a slight lowering of the work function. This gives rise to the phenomenon of field-aided emission from a heated metal surface which is Schottky effect discussed in the previous section.

A different kind of phenomenon is observed when the electric field is of the order of 10^8 V/m. Under the action of such a high field, emission of electrons takes place from a metal surface even without the heating of the metal. This is known as *field-emission* or *cold emission*. At first Schottky and later Fowler and Nordheim developed the quantum mechanical theory of the phenomenon. In this case the potential barrier outside the metal surface becomes very thin (~ 10 Å) and has the appearance as shown in Fig. 18·7b. The effect of the applied field now dominates over that of the image force so that the potential in vacuum outside the metal surface can be written as

$$V(x) = V_0 - Xex$$

where V_0 is a constant. According to quantum mechanics, the electron has now a finite probability of penetrating through the potential barrier, even if its energy is lower than the height of the barrier. This is known as the quantum mechanical tunnel effect (see § 11·4). The barrier penetration probability increases with the increasing electron energy and decreasing barrier thickness. According to Fowler and Nordheim, the field-dependence of the current density for a triangular barrier is given by

$$i_s = \alpha X^2 \exp(-\beta \phi/X)$$

where α and β are constants and ϕ is the work function in volts. Experimental observations are in reasonable agreement with the above formula. (See Prob. 6 at the end of Ch. XII)

18·7 Calculation of the Fermi energy

Using the expression for the Fermi-Dirac distribution function, it is possible to estimate the Fermi energy of the electron gas in a metal in terms of the density n of the free electrons in the metal. We get by the integration of Eq. (18·4-2)

$$n = \int_0^{\varepsilon_f} n(\varepsilon)\, d\varepsilon = \frac{1}{2\pi^2}\left(\frac{2m}{\hbar^2}\right)^{3/2} \int_0^{\varepsilon_f} \frac{\sqrt{\varepsilon}\, d\varepsilon}{\exp(\varepsilon - \varepsilon_f)/kT + 1} \qquad (18\cdot7\text{-}1)$$

For $T = 0$ K, we get, since $\varepsilon < \varepsilon_f$

$$n = \frac{1}{2\pi^2}\left(\frac{2m}{\hbar^2}\right)^{3/2} \int_0^{\varepsilon_f} \sqrt{\varepsilon}\, d\varepsilon = \frac{1}{3\pi^2}\left(\frac{2m}{\hbar^2}\right)^{3/2} \varepsilon_f^{3/2} \qquad (18\cdot7\text{-}1a)$$

This gives
$$\varepsilon_f = \varepsilon_{fo} = \frac{\hbar^2}{2m}(3\pi^2 n)^{2/3} \qquad (18\cdot7\text{-}2)$$

Substituting the numerical values for m and $\hbar$ we get

$$\varepsilon_{fo} = 3\cdot65 \times 10^{-19} n^{2/3} \text{ eV}$$

As an example, the density of silver is $1\cdot05 \times 10^4$ kg/m^3 while its mean atomic weight is about 108. Assuming that there is one free electron per atom in metallic silver, we get

$$n = \frac{6\cdot022 \times 10^{23}}{108} \times 10^3 \times 1\cdot05 \times 10^4$$

$$= 5\cdot85 \times 10^{28} \text{ electrons/m}^3$$

Hence the Fermi energy for silver is

$$\varepsilon_{f0} = 3\cdot65 \times 10^{-19} \times (5\cdot85 \times 10^{28})^{2/3} = 5\cdot5 \text{ eV}$$

The average kinetic energy of the electrons at $T = 0$ K can also be calculated using the Fermi-Dirac distribution function :

Metals and Semi-conductors

$$<\varepsilon>_0 = \frac{1}{n}\int_0^{\varepsilon_{f0}} \varepsilon n(\varepsilon)\, d\varepsilon = \frac{(2m^3)^{1/2}}{\pi^2 \hbar^3 n}\int_0^{\varepsilon_{f0}} \varepsilon^{3/2}\, d\varepsilon$$

$$= \frac{2}{5}\frac{(2m^3)^{1/2}}{\pi^2 \hbar^3 n}\varepsilon_{f0}^{5/2} = \frac{3}{5}\varepsilon_{f0} \qquad (18\cdot7\text{-}3)$$

In the table below are listed the values of the Fermi energy ε_{f0} for $T = 0$ K of a few metals which are found to be of the order of a few electron volts. In the fourth column of the table are listed the experimentally determined values of the work functions ($e\phi$) of these metals. The sum of these two gives the depth of the "potential well".

Table 18·1

Metal	Valency	ε_{f0} (eV)	$e\phi$ (eV)	$T_f\,(10^4$ K)
Na	1	3·1	2·28	3·66
K	1	2·1	2·22	2·37
Cu	1	7·0	4·45	8·12
Ag	1	5·5	4·46	6·36
Ba	2	3·8	2·51	4·24
Al	3	11·7	4·20	13·49

From the above table, it is seen that the work function $e\phi$ for silver is 4·46 eV. Hence at 0 K the free electrons in silver occupy all energy levels starting from the bottom of the potential well upto the highest at an energy $\varepsilon_{f0} = 5\cdot5$ eV which is located 4·46 eV below the top of the potential barrier. Hence the depth of the potential well is about 9·96 volts below the top of the barrier (see Fig. 18·2.).

It may be noted that if the free electrons in metals obeyed Maxwell-Boltzmann statistics, then all the electrons would occupy the lowermost level in the potential well at a depth of 4·46 volts below the top of the barrier at 0 K. So the depth of the well would be 4·46 volts rather than 9·96 volts as deduced above. The depth of the well can also be estimated from electron diffraction experiment (see Ch IX). There is qualitative agreement between these and the values of the well-depth expected on the basis of the Fermi-Dirac statistics as deduced above.

It may be noted that the value of the Fermi energy ε_f depends on the temperature of the metal and decreases somewhat as the temperature increases. However for temperatures upto a few thousand kelvins (*i.e.*, upto the melting points of most metals), ε_f is only slightly less than its value at 0 K which is listed in the table above. In the last column of the table, the values of the Fermi temperature $T_f = 2\varepsilon_{f0}/3k$ are listed which are much higher than the melting points of these metals.

At a temperature $T > 0$ K, one has to use Eq. (18·7-1) rather than (18·7-2) to deduce the Fermi energy. The integral in Eq. (18·7-1) can be evaluated in a series form as was first shown by Sommerfeld. The Fermi energy at a temperature T K is given by

$$\varepsilon_f = \varepsilon_{f0}\left[1 - \frac{\pi^2}{12}\left(\frac{kT}{\varepsilon_{f0}}\right)^2\right] \quad (18\cdot7\text{-}4)$$

where ε_{f0} is the Fermi energy at $T = 0$ K. The average kinetic energy of the electrons at $T > 0$ K is given by

$$<\varepsilon>_T = \frac{1}{n}\int_0^{\varepsilon_f} \varepsilon n(\varepsilon)\, d\varepsilon$$

$$= \frac{(2m^3)^{1/2}}{\pi^2 \hbar^3 n}\int_0^{\varepsilon_f} \frac{\sqrt{\varepsilon^3}\, d\varepsilon}{\exp(\varepsilon - \varepsilon_f)/kT + 1}$$

$$\simeq <\varepsilon>_0\left[1 + \frac{5\pi^2}{12}\left(\frac{kT}{\varepsilon_{f0}}\right)^2\right] \quad (18\cdot7\text{-}5)$$

The above equation show that as the temperature increase, ε_f decreases and $<\varepsilon>$ increases slightly. Thus for silver for which $\varepsilon_{f0} = 5\cdot5$ eV, we get for $T = 300$ K,

$$\frac{\pi^2}{12}\left(\frac{kT}{\varepsilon_{f0}}\right)^2 = \frac{\pi^2}{12}\left(\frac{0\cdot26}{5\cdot5}\right)^2 = 1\cdot84\times 10^{-3}$$

Hence at 3000 K, the value of the Fermi energy is reduced by about 0.18% below it value at 0 K. Correspondingly, the value of the average kinetic energy $<\varepsilon>_T$ is increased by about 1% above its value at 0 K.

Fermi surface :

We can define the Fermi wave vector k_{f0} at $T = 0$ by the equation

$$k_{f0} = \sqrt{2m\,\varepsilon_{f0}/\hbar^2} = (3\pi^2 n)^{1/3}$$

If p_{f0} is the Fermi momentum at $T = 0$, we have $p_{f0} = \hbar k_{f0}$, so that the Fermi velocity at $T = 0$ is

$$v_{f0} = \frac{\hbar}{m}(3\pi^2 n)^{1/3}$$

If k_{f0} is the radius of a sphere in k-space, known as the Fermi sphere, at $T = 0$, the surface of this sphere, known as the Fermi surface, separates

Metals and Semi-conductors

the occupied from the unoccupied states in k-space. The electrons in the metal fill up the Fermi sphere of radius k_f per unit volume.

18·8 Band theory of solids

The free electron theory of metals discussed in the last few sections has been successful in explaining some of the bulk properties of metals, e.g., their electrical and thermal conductivities, thermionic emission etc. However this theory fails completely in explaining those properties which depend on the internal structures of the solids. It does not take into account the importance of the role of the nuclei of the atoms at the lattice sites. It is also unable to explain the classification of solids as conductors, semi-conductors and insulators.

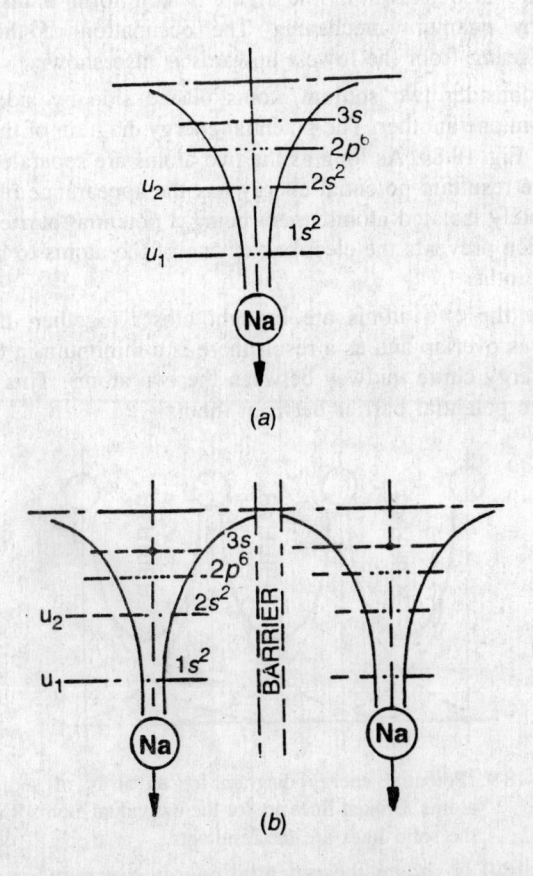

Fig. 18·8. Variation of potential acting upon the electrons with distance : (a) for a single sodium (Na) atom ; (b) for two sodium atoms placed side by side, separated by a large distance.

For a more detailed understanding of the behaviour of the different types of solids, it is necessary to consider the nature of the electronic wave functions and their energy eigenvalues in the periodically varying potential field inside the lattice. The band theory of solids developed for this purpose is actually quite complicated since it must take into account the behaviour of a very large number of interacting particles. In what follows we shall give a simplified version of the theory which will make plausible some of the basic characteristics of crystalline solids.

Consider an isolated atom of a monovalent element, e.g., sodium ($Z = 11$). The variation of the potential acting upon the electrons in this atom due to the Coulomb field of the positively charged nucleus is illustrated in Fig. 18·8a. The various energy levels (e.g., $1s$, $2s$, $2p$, $3s$) of the atom are also shown in the figure which form a discrete set as predicted by quantum mechanics. The occupation of the levels by electrons starting from the lowest upwards is also shown.

Now consider two sodium atoms placed side by side at a large distance from one another. The potential energy diagram of the two atoms is shown in Fig. 18·8b. As long as the two atoms are separated by a large distance, the resultant potential energy has the appearance of that due to two completely isolated atoms, there being a potential barrier separating the two which prevents the electrons of one of the atoms to interact with those of the other.

If now the two atoms are brought closer together, the potential energy curves overlap and as a result there is a minimum in the resultant potential energy curve midway between the two atoms. This reduces the height of the potential barrier between them.

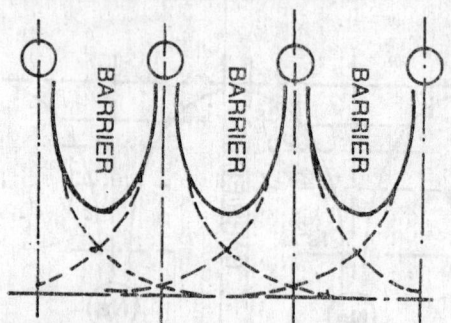

Fig. 18·9. Potential energy diagram for an array of sodium atoms. Broken lines are for the individual atoms while the solid lines are the resultants.

In Fig. 18·9 is shown the potential energy diagram for an array of sodium atoms which is found to change periodically as we go from one lattice ion to the next. Actually the potential energy is infinite at the lattice sites. Midway between two lattice sites, the potential energy attains a minimum value since the attractive forces due to the two neighbouring atoms are equal at these points.

Now if there are n atoms in an array there will be $2n$ electrons in the 1s level for the whole system, since each atom has two 1s electrons with opposite spins. Since according to Pauli's exclusion principle, there cannot be so many electrons in a particular level, we have to assume that the 2s level for the whole system splits up into n close lying levels which can accommodate $2n$ electrons with pairs having oppositely aligned spins in each of the split levels without violation of the exclusion principle. These levels thus constitute an energy band known as a permitted band. The other levels of the individual atoms, e.g., 2s, 2p, 3s etc. are also similarly split up and give rise to energy bands.

The widths of the energy bands depend upon the spacings between the lattice sites (lattice constant) and on the energy spreads of the individual atomic energy levels. They are independent of the total number of atoms in the lattice. The bands associated with the inner orbits of the individual atoms (e.g. 1s, 2s) have usually smaller band-widths because of the very weak influence of the neighbouring atoms on the motion of electrons in the corresponding orbit of the individual atoms. On the other hand the width, of the higher energy bands are much larger since the influence of the neighbouring atoms on the electrons in the upper energy levels (e.g., 2p, 3s) of the individual atoms is much stronger.

In an actual crystal, the spacings between the lattice atoms is a constant and cannot be changed. However if we imagine that it is possible to change these spacings, then we can study the variation of the band-widths with the change of the distance between the lattice atoms which is shown in Fig. 18·10a. When this distance is large, the bands have

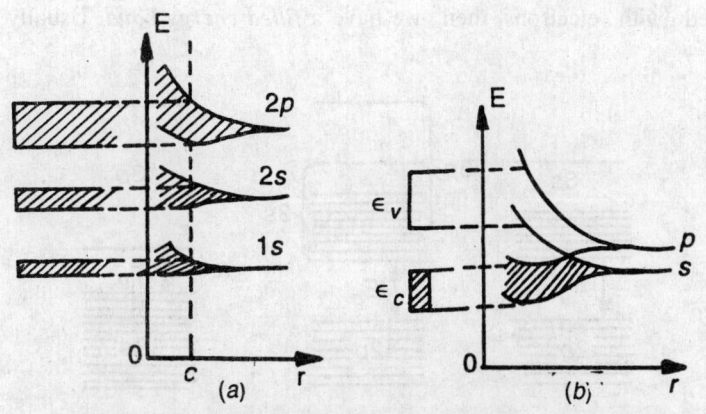

Fig. 18·10 (a) Variation of band-width with distance between lattice atoms. (b) Overlapping between higher energy neighbouring bands for a diamond crystal.

the appearance of the individual atomic levels, having very little spread in energy. As the distance between the lattice atoms is decreased, the splitting of the levels begins to increase which results in an increase of the bandwidth. In the inset of the figure at left are shown the nature of the bands when the distance between the lattice atoms is equal to the actual lattice constant.

Though the band-width does not depend the number of the atoms in the lattice, the number of levels into which a band splits depends upon n. Since n is very large, the separation between the split levels in a band is extremely small. As an example if $n \sim 10^{18}$, then this separation is of the order of 10^{-18} eV for a band-width of ~ 1 eV. This shows that the variation of energy within a band may be taken to be almost continuous.

The gaps between the lower energy bands are usually relatively large. These gaps are known as *forbidden zones*. There can be no electrons in the solid with energies lying in these zones. They can only have energies in the permitted band regions.

The gaps between the higher energy bands are usually much smaller. In some cases, the widths of the bands may be so large that there may be overlapping between neighbouring bands. In Fig. 18·10b. such overlapping bands are shown in the case of a diamond crystal. In this case the bands arising out of the 2s and 2p levels overlap in such a way that they give rise to two other bands as shown. In between these two there is a forbidden zone. The lower band is known as the *valence band* and the upper band as the *conduction band*.

Since the number of electrons in a particular energy level within a band can at most be 2, the maximum number of electrons in the band can be twice the number of levels constituting the band. If all these levels are filled with electrons, then we have a *filled energy band*. Usually the

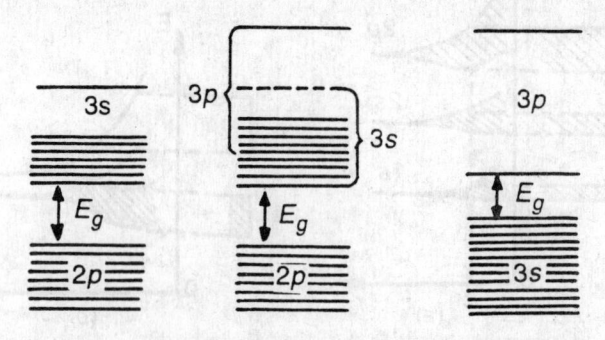

Fig. 18·11. Types of energy bands : (a) apperband fortiaths filled ;
(b) overtapping of filled 3s and partially filled
3p   bands; (c) overlapping of a completely filled and
an   emply band just above it.

lower energy bands are of this type. The higher energy bands may be either completely filled or partially filled. If there is no electron in a band then we have an *empty band*. Partially filled bands may arise either from the atomic levels which are partially filled or due to the overlapping of a completely filled band and an empty band just above it. These different characteristics of the bands are shown in Fig. 18·11.

We have stated earlier that there cannot be any electron in the forbidden zone. However if some impurity atoms are present in a crystal, then there may be permitted energy levels within the forbidden zones. The action of *impurity semi-conductors* depends on the presence of such impurity levels (see later).

18·9 Quantum mechanical theory of the band structure

We have seen that the electrons in a crystalline solid are acted upon by a periodically varying potential field (Fig. 18·12a.). To understand the origin of the energy bands in solids it is necessary to solve the Schrödinger equation in the presence of such a potential. Solving the Schrödinger equation for the three-dimensional motion of an electron considering the actual periodic potential in a solid is an exremely difficult problem. If instead of this we consider a one-dimensional lattice (*i.e.*, a chain of atoms along a line), then assuming a rectangular periodic potential, it is possible to solve exactly the one-dimensional Schrödinger equation which was first done by R. de L. Kronig and W.G. Penny in 1930.

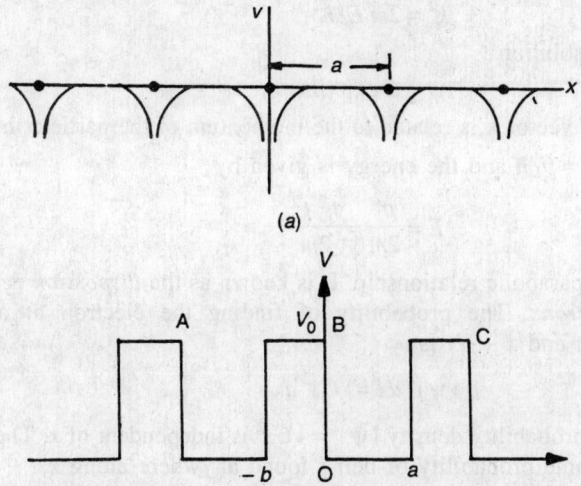

Fig. 18·12. (*a*) Periodically varying potential acting on an electron in a crystalline solid (one dimensional approximation) ; (*b*) Periodic rectangular potential.

The one dimensional rectangular periodic potential shown in Fig. 18·12*b* can be mathematically represented as follows :

$$V = V_0 = \text{constant for } -b < x < 0 \qquad (18\cdot9\text{-}1a)$$
$$V = 0 \text{ for } 0 < x < a \qquad (18\cdot9\text{-}1b)$$
$$V(x+l) = V(x) \qquad (18\cdot9\text{-}1c)$$

where $l = a + b$ is the period of the potential. The Schrödinger equations in the regions of *potential valleys* (v) and *hills* (h) can then be written as

$$\frac{d^2 \psi_v}{dx^2} + \alpha^2 \psi_v = 0 \quad (0 < x < a) \qquad (18\cdot9\text{-}2a)$$

$$\frac{d^2 \psi_h}{dx^2} - \beta^2 \psi_h = 0 \quad (-b < x < 0) \qquad (18\cdot9\text{-}2b)$$

Here $\qquad\qquad\qquad \alpha^2 = 2m E/\hbar^2$
and $\qquad\qquad\qquad \beta^2 = 2m (V_0 - E)/\hbar^2$

$E < V_0$ where V_0 is the height of the potential hills. m and E are the mass and energy of the electron. We now make use of a theorem known as *Bloch's theorem* applicable in the case of a periodic potential. We know that for a free electron moving along the x-axis, the Schrödinger wave equation is of the form

$$\frac{d^2 \psi}{dx^2} + k^2 \psi = 0$$

where $\qquad\qquad\qquad k^2 = 2m E/\hbar^2$

This has solution

$$\psi = C \exp(ikx)$$

The wave vector **k** is related to the momentum of the particle through the relation $k = p/\hbar$ and the energy is given by

$$E = \frac{p^2}{2m} = \frac{\hbar^2 k^2}{2m}$$

This is a parabolic relationship. It is known as the *dispersion relation for free electrons*. The probability of finding the electron at any point between x and $x + dx$ is

$$|\psi|^2 dx = |C|^2 dx$$

Thus the probability density $|\psi|^2 = |C|^2$ is independent of x. The electron has the same probability of being found anywhere along x.

The situation is different when the electron is acted upon by a periodic potential as in Fig. 18·12. In this case the probability of finding the electron at a given point x should change periodically with a period equal to the lattice constant l. Thus the probabilities are the same at the points A, B and C. However within a particular period, *i.e.*, from one lattice site to the next, the probability changes with x. So the amplitude of the wave function should change periodically with a period l. We can write the wave function as

$$\psi(x) = u(x) \exp(ikx) \tag{18.9-3}$$

where the periodic nature of the amplitude $u(x)$ can be expressed by writing

$$u(x+l) = u(x) \tag{18.9-4}$$

Equation (18·9-3) is known as **Bloch's theorem**. The form of the amplitude function $u(x)$ is determined by the nature of the potential $V(x)$.

The solution of the equations (18·9-2) can be written as

$$\psi_v(x) = A \exp(i\alpha x) + B \exp(-i\alpha x) \tag{18.9-5}$$

$$\psi_h(x) = C \exp(\beta x) + D \exp(-\beta x) \tag{18.9-6}$$

Putting $\psi' = \partial \psi/\partial x$ we can write the *boundary conditions* at $x=0$ and $x=-b$ as:

$$\psi_v(0) = \psi_h(0), \; \psi_v'(0) = \psi_h'(0) \tag{18.9-7}$$

$$\psi_h(-b) = \psi_v(-b), \; \psi_h'(-b) = \psi_v'(-b) \tag{18.9-8}$$

From Eqs. (18·9-3) and (18·9-4) we have

$$\psi(x+l) = u(x+l) \exp ik(x+l)$$
$$= u(x) \exp(ikx) \exp(ikl)$$
$$= \psi(x) \exp(ikl)$$

$$\therefore \quad \psi(x) = \exp(-ikl)\, \psi(x+l) \tag{18.9-9}$$

This gives, for $x=-b$

$$\psi_v(-b) = \exp(-ikl)\, \psi_v(l-b) \tag{18.9-10}$$

The boundary conditions (18·9-8) at $x=-b$ then become

$$\psi_h(-b) = \psi_v(-b) = \exp(-ikl)\, \psi_v(l-b) \tag{18.9-11a}$$

$$\psi_h'(-b) = \psi_v'(-b) = \exp(-ikl)\, \psi_v'(l-b) \tag{18.9-11b}$$

Then from the boundary conditions (18·9-7) and (18·9-11) we can get the following two relations:

$$(A+B)\{\cosh \beta b - \exp(-ikl) \cos \alpha a\}$$
$$= i(A-B)\left\{\frac{\alpha}{\beta} \sinh \beta b + \exp(-ikl) \sin \alpha a\right\} \tag{18.9-12}$$

$$(A+B)\left\{\sinh \beta b - \frac{\alpha}{\beta} \exp(-ikl) \sin \alpha a\right\}$$
$$= i(A-B)\left\{\frac{\alpha}{\beta} \cosh \beta b - \frac{\alpha}{\beta} \exp(-ikl) \cos \alpha a\right\} \tag{18.9-13}$$

Writing $A + B = p$ and $A - B = q$, the above two equations can be written as

$$\gamma_{11} p + \gamma_{12} q = 0$$
$$\gamma_{21} p + \gamma_{22} q = 0$$

These two equations can be solved if the determinant of the coefficients of p and q is equal to zero ; i.e.,

$$\gamma_{11} \gamma_{22} - \gamma_{21} \gamma_{12} = 0$$

Substituting the values of $\gamma_{11}, \gamma_{12}, \gamma_{21}$ and γ_{22} from Eqs. (18·9-12) and (18·9-13), we then get the condition :

$$\cos kl = \frac{\beta^2 - \alpha^2}{2\alpha\beta} \sinh \beta b \sin \alpha (l - b)$$
$$+ \cosh \beta b \cos \alpha (l - b) \quad (18\cdot9\text{-}14)$$

Eq. (18·9-14) determines the energy spectrum of the electron in a one-dimensional periodic lattice.

Kronig and Penny considered the limiting case of the potential hills being delta functions such that as $V_0 \to \infty$ and $b \to 0$, the product $V_0 b$ remains finite and constant. In this case we get

$$\beta b = \frac{b}{\hbar} \sqrt{2m (V_0 - E)} = \frac{\sqrt{b}}{\hbar} \sqrt{2m(V_0 b)} = \text{constant} \times \sqrt{b}$$

Thus $\beta b \to 0$ as $b \to 0$ so that $\sinh \beta b \to \beta b$ and $\cosh \beta b \to 1$. Then since $\beta^2 = 2m V_0/\hbar^2 >> \alpha^2$, we get

$$\frac{\beta^2 - \alpha^2}{2\alpha\beta} \sinh \beta b = \frac{(\beta^2 - \alpha^2)\beta b}{2\alpha\beta} = \frac{\beta^2 b}{2\alpha}$$

$$= \frac{\beta^2 lb}{2\alpha l} = \frac{m V_0 lb}{\hbar^2 \alpha l} = \frac{P}{\alpha l}$$

where
$$P = \lim_{b \to 0, V_0 \to \infty} \frac{m(V_0 b) l}{\hbar^2} \quad (18\cdot9\text{-}15)$$

Eq. (18·9-14) then reduces to

$$\cos kl = \frac{P \sin \alpha l}{\alpha l} + \cos \alpha l \quad (18\cdot9\text{-}16)$$

In Eq. 18·9-16), k is a real number. Hence the magnitude of $\cos kl$ cannot be greater than *unity*. So if we write

$$f(E) = \frac{P \sin \alpha l}{\alpha l} + \cos \alpha l \quad (18\cdot9\text{-}17)$$

we have the condition

$$-1 \leq f(E) \leq 1 \quad (18\cdot9\text{-}18)$$

In Fig. 18·13 the value of $f(E)$ has been plotted as a function of αl for the case of $P = 3\pi/2$. Obviously only those energy values are permissible for which the ordinates of the graph lie between ± 1. In the figure, the portions of the abscissa drawn with heavy lines correspond to these permitted energy values. In the remaining portions of the abscissa the values of $f(E)$ are greater than $+1$ or less than -1 and hence correspond to forbidden energy regions.

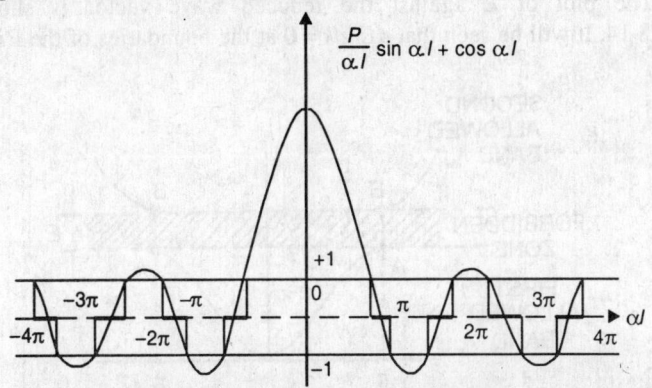

Fig. 18·13. $f(E)$ vs αl curve showing permitted and forbidden energy regions.

The following important conclusions of the theory are to be noted:

(a) Since the energy varies continuously within finite limits in both the permitted and forbidden energy regions, we get a series of allowed energy bands intervened by forbidden gaps.

(b) The width of the allowed energy bands increases with the increasing value of αl, i.e., with increasing energy. This is due to the fact that the first term on the r.h.s. of Eq. (18·9-16) decreases with increasing αl.

(c) The width of an allowed band decreases with increasing P, i.e., with increasing *energy of binding* of the electron. In the limit of $P \to \infty$, the allowed bands become infinitely narrow and we get a line spectrum. In this case $\alpha l = \pm n\pi$ with n = integer so that the energy becomes

$$E_n = \frac{n^2 \pi^2 \hbar^2}{2ml^2}$$

This is equal to the energy of the levels in a potential box with rigid walls.

(d) The plot of E vs. k shows discontinuities at $k = n\pi/l$ where $n = 1, 2, 3, \ldots$ These discontinuities define the boundaries between successive *Brillouin zones*.

(e) Within a given energy band, E is a periodic function of k so that if we replace k by $k + 2n\pi/l$, the l.h.s. of Eq. (18·9-16) remains the same. Thus k is not uniquely determined. The *reduced wave vector* **k** is defined by the relation

$$-\frac{\pi}{a} \le k \le \frac{\pi}{a}$$

The plot of E against the reduced wave vector is shown in Fig. 18·14. It will be seen that $dE/dk = 0$ at the boundaries of the Brillouin zones.

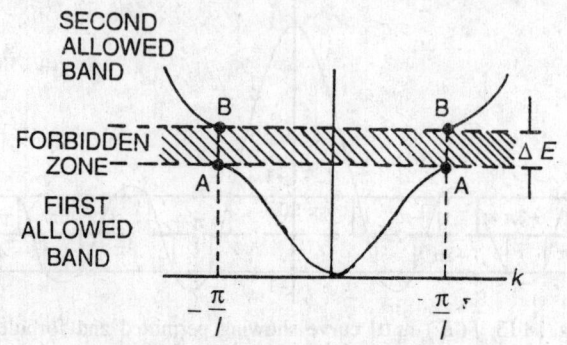

Fig. 18·14. Variation of E with k.

18·10 Motion of electron according to band theory; Effective mass of an electron

According to de Broglie's theory, the velocity of a particle is equal to its group velocity and is given by $\upsilon = \upsilon_g = d\omega/dk$ where ω is the circular frequency of the de Broglie waves so that the energy is $E = \hbar\omega$. We thus have

$$\upsilon = \frac{1}{\hbar}\frac{dE}{dk} \tag{18·10-1}$$

For a free particle, $E = \hbar^2 k^2/2m$ which gives the usual relation $\upsilon = \hbar k/m = p/m$. In the band theory E is not proportional to k^2 in general as can be seen from Fig. 18·14. From such figures it is possible to obtain the curves for the variation of υ with k. Since $dE/dk = 0$ at the top and the bottom of the energy bands, $\upsilon = 0$ at these points. The absolute value of υ reaches a maximum at some value $k = k_0$ beyond which υ *decreases with increasing energy* which is a characteristic feature of the motion of the electron in a band and is different from that of the free electrons.

Consider an external electric field X applied to an electron in a Brillouin zone. The gain in the energy E by the electron will be

$$dE = Xe\upsilon dt = \frac{Xe}{\hbar}\frac{dE}{dk}\,dt = \frac{dE}{dk}dk$$

Hence
$$\frac{dk}{dt} = \frac{Xe}{\hbar} \qquad (18\cdot10\text{-}2)$$

The acceleration of the electron is then given by

$$f = \frac{dv}{dt} = \frac{1}{\hbar}\frac{d}{dt}\left(\frac{dE}{dk}\right) = \frac{1}{\hbar}\frac{d}{dk}\left(\frac{dE}{dk}\right)\frac{dk}{dt}$$

$$= \frac{1}{\hbar}\frac{d^2E}{dk^2} \cdot \frac{dk}{dt}$$

$$\therefore \qquad f = \frac{Xe}{\hbar^2}\frac{d^2E}{dk^2} \qquad (18\cdot10\text{-}3)$$

For a free electron of mass m, the acceleration due to the external field is given by $f = Xe/m$. If we write

$$m^* = \hbar^2 \bigg/ \left(\frac{d^2E}{dk^2}\right) \qquad (18\cdot10\text{-}4)$$

then Eq. (18·10-3) can be rewritten as

$$f = \frac{Xe}{m^*}$$

Thus comparison with the free electron case shows that the electron in the band has an *effective mass m* given by Eq. (18·10-4). It can be obtained as a function of k from the E vs. k curve for the electron in the band. As long as v increases with the increase of k, the effective mass is positive since the application of the external field increases v with time. However beyond the maximum of the v vs. k curve at $k = k_0$. there is decrease of velocity with the increase of k i.e., the same field produces a decreases of v. So the effective mass m^* becomes negative in this region which occurs in the upper part of the band. For the three dimensional case m^* is a tensor given by

$$\left(1/m^*\right)_{\mu v} = \frac{1}{\hbar^2}\left(\frac{\partial^2 E}{\partial k_\mu \partial k_v}\right)$$

The reason for the above strange behaviour of the electron in a crystal can be understood as follows.

For a free electron, the external electric field X performs work which goes to increase the kinetic energy of the electron so that

$$W = E'_k = \frac{\hbar^2 k^2}{2m}$$

The gives $\partial^2 E'_k/dk^2 = \hbar/m$ so that $m^* = m$. Hence the effective mass of the free electron is simply equal to its rest mass.

In a crystal, the electron has both kinetic and potential energies. The work done by the applied field goes to change both of these, so that we can write $W = E_k' + \upsilon$. Thus, a smaller fraction of the work done by the field X is spent in increasing the kinetic energy. As a result there is smaller change in the velocity of the electron in the crystal than in the case of a free electron which can be interpreted by assuming a higher inertial mass for the electron in the crystal. So the effective mass $m^* > m$.

If all the work done by the external field goes to change the potential energy, we have $W = \upsilon$ so that change in the kinetic energy, $E_k' = 0$ which can be interpreted by supposing that the effective mass $m^* \to \infty$.

Finally, if all the work done by the applied field as also a part of the kinetic energy of the electron are together transformed into the potential energy of the electron, then $V = W + E_k'$ so that there is decrease in the electron velocity due to the application of the field which can be interpreted by assuming a negative effective mass for the electron.

The most direct method of determining the effective mass m^* is by cyclotron resonance experiment. (See *Solid State Physics* by Charles Kittel).

18·11 Classification of solids according to band structure

The band theory permits us to classify all crystalline solids into three classes, *viz.*, conductors (metals), insulators and intrinsic semiconductors.

(*a*) *Conductors* : If the uppermost energy band in a solid is partially filled with electrons with the lower energy bands completely filled, then the application of even a very small electric field in the solid causes some of the electrons to make transition to the empty higher energy states within the band. As a result they are able to move about freely within the metal. As explained before such partially filled bands are formed due to partially filled energy levels of the atoms as in the case of alkali metals. They may also be formed due to the overlapping of filled and empty (or partially filled) bands (see Fig. 18·15*a*).

Partially filled bands are characteristics of metals which are good conductors of electricity and heat.

(*b*) *Insulators* : We know that most nonmetals are bad conductors of heat and electricity so that they constitute the insulators. In an insulator, the uppermost energy band (valence band) is completely filled with electrons. So due to the exclusion principle an electron in any level within this band cannot make transition to another level in the same band. They can only go to a level in the upper empty band (conduction band). However in insulators, the gap between the valence band and the empty conduction band is of the order of several electron-volts (see Fig. 18·15 *b*). So even when an electric field is applied to the insulator, the electrons in the uppermost filled band do not acquire sufficient energy to cross the relatively broad forbidden zone and hence cannot make transition to levels in the upper empty conduction band. So the application of an electric field does not induce any electric current to flow in an insulator. Again if an insulator is heated, the electrons in the valence band do not gain sufficient thermal energy to make transitions to the empty conduction band. So they are bad conductors of heat as well.

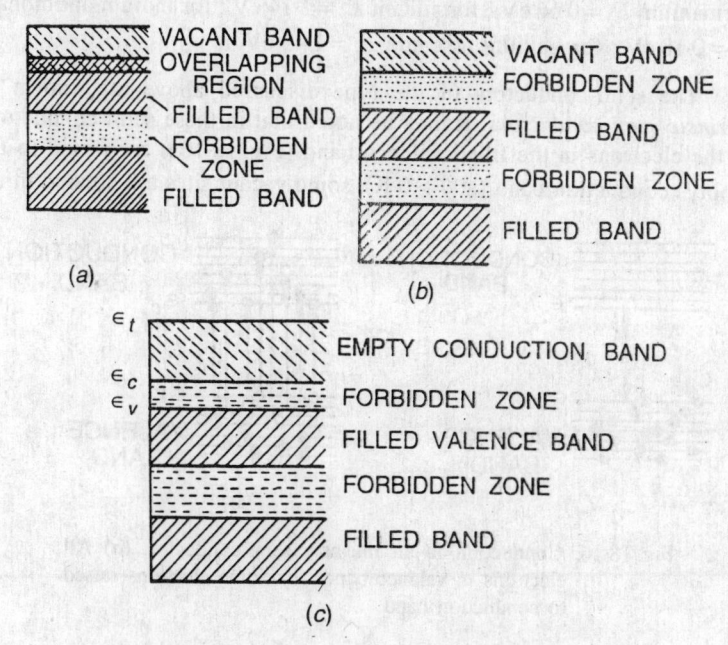

Fig. 18·15. Classification of solids according to band structure; (a) conductors ; (b) insulators ; (c) semi-conductors.

Typical values of the gap-width of some insulator are : for diamond, $E_g = 5\cdot2$ eV ; for Al_2O_3, $E_g = 7$ eV.

(c) *Intrinsic semi-conductors* : In some substances, *e.g.*, silicon, germanium or grey tin in group IV of the periodic table, the gap between the uppermost completely filled valence band and the next higher empty conduction band is quite small (see Fig. 18·15c) At the absolute zero of temperature, they behave like insulators since they do not posses sufficient energy to make transitions to the upper conduction band. However at ordinary room temperature, some of the electrons in the valence band have sufficient thermal energy to make transitions to the conduction band. When an electric field is applied to such a substance, these electrons can move about freely within them. So they exhibit limited conductivity. They are called semi-conductors. When they are heated, more electrons are transferred to the conduction band so that the conductivity of the semi-conductors increases with increase of temperature. This behaviour is just the opposite of what is observed in metals. The conductivity of the latter decreases with increase of temperature. The conductivity of semi-conductors is also found to increase when they are exposed to light.

Typical values of the energy gap of some semi-conductors are : for germanium $E_g = 0.66$ eV ; for silicon $E_g = 1.14$ eV ; for indium antimonate $E_g = 0.18$ eV ; for gallium arsenide $E_g = 1.43$ eV.

The semi-conductors of the type discussed above are known as *intrinsic semi-conductors*. It may be noted that in these semi-conductors, as the electrons in the filled valence band (Fig. 18·16a) are raised to the empty conduction band (at $T > 0$ K), some vacant sites are created in the

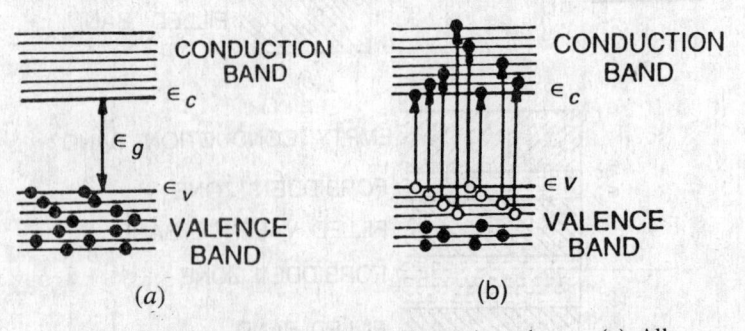

Fig. 18·16. Conduction in an intrinsic semi-conductor. (a) All electrons in valence band. (b) Some electrons raised to conduction band.

valence band shown in Fig. 18·16b, by the small circles. These are capable of accepting electrons (solid points) from amongst those remaining in the valence band. When an electric field is applied, not only are the electrons raised to the conduction band, but also some of the electrons in the valence bands will be able to move about freely under the action of the applied field, thereby giving rise to an electric current.

With very narrow forbidden zone, the number of electrons raised to the conduction band from the valence band as the temperature is raised may be substantial, which will increase the electrical conductivity of the crystal quite appreciably. Thus for germanium with $E_g = 0.66$ eV, the electron density in the conduction band at room temperature is $n_i \sim 10^{19}$ per m^3 and the specific resistance $\rho_i \sim 0.48$ ohm-m. On the other hand for diamond with $E_g = 5.2$ eV, n_i is only about 10^4 per m^3 and correspondingly $\rho_i \sim 10^8$ ohm-m at room temperature. As the temperature is raised to about 600 K, the electron density in diamond however rises by many orders of magnitude and the specific resistance becomes comparable to that of germanium at room temperature.

Hole conduction :

We have seen above that in an intrinsic semi-conductor, a fraction of the electrons from the completely filled valence band at 0 K is raised to the empty conduction band when the temperature is raised. As a result some vacant sites, also known as "holes", lie near the top of the filled band. When there is an electric field, the current associated with all the electrons in a completely filled band is zero so that we can write

$$\mathbf{I} = -e \sum_i \mathbf{v}_i = -e {\sum_i}' \mathbf{v}_i - e\, \mathbf{v}_j = 0 \qquad (18\cdot 11\text{-}1)$$

where $\sum'$ refers to the sum over all electrons except $i = j$; $-e$ is the electronic charge and $\mathbf{v}_i$ is the velocity of the ith electron. If there is one hole in the valence band due to the absence of the jth electron, the current becomes

$$\mathbf{I}' = -e {\sum_i}' \mathbf{v}_i = +e\, \mathbf{v}_j \qquad (18\cdot 11\text{-}2)$$

Thus the current in the valence band due to one "hole" is equivalent to that due to the motion of a single positively charged particle of the same charge as the electron. Such a hole has a positive effective mass equal numerically to the negative effective mass of an electron which initially occupied a state close to the top of the valence band. Only then the current due to the "hole" will be equivalent both in magnitude and direction to that due to the electron of the almost completely filled band.

If an external electric field $\mathbf{X}$ is applied, then the velocity $\mathbf{v}_j$ will be changed so that the current $\mathbf{I}'$ will also change. The rate of charge of $\mathbf{I}'$ with time is given by

$$\frac{d\mathbf{I}'}{dt} = e \frac{d\mathbf{v}_j}{dt} = -\frac{e^2}{m^*} \mathbf{X} \qquad (18\cdot 11\text{-}3)$$

Since the holes tend to occupy the upper parts of the nearly filled band where the effective mass m^* is negative (see above), the r.h.s. of the above equation is positive. So the band in which an electron is missing behaves as a "positive hole" with an effective mass $|m^*|$.

18·12 Extrinsic semi-conductors

The semi-conductors discussed in the previous section are known as intrinsic semi-conductors. It is possible to produce radical change in the conductivity of these materials by the addition of a small quantity of some *impurity* in them. The resulting material is known as *extrinsic* or *impurity semi-conductor*.

Consider the two intrinsic semi-conductors germanium ($Z = 32$) and silicon ($Z = 14$), both belonging to group IV(a) of the periodic table. Their atoms contain two electrons in the outermost $4p$ and $3p$ orbits respectively and are tetravalent.

Now an intrinsic semi-conductor made of Si or Ge always contains some impurity atoms, however pure they may be. These impurity atoms have their own energy levels (*impurity levels*) which may be either in the allowed or in the forbidden bands or in both at different distances above the top of the valence band or below the bottom of the conduction band.

Similarly when impurities are intentionally mixed with an intrinsic semi-conductor, changes in the band structure occur.

n-type semi-conductors :

Suppose some tetravalent germanium atoms in a germanium crystal are replaced by atoms of some pentavalent element *e.g.*, arsenic ($Z = 33$). Germanium has a lattice structure similar to diamond in which each atom of Ge is surrounded by four of its nearest neighbours bound to it by the

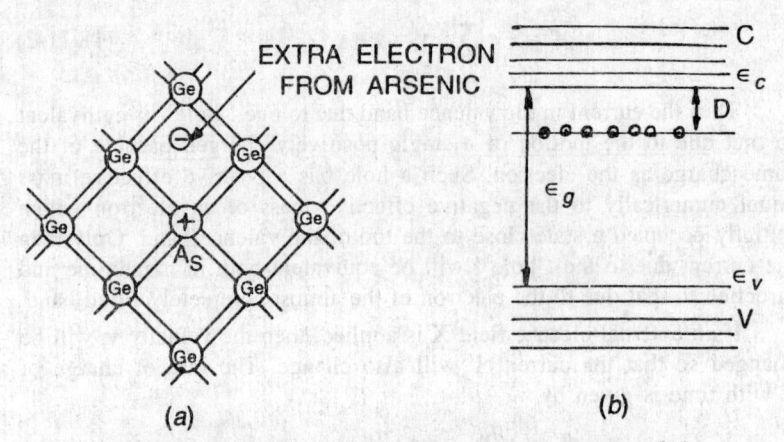

Fig. 18·17. An *n*-type semi-conductor; (*a*) arrangement of the atoms ; (*b*) nature of the bands.

valence forces (see Fig. 18·17a). When an arsenic atom takes the position of a germanium atom in the lattice, it shares four of its valence electrons with the four neighbouring atoms while the fifth electron does not take part in any bonding. It then moves in the field of the positive arsenic ion the field is greatly reduced because of the large permittivity (~ 16) of germanium so that the energy of binding of the electron in the As atom is reduced by a large factor ($\sim 1/256$) compared to that in a free As atom. If the electron gains this amount of energy (~ 0.01 eV) due to thermal motion or otherwise, it will be free to move about inside the lattice and becomes a conduction electron.

In terms of the band theory, we may picture the extra electron as occupying a level (D) in the forbidden zone between the valence (V) and conduction (C) bands as shown in Fig. 15·17b about 0·01 eV below the bottom of the conduction band. If an extra energy amounting to 0·01 eV is given to this electron, it is raised to the conduction band and imparts additional conductivity to the material.

Since the impurity (arsenic) atoms donate electrons to the material, it is called a *donor impurity* and the levels that the additional electrons occupy are called *donor levels*.

The semi-conductors doped with donor type impurity as above are known as *n-type semi-conductors*.

It may be noted that when the electron is raised from the donor level to the conduction band, a hole is created in the immobile arsenic atom which therefore does not take part in the electrical conductivity of the material.

p-type semi-conductors :

If some germanium atoms are replaced by the atoms of a trivalen impurity, *e.g.*, indium ($Z = 49$), then the In atom in the lattice lacks one

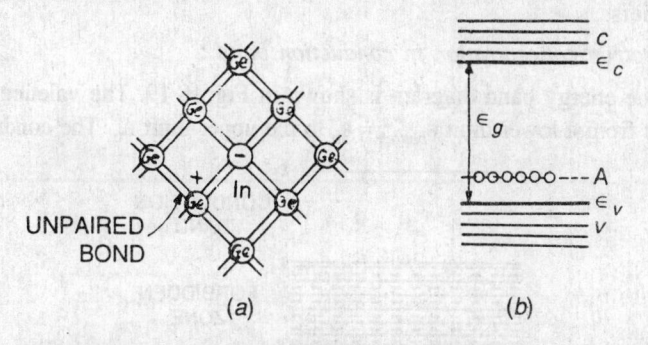

Fig. 18·18. A p-type semi-conductor ; (a) arrangement of the atoms ; (b) nature of the bands.

electron to complete the bonds with its four neighbouring Ge atoms. It can then borrow an electron, so to say, from another Ge atom to complete the bond, causing a rupture in the bond of the Ge atom in the lattice. This corresponds to a hole (Fig. 18·18a). The energy required to form the hole is about 0·01 eV.

The corresponding change in the band structure is shown in Fig. 18·18b which shows the appearance of an impurity level (A) about 0·01 eV above the valence band. Because of the small gap between the impurity level and the valence band, some electrons from the valence band are raised to the impurity level even at room temperature. They complete the bonds of the impurity In atoms and are therefore unable to move in the Ge lattice. So they do not contribute to the conductivity of the material. However, the holes that are created in the valence band due to the absence of electrons now contribute to the conductivity of the materials.

Since the impurity atoms accept electrons from the valence band, they are known as acceptor atoms and the corresponding impurity levels as *acceptor levels*.

Obviously the carriers of charge which contribute to the conduction current in this case are the positive holes. So these semiconductors are known as *p*-type.

18·13 Carrier concentration and positions of Fermi levels in semi-conductors

We have seen that the electrical conductivity of metals is due to the existence of free electrons in a metal which can be treated as a gas (electron gas) obeying Fermi-Dirac statistics.

In a semi-conductor, on the other hand, the conductivity arises due to the mobilities of electrons in the conduction band and holes in the valence band. The electrons and holes are both *carriers* of charge in semi-conductors. We shall first calculate the concentrations of both types of carriers.

(A) *Electron concentration in conduction band* :

The energy band diagram is shown in Fig. 18·19. The valence band extends from a lower limit $\varepsilon_{bottom} = \varepsilon_b$ to the upper limit ε_v. The conduction

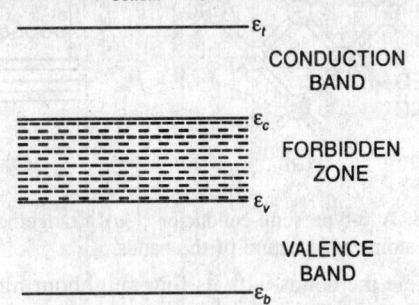

Fig. 18·19. Energy band diagram of an intrinsic semi-conductor.

band, which is above the valence band extends from ε_v to an upper limit $\varepsilon_{top} = \varepsilon_t$. These two bands are separated by the forbidden gap of width $\varepsilon_g = \varepsilon_c - \varepsilon_v$.

According to quantum statistics, the number of electrons in the energy interval ε to $\varepsilon + d\varepsilon$ in the conduction band can be written as (see Ch. XX).

$$n(\varepsilon) = Z(\varepsilon) F(\varepsilon) d\varepsilon \qquad (18\cdot13\text{-}1)$$

so that the total number of electrons in the conduction band is

Metals and Semi-conductors

$$n_e = \int_{\varepsilon_c}^{\varepsilon_t} n(\varepsilon)\, d\varepsilon \qquad (18 \cdot 13\text{-}2)$$

Here $Z(\varepsilon)\, d\varepsilon$ is the density of states in the energy interval ε to $\varepsilon + d\varepsilon$ and is given by (see Eq. 20.7-9)

$$Z(\varepsilon) = \frac{(2m^3)^{1/2}}{\pi^2 \hbar^3}(\varepsilon - \varepsilon_c)^{1/2} \qquad (18 \cdot 13\text{-}3)$$

$F(\varepsilon)$ in Eq. (18·13-1) is the Fermi-Dirac distribution function given by (see Eq. 20.7-8)

$$F(\varepsilon) = \frac{1}{\exp(\varepsilon - \varepsilon_f)/kT + 1} \qquad (18 \cdot 13\text{-}4)$$

where ε_f is the Fermi energy which however does not have the same simple interpretation as in the case of a metal. In fact for insulators and semi-conductors ε_f may even lie in the forbidden zone and has a rather abstract concept.

We then have

$$n_e = \frac{(2m^3)^{1/2}}{\pi^2 \hbar^3} \int_{\varepsilon_c}^{\varepsilon_t} \frac{(\varepsilon - \varepsilon_c)^{1/2}\, d\varepsilon}{\exp(\varepsilon - \varepsilon_f)/kT + 1} \qquad (18 \cdot 13\text{-}5)$$

In the above expression we should take for m the effective mass of the electron m_e^* defined before (see § 18·10).

If we put $y = (\varepsilon - \varepsilon_c)/kT$, then we have $dy = d\varepsilon/kT$ and $\varepsilon/kT = \varepsilon_c/kT + y$. Since ε_f is usually midway between ε_v and ε_c (Fig. 18·19) and ε lies in the conduction band above ε_c, we have

$$(\varepsilon - \varepsilon_f)/kT >> 1$$

so that *unity* in the denominator of the integrand can be neglected and we get after substitution

$$n_e = \frac{(2m_e^{*3})^{1/2}}{\pi^2 \hbar^3}(kT)^{3/2} \exp(\varepsilon_f - \varepsilon_c)/kT$$

$$\times \int_0^\infty y^{1/2} \exp(-y)\, dy \qquad (18 \cdot 13\text{-}6)$$

Here the limits of integration are fixed as follows: For $\varepsilon = \varepsilon_c, y = 0$; for $\varepsilon = \varepsilon_t, y_t = (\varepsilon_t - \varepsilon_c)/kT$. Since $y^{1/2}$ is a slowly rising function of y while $\exp(-y)$ diminishes rapidly with increasing y, we can replace the upper limit by $y_t = \infty$.

The above integral has a standard form and its value is $\sqrt{\pi}/2$. So we get finally

$$n_e = \frac{1}{\sqrt{2}} \left(\frac{m_e^* kT}{\pi \hbar^2}\right)^{3/2} \exp(\varepsilon_f - \varepsilon_c)/kT$$
$$= n_{e0} \exp\{-(\varepsilon_0 - \varepsilon_f)/kT\} \qquad (18\cdot13\text{-}7)$$

where n_{e0} is the value of n_f when $\varepsilon_f = \varepsilon_C$, i.e., the Fermi level rises to the bottom of the conduction band :

$$n_{e0} = \frac{1}{\sqrt{2}} \left(\frac{mk}{\pi \hbar^2}\right)^{3/2} \left(\frac{m_e^*}{m}\right)^{3/2} T^{3/2} = 4\cdot 84 \; 10^2 \left(\frac{m_e^*}{m}\right)^{3/2} T^{3/2} \qquad (18\cdot13\text{-}7a)$$

(B) *Concentration of holes in the valence band* :

Since the holes represent absence of electrons in the valence band, the distribution function in this case measures the probability $1 - F(\varepsilon)$ for a state of energy ε to be unoccupied. Hence for the holes we can write

$$n(\varepsilon) d\varepsilon = Z(\varepsilon) [1 - F(\varepsilon)] d\varepsilon \qquad (18\cdot13\text{-}8)$$

The total number of holes is

$$n_h = \int_{\varepsilon_b}^{\varepsilon_v} n(\varepsilon) d\varepsilon \qquad (18\cdot13\text{-}9)$$

Here $Z(\varepsilon)$ is given by

$$Z(\varepsilon) = \frac{(2m_h^{*3})^{1/2}}{\pi^2 \hbar^3} (\varepsilon_v - \varepsilon)^{1/2} \qquad (18\cdot13\text{-}10)$$

where we use the effective mass m_h^* of the hole. Further ε_v is way below ε_f and $\varepsilon \ll \varepsilon_v$. So we get

$$1 - F(\varepsilon) = 1 - \frac{1}{\exp(\varepsilon - \varepsilon_f)/kT + 1}$$
$$= \frac{\exp(\varepsilon - \varepsilon_f)/kT}{\exp(\varepsilon - \varepsilon_f)/kT + 1} \qquad (18\cdot13\text{-}11)$$

Since $(\varepsilon - \varepsilon_f)/kT$ is a large negative number, the exponential in the denominator can be neglected so that we have

$$1 - F(\varepsilon) = \exp(\varepsilon - \varepsilon_f)/kT \qquad (18\cdot13\text{-}12)$$

Hence we get

$$n_h = \frac{(2m_h^{*3})^{1/2}}{\pi^2 \hbar^3} \int_{\varepsilon_b}^{\varepsilon_v} (\varepsilon_v - \varepsilon)^{1/2} \exp(\varepsilon - \varepsilon_f)/kT \, d\varepsilon \qquad (18\cdot13\text{-}13)$$

As before if we substitute $y = (\varepsilon_v - \varepsilon)/kT$ we can transform the above expression as follows :

$$n_h = \frac{(2m_h^{*3})^{1/2}(kT)^{3/2}}{\pi^2 \hbar^3} \exp(\varepsilon_v - \varepsilon_f)/kT$$

$$\times \int_0^\infty y^{1/2} \exp(-y)\, dy$$

$$= \frac{1}{\sqrt{2}} \left(\frac{m_h^* kT}{\pi \hbar^2}\right)^{3/2} \exp(\varepsilon_v - \varepsilon_f)/kT$$

$$= n_{h0} \exp\{-(\varepsilon_f - \varepsilon_v)/kT\} \qquad (18\cdot13\text{-}14)$$

where $n_h = n_{h0}$ when $\varepsilon_f = \varepsilon_v$. *i.e.*, the Fermi level goes down to the top of the valence band

$$n_{h0} = \frac{1}{\sqrt{2}} \left(\frac{mk}{\pi \hbar^2}\right)^{3/2} \left(\frac{m_e^*}{m}\right)^{3/2} T^{3/2} = 4.84 \times 10^2 \left(\frac{m_h^*}{m}\right)^{3/2} T^{3/2} \quad (18\cdot13\text{-}14a)$$

(C) Calculation of ε_f:

We are now in a position to calculate ε_f for a semi-conductor.

(i) Intrinsic semi-conductor :

The concentrations of the electrons and holes are equal in this case so that $n_e = n_h$.

This gives

$$m_e^{*3/2} \exp(\varepsilon_f - \varepsilon_c)/kT = m_h^{*3/2} \exp(\varepsilon_v - \varepsilon_f)/kT$$

$$\therefore \quad \varepsilon_f = \frac{\varepsilon_v + \varepsilon_c}{2} + \frac{3}{4} kT \ln \frac{m_h^*}{m_e^*} \qquad (18\cdot13\text{-}15)$$

If we assume $m_e^* = m_h^*$, then

$$\varepsilon_f = \frac{\varepsilon_v + \varepsilon_c}{2} \qquad (18\cdot13\text{-}16)$$

i.e., the Fermi level is exactly midway between the top of the valence band (ε_v) and the bottom of the conduction band (ε_c), within the forbidden zone.

In general $m_h^* > m_e^*$ so that ε_f is raised slightly as the temperature T rises.

(ii) Extrinsic semi-conductor of n-type :

For an extrinsic semi-conductor, the sum of the concentrations of the electrons (n_e) and the negatively charged acceptor ions (n_a) must be equal to the sum of the concentrations of the positive holes (n_h) and the positively charged donor ions (n_d). This makes the net charge density under equilibrium to be zero.

$$n_e + n_a = n_h + n_d \qquad (18\cdot13\text{-}17)$$

If we assume that all the acceptor and donor atoms to be completely ionized, n_a and n_d will be equal to the concentration of these atoms. This is a reasonable assumption since the separations of the donor levels from ϵ_c and those of the acceptor levels from ϵ_v are quite small (0·01 eV).

For an *n*-type semi-conductor, we may put $n_h = 0$ and $n_a = 0$ so that we have $n_d = n_e$ which gives

$$n_d = n_{e_0} \exp(\varepsilon_f - \varepsilon_c)/kT \qquad (18\cdot13\text{-}18)$$

where
$$n_{e0} = \frac{1}{\sqrt{2}} \left(\frac{m_e^* kT}{\pi\hbar^2}\right)^{3/2} \qquad (18\cdot13\text{-}19)$$

We then get
$$\varepsilon_f = \varepsilon_c - kT \ln\left(\frac{n_{e0}}{n_d}\right) \qquad (18\cdot13\text{-}20)$$

Usually $n_d < n_{e0}$ which gives $\varepsilon_f < \varepsilon_c$ i.e., the Fermi level lies in the forbidden zone slightly below the conduction band in an *n*-type semi-conductor. As n_d is increased, the Fermi level rises and reaches the bottom of the conduction band (ε_d) when $n_d = n_{e0}$. If n_d becomes larger than n_{e0}, the semi-conductor becomes *degenerate* and behaves like a metal.

(iii) *Extrinsic semi-conductor of the p-type*:

In this case $n_a >> n_d$ and $n_h >> n_e$.

Hence Eqs. (18·13-14) and (18·13-17) give

$$n_a = n_h = n_{h0} \exp(\varepsilon_v - \varepsilon_f)/kT \qquad (15\cdot13\text{-}21)$$

where
$$n_{h0} = \frac{1}{\sqrt{2}} \left(\frac{m_h^* kT}{\pi\hbar^2}\right)^{3/2} \qquad (18\cdot13\text{-}22)$$

This gives
$$\varepsilon_f = \varepsilon_v + \ln\left(\frac{n_{h0}}{n_a}\right) \qquad (18\cdot13\text{-}23)$$

Usually $n_a << n_{h0}$ and ε_f lies slightly above the top of the valence band ε_v and lies in the forbidden zone. If the concentration of the acceptor atoms n_a gradually rises, ε_f goes down towards ε_v and becomes equal to ε_v when $n_a = n_{h0}$.

(D) *Intrinsic carrier concentration* :

For an intrinsic semi-conductor since $n_e = n_h = n_i$ (say), the carrier concentration is

$$n_i = \sqrt{n_e n_h}$$

Using Eqs. (18·13-7) and (18·13-14) we then get

$$n_i = \frac{1}{\sqrt{2}} (m_e^* m_h^*)^{3/4} \left(\frac{kT}{\pi\hbar^2}\right)^{3/2} \exp(-E_g/kT) \qquad (18\cdot13\text{-}24)$$

where $E_g = \epsilon_c - \epsilon_v$ is the band gap between the valence and the conduction bands. The intrinsic carrier concentration is thus independent of the position of the Fermi level. Smaller the band gap, larger is n_i. Similarly higher the temperature, larger is n_i.

18·14 Conductivity of semi-conductors

We have seen that the electrical conductivity of a metal is due to the motion of the free electron gas in them obeying Fermi-Dirac distribution law. In semi-conductors, the conductivity may be due to the motion of negatively charged electrons or positively charged holes or both.

Let v_h and v_e represent the drift velocities of the holes and the electrons respectively and n_h and n_e represent their concentrations, *i.e.*, their number-densities. With the application of the electric field X, the holes move in the field direction while the electrons move opposite to the field direction. The electric current density in the semi-conductor is therefore

$$j = en_h v_h + en_e v_e \qquad (18\cdot14\text{-}1)$$

where e is the charge on the hole and $-e$ that on the electron.

Let u_h and u_e be the mobilities of the holes and electrons respectively, *i.e.*, the velocities gained by them per unit field. Then we can write

$$v_h = Xu_h, \quad v_e = Xu_e \qquad (18\cdot14\text{-}2)$$

So we get $\qquad j = en_h Xu_h + en_e Xu_e \qquad (18\cdot14\text{-}3)$

If σ is the electrical conductivity of the material, then since

$$j = \sigma X$$

we get

$$\sigma = en_h u_h + en_e u_e \qquad (18\cdot14\text{-}4)$$

For intrinsic *semi-conductors*, electron and hole concentrations are equal so that $n_h = n_e = n_i$ (say). Hence in this case we have

$$\sigma_i = en_i(u_h + u_e) \qquad (18\cdot14\text{-}5)$$

In extrinsic semi-conductors, either the hole concentraction is large compared to the electron concentration as in *p*-type semi-conductors ($n_h >> n_e$) or conversely as in *n*-type semi-conductors ($n_e >> n_h$). Hence for these two types we get

$$\sigma_p = en_h u_h, \quad \sigma_n = en_e u_e \qquad (18\cdot14\text{-}6)$$

18·15 Rectifying property of semi-conductors

If an n-type semi-conductor is in contact with a p-type semiconductor, the combination can act as a rectifier. The rectifying action of the combination can be understood by referring to Fig. 18·20.

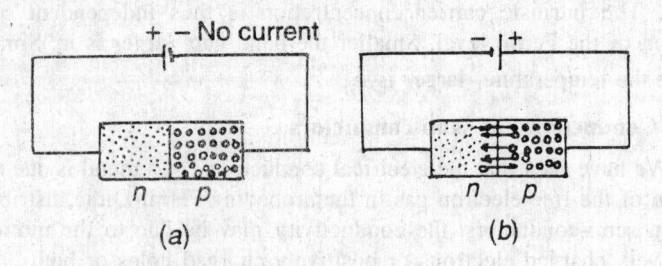

Fig. 18·20. Rectifying property of (a) n-type and (b) p-type semi-conductors.

If the n-type semi-conductor is connected to the positive terminal of a battery while the p-type to the negative terminal as shown in Fig. 18·20 a, then the excess negative carriers in the n-type material and the excess positive carriers (holes) in the p-type material are drawn in opposite directions away from the junction of the two. Hence no electric current can flow across the junction. On the other hand if the n-type semi-conductor is connected to the negative terminal of a battery while the p-type is connected to the positive terminal (Fig. 18·20b), then both negative and positive carriers can flow across the junction so that a current will flow in the external circuit.

Thus such a combination allows electric current to flow in one direction (forward-biased) and prevents the flow of current in the opposite direction (reverse-biased) and hence acts as a rectifier.

If instead of two, three semiconductors in contact are used, the combination, known as a transistor, can be made to work as a triode valve. For example, a very thin layer of a p-type semi-conductor placed between two n-type semi-conductors can be used to construct a transistor which can perform the functions of a triode valve in an amplifier or in an oscillator circuit. Similarly an n-type semi-conductor sandwiched between two p-type semi-conductors can be used to construct a transistor.

Semi-conductor diodes and transistors have largely supplanted the thermionic valves in electronic science and technology. In conjunction with micro-electronics, they have resulted in great reduction in the consumption of electrical power and in the sizes of electronic instruments.

18·16 Hall effect

The concept of holes as positive carriers of current in semi-conductors finds direct experimental support from the phenomenon of Hall effect, first discovered in 1879 by E.H. Hall.

If a magnetic field acts at right angles to the direction of current flow in a material, a voltage is developed across the specimen perpendicular to both the current and the magnetic field. The effect of the magnetic field is to push the moving charges to one side due to the Lorentz force given by

$$\mathbf{F} = e(\mathbf{v} \times \mathbf{B}) \qquad (18\cdot16\text{-}1)$$

where **v** is the velocity of the charge carriers constituting the current and **B** is the magnetic induction field.

As shown in Fig. 18·21, the charge carriers are deflected in a direction at right angles to the direction of the current (e**v**) and the magnetic field and accumulate on the face A or B of the specimen. As a result a potential difference develops between A and B which opposes the motion of the charges towards them by the action of the magnetic field. The effect is known as *Hall effect* and the potential developed between A and B as *Hall potential*.

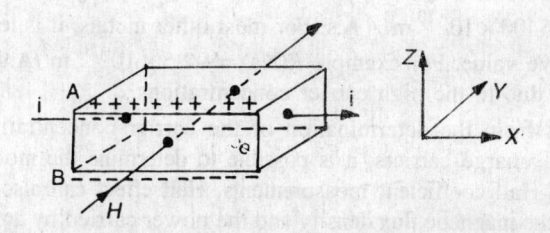

Fig. 18·21. Hall effect.

Hall potential is usually quite small in magnitude. Hall field may be directed from the upper to the lower face of the specimen in Fig. 18·21 in some materials. It may have the opposite direction in some other materials, indicating that the charge carriers in the two types of materials have opposite signs. It is possible to determine the concentration of the charge carriers from experiments on Hall effect.

It is found that the Hall potential V_H depends on the current density j, the applied magnetic induction field B and the distance z between the two faces A and B.

$$V_H = R j B z \qquad (18\cdot16\text{-}2)$$

R is a constant known as the *Hall coefficient*. The Hall field is

$$E_H = \frac{V_H}{z} = R j B \qquad (18\cdot16\text{-}3)$$

If the charge on the carrier is e, then the magnitude of the Lorentz force which is balanced by the force due to the Hall field under equilibrium is given by (see Eq. 18.16–1)

$$F = ev B = eE_H = R jBe \qquad (18 \cdot 16\text{-}4)$$

If the carrier concentration is n per unit volume, then the current density is

$$j = nev$$

So we get from Eq. 18·16-4)

$$R = \frac{ev B}{jBe} = \frac{ev}{ne^2 v}$$

or, $$R = \frac{1}{ne} \qquad (18 \cdot 16\text{-}5)$$

So by measuring the Hall coefficient R, it is possible to determine the carrier concentration n. It is found that R is negative for the metals Li, Na, K, Bi *etc.*, showing that the charge carriers are electrons in them. On the other hand in semi-conductors and some metals like Be, Zn, Cd, R is positive which shows that the charge carriers are the positive holes in them. Measurement shows that Bi has an unusually high Hall coefficient. R (Bi) = -5400×10^{-10} m^3/ A.s. For most other metals, it is less than 1% of the above value. For example R(Na) = $-2 \cdot 5 \times 10^{-10}$ m^3/A.s. Such low values are due to the high carrier concentrations.

Apart from the determination of the carrier concentration and the sign of the charge carriers, it is possible to determine the mobility of the carriers by Hall coefficient measurements, Hall effect can also be used to determine the magnetic flux density and the power carried by an e.m. wave.

18·17 Superconductivity

Superconductivity was discovered by Kamerlingh Onnes in 1911 at Leiden in Holland. Onnes found that the resistivity (ρ) of mercury vanished completely below the temperature 4·2 K (see Fig. 18·22). This is known as the critical temperature (T_c) for superconductivity.

Since $\rho = E/j$ according to Ohm's law, the electric field intensity E must vanish ($E = 0$) for any finite current density j in a superconductor for which $\rho = 0$.

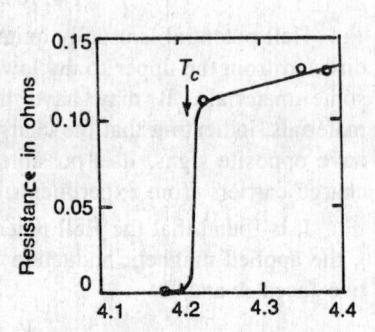

Fig. 18·22. Variation of resistivity (ρ) of mercury with temperature (T).

If a ring of superconductor is cooled in a magnetic field to a temperature below T_c and the magnetic field is switched off, an electric current is induced in the ring. This current is found to persist for hours without any diminution even though no source of e.m.f. is included in the circuit. A group at M.I.T. is reported to have observed that a ring of lead in the superconducting state carried an induced current of several hundred amperes for over a year without any change.

Superconductivity is found to be destroyed by the application of an external magnetic field exceeding a critical value B_c. For the critical temperature T_c, $B_c = 0$ which means that the application of even a very small field destroys superconductivity at $T = T_c$. For $T < T_c$, the value of B_c increases and becomes highest at $T = 0$ K, The variation of B_c with T is shown in Fig. 18·23 and can be represented by the formula $B_c = B_0 [1 - (T/T_c)^2]$. B_0 is the critical field for the destruction of superconductivity at 0 K.

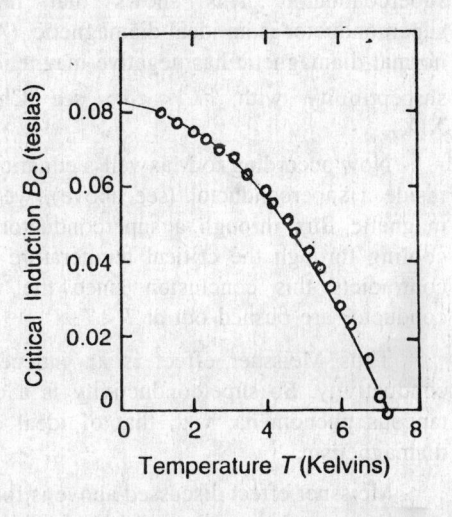

A large number (> 20) of pure chemical elements in the periodic table and few hundred alloys and chemical compounds have been found to be superconducting with T_c ranging from 0 to 20 K. In recent years superconductivity has been discovered in some materials at much higher temperature with T_c upto ~125 K. Concerted efforts are being made to discover materials with still higher values of T_c.

Fig. 18·23. Variation of critical induction (B_c) with temperature (T).

Superconducting metals are usually not good conductors at room temperatures. For example the superconductors titanium, zirconium and hafnium have resistivities of 89×10^{-4}, 45×10^{-4} and 32×10^{-4} ohm-m respectively at room temperature compared to $1·6 \times 10^{-4}$ ohm-m for copper.

We list below some superconductors with their critical temperatures :

Metal	Ti	V	Zn	Nb	Sn	Ta	Hg	Pb
T_c (K)	0.39	5.38	0.88	9.20	3.72	4.48	4.15	7.22

Meissner effect :

In 1933, W. Meissner and R. Ochsenfeld found that in the superconducting state, the lines of magnetic induction are pushed out of a superconducting material (see Fig. 18·24). This is known as Meissner effect which shows that the magnetic induction $B = 0$ inside a superconductor. Now $\mathbf{B} = \mu_0 (1 + \chi)\mathbf{H}$ where $\mu_0 = 4\pi \times 10^{-7}$ henri/m is the permeability of empty space and χ is the magnetic susceptibility. So we must put $\chi = -1$ for a superconductor. This shows that the superconductor is an ideal diamagnetic. (A normal diamagnetic has negative magnetic susceptibility with $|\chi| << 1$; see Ch. XIX).

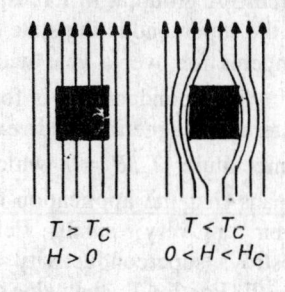

Fig. 18·24 Meissner effect.

Now according to Maxwell's equation $\partial \mathbf{B}/\partial t = -\nabla \times \mathbf{E}$. Since $\mathbf{E} = 0$ inside a superconductor (see above), we must put $\partial \mathbf{B}/\partial t = 0$. Thus the magnetic flux through a superconductor cannot change with time on cooling through the critical temperature T_c. Obviously Meissner effect contradicts this conclusion since the flux lines passing through a conductor are pushed out at $T < T_c$.

Thus Meissner effect is an independent manifestation of superconductivity. So superconductivity is a combination of the two simultaneous phenomena, viz., that of ideal conductivity and that of ideal diamagnetism.

Meissner effect discussed above is for an ideal superconductor. In an actual superconductor in the form of a thin film, there is some penetration of the magnetic field lines near the surface. Two British physicists H. London and F. London have suggested that in this case we can write the current density $\mathbf{j} \propto \mathbf{A}$ where $\mathbf{A}$ is the magnetic vector potential so that $\mathbf{B} = \nabla \times \mathbf{A}$.

Let us write

$$\mathbf{j} = -\frac{1}{\mu_0 \lambda_L^2} \mathbf{A} \qquad (18\cdot17\text{-}1)$$

This is known as London equation. Here λ_L is a constant with the dimension of length known as *penetration depth*.

From Maxwell's equation, we have, neglecting displacement current

$$\nabla \times \mathbf{B} = \mu_0 \mathbf{j}$$

$$\therefore \qquad \nabla \times \nabla \times \mathbf{B} = \mu_0 \nabla \times \mathbf{j}$$

Since $\nabla \cdot \mathbf{B} = 0$, the above equation gives

$$-\nabla^2 \mathbf{B} = \mu_0 (-1/\mu_0 \lambda_L^2) \nabla \times \mathbf{A} = -\frac{1}{\lambda_L^2} \mathbf{B}$$

We thus have
$$\nabla^2 \mathbf{B} = \mathbf{B}/\lambda_L^2 \tag{18·17-2}$$
Taking the x-component, we get
$$\frac{d^2 B_x}{dx^2} = \frac{B_x}{\lambda_L^2} \tag{18·17-3}$$
The solution of this equation is of the form
$$B_x = B_{0x} \exp(-x/\lambda_L) \tag{18·17-4}$$

The positive exponential solution is ruled out since this would make $B_x \to \infty$ as $x \to \infty$. Thus the magnetic flux lines fall off exponentially within a semi-infinite superconductor near the surface. It has been shown that
$$\lambda_L = \left(\frac{m}{\mu_0 n e^2}\right)^{1/2} \tag{18·17-5}$$
where m and e are the electronic mass and charge. n is the concentration of the superconducting electrons. Values of λ_L lie in the range 400 Å to 1000 Å.

Theory of superconductivity :

Theoretical understanding of superconductivity came more than 50 years after the discovery of the phenomenon. The microscopic theory was developed mainly by J. Bardeen, L. N. Cooper and J.R. Schrieffer (1957). The theory involves very difficult concepts and is beyond the scope of the present book. We give below a few salient points.

E. Fröhlich in 1950 proposed a model for an electron moving through a metal lattice continuously emitting and reabsorbing *virtual phonons*. According to him, the electrons perturb the neighbouring atoms causing them to oscillate. These lattice perturbations (phonons) then react on the electron. He pointed out that such *electron-phonon interaction* might produce a ground state of energy lower than that of the completely filled Fermi sea of noninteracting electrons. There will be an energy gap between such superconducting ground state and the normal conducting states of the metal.

Fröhlich's other suggestion was that due to the electron-phonon interaction, T_c should be proportional to T_D, the Debye temperature, which implies that $T_c \propto M^{-1/2}$ where M is the isotopic mass. This dependence of T_c on the isotopic mass (isotope effect) has been confirmed experimentally which provides strong support for the electron-phonon interaction model of superconductivity.*

---

* Actually the variation of T_c with the isotopic mass does not always follow the $M^{-1/2}$ law. The departures are due to the Coulomb interaction between the electrons.

Cooper in 1956 suggested that the *electron-phonon-electron interaction* whould make the Coulomb repulsion between two electrons smaller (or even absent) in the superconducting state than in the normal state. When an electron deforms the lattice in its vicinity a second electron sees the deformation and adjusts itselfs to take advantantage of it to lower its energy. Thus the second electron interacts with the first via the lattice deformation. We may say that due to the deformation of the lattice by the first electron, phonons are created which are propagated through the crystal. The second electron is affected by absorbing this phonon. Since the phonon exchange is *virtual*, the energy is conserved for the crystal but the energy of the interacting electron pair need not be conserved before and after the exchange (c.f. pion exchange between nucleons). Thus Cooper showed that for a pair of electrons just above the Fermi surface a bound state could result if the phonon exchange gave rise to an attractive interaction.

Later Bardeen, Cooper and Schrieffer showed that under certain circumstances, there might be condensation of electrons of opposite spins into bound pairs having opposing values of the wave vectors for zero current density. Such pairing involves electrons in states within the energy kT_D ($k =$ Boltzmann constant) of the Fermi energy. At $T = 0$ with zero current density, the ground state of a superconductor is a highly correlated state. All the states near the Fermi surface are thus filled with pairs of electrons of opposite wave vectors ($\mathbf{k}$) and spins to the fullest possible extent. These are known as *Cooper pairs*. The ground state of all such cooper pairs is represented by a single wave function. A Cooper pair ($\mathbf{k}_1 \uparrow, -\mathbf{k}_1 \downarrow$) by an exchange of a virtual phonon can go over to another unoccupied pair position ($\mathbf{k}_2 \uparrow, -\mathbf{k}_2 \downarrow$). There are strong experimental evidences in support of the existence of such pairs.

There is a significant gap between the energy of a Cooper pair and the energy of two single unpaired electrons which is of the order of millivolts. This is many times larger than the energy required to break up the pair.

Each Cooper pair is symmetric about $k = 0$ for a superconductor with zero current density. If there is a nonvanishing persistent current, the Cooper pairs have non-zero net momenum. As we know, in a normal conductor, the conductivity is prevented from being infinite due to the scattering of the electrons by phonons and by localized defects. This does not happen in the superconducting state because the above energy gap prevents the Cooper pairs against any change of net momentum. Thus the

Metals and Semi-conductors

infinite D.C. conductivity of a suppercondutor emerges as a natural consequence of the BCS theory.

The existence of the energy gap between the superconducting ground state and the states of normal conduction in metals as proposed by the BCS theory has been confirmed experimentally.

The energy gap (ε_g) decreases towards zero as the temperature approaches T_c, since the number of paired electrons decreases with rising temperature (see Fig. 18·25), The ratio ($\varepsilon_{gT}/\varepsilon_{g0}$) of the energy gaps at the temperatures T K and O K is a function of (T/T_c). The functional relation

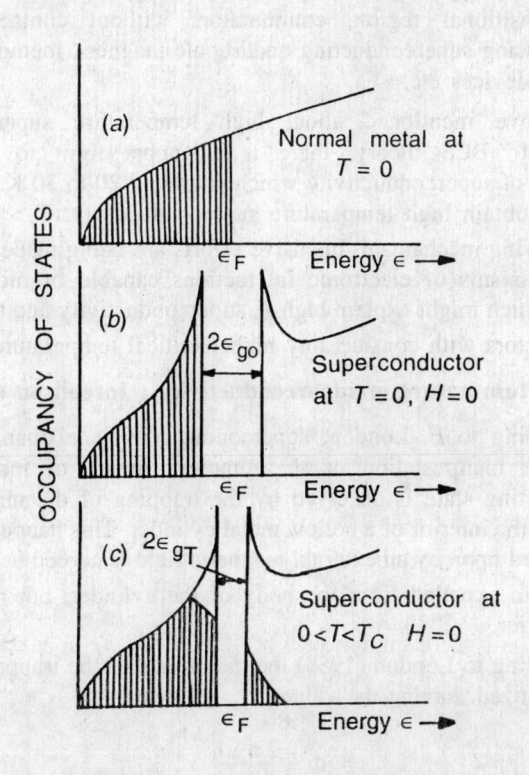

Fig. 18·25 Variation of energy gap between superconducting ground state and normal states in a metal ; (a) nonconducting state ; (b) and (c) superconducting states at $T < T_c$ and $T = $ O K.

between the two expected on theoretical ground has been confirmed experimentally for In and Ta.

The BCS theory is able to explain the ideal diamagnetism of superconductors as also the destruction of superconductivity by an external magnetic field.

There are many important applications of superconductivity. The most notable application is the construction of superconducting magnets to generate extremely high magnetic fields for laboratory and industrial use. The disadvantage of requiring a liquid refrigerant in which the magnet coil must be immersed is often superseded by the advantage of having a magnet coil with negligible resistive loss capable of producing fields upto 15 T ($1\cdot 5 \times 10^5$ gauss) without much difficulty.

Superconductivity is also being increasingly utilized to design audio–frequency modulators, detectors of high frequency modulated signals which utilizes the nonlinearity of the superconductor's resistance in the transitional region, commutators without contact switches, cyclotrons using superconducting quadrupole magnets, memory elements in memory devices etc.

We have mentioned about high temperature superconductors. According to BCS theory, there is an upper limit to the critical temperature of superconductivity which is around 20 to 30 K. So it is not possible to obtain high temperature superconductivity ($T_c > 100$ K) with electron pairing mechanism. Intensive efforts are being made to discover other mechanisms of electronic interactions capable of more efficient attraction which might explain high T_c superconductivity and thus provide superconductors with considerably higher critical temperatures.

18·18 Quantum nature of superconductivity ; Josephson effect

According to F. London, superconductivity is a quantum phenomenon. The manifestation of the quantum nature of matter in the superconducting state is observed by the trapping of the magnetic flux lines within the interior of a hollow metal cylinder. This happens when the cylinder acted upon by an external magnetic field is cooled below T_c. The flux lines are expelled from the body of the cylinder, but not from its hollow interior.

According to London (1950) the magnitude of the trapped magnetic flux is quantized, having the value

$$\Phi = \frac{2\pi s \hbar}{e}$$

where s is an integer and e is the electronic charge,

Measurements by Deaver and Fairbanks at Stanford University in the U.S.A. and by Nabauer in Germany (1961) have confirmed London's prediction. However, the magnitude of the flux quantum was found to be half of that postulated by London *i.e.*,

$$\Phi = \frac{s\pi\hbar}{e} \qquad (18\cdot 18\text{-}1)$$

where the quantized unit of flux is $\Phi_0 = \dfrac{\pi \hbar}{e} = 2 \cdot 06 \times 10^{-15}$ weber.

The above result is in agreement with BCS theory according to which the charge carriers in superconductors are electron pairs (Cooper-pairs) of charge $-2e$ each.

The quantum nature of superconductivity is also manifested by *Josephson effect*, first discovered by B.D. Josephson in 1962. A very thin

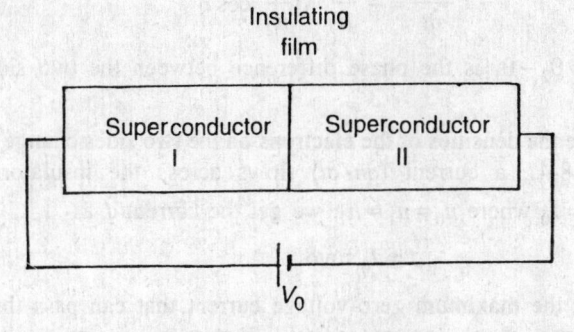

Fig. 18·26. Josephson effect.

layer of an insulator (~ 10 Å thick) between two super-conductors (SIS junction) constitutes a Josephson junction shown in Fig. 18·26. There are three distinct effects.

(a) D.C. Josephson effect : The Cooper pairs which can move freely within a superconductor can make transition from one superconductor to the other through the insulating layer because of its thinness. As a consequence, a finite supercurrent may flow in the circuit containing the Josephson junction even with zero voltage ($V = 0$) across the junction which is known as D.C. Josephson effect. If ψ_1 and ψ_2 represent the common wave functions for all the electrons in the superconductors on the two sides of the insulator, then the Schrödinger equation with $V = 0$ gives.

$$i\hbar \frac{\partial \psi_1}{\partial t} = K \psi_2, \quad i\hbar \frac{\partial \psi_2}{\partial t} = K \psi_1 \qquad (18 \cdot 18\text{-}2)$$

where K is a characteristic of the junction which represents the electron-pair coupling across the insulator. K is a measure of the leakage of ψ_1 into 2 and of ψ_2 into 1. It has the dimensions of energy. For a thick insulator $K = 0$. Let us write

$$\psi_1 = \sqrt{n_1} \exp(i\theta_1), \qquad \psi_2 = \sqrt{n_2} \exp(i\theta_2) \qquad (18 \cdot 18\text{-}3)$$

where n_1 and n_2 are the electron densities and θ_1 and θ_2 are the phases on the two sides of the junction.

Substituting in Eq. (18·18-2) and equating the real and imaginary parts in the two resulting equations, it can be shown that for two identical ideal superconductors on the two sides

$$\frac{\partial n_1}{\partial t} = -\frac{\partial n_2}{\partial t} = \frac{2K}{\hbar}\sqrt{n_1 n_2}\sin\delta \qquad (18\cdot 18\text{-}4)$$

$$\frac{\partial \theta_1}{\partial t} = -\frac{K}{\hbar}\sqrt{\frac{n_2}{n_1}}\cos\delta \qquad (18\cdot 18\text{-}5a)$$

$$\frac{\partial \theta_2}{\partial t} = -\frac{K}{\hbar}\sqrt{\frac{n_1}{n_2}}\cos\delta \qquad (18\cdot 18\text{-}5b)$$

where $\delta = \theta_2 - \theta_1$ is the phase difference between the two sides of the junction.

Since the densities of the electrons on the two sides change with time (Eq. 18·18-4), a current $(\partial n/\partial t)$ flows across the insulator. Writing $2Kn_0/\hbar = J_0$ where $n_0 = n_1 \approx n_2$, we get the current J as

$$J = J_0 \sin\delta \qquad (18\cdot 18\text{-}6)$$

J_0 is the maximum zero-voltage current that can pass through the insulator. Thus with no applied voltage, a D.C. current flows with a value between $+J_0$ and $-J_0$ depending on δ

(b) A.C. Josephson effect : If a D.C. voltage V is applied between the two sides of the insulator then this causes a radio-frequency current to flow across the Josephson junction. This is known as the A.C. Josephson effect. An external r.f. voltage applied with the D.C. voltage can then cause a D.C. current to flow across the junction.

An electron-pair experiences a potential energy difference qV in passing across the junction where $q = -2e$ is the charge of each Copper pair. Assuming that a pair on one side has the potential energy $-eV$ while that on the other side has the p.e. $+eV$ we can write

$$i\hbar\,\frac{\partial \psi_1}{\partial t} = K\psi_2 - eV\psi_1$$

$$i\hbar\,\frac{\partial \psi_2}{dt} = K\psi_1 + eV\psi_2 \qquad (18\cdot 18\text{-}7)$$

As before, substitution for ψ_1 and ψ_2 (Eq. 18·18-3) gives

$$\frac{\partial n_1}{\partial t} = -\frac{\partial n_2}{\partial t} = \frac{2K}{\hbar}\sqrt{n_1 n_2}\sin\delta$$

$$\frac{\partial \theta_1}{\partial t} = \frac{eV}{\hbar} - \frac{K}{\hbar}\sqrt{\frac{n_2}{n_1}}\cos\delta \qquad (18\cdot 18\text{-}8)$$

$$\frac{\partial \theta_2}{\partial t} = -\frac{eV}{\hbar} - \frac{K}{\hbar}\sqrt{\frac{n_1}{n_2}}\cos\delta \qquad (18\cdot 18\text{-}9)$$

Metals and Semi-conductors

For $n_1 \approx n_2$,
$$\frac{\partial \delta}{\partial t} = \frac{\partial}{\partial t}(\theta_2 - \theta_1) = -\frac{2eV}{\hbar} \qquad (18\cdot18\text{-}10)$$

Integration gives

$$\delta(t) = \delta(0) - \frac{2eV}{\hbar}t \qquad (18\cdot18\text{-}11)$$

and
$$J = J_0 \sin\{\delta(0) - \omega_0 t\} \qquad (18\cdot18\text{-}12)$$

where
$$\omega_0 = \frac{2eV}{\hbar} \qquad (18\cdot18\text{-}13)$$

Since $\hbar$ is a small number (compared to ordinary voltages and times), the oscillations are very rapid and the net current is very small.

With $V = 1\ \mu V$, we get the frequency $\nu_0 = \omega_0/2\pi = 483\cdot6$ MHz. When an electron-pair crosses the barrier, a photon of energy $\hbar\omega_0 = 2eV$ is emitted or absorbed.

We have seen that with $V = 0$, there is D.C. current across the junction (Eq. 18·18-6). But if a D.C. voltage V is applied then this goes to zero.

The frequency ν_0 is called the *Josephson frequency*. If e.m. radiation from an outside source having frequency ν falls on the Josephson junction, beats will be produced between the oscillating supercurrent mentioned above and the incident radiation whenever ν is an integral multiple of $\nu_0 : \nu = s\nu_0$ where s is an integer. Under this condition V will be an integral multiple of $\hbar\omega_0/2e = h\nu_0/2e$ so that $V = sh\nu_0/2e$. This shows that the values of V will appear in steps in the current-voltage curve. Such steps were observed by S. Shapiro using a niobium oxide-lead junction cooled to 4·2 K as shown in Fig. 18·27. These steps also exhibit the quantum nature of *super-conductivity*.

Josephson effect was used by Parker, Langenberg, Denenstein and Taylor to determine the ratio e/h very accurately, by using junction of tin-tin oxide-lead at a temperature $\sim 1\cdot 2$ K. The junction was exposed to radiation of frequency 10^4 MHz ($\lambda \sim 3$ cm). The following values of $2e/h$ were obtained from emission and absorption respectively.

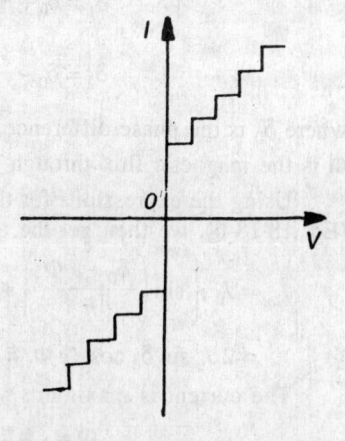

Fig. 18·27. Current-voltage curve for niobium oxide–lead Josephson junction in a microwave field.

$$\frac{2e}{h} = 4 \cdot 835985 \times 10^{14} \text{ coulombs/joule-s}$$

$$\frac{2e}{h} = 4 \cdot 835978 \times 10^{14} \text{ coulombs/joule-s}$$

(c) *Macroscopic quantum interference* : Suppose two Josephson junctions (1 and 2) with $V = 0$ are connected in parallel between two points in a circuit connected to a current measuring device (see Fig. 18·28). If a magnetic field acts in the space between the two junctions then it is found that the phase difference across the Josephson junctions in the two branches are given by

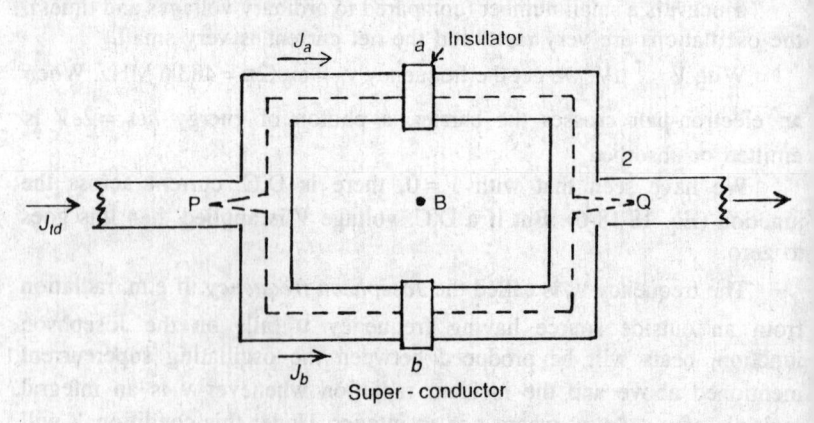

Fig. 18·28. Two Josephson junctions in parallel.

$$\delta_1 = \delta_0 + \frac{e\Phi}{\hbar}$$

$$\delta_2 = \delta_0 - \frac{e\Phi}{\hbar}$$

where δ_0 is the phase difference in the absence of the magnetic field and Φ is the magnetic flux through the closed loop.

Using the expressions for the current across the Josephson junctions (Eq. 18·18-6), we then get the total current as

$$J_{tot} = J_0 \left\{ \sin\left(\frac{\delta_0 + e\Phi}{\hbar}\right) + \sin\left(\delta_0 - \frac{e\Phi}{\hbar}\right) \right\}$$

$$= 2 J_0 \sin \delta_0 \cos (e\Phi/\hbar) \qquad (18\cdot18\text{-}14)$$

The current is maximum when

$$\Phi = s \pi \hbar / e \qquad (18\cdot18\text{-}15)$$

where s is an integer. Thus the current assumes maximum values when the flux linkage has just the quantized values given above (Eq. 18·18-1).

The predicted effect has been confirmed experimentally which shows rapid oscillations of the current with changes in the magnetic field due to the interference term $\cos(e\Phi/\hbar)$.

The superconducting quantum interference effect between two Josephson junctions has been utilized in the construction of extremely sensitive magnetometers which can measure magnetic induction fields of the order of 10^{-11} weber/m^2 (10^{-7} gauss) or less. The sensitivity can be further improved by using a larger number of junctions.

The superconducting quantum interference devices (SQUID) have many applications where extremely feeble magnetic fields are to be measured.

In the absence of any magnetic field, the device would allow a certain amount of electric current to flow. However even an extremely feeble magnetic field reduces this current which can be restored to the previous value by the application of a suitable voltage acoss the junction. Measuring the required voltage, one can then determine the magnetic field with great precision.

This type of magnetometer is used in the analysis of brain waves. The method, known as magneto-encephalography (M.E.G.), is found to be more sensitive than the conventional electro-encephalography (E.E.G.).

The sensitivity of the SQUID type magnetometers have in recent years been considerably improved by the use of multichannel magnetometers using larger number of Josephson-junctions.

Problems

1. Using Fermi-Dirac distribution law show that the highest energy of the free electrons in a metallic lattice is equal to the Fermi energy ε_f at $T = 0$ K. Prove that in this case $n(\varepsilon) \propto \sqrt{\varepsilon}$

2. Prove that for $T > 0$, the F-D. energy distribution is given by $n(\varepsilon) \propto \sqrt{\varepsilon} \exp(\varepsilon - \varepsilon_f)/kT$ upto a few thousand kelvins with $\varepsilon > \varepsilon_f$ and $(\varepsilon - \varepsilon_f)$ of the order of a few electron volts.
   What will be the nature of the change of $n(\varepsilon)$ with ε if $\varepsilon < \varepsilon_f$?

3. Deduce Eqs. (18·7-4) and (18·7-5) for the Fermi energy ε_f and the mean kinetic energy $<\varepsilon>$ of the free electrons in a metal for $T > 0$.

4. Silver has the Fermi energy 5·5 ev at $T = 0$ K. Show that this is reduced by about 0·18% at $T = 3000$ K. Also show that the mean kinetic energy of the free electrons in the silver lattice is increased by about 1% for the above change of temperature.

5. Calculate the reflection coefficient R of the electrons at the surface of platinum during thermionic emission, using the data given in Table 18·1.

6. Assuming $\cos kl = -1$ in Eq. (18·9-6) find the successive positive values of αl by solving the above equation for $P = 3\pi/2$ and find the ranges of αl for the two forbidden zones corresponding to $kl = \pi$ and $kl = 3\pi$. (0·5π ; 0·27π).

7. Using Eq. (18·17-5), calculate the peneration depth λ_L of the magnetic field in a superconductor for which $n_o = 10^{28}$ per m^3 (530 Å)

8. Deduce Eqs. (18·18-4) to (18·18-5b) starting from Eqs. (18·18-1) using the wave functions ψ_1 and ψ_2 given by Eqs. (18·18-3).

19

MAGNETIC PROPERTIES OF SOLIDS

19.1 Introduction

Magnetization under the action of an applied magnetic field is a universal property of all solids. In most of them, the magnetization induced by the external magnetic field is feeble. In some it may be quite high. If M is the magnetization induced by the magnetic field H we can write.

$$M = \chi \cdot H \qquad (19\cdot1\text{-}1)$$

χ is called the *magnetic susceptibility*. In an isotropic substance, M is in the direction of H and χ is a constant. For an anistropic material χ is a tensor of rank 2. In this case M and H are usually not in the same direction. χ is dimensionless in S.I. system.

Magnetic induction **B** is defined as

$$\mathbf{B} = \mu_0 (\mathbf{H} + \mathbf{M}) = \mu_0 \mathbf{H} (1 + \chi) = \mu_0 \mu_r \mathbf{H} \qquad (19\cdot1\text{-}2)$$

$\mu_0 = 4\pi \times 10^{-7}$ henri/m is the magnetic permeability of empty space.

$\mu = \mu_0 (1 + \chi)$ is called the magnetic permeability of the medium.

$\mu_r = (1 + \chi)$ is called the relative permeability.

In an isotropic material. B, H and M are parallel and μ is a scalar.

All solids can be classified according to their magnetic properties into three groups : (*a*) Diamagnetics, (*b*) Paramagnetics and (*c*) Ferromagnetics.

The classification is based on the values and behaviour of χ under different conditions.

(*a*) *Diamagnetics* : In this case $\chi < 0$ and has a very low value, *i.e.*, $|\chi| << 1$. In Fig. 19·1a, the nature of variation of M with H is shown graphically. This is a straight line with a negative slope which shows that χ is a constant independent of H. Some typical values of χ for a few diamagnetic substances are given below :

Material	Bi	Cu	Ge	Si
χ	-15×10^{-5}	$-0\cdot9 \times 10^{-5}$	$-0\cdot8 \times 10^{-5}$	$-0\cdot3 \times 10^{-5}$

(b) *Paramagnetics* : In this case also $|\chi| << 1$. But χ is positive ($\chi > 0$) for these substances. In Fig. 19·1b the variation of M with H for a typical paramagnetic substance is shown graphically. The graph is again a straight line, but with a positive slope. Thus χ is a constant, independent of H in this case also. However unlike in the case of diamagnetics, χ depends on temperature in the case of paramagnetic substances and is given by

$$\chi = C/T \qquad (19\cdot1\text{-}3)$$

where T is the absolute temperature and C is a constant. Eq. (19·1-3) is known as *Curie's law*.

χ for paramagnetics is independent of H only for relatively low magnetic fields. At very high fields and at very low temperatures the magnetization M tends to saturation as H is increased which shows that $\chi \longrightarrow 0$ as H increases (Fig. 19·1b). Values of χ for a few typical paramagnetics at room temperature (300 K) are given below :

Material	CaO	FeCl$_2$	NiSO$_4$	Pt
χ	$+580 \times 10^{-5}$	$+360 \times 10^{-5}$	$+120 \times 10^{-5}$	$+26 \times 10^{-5}$

(c) *Ferromagmetics* : For these substances χ is positive and has usually very high value : $\chi >> 1$. Apart from the first transition group elements iron ($Z = 26$), cobalt ($Z = 27$) and nickel ($Z = 28$), the rare earth elements gadolinium ($Z = 64$), holmium ($Z = 67$) and erbium ($Z = 68$) as also some alloys exhibit ferromagnetism.

For ferromagnetics, the magnetization M at first increases rapidly with the increase of H and ultimately becomes saturated at higher fields (Fig. 19·2a). The values of χ can be obtained from the slopes of the M-H curve at different values of the magnetic field H. When these are plotted against H in a graph, a curve of the type shown in Fig. 19·2b is obtained which shows that χ at first increases rapidly with increasing H, attains a

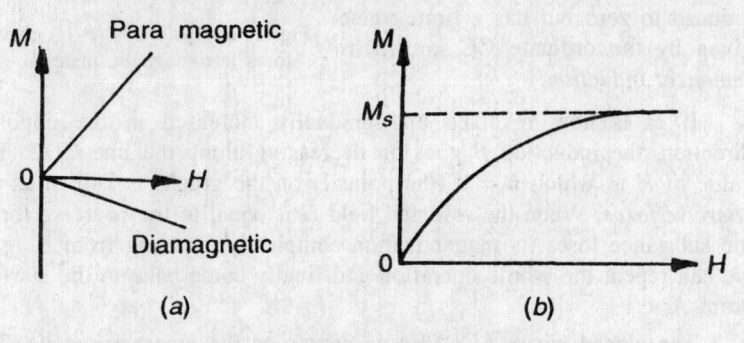

Fig. 19·1 (a) Variation of M with H. (b) Saturation magnetization in a paramagnetic material.

maximum and then decreases rapidly. Ultimately $\chi \to 0$ in the region of saturation of M.

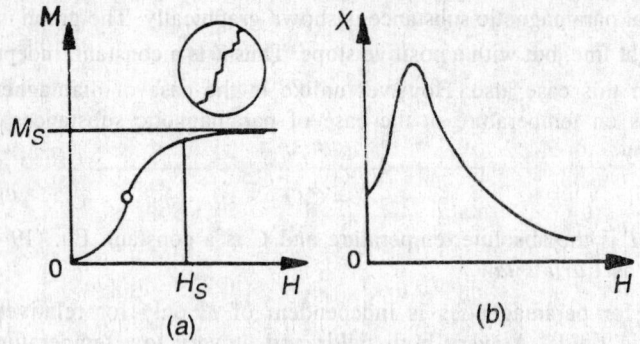

Fig. 19·2. Variation of (a) magnetization M, (b) magnetic susceptibility χ with applied external magnetic field H for a ferromagnetic material.

Careful measurements have shown that M does not increase continuously as H is increased. It increases in short steps, as shown in the inset of Fig. 19·2a. This is known as *Barkhausen effect*.

Another important characteristic of the ferromagnetic substances is the phenomenon of *hysteresis*. In Fig. 19·3 is shown the B vs. H curve for a ferromagnetic. As the magnetic field H is gradually increased from zero, the magnetic induction B increases along the line OA at first. At A the induction B attains almost its saturation value. If now H is gradually reduced, the decrease of B does not follow the line AO, but is along another line AC above OA. When the field H becomes zero, the substance still retains some of its magnetization so that B is not reduced to zero but has a finite value given by the ordinate OC known as *remanent induction*.

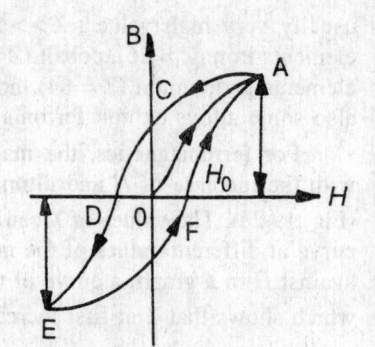

Fig. 19·3. Hysteresis (B-H) curve for a ferromagnetic material.

If H is now reversed and gradually increased in the opposite direction, the induction B goes on decreasing along the line CDE. The value of H at which $B = 0$ (the point D on the graph) is known as the *coercive force*. When the reversed field H is equal to the coercive force, the substance loses its magnetization completely. Starting from E again we can repeat the whole operation and finally come back to the starting point A.

The closed curve ACDEFA is known as the *hysteresis cycle*. The area enclosed by this closed curve is proportional to the work done in taking the ferromagnetic substance through the entire cycle of magnetization which appears as heat and raises the temperature of the substance. If the substance is repeatedly taken through the complete cycle of magnetization, its temperature may rise substantially.

The ferromagnetics for which the coercive force is low (~ 1 to 100 A/m) are known as *magnetically soft*. These include soft iron, permalloy (an alloy of 80% iron, 20% nickel) etc. These are used as cores for transformers, electric motors *etc.* and for the construction of electromagnets. On the other hand, ferromagnetics for which the coercive force is large (10^4 to 10^8 A/m) are known as *magnetically hard*. These usually have large *remanent magnetization* and hence are suitable for making parmament magnets.

If a ferromagnetic is heated, its susceptibility χ and hence its permeability μ decreases. Above a certain critical temperature, it loses its ferromagnetic property and is transformed into a paramagnetic. This critical temperature T_C is known as the *ferromagnetic Curie temperature*. For $T > T_C$, the magnetic susceptibility of the substance (which behaves like a paramagnetic) can be expressed by the formula

$$\chi = \frac{C'}{T - \theta} \qquad (19 \cdot 1\text{-}4)$$

C' is a constant and θ is known as the paramagnetic Curie temperature which is slightly greater than T_C. Eq. (19·1-4) is known as *Curie-Weiss law*.

In Fig. 19·4 is shown the variation of $1/\chi$ against T for a ferromagnetic at $T > \theta$.

The graph is a straight line the slope of which gives C' and the intercept with the T-axis gives θ.

Single crystals of ferromagnetic substances show anisotropic behaviour. There are certain *directions of easy magnetization in such crystals*. Saturation of magnetization is observed in them for a relatively much lower magnetic field applied along these directions. On the other hand, attainment of saturation requires much higher fields along certain other directions, known as *directions of difficult magnetization*. For example in the case of a single crystal of iron which has a cubic structure, the directions of easy magnetization are along

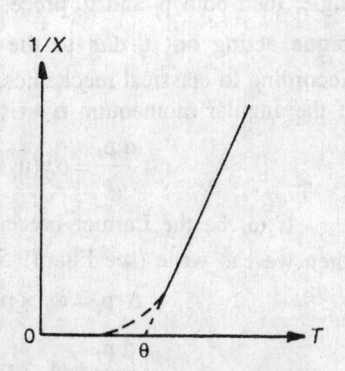

Fig. 19·4. Variation of $(1/\chi)$ with T for a ferromagnetic material.

the three crystal axes. On the other hand for a nickel crystal which also has a cubic structure, the direction of easy magnetization is along its *body diagonal*. In the case of iron, it is quite difficult to magnetize along the body diagonal while for nickel, magnetization is difficult to produce along its crystal axes.

When a ferromagnetic substance is magnetized, there is usually a change in its volume. This is known as *magnetostriction*. This is most prominent in nickel for which there is a decrease in volume with increase of H. For example if the magnetic induction field on a nickel wire is increased from 0.04 to 0.16 T its volume decreases by about 4 parts in 10^5. For iron, the volume at first increases with increasing field and then decreases. For alloys, the change in volume is determined by the relative proportions of their constituents. For most substances the changes in volume are of the order of magnitudes quoted above for nickel.

19·2 Origin of diamagnetism

Diamagnetism of a substance has its origin in the change of the orbital motion of the atomic electrons due to the applied magnetic field. We know that the magnetic moment μ_l of an atomic electron due to its orbital motion is related to its orbital angular momentum p_l by Larmor's formula (see Eq. 6·2-6) :

$$\vec{\mu_l} = -\frac{e}{2m_e} \mathbf{p}_l \qquad (19\cdot2\text{-}1)$$

where e and m_e are the charge (negative) and mass of the electron. $\vec{\mu_l}$ and $\mathbf{p}_l$ are vectors perpendicular to the orbital plane. If these vectors are inclined to the direction of the magnetic induction vector $\mathbf{B}$ at a certain angle, then both $\mathbf{p}_l$ and $\vec{\mu_l}$ precess around $\mathbf{B}$ as shown in Fig. 19·5. The torque acting on μ_l due to the applied magnetic field is $\mu_0(\vec{\mu_l} \times \mathbf{H})$. According to classical mechanics, this is equal to the time rate of change of the angular momentum $\mathbf{p}_l$ of the electron.

$$\frac{d \mathbf{p}_l}{dt} = \mu_0 (\vec{\mu_l} \times \mathbf{H}) = -\frac{\mu_0 e}{2 m_e} \mathbf{p}_l \times \mathbf{H} \qquad (19\cdot2\text{-}2)$$

If ω_L be the Larmor precessional angular velocity of the vector $\mathbf{p}_l$ then we can write (see Fig. 19·5)

$$\Delta \mathbf{p}_l = \omega_L \times \mathbf{p}_l \Delta t \qquad (19\cdot2\text{-}3)$$

$$\therefore \quad \frac{d \mathbf{p}_l}{dt} = \omega_L \times \mathbf{p}_l$$

From Eqs. (19·2-2) and (19·2-3) we get

$$\vec{\omega}_L = \frac{\mu_0 e}{2 m_e} \mathbf{H} \qquad (19\cdot2\text{-}4)$$

So the Larmor precessional frequency $v_L = \omega_L/2\pi$ becomes

$$v_L = \frac{\mu_0 e H}{4\pi m_e}$$

$$= 1.75 \times 10^4 H \qquad (19\cdot2\text{-}5)$$

So for a magnetic field $H = 10^7/4\pi$ amp-turns/m (which corresponds to a magnetic field of 10^4 gauss in C.G.S. system) $v_L \sim 1.4 \times 10^{10}$ Hz. This is small compared to the frequency of rotation of an atomic electron ($\sim 10^{13}$ to 10^{14} Hz) in its orbit.

It was assumed above that p_l is independent of the applied magnetic field. This means that the electron orbit is not affected by the magnetic field which is justified in the first approximation.

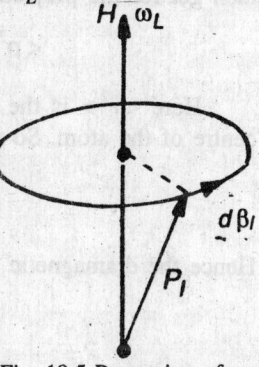

Fig. 19·5 Precession of $\mathbf{p}_l$ about $\mathbf{H}$. Δp_l is perpendicular to the plane of $\mathbf{p}_L$ and H

Due to the precession of the electronic orbit in the applied field, there is an induced magnetic moment $\Delta \mu$ opposite to the direction of B given by

$$\Delta \mu = -\frac{e}{2m_e} \Delta p_l = -\frac{e}{2m_e} m_e \omega_L <\rho^2> = -\frac{e \omega_L}{2} <\rho^2> \qquad (19\cdot2\text{-}6)$$

Here $<\rho^2>$ is the mean squared radius of the projection of the electron orbit in a plane perpendicular to $\mathbf{B}$. The quantity $m_e \omega_L <\rho^2>$ is the change in the angular momentum of the electron due to the precessional motion induced by the magnetic field.

If there are n atoms in a unit volume of a substance and there are Z electrons per atom, then the magnetization induced by the applied field which is equal to the induced magnetic moment per unit volume is

$$M = nZ \Delta \mu = -nZ \frac{e}{2} \frac{\mu_0 e H}{2m_e} <\rho^2>$$

$$= -nZ \frac{\mu_0 e^2 H}{4 m_e} <\rho^2> \qquad (19\cdot2\text{-}7)$$

To calculate $<\rho^2>$, we note that if we assume a spherically symmetric charge distribution in the atom, then

$$<x^2> = <y^2> = <z^2> = \frac{1}{3} <r^2>$$

where x, y, z are the cartesian components of the position vector $\mathbf{r}$ of the charge distribution at any point on the sphere. Taking B to be along z, we then get for the projection of the electron orbit in the x-y plane

$$<\rho^2> = <x^2> + <y^2> = \frac{2}{3}<r^2>$$

Here $<r^2>$ is the mean squared distance of the electron from the centre of the atom. So we get finally.

$$M = -nZ\left(\frac{\mu_0 e^2}{6m_e}\right)<r^2>H \qquad (19\cdot2\text{-}8)$$

Hence the diamagnetic susceptibility will

$$\chi = \frac{M}{H} = -\mu_0 nZ\left(\frac{e^2}{6m_e}\right)<r^2> \qquad (19\cdot2\text{-}9)$$

So χ is negative. Since the mean distance of the orbital electron from the nucleus is of the order of 10^{-10} m, $<r^2> \sim 10^{-20}$ m^2. Also since the classical electron radius $r_0 = \mu_0 e^2/4\pi mc^2 \sim 2\cdot 8 \times 10^{-15}$ m (see § 8·14) and the number density of atoms $n \sim 10^{28}$ per m^3, we get

$$\chi \sim -10^{-7}Z$$

so that $|\chi|$ is of the order of 10^{-5} to 10^{-6} for a solid. It should be noted that diamagnetism is a universal property of all materials, but is usually masked by para or ferro-magnetism.

19·3 Origin of paramagnetism

There are many substances whose atoms have permanent magnetic dipole moments. Paramagnetism owes its origin to the existence of such permament dipole moments. The French physcist Paul Langevin developed the classical theory of paramagnetism based on the above assumption.

(a) *Langevin's theory* : Let μ be the magnetic dipole moment of each atom. Langevin assumed that the magnetic interaction between the dipoles is negligible. Hence in an external magnetic field H the magnetic energy is

$$\varepsilon_m = -\vec{\mu}\cdot B_0 = -\mu B_0 \cos\theta \qquad (19\cdot3\text{-}1)$$

Here $B_0 = \mu_0 H$ is the external induction field. θ is the angle between $\vec{\mu}$ and H. The atomic dipoles which are aligned at random in the absence of the magnetic field tend to align in the direction of the applied field, so that their magnetic energies are lowered. On the other hand, the random thermal motion of the molecules opposes this tendency. Under the influence of these two opposing tendencies, each atomic magnetic dipole is turned to some extent in the direction of the applied field and assumes an equilibrium position. The projection of the dipole moment in the direction of H will be $\mu\cos\theta$ The magnetic moment induced in the substance by the external field will be equal to the algebraic sum of the

Magnetic Properties of Solids

projections of the moments of all the dipoles present in a unit volume.

According to Maxwell-Botzmann classical statistics, the probability of a dipole to align at an angle between θ and $\theta + d\theta$ will be

$$W(\theta)\, d\theta \propto \exp(-\varepsilon_m/kT)\, d\Omega$$

where $d\Omega = 2\pi \sin\theta\, d\theta$ is the solid angle within θ and $\theta + d\theta$ and k is Boltzmann constant.

Using Eq. (19·3-1), we can write

$$W(\theta)\, d\theta = C \exp(\mu B_0 \cos\theta/kT) \sin\theta\, d\theta$$

where C is a constant. The magnetization induced in the substance by the applied field H is

$$M_H = n <\mu \cos\theta> \qquad (19\cdot3\text{-}2)$$

where n is the number of atoms per unit volume and $<\mu \cos\theta>$ is the mean value of the projection of the magnetic moment in the field direction.

$$<\mu \cos\theta> = \int_0^\pi \mu \cos\theta \cdot W(\theta)\, d\theta \Big/ \int_0^\pi W(\theta)\, d\theta$$

$$= \frac{\mu \int_0^\pi \exp(\mu B_0 \cos\theta/kT) \cdot \cos\theta \sin\theta\, d\theta}{\int_0^\pi \exp(\mu B_0 \cos\theta/kT) \cdot \sin\theta\, d\theta} \qquad (19\cdot3\text{-}3)$$

If we substitute

$$x = \cos\theta \text{ and } \alpha = \mu B_0/kT = \mu_0 \mu H/kT,$$

we get
$$<\mu \cos\theta> = \mu \int_{-1}^{1} x \exp(\alpha x)\, dx \Big/ \int_{-1}^{1} \exp(\alpha x)\, dx$$

$$= \mu \left\{ \frac{\exp(\alpha) + \exp(-\alpha)}{\exp(\alpha) - \exp(-\alpha)} - \frac{1}{\alpha} \right\} = \mu \left(\coth\alpha - \frac{1}{\alpha} \right)$$

Hence from Eq (19·3-2) we get

$$M_H = n\mu L(\alpha) = n\mu \left(\coth\alpha - \frac{1}{\alpha} \right) \qquad (19\cdot3\text{-}4)$$

where
$$L(\alpha) = \coth\alpha - \frac{1}{\alpha} \qquad (19\cdot3\text{-}5)$$

is known as Langevin function.

The atomic magnetic moments are usually of the order of the Bohr magneton : $\mu \approx \mu_B \approx 10^{-23}$ J/T. At room temperature ($T \sim 300$ K), $kT \approx 4 \times 10^{-21}$ joule. Hence we get

$$\alpha = \frac{\mu B_0}{kT} \sim 2.5 \times 10^3 B_0$$

So for $B_0 \sim 1$ tesla (i.e., 10^4 gauss), $\alpha << 1$. If we expand $\coth \alpha$ for small α we get

$$\coth \alpha = \frac{1}{\alpha} + \frac{\alpha}{3} - \frac{\alpha^3}{45} \approx \frac{1}{\alpha} + \frac{\alpha}{3}$$

So we have

$$M_H = n\mu \left(\frac{1}{\alpha} + \frac{\alpha}{3} - \frac{1}{\alpha}\right) = \frac{n\mu \alpha}{3}$$

$$= \frac{n\mu}{3} \frac{\mu_0 \mu H}{kT} = \frac{\mu_0 n \mu^2}{3kT} H \qquad (19 \cdot 3\text{-}6)$$

Since $M_H = \chi H$ we get the paramagnetic susceptibility

$$\chi = \frac{\mu_0 n \mu^2}{3kT} \qquad (19 \cdot 3\text{-}7)$$

Eq. (19·3-7) shows that $\chi \propto 1/T$ which agrees with observations (see § 19·1b).

If B_0 is increased to very high values and T is decreased to very low values, then α becomes large. For $\alpha >> 1$, $1/\alpha \to 0$ so that we get in this case

$$\coth \alpha = \frac{\exp(\alpha) + \exp(-\alpha)}{\exp(\alpha) - \exp(-\alpha)} \longrightarrow 1$$

Hence we get in this case $M_H = n\mu$ which shows that at very low temperature, all the atomic dipoles are aligned in the direction of the magnetic field H when a very high field is applied. The magnetization becomes saturated in this case.

(b) *Quantum theory of paramagnetism* :

In Langevin's theory of paramagnetism, it is assumed that the inclination of the atomic dipoles can vary continuously with respect to the direction of the applied magnetic field. However according to Sommerfeld's rules for space quantization, this assumption is not valid (see Ch. VI). According to these rules, the resultant angular momentum vector $J\hbar$ of an atom can align in $(2J+1)$ special directions such that the components of the angular momentum along the field direction can assume the discrete set of values $m_J = J, J-1, J-2 \ldots\ldots -J$. Here m_J is known as the magnetic quantum number.

While discussing about the anomalous Zeeman effect (Ch. VI), we found that the resultant magnetic moment of an atom is given by

$$\mu_J = \sqrt{J(J+1)}\, g\, \mu_B \qquad (19 \cdot 3\text{-}8)$$

Magnetic Properties of Solids

where μ_B is Bohr magneton and g is Landé g-factor given by Eq. (6·8-3)

$$g = 1 + \frac{J(J+1) + S(S+1) - L(L+1)}{2J(J+1)}$$

L and S are the orbital and spin quantum numbers.

The energy of the atomic magnetic dipole in the applied magnetic field is given by

$$\varepsilon_m = -\mu_H B_0 = m_J g \mu B_0 \qquad (19\cdot 3\text{-}9)$$

μ_H is the component of the magnetic moment μ_J along the field direction:

$$\mu_H = -m_J g \mu_B \qquad (19\cdot 3\text{-}10)$$

Since m_J can assume only a finite number of discrete values, the magnetic energy ε_m can also have a discrete set of values. Hence in order to calculate the mean value $<\mu_H>$, we have to replace the integrals in Eq. (19·3-3) by the summation over the discrete set of values of m_J, so that we get

$$<\mu_H> = \frac{\sum_{-J}^{J} m_J g \mu_B \exp(m_J g \mu_B B_0 / kT)}{\sum_{-J}^{J} \exp(m_J g \mu_B B_0 / kT)} \qquad (19\cdot 3\text{-}11)$$

Hence the magnetization of the substance will be

$$M_H = n <\mu_H> = \frac{n \sum_{-J}^{J} m_J g \mu_B \exp(m_J g \mu_B B_0 / kT)}{\sum_{-J}^{J} \exp(m_J g \mu_B B_0 / kT)} \qquad (19\cdot 3\text{-}12)$$

We first consider the special case when the magnetic field is not too high ($B_0 < 1$ T). Then at relatively higher temperatures *e.g.*, at room temperature)

$$\frac{m_J g \mu_B B_0}{kT} << 1$$

In this case the exponential can be expanded as follows:

$$\exp(m_J g \mu_B B_0 / kT) \approx 1 + \frac{m_J g \mu_B B_0}{kT}$$

$$\therefore \quad M_H = \frac{ng\mu_B \sum_{-J}^{J} m_J (1 + m_J g\mu_B B_0/kT)}{\sum_{-J}^{J} (1 + m_J g\mu_B B_0/kT)}$$

$$= \frac{ng\mu_B \left\{ \sum_{-J}^{J} m_J + g\mu_B B_0 \sum_{-J}^{J} (m_J^2)/kT \right\}}{(2J+1) + g\mu_B B_0 \sum_{-J}^{J} m_J \bigg/ kT}$$

Now $\sum_{-J}^{J} m_J = 0$ and $\sum_{-J}^{J} m_J^2 = \frac{J(J+1)(2J+1)}{3}$

$$\therefore \quad M_H = \frac{ng^2 \mu_B^2 B_0}{kT} \frac{J(J+1)(2J+1)}{3(2J+1)}$$

$$= \frac{ng^2 \mu_B^2}{kT} B_0 \frac{J(J+1)}{3} \tag{19.3-13}$$

$$\therefore \quad \chi = \frac{M_H}{H} = \frac{\mu_0 ng^2 \mu_B^2 J(J+1)}{3kT} \tag{19.3-14}$$

If we put $\mu_J^2 = J(J+1) g^2 \mu_B^2$ (see Eq. 19·3-8), then Eq. (19·3-14) assumes the same form as the classical expression (19·3-7). The above equation shows that the quantum mechanical theory is also in agreement with Curie's law.

If we write $\quad \mu_J = p_{eff} \mu_B$

then the *effective number of Bohr magnetons* is given by

$$p_{eff} = g \sqrt{J(J+1)} \tag{19.3-15}$$

p_{eff} can be determined by measuring χ.

Writing $\alpha = \dfrac{Jg\mu_B B_0}{kT}$ the summations in the numerator and denominator on the r.h.s. of Eq. (19·3-11) can be evaluated in the general case to yield the result

$$M_H = ng J \mu_B B_J(\alpha) \tag{19.3-16}$$

where $B_J(\alpha)$, known as a *Brillouin function*, is given by

Magnetic Properties of Solids

$$B_J(\alpha) = \frac{2J+1}{2J} \coth \frac{(2J+1)\alpha}{2J} - \frac{1}{2J} \coth\left(\frac{\alpha}{2J}\right) \qquad (19\cdot3\text{-}17)$$

For high fields and low temperatures, $\alpha \gg 1$. In the limit of $\alpha \longrightarrow \infty$, we have

$$\coth \frac{(2J+1)\alpha}{2J} \longrightarrow 1 \quad \text{and} \quad \coth \frac{\alpha}{2J} \longrightarrow 1$$

Hence $$B_J(\alpha) = \frac{2J+1}{2J} - \frac{1}{2J} = 1$$

The magnetization attains the saturation value $ng J \mu_B$ in this case.

Eq. (19·3-14) can be used to estimate the order of magnitude of paramagnetic susceptibility. Since the atomic magnetic moments have values of the order of the Bohr magneton $\mu_B \sim 10^{-23}$ J/T and $n \sim 10^{28}/\text{m}^3$, we get

$$\chi \sim \frac{10^{28} \times (10^{-23})^2 \times 4\pi \times 10^{-7}}{3 \times 1\cdot38 \times 10^{-23} T} \approx \frac{3}{100\, T}$$

For $T \sim 300$ K (room temperature) we get $\chi \sim 10^{-4}$. On the other hand for $T \sim 1$ K, $\chi \sim 10^{-2}$.

The permanent magnetic dipole moments of the atoms arise from the orbital and spin motions of the electrons. Only those atoms which have partially filled electronic subshells have such permanent magnetic moments. There is no contribution to the magnetic moment of the atoms from the completely filled shells.

Apart from the electronic paramagnetism, the atoms also have *nuclear paramagnetism*. This arises due to the non-vanishing magnetic moments of the nuclei of some atoms. As we shall see in Vol. II, the nuclear magnetic moments have magnitudes less than one thousandth of the atomic magnetic moments. Hence the electronic paramagnetism completely masks the nuclear paramagnetism. However the nuclear paramagnetism is manifested in some diagmanetic substances, *e.g.*, solid hydrogen.

(c) *Paramagnetism of the free electron gas in metals* :

According to the theory of paramagnetism discussed above, the paramagnetic susceptibility is inversely proportiomal to the absolute temperature. However there are some metals which exhibit feeble paramagnetism independent of temperature. Pauli provided an explanation for this on the basis of the free electron theory of metals (Ch. XVIII).

We know that in metals, the free electrons occupy successive energy levels, starting from the bottom of the potential well upto the Fermi level. Each level is occupied by two electrons with their spins aligned in opposite directions. According to Langevin's theory, when a magnetic field is applied, the permanent dipole moments of the atoms in a paramagnetic material tend to align parallel to the field direction, the probability for which exceeds that of antiparallel alignment by a factor of about $\mu B_0/kT$ (μ = magnetic dipole moment). For n atoms, this gives a net magnetic moment $nB_0\mu^2/kT$. However in the case of the free electrons in the metals the probability of turning over to the direction of the magnetic field from antiparallel alignment is zero since the states of parallel spin orientation are already occupied. Only the electrons near the top of the Fermi distribution have a chance to turn over to the field direction from antiparallel orientation. As we have seen that only a fraction (T/T_f) of the total number of electrons which occupy the energy levels in the energy interval kT below the Fermi level $\varepsilon_f = kT_f$ can make such transitions (see §18·1). Hence $\chi \sim (\mu_0 n \mu^2/kT) \times (T/T_f) = n\mu_0 \mu^2/kT_f$ which is independent of temperature and has a value $\sim 10^{-6}$ which is much lower than χ predicted from Langevin's theory. The above result holds for the case $\mu B << kT$ which is true at the room temperature for fields as high as 10^2 T.

The presence of a small temperature-independent paramagnetism in some metals is an additional confirmation of the hypothesis of the existence of free electrons in metals.

19·4 Origin of ferromagnetism

The theory of ferromagnetism was first developed by the French physicist P. Weiss in 1907. It is known as the *theory of molecular field*. Two main assumptions of the theory are :

(*a*) There are some *macroscopic domains* within a ferromagnetic material which are spontaneously magnetized. The magnetization of the whole body is the resultant of the magnetic moments of these domains.

(*b*) The spontaneous magnetization of the macroscopic domains arises due to the existence of molecular fields within the material. Due to this field all the magnetic dipoles within a domain are aligned parallel to one another.

We have seen that paramagnetic substances attain saturation magnetization by the application of very high magnetic fields ($> 10^3$ T). On the other hand, one important characteristic of the ferromagnetic materials is that they attain saturation magnetization by very weak magnetic fields. For example in the case of silicon-steel, saturation is attained by a field as low as 10 A/m. Under this condition magnetization $M \sim 10^5$. For a paramagnetic material the magnetization is about 10^{-4} for the same field (see Eq. 19·3-16) which is only 10^{-9} that of a ferromagnetic substance.

The high value of magnetization produced in a ferromagnetic

Magnetic Properties of Solids

material by a very weak external magnetic field shows that within the macroscopic domains in these materials, all the permanent magnetic dipoles are aligned parallel to one another. Thus there is some kind of mutual interaction between the dipoles within the domain which is responsible for their spontaneous magnetization. We shall discuss about the nature of this interaction later.

Weiss assumed that the total magnetic field acting on the magnetic dipoles within a ferromagnetic material is

$$H_m = H + \gamma M$$

where H is the external field, M the magnetization of the material and γ is a constant, known as the *molecular field constant* or *Weiss constant*. γM predominates over that of the external field H.

Weiss' original theory was based on Langevin's classical theory of paramagnetism. In our discussion below we shall however use the results of the quantum theory of paramagnetism. According to this, the magnetization of a paramagnetic material is given by (see Eq. 19.3-16)

$$M = ngJ\mu_B B_J(\alpha) \qquad (19 \cdot 4\text{-}1)$$

where
$$\alpha = Jg\mu_B B_0/kT$$

Here $B_0 = \mu_0 H$, H being the external field.

For the ferromagnetic material, we have to use the field H_m defined above in place of H so that we have

$$\alpha = Jg\mu_B\mu_0(H + \gamma M)/kT \qquad (19 \cdot 4\text{-}2)$$

For spontaneous magnetisation, $H = 0$ so that we can write

$$\alpha = Jg\mu_B\mu_0\gamma M/kT \qquad (19 \cdot 4\text{-}2a)$$

$$M = \frac{\alpha kT}{\gamma g \mu_0 \mu_B J} \qquad (19 \cdot 4\text{-}3)$$

The two equations (19·4-1) and (16·4-3) can be used to draw two M vs. α curves as shown in Fig. 19.6 (curves 1 and 2).

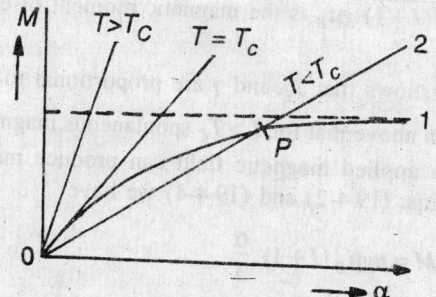

Fig. 19·6. Variation of M with α for a ferromagnetic material.

The graph of M against α given by Eq. (19·4-3) is a straight line (2) passing through the origin with a slope proportional to T.

The point of intersection P of the two curves in Fig. 19·6 gives the spontaneous magnetisation $M(T)$ at the temperature T.

If we draw the tangent to the graph (1) of M vs. α curve determined by Eq. (19·4-1) at the origin ($\alpha = 0$), then the slope of this tangent gives the limiting temperature T_C above which spontaneous magnetization is not possible. All straight lines with higher slopes passing through the origin lie completely above the M vs. α graph (1) and hence there cannot be any point of intersection between them other than the origin. For $T < T_C$, spontaneous magnetization is possible. T_C is the ferromagnetic Curie temperature mentioned in § 19·1c. From the value of T_C, it is possible to determine γ.

From the theory of a paramagnetism, we have from Eq. (19·4-4) for $\alpha << 1$ ($\because \gamma M \sim 0$)

$$M \approx \frac{ng\mu_B(J+1)}{3} \cdot \frac{Jg\mu_B B_0}{kT} = ng\mu_B(J+1)\frac{\alpha}{3} \qquad (19\text{·}4\text{-}4)$$

Thus the slope of the straight line graph (2) for M vs α in Fig. 19·6 is $ng\mu_B(J+1)/3$. On the other hand from Eq. (19·4-3) the slope of the tangent to the M vs α curve (1) at $\alpha = 0$ is found to be $kT_C/\gamma g \mu_0 \mu_B J$ with $T = T_C$. So we get

$$\frac{kT_C}{\gamma \mu_0 \mu_B J} = \frac{ng\mu_B}{3}(J+1)$$

which gives
$$\gamma = \frac{3kT_C}{nJ(J+1)\mu_0 \mu_B^2 g^2} = \frac{3kT_C}{n\mu_0 \mu_J^2} \qquad (19\text{·}4\text{-}5)$$

Thus
$$T_C = \frac{n\mu_0 \mu_J^2 \gamma}{3k} \qquad (19\text{·}4\text{-}6)$$

Here $\mu_J = \sqrt{J(J+1)}\, g\mu_B$ is the magnetic moment of each atom (Eq. (19·3-8).

Eq. (19·4-6) shows that T_C and γ are proportional to each other.

We have seen above that for $T > T_C$ spontaneous magnetization is not possible. Only an applied magnetic field can produce magnetization in this case. From Eqs. (19·4-2) and (19·4-4) we have

$$M = ng\mu_B(J+1)\frac{\alpha}{3}$$

$$= \frac{ng\mu_B(J+1)}{3} \cdot \frac{Jg\mu_B \mu_0(H+\gamma M)}{kT}$$

Magnetic Properties of Solids

$$= nJ(J+1)g^2\mu_B^2\mu_0 \frac{H+\gamma M}{3kT}$$

$$\therefore \quad M\left(1 - \frac{n\mu_J^2\mu_0\gamma}{3kT}\right) = \frac{n\mu_J^2\mu_0 H}{3kT}$$

This gives

$$\chi = \frac{M}{H} = \frac{n\mu_J^2\mu_0/3kT}{1 - n\mu_J^2\mu_0\gamma/3kT}$$

$$= \frac{\mu_0 n\mu_J^2/3k}{T - n\mu_J^2\mu_0\gamma/3k} = \frac{C'}{T-\theta} \qquad (19\cdot4\text{-}7)$$

This is Curie-Weiss law (Eq. 19·1-4). Here

$$C' = \frac{n\mu_J^2\mu_0}{3k}, \quad \theta = \frac{n\mu_J^2\mu_0\gamma}{3k} = \gamma C' \qquad (19\cdot4\text{-}8)$$

We have seen above that it is possible to determine the spontaneous magnetization $M(T)$ at any temperature from the point of intersection of the two graphs (1) and (2) determined by Eqs. (19·4-1) and (19·4-3). If $M(0)$ be the spontaneous magnetization at 0 K, then the graph of $M(T)/M(0)$ against T/T_C is a *universal graph* applicable for all materials (see Fig (19·-7). The above theoretical results agree with the experimental results for different materials if we take $J = \frac{1}{2}$. This indicates that the

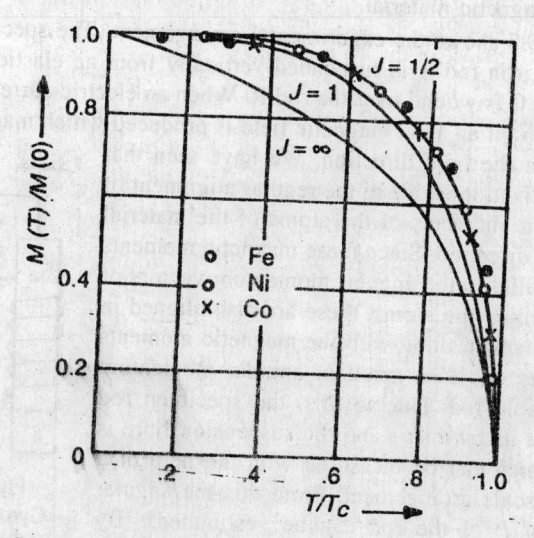

Fig. 19·7. Universal graph of $M(T)/M(0)$ against T/T_c

magnetization of ferromagnetic materials arises from the spin of the electrons. The orbital motion of the electrons has no influence on it. The spin quantum number of the electron is $s = \frac{1}{2}$ so that if the orbital contribution is nil, we have $J = s = \frac{1}{2}$. This conclusion is supported by a series of *gyromagnetic experiments* first by Einstein and de Haas and later by others (see next section).

19·5 Einstein-de Haas experiment.

Magnetization M in a material results from the regular alignment of the magnetic moments μ of the atomic magnets within it. Similarly the regular alignment of the angular momentum vectors $\mathbf{p}$ of the atoms within the material gives rise to a resultant angular momentum $\mathbf{P}$ per unit volume of the whole specimen. Due to the orbital motion of the electrons within the atom the ratio of the magnetic moment to the angular momentum (gyromagnetic ratio) of an electron is given by $\mu_l/p_l = e/2m_e$. On the other hand, the value of this ratio for the spin motion is $\mu_s/p_s = 2 \times e/2m_e = e/m_e$. So by measuring the ratio (M/P) of the magnetization M which is the ratio of the resultant magnetic moment per unit volume of the substance to the resultant angular momentum P per unit volume, it is possible to understand the origin of the magnetization of the substance *viz.*, whether it is due to the orbital or to the spin motion of the electrons or due to both.

To settle the question Einstein and de Haas performed the following 'gyromagnetic experiment' in 1915 to produce *magnetization by rotation* in a ferromagnetic material.

Fig. 19·8 shows the experimental arrangement. The specimen in the form of an iron rod R is suspended vertically from an elastic fibre S. A coil of wire C is wound over the rod R. When an electric current is passed through the coil an axial magnetic field is produced which magnetizes the specimen in the field direction. We have seen that this magnetization is due to the regular alignment of the magnetic moments of the atoms of the material in the field direction. Since these magnetic moments are antiparallel to the angular momentum vectors of the electrons in the atoms, these are also aligned in a regular manner along with the magnetic moments which gives rise to a resultant angular momentum for the whole rod. Due to this, the specimen rod turns about its own axis and the suspension fibre is twisted which can be measured with the help of a lamp and scale arrangement. From this the angular momentum P of the rod can be estimated. By measuring the magnetization M of the rod at the

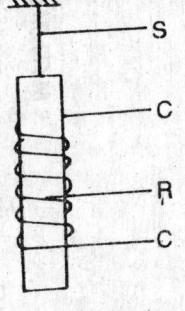

Fig. 19·8. Gyromagnetic experimental set up

same time, the ratio M/P can be found.

Similar experiments were performed by others later. From these and other similar experiments, the gyromagnetic ratio is found to be 2:

i.e. $$\frac{M}{P} = 2 \times \frac{e}{2m} = -\frac{e}{m_e}$$

An experiment which is just the opposite of Einstein-de Haas experiment in principle was performed by S.J. Barnett in which the magnetization produced in a rapidly rotating iron rod was measured. This and similar other experiments on *magnetization by rotation* confirmed the results of Einstein-de Hass experiment.

From all these measurements it is definitely established that the magnetization of a ferromagnetic substance is due to the spin motion of the atomic electrons. Their orbital motion does not contribute to the magnetization.

19·6 Origin of the molecular field

In all atoms and ions having closed shells, the angular momentum vectors due to both orbital and spin motions and the associated magnetic moments of pairs of electrons are oppositely aligned. So the resultant orbital and spin angular momenta of such atoms are zero : $L = 0$, $S = 0$ and hence $J = 0$. The magnetic moments of such atoms also vanish ; $\mu_L = 0$, $\mu_S = 0$ and $\mu_J = 0$.

Materials for which there is complete compensation of the orbital and spin moments and have zero total moments are diamagnetic in nature. Atoms in which such complete compensation does not take place have permanent magnetic moments. This may happen if the atom contains an odd number of electrons so that at least one of them will be unpaired, as in H, Na, K *etc*. If the atom contains an even number of electrons in an incompletely filled inner shell, pairs of electrons may not always have their spin (or orbital) moments oppositely aligned. Some of the pairs may have their moments aligned parallel to each other as in oxygen. When this happens, the atom has a residual permanent dipole moment. If the mutual interaction between such permanent magnetic dipole moments of the atoms in a substance is negligibly small, then the substance exhibits paramagnetism. On the other hand if the mutual interaction between the atoms in a substance is very strong so that the magnetic moments of the neighbouring atoms tend to align parallel to one another then the substance behaves like a ferromagnetic.

We have seen that ferromagnetism occurs in the first transition group elements Fe, Co, Ni and in some rare-earth elements (*e.g.*, Gd and Dy) as also in some alloys and compounds. The above elements have incompletely filled inner subshells. In the first three, the inner $3d$ subshells are partially filled while in the rare-earths the inner $4f$ subshells are partially filled.

Einstein-de Haas experiment and similar other experiments proved that the spin magnetic moments of the electrons are solely responsible for the occurrence of ferromagnetism. Orbital motion of the electrons has no contribution to it.

In the metallic lattice of the transition elements like Fe, Co or Ni, the partially filled 3d subshells are the outermost subshells, the outermost 4s valence electrons forming the free electron gas which has negligible contribution to their magnetic properties. There is strong electrostatic interaction between the 3d electrons and the neighbouring ions in the lattice. Due to this the orbital moments of these electrons are quenched and they do not contribute to the magnetism of these elements. Also the magnetic moments of the closed shells are zero. So ferromagnetism of these elements arises due to the spin moments of the electrons in the partially filled 3d subshells of these atoms.

In 1928, W. Heisenberg proposed a quantum mechanical theory of the molecular fields in ferromagnetic materials.

Since the magnetic energies of the electrons in the molecular field H_m are of the order of $\mu_0 \mu_B H_m \sim kT \sim 10^{-21} J$ where μ_B is Bohr magneton and k is Boltzmann constant, $\mu_0 H_m$ must of the order of 10^2 to 10^3 T. Such a high value of $\mu_0 H_m$ shows that the molecular field cannot originate from the mutual magnetic interaction between the atomic magnetic dipoles. The interaction energy in the latter case is of the order of $U_m \sim (\mu_0 \mu_B^2 / a^3)$ where $a \sim 10^{-10}$ m is the lattice parameter. This gives $U_m \sim 10^{-23}$ J which is much lower than that given above. According to Heisenberg the intense molecular field has its origin in the *exchange interaction* between the electrons. Such exchange interaction is responsible for the binding of the two hydrogen atoms in a H_2 molecule (see § 17.2b).

The basic ideas involved can be understood by considering the case of the H_2 molecule. In Fig. 19.9 the nuclei of the two hydrogen atoms A and B in a H_2 molecule are designated as a and b while their two orbital

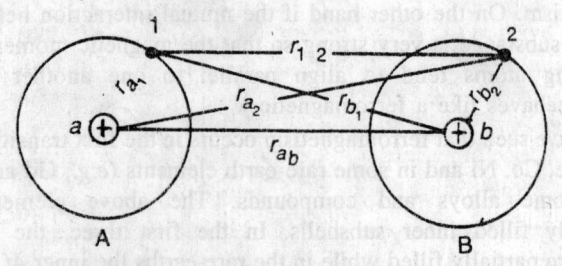

Fig. 19.9. Heisenberg's exchange interaction between two electrons.

electrons are designated as 1 and 2. If r_{ab} denotes the internuclear distance and r_{12} the distance between the two electrons at any instant, then the mutual electrostatic interaction energy of the two atoms is

$$V_{AB} = \frac{e^2}{4\pi\epsilon_0}\left(\frac{1}{r_{ab}} + \frac{1}{r_{12}} - \frac{1}{r_{b1}} - \frac{1}{r_{a2}}\right) \tag{19.6-1}$$

Here r_{a2} is the distance of electron 2 from the nucleus a while r_{b1} is the distance between electron 1 and the nucleus b.

When the two atoms are at a great distance apart, their total energy is $2E_0$ equal to the sum of the energies of the two isolated atoms. When the atoms come close together, then their interaction energy is $V = V_{AB}$. The Schrödinger equation can be written as

$$(\hat{H}_A + \hat{H}_B + V)\psi_{12} = E\psi_{12} \tag{19.6-2}$$

where $\hat{H}_A$ and $\hat{H}_B$ are the Hamiltonian operators for the two seperated hydrogen atoms. An approximate solution of the above equation can be obtained as a linear combination of the products of the wave functions of the two atoms. Let $\psi_a(1)$ denote the wave function of electron 1 in the field of nucleus a alone while $\psi_b(2)$ that of electron 2 in the field of nucleus b alone. Similarly let $\psi_a(2)$ and $\psi_b(1)$ denote the wave functions of the electrons 2 and 1 in the fields of nuclei a and b respectively. Then it can be shown that Eq. (19.6-2) has the following two approximate solutions :

$$\psi_{12}^S = \psi_a(1)\psi_b(2) + \psi_a(2)\psi_b(1) \tag{19.6-3a}$$

$$\psi_{12}^A = \psi_a(1)\psi_b(2) - \psi_a(2)\psi_b(1) \tag{19.6-3b}$$

ψ_{12}^S and ψ_{12}^A are the *symmetric* and *antisymmetric* wave functions respectively. If the electrons 1 and 2 are interchanged ψ_{12}^S remains unchanged, while ψ_{12}^A changes sign. Now the total wave function which is a product of the space part ψ_{12} and a spin part must be antisymmetric in the exchange of all the coordinates (space and spin) of the two electrons because of Pauli principle (see Ch. XIII and XX). So if $\psi_{12} = \psi_{12}^S$ is symmetric, then the spin part must be antisymmetric which happens if the spins of the two electrons are *antiparallel*. On the other hand if $\psi_{12} = \psi_{12}^A$ is antisymmetric then the spin part must by symmetric, which happens if the spins are parallel.

It can be shown that the energies of the system in the above two cases are (see § 13.10)

$$E_S \doteq 2E_0 - W_1 \tag{19.6-4a}$$

$$E_A = 2E_0 + W_2 \qquad (19\cdot6\text{-}4b)$$

where $\quad W_1 = \dfrac{KS + J_e}{1+S} \quad$ and $\quad W_2 = \dfrac{KS - J_e}{1-S} \qquad (19\cdot6\text{-}5)$

$$K = \int\limits_{\tau_1} \int\limits_{\tau_2} |\psi_a(1)|^2 |\psi_b(2)|^2 V d\tau_1 d\tau_2 \qquad (19\cdot6\text{-}6)$$

$$J_e = \int\limits_{\tau_1} \int\limits_{\tau_2} \psi_a(1)\psi_a(2)\, V\psi_b(1)\psi_b(2)\, d\tau_1\, d\tau_2 \qquad (19\cdot6\text{-}7)$$

$$S = \int\limits_{\tau_1} \int\limits_{\tau_2} \psi_a(1)\psi_a(2)\psi_b(1)\psi_b(2)\, d\tau_1 d\tau_2 \qquad (19\cdot6\text{-}8)$$

The integrations in the above expressions are to be carried out over the coordinates of the two electrons 1 and 2. K gives the energy of Coulomb interaction between the electrons in the two atoms.

The second integral J_e is purely quantum mechanical in origin. It is known as the *exchange integral*. It gives the probability of interchange of the two electrons 1 and 2 between the atoms A and B. When evaluated as a function of the internuclear distance r_{ab}, it is found to give an attractive force between the two atoms in the symmetric spatial state of the two electrons while it gives repulsion in the antisymmetric state. Thus for producing a stable H_2 molecule, the spins of the two electrons should be antiparallel.

Heisenberg applied the above ideas to explain the origin of ferromagnetism. He showed that in this case the exchange integral is positive when the spins of the electrons in the neighbouring atoms are aligned parallel to one another under the action of the exchange force. In this case the total energy of the system is less than in the case of antiparallel alignment of the spins. This is the condition under which ferromagnetism arises.

It may be noted that the origin of the exchange energy lies in the electrostatic interaction (V) of the atoms and not in the spin-spin interactions.

In those cases in which the exchange integral J_e is negative, the material becomes non-ferromagnetic. Actually this is a more probable state. Whether J_e will be positive or negative depends on the relative magnitudes of the latice parameter a and diameter d of the partially filled electronic shell. In Fig. 19·10 is shown a plot of J_e against the ratio a/d. It will be seen from the graph that for $a/d < 1\cdot5$, $J_e < 0$. Thus materials for which $a/d < 1\cdot5$ cannot be ferromagnetic. Only if $a/d > 1\cdot5$, $J_e > 0$ and the material is ferromagnetic.

As an example, $a/d < 1\cdot5$ for manganese which is not ferromagnetic.

Magnetic Properties of Solids

If by some means, a/d is made greater than 1·5 for Mn, then it should become ferromagnetic. It is found that by introducing small amount of nitrogen in the lattice of Mn, the lattice parameter a is increased which results in the appearance of ferromagnetism. Similarly ferromagnetism is exhibited by the alloy Mn – Cu – Al (Heusler alloy) and compounds like Mn-Sb and Mn-Bi.

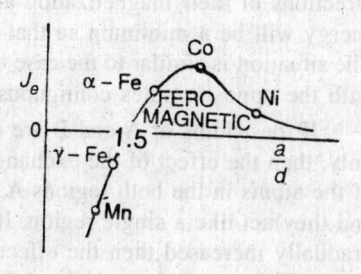

Fig. 19·10. Graph of J_e vs a/d.

The above discussions show that the necessary conditions for the appearance of ferromagnetism are the existence of partially filled internal atomic shells and the positive sign of the exchange integral which cause parallel orientation of the electron spins.

19·7 Domain structure of ferromagnetic materials

We have seen above that the spins of all electrons in the partially filled subshell of an atom in a ferromagnetic substance are aligned parallel to one another. Not only that, due to the influence of the exchange interaction, the spins of the electrons in such partially filled subshells of all neighbouring atoms are aligned parallel to one another. However there is a limit to the number of neighbouring atoms amongst which this effect of the exchange force extends.

To understand the reason for this, consider a small portion (A) of the substance (Fig. 19·11a) in which the spins of the electrons in the partially filled subshells of all the atoms are aligned parallel to one another due to the effect of the exchange force. So the magnetization within A is saturated. Suppose B is another such small portion of the material within which the magnetization is saturated and which is situated just by the side of A. If now the relative orientation of A and B is such that the directions of their magnetization are parallel to each other (which happens if the atomic spins in the two regions are parallel) then the situation is like placing two bar magnets side by side with the like poles contiguous to each other as shown in Fig. 19·11b. The energy of the system is maximum under this condition so that this cannot represent a stable arrangement. On the other hand if A and B are so oriented that the

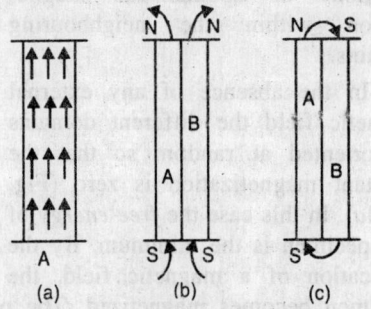

Fig. 19·11. Different orientations of two neighbouring magnetized domains in a ferromagnetic.

directions of their magnetization are antiparallel (Fig. 19·11c). then their energy will be a minimum so that it will represent a stable arrangement. The situation is similar to the case of placing two bar magnets side by side with the opposite poles contiguous to each other.

If the widths of A and B are of the order of a few atomic diameters only, then the effect of the exchange force predominates so that the spins of the atoms in the both regions A and B are alined parallel to each other and they act like a single region. If however the width of the region A is gradually increased then the effect of the magnetic interaction begins to predominate over the exchange force. Ultimately when the width of A reaches a limiting value of ~ 10^4 Å, the directions of magnetization of A and B become oppositely oriented. Thus the regions A and B now behave like two distinct regions.

Thus a ferromagnetic material may be regarded as made up of a large number of regions of very small size. Within each region the magnetization is saturated. They are of macroscopic dimensions and are known as *ferromagnetic domains*. Between two neighbouring domains there is small region within which the directions of orientation of the atomic spin moments gradually change from the direction of magnetization of the one domain to that of the next (see Fig. 19·12). This transition region is called the *Bloch wall*. The width of the Bloch wall is of the order of 1000 Å.

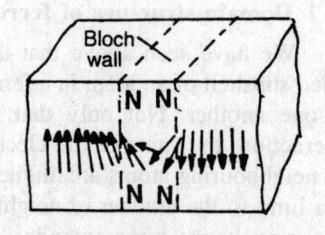

Fig. 19·12. 'Bloch wall' region in a ferromagnetic substance·

In Fig. 19·13, the theoretically expected pattern of the ferromagnetic domains is shown diagrammatically which have been confirmed by photographs of the domain structure. In the figure the arrows indicate the directions of spontaneous magnetization within the neighbouring domains.

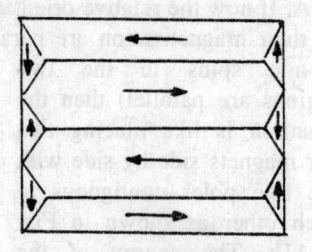

In the absence of any external magnetic field the different domains are oriented at random so that the resultant magnetization is zero (Fig. 19·14a). In this case the *free energy* of the specimen is the minimum. By the application of a magnetic field, the specimen becomes magnetized. The process of **magnetization** may be divided into **three steps** :

Fig. 19·13. Pattern of ferromagnetic domains (theoretical).

Magnetic Properties of Solids 671

(a) Magnetization by displacement of the domain boundaries :

This is shown in Fig. 19·14b. The magnetic moments of the different domains are oriented at different angles with respect to the applied

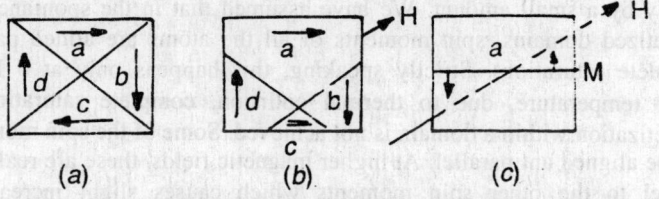

Fig. 19·14. Effect of external magnetic field (H) on domains.

magnetic field H. The size of the domain, the magnetic moment of which subtends the minimum angle with respect to H (e.g., a), increases with the increase of H while the sizes of the domains, the magnetic moments of which are oriented at larger angles with respect to H (e.g., b, c, or d), decrease with increasing H. Such changes in the sizes of the domains cause a diminution in the magnetic energy of the system, which therefore produces a more stable configuration. Since these changes in the domain size are produced by the displacement of domain boundaries, the process is known as the *displacement process of magnetization*. This process actually occurs in the initial stages of magnetization (the portion OP in the magnetization curve of Fig. (19·15). The process ends with the domain a pervading the whole crystal.

At low magnetic fields this process of magnetization takes place continuously and is reversible. If the magnetic field is increased by a small amount and then reduced by the same amount, then the increase and decrease of magnetization take place along the same path, so that at the end of the complete cyclic change, the specimen comes back to the initial state of magnetization.

At higher magnetic field, the magnetization occurs in discontinuous steps and is irreversible. The Barkhausen effect discussed in § 19·1 appears in this case.

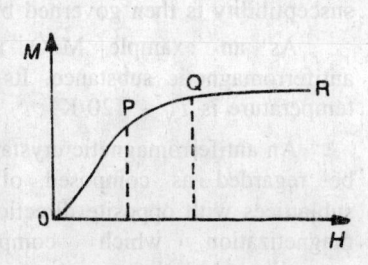

Fig. 19·15. Magnetization (M vs H) in a ferromagnetic.

(b) Magnetization by rotation of the magnetic moment vectors :

After the domain a has expanded to spread over the whole crystal, further increase of H causes its magnetic moment to turn towards the direction of H as shown in Fig. 19·14c. Due to this, the magnetization increases further (the portion PQ in Fig. 19·15). When the magnetic moment of the domain a turns fully in the field direction, the specimen shows what is known as *technical saturation*.

(c) *Para process* :

With further increase of H, the magnetization M increases very 'slowly by a small amount. We have assumed that in the spontaneously magnetized domains, spin moments of all the atoms are alined parallel (complete saturation). Strictly speaking, this happens only at 0 K. At higher temperature, due to thermal collision, complete saturation of magnetization within a domain is not achieved. Some of the spin moments may be aligned antiparallel. At higher magnetic fields, these are realigned parallel to the other spin moments which causes slight increase of magnetization beyond technical saturation. This is known as the para process.

19·8 Anti-ferromagnetism ; Ferri-magnetism and ferrites

We have seen in § 19·6 that if the exchange integral $J_e < 0$, the spin vectors of the atoms at the neighbouring lattice sites are aligned antiparallel. In this case there is no spontaneous magnetization. Such complete demagnetization occurs at 0 K and the phenomenon is known as *anti-ferromagnetism*. As the temperature rises, the regular antiparallel alinement of the magnetic moments of the neighbouring atoms is disturbed and the antiferromagnetic material shows some magnetization. With the rise of temperature, the magnetization increases. Finally above a definite temperature the regular antiparallel alignments of the spin moments of the neighbouring atoms is completely destroyed. This temperature is known as the Neel temperature (see Fig. 19·16). Above this, the substance behaves like an ordinary paramagnetic and its susceptibility is then governed by Curie's law (Eq. 19·1-3).

As an example MnO is an antiferromagnetic substance. Its Neel temperature is $T_N \sim 120$ K.

An antiferromagnetic crystal may be regarded as composed of two sublattices with opposite directions of magnetization which completely neutralise each other.

If in some substance the oppositely aligned magnetizations of the two sublattices are not equal, then they cannot completely compensate each other. Such a substance is said to *exhibit ferrimagnetism*. There can be spontaneous magnetization in them. They are known as *ferrites*.

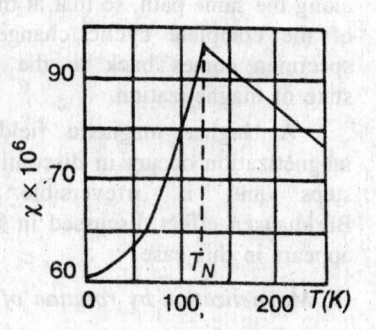

Fig. 19·16. Change of χ with T for an antiferromagnetic material. T_N is the Neel temperature.

Magnetic Properties of Solids 673

As an example if in a magnetite crystal (FeO, Fe_2O_3), the Fe^{++} ions are substituted by some divalent metallic ions (*eg.*, Mg, Ni, Co, Mn, Cu, *etc*) then we get a ferrite. The general formula for them may be written as XO, Fe_2O_3 where X is a divalent metal. The magnetic moments of the trivalent Fe ions in the two sublattices completely compensate each other. The magnetization of the substance is entirely due to the magnetic moments of the divalent metal ions (X^{++}).

One of the important characteristics of the ferrites are their very low electrical conductivity. Their resistivities are about $10^4 - 10^{11}$ times higher than that of iron. They have very high magnetic permeability, very low coercive force and very high saturation magnetization. Due to the combination of the above electrical and magnetic properties the use of ferrites has brought about an almost complete revolution in the field of ultrahigh frequency technology. At very high frequencies, the production of eddy currents is of considerable disadvantage in iron core transformers. Use of ferrite has removed this disadvantage.

In recent years ferrites with higher coercive force have been developed. These are being used in the construction of permanent magnets. Ferrites are also used in the storage elements of computers.

Problems

1. A magnetic induction field $B_0 = \mu_0 H$ acts perpendicular to the circular orbit of an electron of radius r in an atom. Show that to a first approximation the circular frequency of the electron is changed by $\omega_L = eB_0/2m_e$.

2. The diamagnetic susceptibility of helium is 1.05×10^{-9}. Calculate the r.m.s. radius of the electronic charge distribution in the helium atom (0.58Å)

3. Assuming the magnetic dipole moment of the atoms in a paramagnetic substance to be equal to the Bohr magneton μ_B, find the temperature at which its magnetic energy for the field $B_0 = 1$ T is equal to the mean thermal energy kT of the atoms. What is the paramagnetic susceptibility of the substance at 27°C ?

4. Show that for $\alpha << 1$, the Langevin function can be expanded in a power series as
$$L(\alpha) = \frac{\alpha}{3} - \frac{\alpha^3}{45}$$

5. If $\beta = gJ\mu_B\mu_0 H/kT << 1$, prove that in the quantum theory of paramagnetism, the Brillouin function $B_J(\beta) = (J+1)\beta/3J$ and the paramagnetic susceptibility $\chi = n\mu_0 g^2 \mu_B^2 J(J+1)/3kT$. Show that this reduces to the classical expression of Langevin in the limit of large J.

6. The ground state of Fe^{++} ion is $^5D_4$. Calculate the Lande g-factor as also the effective Bohr magneton number $p = g\sqrt{J(J+1)}$ for this state and compare it with the experimental value $p_{ex} = 5.4$. What would be value of g if the magnetism of these ions were solely due to their spin moments?

7. A paramagnetic gas consists of atoms in the $^2S_{1/2}$ state at 300 K. If a magnetic induction $B = 2.5\ T$ acts on the gas, show that the ratio of the number of gas atoms with the two possible projections of the magnetic moments w.r.t. to the field in given by

$$\frac{N_1}{N_2} = \frac{1 + \tanh a}{1 - \tanh a}, \qquad a = \frac{mg\,\mu_B B}{kT} = 0.0056$$

20

STATISTICAL MECHANICS

20·1 Introduction

Matter in bulk is made up of an enormously large number of microscopic particles (atoms and molecules). The behaviour of such a macroscopic system is determined by the behaviour of the ensemble of all the microscopic particles constituting the system. These microscopic particles are in incessant motion. The state of motion of each of these can, in principle, be determined by solving the equations of motion for them if the nature of the forces acting on them is known. Solutions of these equations of motion give us the three position coordinates x, y, z and the components of their momenta p_x, p_y, p_z at a particular instant of time t if the initial conditions are known. The knowledge of these variables of motion for all the microscopic particles would then have to be correlated with the macroscopic state of the system, usually expressed in terms of a set of thermodynamic coordinates, e.g., the pressure (p), the volume (V) and the temperature (T) of the system.

Because of the enormity in the number of the microscopic particles, such a task is prohibitively difficult. Even if it could be achieved in principle it would be of little value, since the states of motion of the individual particles change from one instant to the next due to collisions. Thus such a procedure would be of little practical utility in describing the state of the system.

There is however an alternative statistical method of dealing with such an ensemble of microscopic entities. This approach describes not only the behaviour of systems made up of material particles (atoms or molecules) but also of an enclosure full of radiation which may be regarded as an assembly of energy packets, known as photons. Such application of statistical methods to photons was first carried out by the famous Indian scientist S.N. Bose in 1924 in deriving Planck's radiation formula (see § 20·8).

20·2 Statistical method for describing the state of a system

The thermodynamic coordinates like p, V, T describing the state of an isolated system under equilibrium remain constant in time. On the other hand the motions of the individual particles constituting the system change from one instant to another. However there are certain dynamical parameters connected with the motions of these particles, the averages of which can be shown to remain unchanged under equilibrium condition. The statistical method of describing the macroscopic state of the system involves the correlations of these averages of the dynamical parameters with the observable thermodynamic parameters. Thus the macroscopic system under equilibrium is governed by laws different from those governing the motions of the individual particles. These laws applicable in the case of the whole ensemble are statistical in nature.

As an example if we consider a particular element of volume we cannot predict whether a particular molecule in a volume of gas will enter this element at a given instant t. Inspite of this inability to predict the behaviour of an individual molecule it can be seen that there is a definite regularity in the distribution of the molecules within different elements of volume when the number of molecules is enormously large. The statistical law tells us that on the average, equal elements of volume contain an equal number of molecules under equilibrium, provided the elements are so chosen that the number of molecules in each individual element is also enormously large.

Another example is the velocity distribution of the molecules. Even though it is not possible to predict whether a particular molecule has a specified velocity (say 1 km/s), it is possible to predict the average number molecules within a particular velocity range (say between 1 km/s to 1·1 km/s) from statistical considerations when a very large number of molecules are present.

These facts show that the average behaviour of the microscopic entities constituting the system is governed by the *laws of probability*. We know that when we toss a coin, there is no knowing whether the coin will fall with its head up or down. However if we toss the coin a very large number of times, it will be found that, on the average, the number of times it falls with its head up is approximately $\frac{1}{2}$ the number of throws. The larger the number of throws, more nearly will the above fraction approach the value $\frac{1}{2}$ as predicted by statistical laws.

There will of course be some deviations from the value predicted by the statistical laws. Thus larger the number of throws of the coin in the above example, less will be deviation from the average value of $\frac{1}{2}$ for the number of cases in which the coin falls with its head up. In general the deviation is $\infty/\sqrt{N}$ where N is the total number of cases considered. (see Vol. II, Ch. VII). Thus the percentage deviation is given by $(100 \times \sqrt{N}/N)$ or $100/\sqrt{N}$. In the case of an assembly of say 10^{24} molecules per unit volume of a gas, the percentage deviation of the number contained within the volume will be

$$\frac{100}{\sqrt{N}} = \frac{100}{10^{12}} = 10^{-10} \%$$

which is insignificantly small. So under equilibrium the number of molecules within a unit volume may be taken to be constant.

20·3 Equilibrium distribution ; Maxwell-Boltzmann statistics

We have seen in Ch. X that a quantum mechanical energy level may in some cases be degenerate. *i.e.*, there may be a number of linearly independent wave functions corresponding to a given energy value (see § 10·10). When a particle is confined within a one-dimensional potential box, its energy levels are non-degenerate (§ 11·2). However if a particle is confined within a three dimensional potential box, its energy levels are usually degenerate. The eigenfunction corresponding to a given level is determined by three integers in this case. There may be different combinations of the three integers to give the same energy eigenvalue. Each combination of the three integers gives an independent eigenfunction so that the energy level is degenerate.*

The energy levels for the particle in the box form a discrete spectrum. However these discrete levels may be so closely spaced that there is usually a very large number of levels of different energies $\epsilon_1, \epsilon_2, \epsilon_3 \ldots\ldots$ available for the particles. These levels may be degenerate, as seen above. We denote the degeneracies by $v_1, v_2, v_3, \ldots\ldots$ for the respective levels. If we divide by the total number of states then we get the probabilities of the different levels $g_1, g_2, g_3, \ldots\ldots$ etc.

The problem then reduces to finding the number of ways in which N molecules in a given volume V can be distributed in the different energy levels such that there are N_1 molecules with the energy ϵ_1, N_2 molecules with the energy ϵ_2 etc. In general we say that there are N_i molecules with the energy ϵ_i. The total energy of the system is $E = \Sigma N_i \epsilon_i$ and the total number of molecules is $N = \underset{i}{\Sigma} N_i$.

Now from combinatorial algebra, we know that the number of ways in which N particles can be arranged in the different levels without regard to the order in which the particles occupy the levels or the designation of the particles is

$$\frac{N!}{\underset{i}{\Pi} N_i !} = \frac{N!}{N_1 ! N_2 ! N_3 ! \ldots} \qquad (20\cdot3\text{-}1)$$

Since the probability of occupancy of a state of energy ϵ_i by a

---

* See *Introductory Quantum Mechanics* by the author.

particle is g_i then assuming equal a priori probability for each state the probability of occupancy of the level ϵ_i by N_i particles is $g_i^{N_i}$. Thus the total number of possible ways in which the state can be realised is

$$W = \Pi_i \frac{N! \, g_i^{N_i}}{N_i!} = \frac{N!}{N_1! \, N_2! \, N_3! \ldots} g_1^{N_1} g_2^{N_2} g_3^{N_3} \ldots \qquad (20 \cdot 3\text{-}2)$$

We can illustrate the above by an example. Suppose we have 4 particles which are to be placed in 2 different boxes of equal volume such that there will be 2 particles in each box (see Fig. 20·1). The number of ways in which the 2 particles (α, β) can be put in box 1 is 2! (viz., $\alpha\beta, \beta\alpha$). Similarly the number of ways in which the 2 particles (γ, δ) may be placed in box 2 is 2! (viz. $\gamma\delta, \delta\gamma$). Each of the above for a particular

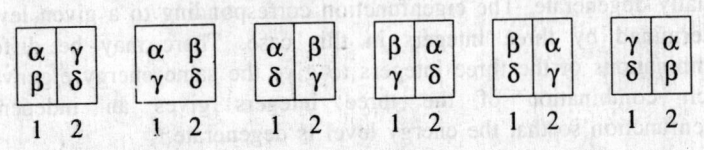

α β	α γ	α δ	β γ	β δ	γ δ
γ δ	β δ	β γ	α δ	α γ	α β
1 2	1 2	1 2	1 2	1 2	1 2

Fig. 20·1. Example of probability of occupancy. (4 particles $(\alpha, \beta, \gamma, \delta)$ are to be placed in two different boxes (1 and 2) of equal volume, each to contain only 2 particles.

box gives identical arrangement. Hence considering the total of 4! ways in which the 4 particles can be arranged between them, the total number of possible arrangements is $4!/(2! \times 2!)$ or 6 as shown in the figure. Since there are 2 boxes the a priori probability for the occupancy of a box by a particle is $(\frac{1}{2})$, by two particles is $(\frac{1}{2})^2$ etc. Hence the total number of possible arrangements is

$$W = \frac{4!}{2! \, 2!} \left(\frac{1}{2}\right)^2 \left(\frac{1}{2}\right)^2$$

which agrees with the expression given above.

Before proceeding further we simplify the expression for W (Eq. 20·3-2) by using *Stirling's formula*, which states that for a large number n, the following approximation can be used for $n!$:

$$n! = (n/e)^n$$

This gives $\quad \ln n! = n \ln n - n$

We then get from Eq. (20·3-2)

$$\ln W = \ln N! + \sum_i N_i \ln g_i - \sum_i \ln N_i!$$

$$= N \ln N - N + \sum_i (N_i \ln g_i - N_i \ln N_i + N_i)$$

$$= N \ln N + \sum_i N_i \ln (g_i/N_i) \qquad (20\cdot3\text{-}3)$$

We shall now find the condition under which $\ln W$ is a maximum which corresponds to the equilibrium (most probable) distribution. To do this we introduce two constraints on the system, viz.,

$$N = \sum_i N_i = \text{constant} \qquad (20\cdot3\text{-}4)$$

$$E = \sum_i N_i \epsilon_i = \text{constant} \qquad (20\cdot3\text{-}5)$$

Differentiating Eq. (20·3-3) and equating it to *zero*, we get the condition for the maximum value for $\ln W$:

$$d \ln W = \sum_i dN_i \left\{ \ln g_i/N_i + N_i \left(\frac{N_i}{g_i}\right)\left(-\frac{g_i}{N_i^2}\right) \right\}$$

$$= \sum_i dN_i (\ln g_i/N_i - 1) = 0 \qquad (20\cdot3\text{-}6)$$

Differentiating Eqs. (20·3-4) and (20·3-5) and equating to *zero* we get

$$\sum_i dN_i = 0 \qquad (20\cdot3\text{-}7)$$

$$\sum_i \epsilon_i \, dN_i = 0 \qquad (20\cdot3\text{-}8)$$

We now use Lagrange's method of undetermined multipliers.*

Multiplying Eq. (20·3-7) by α and (20·3-8) by $(-\beta)$ where α and β are constants and adding to Eq. (20·3-6) we get

$$\sum_i dN_i (\ln g_i/N_i - 1 + \alpha - \beta\epsilon_i) = 0 \qquad (20\cdot3\text{-}9)$$

Equating the coefficient of each dN_i separately to zero, we get

$$\ln g_i/N_i - 1 + \alpha - \beta\epsilon_i = 0$$

or,

$$\ln N_i/g_i = \alpha - 1 - \beta\epsilon_i$$

so that

$$N_i = A g_i \exp(-\beta \epsilon_i) \qquad (20\cdot3\text{-}10)$$

where

$$A = \exp(\alpha - 1) = \text{constant} \qquad (20\cdot3\text{-}11)$$

The total number of particles N is given by

$$N = \sum_i N_i = A \sum_i g_i \exp(-\beta\epsilon_i) \qquad (20\cdot3\text{-}12)$$

The *partition function* Z is defined as

---

* See *Theoretical Physics* by Georg Joos

$$Z = \frac{N}{A} = \sum_i g_i \exp(-\beta \varepsilon_i) \qquad (20\cdot3\text{-}13)$$

$$\therefore \quad A = \frac{N}{Z} \qquad (20\cdot3\text{-}14)$$

and
$$N_i = \frac{N}{Z} g_i \exp(-\beta_i \varepsilon_i) \qquad (20\cdot3\text{-}15)$$

To calculate β, we have to use the thermodynamical method. According to Boltzmann, the entropy $S = k \ln W$ where k is Boltzmann constant. Then it can be shown from thermodynamics* that

$$\beta = \frac{1}{kT} \qquad (20\cdot3\text{-}16)$$

where T is the absolute temperature of the system.

We then get from Eq. (20·3-15)

$$N_i = \frac{N}{Z} g_i \exp(-\varepsilon_i/kT) \qquad (20\cdot3\text{-}17)$$

Eq. (20·3-17) is the Maxwell-Boltzmann energy distribution function. Maxwell's law for the distribution of molecular volocities can be obtained using the above distribution function.

In the case of an ideal gas containing an enormous number of free particles (molecules), the energy levels are so much crowded together that the quantity g_i which is a measure of the number of states for the energy ε_i may be regarded as a continuously varying function of the energy ε and may be written as $g(\varepsilon)\, d\varepsilon$ for the energy range ε to $\varepsilon + d\varepsilon$. The quantity N_i may then be written as $N(\varepsilon)\, d\varepsilon$ equal to the number of particles in the energy range ε to $\varepsilon + d\varepsilon$. Eq. (20·3-17) may then be replaced by

$$N(\varepsilon)\, d\varepsilon = A \exp(-\varepsilon/kT)\, g(\varepsilon) d\varepsilon = f(\varepsilon)\, g(\varepsilon)\, d\varepsilon \qquad (20\cdot3\text{-}18)$$

Here $g(\varepsilon)\, d\varepsilon$ may be regarded as the *density of states* which is calculated in the next section (Eq. 20·4-5). $f(\varepsilon) = A \exp(-\varepsilon/kT)$ is the *distribution function* of the system under consideration. From Eq. (20·4-5) we have

$$g(\varepsilon) d\varepsilon = \frac{4\sqrt{2\pi}\, V}{h^3} m^{3/2} \sqrt{\varepsilon}\, d\varepsilon$$

So from Eq. (20·3-18) we get

$$N(\varepsilon) d\varepsilon = A\, \frac{4\sqrt{2\pi}\, V}{h^3} m^{3/2} \sqrt{\varepsilon} \exp(-\varepsilon/kT)\, d\varepsilon$$

---

* See *ibid*.

$$\therefore\ N = \int_0^\infty N(\varepsilon)\,d\varepsilon = A\,\frac{4\sqrt{2\pi}\,V}{h^3}\,m^{3/2} \int_0^\infty \sqrt{\varepsilon}\,\exp(-\varepsilon/kT)\,d\varepsilon$$

$$= A\,\frac{4\sqrt{2\pi}\,V}{h^3}\,m^{3/2}\,(kT)^{3/2}\,\frac{\sqrt{\pi}}{2}$$

$$\therefore\quad A = \frac{N}{V} \cdot \frac{h^3}{(2\pi\,mkT)^{3/2}} \tag{20·3-19}$$

and
$$N(\varepsilon)\,d\varepsilon = \frac{2\pi N}{(\pi kT)^{3/2}} \sqrt{\varepsilon}\,\exp(-\varepsilon/kT)\,d\varepsilon \tag{20·3-20}$$

This is Maxwell-Boltzmann energy distribution law for the molecules of an ideal gas. Maxwell's law for velocity distribution follows easily. For various applications of M-B statistics, see *A Treatise on Heat* by Saha and Srivastava.

20·4 Calculation of the density of states

To calculate the number of states available for a particle free to move within a volume V in the energy interval ε to $\varepsilon + d\varepsilon$, we note that the total energy ε depends both on the spatial coordinates x, y, z and the momentum components p_x, p_y, p_z of the particle. We define *phase space* as the six dimensional space with the coordinate axes x, y, z, p_x, p_y, p_z. The number of states in the energy interval ε to $\varepsilon + d\varepsilon$ will then be given by the number of *phase points* in the interval x to $x + dx$, y to $y + dy$, z to $z + dz$, p_x to $p_x + dp_x$, p_y to $p_y + dp_y$, p_z to $p_z + dp_z$ which defines an element

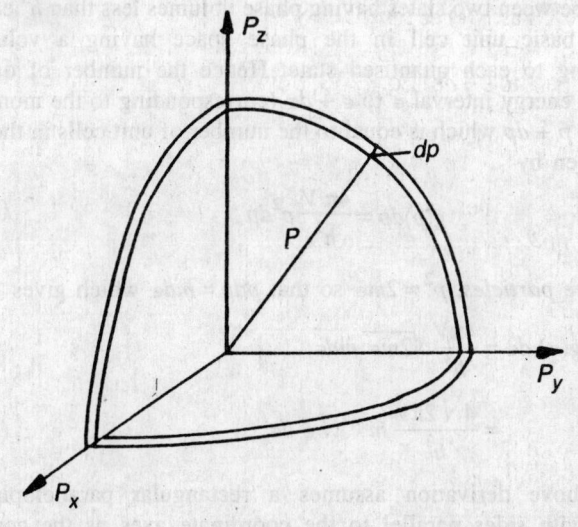

Fig. 20·2. An element of volume in momentum space.

element of phase volume $d\Omega = dV \cdot d\Omega_p$; dV and $d\Omega_p$ represent the elements of volumes in ordinary space and momentum space respectively.

In the case of free particles, the potential energy is zero so that the energy is independent of the coordinates x, y, z. In the case of a container in the shape of a rectangular parallelopiped with sides parallel to the coordinate axes, the entire spatial volume $V = \int dV$ is available for the particles in any given energy state, the energy being determined by the momentum components only. The element of volume $d\Omega_p$ in the momentum space can be calculated by drawing two spherical surfaces of radii p and $p + dp$ in the momentum space with the rectangular axes p_x, p_y, p_z as shown in Fig. 20·2. We get

$$d\Omega_p = 4\pi p^2 \, dp \qquad (20 \cdot 4 \cdot 1)$$

so that the element of volume in the phase space is

$$d\Omega = V \cdot d\Omega_p = 4\pi V p^2 dp \qquad (20 \cdot 4 \cdot 2)$$

Now in quantum mechanics, the position and momentum components are subject to the uncertainty relations

$$\Delta x . \Delta p_x = h, \quad \Delta y . \Delta p_y = h, \quad \Delta z . \Delta p_z = h$$

Hence the product of the intervals of the space coordinates and momentum components is subject to the uncertainly relation

$$\Delta x \, \Delta y \, \Delta z \, \Delta p_x \, \Delta p_y \, \Delta p_z = h^3 \qquad (20 \cdot 4 \cdot 3)$$

Thus for a particle exhibiting wave properties, it is not possible to distinguish between two states having phase volumes less than h^3 each. So there is a basic unit cell in the phase space having a volume h^3 corresponding to each quantised state. Hence the number of quantum states in the energy interval ϵ to $\epsilon + d\epsilon$ (corresponding to the momentum interval p to $p + dp$ which is equal to the number of unit cells in the phase space is given by

$$g(p)dp = \frac{4\pi V}{h^3} p^2 dp \qquad (20 \cdot 4 \cdot 4)$$

For *free particles*, $p^2 = 2m\epsilon$ so that $p dp = m d\epsilon$ which gives

$$g(\epsilon) \, d\epsilon = \frac{4\pi V}{h^3} \sqrt{2m\epsilon} \, m d\epsilon$$

$$= \frac{4\sqrt{2\pi} V}{h^3} m^{3/2} \sqrt{\epsilon} \, d\epsilon \qquad (20 \cdot 4 \cdot 5)$$

The above derivation assumes a rectangular parallelopiped of volume V with sides parallel to the coordinate axes as the container. However it holds for a container of any shape.

20·5 Quantum statistics

In quantum mechanics, the state of a system is described by a wave function. If we consider two identical particles 1 and 2 in the states a and b respectively where a and b denote all the quantum numbers required for the complete specification of the states, then the wave function for the system of the two identical particles can be written as

$$u_{12}(a, b) = u_1(a) u_2(b) \qquad (20\cdot5\text{-}1)$$

Here $u_1(a)$ is the wave function of the particle 1 in the state a and $u_2(b)$ is the wave function for the particle 2 in the state b.

If the two particles are interchanged between the two states, the wave function will be

$$u_{21}(a, b) = u_2(a) u_1(b) \qquad (20\cdot5\text{-}2)$$

As before $u_2(a)$ represents the wave function of the particle 2 in the state a and $u_1(b)$ is the wave function of the particle 1 in the state b.

If $\hat{H}_{12}$ is the Hamiltonian of the system of the two particles, we can write the Schrödinger equation as

$$\hat{H}_{12} u_{12}(a, b) = E u_{12}(a, b) \qquad (20\cdot5\text{-}3)$$

On interchange of the two particles the Hamiltonian becomes $\hat{H}_{21}$ which is equal to $\hat{H}_{12}$ because the two particles are identical ($\hat{H}_{21} = \hat{H}_{12}$). Hence from Eq. (20·5-3) we get by interchanging the particles.

$$\hat{H}_{21} u_{21}(a, b) = \hat{H}_{12} u_{21}(a, b) = E u_{21}(a, b) \qquad (20\cdot5\text{-}4)$$

Thus $u_{12}(a, b)$ and $u_{21}(a, b)$ are both simultaneous eigenfunctions of the Hamiltonian of the system of two identical particles belonging to the same eigenvalue E. So the general solution is a linear combination of the two :

$$\Psi_{12} = A u_{12}(a, b) + B u_{21}(a, b) \qquad (20\cdot5\text{-}5)$$

Interchange of the particles 1 and 2 gives

$$\Psi_{21} = A u_{21}(a, b) + B u_{12}(a, b) \qquad (20\cdot5\text{-}6)$$

Since by the interchange of the two identical particles between the two states, the system remains the same we must have

$$|\Psi_{12}|^2 = |\Psi_{21}|^2 \qquad (20\cdot5\text{-}7)$$

It can be easily seen from Eqs. (20·5-5) and (20·5-6), the condition (20·5-7) is satisfied if and only if

$$A^2 = B^2$$

or, $$A = \pm B \qquad (20\cdot5\text{-}8)$$

Taking the magnitudes of A and B to be unity, we then get the two solutions :

$$\Psi^S = u_{12}(a, b) + u_{21}(a, b) \qquad (20\cdot5\text{-}9)$$

$$\Psi^A = u_{12}(a, b) - u_{21}(a, b) \qquad (20\cdot5\text{-}10)$$

It can be easily seen that by interchanging the particles 1 and 2 between the two states, Ψ^S remains the same (symmetric) ; hence it is called the *symmetric state*. On the other hand due to the interchange of 1 and 2 between the two states Ψ^A changes sign which is thus an antisymmetric wave function giving an antisymmetric state.

In nature all elementary particles or their combinations can be classified into two groups, those with symmetric wave functions known as *bosons* (named after the Indian scientist S. N. Bose) and those with antisymmetric wave functions known as fermions (named after the Italian scientist Enrico Fermi). It is found that particles with *zero or integral* spin angular momenta have symmetric wave functions and hence are bosons. Examples are photons (spin 1), pions (spin 0), deuterons (spin 1), alpha particles (spin 0) *etc*. On the other hand, particles with *half odd integral spins* (1/2, 3/2, 5/2 *etc*.) have antisymmetric wave functions and are fermions. Examples are electrons (spin $\frac{1}{2}$), protons (spin $\frac{1}{2}$), neutrons (spin $\frac{1}{2}$), etc.

In the above discussion we have considered a system consisting of two identical particles. It can be extended to the case of a system made up of more than two identical particles (see *Quantum Mechanics* by L. I. Schiff); (See also Ch. XVIII).

One important characteristic of the two groups of particles can immediately be seen from the Eqs. (20·5-9) and (20·5-10). If the two states a and b are the same, then Eq. (20·5-10) gives

$$\Psi^A = u_{12}(a, a) - u_{21}(a, a)$$
$$= u_1(a) u_2(a) - u_2(a) u_1(a) = 0$$

So the two identical formions cannot occupy the same state which shows that they obey Pauli's exclusion principle.

On the other hand Eq. (20·5-9) gives in this case

$$\Psi^S = u_{12}(a, a) + u_{21}(a, a)$$
$$= u_1(a) u_2(a) + u_2(a) u_1(a) \neq 0$$

So two or more bosons can occupy the same state. In fact the probability of occupancy of the state increases with the increasing number of particles in this case. Thus the bosons do not obey Pauli's exclusion principle.

Depending on whether the particles have symmetric or antisymmetric wave functions, they obey *Bose-Einstein statistics* or

Statistical Mechanics

Fermi-Dirac statistics respectively, both of which are known as *quantum statistics* as distinct from the classical Maxwell-Boltzmann statistics discussed in § 20·3.

Whether or not a group of particles will be governed by classical or quantum statistics depends on whether they are distinguishable or indistinguishable. Quantum statistics applies to particles which are indistinguishable.

If may be noted that in classical physics, two identical particles are distinguishable from the difference in their dynamical states, i.e., from the differences in their position coordinates and momenta at any instant. However this is not possible for two quantum systems since their position coordinates and momenta are governed by Heisenberg's uncertainty principle. Thus two identical particles in quantum mechanics are indistinguishable.

20·6 Exchange operator

To generalize the above considerations to many-particle system, we introduce the particle exchange operator P_{ij} which interchanges the labelling of any two particles i and j of the system. Obviously $P_{ij} = P_{ji}$. In the case of a two-particle system we have $P_{ij} = P_{12}$ and

$$P_{12} u_{12}(a, b) = P_{12} u_1(a) u_2(b)$$
$$= u_2(a) u_1(b) = u_{21}(a, b) \quad (20 \cdot 6\text{-}1)$$

Similarly $\quad P_{12} u_{21}(a, b) = u_{12}(a, b) \quad (20 \cdot 6\text{-}2)$

Taking the linear combination Ψ_{12} defined by Eq. (20·5-5), we get (writing $A = B = 1$)

$$P_{12} \Psi_{12} = P_{12} \{ u_{12}(a, b) + u_{21}(a, b) \}$$
$$= u_{21}(a, b) + u_{12}(a, b) = \Psi_{21} \quad (20 \cdot 6\text{-}3)$$

Operating twice with P_{12} we get

$$P_{12}^2 \Psi_{12} = P_{12} \Psi_{21} = \Psi_{12} \quad (20 \cdot 6\text{-}4)$$

so that P_{12}^2 can have the value 1.

Hence P_{12} can have the two values

$$P_{12} = \pm 1 \quad (20 \cdot 6\text{-}5)$$

Thus for symmetric and antisymmetric combinations Ψ^S and Ψ^A (Eqs. 20·5-9 and 20·5-10), we have

$$P_{12} \Psi^S = +\Psi^S \text{ and } P_{12} \Psi^A = -\Psi^A \quad (20 \cdot 6\text{-}6)$$

Since for two identical particles $H_{12} = H_{21}$ where H is the Hamiltonian, we get

$$P_{12} H_{12} \Psi_{12} = H_{21} \Psi_{21} = H_{12} \Psi_{21} = H_{12} P_{12} \Psi_{12} \qquad (20 \cdot 6\text{-}7)$$

This shows that P_{12} commutes with the Hamiltonian.

$$P_{12} H_{12} = H_{12} P_{12} \qquad (20 \cdot 6\text{-}8)$$

It may be noted that the exchange symmetry of an eigenfunction is constant in time so that if a system is in the symmetric state Ψ^S to start with, it will remain in this state at all subsequent times. Similarly for the antisymmetric state. (See §13·11).

Many particle system :

In this case the Hamiltonian H is invariant under interchange of any pair of particles. For a system of n particles, if we write

$$H = H(1, 2, 3, \ldots n) \qquad (20 \cdot 6\text{-}9)$$

then the indistinguishability of the particles is expressed by the relations of the form.

$$H(1, 2, 3, 4 \ldots n) = H(2, 1, 3, 4, \ldots n)$$
$$= H(2, 1, 4, 3, \ldots n)$$
$$= H(2, n, 4, 3, \ldots 1) \qquad (20 \cdot 6\text{-}10)$$

As in the case of the two particle system, if $\Psi(1, 2, 3, 4, \ldots n)$ represents a solution of the Schrödinger equation, then the functions

$$\Psi(2, 1, 4, 3, \ldots n), \Psi(2, n, 4, 3 \ldots 1) \text{ etc.} \qquad (20 \cdot 6\text{-}11)$$

are also solutions. Additional solutions can be built up by the superposition of the above solutions.

For the two particle system, the antisymmetric wave function can be written as

$$\Psi^A = \frac{1}{\sqrt{2!}} \{ u_{12}(a,b) - u_{21}(a,b) \}$$

$$= \frac{1}{\sqrt{2!}} \{ u_1(a) u_2(b) - u_2(a) u_1(b) \}$$

$$= \frac{1}{\sqrt{2!}} \begin{vmatrix} u_1(a) & u_2(a) \\ u_1(b) & u_2(b) \end{vmatrix} \qquad (20 \cdot 6\text{-}12)$$

For a many-body system of particles obeying Pauli's exclusion principle, we have similarly (see § 13·12)

$$\Psi^A = \frac{1}{\sqrt{n!}} \begin{vmatrix} u_1(a_1) & u_2(a_1) & \ldots & u_n(a_1) \\ u_1(a_2) & u_2(a_2) & \ldots & u_n(a_2) \\ \ldots & \ldots & \ldots & \ldots \\ \ldots & \ldots & \ldots & \ldots \\ u_1(a_n) & u_2(a_n) & \ldots & u_n(a_n) \end{vmatrix} \qquad (20 \cdot 6\text{-}13)$$

where we have replaced the quantum numbers $a, b, \ldots$ etc. by $a_1, a_2 \ldots a_n$.

This is known as the Slater determinant. The determinant vanishes if any two rows are identical which happens if any two of the set of quantum numbers a_1, a_2, a_3 etc. are the same. This is nothing that Pauli's exclusion principle.

The fact that the symmetry characteristics of the wave function is permanent in time shows that no transition is possible between symmetric and antisymmetric states. (See §13·11).

20·7 Fermi-Dirac statistics

We have seen in Ch. XVIII that the free electron gas in metals obeys Fermi-Dirac statistics which is one form of quantum statistics discussed in § 20·5. There are two important characteristics which distinguish this from the classical Maxwell-Boltzmann statistics discussed earlier. Firstly, the particles obeying F-D statistics are *indistinguishable*. Secondly, the particles obeying F-D statistics obey Pauli's exclusion principle. The first of these makes the number of ways of distributing the particles in the different states as derived from combinatorial algebra different from that given by Eq. (20·3-1).

Since the labelling of the particles has no meaning due to their indistinguishability all the possible arrangements of N particles in the different states given by Eq. (20·3-1) are alike and the factor (20·3-1) reduces to 1. This can be seen from the example of the four particles in the two states with two in each state discussed in § 20·3. Since the labellings $\alpha, \beta, \gamma, \delta$ are no longer significant in the present case all the six arrangements in Fig. 20·1 are alike so that there is only one possible arrangement.

However because of Pauli principle, a different kind of arrangement will arise in the present case. Consider the g_i states at the energy ϵ_i. In the present case we take g_i to be the actual number of states and not the probability of the state as discussed in § 20·3. Since each state can contain at most only one particle, the maximum number of particles N_i which can be accommodated in this level is g_i so that $N_i < g_i$. We can select any one of the g_i states in g_i different ways and place one particle in it. Of the remaining $(g_i - 1)$ states, we can chose one in $(g_i - 1)$ ways and place the second particle in it. Proceeding in this way the state for the last state can be chosen in $(g_i - N_i + 1)$ ways. Thus the total number of ways of chosing the states will be

$$g_i(g_i - 1)\ldots(g_i - N_i + 1) = \frac{g_i!}{(g_i - N_i)!}$$

Since the particles are indistinguishable, $N_i!$ ways of permutation of the N_i particles, one in each state, does not give any new arrangement. Hence the number of independent ways of realizing the distribution of the

N_i particles in the g_i states at the energy ϵ_i is

$$\frac{g_i!}{N_i!(g_i-N_i)!} \tag{20.7-1}$$

To illustrate, let us consider the number of ways in which two particles α and β can be placed in four boxes A, B, C or D such that each box can have only one particle or none. The possible arrangements are shown in Fig. 20·3, assuming α and β to be different. The possible number is $4!/2! = 12$.

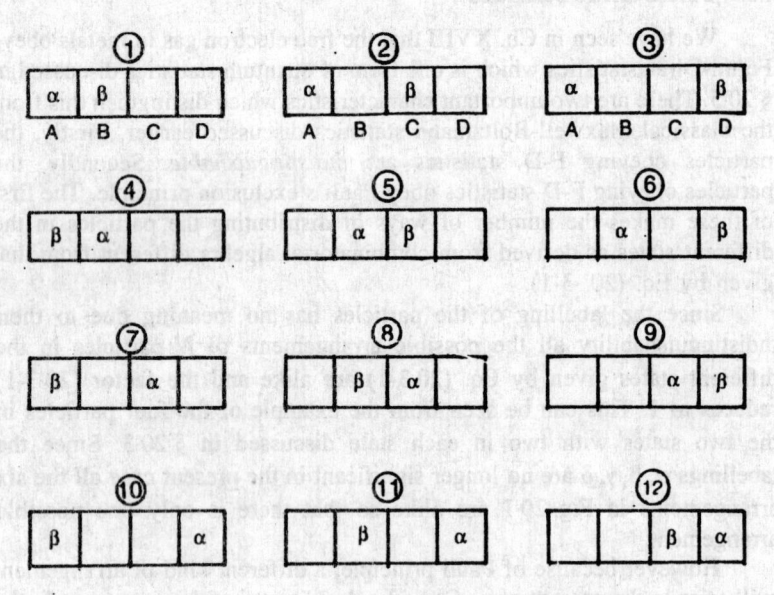

Fig. 20·3. Possible arrangements of two particles α and β in four boxes A, B, C, D, if each box contains 1 or 0 particle.

However if the particles are indistinguishable ($\alpha \equiv \beta$), then the following pairs of arrangements are identical :

① ≡ ④ ; ② ≡ ⑦ ; ③ ≡ ⑩ ;
⑤ ≡ ⑧ ; ⑥ ≡ ⑪ ; ⑨ ≡ ⑫.

So the number of possible arrangements is reduced to

$$\frac{4!}{2!\,2!} = 6$$

which is obtained by dividing the previous number 12 by the possible permutations (2!) between α and β for any arrangement.

Considering all the energy levels $\epsilon_1, \epsilon_2, \epsilon_3 \ldots\ldots$, we get from Eq. (20·7-1) the thermodynamic probability of the state as

$$W = \prod_i \frac{g_i!}{N_i!(g_i - N_i)!} \tag{20.7-2}$$

Taking the logarithm, we have

$$\ln W = \sum_i \{\ln g_i! - \ln N_i! - \ln(g_i - N_i)!\}$$

$$= \sum_i \{g_i \ln g_i - g_i - N_i \ln N_i + N_i - (g_i - N_i)\ln(g_i - N_i) + (g_i - N_i)\}$$

$$= \sum_i \{g_i \ln g_i - N_i \ln N_i - (g_i - N_i)\ln(g_i - N_i)\} \tag{20.7-3}$$

$$d\ln W = \sum_i dN_i \left\{-\ln N_i - \frac{N_i}{N_i} + \ln(g_i - N_i) + \frac{g_i - N_i}{g_i - N_i}\right\}$$

$$= \sum_i dN_i \ln \frac{g_i - N_i}{N_i} = 0 \tag{20.7-4}$$

for maximum probability. The auxiliary conditions (20·3-4) and (20·3-5) remain unchanged so that we have the two further equations (Eq. 20·3-7) and (20·3-8).

$$\sum_i dN_i = 0, \qquad \sum_i \epsilon_i dN_i = 0$$

Using Langrange's method of undermined multipliers we then get as before

$$\sum_i dN_i \left\{\ln \frac{g_i - N_i}{N_i} - \alpha - \beta \epsilon_i\right\} = 0$$

which gives

$$\ln \frac{g_i - N_i}{N_i} = \alpha + \beta \epsilon_i \tag{20.7-5}$$

or,

$$\frac{g_i - N_i}{N_i} = \exp(\alpha + \beta \epsilon_i)$$

∴

$$\frac{g_i}{N_i} = 1 + \exp(\alpha + \beta \epsilon_i)$$

or,

$$N_i = \frac{g_i}{1 + \exp(\alpha + \beta \epsilon_i)} \tag{20.7-6}$$

This is the Fermi-Dirac distribution function. As in §20·3 we shall take $\beta = 1/kT$. If we write $\alpha = -\epsilon_f/kT$ where ϵ_f is the *Fermi energy*, we get

$$N_i = \frac{g_i}{1 + \exp(\epsilon_i - \epsilon_f)/kT} \tag{20.7-7}$$

If the levels are crowded together to form almost a continuum we get

$$N(\epsilon)\, d\epsilon = g(\epsilon) f(\epsilon)\, d\epsilon = \frac{g(\epsilon)\, d\epsilon}{1 + \exp(\epsilon - \epsilon_f)/kT} \qquad (20\cdot 7\text{-}8)$$

For a gas of free particles as in the case of the free electron gas, we can get $g(\epsilon)\, d\epsilon$ with the help of Eq. (20·4-5). However in the case of an electron gas, it is possible to place two electrons with spins *up* and *down* in each state without violating Pauli's exclusion principle. So the expression for $g(\epsilon)\, d\epsilon$ should be multiplied by 2. Thus we get finally

$$N(\epsilon)\, d\epsilon = \frac{8\sqrt{2}\,\pi V}{h^3} \cdot \frac{m^{3/2}\sqrt{\epsilon}\, d\epsilon}{1 + \exp(\epsilon - \epsilon_f)/kT}$$

$$= 4\pi V \left(\frac{2m}{h^2}\right)^{3/2} \frac{\sqrt{\epsilon}\, d\epsilon}{1 + \exp(\epsilon - \epsilon_f)/kT}$$

$$= \frac{V}{2\pi^2}\left(\frac{2m}{\hbar^2}\right)^{3/2} \frac{\sqrt{\epsilon}\, d\epsilon}{1 + \exp(\epsilon - \epsilon_f)/kT} \qquad (20\cdot 7\text{-}9)$$

The quantity $A = \exp(-\alpha)$ is called the *degeneracy parameter*. The denominator of the distribution function (20·7-6) is always greater than 1, which means that α can have any value from $-\infty$ to $+\infty$ so that the limits of A are 0 and ∞. For large $\alpha, A \ll 1$. In this case we can get an approximate expression for A which is the half of that given by Eq. (20·3-19). F-D distribution law passes over to the classical statistics in this case. For $\alpha \to -\infty, A \gg 1$. An approximate expression for A in this case gives

$$\ln A = -\alpha = \frac{h^2}{2mkT}\left(\frac{3n}{8\pi}\right)^{2/3} + \ldots\ldots$$

Assuming a free electron density $10^{28}/m^3$, A is found to be large compared to 1 in most metals at room temperature which shows that the electron gas in metals is degenerate.

For an application of the F-D statistics to the electron gas in metals and semi-conductors, see Chapter XVIII.

20·8 Bose-Einstein statistics

This is the second type of quantum statistics obeyed by particles of integral or zero spin. As in the case of F-D statistics, the particles obeying Bose-Einstein statistics are indistinguishable. Hence the statistical factor given by Eq. (20·3-1) reduces to 1 (see § 20·7). However, in the present case there is no restriction to the number of particles which can occupy a particular state, since the particles obeying B-E statistics do not obey Pauli's exclusion principle.

The enumeration of the number of possible ways in which N_i particles can be placed in g_i states at the energy ϵ_i can be carried out by

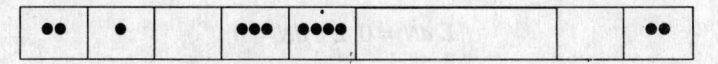

Fig. 20·4. Placing of N_i particles in g_i states at the energy ϵ_i according to B-E statistics.

referring to Fig. 20·4. in which the different cells are represented by the different boxes and the particles by the dots. As can be seen the boxes and the particles are arranged in an entirely arbitrary manner, the only restriction being that we start with the first box at the left and end with box number g_i on the right. The first box can be chosen in g_i different ways. Then the remaining $(g_i + N_i - 1)$ entities (the partitions between the boxes plus the particles) can be arranged in $(g_i + N_i - 1)!$ ways. However the permutations of the g_i boxes or of the N_i particles amongst themselves do not give any new arrangement. So we have divide by $(g_i! \times N_i!)$ to get the total number of ways in which the N_i particles can be arranged between the g_i boxes :

$$\frac{g_i(g_i + N_i - 1)!}{g_i! N_i!} = \frac{(g_i + N_i - 1)!}{(g_i - 1)! N_i!} \qquad (20 \cdot 8\text{-}1)$$

To get the thermodynamic probability we have to multiply the above quantities for all the energy levels ϵ_i.

$$W = \prod_i \frac{(g_i + N_i - 1)!}{(g_i - 1)! N_i!} \qquad (20 \cdot 8\text{-}2)$$

Taking the logarithm, we get

$$\ln W = \sum_i \{\ln(g_i + N_i - 1)! - \ln(g_i - 1)! - \ln N_i!\}$$

$$= \sum_i \{(g_i + N_i - 1) \ln(g_i + N_i - 1) - (g_i + N_i - 1) - (g_i - 1) \ln(g_i - 1)$$
$$+ (g_i - 1) - N_i \ln N_i + N_i\}$$

$$= \sum_i \{(g_i + N_i - 1) \ln(g_i + N_i - 1) - (g_i - 1) \ln(g_i - 1) - N_i \ln N_i\}$$

$$(20 \cdot 8\text{-}3)$$

Differentitating and equating to zero we get ln W for the most probable arrangement :

$$d \ln W = \sum_i dN_i \left\{ \ln(g_i + N_i - 1) + \frac{g_i + N_i - 1}{g_i + N_i - 1} - \ln N_i - \frac{N_i}{N_i} \right\}$$

$$= \sum_i dN_i \ln \frac{g_i + N_i - 1}{N_i} = 0 \qquad (20 \cdot 8\text{-}4)$$

We use the auxiliary conditions (20·3-4) and (20·3-5) for the constancy of the number of particles and the total energy of the system. These give two further equations (20·3-7) and (20·3-8) :

$$\Sigma_i dN_i = 0, \Sigma_i \epsilon_i dN_i = 0$$

Using Lagrange's method of undetermined multipliers we get as before

$$\Sigma_i dN_i \left\{ \ln \frac{g_i + N_i - 1}{N_i} - \alpha - \beta \epsilon_i \right\} = 0$$

which gives
$$\ln \frac{g_i + N_i - 1}{N_i} = \alpha + \beta \epsilon_i \qquad (20 \cdot 8\text{-}5)$$

Since g_i and N_i are both very large numbers, we can neglect 1 in the numerator on the left side of the above equation. We then get

$$\frac{g_i + N_i}{N_i} = \exp(\alpha + \beta \epsilon_i)$$

$$\therefore \qquad N_i = \frac{g_i}{\exp(\alpha + \beta \epsilon_i) - 1} \qquad (20 \cdot 8\text{-}6)$$

As before for very closely spaced levels, this can be written in the from

$$N(\epsilon)d\epsilon = \frac{g(\epsilon)d\epsilon}{\exp(\alpha + \epsilon/kT) - 1} \qquad (20 \cdot 8\text{-}7)$$

where we have put $\beta = 1/kT$. Eq. (20·8-7) gives the Bose-Einstein distribution function.

Substituting for $g(\epsilon)d\epsilon$ from Eq. (20·4-5), we get

$$N(\epsilon)d\epsilon = f(\epsilon)g(\epsilon)d\epsilon = \frac{4\sqrt{2}\pi V}{h^3} m^{3/2} \frac{\sqrt{\epsilon} d\epsilon}{\exp(\alpha + \epsilon/kT) - 1} \qquad (20 \cdot 8\text{-}8)$$

If we compare Eq. (20·8-7) with Eq. (20·7-7) for the Fermi-Dirac distribution law, we see that except for the sign before 1 in the denominator the two formulas are the same. This difference in the sign, however, is of profound significance.

In the present case $\alpha > 0$. It cannot be negative since in that case, the denominator vanishes for $\alpha = -\beta \epsilon$ which makes $N = \infty$. If α is very large, the degeneracy parameter $A = \exp(-\alpha) << 1$, so that $\exp(\alpha + \beta \epsilon) = \exp(\beta \epsilon)/A >> 1$ which means that we can neglect 1 in the denominator of Eq. (20·8-8). This gives the classical M-B distribution law :

$$N(\epsilon)d\epsilon = \frac{4\sqrt{2}\pi V}{h^3} m^{3/2} \sqrt{\epsilon} A \exp(-\epsilon/kT) d\epsilon \qquad (20 \cdot 8\text{-}9)$$

A is given by Eq. (20·3-19) in this case. Thus when A is very small. Bose-Einstein distribution law passes over to classical distribution law. The limiting value of $A = 1$ when $\alpha = 0$. In this case deviations from classical distribution law are observed and the gas is said to be

Statistical Mechanics

degenerate. A increases as $n = N/V$ becomes larger, *i.e.*, as the pressure of the gas increases. It also increases as the temperature T becomes lower and as the atomic weight decreases.

For hydrogen gas at $T = 300$ K, $(n \sim 3 \times 10^{27}\,\text{m}^{-3})$, $A \sim 3 \times 10^{-5} \ll 1$ so that the gas is non-degenerate. Degeneracy would set in only at extremely low temperature or at very high pressures.

20·9 Derivation of Planck's law of radiation using Bose-Einstein statistics

In the case of a photon gas, the restriction regarding the constancy of the number of particles (Eq. 20·3-5) does not apply. So the auxiliary condition (20·3-7) need not be taken into account. Putting $\alpha = 0$, we get from Eq. (20·8-6), for a photon gas

$$N_i = \frac{g_i}{\exp(\beta \epsilon_i) - 1} = \frac{g_i}{\exp(\epsilon_i / kT) - 1} \qquad (20\cdot9\text{-}1)$$

So in the limit of very close levels

$$N(\epsilon)d\epsilon = \frac{g(\epsilon)d\epsilon}{\exp(\epsilon / kT) - 1} \qquad (20\cdot9\text{-}2)$$

Since for photons, the momentum $p = h\nu/c$ where ν is the frequency and c is the velocity of light, we get from Eq. (20·4-4)

$$g(\nu)d\nu = g(\epsilon)d\epsilon = \frac{4\pi V p^2 dp}{h^3} = \frac{4\pi V}{h^3}\left(\frac{h}{c}\right)^3 \nu^2 d\nu = \frac{4\pi V}{c^3}\nu^2 d\nu \qquad (20\cdot9\text{-}3)$$

This gives

$$N(\nu)\,d\nu = \frac{4\pi V}{c^3}\frac{\nu^2 d\nu}{\exp(h\nu/kT) - 1} \qquad (20\cdot9\text{-}4)$$

Since each photon carries an energy $h\nu$ and has two directions for polarization, the energy density of the radiation in the spectral range ν to $\nu + d\nu$ comes out to

$$u_\nu\,d\nu = 2 \times \frac{N(\nu)d\nu}{V} h\nu = \frac{8\pi}{c^3}\frac{h\nu^3 d\nu}{\exp(h\nu/kT) - 1} \qquad (20\cdot9\text{-}5)$$

Eq. (20·9-5) is the same as Planck's radiation formula Eq. (3·3-4). As stated earlier, S.N. Bose was the first to deduce this formula by statistical method and thereby laid the foundation of quantum statistics (§ 20·1).

Problems

1. Using the M–B energy distribution function obtain the velocity distribution of the molecules of an ideal gas.

   Hence calculate the mean velocity and the mean squared velocity of the gas molecules.

2. Using the M–B energy distribution function, show that the number of molecules in an ideal gas having the x-component of the velocity lying between v_x and $v_x + dv_x$ regardless of the values of the y and z components of the velocity is given by

$$N(v_x)\,dv_x = N\sqrt{\frac{m}{2\pi kT}}\,\exp(-mv_x^2/2kT)dv_x$$

3. Using the result of Prob. 2, show that if the electron gas in a metal obeys M–B statistics, then the thermionic emission current from a metal surface perpendicular to the x-axis is given by (Richardson equation 5·9-1)

$$i_s = AT^{1/2}\exp(-mv_0^2/2kT)$$

where $mv_0^2/2$ is minimum energy required for the emission.

4. Using Eq. (20·3-19), calculate the degeneracy parameter A for hydrogen gas at 273 K at 1 atmospheric pressure and show that the gas is non-degenerate in this case. Density of hydrogen = 0.0899 kg/m³. (1.11×10^{-5})

5. Using the approximate expression for the degeneracy parameter $A = \exp(-\alpha)$ given by the classical expression (20·3-19) multiplied by 1/2, calculate A for the free electron gas in copper at $T = 300$ K, assuming one conduction electron per atom in the lattice of the copper crystal. Density of copper is $\rho = 8.96 \times 10^3$ kg/m³. Atomic weight of copper $M = 63.54$. Hence show that the electron gas is degenerate. (3·38)

6. In a nucleus with mass number A with equal numbers of protons and neutrons ($Z = N = A/2$), both types of nucleons behave like a Fermi gas. Show that the Fermi energy of each type of nucleon is ~ 21 MeV.

References for Further Reading

1. Atomic Spectra and Atomic Structure by G. Hertzberg, Dover Publications (1944)
2. Introduction to Atomic Spectra by H. White, McGraw-Hill Book Co. Inc. (1934).
3. Atomic Structure and Spectral Lines by A. Sommerfeld, E.P. Dutton & Co. Inc.
4. Modern Physics and Quantum Mechanics by E.E. Anderson, Mc Millan Co. Ltd. (1979).
5. Quantum Mechanics by L.I. Schiff, McGraw-Hill Book Co. Inc. (1949)
6. Physics of Atoms and Molecules by B.H. Bransden and C.J. Joachain, Longman (1983).
7. Spectra of Diatomic Molecules by G. Hertzberg, Van-Norstand Reinhold Co. (Copy right 1950) by Litton Publishing Inc.
8. Introduction to Modern Physics by F.K. Richtmeyer, H. Kenard and Lauritsen, McGraw-Hill Book Co. Inc. (1955).
9. Quantum Mechanics by P.A.M. Dirac, Oxford at Clarendon Press (1958).
10. Quantum Mechanics by J. L. Powell and B. Craseman, Narosa Publishing House (1988).
11. Quantum Mechanics by V. Fock, Mir Publisher, Moscow (1986).
12. Introductory Quantum Mechanics by L. Pauling and E.B. Wilson, McGraw-Hill Book Co. Inc. (1935).
13. Introduction to Solid State Physics by Charles Kittel, Wiley–Eastern.
14. Solid State and Semi Conductor Physics by J.P.McKelvy, Harper and Row (1966).
15. X-rays in Theory and Experiment by A.H. Compton and S.K. Allison, D.Van Norstand Co. Inc. (1947).
16. Quantum Theory of Atomic Structure Vol-II, by J.C. Slater, McGraw-Hill Book Co.Inc. (1960).
17. Quantum Theory of Molecules and Solids, Vol-I, by J.C. Slater, McGraw-Hill Book Co.Inc. (1963).

APPENDIX A–I

HERMITE POLYNOMIALS

(i) The Hermite polynomials have the general form
$$H_n(q) = (-1)^n \exp(q^2) \left(\frac{d}{dq}\right)^n \exp(-q^2), \quad n = 0, 1, 2, 3...$$

(ii) They obey the differential equation
$$H''_n(q) - 2qH'_n(q) + 2n H_n(q) = 0$$

(iii) Generating function, $\exp(-z^2 + 2qz) = \sum_{n=0}^{\infty} H_n(q) \frac{z^n}{n!}$

(iv) Recurrence relations between Hermite polynomials of different orders:
  (a) $H'_n(q) = 2n H_{n-1}(q)$
  (b) $H''_n(q) = 4n(n-1) H_{n-2}(q)$
  (c) $H_{n+1}(q) - 2q H_n(q) + 2n H_{n-1}(q) = 0$

(v) $H_n(-q) = (-1)^n H_n(q)$

(vi) $\int_{-\infty}^{\infty} H_n(q) H_m(q) \exp(-q^2) dq = 0$ for $m \neq n$

$\qquad\qquad\qquad\qquad\qquad = 2^n n! \sqrt{\pi}$ for $m = n$

(vii) $\int_{-\infty}^{\infty} H_n(q) q H_m(q) \exp(-q^2) dq = 2^{n-1} n! \sqrt{\pi}$ for $m = n-1$

$\qquad\qquad\qquad\qquad\qquad = 2^n (n+1)! \sqrt{\pi}$ for $m = n+1$
$\qquad\qquad\qquad\qquad\qquad = 0$ for $m = n$

(viii) $\int_{-\infty}^{\infty} H_n(q) q^2 H_m(q) \exp(-q^2) dq = (2n+1) 2^{n-1} n! \sqrt{\pi}$ for $m = n$

APPENDIX A–II

LEGENDRE POLYNOMIALS AND ASSOCIATED LEGENDRE FUNCTIONS

(i) Lengendre polynomials are defined as follows:

$$P_l(\mu) = \frac{1}{2^l l!} \left(\frac{d}{d\mu} \right)^l (\mu^2 - 1)^l \qquad \ldots (A2 \cdot 1)$$

with $l = 0, 1, 2, 3, \ldots$

(ii) Associated Legendre functions are defined as follows:

$$P_l^m(\mu) = \frac{1}{2^l l!} (1 - \mu^2)^{m/2} \left(\frac{d}{d\mu} \right)^{l+m} (\mu^2 - 1)^l$$

$$= (1 - \mu^2)^{m/2} \left(\frac{d}{d\mu} \right)^m P_l(\mu) \qquad \ldots (A2 \cdot 2)$$

with $m = 0, 1, 2, \ldots l$. Note that this is defined for positive $m \leq l$.

For $m = 0$, $P_l^0(\mu) = P_l(\mu)$

(iii) Generating functions:

$$(1 - 2z\mu + z^2)^{-1/2} = \sum_{l=0}^{\infty} P_l(\mu) z^l, \quad |z| < 1 \qquad \ldots (A2 \cdot 3)$$

$$\frac{(2m)! (1 - \mu^2)^{m/2} z^m}{2^m m! (1 - 2z\mu + z^2)^{m+1/2}} = \sum_{l=m}^{\infty} P_l^m(\mu) z^l \qquad (A2 \cdot 4)$$

(iv) Differential equations for $P_l(\mu)$ and $P_l^m(\mu)$:

$$(1 - \mu^2) \frac{d^2 P_l}{d\mu^2} - 2\mu \frac{dP_l}{d\mu} + l(l+1) P_l = 0 \qquad \ldots (A2 \cdot 5)$$

$$(1 - \mu^2) \frac{d^2 P_l^m}{d\mu^2} - 2\mu \frac{dP_l^m}{d\mu} + \left\{ l(l+1) - \frac{m^2}{1 - \mu^2} \right\} P_l^m = 0 \quad \ldots (A2 \cdot 6)$$

Note that the differential equation (11·14-2) satisfied by Θ is the same as Eq. (A2·6) with $\lambda = l(l+1)$. Eq. (A2·5) is known as the Legendre equation.

(v) $P_l(-\mu) = (-1)^l P_l(\mu)$

$P_l^m(-\mu) = (-1)^{l+m} P_l^m(\mu)$

(vi) $\int_{-1}^{1} P_l(\mu) P_{l'}(\mu) d\mu = \frac{2}{2l+1} \delta_{ll'} \qquad \ldots (A2 \cdot 7)$

where $\delta_{ll'} = 1$ for $l = l'$ and $\delta_{ll'} = 0$ for $l \neq l'$.

$$\int_{-1}^{1} P_l^m(\mu) P_{l'}^m(\mu) \, d\mu = \frac{2}{2l+1} \cdot \frac{(l+m)!}{(l-m)!} \delta_{ll'} \quad \ldots (A2\cdot 8)$$

(vii) *Recurrence relations*

$$(l+1) P_{l+1}(\mu) = (2l+1) \mu P_l(\mu) - l P_{l-1}(\mu) \quad \ldots (A2\cdot 9)$$

$$\mu P'_l(\mu) - P'_{l-1}(\mu) = l P_l(\mu) \quad \ldots (A2\cdot 10)$$

$$P'_{l+1}(\mu) - P'_{l-1}(\mu) = (2l+1) P_l(\mu) \quad \ldots (A2\cdot 11)$$

$$(\mu^2 - 1) P'_l(\mu) = l\mu P_l(\mu) - l P_{l-1}(\mu) \quad \ldots (A2\cdot 12)$$

$$P'_{l+1}(\mu) - \mu P'_l(\mu) = (l+1) P_l(\mu) \quad \ldots (A2\cdot 13)$$

where the primes (′) indicate the first differential coefficients with respect to μ.

(viii) A few values of the Legendre polynomials are :

$$P_0(\mu) = 1 \qquad P_3(\mu) = \tfrac{1}{2}(5\mu^3 - 3\mu) \quad \ldots (A2\cdot 14)$$

$$P_1(\mu) = \mu \qquad P_4(\mu) = \tfrac{1}{8}(35\mu^4 - 30\mu^2 + 3)$$

$$P_2(\mu) = \tfrac{1}{2}(3\mu^2 - 1) \qquad P_5(\mu) = \tfrac{1}{8}(63\mu^5 - 70\mu^3 + 15\mu)$$

For any l, $P_l(+1) = 1$, $P_l(-1) = (-1)^l$ $\quad \ldots (A2\cdot 15)$

$$P_l^m(+1) = P_l^m(-1) = 0 \text{ for } m \neq 0 \quad \ldots (A2\cdot 16)$$

APPENDIX A-III

SPHERICAL HARMONICS

(i) Defining relation

$$Y_l^m(\theta, \phi) = \sqrt{\frac{2l+1}{4\pi} \cdot \frac{(l-m)!}{(l+m)!}} \, (-1)^m P_l^m(\cos\theta) \, \exp(im\phi)$$

.... (A3·1)

The above definition is for positive $m \leq l$.
For negative m (with the restriction $|m| < l$), we have

$$Y_l^{-m}(\theta, \phi) = (-1)^m Y_l^{m*}(\theta, \phi) \qquad \text{.... (A3·2)}$$

where Y_l^{m*} is the complex conjugate of $Y_l^m(\theta, \phi)$.

(ii) Differential equation for $Y_l^m(\theta, \phi)$:

$$\frac{l}{\sin\theta} \frac{\partial}{\partial\theta}\left\{\sin\theta \frac{\partial}{\partial\theta} Y_l^m(\theta, \phi)\right\} + \frac{1}{\sin^2\theta} \frac{\partial^2}{\partial\phi^2} Y_l^m(\theta, \phi)$$

$$= -l(l+1) Y_l^m(\theta, \phi) \qquad \text{.... (A3·3)}$$

Note that for a central potential, $Y_l^m(\theta, \phi)$ gives the angular part of the wave function. Eq. (11·12-9) for the angular part is the same as Eq. (A3·3) with $\lambda = l(l+1)$.

(iii) $\int_0^{2\pi} d\phi \int_0^{\pi} Y_l^{m*} Y_{l'}^{m'} \sin\theta \, d\theta = \delta_{ll'} \delta_{mm'}$ .... (A3·4)

where $\delta_{ll'} = 1$ for $l = l'$ and $\delta_{ll'} = 0$ for $l \neq l'$

$\delta_{mm'} = 1$ for $m = m'$ and $\delta_{mm'} = 0$ for $m \neq m'$

APPENDIX A–IV

LAGUERRE POLYNOMIALS AND ASSOCIATED LAGUERRE POLYNOMIALS

(i) Laguerre polynomials are defined by the relation

$$L_n(\rho) = \exp(\rho) \left(\frac{d}{d\rho}\right)^n \left\{ \exp(-\rho) \rho^n \right\} \quad \ldots \text{(A4·1)}$$

$$= (-1)^n \left\{ \rho^n - \frac{n^2}{1!} \rho^{n-1} + \frac{n^2(n-1)^2}{2!} \rho^{n-2} + \ldots + (-1)^n n! \right\}$$

where n is an integer.

(ii) Associated Laguerre polynomial of degree $(n-k)$ is defined by the relation

$$L_n^k(\rho) = \left(\frac{d}{d\rho}\right)^k L_n(\rho) \quad \ldots \text{(A4·2)}$$

where $k > 0$ is an integer.

(iii) Generating functions :

$$\frac{\exp\{-\rho z/(1-z)\}}{1-z} = \sum_{n=0}^{\infty} \frac{L_n(\rho)}{n!} z^n, \, |z| < 1 \quad \ldots \text{(A4·3)}$$

$$\frac{(-z)^k \exp\{-\rho z/(1-z)\}}{(1-z)^{k+1}} = \sum_{n=k}^{\infty} \frac{L_n^k(\rho)}{n!} z^n \quad \ldots \text{(A4·4)}$$

(iv) Differential equations for $L_n(\rho)$ and $L_n^k(\rho)$:

$$\rho \frac{d^2 L_n}{d\rho^2} + (1-\rho) \frac{dL_n}{d\rho} + nL_n = 0 \quad \ldots \text{(A4·5)}$$

$$\rho \frac{d^2 L_n^k}{d\rho^2} + (k+1-\rho) \frac{dL_n^k}{d\rho} + (n-k) L_n^k = 0 \quad \ldots \text{(A4·6)}$$

(v) $$\int_0^{\infty} \exp(-\rho) \rho^{2l} \{L_{n+l}^{2l+1}(\rho)\}^2 \rho^2 d\rho = \frac{2n\{(n+l)!\}^3}{(n-l-1)!} \quad \ldots \text{(A4·7)}$$

(vi) $L_{n+l}^{2l+1}(\rho)$ is given explicitly by

$$L_{n+l}^{2l+1}(\rho) = \sum_{k=0}^{n-l-1} \frac{(-1)^{k+1}\{(n+l)!\}^2 \rho^k}{(n-l-1-k)!(2l+1+k)!k!} \quad \ldots \text{(A4·8)}$$

APPENDIX A–V

List of hydrogenic total wave functions

$$\psi_{100} = \frac{1}{\sqrt{\pi}} \left(\frac{Z}{a_0}\right)^{3/2} \exp(-\sigma)$$

$$\psi_{200} = \frac{1}{4\sqrt{2\pi}} \left(\frac{Z}{a_0}\right)^{3/2} (2-\sigma) \exp(-\sigma/2)$$

$$\psi_{210} = \frac{1}{4\sqrt{2\pi}} \left(\frac{Z}{a_0}\right)^{3/2} \sigma \exp(-\sigma/2) \cdot \cos\theta$$

$$\psi_{21\pm 1} = \mp \frac{1}{8\sqrt{\pi}} \left(\frac{Z}{a_0}\right)^{3/2} \sigma \exp(-\sigma/2) \cdot \sin\theta \exp(\pm i\phi)$$

$$\psi_{300} = \frac{1}{81\sqrt{3\pi}} \left(\frac{Z}{a_0}\right)^{3/2} (27 - 18\sigma + 2\sigma^2) \exp(-\sigma/3)$$

$$\psi_{310} = \frac{1}{81} \sqrt{\frac{2}{\pi}} \left(\frac{Z}{a_0}\right)^{3/2} \sigma(6-\sigma) \exp(-\sigma/3) \cdot \cos\theta$$

$$\psi_{31,\pm 1} \pm \frac{1}{81\sqrt{\pi}} \left(\frac{Z}{a_0}\right)^{3/2} \sigma(6-\sigma) \exp(-\sigma/3) \cdot \sin\theta \exp(\mp i\phi)$$

$$\psi_{320} = \frac{1}{81\sqrt{6\pi}} \left(\frac{Z}{a_0}\right)^{3/2} \sigma^2 \exp(-\sigma/3) \cdot (3\cos^2\theta - 1)$$

$$\psi_{32\pm 1} = \pm \frac{1}{81\sqrt{\pi}} \left(\frac{Z}{a_0}\right)^{3/2} \sigma^2 \exp(-\sigma/3) \cdot \sin\theta \cos\theta \exp(\pm i\phi)$$

$$\psi_{32,\pm 2} = \frac{1}{162\sqrt{\pi}} \left(\frac{Z}{a_0}\right)^{3/2} \sigma^2 \exp(-\sigma/3) \cdot \sin^2\theta \exp(\pm 2i\phi)$$

Here $\sigma = Zr/a_0$; $a_0 = 4\pi\hbar^2\varepsilon_0/me^2$.

APPENDIX A–VI

TABLE OF IMPORTANT PHYSICAL CONSTANTS

Atomic mass unit and its equivalent rest energy
$1\,u = 1.66057 \times 10^{-27}$ kg
$= 931.502$ MeV

Avogadro number
$N_0 = 6.02205 \times 10^{23}$ molecules/mole

Bohr magneton
$\mu_B = 9.2741 \times 10^{-24}$ joule/tesla

Bohr radius
$a_0 = 0.529177 \times 10^{-10}$ m

Boltzmann costant
$k = 1.3807 \times 10^{-23}$ joule/K

Compton wavelength for the electron
$\lambda_c = 2.42631 \times 10^{-12}$ m

Electronic charge
$e = 1.60219 \times 10^{-19}$ Coulomb

Electron rest mass and rest energy
$m_e = 9.10953 \times 10^{-31}$ kg
$m_e c^2 = 0.511003$ MeV

Electron-volt
$1\,\text{eV} = 1.60219 \times 10^{-19}$ joule

Fine structure constant
$\alpha = 7.29735 \times 10^{-3}$
$= \dfrac{1}{137.03605}$

Planck's constant
$h = 6.62618 \times 10^{-34}$ joule-second.
$\hbar = h/2\pi = 1.05459 \times 10^{-34}$ joule-second.

Ratio of proton to electron mass
$m_p/m_e = 1836.15$

Rydberg constant for infinite mass
$R_\infty = 1.0973732 \times 10^7\,m^{-1}$

Universal gravitational constant
$G = 6.6720 \times 10^{-11}$ newton m^2/kg^2

Velocity of light in vaccum
$c = 2.997925 \times 10^8$ m/sec

Permeabity of free space
$\mu_0 = 4\pi \times 10^{-7}$ H/m
$= 1.25664 \times 10^{-6}$ H/m

Permittivity of free space
$\epsilon_0 = 8.85419 \times 10^{-12}$ F/m

Classical electron radius
$r_0 = 2.81794 \times 10^{-15}$ m

APPENDIX A–VII. PERIODIC TABLE OF ELEMENTS

I	II	III	IV	V	VI	VII	VIII		
1 H 1.0080							2 He 4.0026		
3 Li 6.939	4 Be 9.0122	5 B 10.811	6 C 12.011	7 N 14.007	8 O 15.999	9 F 18.998	10 Ne 20.183		
11 Na 22.990	12 Mg 24.312	13 Al 26.981	14 Si 28.086	15 P 30.974	16 S 32.064	17 Cl 35.453	18 Ar 39.948		
19 K 39.102	20 Ca 40.08	21 Sc 44.956	22 Ti 47.90	23 V 50.942	24 Cr 51.996	25 Mn 54.938	26 Fe 55.847	27 Co 58.933	28 Ni 58.71
29 Cu 63.54	30 Zn 65.37	31 Ga 69.72	32 Ge 72.59	33 As 74.922	34 Se 78.96	35 Br 79.909	36 Kr 83.80		
37 Rb 85.47	38 Sr 87.62	39 Yt 88.905	40 Zr 91.22	41 Nb 92.906	42 Mo 95.94	43 Tc 99	44 Ru 101.07	45 Rh 102.905	46 Pd 106.4
47 Ag 107.870	48 Cd 112.40	49 In 114.82	50 Sn 118.69	51 Sb 121.75	52 Te 127.60	53 I 126.904	54 Xe 131.30		
55 Cs 132.905	56 Ba 137.34	57§ La 138.91	72 Hf 178.49	73 Ta 180.95	74 W 183.85	75 Re 186.2	76 Os 190.2	77 Ir 192.2	78 Pt 195.09
79 Au 196.97	80 Hg 200.59	81 Tl 204.37	82 Pb 207.19	83 Bi 208.98	84 Po 210*	85 At 210	86 Rn 222		
87 Fr 223	88 Ra 226	89§ Ac 227	104 Ku	105 Ha					

* In case of elements having no stable or very long-lived isotopes, the mass numbers given are those for the most well-known isotope having the longest half-life

§ For the elements in the lanthanide group and in the actinide group see next page.

§ 58 Ce 140.12	59 Pr 140.91	60 Nd 144.24	61 Pm 251	62 Sm 150.35	63 Eu 151.96	64 Gd 157.25	65 Tb 158.92	66 Dy 162.50	67 Ho 164.93	68 Er 167.26	69 Tm 168.93	70 Yb 173.04	71 Lu 174.97
§§ 90 Th 232.04	91 Pa 231	92 U 238.03	93 Np 237	94 Pu 242	95 Am 243	96 Cm 247	97 Bk 249	98 Cf 251	99 Es 254	100 Fm 253	101 Md 257	102 No 255	103 Lw 257

Atomic weights are given on the carbon 12 scale

§ These 14 elements belong to the lanthanide (rare earth) group.
§§ These 14 elements belong to the actinide group.

APPENDIX A–VIII

EVALUATION OF THE INTEGRALS IN THE CALCULATION OF THE GROUND STATE ENERGY OF HELIUM

(a) We evaluate the integral in Eq. (13·9-10) of Ch. XIII for the calculation of the perturbation in energy $W_0^{(1)}$ of the ground state of the two electron system. We rewrite Eq. (13·9-10) as

$$W_0^{(1)} = H'_{00} = <u_0 | H' | u_0>$$

$$= \int u_0^* \, H' \, u_0 \, d\tau_1 \, d\tau_2 \quad (A8 \cdot 1)$$

where u_0 is the unperturbed wave function for the 1s state of the two–electron atom (see Eq. 13·9-19) given by

$$u_0 = \Psi_{100} \, \Psi_{100} = \frac{Z^3}{\pi a_0^3} \exp\left(-\frac{\rho_1 + \rho_2}{2}\right) \quad (A8 \cdot 2)$$

$$\Psi_{100} = \Psi_{1s} = \left(\frac{Z^3}{\pi a_0^3}\right)^{1/2} \exp(-\rho/2) \quad (A8 \cdot 3)$$

for each electron in the 1s state of the two-electron atom of nuclear charge Ze; $\rho = 2Zr/a_0$ and $a_0 = 4\pi\epsilon_0 \hbar^2/me^2$ is the Bohr radius. The volume elements available for the two electrons are

$$d\tau_1 = r_1^2 \sin\theta_1 \, dr_1 \, d\theta_1 \, d\varphi_1 \quad (A8 \cdot 4)$$

$$d\tau_2 = r_2^2 \sin\theta_2 \, dr_2 \, d\theta_2 \, d\varphi_2 \quad (A8 \cdot 5)$$

Since $r = (a_0/2Z)\rho$, we have

$$r_1^2 \, dr_1 = \left(\frac{a_0}{2Z}\right)^3 \rho_1^2 \, d\rho_1 \text{ and } r_2^2 \, dr_2 = \left(\frac{a_0}{2Z}\right)^3 \rho_2^2 \, d\rho_2 \quad (A8 \cdot 6)$$

Since the perturbation $H' = e^2/4\pi\epsilon_0 r_{12}$ we get

$$W_0^{(1)} = \left(\frac{Z^3}{\pi a_0^3}\right)^2 \int \exp\{-(\rho_1 + \rho_2)\} \frac{e^2}{4\pi\epsilon_0 r_{12}}$$

$$\times r_1^2 \sin\theta_1 \, dr_1 \, d\theta_1 \, d\varphi_1 \cdot r_2^2 \sin\theta_2 \, dr_2 \, d\theta_2 \, d\varphi_2$$

$$= \left(\frac{Z^3}{\pi a_0^3}\right)^2 \left(\frac{a_0}{2Z}\right)^6 \frac{2Z}{a_0} \int\int\int\int\int\int \frac{\exp\{-(\rho_1 + \rho_2)\}}{4\pi\epsilon_0 \rho_{12}}$$

$$\times \rho_1^2 \sin\theta_1 \, d\rho_1 \, d\theta_1 \, d\varphi_1 \cdot \rho_2^2 \sin\theta_2 \, d\rho_2 \, d\theta_2 \, d\varphi_2 \quad (A8 \cdot 7)$$

The ρ integrals represent the electrostatic energy of two spherically symmetric charge distributions with the density functions $\exp(-\rho_1)$ and $\exp(-\rho_2)$.

The potential due to a spherical shell of radius ρ and thickness $d\rho$ containing the charge $4\pi\rho^2 \exp(-\rho) d\rho$ at a distance r from the centre is

$$V_r = \frac{4\pi\rho^2 \exp(-\rho) d\rho}{4\pi\epsilon_0 \rho} \text{ for } r < \rho$$

and $\quad V_r = \dfrac{4\pi\rho^2 \exp(-\rho) d\rho}{4\pi\epsilon_0 r}$ for $r > \rho$

So the potential for the complete distribution is

$$V(r) = \frac{4\pi}{4\pi\epsilon_0 r} \int_0^r \exp(-\rho)\rho^2 d\rho + \frac{4\pi}{4\pi\epsilon_0} \int_r^\infty \frac{\exp(-\rho)\rho^2}{\rho} d\rho$$

$$= \frac{1}{\epsilon_0 r} \int_0^r \exp(-\rho)\rho^2 d\rho + \frac{1}{\epsilon_0} \int_r^\infty \exp(-\rho)\cdot\rho\, d\rho \quad (A8\cdot 8)$$

These integrals can be easily evaluated:

$$\int_r^\infty \rho \exp(-\rho) d\rho = [-\rho \exp(-\rho)]_r^\infty + \int_r^\infty \exp(-\rho) d\rho$$

$$= r\exp(-r) + \exp(-r) = (r+1)\exp(-r)$$

$$\int_0^r \rho^2 \exp(-\rho) d\rho = [-\rho^2 \exp(-\rho)]_0^r + 2\int_0^r \rho \exp(-\rho) d\rho$$

$$= -r^2 \exp(-r) + 2[-\rho \exp(-\rho) - \exp(-\rho)]_0^r$$

$$= -r^2 \exp(-r) - 2r \exp(-r) - 2 \exp(-r) + 2$$

$$= -(r^2 + 2r + 2)\exp(-r) + 2$$

$$\therefore V(r) = \frac{1}{\epsilon_0}\left[\left(-r - 2 - \frac{2}{r} + r + 1\right)\exp(-r) + \frac{2}{r}\right]$$

$$= \frac{1}{\epsilon_0}\left[\frac{2}{r} - \left(1 + \frac{2}{r}\right)\exp(-r)\right] \quad (A8\cdot 9)$$

If we put $r = \rho_2$, we get from Eq. (A8·9) the potential due to a spherically symmetric charge distribution of density $\exp(-\rho_1)$ at a distance $r = \rho_2$ from the centre. The Coulomb energy of a second charge distribution of density $\exp(-\rho_2)$ in the field of the first distribution can then be obtained by considering a spherical shell of radius ρ_2 and thickness $d\rho_2$ containing an amount of charge $4\pi\rho_2^2 \exp(-\rho_2) d\rho_2$ in the

Appendix

Coulomb field of the first distribution. The potential at $r = \rho_2$ due to the latter is

$$V_1(\rho_2) = \frac{1}{\epsilon_0}\left[\frac{2}{\rho_2} - \left(1 + \frac{2}{\rho_2}\right)\exp(-\rho_2)\right] \quad (A8\cdot10)$$

Then the Coulomb energy of the second charge distribution of density $\exp(-\rho_2)$ in this field will be

$$V = \int_0^\infty V_1(\rho_2)\exp(-\rho_2)\, 4\pi\rho_2^2\, d\rho_2$$

$$= \frac{4\pi}{\epsilon_0}\int_0^\infty \left[\frac{2}{\rho_2} - \left(1 + \frac{2}{\rho_2}\right)\exp(-\rho_2)\right]\exp(-\rho_2)\rho_2^2\, d\rho_2$$

$$= \frac{4\pi}{\epsilon_0}\left[2\int_0^\infty \rho_2\exp(-\rho_2)\, d\rho_2 - \int_0^\infty \rho_2^2 \exp(-2\rho_2)\, d\rho_2\right.$$

$$\left. - 2\int_0^\infty \rho^2 \exp(-2\rho_2)\, d\rho_2\right]$$

$$= \frac{4\pi}{\epsilon_0}\left(2 \times 1 - \frac{2}{4} - \frac{1}{4}\right) = \frac{5\pi}{\epsilon_0} \quad (A8\cdot11)$$

$$\therefore\quad W_0^{(1)} = \frac{2Ze^2}{64\pi^2 a_0}\cdot\frac{5\pi}{\epsilon_0} = \frac{5Ze^2}{32\pi\epsilon_0 a_0} = -\frac{5}{4}Z E_H \quad (A8\cdot12)$$

where $\quad E_H = -\dfrac{e^2}{8\pi\epsilon_0 a_0} = -13\cdot6\text{ eV} \quad (A8\cdot13)$

E_H is the ground state energy of the hydrogen atom. Eq. (A8·12) gives the expression for the first order perturbation in the energy of the ground state of a two electron atom.

(See Eq. 13·9-11 in Ch. XIII).

(b) We now evaluate the integrals appearing in Eq. (13·9-21) in Ch. XIII for the calculation of the ground state energy of the two-electron system by variational method. We rewrite Eq. (13·9-21) below:

$$<H> = \int |\Psi_{100}(\mathbf{r}_2)|^2 d\tau_2 \int \Psi_{100}^*(\mathbf{r}_1)\left(-\frac{\hbar^2}{2m}\nabla_1^2 - \frac{Z'e^2}{4\pi\epsilon_0 r_1}\right)$$

$$\times \Psi_{100}(\mathbf{r}_1)\, d\tau_1 + \int |\Psi_{100}(\mathbf{r}_1)|^2 d\tau_1 \int \Psi_{100}^*(\mathbf{r}_2) \times$$

$$\left(-\frac{\hbar^2}{2m}\nabla_2^2 - \frac{Z'e^2}{4\pi\epsilon_0 r_2}\right)\Psi_{100}(\mathbf{r}_2)\, d\tau_2$$

$$+ (Z' - Z) \frac{e^2}{4\pi \epsilon_0} \int\int_{\tau_1 \tau_2} \Psi^* \left(\frac{1}{r_1} + \frac{1}{r_2}\right) \Psi \, d\tau_1 \, d\tau_2$$

$$+ \int\int_{\tau_1 \tau_2} \Psi^* \frac{e^2}{4\pi \epsilon_0 r_{12}} \Psi \, d\tau_1 \, d\tau_2 \qquad (A8\cdot 14)$$

Here $\Psi = \Psi_{100}(r_1) \Psi_{100}(r_2)$. (see Eq. 139–15).

Since the wave functions $\Psi_{100}(r_1)$ and $\Psi_{100}(r_2)$ are normalized, the first integrals in the first and second terms are unity each:

$$\int_{\tau_1} |\Psi_{100}(r_1)|^2 \, d\tau_1 = \int |\Psi_{100}(r_2)|^2 \, d\tau_2 = 1$$

The second integrals in these terms are the expectation values $<H>_1$, and $<H>_2$ respectively of the Hamiltonians for the two electrons in the ground state of the hydrogen–like atom of nuclear charge $Z'e$. These are given by

$$<H>_1 = \int_{\tau_1} \Psi_{100}^*(r_1) \left(-\frac{\hbar^2}{2m} \nabla_1^2 - \frac{Z' e^2}{4\pi \epsilon_0 r_1}\right) \Psi_{100}(r_1) \, d\tau_1 = \epsilon_1 \qquad (A8\cdot 15a)$$

$$<H>_2 = \int_{\tau_2} \Psi_{100}^*(r_2) \left(-\frac{\hbar^2}{2m} \nabla_2^2 - \frac{Z' e^2}{4\pi \epsilon_0 r_2}\right) \Psi_{100}(r_2) \, d\tau_2 = \epsilon_2 \qquad (A8\cdot 15b)$$

Here $\epsilon_1 = \epsilon_2 = Z'^2 E_H$.

We now calculate the third term in Eq. (A 8-14). We take the nuclear charge to be Z'.

$$\int\int_{\tau_1 \tau_2} \Psi^* \frac{1}{r_1} \Psi \, d\tau_1 \, d\tau_2$$

$$= \int\int_{\tau_1 \tau_2} \Psi_{100}^*(r_1) \Psi_{100}^*(r_2) \Psi_{100}(r_1) \Psi_{100}(r_2) \frac{d\tau_1 \, d\tau_2}{r_1}$$

$$= \int_{\tau_2} |\Psi_{100}(r_2)|^2 \, d\tau_2 \int_{\tau_1} \frac{|\Psi_{100}(r_1)|^2}{r_1} \, d\tau_1$$

$$= \int_{\tau_1} \frac{|\Psi_{100}(r_1)|^2}{r_1} \, d\tau_1 = \frac{1}{\pi} \left(\frac{Z'}{a_0}\right)^3 \int_0^\infty \exp\left(-\frac{2Z' r_1}{a_0}\right) \frac{4\pi r_1^2 \, dr_1}{r_1}$$

$$= \frac{4\pi}{\pi} \left(\frac{Z'}{a_0}\right)^3 \int_0^\infty \exp\left(-\frac{2Z' r_1}{a_0}\right) r_1 \, dr_1$$

$$= 4\left(\frac{Z'}{a_0}\right)^3 \times \left(\frac{a_0}{2Z'}\right)^2 = \frac{Z'}{a_0}$$

Similarly $\int\int_{\tau_1 \tau_2} \Psi^* \frac{1}{r_2} \Psi \, d\tau_1 d\tau_2 = \frac{Z'}{a_0}$

So the third term is

$$(Z'-Z)\frac{e^2}{4\pi\epsilon_0} \int\int_{\tau_1 \tau_2} \Psi^* \left(\frac{1}{r_1}+\frac{1}{r_2}\right)\Psi \, d\tau_1 \, d\tau_2$$

$$= (Z'-Z)\frac{e^2}{4\pi\epsilon_0} \cdot \frac{2Z'}{a_0} = 4Z'(Z'-Z)\frac{e^2}{8\pi a_0 \epsilon_0}$$

$$= -4Z'(Z'-Z)E_H \tag{A8.16}$$

The last term in Eq. (A 8-14) has already been evaluated (see Eq. A8·12). Here we write Z' in place of Z. We have

$$\int\int_{\tau_1 \tau_2} \Psi^* \frac{e^2}{4\pi\epsilon_0 r_{12}} \Psi \, d\tau_1 \, d\tau_2 = -\frac{5}{4} Z' E_H \tag{A8.17}$$

Combining the result expressed by Eqs. (A8·15a), (A8·15b), (A8·16) and (A8·17), we then get

$$<H> = 2Z'^2 E_H - 4Z'(Z'-Z) E_H - \frac{5}{4} Z' E_H$$

$$= \left\{ 2Z'^2 - 4Z'(Z'-Z) - \frac{5}{4} Z' \right\} E_H \tag{A8.18}$$

This is the same as Eq. (13·9-22) in § 13·9.

APPENDIX A–IX

HAMILTONIAN OF A CHARGED PARTICLE IN AN ELECTROMAGNETIC FIELD

The Lorentz force acting on an electron is

$$\mathbf{F} = e\,(\mathbf{X} + \mathbf{v} \times \mathbf{B}) \tag{A9.1}$$

where $\mathbf{X}$ is the electric field and $\mathbf{B}$ is the magnetic induction field. $\mathbf{v}$ is the velocity of the particle. We can then write

$$m\frac{d\mathbf{v}}{dt} = e\,(\mathbf{X} + \mathbf{v} \times \mathbf{B}) \tag{A9.2}$$

where $\mathbf{X} = -\nabla \phi - \partial \mathbf{A}/\partial t$ and $\mathbf{B} = \nabla \times \mathbf{A}$.

$\mathbf{A}$ is magnetic vector potential and ϕ is the scalar potential.

Eq. (A9.2) can be written in Lagrangian form of classical mechanics:

$$\frac{d}{dt}\frac{\partial L}{\partial v_i} - \frac{\partial L}{\partial x_i} = 0 \tag{A9.3}$$

with $i = 1, 2$ or 3 v_i denotes the Cartesian components of the velocity vector $\mathbf{v}$: $v_1 = v_x$, $v_2 = v_y$ and $v_3 = v_z$. Similarly $x_1 = x, x_2 = y, x_3 = z$ are the three components of the position vector $\mathbf{r}$. $L(x_i, v_i)$ is the classical Lagrangian given by $L = T - V$ where T is the kinetic energy and V is the potential energy.

We make use of the following vector identity

$$\nabla(\mathbf{v} \cdot \mathbf{A}) = \mathbf{v} \times \nabla \times \mathbf{A} + \mathbf{A} \times \nabla \times \mathbf{v} + (\mathbf{A} \cdot \nabla)\mathbf{v} + (\mathbf{v} \cdot \nabla)\mathbf{A}$$

$$= \mathbf{v} \times \nabla \times \mathbf{A} + (\mathbf{v} \cdot \nabla)\mathbf{A} \tag{A9.4}$$

since $\mathbf{v}$ is independent of the coordinates. Then

$$m\frac{d\mathbf{v}}{dt} = -e\nabla \phi - e\frac{\partial \mathbf{A}}{\partial t} + e\mathbf{v} \times \nabla \times \mathbf{A}$$

$$= -e\nabla \phi - e\left\{\frac{\partial \mathbf{A}}{\partial t} + (\mathbf{v} \cdot \nabla)\mathbf{A}\right\} + e\nabla(\mathbf{v} \cdot \mathbf{A}) \tag{A9.5}$$

But we can write

$$\frac{d\mathbf{A}}{dt} = \frac{\partial \mathbf{A}}{\partial t} + (\mathbf{v} \cdot \nabla)\mathbf{A} \tag{A9.6}$$

which is the time rate of change of $\mathbf{A}$ as observed by the moving particle. So we get

$$\frac{d}{dt}(m\mathbf{v} + e\mathbf{A}) = -e\nabla \phi + e\nabla(\mathbf{v} \cdot \mathbf{A}) \tag{A9.7}$$

We can write the last term $e\nabla(\mathbf{v} \cdot \mathbf{A})$ in the above equation as

$$e\nabla(\mathbf{v} \cdot \mathbf{A}) = \frac{1}{2m}\nabla\{(m\mathbf{v} + e\mathbf{A})^2 - e^2 A^2\} \tag{A9.8}$$

Appendix

Here we have made use of the fact that $\mathbf{v}$ is an independent variable. Eq. (A9·7) now takes the form

$$\frac{d}{dt}(m\mathbf{v}+e\mathbf{A}) - \nabla\left\{\frac{1}{2m}(m\mathbf{v}+e\mathbf{A})^2 - \frac{e^2}{2m}A^2 - e\phi\right\} = 0 \qquad (A9·9)$$

If we write

$$L = \frac{1}{2m}(m\mathbf{v}+e\mathbf{A})^2 - \frac{e^2}{2m}A^2 - e\phi \qquad (A9·10)$$

then

$$\frac{\partial L}{\partial v_i} = \frac{d}{dt}(m\mathbf{v}+e\mathbf{A})_i \qquad (A9·11)$$

$$\frac{\partial L}{\partial x_i} = \frac{\partial}{\partial x_i}\left\{\frac{1}{2m}(m\mathbf{v}+e\mathbf{A})^2 - \frac{e^2}{2m}A^2 - e\phi\right\} \qquad (A9·12)$$

So Eq. (A9·9) has the form of the Lagrangian equation of motion (Eq. A9·3). Hence the momentum conjugate to x_i is

$$p_i = \left(\frac{\partial L}{\partial v_i}\right)_\mathbf{p} = (m\mathbf{v}+e\mathbf{A})_i$$

and

$$\mathbf{p} = m\mathbf{v} + e\mathbf{A} \qquad (A9·13)$$

The Hamiltonian of the system is given by

$$H = \mathbf{p}\cdot\mathbf{v} - L$$

Since

$$\mathbf{v} = \frac{1}{2m}(p - e\mathbf{A})$$

we then get for a charged particle in an e.m. field the Hamiltonian function (non-relativistic)

$$H(\mathbf{p},\mathbf{r}) = \frac{1}{2m}(\mathbf{p}-e\mathbf{A})^2 + e\phi \qquad (A9·14)$$

In quantum mechanics we have to take the operator representation for H and write

$$\hat{H} = \frac{1}{2m}(\hat{\mathbf{p}} - e\mathbf{A})^2 + e\phi \qquad (A9·15)$$

where

$$\hat{\mathbf{p}} = \frac{\hbar}{i}\nabla, \quad \hat{\mathbf{r}} = \mathbf{r}$$

Then

$$\hat{H} = \frac{1}{2m}\left(\frac{\hbar}{i}\nabla - e\mathbf{A}\right)^2 + e\phi \qquad (A9·16)$$

The Schrödinger equation then becomes

$$\frac{1}{2m}\left(\frac{\hbar}{i}\nabla - e\mathbf{A}\right)^2\Psi + e\phi\Psi = E\Psi \qquad (A9·17)$$

where Ψ is the wave function.

APPENDIX A-X
PAULI'S THEORY OF SPIN

The spin of a spin 1/2 particle like the electron moving non-relativistically can be treated by Pauli's theory. Pauli introduced the spin operator $\vec{\sigma}$ related to the spin vector through the relation $s = (\hbar/2)\vec{\sigma}$.

The vector operator $\vec{\sigma}$ has the three components $\sigma_x, \sigma_y, \sigma_z$ which are 2×2 matrices (Pauli matrices) as given below (See *Introductory Quantum Mechanics*, Second Edition by the author):

$$\sigma_x = \begin{pmatrix} 0 & 1 \\ 1 & 0 \end{pmatrix}, \sigma_y = \begin{pmatrix} 0 & -i \\ i & 0 \end{pmatrix}, \sigma_z = \begin{pmatrix} 1 & 0 \\ 0 & -1 \end{pmatrix} \quad \text{(A10·1)}$$

Then $\sigma_x^2 = \sigma_y^2 = \sigma_z^2 = 1$ which is the 2×2 unit matrix

$$I = \begin{pmatrix} 1 & 0 \\ 0 & 1 \end{pmatrix} \quad \text{(A10·2)}$$

The two states of the particle ('spin up' and 'spin down') are the two component Pauli spinors

$$\alpha = \begin{pmatrix} 1 \\ 0 \end{pmatrix} \uparrow \, ; \, \beta = \begin{pmatrix} 0 \\ 1 \end{pmatrix} \downarrow \quad \text{(A10·3)}$$

Operation of α and β by the Pauli matrices gives the following results as can be easily verified by direct matrix multiplication:

$$\sigma_x \alpha = \beta \qquad \sigma_x \beta = \alpha$$
$$\sigma_y \alpha = i\beta \qquad \sigma_y \beta = -i\alpha$$
$$\sigma_z \alpha = \alpha \qquad \sigma_z \beta = -\beta \quad \text{(A10·4)}$$

We also have

$$\sigma^2 \alpha = 3\alpha \qquad \sigma^2 \beta = 3\beta \quad \text{(A10·5)}$$

We then get

$$s^2 \alpha = \frac{3\hbar^2}{4}\alpha = \frac{1}{2}\left(\frac{1}{2}+1\right)\hbar^2 \alpha = s(s+1)\hbar^2 \alpha$$

$$s^2 \beta = \frac{3\hbar^2}{4}\beta = \frac{1}{2}\left(\frac{1}{2}+1\right)\hbar^2 \beta = s(s+1)\hbar^2 \beta \quad \text{(A10·6)}$$

Also $\qquad s_z \alpha = \dfrac{\hbar}{2}\alpha, \, s_z \beta = -\dfrac{\hbar}{2}\beta \quad \text{(A10·7)}$

Thus α and β are the simultaneous eigenvectors of s^2 and s_z belonging to the eigenvalues $3\hbar^2/4$ and $\pm \hbar/2$ respectively, the plus sign corresponding to the 'spin up' ($\uparrow$) state and the minus sign corresponds to

Appendix

the 'spin down' ($\downarrow$) state.

The three components of σ anticommute :

$$\sigma_x \sigma_y + \sigma_y \sigma_x = 0, \ \sigma_y \sigma_z + \sigma_z \sigma_y = 0, \ \sigma_z \sigma_x + \sigma_x \sigma_z = 0 \qquad (A10\cdot 8)$$

We further have

$$\sigma_x \sigma_y - \sigma_y \sigma_x = 2i \sigma_z, \ \sigma_y \sigma_z - \sigma_z \sigma_y = 2i \sigma_x$$
$$\sigma_z \sigma_x - \sigma_x \sigma_z = 2i \sigma_y \qquad (A10\cdot 9)$$

These relations give

$$\sigma_x \sigma_y = i \sigma_z, \ \sigma_y \sigma_z = i \sigma_x, \sigma_z \sigma_x = i \sigma_y \qquad (A10\cdot 10)$$

APPENDIX A–XI
DIRAC'S THEORY OF THE RELATIVISTIC ELECTRON

Schrödinger wave equation discussed in Ch. XI is nonrelativistic and applies to particles for which relativistic variation of mass with velocity can be neglected. However for electrons this relativistic effect becomes important even at a kinetic energy above about 10 keV. This points to the obvious necessity of formulating a relativistic wave equation for the electron.

Klein-Gordon equation :

The simplest way to approach the problem would be to use the Schödinger operators for the energy and momentum :

$$\hat{E} \to i\hbar \frac{\partial}{\partial t}, \quad \hat{p} \to -i\hbar \nabla \qquad (A11 \cdot 1)$$

For a free particle we have

$$E^2 = p^2 c^2 + m^2 c^4 \qquad (A11 \cdot 2)$$

Writing $\hat{E}$ and $\hat{p}$ in operator forms (A11·1) and introducing the wave function Ψ, we then get

$$\left(i\hbar \frac{\partial}{\partial t} \right)^2 \Psi = \left\{ c^2 (-i\hbar \nabla)^2 + m^2 c^4 \right\} \Psi$$

or,

$$-\hbar^2 \frac{\partial^2 \Psi}{\partial t^2} = -c^2 \hbar^2 \nabla^2 \Psi + m^2 c^4 \Psi$$

so that

$$\nabla^2 \Psi - \frac{1}{c^2} \frac{\partial^2 \Psi}{\partial t^2} - \frac{m^2 c^2}{\hbar^2} \Psi = 0 \qquad (A11 \cdot 3)$$

This is known as the Kelin–Gordon equation. It is a second order differential equation w.r.t. time. It can be shown that it holds for particles without spin (e.g. for pions and K–mesons). It cannot be applied to the case of an electron which has spin 1/2.

Dirac equation for a free particle :

In special theory of relativity space coordinates (x, y, z) and time (t) appear symmetrically and hence the wave equation for the relativistic particle must also be symmetric in the space and time coordinates.

In Ch. X we saw that the wave equation for the particle must be first order in time so that if Ψ is known at some initial time t_0, then at any subsequent time t, $\Psi(\mathbf{r}, t)$ is uniquely determined. So according to the symmetry requirement mentioned in the previous paragraph, the wave equation must also be first order in the space coordinates i.e. linear in the components of the momentum operator $\hat{\mathbf{p}} = -i\hbar \nabla$.

The simplest Hamiltonian that Dirac chose which is linear in the momentum components as also in energy mc^2 is

$$\hat{H} = -c\,\vec{\alpha}\cdot\mathbf{p} - \beta mc^2$$
$$= -c\,(\alpha_1 p_1 + \alpha_2 p_2 + \alpha_3 p_3) - \beta mc^2 \qquad (A11\cdot 4)$$

$\alpha_1, \alpha_2, \alpha_3$ are the components of $\vec{\alpha}$ while $\hat{p}_1, \hat{p}_2, \hat{p}_3$ are the components of $\hat{\mathbf{p}}$.

Writing the operator representations (A11·1) for $\hat{\mathbf{p}}$ and the energy $\hat{E}$, we get, since $H\Psi = E\Psi$

$$(\hat{E} + c\,\vec{\alpha}\cdot\hat{\mathbf{p}} + \beta mc^2)\Psi = 0 \qquad (A11\cdot 5)$$

or,

$$\left(i\hbar\frac{\partial}{\partial t} - i\hbar c\,\vec{\alpha}\cdot\nabla + \beta mc^2\right)\psi = 0 \qquad (A11\cdot 6)$$

The operators $\vec{\alpha}$ and β must be independent of $\mathbf{r}$, t, $\mathbf{p}$ and E. So $\vec{\alpha}$ and β must commute with $\mathbf{r}$, t, $\mathbf{p}$ and E. However, $\vec{\alpha}$ and β may not commute with each other and hence may not be numbers.

It may be noted that any solution Ψ of the wave equation (A11·6) must also be a solution of Klein-Gordon equation (A11·3) though the converse need not be true.

Multiplying the wave equation (A11·5) on the left by the operator

$$E - c\,\vec{\alpha}\cdot\mathbf{p} - \beta mc^2$$

we get

$$(E - c\,\vec{\alpha}\cdot\mathbf{p} - \beta mc^2)(E + c\,\vec{\alpha}\cdot\mathbf{p} + \beta mc^2)\Psi = 0$$

This can be expanded as follows :

$$\left\{ E^2 - c^2\left[\sum_{j=1}^{3}\alpha_j^2 p_j^2 + \sum_{j=1}^{3}\sum_{k\neq j}^{\prime}(\alpha_j\alpha_k + \alpha_k\alpha_j)p_j p_k\right.\right.$$
$$\left.\left. - mc^3\sum_{j=1}^{3}(\alpha_j\beta + \beta\alpha_j)p_j - m^2 c^4 \beta^2\right]\right\}\Psi = 0 \qquad (A11\cdot 7)$$

In order that Eqs. (A11·2) and (A11·7) may be compatible, the following conditions must be satisfied :

$$\alpha_1^2 = \alpha_2^2 = \alpha_3^2 = \beta^2 = 1$$
$$\alpha_j\alpha_k + \alpha_k\alpha_j = 0 \text{ for } j \neq k\ (j, k = 1, 2, 3)$$
$$\alpha_j\beta + \beta\alpha_j = 0\ (j = 1, 2, 3) \qquad (A11\cdot 8)$$

The components of α and β anticommute. So they cannot be numbers. Quantities of this type can be written as matrices. So we shall represent them as matrices.

Since H must be Hermitian (i.e. $H = H^\dagger$), each of the four matrices $\alpha_1, \alpha_2, \alpha_3$ and β must also be Hermitian :

$$\alpha_1 = \alpha_1^\dagger, \alpha_2 = \alpha_2^\dagger, \alpha_3 = \alpha_3^\dagger, \beta = \beta^\dagger \tag{A11.9}$$

Hence they must be square matrices one of which may be chosen diagonal. The other three then cannot be diagonal. It can be shown that the minimum dimensions for these matrices are 4×4 in order that they may satisfy the conditions (A11.8) and (A11.9). Correspondingly the wave function Ψ must have four components given by the following column matrix :

$$\Psi = \begin{pmatrix} \Psi_1 \\ \Psi_2 \\ \Psi_3 \\ \Psi_4 \end{pmatrix} \tag{A11.10}$$

It has been shown that a representation of the matrices $\alpha_1, \alpha_2, \alpha_3$ and β which is particularly useful in the non-relativistic limit of the Dirac equation is given by

$$\vec{\alpha} = \begin{pmatrix} 0 & \sigma \\ \sigma & 0 \end{pmatrix} \qquad \beta = \begin{pmatrix} \mathbf{I} & 0 \\ 0 & -\mathbf{I} \end{pmatrix} \tag{A11.11}$$

where $\mathbf{I}$ is the 2×2 unit matrix while the components $\sigma_1, \sigma_2, \sigma_3$ of $\vec{\sigma}$ are the Pauli matrices (See A-X) :

$$\sigma_1 = \begin{pmatrix} 0 & 1 \\ 1 & 0 \end{pmatrix}, \quad \sigma_2 = \begin{pmatrix} 0 & -i \\ i & 0 \end{pmatrix}, \quad \sigma_3 = \begin{pmatrix} 1 & 0 \\ 0 & -1 \end{pmatrix} \tag{A11.12a}$$

$$\mathbf{I} = \begin{pmatrix} 1 & 0 \\ 0 & 1 \end{pmatrix} \tag{A11.12b}$$

Using the properties of the Pauli matrices given in A-X, it is easy to see that the relations (A11.7) and (A11.9) are satisfied.

FREE PARTICLE SOLUTIONS :

The four components of the wave functions Ψ_i are the eigenfunctions of the energy and momentum operators with the eigenvalues $\hbar \omega$ and $\hbar \mathbf{k}$ respectively. Substitution in the wave equation (A11.5) gives with the help of the explicit forms of the matrices α, β four linear homogeneous algebraic equations in Ψ_1, Ψ_2, Ψ_3 and Ψ_4 :

Written explicity, we have

$$(E + \beta mc^2 + c\vec{\alpha} \cdot \mathbf{p}) \Psi$$

$$= \begin{pmatrix} E+mc^2 & 0 & cp_z & c(c_x-ip_y) \\ 0 & E+mc^2 & c(p_x+ip_y) & -cp_z \\ cp_z & c(p_x-ip_y) & E-mc^2 & 0 \\ c(p_x+ip_y) & -cp_z & 0 & E-mc^2 \end{pmatrix} \begin{pmatrix} \Psi_1 \\ \Psi_2 \\ \Psi_3 \\ \Psi_4 \end{pmatrix} = 0$$

(A11·13)

We thus get the following four linear homogeneous algebraic equations in Ψ_1, Ψ_2, Ψ_3 and Ψ_4:

$$(E+mc^2)\Psi_1 + cp_z\Psi_3 + c(p_x-ip_y)\Psi_4 = 0$$

$$(E+mc^2)\Psi_2 + c(p_x+ip_y)\Psi_3 - cp_z\Psi_4 = 0$$

$$cp_z\Psi_1 + c(p_x-ip_y)\Psi_2 + (E-mc^2)\Psi_3 = 0$$

$$c(p_x+ip_y)\Psi_1 - cp_z\Psi_2 + (E-mc^2)\Psi_4 = 0 \qquad (A11·14)$$

The above equations apply to a free particle. In the case of a charged particle acted upon by a potential V, the Hamiltonian H should include V on the r.h.s. of Eq. (A11·4) with consequent changes in the system of Eqs. (A11·14).

The system of Eqs. (A11·14) can have non-trival solutions only if the determinant of the coefficients of Ψ_1, Ψ_2, Ψ_3 and Ψ_4 vanishes. When evaluated, the determinant comes out to be

$$D = (E^2 - m^2c^4 - p^2c^2)^2 = 0 \qquad (A11·15)$$

This is in agreement with the relativistic relation (A11·2). Solution of Eq. (A11·15) gives two possible energy values for the electron:

$$E_+ = +(m^2c^4 + p^2c^2)^{1/2} \qquad (A11·16)$$

and

$$E_- = -(m^2c^4 + p^2c^2)^{1/2} \qquad (A11·17)$$

For a given momentum **p**. each of the above energy values gives two linearly independent solutions.

If we write $\Psi_j = u_j \exp\{i(\mathbf{k}\cdot\mathbf{r} - \omega t)\}$

we get for $\quad E = E_+$

$$u_1 = -\frac{cp_z}{E_+ + mc^2},\ u_2 = -\frac{c(p_x+ip_y)}{E_+ + mc^2},\ u_3 = 1,\ u_4 = 0$$

$$u_1 = -\frac{c(p_x-ip_y)}{E_+ + mc^2},\ u_2 = \frac{cp_z}{E_+ + mc^2},\ u_3 = 0,\ u_4 = 1 \qquad (A11·18)$$

For $E = E_-$, we get

$$u_1 = 1,\ u_2 = 0,\ u_3\frac{cp_z}{-(E_- - mc^2)},\ u_4 = \frac{c(p_x+ip_y)}{-(E_- - mc^2)}$$

$$u_1 = 0, \; u_2 = 1, \; u_3 = \frac{c\,(p_x - ip_y)}{-(E_- - mc^2)}, \; u_4 = -\frac{cp_z}{-(E_- - mc^2)} \qquad (A11 \cdot 19)$$

The above solutions can be normalized using the relation $\Psi^\dagger \Psi = 1$ where $\Psi^\dagger$ is the Hermitian adjoint of Ψ given by the row matrix

$$\Psi^\dagger = \left(\Psi_1^*, \Psi_2^*, \Psi_3^*, \Psi_4^* \right) \qquad (A11 \cdot 20)$$

EXISTENCE OF ELECTRON SPIN :

A remarkable feature of Dirac equation is that it predicts the existence of the spin angular momentum for the electron.

We consider the Dirac Hamiltonian in the presence of an electrostatic potential $V(r)$ which is central :

$$\hat{H} = V(r) - c\,\vec{\alpha} \cdot \mathbf{p} - \beta' mc^2 \qquad (A11 \cdot 21)$$

Classically the x-component of the orbital angular momentum L_x is a constant of motion in a central field. As we shall see this is no longer true for the case of the Dirac Hamiltonian. Now $\hat{L}_x = (\mathbf{r} \times \hat{\mathbf{p}})_x = y\hat{p}_z - z\hat{p}_y$ commutes with every quantity in (A11·21) except $\hat{p}_y$ and $\hat{p}_z$. We have

$$\hat{L}_x \hat{H} - \hat{H} \hat{L}_x = -c\,\vec{\alpha} \cdot \{ (y\hat{p}_z - z\hat{p}_y)\hat{\mathbf{p}} - \hat{\mathbf{p}}(y\hat{p}_z - z\hat{p}_y) \}$$

So we get from the equation of motion (10.17-8)

$$i\hbar \frac{d\hat{L}_x}{dt} = \hat{L}_x \hat{H} - \hat{H} \hat{L}_x$$

$$= i\hbar c\,(\alpha_3 p_y - \alpha_2 p_z) \qquad (A11 \cdot 22)$$

Thus $\hat{\mathbf{L}}$ does not commute with $\hat{H}$ and hence is not a constant of motion.

We now seek another operator such that the commutator of its x-component with $\hat{H}$ just cancels out the commutator on the r.h.s. of Eq. (A11·22).

We introduce the operator

$$\vec{\sigma}' = \begin{pmatrix} \vec{\sigma} & 0 \\ 0 & \vec{\sigma} \end{pmatrix} \qquad (A11 \cdot 23)$$

The '0' in Eq. (A11·23) represents the 2×2 null matrix :

$$0 = \begin{pmatrix} 0 & 0 \\ 0 & 0 \end{pmatrix}$$

$\vec{\sigma}$ has the three components $\sigma_1, \sigma_2, \sigma_3$ which are 2×2 Pauli spin matrices defined in Eq. (A11·11). Thus the components $\sigma_1', \sigma_2', \sigma_3'$ of σ' are the 4×4 matrices given below :

$$\sigma_1' = \begin{pmatrix} \sigma_x & 0 \\ 0 & \sigma_x \end{pmatrix}, \quad \sigma_2' = \begin{pmatrix} \sigma_y & 0 \\ 0 & \sigma_y \end{pmatrix}, \quad \sigma_3' = \begin{pmatrix} \sigma_z & 0 \\ 0 & \sigma_z \end{pmatrix}$$

It can be easily seen that σ_1' commutes with α_1 and β (see Eq. A11·11), though it does not commute with α_2 or α_3.

$$[\sigma_1', \alpha_1] = [\sigma_1', \beta] = 0 \tag{A11·24}$$

$$[\sigma_1', \alpha_2] = 2i\alpha_3, \quad [\sigma_1', \alpha_3] = 2i\alpha_2 \tag{A11·25}$$

We therefore get

$$i\hbar \frac{d\sigma_1'}{dt} = \sigma_1' H - H \sigma_1' = -2ic (\alpha_3 p_y - \alpha_2 p_z) \tag{A11·26}$$

Comparing Eqs. (A11·22) and (A11·25), we then get

$$i\hbar \frac{d}{dt}\left(L_x + \frac{\sigma_1'}{2}\right) = 0 \tag{A11·27}$$

Eq. (A11·27) shows that if we add a term $\sigma_1'/2$ with $\hat{L}_x$, the resultant $\hat{J}_x = \hat{L}_x + \frac{\sigma'_1}{2}$ commutes with the Hamiltonian $\hat{H}$. We therefore interpret J_x as the total angular momentum component along x which is obtained by adding to the x-component of the orbital angular momentum L_x the component $(\sigma_1'/2)$ of $(\vec{\sigma}/2)$. We can thus interpret

$$s = \frac{1}{2}\hbar \sigma_1'$$

as the intrinsic spin angular momentum of the electron.

Negative energy states :

The relativistic expression (Eq. A11·2) for the energy of the electron shows that the energy can be both positive and negative. Thus both Schrödinger theory and Dirac theory which start from this equation admit of solutions for electrons having negative kinetic energy and negative rest mass. These negative energy solutions are important in quantum mechanics (though they are ignored in classical physics) since the electron can make a transition from the state of positive energy to a state of negative energy.

To prevent all the electrons of positive energy making transitions of negative energy states, Dirac assumed that all negative energy states are filled with electrons so that Pauli's exclusion principle prevents an electron from the positive energy state making a transition to a state of negative energy. The electrons in the negative energy state (which have infinite density) have no gravitational or electromagnetic effects. However if somehow an electron in the negative energy state is dislodged (which has negative mass and negative kinetic energy) so that it makes a transition to an empty positive energy state (which can be done by giving it energy $E > 2mc^2$), then a *hole* is created in the sea of negative energy states. Such a hole, called the positron, manifests itself as a positively charged particle of positive mass and positive energy.

The above hole theory is not a one-particle theory. Since it leads to an infinite sea of negative energy electrons, a many particle theory based on Dirac equation in accordance with formalism of quantum field theory is appropriate for the description of the positron.

Index

Aberration, stellar 442, 458
Absorption edges 181, 182
Acceptor levels 628
Action integral 68
Alkali spectra 100
    doublet structure 103, 116
Anharmonic oscillators, 505
Angular momentum quantization of 51
Angular momentum operator 249, 276, 309
    commutation relations 310
    eigenfunctions of 314
    eigenvalues of z component of 314
    eigenvalues of square of 313
Anisotropy in solids 570
Antiferromagnetism 672
Approximate methods 334
    degenerate case 339
    first order perturbation 335, 381
    second order perturbation 337
    stationary perturbation 334
    time-dependent perturbation 341, 344, 421
    variational methods, 347, 384
    WKB approximation 349
Atomic and molecular radii 9
    Atomic orbitals 322, 528
Atomic scattering factor 559
Atomic theory, Dalton's 1
Auger effect 175
Avogadro hypothesis 3
Avogadro number 3
    determination of 5

Balmer series 54
Band spectrum 492
    origin of 494
Band, electronic 510
    pure rotational 497
    rotation-vibration 501
Band theory of solids 611, 615
    forbidden zones 614, 619
    Kronig-Penny model 615
    permitted bands 613, 619
Barkhausen effect 650, 671
Barkla's experiment 188
Basis 544
Blackbody radiation 33
    Kirchhoff's theorem 33
    Planck's law 41
    Rayleigh-Jeans law 36, 43
    Stefan-Boltzmann law 35, 44
    Wien's displacement law 35, 36, 44
    Wien's distribution law 35, 43
Bloch wall 670

Bohr magneton 105, 657
    effective number of 658
Bohr's postulates 48
Bohr's theory of atomic spectra 50
    spectral series 53
Boundary conditions 280
Box normalization 308
Bragg equation 199, 206
    from Laue equations 559
Bragg's experiment 196, 561
Brillouin zones 619

Cathode rays 13
    e/m of 16
    Duninngton's experiment 19
    properties of 13
    Thomson's experiment 16
Central field approximation, 397, 407
    Hartree-Fock method 403, 405, 407
    Thomas-Fermi Model 397, 400
Centre of mass system, motion in 60, 329
Classical electron radius 188
Classification of solids 538, 622
Coherent scattering of x-rays 560
Commutator of operators 246
Complementarity principle 236
Compton effect 189
    discovery of 189
    experimental study 193
    theory of 190
Conduction band 614
Conservation of mass, Law of 1
Constants of motion in
    quantum mechanics 323
Cooper pairs 640
Correlation energy 409
Correspondence principle 66, 242, 262
Covalent bond 541
Crystal structure determination
    Bragg's method 567
    Debye-Scherrer method 568
    Laue's method 566
    Rotation photograph method 569
Crystal system 546
Cubic lattice 555

De Broglie wavelength 218, 222
Definite proportions, Law of 2
Degeneracy of state 259
Degenerate gas 690, 693
Density of states 681
Diamagnetism 648
    theory of 652

Diatomic molecules 306, 495, 528
 molecular orbital thory 528
 spectra of 494, 497, 501
Dirac notations 328
Dirac's theory of relativistic electron 714
Donor levels 627
Doppler effect 148, 471
 relativistic 471
Double slit, interference by 229

Effective mass 620
Ehrenfest's theorem 262
Eigenfunctions 254
 completeness of 256
 orthogonality of 256, 272
 well-behaved 254
Eigenvalues 254
 degeneracy of 259
 reality of 272
Eigenvalue equation 256, 277
Einstein-de Hass experiment 664
Einstein's A and B coeffieients 143, 427
Einstein's general theory of relativity 484
Einstein's special theory of relativity 447
Einstein's velocity addition
 theorem 456
Electrical discharge in gases 11
Electromagnetic field
 Hamiltonian of a charged particle 422, 710
Electromagnetic radiation 421
 dipole approximation 426
 forbidden transition 431
 interaction with atoms 421
 relations with A, B coefficients 427
 resonance absorption 422
 selection rules 428, 432, 433
 spin of photon 430
 spontaneous transition 427
 stimulated emission 422
 transition probability 425
 width of spectral lines 432
Electron cloud 322
Electron microscope 223
Elliptic orbits 66, 69
 relativity correction 73
 Sommerfeld's theory of 75
Energy levels of complex atoms 130
Energy levels of hydrogen 55
Energy level diagrams for
 x-ray emission 170, 172
Equilibrium distribution 677
Ether hypothesis 441
Exchange integral 668
Exchange operator 685
Expectation value 259, 262

Fermi energy 600, 608

Ferrimagnetism 672
Ferromagnetism 649
 Curie-Weiss law 651
 domain structure 660, 669, 670
 exchange integral 668
 Heisenberg's theory 666
 molecular field 665
 Weiss theory 661
Field emission 607
Fine structure constant 75
Fine structure of hydrogen spectra 76, 118
 quantum mechanical theory 367
Fine structure of x-ray levels 171, 173
Fizeau's experiment 460
 explanation of 461
Fluctuation 5
Fluorescence 81
Forbidden zone 614, 619
Fortrat diagram 515
Four vector 477
Fourier integral 211
Fourier series 256
Frames of reference 437
Franck-Condon principle 511
Franck and Hertz experiment 77
 resonance potentials 79
Free particle eigenfunctions 307

Galilean transformation 441
Graphical represenation 467
Group velocity 211, 214
Gyromagnetic experiments 664
Gyromagnetic ratio 105, 109

Hall effect 634
Hamiltonian operator 245, 248, 249, 253
Hermitian character 268
Hermitian operators 266, 268
 orthogonality of eigenfunctions 272
 reality of eigenvalues 272
Harmonic oscillator 298
 eigenfunetions of 301
 eigenvalues of 301
 parity of wave functions 303
Heavy hydrogen, discovery of 63
Helium atom 376
 excited states 387
 ground state 381
 term scheme 380
Hermite polynomials 302, 696
 differential equation for 696
Hermitian operators 266, 268
 orthogonality of eigenfunctions 272
 reality of eigenvalues 272
Hole conduction 625
Hund's rule 412
Hydrogen atom
 Bohr's theory 50
 elliptic orbits 69
 energy levels 55
 fine structure of spectra 75, 118, 362

Index

relativity correction 75
term values 56
Hydrogen atom; quantum
  mechanical theory 310
  degeneracy of eigenvalues 319
  energy levels 318
  θ—equation 312
  radial equation 314
  wave functions of 319, 701
Hydrogen bond 543
Hydrogen molecular ion 528
Hydrogen molecule 532, 666
Hysteresis 650

Identical particles 390
  exchange degeneracy 391
  symmetry of wave function 391
Incoherent scattering of x-rays 560
Indistinguishability 687, 690
Inertial frames 440
Infra-red bands, rotational structure 507
Intensity of x-rays 164
Ionic bond 540
Ionization potential 57, 81, 386
Ionized helium, spectrum of 60
Isotopes 29

j-j coupling 115, 415
Josephson effect 642

Kinetic theory of matter 7
Kossel diagram 170

L–S coupling 115, 409, 417
Laguerre differential equation 700
Laguerre polymials 318, 319, 700
Lamb–Retherfor experiment 370
Lande interval rule 413
Larmor frequency 106
Laser 144
  applications of 157
  experimental arrangement 149
  gain constant 147
  helium-neon 154
  optical pumping 150, 151
  population inversion 145, 147, 150
  ruby 153
  stimulated emission 143, 144
Lattice, Bravais 546
Lattice energy 550
  Madelung constant 551
Lattice structure 201, 544
  Miller indices 552
Laue equations 556
LCAO approximation 528
Legendre differential equation 697
Legendre functions, associated 314, 697

Legendre polynomials 697
Limitations of quantum theory 77
Lorentz-Fitzgerald contraction 447
Lorentz transformation equations 350, 468

Magnetic moment
  of an atom 132, 139, 654, 662
  of an electron 109
Many–electron atoms 393
  Slater determinants 395
Mass change with velocity 461
Mass-energy equivalence 463
Mass of atoms 9, 30
Matter wave, nature of 229
Meissner effect 637
Metallic bond 542
Metallic conductivity, free electron theory 591
Metastable states 82, 151, 153, 155
Michelson-Morley experiment 443
Millikan's oil drop experiment 21
Minkowski diagram 468
Molecular spectra 492
Momentum eigenfunctions 307
Momentum operator 247, 249, 268
Morse potential 505
Multiple proportion, Law of 1
Multiplicity of term values 116

Neel temperature 672
Newtonian relativity 437, 440
Normal modes 576, 584

Operator 245
  angular momentum 249, 276, 309
  Hamiltonian 245, 248, 276
  Hermitian 266, 272
  linear 245
  momentum 247, 249, 276
  quantum mechanical 247

Paramagnetism 649
  Curie's law 649
  Langevin's theory 654
  nuclear 659
  of free electron gas 659
  quantum theory of 656
Parity operator 326
Parity of wave function 296, 303, 326
Particle in a potential box 282
Paschen-Back effect 136, 375
Pattern unit 544
Pauli's exclusion principle 120, 392
Pauli's theory of spin 712
Periodic classification of elements 120, 396
  quantum mechanical theory 396
Periodic potential 615

Periodic table 120, 703 704
Phase velocity 211, 214
Phonons 584, 639
Phosphorescence 81
Photoelectric effect 84
    applications of 92
    Einstein's light quantum hypothesis 90
    Lenard's experiment 84
    Millikan's experiment 86
Photoelectric equation 90
Planck's law of radiation from Bose statistics 693
Positive rays 25
Thomson's parabola method 24
Postulates of quantum mechanics 275
Potential barrier, rectangular 289
    explanation of alpha decay 292
Potential well, rectangular 294
Principal series 102, 117
Primitive cell 545
Primitive translation 546
Probability
    conservation of 251
    current density 251
    density 249
    interpretation 231, 249

Quantum concepts, applicability of 237
Quantum hypothesis 41
Quantum mechanics
    equations of motion 273
    fundamental postulates 275
Quantum number
    azimuthal 68, 101, 313
    effective 100
    magnetic 104, 312
    principal 70, 317
    radial 68
    rotational 498
    spin 109
    total 112
    vibrational 307, 503
Quantum theory, limitations of 77

Radial probability density 322
Raman effect 515, 517
    applications 527
    classical theory of 518
    experimental techniques 525
    quantum theory of 519
Raman spectrum 516
    rotational structure of 522
Rare earth elements 130
Rayleigh-Jeans law 46, 43
Reduced mass correction 62
Reflection coefficient 287, 289
Refraction of electron wave 220

Relativistic acceleration and force 485
Reactivity of length measurement 453
Relativity of time measurement 454
Relaxation time 593
Resonance potentials 79
Ritz combination principle 59
Russel-Saunders coupling 115, 409, 417
Rutherford model of atom 47

Schottky effect 605
Schrodinger equation 242
    one dimensional 242, 280
    three dimensional 244, 308
    time-independent 253
Selection rules 76, 101, 107, 117, 449 503, 520, 522
Semi-conductor diodes 634
Semi-conductor 623
    conductivity of 633
    carrier concentration of 628, 635, 636
    extrinsic 625
    intrinsic 623
    Fermi levels of 631
    n-type 626
    p-type 627
    rectifying property of 634
Shell 121
Simultaneity 450
Slater determinants 305
Space-quantization 103
Space-time diagram
    in Newtonian relativity 466
Specific heat of metals 588
Specific heat of solids 571
    Debye's theory 575
    Dulong–Petit's law 573
    Einstein's theory 573
Spectroscopic terms 56
    of equivalent electrons 130
    of non-equivalent electrons 130
Spherical harmonics 314, 699
Spherically symmetric potential 310
Spin of electron 76, 108
Spin-orbit interaction 115, 319, 409, 417
    equivalent electrons 411, 413
    $j-j$ coupling 415
    $L-S$ coupling 409, 417
Stationary states 254, 257
    probability of 257
Statistics
    Bose-Einstein 690
    Fermi-Dirac 687
    Maxwell-Boltzmann 681
Stefan-Boltzmann law 35, 44
Step potential 285
Stern-Gerlach experiment 138
Stimulated emission 143
    Einstein's theory 143

Index

Stirling's formula 678
Structure factor 559
Superconductivity 636
    BCS theory 639
Superposition principle
    in quantum mechanics 238
Symmetry of crystals
    rotational 548
    translational 545
Symmetry of wave functions 391, 684

Thermionic emission 94, 599
    effect of temperature 96
    Richardson-Dushman
    equation 98, 602
Thermionic work function 98, 599
Thomson model of the atom 46
Time dilatation 454
Transistors 634
Transition between states
    Einstein's theory 143
Transition elements 129, 130
Transmission coefficient 287, 292
Trouton and Noble's experiment 443
Twin paradox in relativity 482

Uncertainty principle 231
    formal proof 264
    gamma-ray microscope
    experiment 235
Unit of atom mass 30
Valence band 614
Van der Waal force 543
Vector model of the atom 108

Wave function 242
    normalization 250
    probability interpretation 249
    square integrability 250

Wave packet 211, 229 242
    and Schrodinger equation 242
Wave nature of matter 210, 229
    Davisson and Germer's
    experiments 216
    de Broglie's theory 213
    G.P. Thomson's experiment 218
    relation between phase and group
    velocity 214
Wiedemann-Franz law 597
Wien's displacement law 33, 35, 44
Wilson-Sommerfeld quantization rules, 66
World line 468
World points 467

X-rays
    absorption edges 181, 182
    absorption of 178
    characteristic 166, 171
    continuous 167
    diffraction of 194
    discovery of 160
    fine structure 173
    Moseley's law 176
    polarization of 207
    production of 161
    properties of 163
    reflection of 198
    refraction of 204
    scattering of 183, 188
    Thomson's theory of scattering 183
    wavelength measurement 201, 203, 205

Zeeman effect
    anomalous 131, 135
    normal 103, 108, 136
    quantum mechanical theory 372
Zero-point energy 301, 503

Win Prizes!

Attention: Students

We request you, for your frank assessment, regarding some of the aspects of the book, given as under:

16 173   **Atomic and Nuclear Physics**
S.N. Ghoshal   **Reprint 2011**

Please fill up the given space in neat capital letters. Add additional sheet(s) if the space provided is not sufficient, and if so required.

(i) What topic(s) of your syllabus that are important from your examination point of view are not covered in the book?

....
....
....
....

(ii) What are the chapters and/or topics, wherein the treatment of the subject-matter is not systematic or organised or updated?

....
....
....

(iii) Have you come across misprints/mistakes/factual inaccuracies in the book? Please specify the chapters, topics and the page numbers.

....
....
....
....

(iv) Name top three books on the same subject (in order of your preference - 1, 2, 3) that you have found/heard better than the present book? Please specify in terms of quality (in all aspects).

1 ....
....
2 ....
....
3 ....
....

(v) Further suggestions and comments for the improvement of the book:

....
....
....
....

Other Details:

(i) Who recommended you the book? (Please tick in the box near the option relevant to you.)
☐ Teacher   ☐ Friends   ☐ Bookseller

(ii) Name of the recommending teacher, his designation and address:

....
....
....

(iii) Name and address of the bookseller you purchased the book from:

....
....

(iv) Name and address of your institution (Please mention the University or Board, as the case may be)

....
....
....

(v) Your name and complete postal address:

....
....
....

(vi) Write your preferences of our publications (1, 2, 3) you would like to have ....

....

The best assessment will be awarded half-yearly. The award will be in the form of our publications, as decided by the Editorial Board, amounting to Rs. 300 (total).

Please mail the filled up coupon at your earliest to:
Editorial Department
S. CHAND & COMPANY LTD.,
Post Box No. 5733, Ram Nagar, New Delhi 110 055

OTHER IMPORTANT BOOKS ON PHYSICS

MODERN ENGINEERING PHYSICS
A.S. Vasudeva

The book includes a wide spectrum of topics covering all the important sections of the subject. The stress has been put to give an insight into the topics that form the syllabi of various technical institutes and universities. The fundamentals are explained in a way that the students grasp the subject and in fact enjoy the various aspects of the subject.

10 182 pp. 320

THEORY OF SPACE, TIME AND GRAVITATION
S.G. Pimpale

The book is designed for B.Sc. (Pass and Hons.) and postgraduate courses in Physics. The main emphasis is on the Special relativity, with a number of exceptional features, such as detailed analysis of the 'Paradoxical' experiments and the decomposition of a Lorentz trasformation into a rotation. The use of Tensers, in the gravitation theory of relativity has been added as an appendix. University questions and bibliography, at the end of each chapter, add to the utility of the book, in its own way.

16 227 pp. 192

OPTICAL FIBRES AND FIBRE OPTIC COMMUNICATION SYSTEMS
Subir Kumar Sarkar

This book deals with the fundamentals of 'fibre optics' with a mathematical approach. It fully covers the course for two semester undergraduate course in "Electronics and Telecommu-nication Engineering" and "Applied Physics", for Indian universities. It contains numerical examples, both solved and unsolved, to meet the trends of the latest competitive examinations. A glossary of terms and abbreviations, appendices of design-data and specifications, etc., a bibliography and the index at the end of book make the book compact and complete in itself.

16 205 pp 320

SIGNALS, SYSTEMS AND SIGNAL PROCESSING
S.P. Eugene Xavier

The book presents an introductory yet comprehensive treatment of Signals and Systems, with a strong emphasis on numerical examples. A lot of emphasis has been placed on the meaning and significance of the various concepts and how these concepts fit together to give an overall picture. There are over 120 illustrative examples in the book and chapter 6 in particular deals with the mathematical tools necessary for the study of signals and systems. In this revised edition a new chapter on 'Laplace Trans-form Techniques', and a large number of miscellaneous examples have been added in different chapters. The book will be helpful for students, teachers and working engineers.

10 199 pp. 496

SOLID STATE PHYSICS AND ELECTRONICS
R.K. Puri & V.K. Babbar

The book has been written to meet the requirements of the students, studying Solid State Physics and Electronics, as a subject under the new syllabi introduced in a number of Indian Universities. A summary of each chapter is given for a quick review of the topics. A set of questions & unsolved problems are given at the end of each chapter to help the students in comprehending these topics. There are also solved examples at the end of the text to help the students with the applications of various principles and formula.

10 176 pp. 576

ELEMENTS OF PROPERTIES OF MATTER
D.S. Mathur

The revised edition of the book gives a lucid and comprehensive treatment of the subject matter, beside many typical problems with solutions given at the end of each chapter. A new chapter on "Special Theory of Relativity" has been added.

16 062 pp. 592

HEAT AND THERMODYNAMICS
Brij Lal & Subrahmanyam

The present revised edition includes many new topics of great importance viz. "Statistical Thermodynamics". The old chapter on thermodynamics has been thoroughly reoriented. Many important questions from recent university papers have also been added.

16 060 pp. 528
16 061 (Hindi) pp. 470

A TEXTBOOK OF OPTICS
N. Subrahmanyam & Brij Lal

The book has been revised and up-dated to suit the needs of B.Sc. and Engineering students of Indian universities. Worked examples and University questions have also been given at the end of each chapter.

16 056 pp. 736

OPTICS AND SPECTROSCOPY
R. Murugeshan

The present book, written for students of B.Sc. Physics, deals with the subject with a modern outlook. S.I. System of Units has been used throughout the text. A large number of questions and problems have been given at the end of each chapter for the students to attempt them and gain better insight into the subject.

16 206 pp. 168

ELECTRICITY AND MAGNETISM
K.K. Tewari

Written purely in S.I. units with a complete vector treatment, the book emphasizes on the basic Physics with some instructive, stimulating and useful applications; and covers the syllabi of B.Sc. and Engineering students of various Universities in India.

16 175   pp. 816

FUNDAMENTALS OF MAGNETISM AND ELECTRICITY
D.N. Vasudeva

This revised and updated edition is designed to be used as a textbook by the students of B.Sc. degree course and Engineering students. The book gives logical and systematic treatment to fundamental principles of the subject covered.

16 058   pp. 952

ATOMIC AND NUCLEAR PHYSICS
N. Subrahmanyam & Brij Lal

This book is suitable for B.Sc. (Pass and Honours) and Engineering students of Indian Universities. The mathematical treatment is explanatory. New topics *viz.*, Wave and Probability, Group Velocity, Wave Function, Transmission and Reflection, etc., have been added in the book.

16 132   pp. 568

ATOMIC AND NUCLEAR PHYSICS
(In Two Vols.)
S.N. Ghoshal

The subject, which has gained immense speculation and has also grown tremendously over the last quarter of the century, is treated exclusively in this edition dealing mainly with the Atomic Physics, comprising the extra-nuclear part of the atom. The books are designed to cover the B.Sc. (Pass and Hons.) course.

16 173   Vol. I   pp. 752
16 187   Vol. II   pp. 1068

MODERN PHYSICS
R. Murugeshan

The revised edition of this book gives an account of the basic principles of Atomic, Nuclear and Solid State Physics. It is meant to serve as a textbook for the B.Sc. Physics students of Indian Universities. Seven new chapters and there news appendices have been added.

16 089   pp. 648

QUANTUM MECHANICS, STATISTICAL MECHANICS AND SOLID STATE PHYSICS: AN INTRODUCTION
D. Chattopadhyay & P.C. Rakshit

The book provides the subject matter in a clear, concise and easy-to-read manner. Emphasis has been given to the clarification of Physical principles. Mathematical details have also been worked out to the extent suitable for the undergraduate students.

16 155 pp. 352

QUANTUM MECHANICS
S.P. Singh, M.K. Bagde & Kamal Singh

The revised edition of the book covers the basic principles and methods of quantum mechanics which is useful for Physics students at the undergraduate level. A number of new topics *viz.*, Eigen Functions, Square Potential Well, and Rigid Rotator, etc., have been added at the proper places to update the subject. At the end of each chapter a set of solved examples and questions are also given.

CONTENTS: Development of Quantum Mechanics • Wave Properties of Matter • Heisenberg's Uncertainty Principle • Schrodinger's Wave Equation • Free States • Bound States • The Hydrogen Atom and the Rigid Rotator • Appendices.

16 129 pp. 224

ELECTRICITY AND MAGNETISM WITH ELECTRONICS
K.K. Tewari

This book covers the syllabi of B.Sc. (Pass & Honours) and Engineering students of various Universities in India and is written in S.I. units with a complete vector treatment. There are a large number of worked examples and objective questions given along with comparative study of valves and semi-conductor devices.

16 150 pp. 956

BASIC ELECTRONICS (SOLID STATE)
B.L. Theraja

This revised and updated edition of the book has been written expressively to meet the syllabus requirements of students preparing for a diploma and degree in Electronics & Communications. Also, it could be used profitably by the postgraduate students of Physics. A new chapter entitled 'Fibre Optics' has been added to this new edition.

10 051 pp. 656

PRINCIPLES OF ELECTRONICS
V.K. Mehta

The revised and updated edition of the book is immensely useful for the students of Diploma, AMIE Section B, Degree and other Engineering examinations. A number of important topics like "Equivalent Circuits, Class - C Amplifier; Semi conductor Physics, etc. have been added alongwith many practical problems.

10 009 pp. 496